Towards the First Silicon Laser

NATO Science Series

A Series presenting the results of scientific meetings supported under the NATO Science Programme.

The Series is published by IOS Press, Amsterdam, and Kluwer Academic Publishers in conjunction with the NATO Scientific Affairs Division

Sub-Series

I. Life and Behavioural Sciences	IOS Press
II. Mathematics, Physics and Chemistry	Kluwer Academic Publishers
III. Computer and Systems Science	IOS Press
IV. Earth and Environmental Sciences	Kluwer Academic Publishers
V. Science and Technology Policy	IOS Press

The NATO Science Series continues the series of books published formerly as the NATO ASI Series.

The NATO Science Programme offers support for collaboration in civil science between scientists of countries of the Euro-Atlantic Partnership Council. The types of scientific meeting generally supported are "Advanced Study Institutes" and "Advanced Research Workshops", although other types of meeting are supported from time to time. The NATO Science Series collects together the results of these meetings. The meetings are co-organized bij scientists from NATO countries and scientists from NATO's Partner countries – countries of the CIS and Central and Eastern Europe.

Advanced Study Institutes are high-level tutorial courses offering in-depth study of latest advances in a field.
Advanced Research Workshops are expert meetings aimed at critical assessment of a field, and identification of directions for future action.

As a consequence of the restructuring of the NATO Science Programme in 1999, the NATO Science Series has been re-organised and there are currently Five Sub-series as noted above. Please consult the following web sites for information on previous volumes published in the Series, as well as details of earlier Sub-series.

http://www.nato.int/science
http://www.wkap.nl
http://www.iospress.nl
http://www.wtv-books.de/nato-pco.htm

Series II: Mathematics, Physics and Chemistry – Vol. 93

Towards the First Silicon Laser

edited by

Lorenzo Pavesi
INFM and Department of Physics,
University of Trento, Italy

Sergey Gaponenko
National Academy of Sciences of Belarus,
Minsk, Belarus

and

Luca Dal Negro
INFM and Department of Physics,
University of Trento, Italy

Kluwer Academic Publishers

Dordrecht / Boston / London

Published in cooperation with NATO Scientific Affairs Division

Proceedings of the NATO Advanced Research Workshop on
Towards the First Silicon Laser
Trento, Italy
21–26 September 2002

A C.I.P. Catalogue record for this book is available from the Library of Congress.

ISBN 1-4020-1193-8 (HB)
ISBN 1-4020-1194-6 (PB)

Published by Kluwer Academic Publishers,
P.O. Box 17, 3300 AA Dordrecht, The Netherlands.

Sold and distributed in North, Central and South America
by Kluwer Academic Publishers,
101 Philip Drive, Norwell, MA 02061, U.S.A.

In all other countries, sold and distributed
by Kluwer Academic Publishers,
P.O. Box 322, 3300 AH Dordrecht, The Netherlands.

Printed on acid-free paper

Printed in the Netherlands.

TABLE OF CONTENTS

Part II. Silicon nanocrystals

Part III. Optical gain in silicon nanocrystals

Part IV. Theory of silicon nanocrystals

Part V. Silicon/Germanium quantum dots and quantum cascade structures

Part VI. Terahertz silicon laser

PREFACE

Silicon, the leading material in microelectronics during the last four decades, promises to be the key material also in the future. Despite of many claims that silicon technology has reached fundamental limits (see e. g. Nature 399 (2001) 729), the performances of silicon microelectronics steadily improve: for example a 15 nm wide transistors operating at 2.63 THz and at 0.8 V have been demonstrated by INTEL (www.intel.com/research/silicon). The same is true for almost all the applications where Si was considered unsuitable. For example, front-end circuits for OC768 (40Gbit/s) have been demonstrated by IBM, which makes Si a player in high frequency applications. Another example is silicon photonics, which is considered now a viable technology. The main exception to this positive trend is that of the silicon laser, for which no demonstrator exists up to now. The main reason for such a stumbling situation comes from a fundamental limitation related to the indirect nature of the Si band-gap. Demonstration of a silicon laser is still a very challenging research activity because it is trying to change the material properties of silicon and because of the opportunity to integrate a silicon laser within the available silicon technology. In the recent past, many different approaches have been undertaken to achieve this goal. All of them are abundantly illustrated in the present book. The interest in a silicon laser is now very widespread as revealed by the many articles published on advances on newspapers. The last in the series is about a 10% efficient Er based light emitting diode developed by ST Microelectronics (New York Times 27 October 2002).

A meeting to discuss, weight out and try to make the point on the silicon laser was organized by ourself in September 2002 in Trento as a NATO Advanced Research Workshop. This book contains the reports presented at the workshop on such different approaches to a silicon laser as dislocated silicon, extremely pure silicon, silicon nanocrystals, porous silicon, Er doped Si-Ge, SiGe alloys and multiquantum wells, SiGe quantum dots, SiGe quantum cascade structures, shallow impurity center in silicon and Er doped silicon.

It should be noted that the success of our meeting and of this Book was guaranteed in advanced by the enthusiastic work of the organizing committee: H. Atwater, U. Gösele, L. Kimerling and L. Ledentsov. We thank the staff of the Ufficio Congressi of Trento University, of Opera Universitaria of Trento and the members of the research group on Silicon Photonics of Trento for the help in organizing and managing the workshop. We thank all the authors who responded eagerly to this demanding task and promptly delivered high quality manuscripts. We are most grateful to them for all their efforts. We gratefully acknowledge the generous support of the NATO Scientific Affairs Division, of the University of Trento, of Istituto Trentino di Cultura ITC-irst, of Provincia Autonoma di Trento, of the National Institute for the Physics of the Matter INFM and of Laser Optronic SPA. We also express our gratitude to Kluwer for efficient publication.

Lorenzo Pavesi and Luca Dal Negro
Trento, Italy, December 2002

Sergey Gaponenko
Minsk, Belarus, December 2002

Conference Photograph September 24th, 2002 © Jan Valenta, 2002

INTRODUCTION TO *TOWARDS THE FIRST SILICON LASER*

S. GAPONENKO
Institute of Molecular and Atomic Physics, National Academy of Science, Minsk, Belarus
L. PAVESI L. DAL NEGRO
INFM and Dipartimento di Fisica, Università di Trento, Italy

Silicon is the leading material for microelectronics. Integration and economy of scale are the two keys ingredients for technological success of silicon. Silicon has a band-gap (1.12eV) that is ideal for room temperature operation and an oxide (SiO_2) that allows the processing flexibility to place today more than 10^8 devices on a single chip. The continuous improvements of the silicon technology has made possible to grow routinely 200 mm single silicon crystals at low cost and even larger crystals are now commercially available. The high integration levels reached by the silicon microelectronics industry have permitted high-speed device performances, and unprecedented interconnection levels. The present interconnection degree is sufficient to cause interconnect propagation delays, overheating and information latency between the single devices. Overcoming of this "Interconnection bottleneck" is the main motivation and opportunity for present-day silicon microphotonics. Attempts to combine photonics and electronic components on a single Si chip or wafer are made. In addition, microphotonics aim also to combine the power of silicon microelectronics with the advantages of photonics. The main limitation of this technology when it is implemented in silicon is the lack of any practical silicon light sources: either an efficient LED or a Si laser.

Silicon is an indirect band-gap material. Light emission in Si is naturally a phonon-mediated process with low probability (spontaneous recombination lifetimes in the millisecond range). In standard bulk silicon, competitive non-radiative recombination rates are much higher than the radiative ones and most of the excited e-h pairs recombine non-radiatively. This usually yields very low internal quantum efficiency ($\eta_i \approx 10^{-6}$) for bulk silicon luminescence. In addition, fast non-radiative processes such as *Auger* or *free carrier* absorption severely prevent population inversion from silicon optical transitions at the high pumping rates needed to achieve optical amplification. Despite of all, during the last 10 years many different strategies have been employed to overcome these material limitations and the best silicon LED efficiencies are now only a factor of ten below what is required for actual applications. The main future challenge for silicon microphotonics is the demonstration of a silicon-based laser action. During the nineties many research efforts have been focused towards this goal. A steady improvement in silicon LED performance has been achieved. However it was only at the end of 2000 and during 2001 that major advances demonstrated that this field is very active and promising.

Silicon photonics today is a challenging technological field which involves a number of basic scientific issues and promises an application breakthrough in optoelectronics and communications industry. The last decade showed that silicon should not be ignored by optoelectronics. Visible photo-and electroluminescence have been reported for silicon nanostructures. The communication by Pavesi et al. in 2000 on optical gain in silica containing silicon nanocrystals have stimulated extensive research which span from technological advances in silicon nanostructure synthesis to the detailed studies of their optical properties under condition of optical and electrical excitation. Will silicon lasers become a reality in the near future? If this happens, which impact can be expected on optoelectronic industry and information technology? To discuss the state-of-art in the field and to answer the above questions scientists from 19 countries gathered at Trento in the period of September 21-26, 2002. This book contains results presented at the NATO Advanced Workshop "Towards the first silicon laser".

An overview of photonic components and potential impact of photonics on information processing, storage and transfer was presented in the introductory talk by L. Kimerling (MIT, USA). Basic properties of silicon nanocrystals and their potential applications in electronic circuitry and optical devices were the subject of the talk by H. Atwater (CalTech, USA). Novel concepts in photonic engineering based on 1-, 2-, and 3-dimensional photonic crystals for spontaneous emission control, waveguiding, and optical feedback in microlasers were discussed by S. Gaponenko (NAS, Belarus).

The most challenging and promising trend towards silicon lasers in the near-IR – visible range is making use of quantum confinement effects on the energy spectrum and optical transition probabilities. The first report on optical gain in Si nanocrystals by Pavesi and co-workers in 2000 did stimulate extensive experiments on synthesis and optical research of nanometer-size silicon nanocrystals/clusters in a silica matrix. A comprehensive overview of the state of the art in this field comprises the principal part of the present volume. The section includes theory of electron states in silicon nanocrystals (Delerue et al., Ossicini et al.) including isolated nanocrystals and nanocrystals in a silica matrix with an evaluation of processes resulting in optical gain; new techniques of fabrication and processing of silicon quantum dots/clusters (Zacharias, Miu) optical studies of luminescent silicon nanocrystals in various matrices including fine experiments with detectability down to a single Si nanocrystal (Kanemitsu, Linnros et al., Kovalev). Extensive analysis of experiments on observation of optical gain in silicon nanostructures was presented in several contributions by different groups. Evidences of optical gain were reported by Dal Negro et al., Nayfeh, Fauchet et al., Khriachtev et al, Ivanda et al. Technological and experimental obstacles to wards the achievement of stimulated emission and optical gain in existing silicon based nanostructures were the subject of a detailed analysis by Polman et al. and Valenta et al. Pelant/Luterova and Ryabtsev outlined peculiar aspects of the evaluation of internal parameters (net gain coefficient, radiative to non-radiative rate ratio) in experiments performed on thin film silicon-based structures. Quantum confinement effects on energy spectrum and optical transition probabilities result in drastic modifications of optical properties. Stimulated emission and optical gain in Si which seemed unreal a decade ago have became the subject of systematic research in many laboratories all over the world. Quantum confinement of electron states in Si can be further combined with another powerful modern trend in photonic engineering, namely photonic crystal concepts allowing efficient waveguiding, spontaneous emission control and positive optical feedback in microphotonic devices and optical circuitry components.

Potential applications of a Si laser will need electric rather than optical pumping. The problem of efficient light generation and extraction from silicon electroluminescent diodes comprise another section of this book. The topics here are divided into optimisation of traditional near- IR silicon diodes using a number of technical solutions developed for solar cell devices (Green et al.), by means of special material processing to reduce non-radiative recombination rate (Gardner et al.) and the new approach to silicon diode fabrication based on silicon nanocrystals (Iacona et al., Garrido et al., Heikila) The use of dislocations and defect states engineered to get electroluminescence from Si-based structures were reported by Homewood, Gusev and Rebohle and coworkers. Electroluminescence of porous silicon light emitting diode was reviewed by Lazarouk and a visibly emitting microdisplay panel was demonstrated at the workshop.

In spite of the impressive progress in laser science and technology there is a practical need to fill the gap of far-infrared and terahertz (1-10 THz) frequency range with a tuneable laser source. This range lies above the present upper frequency limit of millimetre wave electronic oscillators and below the range of solid state lasers. In spite of the valuable practical need for the purpose of sub-millimetre imaging in various applications this range still remains inaccessible for science and engineering applications. Silicon-based structures do offer a solution of this problem by means of lasing in quantum well cascade structures and in doped silicon. This field was another focus of the workshop and the contributions on THz-lasing in silicon based structures comprise a separate section in the book. Several group reported on light emission, optical gain and lasing in Si/Ge nanostructures (Sun et al., Cirlin, Gruetzmacher et al., Kelsall et al., Yassievich et al). Optical gain and lasing in Si doped with shallow donor centres (P, Bi, Sb) were discussed by Shastin et al., Pavlov et al., and Kagan.

The problem of using Si in laser devices and development of light emitting structures for optical interconnect compatible with silicon based electronics should not be reduced to a search for intrinsic silicon emission but may involve other light emitting materials as well. Alternative Si/Ge and Si—III-V heterostructures were considered by several authors. Dvurechenskii et al. provide an overview of growth and optical properties of Ge quantum dots on silicon facets, Pchelyakov et al. reported on MBE growth of smaller Ge nanoparticles on silicon. Yablonskii and Heuken reviewed the state-of-art of InGaN/GaN/Si lasers emitting in the blue and near-UV range under optical pumping.

In modern optical communication technology Erbium plays a major role providing optical gain in the range of superior waveguides transparency around 1.55 μm. Er-doped silicon and Si-containing Er-doped silica, their optical properties, optical gain and lasing form the subject of the concluding section of the book. Several authors reported on the sensitising effect on Er by silicon nanocrystals with application to optical gain enhancement. Very large absorption/emission cross section for Er transitions were observed by several authors (Shin et al., Kik and Polman, Pacifici et al., Krasilnik et al., Yassievich et al.).

In conclusion we outlined trends, prospects and problems in silicon laser development. First, a development of visible—near infrared laser remains a major challenge for scientists. Though quantum confinement effects offer good perspectives in visible-near IR light emitting devices for practical lasing devices the evident obstacles are low optical gain and low radiative recombination rate. However, if one takes into account that modern gas lasers operate with even lower gain and modern solid state lasers possess low radiation decay rates, then the prospects of visible silicon lasers become more optimistic. The optimism is further enhanced by the fact that advances and facilities of silicon technology can be used to develop silicon nanocrystals with desirable optical properties. There is no other field in laser technology with such a solid technological/material background like this is given for the development effort on silicon lasers just now. Nevertheless, the judge is still out whether a visible-near infrared silicon based laser may actually be realized in the foreseeable future. Second, silicon-based devices can competitively fill the THz-gap in the electromagnetic spectrum. Here silicon photonics may offer a reliable practical solution in the near future. Third, Si-Erbium composite structures will definitely contribute to optical communication technology as well.

HIGH EFFICIENCY SILICON LIGHT EMITTING DIODES

MARTIN A. GREEN, JIANHUA ZHAO, AIHUA WANG AND THORSTEN TRUPKE
Centre for Third Generation Photovoltaics
University of New South Wales, Sydney, Australia, 2052

1. Introduction

Efficient silicon light emitting diodes (LEDs) could allow "super-integration" of optical and electronic functions in high density silicon microelectronic circuits [1] and are consequently of considerable interest. However, performance with standard p-n junction emitters has been modest, with devices with electrical to light conversion efficiency in the 0.01-0.1% range recently described as "high performance" [2]. Slightly better results have been obtained using approaches based on porous silicon [3,4], although this material is fragile and not fully compatible with standard microelectronics processing.

Silicon's poor performance is usually attributed to the fact that silicon is an indirect bandgap material and hence optical processes are weak, involving at least one phonon. However, there is a fundamental reciprocal relation between light absorption and emission that shows that similar performance to direct bandgap material intrinsically is feasible.

The poorer performance most fundamentally arises from the relative enhancement of competing "parasitic" processes by the length and time scaling mentioned above. An example is non-radiative recombination through defects. As an elemental semiconductor with well developed materials technology, the background level of bulk defects in silicon is lower than in compound semiconductors. Oxides and amorphous silicon interfaces also produce surface or interfacial recombination properties comparable to those offered by III-V heterojunctions. However, defect recombination is not slowed to the same extent as radiative recombination by silicon's indirect bandgap. Since defect sites are fixed in space, the uncertainty relationship effectively removes the momentum conservation requirement for recombination through defects, giving recombination rates per defect not fundamentally different from direct gap materials.

Similarly, free carrier absorption can be better controlled in silicon than compound semiconductors due to better control over doping that is possible for several reasons. However, free carrier absorption rates are fundamentally no different in direct and indirect bandgap material.

The result is that, on a relative basis, parasitic non-radiative recombination and absorption are more significant. To demonstrate high light emission efficiency in silicon, both bulk and surface non-radiative defect recombination and parasitic absorption processes have to be reduced to very low levels.

L. Pavesi et al. (eds.), Towards the First Silicon Laser, 1–10.

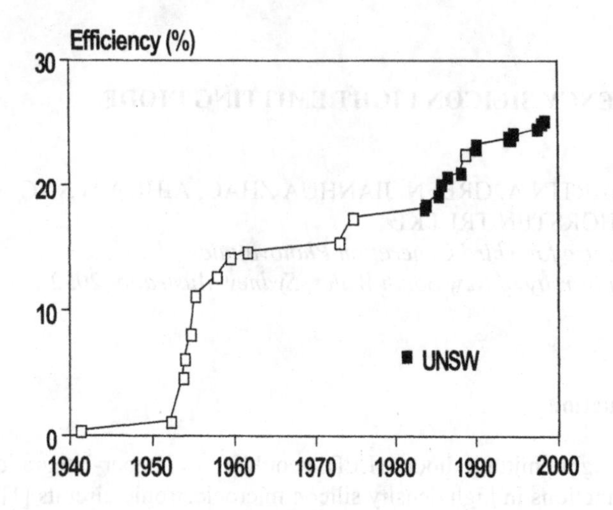

Figure 1: Evolution of silicon solar cell efficiency.

2. High Efficiency Silicon Solar Cells

Similar requirements also apply to improving the performance of silicon solar cells since time reversal symmetry provides a fundamental link between conversion of light into electricity and electricity into light. The last 20 years has been a period of sustained refinement of silicon cell design (Figure 1), most of which is directly relevant to improved light emission performance. Solar cell ideas introduced into the III-V LED area also produced striking results in the mid-1990's [5].

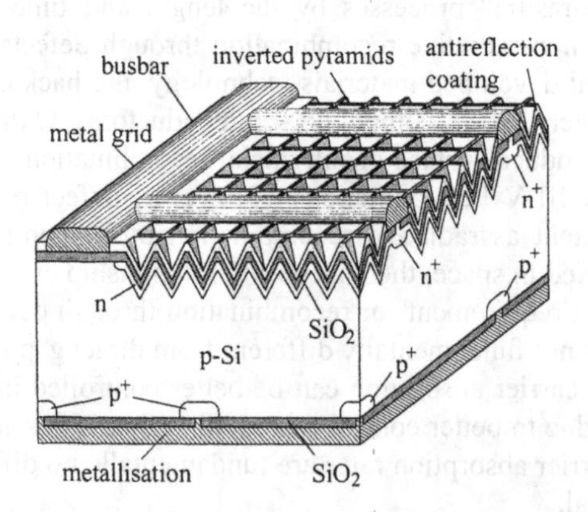

Figure 2: High efficiency UNSW PERL cell (passivated emitter rear locally-diffused cell) and light emitting diode.

As Figure 1 shows, there has been over 50% relative improvement in silicon cell performance since 1980, with the best cells now around 25% energy conversion efficiency. A schematic of such a device is shown in Figure 2 (not to scale). Several features that contribute to its high solar conversion efficiency are even more important to its performance as a light emitter, as discussed below.

There are three optical features worthy of note. The most obvious is the inverted pyramid structure on the top surface of the device (base of pyramids ~ 10 μm). These pyramids are formed by intersecting (111) equivalent crystallographic planes [6], which are exposed by anisotropic surface etching. The pyramids have one obvious role in reducing reflection, since light reflected from a pyramid side has at least one more chance to be coupled in. A second role is as part of a "light trapping" scheme that enhances the absorption of weakly absorbed long wavelengths by making it difficult for this light to escape once within the cell. After rear surface reflection, this light impinges on the pyramids from underneath. About half will strike a side of a pyramid of opposite orientation to that coupling it into the cell. Symmetry dictates this will be coupled straight out. However, the rest will strike oblique pyramid faces and be totally internally reflected and skewed in a way that makes it difficult for it to escape on future encounters with the top surface.

The second optical feature is the rear reflector itself. The low index dielectric layer between it and the semiconductor greatly increases reflectance. The optimum thickness of this layer is thicker than for an interference layer since much of the light striking this layer from above strikes obliquely. A thick layer allows total internal reflection of such light with minimal absorption of the evanescent wave in the rear metal.

The absorption that does occur in the rear metal (or in the top metal contact from underneath) is by free carriers. In future discussion, since small, it will be lumped with free carrier absorption in the semiconductor bulk by converting to an effective value (α_{eff}) distributed across the semiconductor bulk (in the weak absorption limit, $\alpha_{eff} \approx (1 - R)/2W$ for a planar top surface, where R is the reflectance and W is the device thickness).

The third important optical feature arises from the localised heavily diffused regions in Figure 2. By keeping these areas small, their contribution to free carrier absorption in the semiconductor bulk is reduced. Even so, free carrier absorption in diffused regions accounts for about half the total free carrier absorption in the semiconductor.

There are also three key features, one common to the optical features that reduce non-radiative recombination in the device under operating voltages. One is the use of high carrier lifetime wafers such as prepared by the floatzone (FZ) process or magnetically confined Czochralski growth (MCZ) and, importantly, the selection of cell processing conditions to maintain (or enhance) these lifetimes during device fabrication.

The second non-radiative recombination control feature is the passivation of surface recombination by almost complete enshroudment of the device by a high quality thermal oxide and the selection of surface carrier concentrations to minimise recombination rates at these passivated surfaces. Silicon seems to prefer moderate n-type properties at surfaces for minimum recombination [6], as ensured by a moderate diffusion at the top surface and electrostatic induction of an n-type inversion layer at the rear by the low work function Al rear contact/reflector metal.

The third feature controls recombination at the metal contacts by restricting contacts to small areas as shown, and by having these contact areas heavily doped. This ensures low contact resistance while screening the contact from minority carriers.

The refinement of these features and the way they subtly reinforce one another is the reason for greatly improved LED performance, by one or two orders of magnitude compared to earlier silicon devices. Additional optimisation specifically for light emission should allow further improvement.

3. LEDs as Blackbodies

Solar cell researchers think of p-n junction light emission largely as an external property of a diode, due to the seminal work of Shockley and Queisser [7] on limiting solar cell conversion efficiency. Realising that this limit would occur when all recombination in the cell was radiative, these authors found a simple solution to the problem of keeping track of internal radiative recombination events, with the complications of photon recycling and total internal reflection of emitted photons that occurs. They merely restricted their attention to outside the device.

A solar cell must be a good absorber of light of energy above its bandgap. Hence, in thermal equilibrium with the ambient, say at 300K, it must be emitting close to blackbody radiation at this temperature, at least for energies above the bandgap. The source of the photons emitted at this energy is overwhelmingly band to band recombination. These are the photons that are not reabsorbed by photon recycling and not totally internally reflected.

If a voltage is applied to such a p-n junction, radiative recombination rates increase exponentially throughout the cell volume and hence externally, since these rates depend on the electron-hole product. The latter product depends exponentially on the difference between electron and hole quasi-fermi levels. For a good cell, carrier diffusion lengths and hence carrier mobilities must be large, ensured by large carrier mobilities. Since quasi-Fermi level gradients approach zero as the mobility approaches infinity, quasi-Fermi levels must be approximately constant throughout a good device and, in the absence of parasitic resistance, equal to the applied voltage. This argument has been given a solid basis by work that shows this is the condition required for full photogenerated carrier collection [8].

The result of this is that under bias, a good solar cell emits blackbody radiation at the cell's temperature, for photon energies above its bandgap, exponentially enhanced by the applied voltage. In the following, it will be shown that good quality silicon p-n junctions do almost exactly this.

4. Silicon Light Emission

The light emission spectrum of the device of Figure 2 (uppermost curve) is compared with that of two other devices in Figure 3 [9]. The lowermost curve is that of a high quality silicon space cell (also shown magnified 100X) used as a baseline in this work. Its performance, with 0.004% peak quantum efficiency at room temperature, is representative of good results previously reported for silicon. The next lowest curve marked "planar" (also shown magnified 10X) is for a device as in Figure 2, but with a planar top surface, rather than with inverted pyramids. Its emission is about 10 times stronger than the baseline device, due primarily to the suppression of non-radiative recombination by the three techniques previously described. Texturing the top surface (uppermost curve) brings the "light trapping" features previously described into play, increasing output by a further factor of 10.

The textured device has a "wall plug" electricity to light conversion efficiency close to 1% at room temperature increasing slightly to a peak at 200K, before decreasing at lower temperatures as discussed elsewhere [9]. This is well within the range of commercial III-V devices and not far below the best readily available. For example, we have compared the power efficiency of the silicon LED with a commercial "high efficiency" GaAlAs LED (TO39 from Opto-Diode Co., USA) with a peak emission wavelength of 880 nm and a peak efficiency of 11%. The silicon devices outperform it at power densities below 1 mW/cm^2.

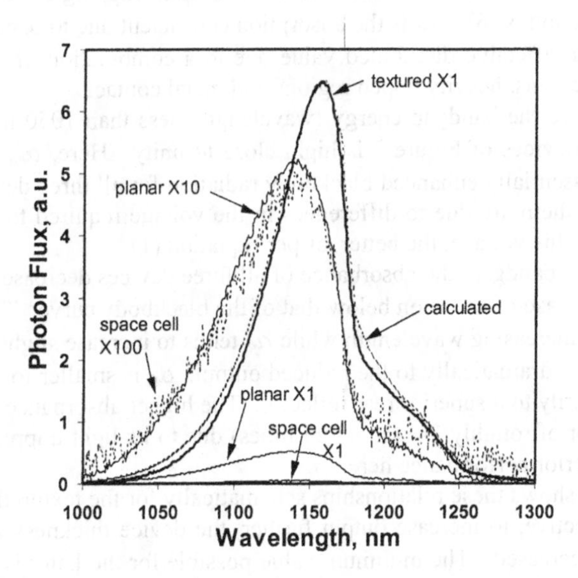

Figure 3: Photon flux versus wavelength emitted by three silicon diodes at 32.5 mA/cm^2 current bias. The lowest curve (also shown multiplied 100 times) is for a high performance silicon space cell, representative of the best results previously reported. Planar PERL cells perform 10 times better, due to reduction of parasitic non-radiative recombination. Textured PERL cells (Figure 2) perform 10 times higher again due to better optical properties, particularly much higher absorbance and hence emittance at the wavelengths shown.

5. Detailed Output Spectrum

The output of the device is compared to the prediction of Shockley/Queisser theory in Figure 4. At short wavelengths (high photon energies), the device does indeed emit enhanced blackbody radiation. However, the device emits considerable sub-bandgap radiation, rather than ceasing to emit at the band edge. The difference is that the Shockley/Queisser theory assumes an abrupt transition at the band edge, from strong absorption to zero absorption. This turns out to be consistent with these author's aims in calculating the minimum emission from a good absorber. In fact, band to band excitations are possible with sub-bandgap photons when phonon assistance is taken into account. Actual devices show sub-bandgap emission described by the equation [9]:

$$Emission = Blackbody \times exp\ (qV/kT) \times A \times \alpha_{bb}/(\alpha_{bb} + \alpha_{fc}) \qquad (1)$$

where (kT/q) is the thermal voltage and A is the device absorbance approximately given by [9,10]:

$$A \approx (1 - R_{ext})\ P\ (\alpha_{bb} + \alpha_{fc})/[1 + P\ (\alpha_{bb} + \alpha_{fc})] \qquad (2)$$

R_{ext} is reflection from the top surface while P is an effective pathlength, equal to $2W$ for a planar device and $4n^2W$ for a device with a good light trapping scheme, where n is silicon's refractive index [6]. α_{bb} is the absorption coefficient due to band to band processes and α_{fc} is an effective distributed value due to a combination of free carrier absorption in bulk regions, heavily doped regions and metal contacts.

Well above the bandgap energy (wavelengths less than 1050 nm) the absorbance of all three devices of Figure 3 is high, close to unity. Here, α_{bb} is much larger than α_{fc}, giving essentially enhanced blackbody radiation for all three devices. The differences between them are due to differences in the voltage required to pass the given current; the larger this voltage, the better, as per Equation (1).

Below the bandgap, the absorbance of all three devices decreases, which is one reason for the decrease of emission below that of the blackbody curve. The value of α_{bb} falls rapidly with increasing wavelength while α_{fc} tends to increase slightly contributing further and, finally, dramatically to the reduced output. α_{fc} is smaller for the two PERL devices due primarily to a superior rear reflector. The higher absorbance of the textured device, by a factor of roughly $2n^2$ (about 25 times) due to its light trapping scheme, accounts for its superior performance here.

Figure 4 shows these relationships schematically for the textured device. From the optical perspective, to increase output further, the device thickness W could be increased and α_{fc} decreased. The minimum value possible for the latter is determined by the carrier concentrations induced in the bulk of a lightly doped device at a given operating voltage. Noting the good performance demonstrated by present devices at low current levels, quite low values would be feasible. At such current levels, it would be feasible to reduce the amount of heavily doped material near the contacts as well. Increased rear reflectance is also possible if material of lower refractive index is placed under the rear metal, and/or a more reflective metal such as Ag, Au, or Cu is used.

These optical changes would need to be made without affecting the electronic properties. For example, increasing W will increase bulk recombination and would eventually lead to difficulties with maintaining constant quasi-Fermi levels throughout the cell volume, reducing performance. One way of overcoming the latter difficulty is to design devices for edge emission (Figure 5). In this case, W in the former equations is replaced by L, the lateral cell dimensions which can be very much larger. This design also decouples the contacting and light-emission surfaces, giving scope for designs with reduced surface recombination rates, critical for high efficiency with such edge emitters.

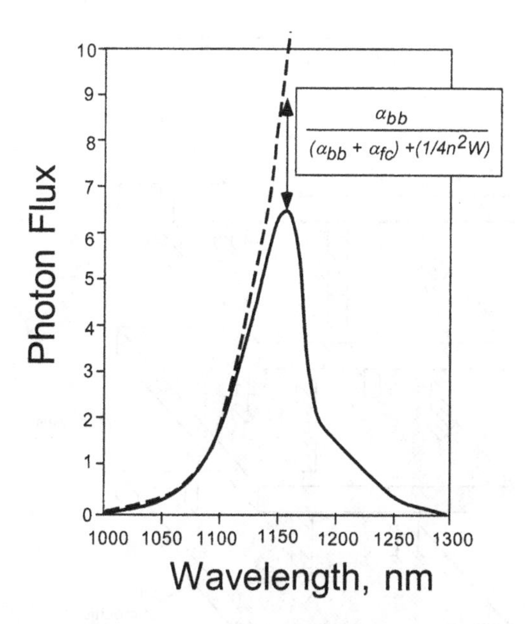

Figure 4: The device output compared to the blackbody emission (dashed) predicted by Shockley/Queisser theory for energies above the silicon bandgap (wavelengths shorter than 1102 nm). The ratio is given in the inset.

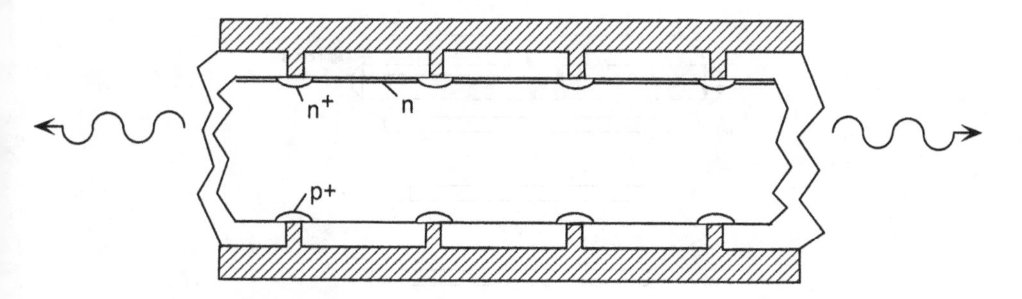

Figure 5: One way of overcoming the latter difficulty is to design devices for edge emission.

6. Applications

One application of a high performance silicon light emitter might be for intra- and inter-chip communication, with the latter indicated in Figure 6. High emission efficiency in specific directions at low power levels and high modulation frequencies would be essential. Directly modulating the device drive current is too slow a modulation approach, even for cutting-edge III-V devices [11], with the time scaling factor making this a hopeless strategy for the present high efficiency silicon emitters at frequencies above kilohertz. An innovative strategy is subsequently discussed, with present attention focussing on output directionality.

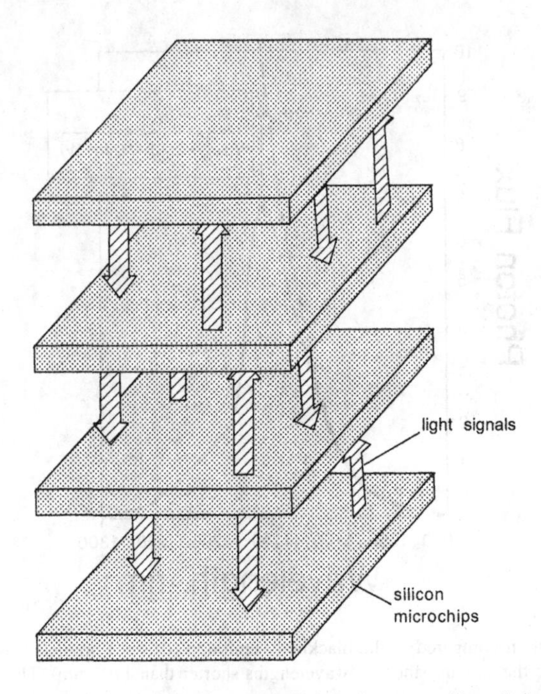

Figure 6: Interchip communication using high performance silicon light emitters.

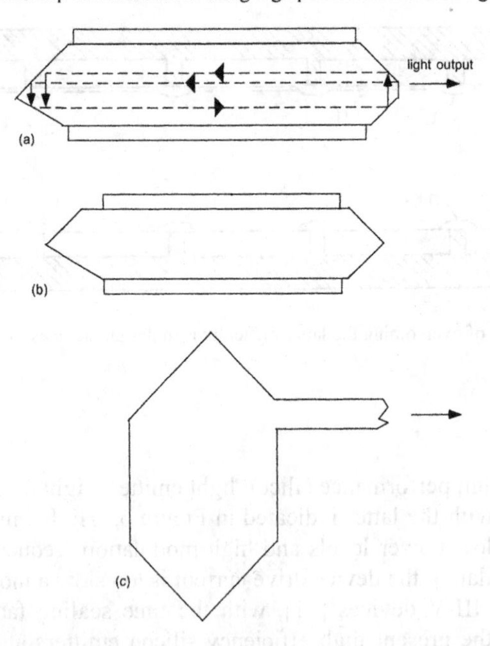

Figure 7: Geometrically based schemes for achieving high emission directionality: (A) Edge bevel used to increase light output in edge emitting device; (b) in surface emitting device; (c) plan view of scheme where diode geometry is used in a similar way to enhance power output in edge emitting device.

The previous theory shows that optical output in specific directions can be greatly enhanced by reducing output in unwanted directions. In principle, the effective optical path length, P, can be increased to values as high as $4n^2W/\sin^2\theta$ where θ is the angle over which the enhanced output is required. This can greatly reduce the device dimensions required for high output at a given applied voltage. Working from the previous theory, the design of device structures with high emission in specific directions reverts to the simpler problem of designing for high absorbance from these directions. Some simple geometrical structures have highly directional absorbance, such as for the devices of Figure 7.

At least some experienced commentators believe that microelectronics will make a full transition to Silicon on Insulator (SOI) technology over the coming decade in order to meet increasingly stringent criteria required to maintain the historically impressive rate of evolution of this technology [12,13]. This may be fortuitous as this also is the approach most conducive to the integration of silicon LEDs (Figure 8). Either device sized to the wavelength in silicon or rib structures would be appropriate. The length of these devices would depend on the directionality in emission that was required for the given application. Note that waveguides could be active with this approach.

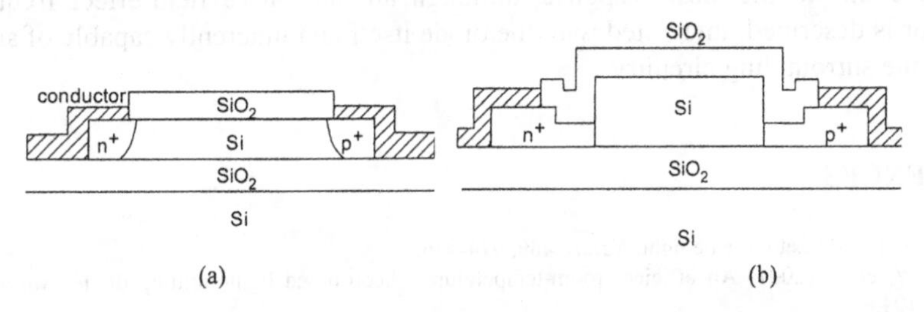

Figure 8: Side view of possible geometries for integration of silicon light emitting diodes onto microchips: (a) slab; (b) rib.

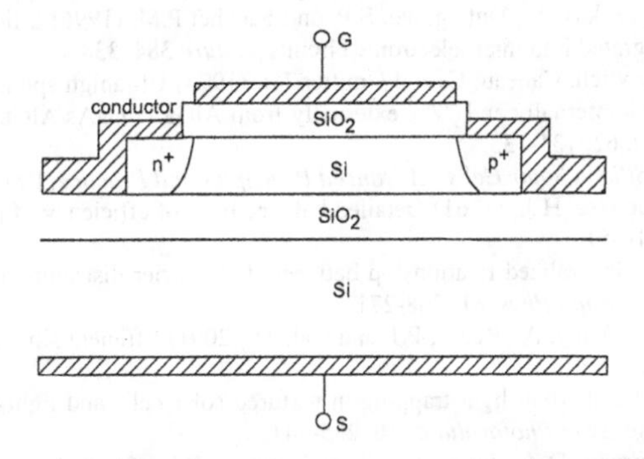

Figure 9: Super integrated light emitting diode and field-effect wavelength modulator [14].

External modulators are considered necessary even for the highest speed III-V emitters [11]. The Stark effect modulators favoured in this case could also be applied to silicon, particularly in conjunction with thin SOI layers as suggested in Figure 8(a). However, the weaker optical processes would mean additional real estate devoted to this function. An elegant solution [14] is to integrate both light generation and modulation functions as suggested in Figure 9. Similarly to fully depleted CMOS, voltages applied to the gate modulate the electric field strength across the silicon layer. This in turn modifies silicon's absorption coefficient, such as by the Stark effect in devices sufficiently thin that quantum confinement effects are strong or by the Franz-Keldysh effect in thicker devices [15]. In this way, the frequency of light emission can be modulated by the gate voltage. Given the geometrical similarity to a CMOS circuit element, modulation speeds automatically should be compatible with sub-micron CMOS circuitry.

7. Conclusions

By careful attention to the reduction of parasitic non-radiative recombination and optical absorption losses, greatly improved silicon LED efficiency has been demonstrated. Although already within the range of III-V performance and superior to even the best III-V emitters at low power densities, there is scope for further improvement by continuing to reduce these parasitics. SOI technology seems best suited for the development of integrated circuit versions of these emitters. Drive current modulation is unlikely to be feasible due to the slow response, although an innovative field-effect frequency modulator is described, integrated with the diode itself and inherently capable of similar speed to the surrounding circuitry.

REFERENCES

[1] Bell, P. (2001) Let there be light, *Nature* **409**, 974-976.

[2] Ng, W. et al. (2001) An efficient room-temperature silicon-based light-emitting diode, *Nature* **410**, 192-194.

[3] Gelloz, B. and Koshida N. (2000) Electroluminescence with high and stable quantum efficiency and low threshold voltage from anodically oxidized thin porous silicon diode, *J. Appl. Phys.* **88**, 4319.

[4] Hirschman K.D., Tsybeskov, L., Duttagupta, S.P. and Fauchet P.M. (1996) Silicon-based visible light-emitting devices integrated into microelectronic circuits, *Nature* **384**, 338.

[5] Schnitzer, I., Yablonovitch, Caneau, C. and Gmitter,T.J. (1993) Ultrahigh spontaneous emission quantum dfficiency, 99.7% internally and 72% externally from AlGaAs/GaAs/AlGaAs double heterostructures, *Appl. Phys. Lett.* **62**, 131-133.

[6] Green, M.A. (1995) *Silicon Solar Cells: Advanced Principles and Practice*, Bridge Printery, Sydney.

[7] Shockley, W. and Queisser, H.J., (1961) Detailed balance limit of efficiency of p-n junction solar cells, *J. Appl. Phys.* **332**, 510-519.

[8] Green, M.A., (1997) Generalized relationship between dark carrier distribution and photo-carrier collection in solar cells, *J. Appl. Phys.* **81**, 268-271.

[9] Green, M.A., Zhao, J., Wang, A., Reece, P.J. and Gal, M. (2001) Efficient silicon light emitting diodes, *Nature* **412**, 805-808.

[10] Green, M.A. (2002) Lambertian light trapping in textured solar cells and light-emitting Diodes: analytical solutions, *Progress in Photovoltaics* **10**, 235-241.

[11] Thomas, G.A., Ackerman, D.A., Prucnal, P.R. and cooper, S.L. (2000) Physics in the whirlwind of optical communications, *Physics Today*, 300-365.

[12] Sze, S.M. (2001), Four decades of developments in microelectronics: achievements and challenges", 2001 Advanced Research Workshop on Future Trends in Microelectronics, Ile de Bendor, France.

[13] Soloman, P.M. (2001) "Strategies at the end of CMOS scaling", 2001 Advanced Research Workshop on Future Trends in Microelectronics, Ile de Bendor, France.

[14] Green, M.A. and Zhao, J. (2000) Silicon light emitters", Provisional Patent specification, initial filing.

[15] Vasko, F.T. and A.V. Kuznetsov, (1999) *Electronic States and Optical Transitions in Semiconductor Heterostructures*, Springer-Verlag, New York.

DISLOCATION-BASED SILICON LIGHT EMITTING DEVICES

M. A. LOURENÇO[1], M. S. A. SIDDIQUI[1], G. SHAO[2], R. M. GWILLIAM[1] and K. P. HOMEWOOD[1]
[1]School of Electronics and Physical Sciences, [2]School of Engineering, University of Surrey, Guildford, Surrey, GU2 7XH, United Kingdom

1. Introduction

Silicon is by far the most commonly used and preferred semiconductor for the electronics industry, not least because of the much lower cost that results from, for example, cheaper substrates and the savings involved by the use of ultra large scale integration (ULSI) technology. Silicon's main disadvantage is its inability to act as an efficient light emitter due to the indirect nature of the band gap.

The niche of light emitters in the infra red and visible has been taken by devices largely based on the use of III-V and II-VI direct gap semiconductors. Such materials are by comparison with silicon technology very costly and difficult to produce and are also associated with more toxic and hazardous processing. Also longer term there are concerning issues about the sustainability of such technologies given the limited known resources of some of the key elements for example indium [1]. There are therefore clearly very significant advantages in to moving to all silicon electronics and optoelectronics. Indeed in the microelectronics sector it is now realised that optical data transfer on chip or at the least optical data transfer for critical parts of the interconnection paths are required for the continuing improvement of microprocessor and related chips by around 2010 [2]. The provision of this functionality by hybrid technologies is both expensive and may well be technologically impossible at the specification that ULSI technology demands. Integrable silicon based light source that is truly ULSI compatible would be invaluable.

Another large application for a silicon optical emitter is the provision of transceiver modules for domestic use to enable cheap high bandwidth connectivity to optical fibre based networks.

A number of methods have been attempted to obtain light emission from silicon [3-10] but in general have not been either efficient enough and not ULSI compatible. A major problem in the systems based on crystalline silicon is obtaining good high temperature performance – several approaches have looked interesting at low (liquid nitrogen) temperatures but strong thermal quenching due to the competing non radiative recombination routes has prevented the required high temperature operation.

Here we describe a new approach, dislocation engineering [11], and its application to light emitters in silicon. The new technology is based entirely on ULSI technology

L. Pavesi et al. (eds.), Towards the First Silicon Laser, 11–20.

and solves the key problem of thermal quenching providing efficient light emission at room temperature.

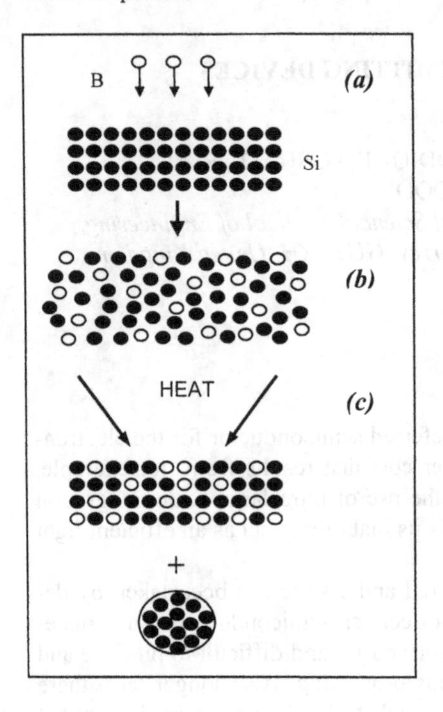

Figure 1. A schematic showing the formation of dislocation loops by ion implantation: *(a)* the implant process, *(b)* the disordered lattice after implantation and *(c)* the annealing step showing activation of the implant and formation of the loop from the excess interstitials.

2. Theory

The implantation of boron is made in to a device grade n-type silicon wafer to form the p-type top layer of the diode and to form dislocation loops. After the initial implant the silicon is disordered. A subsequent annealing step is undertaken to electrically activate the implanted boron dopants. The activation involves placing the boron atom on to a silicon lattice site. The consequence of this is that the silicon atom that was on that lattice site is now present as an interstitial atom. Within the implanted region there is now present an excess population of interstitial silicon atoms equal to the original implant dose of the boron. If a suitable anneal schedule is applied then the silicon atoms can diffuse, aggregate and form into small crystalline planes of silicon atoms, one atom thick, known as dislocation loops. The size and distribution of these loops is a function of the implant and anneal parameters and so can be engineered. If we wish to form dislocation loops independently of doping the material, we can use silicon to implant silicon. The basic steps of dislocation loop formation are shown in Figure 1.

A dislocation loop is just an island, a single atom thick, of bulk crystalline silicon whose rim is formed of a simple single edge dislocation turned around on its self. Figure 2 is a schematic diagram showing how the dislocation loop leads to modification of the band gap energies.

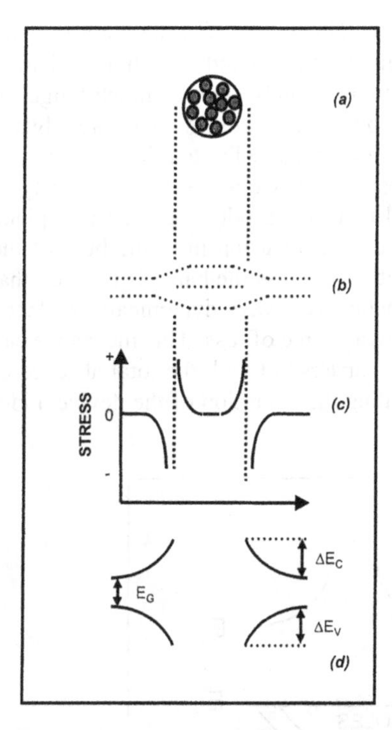

Figure 2. A schematic diagram showing how the dislocation loop leads to modification of the band gap energies: *(a)* looking down on a loop; *(b)* a loop inserted into the silicon lattice - just outside the loop the lattice silicon atoms are forced further apart than the bulk value; *(c)* the stress distribution across the loop; *(d)* illustration of the band gap modification.

Inspection of figure 2(b), showing a dislocation loop inserted into the silicon lattice, reveals that just outside the loop the silicon interatomic spacing is increased and just inside it is decreased. So, just outside the loop the silicon lattice is placed under negative hydrostatic pressure and just inside it is placed under positive hydrostatic pressure. The stress distribution across a diameter of the loop is shown schematically in Figure 2(c). The magnitude and form of the stress field can be calculated using standard elastic theory of dislocations [12] which shows that the stress field decays inversely with distance and reaches a maximum stress of 25 to 50 GPa. This uses the known values of Poisson's ratio and Young's modulus for silicon, 0.42 and 113 GPa, respectively. The importance of this stress field for our applications is that the band gaps of semiconductors are pressure dependent and, in the case of silicon, the band gap decreases with increasing pressure and increases under negative pressure. A schematic of the dependence of the conduction and valence band energies is shown in Figure 2(d). The total change in the band gap, $\Delta E_G = \Delta E_C + \Delta E_V$, just outside the loop, can be easily calculated using the known pressure coefficient of the band gap for silicon. However, the distribution of this band gap change between the conduction and valence bands is still uncertain. The total band gap change is from 0.325 to 0.75 eV, a significant fraction of the band gap itself which is 1.1 eV. The key to our device operation is the utilisation of this band gap change to modify the carrier flow after injection. In a conventional silicon diode, after injection the carriers diffuse and eventually recombine. The average linear distance before recombination occurs is just the diffusion length and in silicon this can be tens of microns although often the surface is reached first and recombination occurs there. The long diffusion lengths in silicon are the result of the low density of recombination centres in silicon and the fact that the material is indirect gap. This means that the band to band recombination is forbidden or in practice the radiative band to band transition is characterised by a long lifetime in the range of tens of microseconds, or even much longer in very pure material. The faster non-radiative recombination routes therefore always dominate. However, if the carrier diffusion can be prevented or limited then the carriers will eventually recombine radiatively across the band. Here we make use of an array of dislocation loops, or rather the blocking potential that they produce, placed in a plane just beyond the depletion region edge to prevent carrier diffusion in to the bulk of the device. The injected carriers then recombine radiatively across the band. It is noted that any injected carrier diffusing back across the junction is always dynamically replaced by a carrier moving in the other direction and is not a source of loss. This means we can have tight effective spatial localisation of injected carriers but with the total absence of any quantum confinement. A band diagram indicating the operation of the device under zero and forward bias is shown in Figure 3.

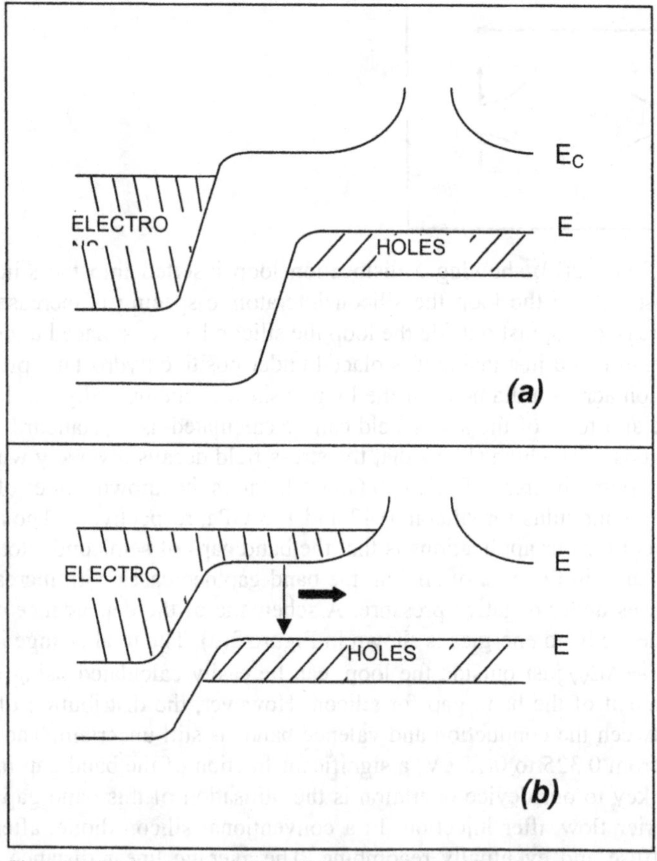

Figure 3. A schematic of the band diagram of the dislocation engineered diode. The potential change is due to an array of dislocation loops formed parallel to the depletion region edge at that point. *(a)* The device at zero bias and *(b)* under forward bias.

3. Experimental Details

To demonstrate how the dislocation engineering method can be used to fabricate efficient silicon light emitting devices operating at room temperature, and also to show how this method can be employed to reduce the luminescence thermal quenching in other systems, several types of devices were fabricated. Initially, boron was implanted into [100] n-type silicon substrates, followed by a high temperature anneal, at several temperatures and times, in nitrogen ambient. The boron implantation served as a means of introducing the dislocation loops as well as the p-type dopant to form a p-n junction. The implants were performed in the range 20 – 70 keV to a constant peak dopant concentration of 10^{20} cm^{-3}. The dislocation loops were formed by annealing between 900 and 1100 °C, from 5 seconds to 60 minutes.

To evaluate the influence of the controlled introduction of dislocation loops on the luminescence thermal quenching for other material systems, dislocation engineered β-$FeSi_2$ and Er light emitting devices were fabricated.

Ion beam synthesised β-$FeSi_2$ light emitting diodes (LEDs) were fabricated by a conventional procedure, as described in Ref. [13], and by the dislocation engineered method. Initially, Fe was implanted (1.5×10^{16} to 1.5×10^{17} cm^{-2} at 180 keV) into epitaxial chemical vapour deposition grown silicon p-n^+ junctions, in the recombination region adjacent to the p side of the depletion region. Following the Fe implantation, all samples were annealed at 900 °C for 18 hours in nitrogen ambient [13]. Subsequently, the processed epilayers were cut into two pieces. The first one (conventional devices) was not processed any further; the second one (dislocation engineered devices) was further implanted with 30 keV boron at a dose of 10^{15} cm^{-2}, and then annealed at 900 °C for 1 minute.

Two different sets of erbium/silicon devices were fabricated. For the first one Er ions were implanted into a standard [100] n-type silicon substrate (5×10^{12} to 2×10^{14} Er/cm^2 at 0.4 MeV). The second batch was fabricated by Er implantation (same doses and energies as the previous batch) into wafers previously implanted with boron (10^{15} B cm^{-2} at 30 keV and annealed at 950 °C for 20 minutes). The Er implants were followed by a rapid thermal anneal at 950 °C for 1 minute in nitrogen ambient to activate the Er, for both batches. Ohmic contacts, when required, were formed by vacuum evaporation of Al and AuSb eutectic on the p-type region and n-type substrate, respectively and sintered for 360 °C for 2 minutes. Subsequently, the samples were mesa etched to isolate the p-n junction; the device area was 8×10^{-3} cm^{-2}.

The devices were characterised by electroluminescence (EL) and photoluminescence (PL) measurements. The EL and PL experiments were performed in the temperature range 80 – 300 K. In both cases the samples and devices were mounted in a continuous-flow liquid nitrogen cryostat placed in front of a conventional half-metre spectrometer. A liquid nitrogen cooled germanium p-i-n diode was used for detection of the EL and PL. The PL experiments were performed at 150 mW excitation laser power using a wavelength of 514 nm. The EL measurements were performed under forward bias extracting the light through a window at the back of the samples. Typical current densities were of order of 15 A/cm^2.

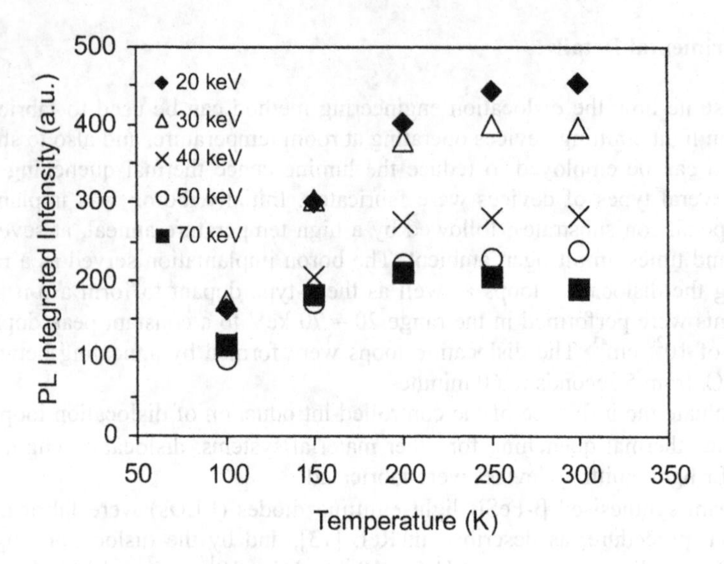

Figure 4. Photoluminescence integrated intensity as a function of measurement temperature for devices implanted with boron at several energies.

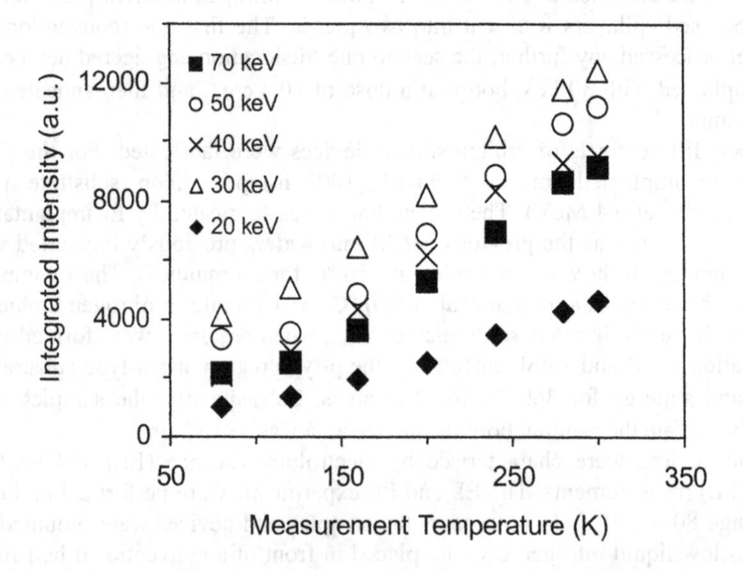

Figure 5. Electroluminescence integrated intensity as a function of measurement temperature for devices implanted with boron at several energies.

4. Results and Discussion

The samples implanted with boron only showed similar PL and EL emissions, peaking, at room temperature, at ~1154 nm. The post-implant conditions were critical to determine device performance; the maximum integrated luminescence intensity was obtained from devices annealed at 975 °C for 5 seconds [14]. For crystalline silicon, annealing at 600 °C restores the crystallinity from the damage induced by ion implantation. However, annealing between 700 °C and 1100 °C can induce the formation of defects such as dislocation loops. Figures 4 and 5 shows the PL and EL integrated intensity plotted against measurement temperature for samples implanted in the range 20 to 70 keV and annealed at 950 °C for 20 minutes. No temperature quenching is observed; indeed, the integrated intensity increases with increasing temperature. The PL integrated intensity increases with decreasing implant energy. At higher energies the dislocation loops are formed further from the surface so a greater volume of non-spatially confined silicon is available for absorption of photons before reaching the loops. The EL emission did not show any dependence upon the energy range and doses studied here.

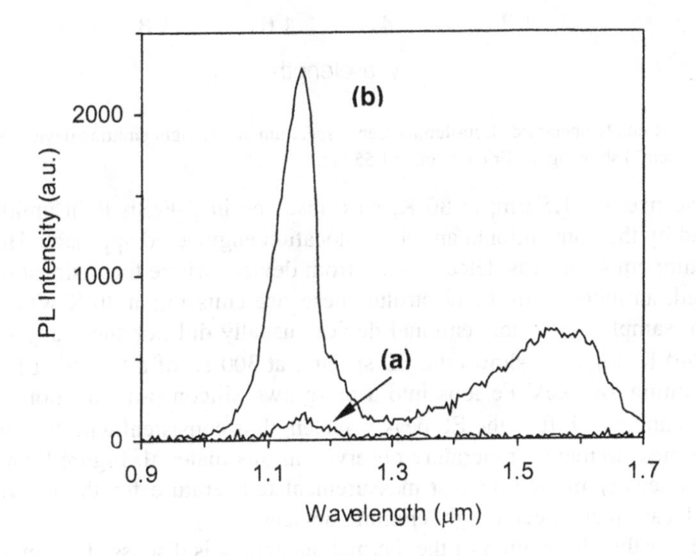

Figure 6. Room temperature photoluminescence spectra of Er samples: structures fabricated by *(a)* implanting only Er ions into a silicon substrate (2×10^{13} Er cm^{-2}) and *(b)* initial B implantation and high temperature anneal to form the dislocation loops followed by the Er implant (2×10^{13} Er cm^{-2}).

The room temperature PL spectrum of the Er implanted structures is shown in Figure 6. Spectra obtained from samples implanted with Er only show only a weak Si band edge emission; while for samples incorporating the dislocation loops a broad peak centred at ~1.5 μm can be observed, in addition to a strong Si band edge emission. This long wavelength peak is related to the Er centre. Indeed, room temperature EL spectra obtained from light emitting devices fabricated from substrate implanted with B and Er showed, in addition to the Si band edge, a clear peak at 1.55 μm, corresponding to an Er internal transition, as shown in Figure 7.

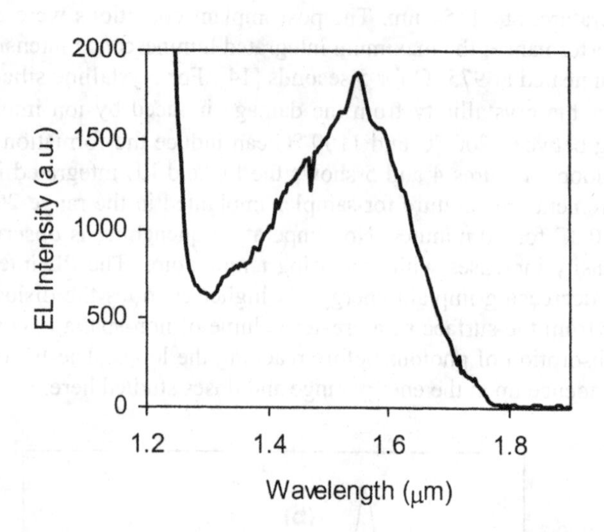

Figure 7. Room temperature electroluminescence spectrum of a Er light emitting device (5×10^{13} Er cm^{-2}) showing the Er emission at 1.55 μm.

Electroluminescence at ~1.5 μm, at 80 K, was observed in β-$FeSi_2$ light emitting devices fabricated by the conventional and the dislocation engineered approach. However, room temperature emission was detected only from devices where the dislocation loops were introduced; an increase in the electroluminescence emission at 80 K was also observed in these samples. The conventional devices usually did not show any luminescence above 240 K. Figure 8 shows the EL spectra, at 300 K, of a β-$FeSi_2$ LED fabricated by implanting 180 keV Fe ions into a pre-grown silicon p-n^+ junction. The EL peaks at 1.6 μm and the shift of the EL peak wavelength is consistent with the change in band gap with measurement temperature observed in this material. Figure 9 shows the EL integrated intensity plotted against measurement temperature for the conventional and for the dislocation engineered light emitting devices.

The model for the elimination of the thermal quenching is discussed in detail in reference [11]. In summary, the dislocation loop edge distorts the silicon lattice by applying a negative hydrostatic pressure to the adjacent silicon lattice just outside the loop. The negative pressure dependence of the band gap of silicon leads to an increase of band gap of around 0.35 to 0.7 eV. A closely packed array of dislocation loops placed in a plane parallel and just beyond the depletion edge on the p side forms a potential blocking barrier for carrier diffusion into the bulk and to the silicon surface. Non radiative defect centres in the bulk and at the surface are normally the dominant recombination route and are responsible for the thermal quenching of the band to band radiative route in silicon which is otherwise to first order temperature independent. The thermal quenching in iron disilicide in silicon and erbium doped silicon is also the result of the

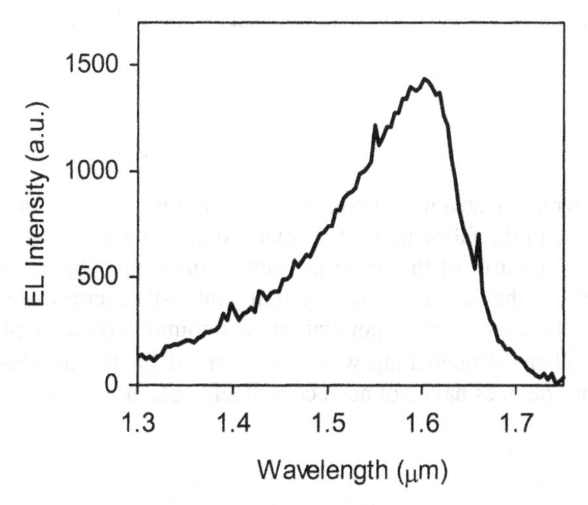

Figure 8. Room temperature electroluminescence spectrum of a β-$FeSi_2$ light emitting device (1.5×10^{17} Fe cm^{-2}) showing the β-$FeSi_2$ emission at ~ 1.6 μm.

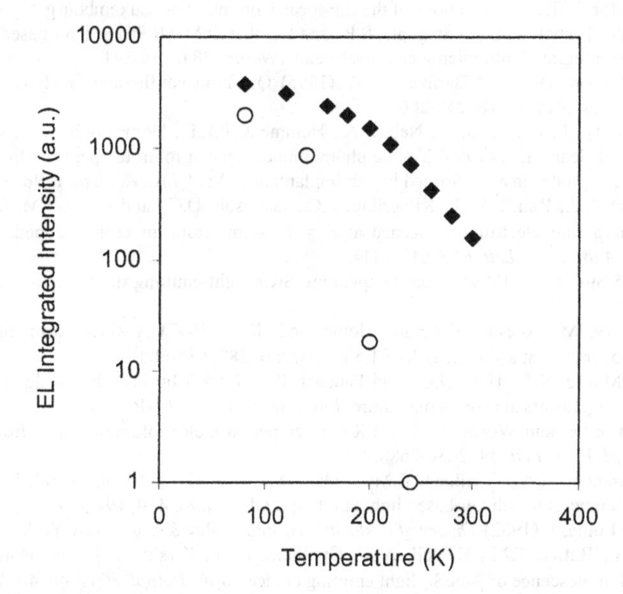

Figure 9. Electroluminescence integrated intensity as a function of measurement temperature for a conventional (°) and a dislocation engineered (♦) β-$FeSi_2$ light emitting devices.

same non-radiative recombination after back emission to the silicon edge from the iron disilicide and erbium levels respectively. The provision of the dislocation barrier between the radiative centres and the bulk and surface non-radiative regions enables recapture of the blocked carriers enhancing the radiative routes and reducing significantly the thermal quenching of the EL.

5. Conclusions

Dislocation engineering, which makes use of conventional ion implantation and thermal processing, is a viable route to the fabrication of efficient room temperature silicon light emitting devices and to the control of the luminescence thermal quenching. Indeed, in the case of the silicon LED's, the luminescence integrated intensity increased with temperature showing that the radiative carrier population was completely decoupled from the non-radiative routes. Thermal quenching was still observed for the β-$FeSi_2$ and Er structures – however, those devices have not been completely optimised yet.

References

[1] Makita, Y. (1997), *Proc. First NREL Conf.*, 3.

[2] European Commission (1998), *Technology Roadmap, Optoelectronic interconnects for integrated circuits*, Office for Official Publications of the European Communities, Luxembourg.

[3] Hirschman, K.D., Tysbekov, L., Duttagupta, S.P., and Fauchet, P.M. (1996) Silicon-based visible light-emitting devices integrated into microelectronic circuits, *Nature* **384**, 338-341.

[4] Lu, Z.H., Lockwood, D.J., and Baribeau, J.M. (1995) Quantum confinement and light emission in SiO_2/Si superlatices, *Nature* **378**, 258-260.

[5] Komoda, T., Kelly, J., Cristiano, F., Nejim, A., Hemment, P.L.F., Homewood, K.P., Gwilliam, R., Mynard, J.E., and Sealy, B.J. (1995) Visible photoluminescence at room temperature from microcrystalline silicon precipitates in SiO_2 formed by ion implantation, *Nucl. Inst. & Meth.* B **96**, 387-391.

[6] Zheng, B., Michel, J., Ren, F.Y.G., Kimerling, L.C., Jacobson, D.C., and Poate, J.M. (1994) Room-temperature sharp line electroluminescence at $\lambda = 1.54$ μm from an erbium-doped, silicon light-emitting diode, *Appl. Phys. Lett.* **64**, 2842-2844.

[7] Vescan, L., and Stoica, T. (1999) Room-temperature SiGe light-emitting diodes, *Journal of Luminescence* **80**, 485-489.

[8] Leong, D., Harry, M., Reeson, K.J., and Homewood, K.P. (1997) A silicon/iron disilicide light-emitting diode operating at a wavelength of 1.5 μm, *Nature* **387**, 686-688.

[9] Tybeskov, L., Moore, K.L., Hall, D.G., and Fauchet, P.M. (1996) Intrinsic band-edge photoluminescence from silicon clusters at room temperature, *Phys. Rev B* **54**, R8361-R8364.

[10] Sveinbjörnsson, E.O., and Weber, J. (1996) Room temperature electroluminescence from dislocation rich silicon, *Appl. Phys. Lett.* **69**, 2686-2688.

[11] Ng, W.L., Lourenço, M.A., Gwilliam, R.M., Ledain, S., Shao, G., and Homewood, K.P. (2001) An efficient room-temperature silicon-based light emitting diode, *Nature* **410**, 192-194.

[12] Hirth, J.P., and Lothe, J. (1982) *Theory of Dislocations*, John Wiley & Sons, New York.

[13] Lourenço, M.A., Butler, T.M., Kewell, A.K., Gwilliam, R.M., Kirkby, K.J., and Homewood, K.P. (2001) Electroluminescence of β-$FeSi_2$ light emitting devices, *Jpn. J. Appl. Phys.* **40**, 4041-4044.

[14] Lourenço, M.A., Ng, W.L., Shao, G., Gwilliam, R.M., and Homewood, K.P. (2002) Dislocation engineered silicon light emitting diodes, *Proc. SPIE* **4654**, 138-141.

EFFICIENT ELECTROLUMINESCENCE IN ALLOYED SILICON DIODES

O.B. GUSEV, M.S. BRESLER, I.N. YASSIEVICH,
B.P. ZAKHARCHENYA
A F Ioffe Physico-Technical Institute
Politekhnicheskaya 26, 194021 St. Petersburg, Russia

1. Introduction

Recently the interest to electroluminescence (EL) of silicon at room temperature has considerably increased. This is caused by the need to develop an optoelectronics compatible with conventional silicon technology, in particular, with ultra-large-scale integration (ULSI) technology. The main progress was achieved in three fields of studies: application of silicon-based quantum structures [1,2,3], doping of silicon with rare earths (mostly, erbium) [4,5] and renewed attempts to increase the efficiency of band edge silicon luminescence [6,7]. The latter presents probably the simplest and most straightforward approach to silicon optoelectronics, needing however considerable efforts to increase internal and external quantum efficiencies. We note that high external quantum efficiency is not crucial if the light-emitting element is integrated into a chip.

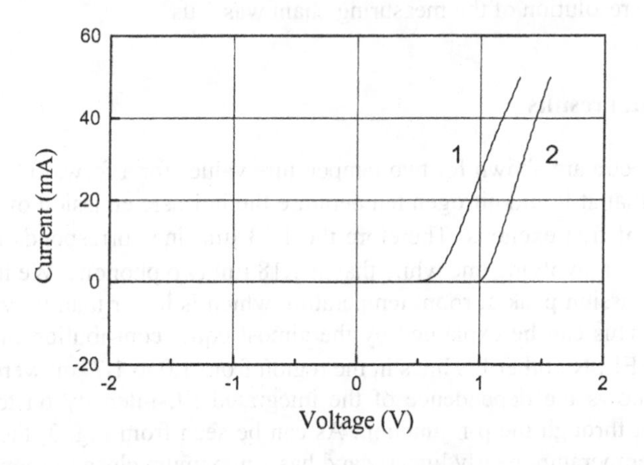

Figure 1. Current - voltage characteristics at T = 300 K (1) and T = 77 K (2).

A significant increase of light-emitting diode (LED) emission intensity at the silicon band edge emission was obtained in [6,7] where optimization of light extraction

L. Pavesi et al. (eds.), Towards the First Silicon Laser, 21–28.

efficiency [6] or on internal quantum efficiency of bare p-n junction [7] were done. In Ref. 7 it is claimed that the dislocation loops generated by boron implantation lead to confinement of minority carriers close to the p-n junction which in turn generates an increase of electroluminescence (EL) internal quantum efficiency at room temperature. Unfortunately, no quantitative estimates supporting the mechanism proposed were presented.

In the present paper we have studied the band edge EL for a commercial alloyed silicon diode operating in forward bias. Our experimental results confirm those of [7]. We have developed a simple model of recombination processes in the forward-biased p-n junction, which permitted to explain most of the experimental data, including the temperature enhancement of the EL intensity from liquid nitrogen to room temperature, observed both on our structure and that studied in [7]. We have determined also the optimal parameters of a p-n junction necessary to achieve the maximum quantum efficiency and highest modulation frequency.

2. Experimental procedure

We have used a commercial alloyed silicon diode with its metal package partly removed to permit registration of the emission. I-V characteristics of the LED at liquid nitrogen and room temperatures are shown in Fig. 1.
The EL was observed when square shaped current pulses up to 100 mA with a 1:1 duty cycle were passed through the diode. The emission was collected from the edge of the diode and analyzed using of a grating spectrometer with a focal length of 822 mm interfaced with a liquid nitrogen-cooled germanium detector. The time constant of EL decay after the switch-off of the current through the diode was measured with a digital oscilloscope. The time resolution of the measuring chain was 3 μs.

3. Experimental results

In Fig. 2 EL spectra are shown for two temperature values for a forward biased diode. It is well known that at liquid nitrogen temperature the intrinsic emission of silicon is due to annihilation of free excitons. Therefore the 1.13 μm line corresponds to an indirect radiative transition involving one while that at 1.18 μm two phonons. We have observed a shift of the emission peak at room temperature which is lower than the silicon energy gap reduction. This can be explained by the almost equal contribution of free charges and excitons to EL. No other EL lines in the region from 1.0 to 1.7 μm were observed.

Figure 3 shows the dependence of the integrated EL intensity on temperature at constant current through the p-n junction. As can be seen from Fig. 3, the EL intensity rises with the temperature nearly linearly and has a maximum close to room temperature similarly to results of Ref. 7. In Fig. 4 we present the EL-I characteristics for liquid nitrogen and room temperatures. It is close to linear for both values of the temperature. The decay time of the EL intensity is about 20 μs at room temperature and less than 3 μs at liquid nitrogen temperature. These results are similar to those reported in Ref. 7. Also the doping levels are nearly the same for this and the LED produced in [7].

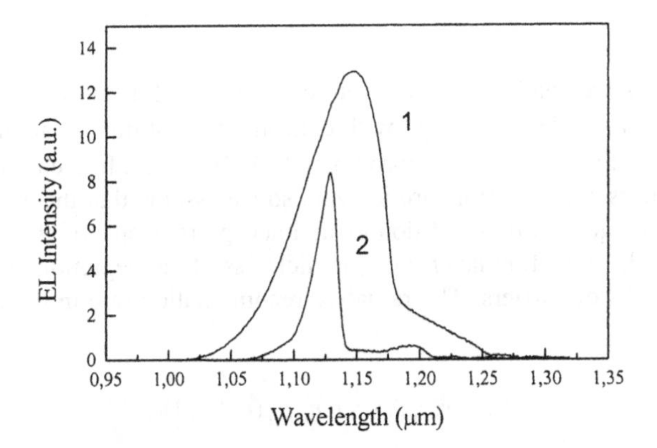

Figure 2. EL spectra of silicon diode structure at T = 300 K (1) and T = 77 K (2) and constant current through the structure of 20 mA.

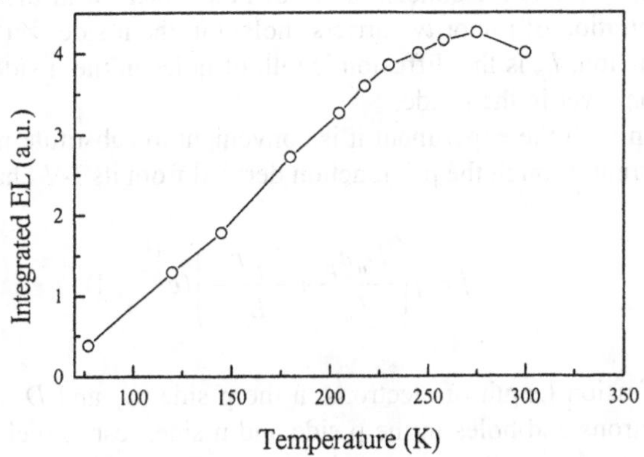

Figure 3. Integrated EL intensity as a function of temperature. Points represent experimental data; the solid line is a guide for eye. The current through the structure is 20 mA.

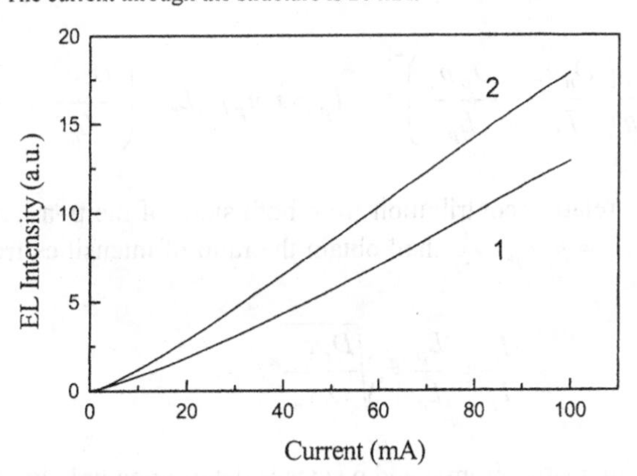

Figure 4. EL intensity of the diode structure as a function of current through the p–n junction. T = 100 K (1), T = 300 K (2).

4. Discussion

To treat our results we shall use the conventional theory of a thin abrupt p-n junction. The substrate of our diode was n-type with concentration of donors about 1×10^{15} cm^{-3}. The concentration of acceptors was approximately 1×10^{19} cm^{-3}. In the temperature range of interest all donors and acceptors are ionized, so we assume that the electron and hole concentrations are equal to those of donors and acceptors, respectively. We assume no recombination in the depletion layer, i.e. the thickness of the depletion layer is less than the diffusion lengths of carriers. The radiative recombination rate in the n side is given by the expression:

$$R_n = r_r n_n \Delta p = r_r n_n p_n (e^{\frac{qV}{kT}} - 1)e^{-\frac{x-l_n}{L_p}} \quad , \qquad (1)$$

where r_r is the radiation recombination coefficient, n_n is the concentration of the majority carriers (electrons), Δp is the concentration of holes injected in the n side, p_n is the equilibrium concentration of minority carriers (holes) in the n side, V is the voltage applied to the p-n junction, L_p is the diffusion length of holes in the n side, l_n is the thickness of the depletion layer in the n side.

For comparison with the experiment it is convenient to substitute in Eq. (1) the expression for the current through the p-n junction derived from its I-V characteristic

$$j = q\left(\frac{D_n n_p}{L_n} + \frac{D_p p_n}{L_p}\right)(e^{\frac{qV}{kT}} - 1) \quad , \qquad (2)$$

where L_n is the diffusion length of electrons in the p side, D_n and D_p are the diffusion coefficients of electrons and holes in the p side and n side, respectively; n_p is the equilibrium concentration of minority carriers (electrons) in the p side.

Integrating Eq. (1) over x, we obtain the density of the photon flux from the n side or the p side emitted in the direction perpendicular to the plane of the p-n junction

$$I_n = r_r n_n p_n L_p \frac{j}{q}\left(\frac{D_n n_p}{L_n} + \frac{D_p p_n}{L_p}\right)^{-1} \qquad I_p = r_r n_p p_p L_n \frac{j}{q}\left(\frac{D_n n_p}{L_n} + \frac{D_p p_n}{L_p}\right)^{-1} \qquad (3)$$

Let us compare the relative contribution from both sides of the p–n junction into luminescence. Since $n_n p_n = p_p n_p$, we shall obtain the ratio of intensities from both sides in the form:

$$\frac{I_n}{I_p} = \frac{L_p}{L_n} = \sqrt{\frac{D_p \tau_p}{D_n \tau_n}} \quad , \qquad (4)$$

i.e. the ratio of EL intensities from n and p sides is determined only by the ratio of diffusion lengths of minority carriers injected in these sides. In Fig. 5, typical literature data for the minority carrier lifetimes are shown [8]. For our structure the lifetime of holes in the n side is 2×10^{-5} s (the concentration of donors is 10^{15} cm^{-3}), while for electrons in the p side 10^{-7} s (the concentration of acceptors is 10^{19} cm^{-3}). To estimate the diffusion coefficients we used the data for mobilities taken from [9]. For the doping levels involved, the diffusion coefficient of holes is approximately five times higher than that of the electrons. Therefore, the diffusion length of holes L_p is several orders of magnitude larger than that of electrons L_n. Consequently, EL is mainly due to recombination in the n-side of the junction. Taking these consdiertions and the ratio of the minority carriers concentrations, $p_n/n_p \sim 10^4$, we obtain from Eq. (3):

$$I_n = r_r n_n p_n L_p \frac{j}{q}\left(\frac{D_n n_p}{L_n} + \frac{D_p p_n}{L_p}\right)^{-1} \approx r_r n_n \tau_p \frac{j}{q} \tag{5}$$

The EL intensity is linear in the current density, which is in agreement with the data of Fig. 4. According to Eq. (5) the temperature dependence of the EL integrated intensity is determined by the temperature dependence of the hole lifetime in the n side and of the radiative recombination coefficient r_r.

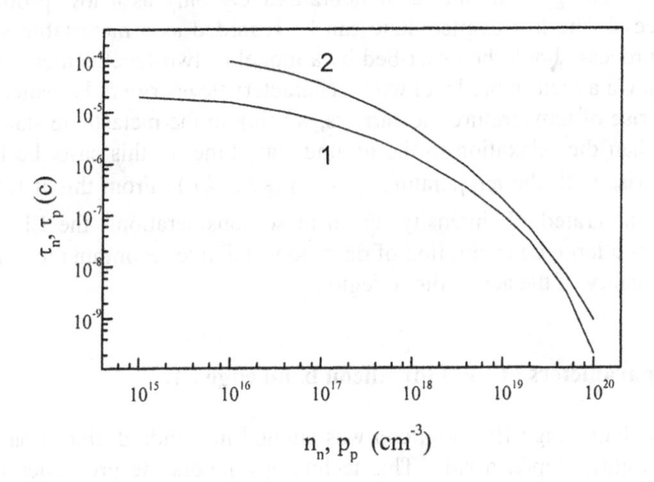

Figure 5. Lifetime of minority carriers as a function of the majority carrier concentration τ_p (1), τ_n (2). T = 300 K.

In Ref. 10 it was shown that for indirect band-to-band and exciton recombinations the radiative recombination coefficient decreases with temperature:

$$r_r \propto \sqrt{\frac{E_x}{kT}}\left[1+2\frac{E_x}{kT}\exp\left(\frac{E_x}{kT}\right)\right]\coth\left(\frac{\theta}{2T}\right), \tag{6}$$

where E_x is the exciton binding energy, θ is the Debye temperature for the transverse optical phonon. Due to a high value of θ the last factor in Eq. (6) is temperature independent and close to unit. The factor in square brackets refers to ionized (free electron and hole) and free excitons recombinations. It follows from Eq. (6) that the temperature dependence of r_r is determined primarily by thermal dissociation of excitons which decreases the luminescence intensity at high temperatures. On the contrary, EL data demonstrate an intensity increase together with a longer decay time. To explain these results we should assume that the total carrier lifetime increases with temperature faster than the r_r reductions. We suggest that in the n side deep recombination centers are present whose concentration is lower than that of the majority carriers and whose origin could be related to the heavy doping of the p side. This assumption is confirmed by the observation that in a n-type substrate with the same doping level as that of the n side of our p-n junction, the photoluminescence intensity decreases with the temperature as predicted by Eq. (6). We can expect that at liquid nitrogen temperature the hole lifetime τ_p is controlled by hole capture in deep recombination centers. The capture process can be split into two steps: first, the holes are captured on the upper Coulomb states and quickly fall down in energy (cascade capture mechanism). However, as a rule, the dense spectrum of highly excited states having the depth of Bohr binding energy for shallow donors, i.e. 45-50 meV in silicon is separated from the ground state by a large energy gap. Therefore, the last steps in the capture process, i.e. the transition to the ground state, needs a large transfer of energy and can occur nonradiatively only as a low probability multiphonon process. The last excited state can be regarded as a metastable state and recombination process should be described by a model of two-level center. The recombination process via a metastable level with a characteristic energy E_t is treated in [8].

With the rise of temperature the carriers, waiting in the metastable state, can be released faster than the relaxation to the ground state time. In this case the lifetime will exponentially rise with the temperature $\tau_T \propto \exp(-E_t / kT)$. From this it follows an increase of the integrated EL intensity. From these considerations, the EL intensity increase indicates a large concentration of deep non-radiative recombination centers, i.e. a bad material quality of the active diode region.

5. Optimal parameters of LED for silicon band edge EL

Up to now the band edge EL in silicon was studied in standard diodes having heavily doped p and lightly doped n side. This technology favors the properties of rectifiers, since in this case large breakdown reverse voltage is necessary. However, in the case of LED the reverse characteristic is not important. Therefore the parameters of the diode can be optimized by variation of doping level both in n and p side. We have calculated the optimal light emitting structure. The internal quantum efficiency is given by the formula:

$$\eta = \frac{h\nu(I_n + I_p)}{jV} \equiv \frac{h\nu(j/q)r_r A_I}{jV}, \tag{7}$$

where the numerator determines the power of band edge EL and the denominator the electric power dissipated in the diode. Assuming $V \cong E_g / q$ and $h\nu \cong E_g$, the internal quantum efficiency in the form:

$$\eta = r_r A_I \equiv r_r n_n p_p \left(L_n + L_p\right)\left(\frac{D_n n_p}{L_n} + \frac{D_p p_n}{L_p}\right)^{-1} \quad (8)$$

The physical meaning of this expression is very simple. If for example the main contribution to EL comes from the p side of the junction, η is determined by the ratio of the total lifetime of minority carriers (electrons) to the radiative lifetime: $\eta \approx r_r p_p \tau_n \equiv \tau_n / \tau_r$. The results of calculations of η for different doping levels are shown in Fig. 6, where we used $r_r = 10^{-14}$ cm^{3}s^{-1} [10].

According to these simulations, the LED studied up to now are far from the optimal regime. In Ref. 6, $n_n = 10^{19}$ cm^{-3} and $p_p = 1.4 \times 10^{16}$ cm^{-3} yielding a η of 1 percent in agreement with our calculations. In [7] and in our diode, the doping levels were of 10^{18} cm^{-3} and 10^{15} cm^{-3} for the p and n sides, respectively. For this LED we expect a η of 0.01 %. It should be noted that the decay time of the edge EL at room temperature in our work and in [7] is about 20 μs, a value consistent with the data of Fig. 5. However, the internal quantum efficiency measured in Ref. 7 reaches approximately 0.5 % in evident contradiction to our calculations. The reasons for such a difference can be either an insufficient accuracy in the measurements of the quantum efficiency in [7] or an underestimate of the radiative recombination coefficient calculated in [9].

It is clear from Fig. 6 that the optimal value of the internal quantum efficiency is obtained for doping level of 10^{19} cm^{-3} for the n side and that of 10^{18} cm^{-3} for the p side. In this case the main contribution to the EL comes from the p side and the efficiency is due to the comparatively high diffusion length of electrons. At higher doping levels the quantum efficiency drops due to strong Auger recombination of the minority carriers while at lower doping levels the radiative recombination rate decreases because it is proportional to $n_n p_p$. Since higher concentrations are usually accompanied by higher densities of defects, it is possible that a concentration of 10^{17} cm^{-3} in the p side would be more favorable. It follows from the data of Fig. 6 that it is possible to achieve internal quantum efficiency as high as 3% at doping level of 10^{18} cm^{-3} on the n side and of 10^{17} cm^{-3} on the p side. In this case the modulation frequency of the LED according to the data on the lifetime of the minority carriers should be on the order of 50 kHz.

5. Conclusions

We have studied the intrinsic EL for a forward-biased silicon p-n junction. Our theoretical consideration describes consistently the full set of experimental data. The considerable increase of EL intensity and the lengthening of the EL decay time with the temperature can be explained by a thermal suppression of nonradiative recombination channel due to deep recombination centers. We have discussed quantitatively the radiative properties of a p-n junction and shown that a significant increase of its quantum efficiency can be obtained on optimization of doping levels of both sides of a p-n junction. The calculations show that the maximum internal quantum efficiency of a diode can reach several percent with a modulation frequency of 50 kHz. It should be noted that optimization of the diode parameters demands high doping levels of both sides of the junciton. The increase of a doping level can lead simultaneously to introduction of large concentration of defects which should be faced technologically.

This work was supported by the grants from Russian Foundation of Basic Research, Nederlandse Organisatie voor Wetenschappelijk Onderzoek (NWO), Russian Ministry of Science and Technology, and Russian Academy of Sciences. We want to thank V.B. Shuman for useful discussions. I.N.Y. is grateful to the Wenner-Gren foundations for a scholarship.

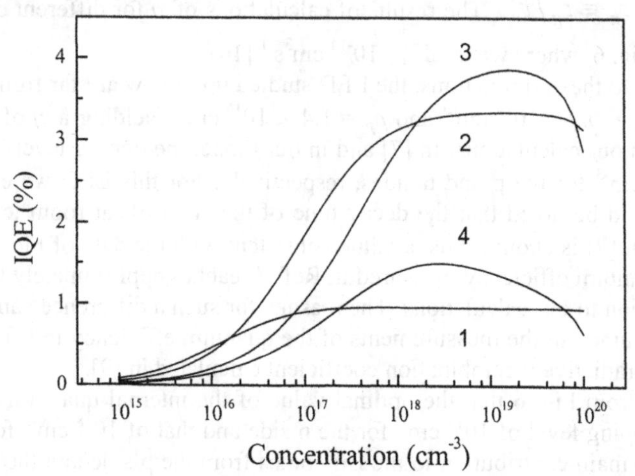

Figure 6. Internal quantum efficiency (IQE) as a function of electron concentration n_n in the n side of the p-n junction for different hole concentrations in the p side: $p_p = 10^{16}$ cm^{-3} (1), $p_p = 10^{17}$ cm^{-3} (2), $p_p = 10^{18}$ cm^{-3} (3), $p_p = 10^{19}$ cm^{-3} (4).

6. References

1. Lu Z.H., Lockwood D.J., and Baribeau J.M. (1995) Quantum confinement and light emission in SiO_2/Si superlattices, *Nature* **378**, 258-261.

2. Hirschman K.D., Tsybeskov L., Duttagupta S.P., and Fauchet P.M. (1996) Silicon-based visible light emitting devices integrated into microelectronic circuits, *Nature* **384**, 338-341.

3. Franzò G., Irrera A., Moreira E.C., Miritello M., Iacona F., Sanfilippo D., DiStefano G., Fallica P.G., and Priolo F. (2002) Electroluminescence of silicon nanocrystals in MOS structures, *Appl. Phys., A* **74**, 1-5.

4. Franzò G., Priolo F., Coffa S., Polman A., and Carnera A. (1994) Room-temperature electroluminescence from Er-doped crystalline Si, *Appl. Phys. Lett.* **64**, 2235-2237.

5. Zheng B., Michel J., Ren F.Y.G., Kimerling L.C., Jacobson D.C., and Poate J.M.. (1994) Room temperature sharp line electroluminescence at $\lambda = 1.54$ μm from an erbium doped, silicon light-emitting diode, *Appl. Phys. Lett.* **64**, 2942-2944.

6. Green M.A., Zhao J., Wang A., Reece P.J., and Gal M.. (2001) Efficient silicon light-emitting diodes, *Nature* **412**, 805-808.

7. Ng.W.L., Lourenço M.A., Gwilliam R.M., Ledain S., Shao G., and Homewood K.P. (2001) An efficient room-temperature silicon-based light-emitting diode, *Nature* **410**, 192-194.

8. Abakumov V.N., Perel V.I., and Yassievich I.N. (1991) *Nonradiative recombination in semiconductors*, North Holland, Amsterdam.

9. Jacoboni C. , Canali C., Ottaviani G., Alberigi Quaranta A. (1977) A review of some charge transport properties of silicon, *Solid State Electronics* **20**, 77-89.

10 Schlangenotto H., Maeder H., and Gerlach W. (1974) Temperature dependence of the radiative recombination coefficient in silicon, *Phys. stat. sol. (a)* **21**, 357-367.

LIGHT EMITTING DEVICES BASED ON SILICON NANOCRYSTALS

A. IRRERA, D. PACIFICI, M. MIRITELLO, G. FRANZÒ, AND F. PRIOLO
INFM and Dipartimento di Fisica e Astronomia,
Università di Catania, Corso Italia 57, I-95129 Catania, Italy
F. IACONA
CNR-IMM, Sezione di Catania,
Stradale Primosole 50, I-95121 Catania, Italy
D. SANFILIPPO, G. DI STEFANO, P.G. FALLICA
STMicroelectronics,
Stradale Primosole 50, I-95121 Catania, Italy

1. Introduction

Silicon has for a long time been considered unsuitable for optoelectronic applications. Due to the indirect nature of its energy band gap, bulk silicon is indeed a highly inefficient light source. Many efforts have been devoted towards the development of Si-based materials able to act as light emitters [1]. Quantum confinement in Si nanostructures, including porous Si [2], Si nanocrystals (nc) embedded in SiO_2 [3-11], and Si/SiO_2 superlattices [12-14], and rare earth doping of silicon have dominated the scientific scenario of silicon-based microphotonics.In particular, Si nc embedded in SiO_2 have attracted a great attention, due to their high stability and to their full compatibility with Si technology. Si nc are characterized by an energy band gap which is enlarged with respect to bulk Si, and exhibit an intense and tunable room temperature photoluminescence (PL) in the visible-near IR range [15]. Si nc embedded in SiO_2 are produced with several different techniques, such as ion implantation [4, 6], chemical vapor deposition [11, 15], sputtering and laser ablation [16, 17]. Recently, the interest towards this material is greatly increased due to the first observation of light amplification in Si nanostructures [18].

On the other hand, Er-doped crystalline Si has been extensively studied to take advantage of the radiative intra-4f shell Er transition and room temperature operating devices have been achieved [19]. Recently, the coupling of the two approaches has been demonstrated to produce quite interesting effects. Indeed, Er doping of Si nc results in a very efficient light emission at 1.54 μm [20-22] since Si nanoclusters act as efficient sensitizers for the Er ions [15, 23]. In particular, the nc, once excited, promptly transfer its energy to the nearby Er ions, which can then decay radiatively emitting a 1.54 μm photon. In spite of the large amount of experimental data present in the literature on the PL properties of Si nc, only a few papers have reported their electroluminescence (EL) characteristics, mainly due to the difficulties of carrier injection in a semi-insulating ma-

L. Pavesi et al. (eds.), Towards the First Silicon Laser, 29–43.

terial. In this work we report the electrical, optical and structural properties of light emitting devices based on Si nc embedded in SiO_2 with and without Er doping. We will demonstrate that by properly tailoring few relevant parameters of the SiO_x layer acting as the active medium of the devices (Si concentration, annealing temperature, thickness and Er concentration) it is possible to obtain intense and stable EL signals, in the visible or in the infrared region, from devices operating at room temperature and at low voltages. Also a detailed study of the EL properties as a function of the current density and of the temperature, including the calculation of the excitation cross section under electrical pumping, will be presented.

2. Light Emitting Devices Based on Si nc

A substoichiometric SiO_x ($x < 2$) film, about 80 nm thick, was deposited on a p-type Si substrate (0.01-0.02 Ω cm) by plasma enhanced chemical vapor deposition (PECVD). Deposition processes were performed by using 50 W of input power and the source gases used were high purity SiH_4 and N_2O. By changing the N_2O/SiH_4 flow ratioγ, three different total Si concentrations of 39, 42 and 46 at.% were obtained. After deposition, the SiO_x films were annealed at high temperature (in the range 1100 - 1250 °C) for 1 h in N_2 atmosphere. The annealing process induces the separation of the Si and SiO_2 phases with the formation of Si nc embedded in SiO_2 whose size depends on the excess Si amount as well as on the annealing temperature [11]. Then a n-type polysilicon layer (resistivity 0.01 Ω cm), about 100 nm thick, was deposited on the SiO_x film to promote the carrier injection into the Si nc. Finally, the active area of the device was defined with photolithographic processes. Aluminium-based contacts, defined as circular rings to allow a free central area for the exit of the light, were made to the n-type polysilicon film and the p-type substrate. The structure of the MOS device is shown in Figure 1, displaying a dark field cross-sectional micrograph obtained by transmission electron microscopy (TEM). From the top, the metallization layer, the poly-Si film, the SiO_x active region (having a Si content of 46 at.% and annealed at 1100 °C) and the Si substrate are distinguishable. The Si nc present in the SiO_x layer appear as white dots on the dark SiO_2 background.

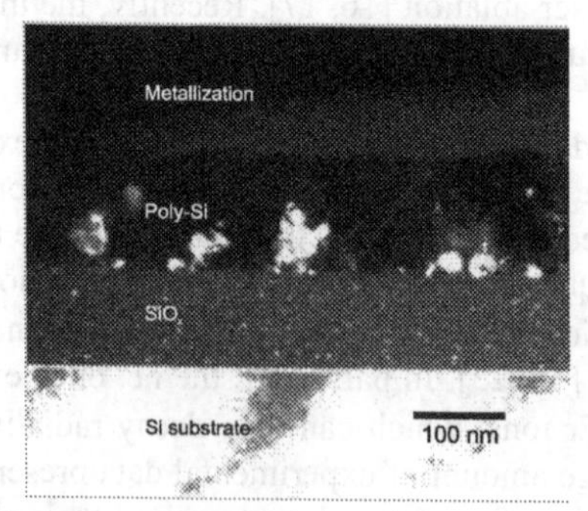

Figure 1. Dark-field cross-sectional TEM micrograph of a device having a Si content of 46 at.% in the SiO_x layer. Starting from the bottom the Si substrate, the SiO_x active region, the poly-Si film and the metallization layer are clearly visible. In the SiO_x layer Si nc appear as bright spots on a dark background.

From a statistical analysis of plan view TEM micrographs taken in dark field conditions from the same sample it is possible to extract a Si nc mean radius of about 1.0 nm. By changing the Si concentration or the annealing temperature mean sizes ranging from 1 to 2 nm have been obtained.

In figure 2 we report the current density-voltage (J-V) curves of different devices under both forward and reverse bias conditions, recorded with a parameter analyser HP 4156.

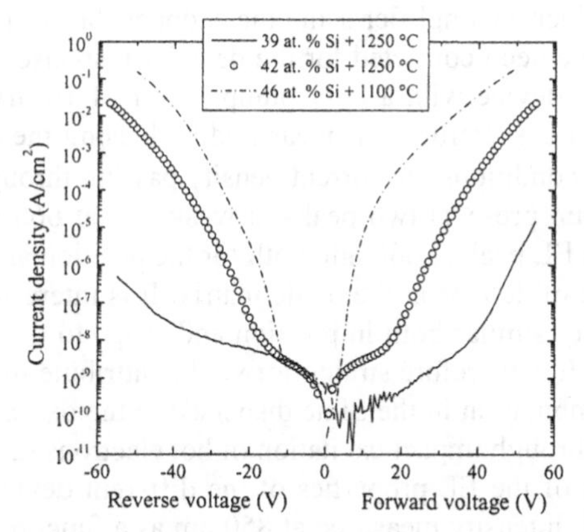

Figure 2. Current density as a function of voltage for the Si nc devices having in the SiO_x layer a Si content of 39 at.% (—) and 42 at.% (o) annealed at 1250 °C, and 46 at.% annealed at 1100 °C (----).

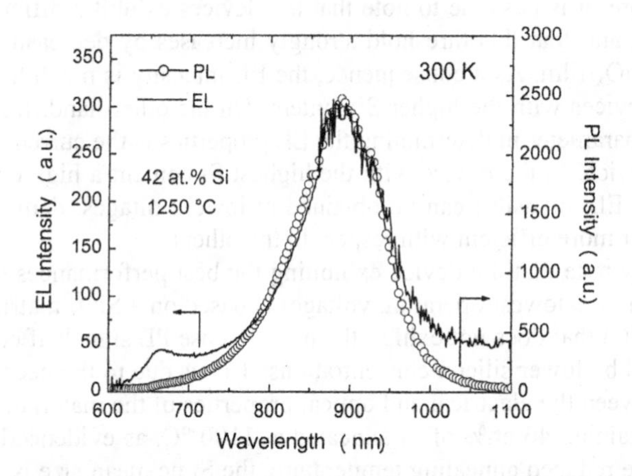

Figure 3. Comparison between the PL and the EL spectra of the device with a Si content of 42 at.% in the SiO_x layer. The PL spectrum (o) was measured with a laser pump power of 10 mW. The EL spectrum (—) was measured with a voltage of 48 V and a current density of 4 mA/cm^2.

From the comparison of the J-V curves we can conclude that by increasing the Si concentration in the SiO_x layer there is a very strong increase in the current that can pass through the device for a fixed applied voltage. The carrier injection through the device is due to direct tunneling between Si nc.

Figure 3 shows the room temperature PL and EL spectra measured in a device based on a SiO_x layer with a Si content of 42 at.% and annealed at 1250 °C. PL measurements were performed by pumping with the 488 nm line of an Ar-Kr ion laser. The laser beam was mechanically chopped at a frequency of 55 Hz. EL spectra were taken by biasing the device with a square pulse at a frequency of 55 Hz, using a fast Agilent Pulse Generator. EL and PL signals were analyzed by a single grating monochromator and detected by a photomultiplier tube or by a liquid nitrogen-cooled Ge detector. Spectra were recorded with a lock-in amplifier using the chopper (or the TTL) frequency as a reference. All spectra have been corrected for the detector response. The PL spectrum, taken by illuminating the device with a laser pump power of 10 mW, presents a peak centered at 890 nm. The EL spectrum was measured by biasing the device with a voltage of 48 V; under these conditions the current density passing through the device is ~4 mA/cm^2. The EL spectrum presents two peaks: a weak one at 660 nm and a most intense one at 890 nm. The EL peak at 660 nm, both for the position and the shape, can be attributed to the presence of defects in the oxide matrix. It is interesting to note that the EL peak at 890 nm is very similar both in position and shape to the PL peak measured in the very same sample. It is therefore straightforward to attribute this emission to electron-hole (e-h) pair recombination in the Si nc dispersed in the oxide layer. The e-h pair is generated within a nc through impact excitation of hot electrons [24-25].

The comparison of the EL properties of the different devices can be found in figure 4, reporting the EL intensity measured at 850 nm as a function of voltage (a) and of current density (b) for the samples with 39 and 42 at.% Si annealed at 1250 °C and with 46 at.% Si annealed at 1100 °C under forward bias conditions. Figure 4(a) demonstrates that for each device the EL intensity strongly increases by increasing the applied voltage; furthermore, it is possible to note that the devices exhibit a different threshold for light emission, and that this threshold strongly increases by decreasing the Si concentration in the SiO_x film. As a consequence, the EL intensity is much higher at lower voltages for the devices with the higher Si content. On the other hand, figure 4(b) suggests that the key parameter in determining the EL properties is the current density passing through the device. In the device with the highest Si content a high current density (and hence a high EL intensity) can be obtained at lower voltages, demonstrating that this device is by far more efficient with respect to the others.

We finally note that the device exhibiting the best performances (i.e. the highest EL intensity and the lowest operating voltage) is based on a SiO_x matrix (46 at.% Si annealed at 1100 °C) that does not exhibit the most intense PL signals if compared with films characterized by lower silicon concentrations. This is due to the necessity to reach a compromise between the electrical and optical properties of the material. In particular, in the sample containing 46 at.% of Si annealed at 1100 °C, as evidenced by the TEM analysis, due to the reduced annealing temperature, the Si nc mean size is quite small (r ~ 1 nm). This careful balance between Si concentration and annealing temperature implies the formation of a very dense distribution of Si nc having the proper size for light emission and also the fact that a relevant fraction of the silicon excess remains dissolved in the oxide matrix; both effects are expected to strongly contribute to a high electrical conduction, and therefore to a good electrical injection in Si nc.

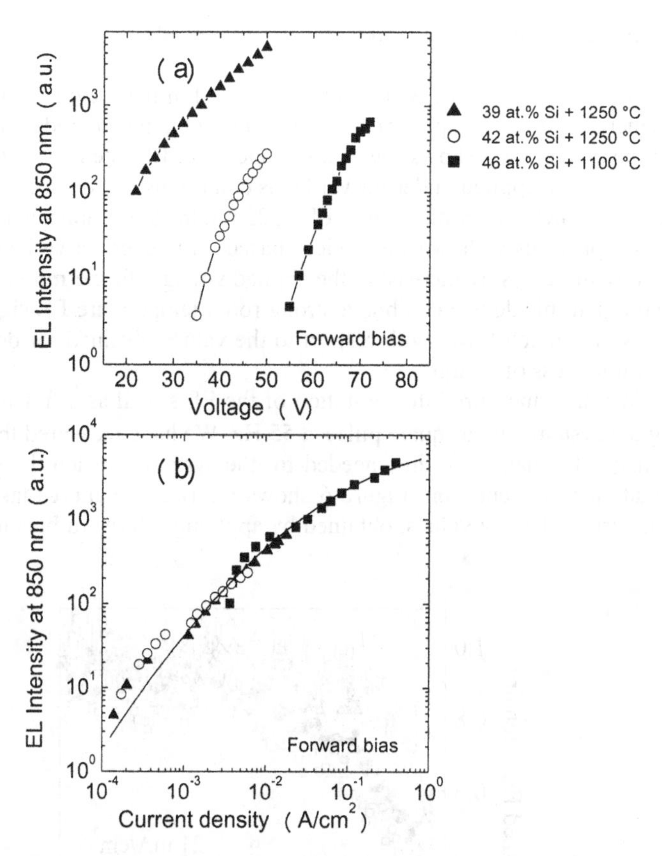

Figure 4. EL intensity at 850 nm as a function of the applied voltage *(a)* and of current density *(b)* for the samples with 39 at.% Si (■) and 42 at.% Si (○) annealed at 1250 °C and for the one with 46 at. % Si annealed at 1100 °C (▲) under forward bias conditions.

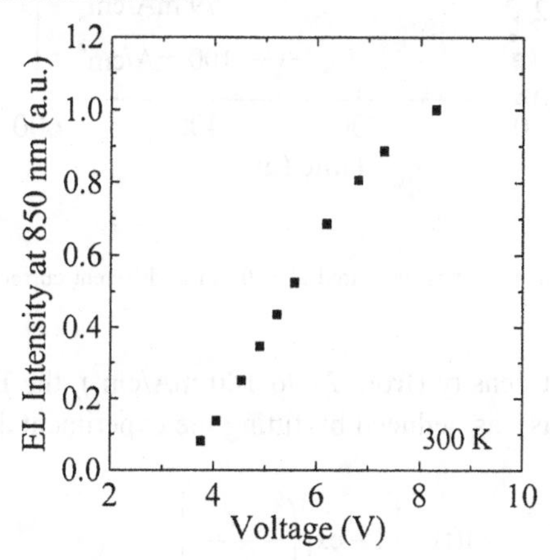

Figure 5. Room temperature EL intensity at 850 nm as a function of the voltage under forward bias conditions for Si nc prepared by annealing at 1100 °C a SiO_x film, 25 nm thick, having a Si content of 46 at.%.

3. Radiative and non- radiative decay processes of Si nc

After the optimization of the SiO_x matrix discussed in the previous paragraph, a further improvement of the device performances has been obtained by reducing the thickness of the SiO_x layer. Figure 5 shows the room temperature EL intensity at 850 nm as a function of the voltage applied under forward bias conditions to a device with an active layer consisting of a SiO_x film with 46 at.% of Si, 25 nm thick and annealed at 1100 °C.

As previously shown for devices based on thicker active layers, the EL intensity strongly increases by increasing the applied voltage; furthermore, it is very interesting to note that this device exhibits a strong room temperature EL signal at voltages as low as 4 V, i.e. much lower with respect to the values obtained for devices with an active layer thickness of 80 nm.

We have measured the evolution of the EL signal as a function of the time by exciting the system with a square pulse at 55 Hz. We have measured the risetime (τ_{on}) of the EL intensity, that is the time needed for the system to reach the steady state when the excitation is switched on. Figure 6 shows the risetime curves taken at 850 nm for different current density values, obtained by applying a forward bias in the range of 4-6 V.

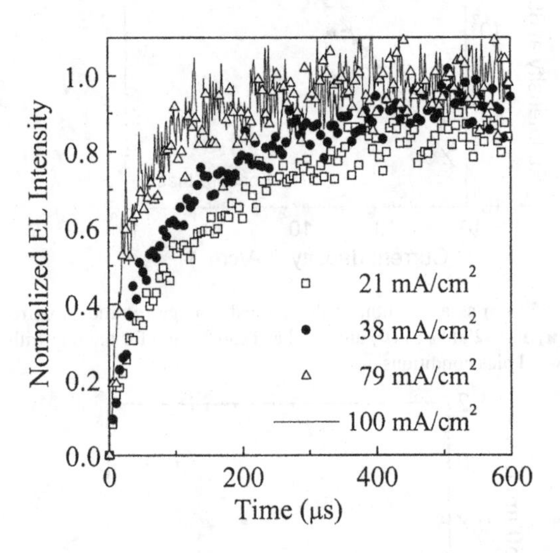

Figure 6. Room temperature EL risetimes measured at 850 nm at different current densities under forward bias conditions for Si nc.

By increasing the current density (from 21 to 100 mA/cm^2), the EL risetime becomes shorter (from 175 to 36 μs), as deduced by fitting the experimental curves with the following equation:

$$I(t) = \left[1 - \exp\left(-\frac{t}{\tau_{on}}\right)\right] \qquad (1)$$

where I(t) is the normalized EL intensity as a function of time. It can be demonstrated [15] that the risetime τ_{on} is given by the equation:

$$\frac{1}{\tau_{on}} = \sigma\phi + \frac{1}{\tau} \tag{2}$$

where σ is the excitation cross section, ϕ is the electron flux and τ the luminescence decay time. By plotting the reciprocal of τ_{on} versus the electrons flux the slope of the straight line that fits the experimental data gives for the excitation cross section a value of $\sim 4.7\times10^{-14}$ cm^2, as shown in figure 7. This value is almost temperature independent, being $\sim 6.5\times10^{-14}$ cm^2 at 12 K [25].

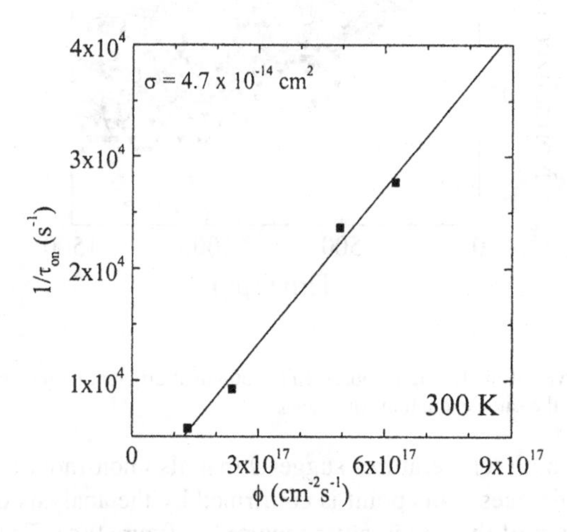

Figure 7. Reciprocal of the EL risetime (τ_{on}) as a function of the carrier flux (ϕ) passing through the device. The excitation cross section of Si nc embedded in SiO_2 under electrical pumping is represented by the curve slope.

We have calculated for the same system the excitation cross section at 850 nm for optical pumping under 488 nm laser beam excitation and we have found a value of $\sim 1\times10^{-16}$ cm^2 at 300 K, in agreement with recent results [15]. Therefore the excitation cross section under electrical pumping, i.e. due to the impact of hot electrons on a Si nc, is almost two orders of magnitude higher with respect to that relative to optical pumping.
Interesting information on the de-excitation mechanisms of Si nc under electrical pumping can be obtained by studying the time evolution of the EL signal at the device switch off. In figure 8 the time-decay curves of the EL signal at 850 nm measured at different temperatures under the same excitation conditions (J = 40 mA/cm^2) are reported. The lifetime values of the EL signals have been extracted by fitting the experimental curves reported in figure 8 with the following equation:

$$I(t) = I_o \exp\left[-\left(\frac{t}{\tau}\right)^{\beta}\right] \tag{3}$$

where I_0 is the EL intensity at t = 0, τ is the EL decay time, and β is a dispersion factor ≤ 1. The situation in which β is < 1 ("stretched" exponential behaviour) is characteristic of systems of interacting Si nc [10]. For the curves reported in figure 8, β is constant and the value is ~0.6, while τ decreases from 155 to 43 μs by increasing the temperature from 12 to 300 K.

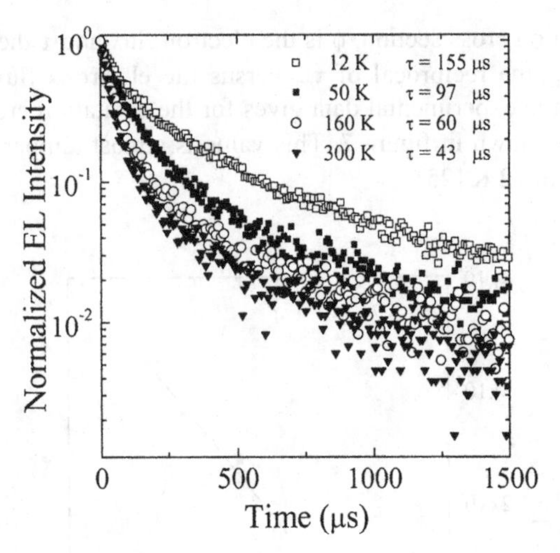

Figure 8. Time-decay curves of the EL signal observed at 850 nm at different temperatures for Si nc. All data have been collected under the same excitation conditions.

The decrease of τ at high temperatures suggests that also non-radiative decay processes are operating in our devices. This point is confirmed by the analysis of the EL intensity at 850 nm as a function of the temperature reported in figure 9(a). The figure shows that the EL intensity increases in the range 12-150 K, has a maximum at ~150 K and then decreases with increasing the temperature up to 320 K. However, the temperature dependence of the EL intensity is weak, the difference between 12 K and 150 K being only about a factor of 2. This behaviour of the temperature dependence of the EL resembles that reported for the PL of Si nc [13].

At this point, the temperature dependence of the radiative rate R_R, defined as the reciprocal of the radiative lifetime (τ_R), can be extracted from the ratio between the EL intensities reported in figure 9(a) and the decay times extracted from the data of figure 8 and reported in figure 9(b). The result of this calculation is shown in figure 9(c). The radiative rate increases by a factor of 4 on going from 12 to 300 K. This behaviour can be explained with a model proposed by Calcott et al. [26] for porous silicon and recently applied also to Si nc [27, 28]. According to this model the exchange electron-hole interaction splits the excitonic levels by an energy Δ. The lowest level in this splitting is a triplet state and the upper level is a singlet state. The triplet state (threefold degenerate) has a radiative decay rate R_T much smaller than the radiative decay rate R_S of the singlet one. Once excited the excitonic population will be distributed according to thermal equilibrium law. Hence, at a temperature T the radiative decay rate will be:

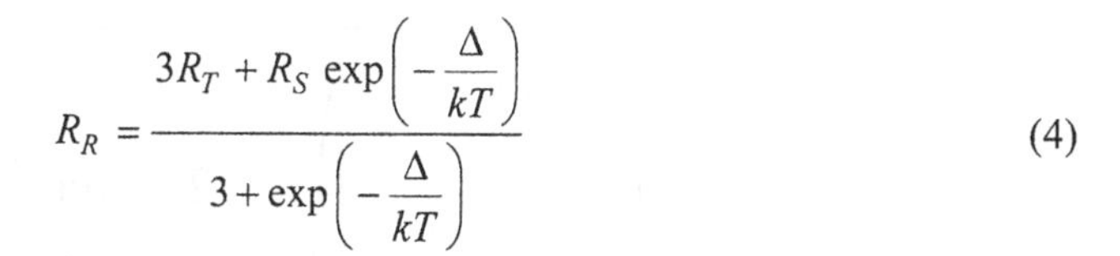

$$R_R = \frac{3R_T + R_S \exp\left(-\frac{\Delta}{kT}\right)}{3 + \exp\left(-\frac{\Delta}{kT}\right)} \tag{4}$$

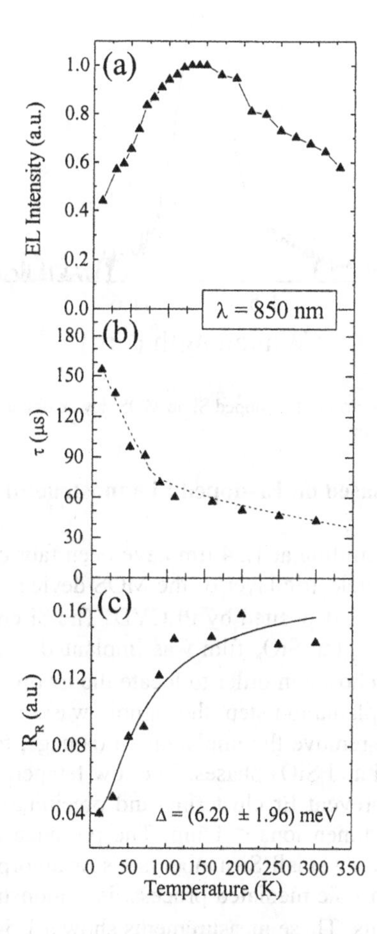

Figure 9 (a) EL intensity at 850 nm, *(b)* decay lifetime at 850 nm, and *(c)* radiative rate as a function of the temperature for Si nc. The lines in *(a)* and *(b)* are drawn to guide the eye. The radiative rate data have been evaluated by dividing the EL intensities of *figure* 9 *(a)* by the lifetime values of *figure 9 (b)*, and have been fitted according to a model accounting for the different transition rates from the singlet or triplet states to the ground one. The resulting splitting between the two excited levels is 6.20 meV.

This rate tells us that, by increasing temperature, the relative population of the singlet state will increase and, being the radiative rate of the singlet state much higher than that of the triplet state, also the total radiative rate will consequently increase. We have used equation (4) to fit the data in figure 9(c); from the fit (shown as a continuous line) experimental values for the level splitting ($\Delta = 6.20 \pm 1.96$ meV) and for the ratio between the radiative rates of the two levels ($R_T/R_S = 0.07 \pm 0.02$) have been obtained.

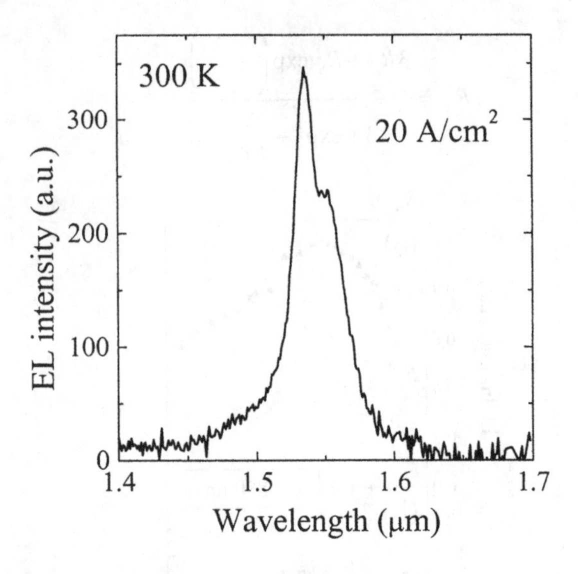

Figure 10. Room temperature EL spectrum of Er doped Si nc MOS device obtained under forward bias conditions with a current density of 20 A/cm^2.

4. Light Emitting Devices Based on Er-doped Si nanostructures

Electroluminescent devices emitting at 1.54 μm have been fabricated by Er ion implantation in a SiO_x matrix. The dielectric layer of the MOS device was a substoichiometric SiO_x ($x < 2$) film, 70 nm thick, deposited by PECVD. The Si concentration in the film was 46 at.%. After deposition, the SiO_x film was implanted with Er ions to a dose of 7×10^{14} cm^{-2}; the energy was chosen in order to locate the Er profile in the middle of the dielectric layer. After the implantation step, the sample was annealed at 900 °C for 1 h in N_2 atmosphere in order to remove the implantation damage, to activate Er and to induce the separation of the Si and SiO_2 phases. The low temperature chosen for the annealing process, needed to prevent Er clustering and precipitation, determines the nucleation of Si clusters with dimensions < 1 nm. The presence of these Si aggregates, possibly under the form of very small Si nanocrystals or amorphous clusters, that can excite Er through an electron-hole mediated process, is demonstrated by photoluminescence excitation measurements. These measurements show a 1.54 μm signal that monotonically increases with decreasing the excitation wavelength without the oscillations typical of resonant photon absorption by Er [15, 21, 22]. The presence of this large amount of closely spaced and very small Si nc dispersed in the SiO_2 matrix determines an efficient carrier injection through the insulating layer. Figure 10 reports the room temperature EL spectrum obtained by forward biasing the device at 32 V, with a current density of 20 A/cm^2. The spectrum shows the typical features of Er emission with a main peak centered at 1.54 μm and a shoulder at a slightly higher wavelength, both of them due to the transitions from the Er first excited multiplet $^4I_{13/2}$ to the ground state $^4I_{15/2}$. No relevant signals coming from Si nc or due to the matrix have been observed after Er implantation.

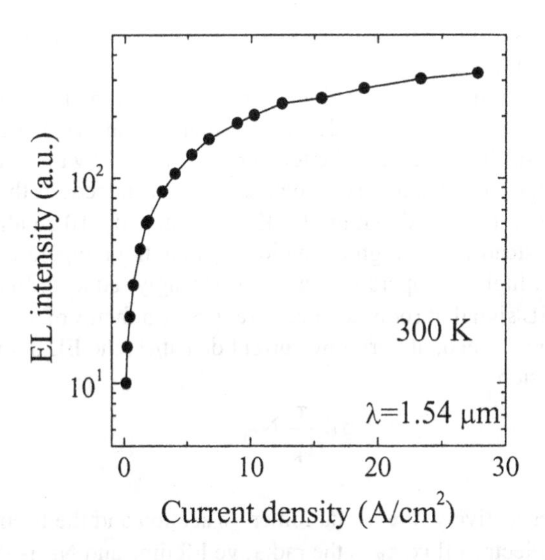

Figure 11. EL intensity at 1.54 μm as a function of the current density at 300 K under forward bias conditions.

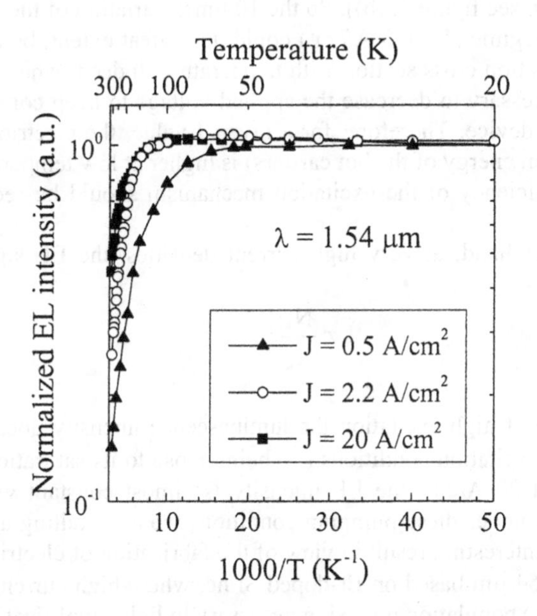

Figure 12. Normalized EL intensity at 1.54 μm as a function of the temperature for three different current density values of 0.5 A/cm^2 (▲), 2.2 A/cm^2 (○) and 20 A/cm^2 (■).

Figure 11 shows the EL intensity at 1.54 μm as a function of the current density. The EL intensity at 1.54 μm increases less than linearly and tends to saturate with increasing current. This result demonstrates that a saturation of the Er emitting centers is possible under electrical pumping.

We have also studied the behaviour of the EL signal at 1.54 μm as a function of the temperature for different current densities. Figure 12 shows that the EL intensity monotonically decreases by increasing the temperature, for all J values, in the 12-300 K range. The figure evidences also a clear dependence on J. Indeed, in the low excitation regime (J = 0.5 A/cm^2), the EL signal at 300 K is a factor of ~10 smaller than the low temperature value. Instead in the high excitation regime, the temperature quenching of the EL signal starts at higher temperatures and it is strongly reduced. In particular, at 20 A/cm^2 the 1.54 μm EL signal at room temperature is only a factor of ~2 smaller than the low temperature value. Indeed, at very low current densities, the EL intensity can be described by the expression

$$I \propto \sigma \phi \frac{\tau}{\tau_R} N_{Er} \tag{5}$$

where σ and τ are, respectively, the excitation cross section and the lifetime of the emitting centers, ϕ is the electron flux, τ_R is the radiative lifetime and N_{Er} is the total concentration of excitable Er ions. The lifetime of Er has a weak dependence on temperature. In fact in the range 12-300 K the lifetime variation is less than a factor of 2 ($\tau_{12\,K}$ ~1120 μs and $\tau_{300\,K}$ ~660 μs, see figure 13(b)). So the 10 times variation of the EL signal in the very low excitation regime (J = 0.5 A/cm^2) could, to a great extent, be attributed to the variation of the excitation cross section with temperature. Indeed, when the temperature is increased, it is necessary to decrease the applied voltage to keep constant the current flowing through the device. Therefore, for a given J value, the electric field (and as a consequence the mean energy of the hot carriers) is higher at low temperatures, and, as a consequence, the efficiency of the excitation mechanism should be reduced at 300 K with respect to 12 K.

On the other hand, at very high current densities, the EL signal can be expressed by

$$I \propto \frac{N_{Er}}{\tau_R} \tag{6}.$$

Under this condition of high excitation the luminescence intensity does not depend on excitation (σ) or de-excitation conditions (τ), being close to its saturation value. Therefore, the fact that at 20 A/cm^2 the EL intensity is almost constant with temperature clearly indicates that under these pumping conditions we are exciting almost all the Er ions. This is a quite interesting result in view of the fabrication of electrically driven optical amplifiers at 1.54 μm based on Er-doped Si nc, where high current densities are a must for obtaining the population inversion necessary to light amplification.

A more detailed analysis of the excitation and de-excitation properties of this system has been done through time resolved EL measurements. We have performed the measurements, showed in figure 13, by electrically pumping the system with a square pulse at 11 Hz. The τ_{on} and τ values, obtained by fitting the curves with the equations (1) and (3), are respectively ~77 μs and ~660 μs. In fig. 13b it is also showed that τ does not depend on J. Moreover we have compared it with the value of lifetime obtained under optical pumping by exciting the system with the 488 nm line of an Ar laser. Also under this different excitation condition the lifetime is ~660 μs, suggesting that the excited states and the de-excitation mechanisms are independent of the type of excitation [15].

Through these experimentally determined values, for a current density of 0.15 A/cm^2, a value for the excitation rate $R = \tau_{on}^{-1} - \tau^{-1}$ of ~11470 s^{-1} can then be estimated. Since $R = \sigma\phi$, a value of the excitation cross section σ of ~1×10^{-14} cm^2 can be estimated. The obtained value is comparable to the value we have found for the electrical excitation of undoped Si nc (~4×10^{-14} cm^2).

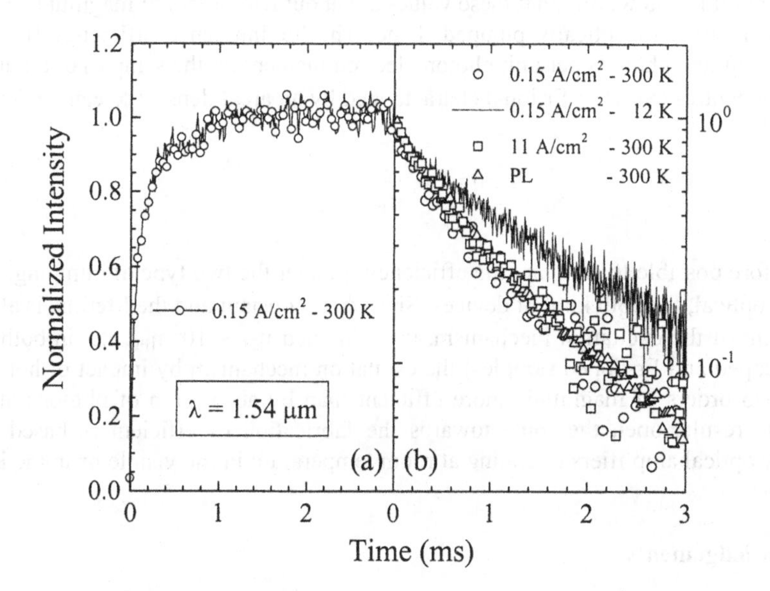

Figure 13. (a) Time evolution of the room temperature EL signal at 1.54 μm at the device switch on for a current density of 0.15 A/cm^2. *(b)* Room temperature (○) and low temperature (—) EL time decay curves at 1.54 μm for an Er-doped Si nc device measured at a low current density of ~0.15 A/cm^2. For comparison, the time decay curve under high current density, 11 A/cm^2 (□) and under low optical pumping at 488 nm, 10 mW (Δ) are also reported.

5. Conclusions

In conclusion, we have studied the structural and optical properties of MOS devices where the dielectric layer consists of a substoichiometric SiO_x ($x < 2$) thin film grown by plasma enhanced chemical vapor deposition. We have demonstrated that the devices with silicon nanocrystals are efficient light sources at room temperature. Samples with different quantum dot mean sizes were used. Excitation occurs by impact of hot electrons and subsequent electron-hole generation and recombination. We observed that by increasing the Si concentration in the SiO_x layer there is a strong increase of the current that can pass through the device for a given voltage. It is in fact necessary to reach a compromise between a high silicon content, which is favorable for the electrical properties, and the silicon nanocrystals mean size, which influences the optical properties. Devices with very high densities of small nc (i.e. high Si content annealed at temperatures not too high, namely 1100 °C) have demonstrated to reach the best compromise. We have also demonstrated that by reducing the thickness of the SiO_x layer it is possible to obtain a very strong reduction of the operating voltage. Moreover, we have studied the EL properties of light emitting devices based on Er-doped Si nc. The devices are very stable, exhibit a high EL efficiency at room temperature and a weak temperature dependence of the EL signal.

We have calculated for both kind of systems the cross section under electrical pumping and it is noteworthy that these values are about two orders of magnitude higher than those relative to optically pumped Si nc. The luminescence efficiency (i.e., the number of emitted photons per unit photon/electron incident on the sample) of a number of emitting centres (Si nc or Er ions) characterized by an areal density ϕ_c can be demonstrated to be:

$$\eta = \sigma\phi_c \frac{\tau}{\tau_R} \qquad (7)$$

It is therefore possible to compare the efficiency η under the two type of pumping, electrical and optical, for both kind of devices. Since ϕ_c is constant and the lifetime is almost independent of the excitation mechanism, it is obtained $\eta_{EL} \sim 10^2\ \eta_{PL}$, i.e. in both systems (undoped and Er-doped samples) the excitation mechanism by impact of hot electrons is two orders of magnitude more efficient than by absorption of photons at 488 nm. These results open the route towards the fabrication of efficient Si-based light sources or optical amplifiers operating at room temperature in the visible or in the infrared region.

6. Acknowledgements

The authors wish to thank S. Pannitteri, N. Marino, A. Marino, and A. Spada for their expert technical assistance and A. La Porta for collaboration. This work has been partially supported by the EU IST-SINERGIA project, by the INFM project RAMSES, and by the project FIRB financed by MIUR.

REFERENCES

1. Iyer, S.S., and H. Xie, Y., (1993) Light-emission from silicon, *Science* **260**, 40-46.
2. Cullis, A. G., T. Canham, L., Calcott, P. D. J., (1997) The structural and luminescence properties of porous silicon, *J. Appl. Phys.* **82**, 909-965.
3. Kanemitsu, Y., Ogawa, T., Shiraishi, K., and Takeda, K., (1993) Visible photoluminescence from oxidized Si nanometer-sized spheres: Exciton confinement on a spherical shell, *Phys. Rev. B* **48**, 4883.
4. Shimizu-Iwayama, T., Fujita, K., Nakao, S., Saitoh, K., Fujita, T., and Itoh, N., (1994) Visible photoluminescence in Si^+-implanted silica glass, *J. Appl. Phys.* **75**, 7779-7783.
5. Zhu, J.G., White, C.W., Budai, J.D., Withrow, S.P., and Chen, Y., (1995) Growth of Ge, Si, and SiGe nanocrystals in SiO_2 matrices, *J. Appl. Phys.* **78**, 4386-4389.

6. Min, K.S., Shcheglov, K.V., Yang, C.M., Atwater, H.A., Brongersma, M.L., and Polman, A., (1996) Defect-related versus excitonic visible light emission from ion beam synthesized Si nanocrystals in SiO_2, *Appl. Phys. Lett.* **69**, 2033-2035.
7. Brongersma, M.L., Polman, A., Min, K.S., Boer, E., Tambo, T., and Atwater, H.A., (1998) Tuning the emission wavelength of Si nanocrystals in SiO_2 by oxidation, *Appl. Phys. Lett.* **72**, 2577-2579.
8. Klimov, V.I., Schwarz, Ch., McBranch, D., and White, C.W., (1998) Initial carrier relaxation dynamics in ion-implanted Si nanocrystals: Femtosecond transient absorption study, *Appl. Phys. Lett.* **73**, 2603-2605.
9. Shimizu-Iwayama, T., Kunumado, N., Hole, D.E., and Townsend, P., (1998) Optical properties of silicon nanoclusters fabricated by ion implantation, *J. Appl. Phys.* **83**, 6018-6022.
10. Linnros, J., Lalic, N., Galeckas, A., and Grivickas, V., (1999) Analysis of the stretched exponential photoluminescence decay from nanometer-sized silicon crystals in SiO_2, *J. Appl. Phys.* **86**, 6128-6134.
11. Iacona, F., Franzò, G., and Spinella, C., (2000), Correlation between luminescence and structural properties of Si nanocrystals, *J. Appl. Phys.* **87**, 1295-1303.
12. Lu, Z.H., Lockwood, D.J., and Baribeau, J.M., (1995), Quantum confinement and light-emission in SiO_2/Si superlattices, *Nature* **378**, 258-260.
13. Vinciguerra, V., Franzò, G., Priolo, F., Iacona, F., and Spinella, C., (2000) Quantum confinement and recombination dynamics in silicon nanocrystals embedded in Si/SiO_2 superlattices, *J. Appl. Phys.* **87**, 8165-8173.
14. Photopoulos, P., Nassiopoulou, A.G., Kouvatsos, D.N., and Travlos, A., (2000) Photoluminescence from nanocrystalline silicon in Si/SiO_2 superlattices, *J. Appl. Phys.* **76**, 3588-3590.
15. Priolo, F., Franzò, G., Pacifici, D., Vinciguerra, V., Iacona, F., and Irrera, A., (2001) Role of the energy transfer in the optical properties of undoped and Er-doped interacting Si nanocrystals, *J. Appl. Phys.* **89**, 264-272.
16. Hayashi, S., Nagareda, T., Kanzawa, Y., and Yamamoto, K., (1993) Photoluminescence of Si-rich SiO_2-films - Si clusters as luminescent centers, *Jpn. J. Appl. Phys.* **32**, 3840-3845.
17. Werwa, E., Seraphin, A.A., Chin, L.A., Zhou, C., and Kolenbrander, K.D., (1994) Synthesis and processing of silicon nanocrystallites using a pulsed laser ablation supersonic expansion method, *Appl. Phys. Lett.* **64**, 1821-1823.
18 Pavesi, L, Dal Negro, L, Mazzoleni, C, Franzò, G, Priolo, F, (2000) Optical gain in silicon nanocrystals, *Nature* **408**, 440-444.
19 Franzò, G., Priolo, F., Coffa, S., Polman, A., Carnera, A., (1994) Room-temperature electroluminescence from Er-doped crystalline Si, *Appl. Phys. Lett.* **64**, 2235-2237.
20. Kenyon, A.J., Trwoga, P.F., M. Federighi, and C.W. Pitt, (1994) Optical-properties of PECVD erbium-doped silicon-rich silica - evidence for energy-transfer between silicon microclusters and erbium ions, *J. Phys.: Condens. Matter* **6**, L319- L324.
21 Fujii, M., Yoshida, M., Kanzawa, Y., Hayashi, S., and Yamamoto, K., (1997) 1.54 μm photoluminescence of Er^{3+} doped into SiO_2 films, containing Si nanocrystals: Evidence for energy transfer from Si nanocrystals to Er^{3+}, *Appl. Phys. Lett.* **71**, 1198-1200.
22. Franzò, G., Vinciguerra, V., and Priolo, F., (1999) The excitation mechanism of rare-earth ions in silicon nanocrystals, *Appl. Phys. A* **69**, 3-12.
23. Franzò, G., Pacifici, D., Vinciguerra, V., Iacona, F., and Priolo, F., (2000) Er^{3+} ions-Si nanocrystals interactions and their effects on the luminescence properties, *Appl. Phys. Lett.* **76**, 2167-2169.
24. Franzò, G., Irrera, A., C. Moreira, E., Miritello, M., Iacona, F., Sanfilippo, D., Di Stefano, G., Fallica, P.G., and Priolo, F., (2002) Electroluminescence of silicon nanocrystals in MOS structures, *Appl. Phys. A* **74**, 1-5.
25. Irrera, A., Pacifici, D., Miritello, M., Franzò, G., Priolo, F., Iacona, F., Sanfilippo, D., Di Stefano, G., and Fallica, P.G., (2002) Excitation and de-excitation properties of silicon quantum dots under electrical pumping, *Appl. Phys. Lett.* **81**, 1866-68.
26. Calcott, P.D.J., Nash, K.J., Canham, L.T., Kane, M. J., and Brumhead, D., (1993) Identification of radiative transitions in highly porous silicon, *J. Phys. Condens. Matter* **5**, L91-L98.
27. Brongersma, M.L., Kik, P.G., Polman, A., Min, K.S., and Atwater, H.A., (2000) Size-dependent electron-hole exchange interaction in Si nanocrystals, *Appl. Phys. Lett.* **76**, 351-353.
28. Yu. Kobitski, A., Zhuravlev, K.S., Wagner, H.P., and Zahn, D.R.T., (2001) Self-trapped exciton recombination in silicon nanocrystals, *Phys. Rev. B* **63**, 115423-1-115423-5.

OPTICAL AND ELECTRICAL CHARACTERISTICS OF LEDs FABRICATED FROM Si-NANOCRYSTALS EMBEDDED IN SIO_2

B. GARRIDO, O. GONZÁLEZ, S. CHEYLAN, M. LÓPEZ, A. PÉREZ-RODRÍGUEZ, C. GARCÍA, P. PELLEGRINO, R. FERRER, J.R. MORANTE
EME, Department d'Electrònica, Universitat de Barcelona, Martí i Franquès 1, 08028 Barcelona, Spain
J. DE LA TORRE, A. SOUIFI, A. PONCET, C. BUSSERET, M. LEMITI, G. BREMOND, G. GUILLOT
Laboratoire de Physique de la Matière, UMR-CNRS 5511, INSA de Lyon, Bât.502, 20 Av. Albert Einstein, 69621 Villeurbanne cedex, France

1. Introduction

Structures containing Si nanocrystals embedded in SiO_2 are candidates for optoelectronic and photonic applications due to their intense visible emission at room temperature, high thermal and chemical stability and compatibility with CMOS technology. Extensive investigations have demonstrated strong tunable photo- and electroluminescence (PL, EL), and even optical amplification [1-6]. The photon absorption in such systems is a fundamental transition modulated by quantum confinement effects and takes place in the core of the nanocrystal, while the PL consists of an intense and wide emission peaking in the near infrared or visible spectrum [7]. The Si-SiO_2 interface plays a dominant role in the emission mechanism. This is possible either through the existence of surface radiative states [8] or because of vibronic interactions among electron-hole pairs created in the nanocrystals and the polarizable material surrounding them [9,10]. The suppression of non-radiative defects at the Si-SiO_2 interface is a process required in order to increase the radiative yield without affecting the emission mechanism. It has been reported that an increase in radiative efficiency is possible by a suitable thermal treatment in hydrogen atmosphere [11-12]. However, the microscopic nature of the H-passivated defects was not fully understood. This research topic has been addressed by us extensively in recent papers [13-14].

Nevertheless, it remains quite difficult to obtain radiative recombination by means of electrical pumping in Si-nc/SiO_2 systems, and thus produce efficient and reliable LEDs. EL visible emission of Si implanted SiO_2 was first reported in 1984 by DiMaria et al. [15]. The purpose at that time was not producing LEDs, the mechanism of luminescence was not understood and the subject remained into oblivion until the rush of reports in the 90's following the discovering of PL of nanostructured silicon [16]. Recently, several attempts to obtain stable and reliable Si-based LEDs have been rather successful. The approaches were based either on excitation of Si-nc embedded in SiO_2

L. Pavesi et al. (eds.), Towards the First Silicon Laser, 45–54.

[5,17-18] or defects in SiO_2 created by ion implantation [19-21]. Nevertheless, the carrier injection remained in the Fowler-Nordheim regime which brings about device degradation due to hot carrier injection. Very recently, we have reported the production of a stable and efficient LEDs based on Si-nc in SiO_2 which were optimised to allow both electron and tunnel injection into the nanocrystals [22].

In this paper we report the development and optimization of highly luminescent Si-nc/SiO_2 systems with the aim of fabricating efficient LEDs. We present a systematic study of the evolution of PL emission and absorption features of nanocrystals versus implantation dose and annealing conditions and find a correlation between the structural and optical properties. We address as well the issue of identifying the non-radiative defects that quench PL and establish a quantitative relationship between their concentration and PL intensity. After a description of the LED process, we model charge injection and I-V characteristics in the forward and reverse bias conditions. Finally, EL measurements are described and the results are related to the PL and the structural characterisation.

2. Experimental

The thickness of the SiO_2 was about 400 nm for the optical and structural studies. The samples were implanted with 150 keV Si^+ single implants to doses up to $3x10^{17}$ ions/cm^2. The peak of Si excess ranged from 1 atomic percentage (1 %) to 30 %. Afterwards, the samples were annealed at 1100 °C in N_2 atmosphere for different durations, ranging from 1 min to 16 h. In addition, some selected samples underwent forming gas (95% N_2+ 5% H_2) post-annealing at 450 °C for duration from 15 to 120 min. SiO_2 films of 20-40 nm thick were grown for the devices. Single implantations at 15 KeV were performed for the studies of charge injection into the Si-nc. Double implantations at 15 KeV and 25 KeV were performed to obtain planar profiles of Si-nc for the LEDs. The implanted doses correspond to Si excess between 1 and 15%. Afterwards the samples were annealed at 1100 °C under N_2 for 4 h. The LEDs were fabricated by etching the superficial unimplanted layer followed by the evaporation of a thin Al film (15nm). The contact at the back side was made with InGa and silver lack.

The PL measurements were performed by exciting the samples at room temperature with an He-Cd laser (325 nm,) and by detecting the emission with a photomultiplier. Electron spin resonance (ESR) experiments were performed at room temperature using conventional absorption ESR at 9.1 GHz and employing a 100 kHz field modulation when detecting the first derivative of absorption. High-resolution electron microscopy (HREM) was used for measuring the size distribution of nanocrystals. Full details on the TEM, ESR and PL characterization are provided in [14]. I-V curves were recorded with a Keithley 4200 data acquisition system.

3. Optical properties of Si-nc embedded in SiO_2

Figure 1a shows the Si-nc mean diameter as a function of annealing time at 1100 °C. The results are shown only for the 10 to the 30% of Si excess because visibility reasons prevented us to build reliable size histograms for lower excess. The mean diameter increases very slowly when the annealing time changes from 1 min to 16 h (less than 20% of relative increase). This is consistent with an asymptotic Ostwald ripening growth stage of the nanocrystals [14] and with the small values of the diffusion coefficient of Si in SiO_2 (10^{-17} cm^2/s at 1100 °C). On the contrary, for a fixed annealing time such as 16 h, the mean diameter increases roughly from 3 to 5.5 nm when the excess changes from the 10 to the 30%. This remarkable increase of diameter is in contrast with the standard Ostwald ripening theory (dilute systems), which predicts a final diameter that is almost independent of the initial excess. These results can be reconciled with theory if one considers proximity effects by taking into account overlapping gradients of Si excess for neighbor nanocrystals. The curves in Fig. 1a display the results of the simulations. The main conclusion of such a study is that the average diameter of Si-nc in SiO_2 is between 2-3 nm for Si excess up to 10-15% (weak dependence with Si excess) and starts to grow fast afterwards.

All the annealed samples are characterized by a strong and wide (about 0.28 eV at half maximum) PL emission extended through the red and near infrared. The PL emission (25 mW excitation @ 325 nm) is always visible with the naked eye under ordinary laboratory illumination and is comparable with the PL intensity of porous Si. We have represented in Fig. 1b the evolution of PL intensity with Si excess for samples annealed 8 h at 1100 °C in N_2, that is to say, well into the ripening stage. It is remarkable the linear dependence of PL intensity versus Si excess in the range from 1 to 10% ("dilute system"). Nevertheless, for the 20 and 30% of Si excess ("non-diluted system") there is a substantial decrease of PL intensity. This decrease is clearly linked to: i) An increase in the density of Si-nc and thus self-absorption; ii) A decrease in the oscillator strength per Si-nc when the size increases (see below) and iii) Loss of confinement due to proximity effects. It is then clear that the highest efficiency is obtained for a Si excess somewhere between 10-20%. Indeed, we obtained the highest efficiency for the 15% of Si excess in the set of samples prepared for the LEDs. As the average diameter is 3 nm ($\approx 10^3$ atoms), the optimum density is about 10^{19} Si-nc/cm^3.

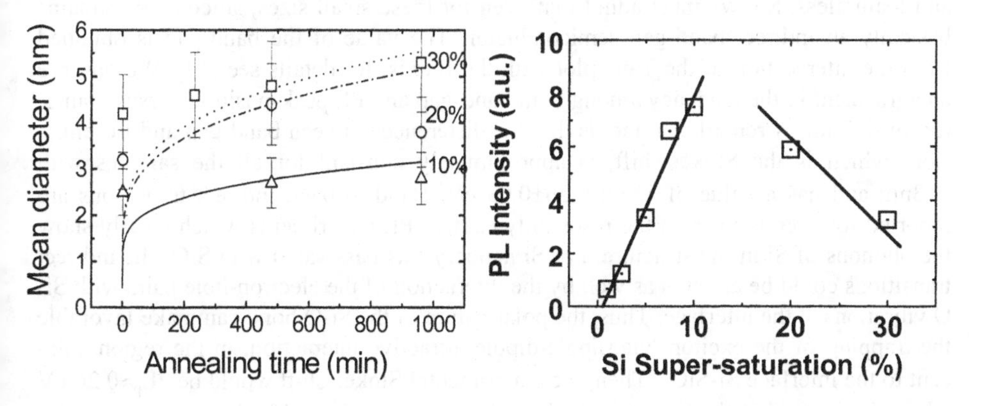

Figure 1. a) Left: Experimental values (symbols) and simulated curves of mean diameter vs. annealing time; b) Right panel: PL intensity vs. Si excess.

Figure 2a shows the peak-position of PL vs. Si excess. The peak-position is pinned at about 1.75 eV for Si excess of less than 10% (d≈2-3 nm), while it steadily shifts to the infrared with increasing Si excess, down to 1.4 eV for the 30% (d≈5.5 nm). So, recalling the increase of average size displayed in Fig. 1a, this is a clear hint that quantum confinement effects occur for sizes larger than 3 nm. On the contrary, the emission is not wavelength dependent for smaller Si-nc, in agreement with the surface states postulated by Wolkin et al. [8]. Although a certain degree of tunability of the emission in the red is shown for Si-nc passivated by SiO_2, this is far from that shown by porous silicon and other nanostructures passivated with H.

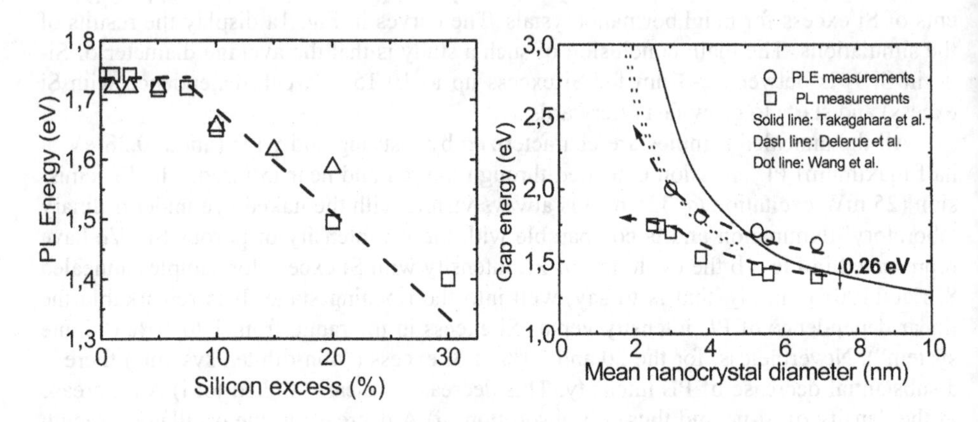

Figure 2. a) Left: Evolution of PL energy vs. Si excess; b) Righ: Theoretical data for the band-gap energy. The symbols are the experimental results of this work.

Figure 2b shows the band gap energy obtained from the PLE spectra, the PL peak energy and the results of the theoretical calculation of band-gap vs. size by different authors [23-25]. All the PLE spectra obtained are smooth, almost quadratic with energy and featureless. So, we must admit that even for these small sizes, silicon still remains basically an indirect band-gap semiconductor. The value of the band-gap is obtained from the intersection of the Tauc plot with the E axis; for details see [14]. We observe an agreement in the tendency among both band-gap and PL peak-position versus diameter for d>3nm. A remarkable fact is that the difference between band-gap and PL emission, which is the Stokes-shift, is approximately constant for all the samples with d>3nm and has a value of about 0.26±0.03 eV. Band to band indirect transitions are reported for porous Si from the resonantly excited PL experiments which clearly show the phonons of Si in the structure. For Si nanocrystals passivated with SiO_2 the indirect transitions could be assisted as well by the interaction of the electron-hole pairs with Si-O vibrations at the interface. Thus, the polar nature of the Si-O bond can make favorable the trapping of the exciton, via dipole-dipole attractive interaction, in the region adjacent to the interface Si-SiO_2. Then, the experimental Stokes-shift would be $2E_p \approx 0.26$ eV where E_p is a vibronic excitation and, in consequence, $E_p \approx 0.13$ eV, which coincides with the energy of the Si-O vibration. Kanemitsu [9] has reported certain steps in the resonantly excited spectra that can account for this interaction with the SiO_2 vibrations. Nevertheless, this recombination channel does not exclude the direct recombination or indirect assisted by Si phonons or recombination via surface states. In fact, all our results are consistent with the existence of surface states for small Si-nc.

We have also investigated the ability of surface passivation to enhance the PL emission. As reported above, Si precipitation in implanted samples takes place in a time scale of few minutes at 1100 °C. For longer annealing at the same temperature, the PL intensity of the Si nanocrystals increases and eventually reaches saturation, while it correlates inversely with the amount of Si dangling bonds at the Si-SiO_2 interface (P_b centers), as measured by electron spin resonance. This combined behaviour is shown for a particular sample in Fig. 3a, but is independent on the silica matrix properties, as reported in [13,26]. The observation that the light emission enhancement is directly related to the annealing of P_b centres is confirmed by treatment in forming gas. This mild hydrogenation at much lower temperature (450 °C) leads to a complete passivation of the P_b defects, increasing at the same time the PL yield and the lifetime.

We have also measured the absorption cross-sections and the PL lifetimes as a function of size [27]. The lifetimes span from 50 μs for the smallest Si-nc (d≈2.5 nm) to more than 200 μs for the largest Si-nc (d≈5.5 nm). Once known the PL decay lifetimes, one can calculate the absorption cross-sections per nanocrystal by fitting the power dependence of the PL intensity [27]. The results as a function of Si-nc size are reported in Fig. 3b, for 457 nm excitation. The cross-section can be expressed as $\sigma(E) \propto g(E)f(E)$ where g(E) is the density of states and f(E) is the oscillator strength of the optical transition. We find that the absorption cross-section increases for increasing size up to roughly 3 nm of diameter. This initial behaviour is easily understood if we consider that the excitation at 457 nm is not able to excite efficiently the smallest Si-nc due to the small density of states of them at that energy. On the contrary, we find a diminution of the cross-section for larger sizes, which can be clearly interpreted as a reduction of the oscillator strength for the optical transition in large nanocrystals. This result is in perfect agreement with the PL intensity dependence reported above (see Fig. 1b).

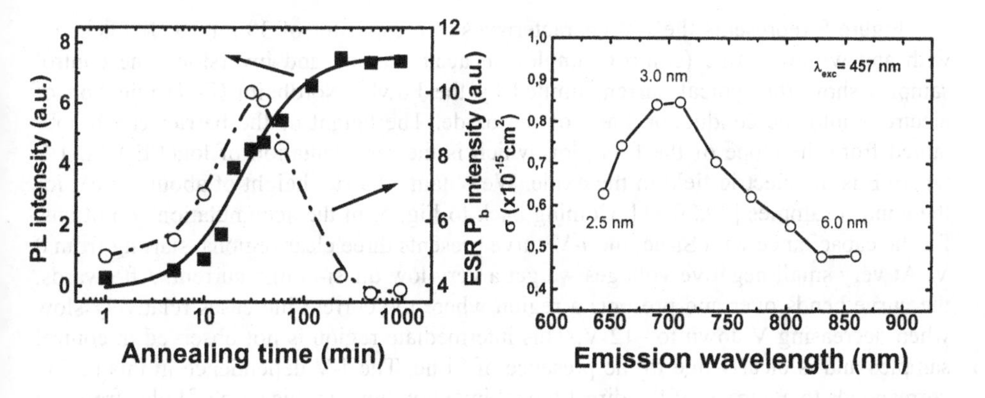

Figure 3. a) Correlation of PL int. with P_b centers; in the first min of annealing, the interfaces form and number of P_b increases; b) Righ: absorption cross-section vs. size.

4. Electro-optical characteristics of Si-nc based LEDs

We have measured the electro-optical characteristics in the MOS capacitors depicted in Fig. 4. The thickness of the gate oxide was 20, 30 and 40 nm while a thick field oxide of 800 nm was grown for isolation. For simultaneous injection of electron and holes in "direct" polarisation we engineered the gate contact as an electron injector with either n-polysilicon or Al. For hole injection the substrate was in all cases p-doped silicon. In a first stage, a single implantation of 1-2x10^{16} cm^{-2} Si atoms at 15 keV was done for the three thickness to study electron and hole injection from the substrate into the Si-nc. These doses correspond to a 5-10% of Si excess at the projected range (20 nm). Recent publications have shown that for low energy implantation a denuded zone of Si-nc is always present close to the interface, regardless of the implantation profile because the Si-SiO_2 interface acts as a sink of Si atoms [28]. An estimation of the tunnelling distance from the nanocrystals to the substrate is in Table I. The control oxide thickness is of about 15 nm and prevents electron injection from the gate for low voltage.

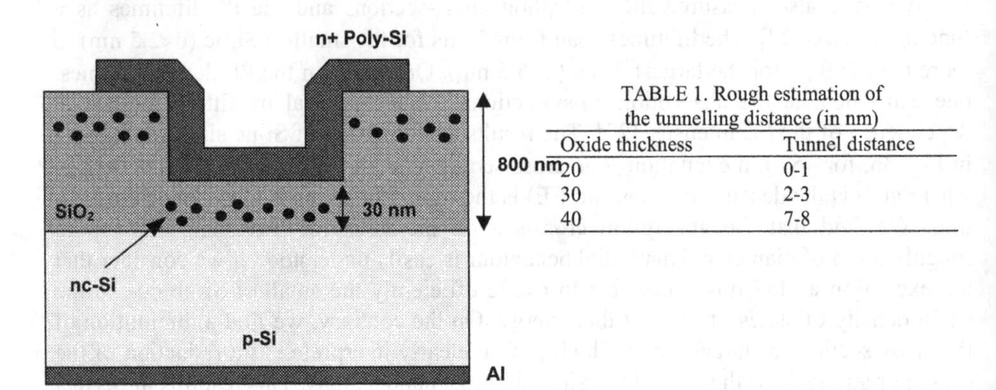

TABLE 1. Rough estimation of the tunnelling distance (in nm)

Oxide thickness	Tunnel distance
20	0-1
30	2-3
40	7-8

Figure 4. Cross-section of the devices

Figure 5 represents the J-V characteristics of a capacitor of 30 nm oxide thickness with and without Si-nc (control sample), in accumulation and inversion. The control samples show the typical current limited by the Fowler-Nordheim (F-N) injection of electrons into the conduction band of the oxide. The height of the barrier can be obtained from the slope of the F-N plot, which is the representation of log(J/E^2) vs. 1/E where E is the electric field in the oxide. We obtain a barrier height of about 2.5 eV for the control samples [29,30,31]. Coming back to Fig. 5, in the accumulation conditions, for the capacitance with Si-nc, the J-V curve presents three clear regions, starting from 0 V. At very small negative voltages we get a very low quasi-ohmic current. Afterwards, the curve bends over and presents a region where the current increases relatively slow when decreasing V down to –12 V. This intermediate region is not observed in control samples and is clearly due to the presence of Si-nc. The J-V dependence in this region corresponds to a current of the direct tunnel injection type, in this case of holes from the Si substrate to the Si nanocrystals, as can be seen from the linear slope of the representation of log(J/E) vs. E. For higher voltages, the current density is purely of the F-N type, but with a lower slope compared to the non-implanted sample. The currents for inversion (positive V applied to the gate) can be interpreted similarly as a tunnel electron injection from the substrate to the Si-nc for low V and a F-N type for higher V. The saturation of the current in the inversion regime is due to exhaustion of minorities in the substrate (electrons in a p-type substrate).

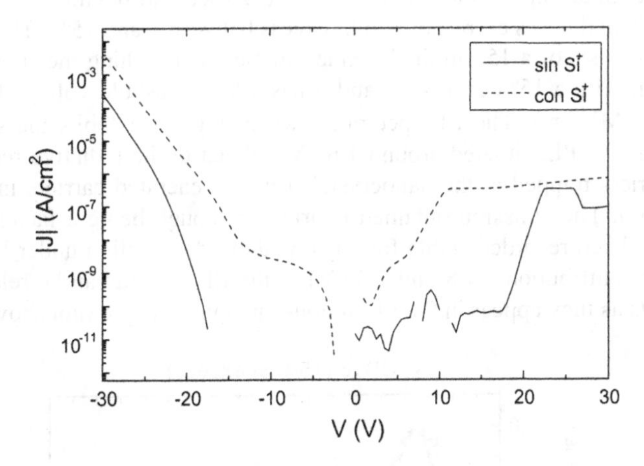

Figure 5. J(V) characteristics of samples with and without Si-nc.

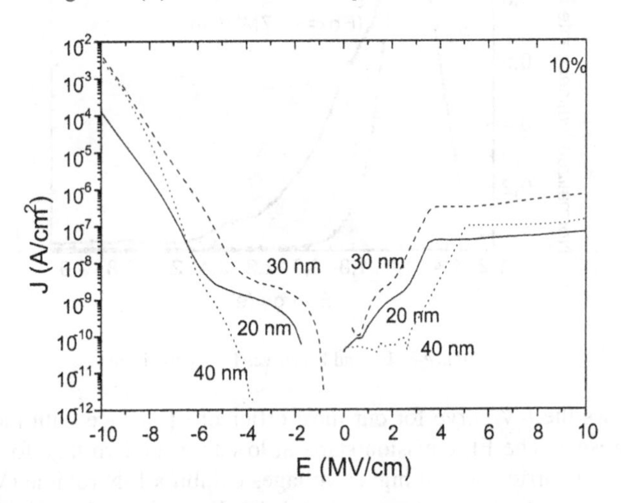

Figure 6. J-V characteristics for different tunnel oxides.

The effect of the different tunnel oxide thickness (see Table I) is displayed in Fig. 6. In the oxides of 20 and 30 nm, the effect of the nanocrystals is clearly visible at low voltages in both accumulation and inversion. The difference observed between the 20 and 30 nm thick oxide case is due to the position of the Si-nc distribution in the oxide and the difference in their density. For the same electric field applied, the 30 nm oxide presents a distribution wider and with more density of Si-nc, and this does favour the injection of carriers at lower voltages. Nevertheless, the sample 40 nm thick (7-8 nm of tunnel oxide) behaves almost identically than the non-implanted sample. In conclusion, if the nanocrystals are far from the interface, the injection will require higher voltages to reach the first nanocrystals, being dominant the F-N injection. The tunnel oxide thickness for reasonable voltage is about 3 nm, corresponding to the 30 nm thick oxide.

For maximising the EL in the LED devices we used only 40 nm thick oxides and we performed double implantation at 15 KeV and 25 KeV to obtain planar profiles of Si-nc. The implanted doses correspond to Si excess between 1 and 15%. The best device so far for EL consists of a 15 nm thick oxide sample, after etching the superficial non-implanted layer, with a 15% excess Si, and it has a low threshold voltage EL emission (≈7 V, i.e. E=5 MV/cm). The EL spectra shown in Fig. 7a exhibits the same energy peak position as the PL, situated around 1.6 eV, related to the radiative recombination of injected carriers trapped at the nanocrystals (photo-generated carriers in the case of the PL emission). The emission obtained is bright, although being a factor 7 less than the PL, and has been recorded stable for days, without degradation under low polarisation. The other contributions (1.8 and 2.3 eV) to the EL spectra can be related to SiO_2 radiative defects as they appear in the PL of non-implanted sample (not shown).

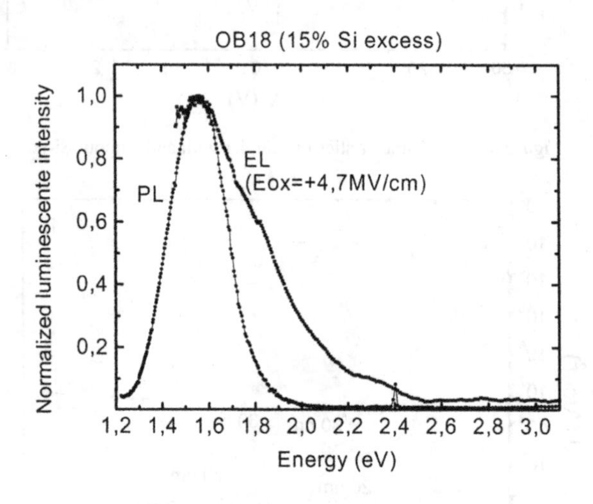

Figure 7. EL and PL curves for comparison

Figure 8 shows the I-V curve for our most efficient EL device with the current injection regime present. The EL emission starts at low threshold voltage for a regime of tunnel conduction of carriers and at higher voltages exhibit a F-N regime (V> 10-12V), for which the EL intensity does not increase and the device degrades. The EL intensity increases with current and saturates eventually. A linear variation of ln(J/E) as a function of E is obtained. This result shows that the conduction is indeed a direct tunneling between nanocrystals and electrodes. The theoretical tunnel injection model fits the experimental curve. From the slope of the plot, the average distance between tunneling centres is in the range of 1.4–1.9 nm. This coincides with the average distance between surfaces of Si-nc of about 3 nm of diameter for 10-15% excess. Then, one can conclude that the conduction is possible either from dot to dot, or assisted by tunnelling centres between dots. The important point is that the current is not thermally activated and is mainly due to cold carriers. Further evidence of such results is the very weak temperature dependence of I-V in accordance with a tunnelling conduction process.

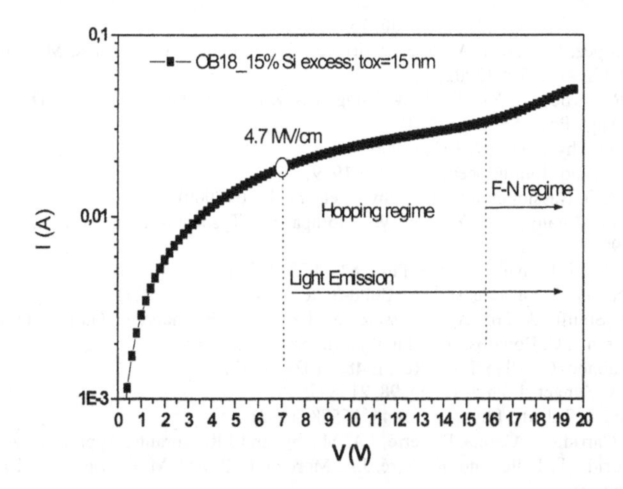

Figure 8. I-V curves and EL

5. Conclusions

We have studied and optimized the emission properties of Si nanocrystals embedded in SiO_2 synthesized by ion implantation plus annealing. The most efficient structures have Si-nc with average size of 3 nm and densities of 10^{19} cm^{-3}. Moreover, we have studied and optimized the injection of electron and holes from the electrodes to Si-nc situated at different depths. We have obtained red EL at room temperature and the I-V characteristics prove that the current is related to a pure tunneling process. Fowler-Nordheim injection is not observed during light emission for electric fields below 5 MV/cm. Thus, hot carrier injection is avoided and efficient and reliable devices are obtained.

References

[1] T.S. Iwayama, S. Nakao and K. Saitoh, Appl. Phys. Lett. **65**, 1814 (1994).

[2] K.S. Min, K.V. Sheheglov, C.M. Young, H.A. Atwater, M.L. Brongersma and A. Polman, Appl. Phys. Lett. **69**, 2033 (1996).

[3] P.F.Trowga, A.J. Kenyon and C.W. Pitt, J. Appl. Phys. **83**, 3789 (1998).

[4] V. Vicinguerra, G. Franzò, F. Priolo, F. Iacona and C. Spinella, J. Appl. Phys. **87**, 8165 (2000).

[5] G. Franzò, A. Riera, E.C. Moreira, M. Miritello, F. Iacona, D. Sanfilippo, G. Di Stefano, P.G. Fallica and F. Priolo, Appl. Phys. A **74**, 1 (2002).

[6] L.Pavesi, L.Dal Negro, C.Mazzoleni, G.Franzò and F.Priolo, Nature **408**, 440 (2000).

[7] T. S. Iwayama, N. Kurumado, D.E. Hole and D.E. Townsend, J. Appl. Phys. **83**, 6018 (1998).

[8] M.V. Wolkin, J. Jorne, P. Fauchet, G. Allan, and C. Delerue, Phys. Rev. Lett. **82**, 197 (1999).

[9] Y. Kanemitsu, Phys. Rev. B **53**, 13515 (1996).

[10] B. Garrido, M. López, O. González, A. Pérez-Rodríguez, J.R. Morante and C. Bonafos, Appl. Phys. Lett. **77**, 3143 (2000).
[11] S. Guha, M.D. Pace, D.N. Dunn, and I.L. Singer, Appl. Phys. Lett. **70**, 1207 (1997).
[12] S. Cheylan, R.G. Elliman, Appl. Phys. Lett. **78**, 1225 (2001).
[13] M. Lopez, B. Garrido, C. Garcıa, P. Pellegrino, A. Perez-Rodrıguez, J. R. Morante, C. Bonafos, M. Carrada, and A. Claverie, Appl. Phys. Lett. **80**, 1637 (2002).
[14] B. Garrido, M. Lopez, C. Garcıa, A. Perez-Rodrıguez, J. R. Morante, C. Bonafos, M. Carrada, and A. Claverie, J. Appl. Phys. **91**, 798 (2002).
[15] D.J. DiMaria, J.R. Kirtley, B. Pakulis, D.W. Dong, T.S. Kuan, F.L. Pesavento, T.N. Theis, J.A. Cutro, S.D. Brorson, J. Appl. Phys. **55**, 401 (1984).
[16] L.T. Canham, Appl. Phys. Lett. **57**, 1046 (1990).
[17] N. Lalic, J. Linnros, J. of Luminescence **80**, 263 (1999).
[18] P. Photopoulos , A.G. Nassiopoulou, Appl. Phys. Lett. **77**, 1816 (2000).
[19] L. Rebohle, J. Von Borany, R.A. Yankov, W. Skorupa, I.E. Tyshenko, H. Frob, K. Leo, Appl. Phys. Lett. **71**, 2809 (1997).
[20] H. Song, X. Bao, N. Li, J. Zhang, J. Appl. Phys. **82**, 4028 (1997).
[21] T. Matsuda, M. Nishio, T. Ohzone, H. Hori, Solid-State Electr. **41**, 887 (1997).
[22] J. de la Torre , A. Souifi , A. Poncet, C. Busseret, M. Lemiti , G. Bremond, G. Guillot., O. Gonzalez, B. Garrido, J. R. Morante, C. Bonafos, Appl.Phys. A, (in press).
[23] C. Delerue, M. Lannoo, G. Allan, Phys. Rev. B **48**, 11024 (1993).
[24] L.W. Wang and A. Zunger, J. Phys. Chem. **98**, 2158 (1994).
[25] T. Takagahara and K. Takeda, Phys. Rev. B **46**, 15578 (1992).
[26] P. Pellegrino, B. Garrido, C. García, R. Ferré, J.A. Moreno and J.R. Morante, Appl. Phys. A (in press).
[27] C. García, B. Garrido, P. Pellegrino, R. Ferré, J.A. Moreno, L. Pavesi, M. Cazzanelli and J.R. Morante, Appl. Phys. A (in press).
[28] T. Müller, K.H. Heinig, W. Möller, Appl. Phys. Lett. 2002 (in press).
[29] S.M. Sze, "Physics of Semiconductor Devices", John Wiley & Sons, 1981.
[30] E. Kameda, T. Matsuda, Y. Emura and T. Ohzone T., Solid-State Electron. **42**, 2105 (1998).
[31] D.R. Wolters and H.L. Peek, Solid-State Electron. **30**, 835 (1987).

ELECTROLUMINESCENCE IN SI/SIO$_2$ LAYERS

L. HEIKKILÄ, R. PUNKKINEN, and H.-P. HEDMAN
Laboratory of Microelectronics, Department of Information Technology
University of Turku, 20014 Turku, Finland

1. Introduction

Silicon nanocrystals (NCs) and artificial nanostructures are potential building blocks of future nanoelectronics and nanophotonics. Electronics is mainly fabricated by silicon technology, but there has been a lack of effective light sources, which could be easily coupled to silicon-based electronics. Among other materials, nanocrystalline silicon has attracted much interest after the discovery of strong photoluminescence (PL) emission from porous Si [1]. Electroluminescence (EL) was detected in 1984 from silicon chopped into tiny pieces like nanoparticles embedded in a silicon dioxide layer [2]. EL was attributed to quantum confinement effect and inter-band transition in the tiny Si islands. EL has also been studied from native oxide film and from extra-thin Si-rich silica film on p-Si structure and EL was attributed to radiative recombination of electron-hole pairs via the luminescence centers (LCs) in films rather than in nanometer scale Si particles in silicon oxide [3, 4]. The EL mechanism in silicon NCs and nanolayers has been intensively debated and several models proposed, but a comprehensive theory has not been found yet. Recent discoveries [5, 6, 7] of light amplification in silicon dioxide containing silicon nanoparticles suggests that a certain pump power level is needed before the amplification starts. This has raised the idea that in EL samples might have similar situation: the amplification starts when the intensity is high enough. In our experiments electroluminescence spectra are broad, typically 100 - 150 n wide, which can be related to the presence of many distinct and differently coloured spots. By reducing the effective EL area it is believed that the current density will be increased and thus possible narrowing of the spectrum [8] at certain wavelengths could be detected.

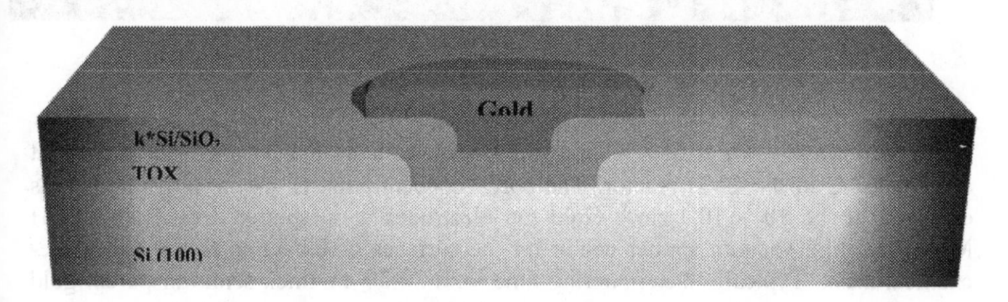

Figure 1. A structural layout of the processed EL samples on a silicon wafer.

L. Pavesi et al. (eds.), Towards the First Silicon Laser, 55–60.

In this paper we present results from two different sample types. The first one is a Si/SiO_2 structure grown over a thermal oxide layer with small openings in the silicon wafer. The second one is a p-type polysilicon layer prepared on quartz substrate. We have studied morphology and EL from these structures by using AFM, optical microscopy, spectrometry and TEM. Sample areas will vary from 100 μm^2 to 1 mm^2 and the maximum current densities are up to several kA/cm^2. The density, size and colors of EL spots are studied by using optical microscopy.

2. Preparation of samples

Most of the samples were processed on 2-5 Ωcm p-type (100)-oriented silicon wafers and one quartz wafer. The wafers were carefully washed with a RCA treatment and finally dipped in HF-solution (5%) and dried before inserting into the processing equipment or into the oxidation oven. In older experiments [8] the current is passing through whole gold electrode producing EL from an area of several square millimeters. In this way only small current densities can be achieved which does not favor any LED or laser emission. In order to reduce this area, silicon wafers were oxidized thermally (70 or 130 nm) and small openings were lithographically patterned into the oxide layer. The wafers were then wet etched in HF (5%) or dry etched in CF_4/H_2 -plasma at 0.1 mbar.

Polysilicon and SiO_2 layers were grown on the top of the oxide by CVD. Silicon layers were grown at 0.064 mbar pressure and 650 - 700 °C temperature (silane flow rate 15 $cm^3 min^{-1}$) and typical growth time was 5-7 min. For SiO_2 layers O_2 flow (rate 50 cm^3 min^{-1}) was added (total pressure 0.1 mbar) and the time was varied from 0.5 to 2 min depending on the aimed thickness. For multilayer structures ($k*Si/SiO_2$) these process phases were repeated.

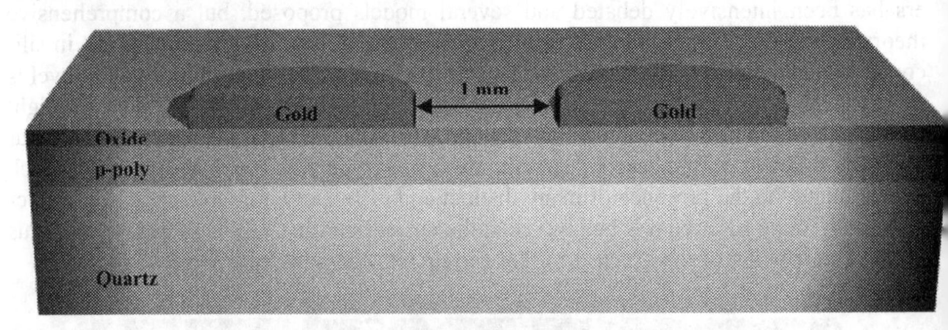

Figure 2. Layout of the structure on a quartz wafer.

On the quartz wafer a boron doped polysilicon (200-300 nm) and a thin oxide layer (some nm) was grown at 650 °C and the wafer was annealed 20 min in hydrogen at about 800 °C in the CVD reactor. The boron concentration of p-type polysilicon was estimated to be 10^{16}- 10^{17} cm^{-3}. Gold top electrodes were sputtered by BAL-TECH MED 020 high vacuum sputter coater on the surfaces of the samples through a mechanical mask. Typically the sputtering time for a 20-30 nm thick semitransparent gold film was 270 s when the argon pressure was $2x10^{-2}$ mbar and the current 40 mA.

3. Measurements

All spectral measurements were made by using a commercial spectrometer (Oriel 77250 1/8m Monochromator) with a cooled (-20 °C) photomultiplier tube with spectral range 160 -900 nm. In the spectrometer the grating response is changed approx. at 600 nm which yields a sharp notch in the spectrum. Typical wavelength window in the monochromator was 15 nm and the spectrum was taken at 5 nm steps in about 12 minutes. The PMT was used in current mode. The represented spectra are not corrected for the sensitivity of the spectrometer.

The samples were studied by AFM (Park Scientific), TEM (Jeol) and optical microscope. Photographs were taken in two ways: In normal microphotography (magnifications typically 200–500) typical exposure times were around 5 minutes. The microscope was also equipped with a cooled CCD cell (KAF 1600-2, 1.2 million pixels) to ensure fast shots. Usual exposure times were 1–2 s. Oil immersion objectives were used for the highest magnification (100x, NA 1.3). Mathematically the magnification was such that one pixel in CCD cell corresponded to 92 nm in real sample.

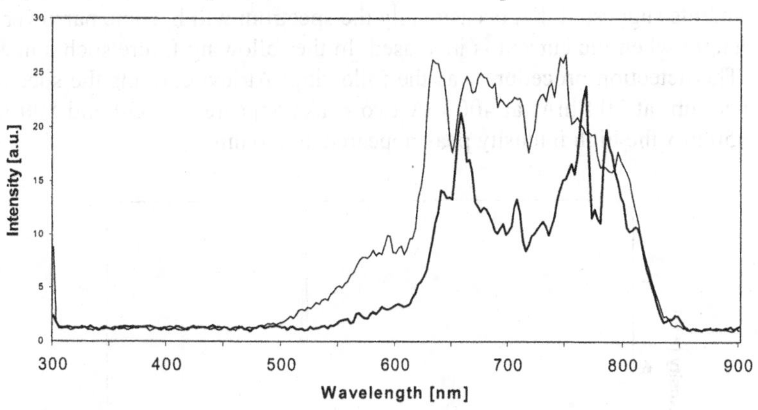

Figure 3. Electroluminescence spectra from Si/SiO_2 layers on the quartz substrate are shown at 5 mA (thin line) and 40 mA (thick line).

4. Electroluminescence from Si/SiO_2 layers on quartz substrate

Here we report EL from Si/SiO_2 structure on quartz substrate. In the experiments a DC voltage was applied to the gold electrodes locating on top of the structure (Fig. 2). Typically EL is seen from electrodes, just very close to the edge of the electrode (location of bright spots about 200 μm from the edge of the gold layer). This might be due to the fact that electric field is strongest at edges. EL is seen simultaneously from both electrodes but normally more luminating spots and stronger intensity was found from the negative electrode. The resistance between the electrodes was around 10 kΩ. Typically, visible EL started at currents around 2 mA. Increasing the current EL was increased simultaneously and the maximum intensity was reached at 5 mA. Further current increase reduced the intensity. Two EL spectra from the sample on quartz with 5 and 40 mA driving currents are shown in figure 3. The spectra revealed the red spots seen by the microscope. The number of luminating spots was only some tens, which clearly differs from samples on silicon substrate. Thus one might suggest that the sharp peaks in the spectra are due to the most luminating spots.

5. Electroluminescence from Si/SiO_2 layers on silicon substrate

Several luminating samples were studied, photographed and their spectra recorded. The typical sample size was about 1 cm^2 and the effective openings in the samples varied from 1 mm^2 to 100 μm^2 (Fig. 1). For the detection of the spectrum a certain "base" intensity has to be supplied, i.e. the sample is usually seen by naked eye in a dark room. This boundary was reached by having current densities from 10Acm^{-2} (1 mm^2) to 4.5 $kAcm^{-2}$ (100 μm^2). In the case of the smallest samples this is about the current density the gold electrode barely holds. At higher densities the gold electrode is possibly damaged. In order to prevent overheating the samples are glued with a silver paste to a copper sheet, which is then screwed to a cooled aluminum frame. Earlier studies [8] and new experiments suggested that occasionally the spectrum will become narrower at certain wavelength when the current is increased. In the following figure such a new event is shown. The detection procedure was the following: At low currents the spectra were like the spectrum at 300 mA, at 400 mA two peaks appeared at 330 and 800 nm and finally at 450 mA the high intensity peak appeared at 670 nm.

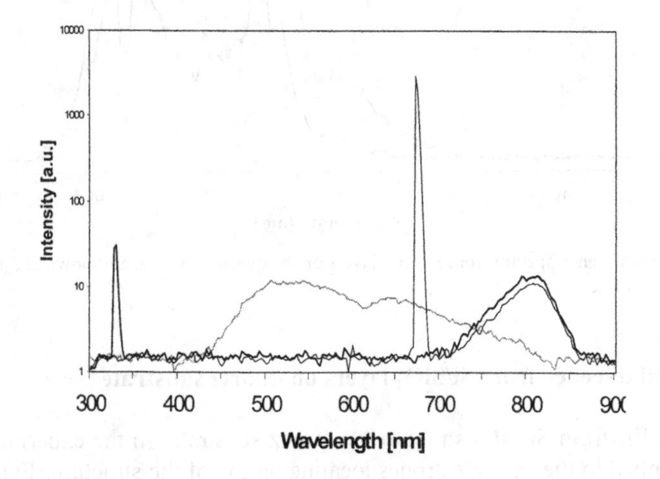

Figure 4. Electroluminescence from a sample having 1 mm^2 surface area. The spectrum drawn by thick gray line is detected with 300 mA sample current, thick black line shows the spectrum with 400 mA and thin line at 450 mA, respectively.

The TEM image (Fig. 5) of the thin CVD oxide layer on silicon shows that there are silicon nanocrystals in the oxide which might be the origin of light emission. On the other hand the AFM picture (Fig. 5) shows that the structure is uneven having possible holes up to the silicon substrate. Partially the current may pass directly to the substrate.

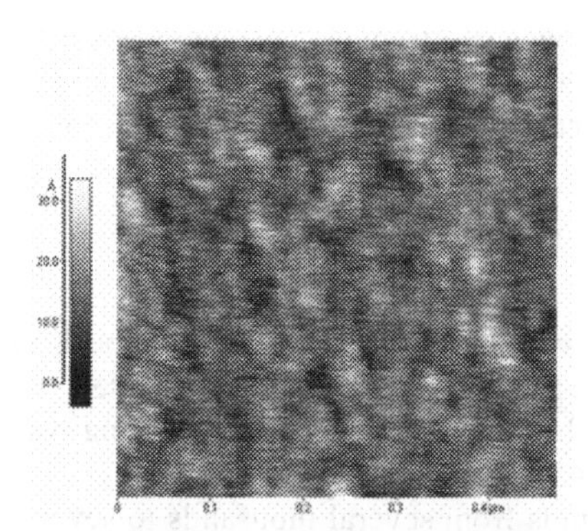

Figure 5. TEM image of thin CVD oxide layer on silicon, where silicon nanocrystals are seen as darker spots.

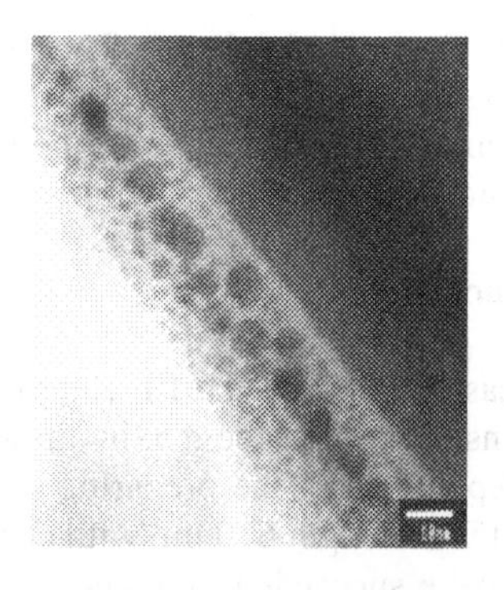

Figure 6. AFM image of the top of the same thin CVD oxide layer on silicon that is shown in fig. 5.

6. The size and color of light emitting EL spots

An important theme is the size and color of the light spot. In various experiments the size (and density) of the lighting spots was measured by using a microscope equipped with a CCD cell (1.2 Million pixels). In the series of experiments (Fig. 8) through the 15x15 μm^2 square sample the activating currents of 4.2, 5.4, 8.0 and 10.0 mA were used. Simultaneously the light from the sample was microphotographed with exposure times 2 s. It seems that the same light spots are luminating, only the intensity increases with raising current. Inside the square there are three areas without spots. This is expected to be resulting from the fact that the gold foil is not tightly bonded to the surface..

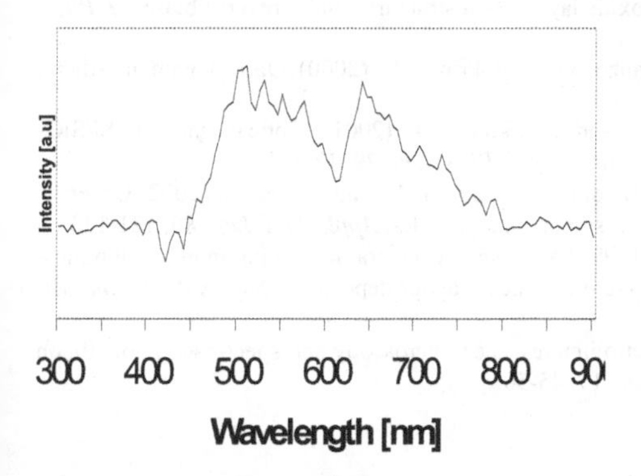

Figure 7. Electroluminescence spectrum measured from 15*15 μm^2 square (shown in Fig. 8) with 10.0 mA current.

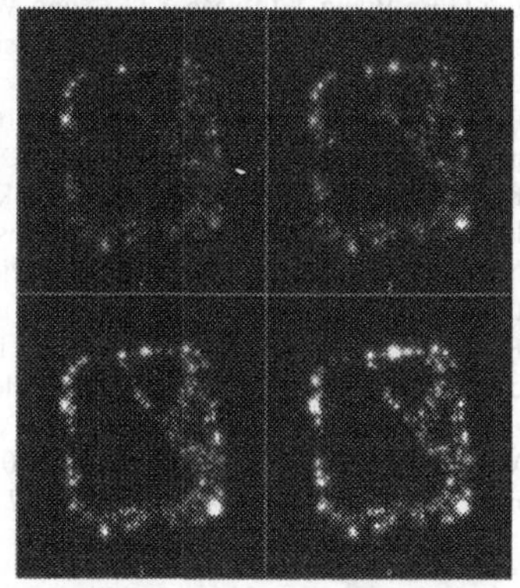

Figure 8. Microscope photographs taken from 15·15 μm^2 square sample with four different current values 4.2 mA, 5.4 mA, 8.0 mA and 10.0 mA in inserts 1 to 4.

We have succeeded to lower the number of spots from several thousands in EL samples with 5-7 mm^2 area of gold electrode to several hundreds in the samples having active electrode areas of 15x15 μm^2. The electroluminescence spectrum (Fig. 7) was measured from the sample with 10.0 mA current. Several measurements were carried out to measure the approximate diameter of the EL spot. Typically the diameter of the single spot was found to be less than 600 nm

7. Conclusions

The measurements of the EL samples showed a wide (100 - 150 nm FWHM) peak in the intensity spectra, the maximum locating in the range of 500 - 700 nm. This seemed to be depending on the preparing of the sample. In one case a spectral narrowing was found at wavelength 670 nm with current density 45 A cm^{-2}.

We have succeeded to lower the number of spots from several thousands to several hundreds by reducing the active area of the electrodes. Our measurements suggested that the size of a luminescent spot is less than 600 nm, which is in good agreement with another founding [9].

EL is seen in both electrodes independently from the polarity of the voltage, but EL from negative electrode is clearly stronger than from positive electrode.

References

1. Cullis, A.G., Canham, L.T., and Calcott, P.D.J. (1997) The structural and luminescence properties of porous silicon, *J. Appl. Phys.* **82**, 909-965.
2. DiMaria D.J., Kirtley J.R., Pakulis E.J., Dong D.W., Kuan T.S., Pesavento F.L., Theis T.N., Cutro J.A., and Brorson S.D. (1984) Electroluminescence studies in silicon dioxide films containing tiny silicon islands, *J. Appl. Phys.* **56**, 401-416.
3. Qin G.G., Li A.P., Zhang B. R., and Li B.C. (1995) Visible electroluminescence from semitransparent Au film/extra thin Si-rich silicon oxide film/p-Si structure, , *J. Appl. Phys.* **78**, 2006-2009.
4. Bai G.F., Wang Y.Q., Ma Z.C., Zong W.H. and Qin G.G. (1998) Electroluminescence from Au/native silicon oxide layer/p^+-Si and Au/native silicon oxide layer/n+-Si structures under reverse biases, *J. Phys. Condens. Matter* **10**, L717-L721.
5. Pavesi L., Dal Negro, L., Mazzoleni, G., Franzó, G., and Priolo, F. (2000) Optical gain in silicon nanocrystals, *Nature* **408**, 440-444.
6. Khriachtchev, L. Räsänen, M., Novikov, S., and Sinkkonen, J. (2001) Optical gain in Si/SiO_2 lattice:Experimental evidence with nanosecond pulses, *Appl. Phys. Lett.* **79**, 1249-1251.
7. Nayfeh, M.H., Rao, S., Barry, N., Therrien, J., Belomoin, G., Smith, A., and Chaieb, S. (2002) Observation of laser oscillation in aggregates of ultrasmall silicon nanoparticles, *Appl. Phys. Lett.* **80**, 121-123.
8. Heikkilä, L., Kuusela, T., and Hedman, H.-P. (1999) Laser type of spectral narrowing in electroluminescent Si/SiO2 superlattices prepared by low-pressure chemical vapour deposition, *Superlattices and Microstructures* **26**, 159-169.
9. Valenta J., Lalic N., and Linnros J. (2001) Electroluminescence microscopy and spectroscopy of silicon nanocrystals in thin SiO_2 layers, *Optical materials* **17**, 45-50.

REVERSE BIASED POROUS SILICON LIGHT EMITTING DIODES

S. LAZAROUK
Belarusian State University of Informatics and Radioelectronics, P. Brovka 6, Minsk 220027, Belarus

1. Introduction

Silicon is the basic material in microelectronics. However, bulk silicon is an indirect gap semiconductor which makes it inefficient in emitting light. Therefore, silicon has limited optoelectronic applications, as compared to direct gap semiconductors. In the last years, porous silicon (PS) based light emitting diodes have been extensively studied because of room temperature electroluminescence, which raises hopes of an all silicon based optoelectronics. The first PS LED based on a Schottky junction was reported in 1991 [1]. It emitted light at both forward and reverse biases. At present, more than 200 papers dealing with PS LEDs have been published. They are mostly dedicated to light emission at forward bias. However the best efficiency and frequency parameters have been observed for reverse biased devices [2,3], while published reviews [4] deal mainly with forward biased PS LEDs. The state of the art for reverse biased PS LEDs is reviewed in this paper.

2. State of the Art

The first reverse biased PS LED was demonstrated by Richter et al. [1] in 1991. A PS layer was formed on an n-type silicon substrate. Then, a semitransparent gold electrode was deposited upon the PS layer in order to form the Schottky barrier between the gold electrode and the PS layer. Light emission was observed in the visible range with the peak at 650 nm. The efficiency was in the range of $10^{-5} - 10^{-6}$ [1,5]. The lifetime of such reverse biased PS LEDs varied from 45 min to 100 h, after which the parameter degradation and emission attenuation took place [1,5,6]. A significant improvement in the efficiency and stability was made by Lazarouk et al. [2] in 1995 through the formation of an oxidized PS layer protected from atmospheric oxygen by an additional passivation layer. The oxidized PS layer was formed on low resistivity n-type silicon by anodization in the transition regime [7], providing a continuous anodic oxide on the surface [8]. Moreover, an additional passivation layer of transparent anodic alumina was formed on the PS layer by a selective aluminum anodization in an oxalic electrolyte during formation of the aluminum Schottky electrode. The passivation ensured the stability of continuous PS LED operation during 1000 h without degradation effects. The quantum efficiency for oxidized PS LEDs was in the range of $10^{-4} - 10^{-3}$ [2,9].

L. Pavesi et al. (eds.), Towards the First Silicon Laser, 61–68.

In 1998, Kuznetsov et al. [10] improved the PS LED design in order to enhance the device efficiency. In particular, they replaced the opaque aluminum electrode with a semi-transparent silver electrode. The efficiency of this reverse biased PS LED was about 0.5×10^{-2} [8].

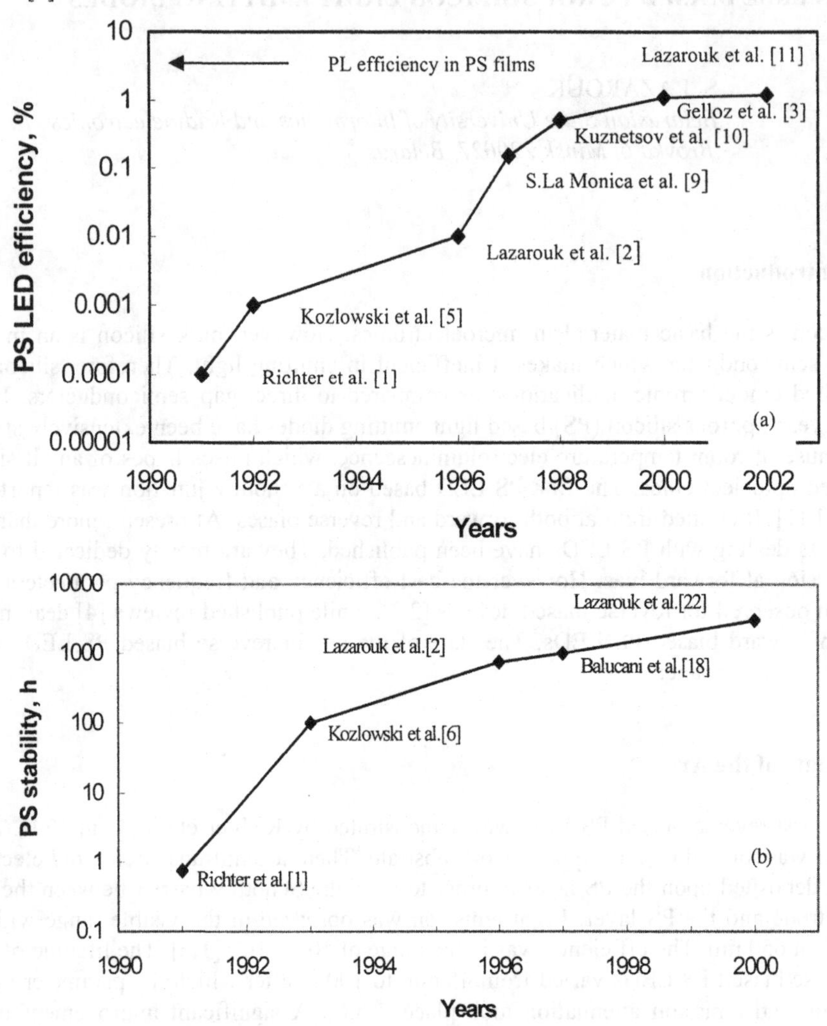

Figure 1. Efficiency (a) and stability (b) vs years for reverse biased PS LEDs.

In 2000, the more efficient reverse biased PS LEDs have been reported by Gelloz et al. [3]. The quantum efficiency of about 10^{-2} has been obtained by using oxidized PS. Porous layer was formed on n^+-type silicon at 0 °C. Then, the PS layer without drying was electrochemically oxidized by anodization in an aqueous solution of sulfuric acid. A Schottky barrier was formed by a transparent indium – tin – oxide deposition. The advancement of the anodization process by this way decreased the size of nonconfined silicon nanocrystals in PS. The enhanced quantum efficiency can be explained by the reduction of leakage carrier flow through the non-confined silicon nanocrystals. Recently the highest quantum efficiency of about 1.2×10^{-2} has been obtained by pulsed excitation [11]. Pulse LED operation allows one to reach the highest current density through a LED structure, which corresponds to maximum efficiency values [3,11].

Some common design ideas are present in all these LEDs. All are formed on n-type silicon substrates due to the higher Schottky barrier for n-type silicon compared to p-type material [12]. In addition, n^+-type silicon substrates are preferable because of minimum series resistance of PS LEDs. PS must have homogeneous size distribution of nanocrystals over all the thickness and down to 1 μm [2,3,10]. For this reason, oxidized PS was used in the device fabrication and the temperature of the PS formation was chosen to be 0 °C [3]. These reverse biased LEDs have a non-linear EL-I characteristics with an efficiency which increases with increasing current [2,3]. For this reason optimized LED geometry and pulsed LED bias provide maximum LED current with the record efficiency value of 1.2×10^{-2} [11].

Other approaches to form PS LED have been also used. Though the performances are lower than the one reviewed here, they demonstrate that the LEDs can be fabricated on different silicon substrates. Thus, reverse biased PS LEDs were formed on polysilicon [13] and amorphous silicon [14] layers. Also, PS layers for reverse biased LEDs can be fabricated by stain chemical etching [15]. Furthermore, reverse biased PS LEDs were formed on transparent substrates such as sapphire [16] and glass [14]. The wide variety of initial substrates and silicon layers are especially important for practical applications, which will be considered in the next sections.

3. Reverse Biased PS LEDs for Optical Intra-chip Interconnects

Since the discovery of efficient light emission from PS, this material is considered to be promising for integrated silicon based optoelectronic systems able to emit, transmit, and detect light in the visible range. The analysis of LED parameters for optical interconnects has been carried out in [4]. Most of these are already reached in PS LEDs: for example, threshold voltage of 4 V and on-set current density of 0.02 mA/cm^2 [10,13], emission optical power density of 1 W/cm^2 at pulse excitation [17], operational frequency of 200 MHz [18]. However, nowadays the main challenge for implementing PS LEDs in optical interconnects is their low efficiency. As noted in [4], an external LED efficiency of about 10^{-1} is the required level for optical interconnects. On the other hand, intra-chip optical interconnects as opposed to inter-chip interconnects are operated at distance usually shorter than 1-2 cm, thus the LED efficiency for this application can be at least one order of magnitude lower. Thus, reverse biased PS LEDs are more attractive for intra-chip optical interconnects

A schematic cross-section of a developed optoelectronic unit based on reverse biased PS LEDs is shown in Fig. 2a. The main technological steps are described elsewhere [2,11,19,20]. Figure 2b presents the equivalent circuit. It is composed of two Al/PS Schottky junctions with area of 0.1 mm^2 and an alumina layer between them. One of the junctions operates as a LED, the other as a photodetector (PD). The distance between them is 5 μm. The anodic alumina protects the PS surface from atmospheric oxygen. Moreover, it plays one more important role in the device. The light emitted by one of the Schottky junctions is transmitted within the alumina layer as in an optical waveguide. As far as the refractive index of porous silicon (1.3-1.6) is lower than that of alumina (1.6-1.65) or that of alumina doped with titanium oxide (1.7-1.8), the anodic alumina layer provides an appropriate light guiding effect [19]. The niobium film acts as a reflector, which assists light spreading within the anodic alumina layer, as it illustrates in Fig. 2a.

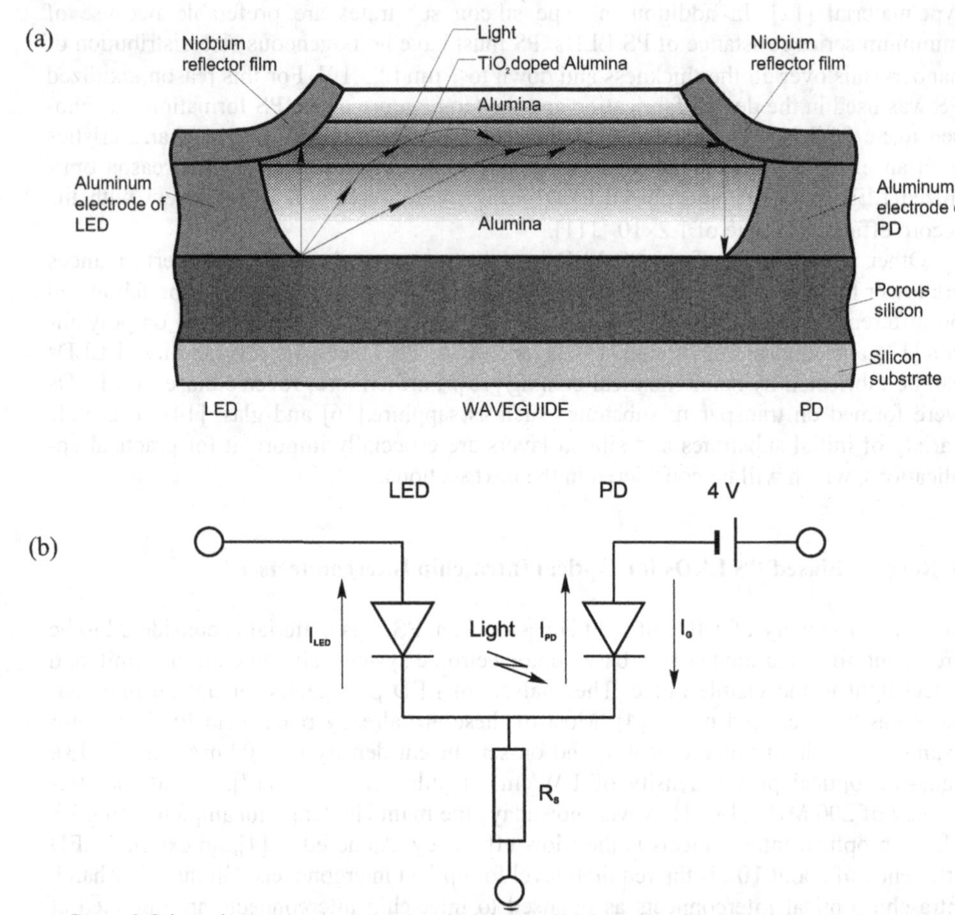

Figure 2. Schematic cross-section (a) and equivalent electric scheme (b) of the integrated optoelectronic unit based on reverse biased PS LEDs.

When the left Schottky junction is biased near to avalanche breakdown, i.e. more than 5 V, a reverse current (I_{LED}) passes through it and light emission can be seen around the aluminum electrode. At the same time, a reverse current appears in the right Schottky junction operating in a photodetector mode with the reverse bias less than 5 V. The current through this junction is increased with an increase of I_{LED} as it is shown in Figure 3. The similar behavior is observed at increasing of external light intensity. The relationship between I_{PD} and I_{LED} is close to quadratic dependence. Meantime, the relationship between LED electroluminescence and I_{LED} is also quadratic [21] So, we conclude that the measured current is a photo response of the right Schottky junction.

There is a galvanic link between the LED and the PD (see Figure 2b). However, the direction of the galvanic current (I_G) is opposite to the measured PD current (I_{PD}). To reduce the influence of the galvanic current, we used an additional 4 V battery connected with the PD, as shown in Figure 2b. The 4 V battery provides an additional reverse bias for the PD and precludes the changing of the net PD bias as it was observed in case of the 1.5 V battery [22].

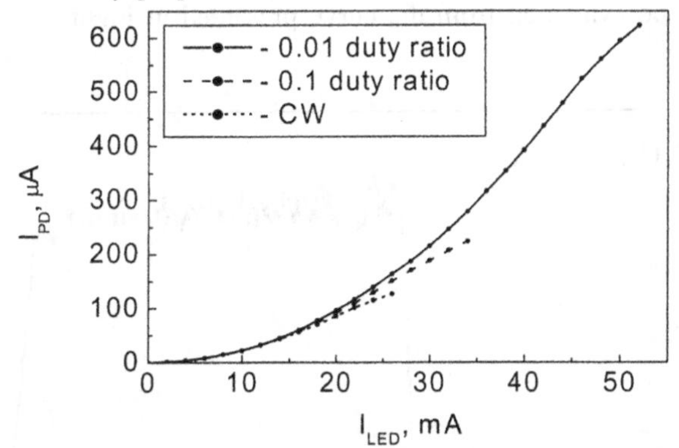

Figure 3. PD current versus LED current in the porous silicon based optoelectronic unit with 5 µm alumina waveguide between LED and PD for different LED bias.

Fig. 3 shows the PD current versus the LED currents in the PS optoelectronic unit for different bias. The relation between I_{PD} and I_{LED} is close to a quadratic dependence with the transition to a linear one at high current values. It is important to note, that the transition to a linear region is observed at different I_{LED} values for different LED bias regimes. In particular, in CW the linear dependence appears at lower I_{LED} values than with pulsed LED operation. This phenomenon can be explained by the LED thermal heating, which increases the probability of nonradiative recombination processes due to Auger effects. The subsequent thermal breakdown supports this suggestion. Under pulsed LED operation the produced heat dissipates in the time intervals between pulses safeguarding the LED structures against thermal heating. Therefore, the transition to the linear region and the irreversible thermal breakdown can be observed at higher amplitudes of the input LED currents. It is obvious, that the lower is the duty ratio, the lower is the thermal heating. So with the decreasing the duty ratio, maximum efficiency values can be obtained [3,11]. The LED quantum efficiency can be calculated as the ratio of the PD output current to the LED input current [11,22]. As Fig. 3 illustrates, a maximum efficiency of 1.2 % is observed at the duty ratio of 0.01 where the LED is operated with the largest input currents. But the increase of the quantum efficiency with the operating current is limited, the reaching of the saturation stage being connected with the thermal heating of the diode. The increased quantum efficiency of the reverse biased PS LEDs as compared to our previous results has been achieved due to the improved heat sinking. In addition, a threefold reduction in the aluminum electrode area has made possible to measure an efficiency of 0.5 % in CW, which is more than twice as large as the value obtained earlier [22]. The employment of pulsed operation has enabled the heat sinking to be improved and the light emission efficiency of 1.2 % - the best value for such LEDs – to be gained [11].

The special attention has been paid to the time response of the LED [22]. The transient electroluminescence waveform with the minimized time response is shown in Figure 4. This curve corresponds to the lowest of the series resistance and capacitance in the developed LED. The basic feature of the transient electroluminescence can be characterized by the delay time (time between the application of the drive pulse and the start of the light response) and the rise time. The delay time of 1.2 ns and the rise time of 1.5 ns can be evaluated from the curve presented in Figure 4 for the voltage pulse of 12 V.

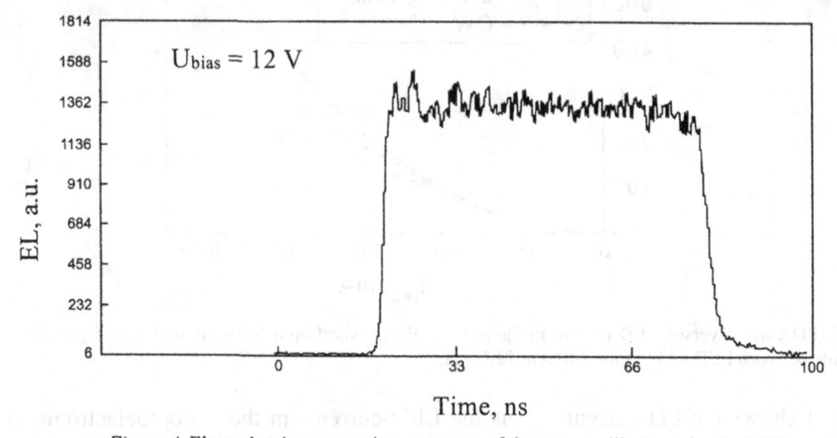

Figure 4. Electroluminescence time response of the porous silicon avalanche LED.

The shortest time response is observed for the maximum bias applied, as it was already described in [23] for forward biased porous silicon LEDs. However, our electroluminescence devices are faster, because they have no diffusion capacitance, which limits the time response of the light emission. The main mechanism of minor carrier generation in the light emission from reverse biased junctions is impact ionization at avalanche breakdown. For the avalanche breakdown to take place, a high value of electric field is necessary. A regular columnar structure of porous silicon promotes the avalanche breakdown due to non-uniform electric field distribution inside the porous layer [7]. The effect of the impact ionization at the avalanche breakdown is very fast. For example, the time of the avalanche response is estimated to be about 1 ps [12]. Thus, we have shown that the developed LED can operate in the nanosecond range. It should be noted that these values are not limited for these devices. By further technology optimization, we hope to reach the sub-nanosecond range, which is promising for LED applications in optical intra-chip interconnect.

4. Reverse Biased PS LEDs for Microdisplay Applications

Porous silicon is also considered as a candidate for display technologies. Such devices can be fabricated on silicon substrates that are especially attractive for microdisplay technologies. The advantages of porous silicon LEDs for microdisplay application are:

i) possible integration of driver IC with a PS microdisplay; ii) high resolution of such microdisplay devices because the size of a light emitting pixel can be few microns; iii) low cost and simplicity of PS fabrication. The main disadvantage of such LEDs is related to the low efficiency. Taking into account the achieved quantum efficiency of about 1 % (or power efficiency of about 0,3 %), the estimate thermal load for a PS LEDs with brightness 100 Cd/m^2 is about 0.3 W/cm^2. It is obvious that in this case the heat should be removed to prevent overheating effects. However, if the light emission brightness is limited at the level of 20 Cd/m^2 (a usual level for head mounted microdisplay devices operated in the dark), the heat dissipation will not result in catastrophic overheating effects. Such PS microdisplay devices can operate in the continuous regime for more than 1000 h without any considerable degradation [22]. Moreover, PS microdisplays can contain more than one million pixels over an area of 1 cm^2 which cannot be achieved by other existing microdisplay technologies. In this case, the operating current for the single pixel is about 10 μA, which corresponds to the operating current of silicon VLSI components. Depending on PS anodizing regimes the emission peak can be both in the blue and in the red [4]. But the emission spectra are very broad, with example which covers the whole visible range [21]. Of course, such light emission is suitable for black and white displays, but for a color display other approaches are to be used to get a narrow light emission spectrum. The simplest approach is the employment of light filter. An alternative method of providing a narrow light emission spectrum reported recently in [20] is the integration of PS LED with PS microcavities. Thus, PS LEDs can solve miniaturization problems for microdisplay technologies.

5. Conclusion

The analysis of reverse biased PS LED developments for the last ten years has shown considerable parameter improvement towards practical implementations of these devices in optoelectronics. The only unresolved problem is insufficient LED efficiency. Nevertheless, the achieved efficiency level of about 1 % allows us consider some special applications. In particular, reverse biased PS LEDs could be used for optical intrachip interconnects. The fabricated prototype of optoelectronic unit based on these LEDs has demonstrated the possibility of using photons for communications inside silicon chips. Also the attained LED efficiency is close to the value corresponding to the required level for microdisplay devices. In this case, the high resolution could be afforded that could not be attained by other methods. The development of reverse biased PS LEDs allows us to continue consideration of these devices as candidates for Si based optoelectronics in the near future.

This work is supported by EC within INCO-COPERNICUS, project 977037, as well as by the Information Fund of Belarus and by the National Basic Research Foundation. The author would like to thank Professor V. Borisenko for fruitful discussions. The hard work of the group members has permitted to reach the results presented in this paper: P. Jaguiro, A. Leshok, S. Katsouba are gratefully thanked.

References

1. Richter, A., Steiner, P., Kozlowski, F., Lang, W. (1991) Current-induced light emission from a porous silicon device, *IEEE Elec. Dev. Lett.* **12**, 691-693.
2. Lazarouk, S., Jaguiro, P., Katsouba, S., Masini, G., Monica, S.La, Maiello, G., Ferrari, A. (1996) Stable electroluminescence from reverse biased n-type porous silicon-aluminum Schottky junction device, *Applied Physics Letters* **68**, 2108-2110.
3. Gelloz, B., Koshida, N. (2000) Electroluminescence with high and stable quantum efficiency and low threshold voltage from anodically oxidized thin porous silicon diode, *J. Appl. Phys.* **88**, 4319-4322.
4. Cullis, A. G., Canham, L. T., Calcott, P. D. J. (1997) The structural and luminescence properties of porous silicon, *J. Appl. Phys.* **82** 909-965; Canham, L. T. (1997), *Properties of Porous Silicon* INSPEC, The Institution of Electrical Engineers, London; Bisi, O., Ossicini, S., Pavesi, L. (2000) Porous silicon: a quantum sponge structure for silicon based optoelectronics, *Surf. Sci. Rep.*, **264**, 1-126.
5. Kozlowski, F., Sauter, M., Steiner, P., Richter, A., Sandmaier, H., Lang, W. (1992) Electroluminescent performance of porous silicon, *Thin Solid Films* **222**, 196-199.
6. Kozlowski, F., Steiner, P., Lang, W. (1993) Current-induced light emission from nanocrystalline silicon structures, *Proc. NATO ARW Series E: Applied Sciences* **244**, 123-133.
7. Bertoloti, M., Carassiti, F., Fazio, E., Ferrari, A., Monica, S.La , Lazarouk, S., Liakhou, G., Maello, G., Proverbio, E., Schirone, L. (1995) Porous silicon obtained by anodization in the transition regime, *Thin Solid Films*, **255**, 152-154.
8. Zhang, X. G., Collins, S.D., and Smith, R.L. (1989) Porous silicon formation and electropolishing of silicon by anodic polarization in HF solution, *J. Electrochem. Soc.* **136**, 1561-1565.
9. La Monica, S., Maiello, G., Ferrari, A., Masini, G., Lazarouk, S., Jaguiro, P., Katsouba, S. (1997) Progress in the field of integrated optoelectronics based on porous silicon, *Thin Solid Films* **297**, 261-264.
10. Kuznetsov, V., Andrienko, I., Haneman, D. (1998) High efficiency blue-green electroluminescence and scanning tunneling microscopy studies of porous silicon, *Appl. Phys. Lett.* **72**, 3323-3325.
11. Lazarouk, S., Jaguiro, P. ,.Leshok A. (2002) Increasing of porous silicon LED efficiency at pulsed bias, *Izvestia Belorusskoi Injenernoi Academii* **11**, 19-21 (in russian)..
12. Sze, S.M. (1985) *Semiconductor Devices: Physics and Technology*, Bell Lab., A Wiley-Interscience publication, New York.
13. Lazarouk, S., Bondarenko, V., La Monica, S., Maello, G., Masini, G., Pershukevich, P., Ferrari, A. (1996) Electroluminescence from aluminum-porous silicon reverse bias Schottky diodes formed on the base of highly doped n-type polysilicon, *Thin Solid Films* **276**, 296-299.
14. Toyama, T., Matsui, T., Kurokawa, Y., Okamoto, H., Hamakawa, Y. (1996) Visible photo- and electroluminescence from electrochemically formed nanocrystalline Si thin film, *Appl. Phys. Lett.* **69**, 1261-1263.
15. Sercel, P., Kwon, D., Vilbrandt, T., Yang, W., Hautala, J., Cohen, J., Lee, H. (1996) Visible electroluminescence from porous silicon/hydrogenated amorphous silicon pn-heterojunction devices, *Appl. Phys. Lett.* **68**, 684-686.
16. Lazarouk, S. (1998) Light emitting devices based on Al/porous silicon Schottky junctions on supphire substrates for display applications, *Proc. of the 7-th International Symposium Advanced Display Technologies* Minsk, Belarus, 193-196.
17. La Monica, S., Balucani, M., Lazarouk, S., Maiello, G., Masini, G., Jaguiro, P., Ferrari, A. (1997) Characterization of Porous Silicon Light Emitting Diodes in High Current Density Conditions, *Solid State Phenomena* **54**, 21-26.
18. Balucani, M., La Monica, S., Lazarouk, S., Maiello, G., Masini, G., Ferrari, A. (1997) Silicon Emitting Device Will Knock Down Communication Bottleneck?, *Solid State Phenomena* **54**, 8-12.
19. Lazarouk, S. K., Jaguiro, P. V., Leshok, A. A., Borisenko, V. E. (1999) Porous Silicon Light Emitting Diode and Photodetector Integrated with a Multilayer Alumina Waveguide *Physics, Chemistry and Application of Nanostructures* World Scientific, Singapore, 370-373.
20. Lazarouk, S., Leshok, A., Borisenko, V.E., Mazzoleni, C., Pavesi, L. (2000) On the Route Towards Si-based Optical Interconnects, *Microelectronic Engineering* **50**, 81-86.
21. Lazarouk, S., Katsouba, S., Tomlinson, A., Benedetti, S., Mazzoleni, C., Mulloni, V., Mariotto, G., Pavesi, L. (2000) Optical Characterization of reverse biased porous silicon light emitting diode, *Materials Science and Engineering* **B69-70**, 114-117.
22. Lazarouk, S., Jaguiro, P., Melnikov, S., Prohorenko, A. (2000) LEDs based on porous silicon for intrachip optical interconnects, *Izvestia Belorusskoi Injenernoi Academii* **9**, 67-69 (in russian).
23. Cox, T.I., Simons, A.J., Loni A., Calcott, P.D.J., Canham, L.T., Uren, M.J. and Nash K.J. (1999) Modulation speed of an efficient porous silicon light emitting device, Journal of Applied Physics **86**, 2764-2773.

STRONG BLUE LIGHT EMISSION FROM ION IMPLANTED Si/SiO_2 STRUCTURES

W. SKORUPA, L. REBOHLE*, T. GEBEL, M. HELM
nanoparc GmbH, Bautzner Landstr.45,
D-01454 Dresden, Germany, and
Institute of Ion Beam Physics and Materials Research
Forschungszentrum Rossendorf e.V., POB 510119
D-01314 Dresden, Germany
**now with Institut für Festkörperelektronik, TU Wien, Austria*

1. Introduction

The enormous development of information technology has created an ever increasing demand for optoelectronic devices able to generate, modulate and process optical signals. Unfortunately, silicon is badly suited to operate as a light emitter due to its indirect band gap of about 1.1 eV. To date, mostly compound semiconductors are used as discrete devices in optoelectronics, rendering impossible complete Si-based optoelectronic integrated circuits. Various Si-based materials for light emission have been studied so far, among them Er-doped SiO_2/Si and porous Si being the prime examples with emission in the near-infrared and red spectral region, respectively. For further details about the different approaches to extract light from silicon we refer to Refs.[1-3] and the references therein. In this article the number of references is kept intentionally low since other relevant works on Si-based light emission are extensively treated in this volume.

The advantages of ion-implanted SiO_2 layers are based on the excellent mechanical, chemical and electrical properties of SiO_2 as well as the well-known reproducibility of ion beam processing and its compatibility to standard Si technology. Red electroluminescence (EL) has been observed in Si- and Ge-implanted SiO_2 layers, but the power efficiency was either not given or below 10^{-4}. Our recent investigations revealed the possibility to extract blue-violet EL from Ge-implanted SiO_2 films and blue EL from Si-implanted SiO_2 films with a power efficiency between values $\leq 10^{-4}$ and more than 1×10^{-3} [2,4]. More specifically, the present-state optimised Rossendorf light source based on Ge-implanted SiO_2 layers exhibits bright blue-violet EL with a record wall-plug efficiency of 0.5%!

This article will briefly summarise the work performed at the FZR during the past years in producing nanoclusters in silicon dioxide layers enriched with Si, Ge and Sn by ion beam synthesis (IBS) for silicon based light emission, see [1-11] and references therein. The focus is put on the luminescence and electrical properties; for the results of our microstructural investigations we refer the reader to Refs. [2,5,7,9].

L. Pavesi et al. (eds.), Towards the First Silicon Laser, 69–78.

2. Experimental Details

For the luminescence investigations SiO_2 films with a thickness between 80 and 500 nm on [100]-oriented, n-type Si substrates were thermally grown at 1000°C. For 500 nm oxides a double implant was performed (for details, see [2]). The thinner oxide films were single implanted with Ge ions resulting in a Gaussian implant profile with a peak Ge concentration ranging between 0.3 and 3 %. After implantation, furnace annealing (FA) at 400-1200°C for 30 min or rapid thermal annealing (RTA) at 1000°C for 1-150 s was applied. Metal-oxide-semiconductor (MOS) capacitor structures for EL studies were prepared using sputtered 80 nm thick layers made of indium tin oxide (ITO) and Al as top and bottom electrodes, respectively. The transmission of ITO is higher than 80 % in the relevant wavelength region. EL and PL measurements were performed at room temperature in a Spex Fluoromax spectrometer with a R298 Hamamatsu photomultiplier. Electrical properties of the MOS capacitors were studied through IV (dc-current-voltage) and high frequency CV (capacitance-voltage at 1 MHz) measurements.

3. Photoluminescence

Fig. 1 shows the PL spectra of Ar-, Si-, Ge- and Sn-implanted as well as unimplanted SiO_2 layers. The spectra were excited at 4.96 eV (Si, Sn) or 5.17 eV (Ge) consisting of an UV peak between 4.1 and 4.3 eV as well as a PL peak in the blue-violet spectral region. The position of the UV peak does not change, but the position of the blue-violet peak shifts from 2.7 eV (Si) to 3.2 eV (Ge) and 3.1 eV (Sn), respectively. The non-implanted as well as the Ar-implanted thermally grown SiO_2 layers show very weak PL. In both cases the low PL intensity decreases or vanishes after moderate temperature annealing. In contrast to that, SiO_2 films implanted with Si, Ge or Sn achieve a PL

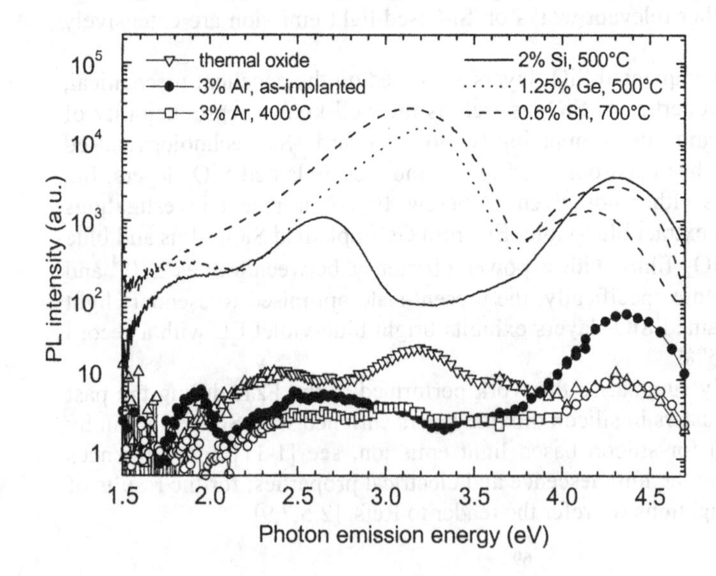

Fig 1: PL spectra of thermally grown SiO_2 layers in comparison with those of Ar-, Si-, Ge- and Sn-double-implanted SiO_2 layers after furnace annealing at different temperatures [2].

intensity up to three orders of magnitude (!) higher, in which the maximum appears after an annealing step of at least 400°C only. Based on these results intrinsic defects of the SiO_2 network and the pure radiation damage simulated by the Ar implant can be excluded as a source of the intense PL of Si-, Ge- or Sn-implanted SiO_2 layers.

In our TEM studies we have observed the formation of Ge and Sn nanoclusters for high anneal temperatures (>800°C) and high concentrations (>2%) of Ge and Sn only. However, strong PL has been found also for implanted oxides annealed at moderate temperatures (<800°C) and with a low concentration of Ge or Sn (<2%). Moreover, the emission and excitation energies are nearly independent of the anneal temperature, of the ion concentration and of the size of the nanoclusters, if they are present. Undoubtedly, these results show that the observed PL is definitely not caused by the effects of quantum confinement in Si, Ge or Sn nanocrystals.

Another crucial question to be answered is, whether ion implantation has a certain advantage compared to other methods used to synthesize Si-, Ge- or Sn-rich SiO_2. We investigated the PL of Si- and Ge-rich SiO_2 layers produced by magnetron sputtering in comparison with the corresponding implanted films. Both sputter-deposited layer types show the blue-violet and the UV peak at the same position and with the same relative intensity ratio as in the case of the implanted species. Although the direct comparison of the PL intensity is difficult due to the partly different processing conditions, it can be stated that implanted and sputtered SiO_2 films with small Si- and Ge excess contain the same luminescence centers, but that the PL intensity of the implanted SiO_2 layers is more than one order of magnitude higher. The main reason for this difference are the synergy effects between the energy deposition during implantation and the excess of Si and Ge. First, the excess Si- or Ge-atoms are more dispersed in the case of ion implantation than in the case of sputtering, where the first clustering already occurs during the deposition. Secondly, the network of the sputter-deposited SiO_2 layers is characterized by a smaller degree of damage than that after a high dose implantation. Due to the complete destruction of the network bonds during implantation the precursors of luminescence centers, in particular E' centers, have better chances to transform into luminescence centers during annealing. These possibilities are more restricted in the case of sputter-deposited materials.

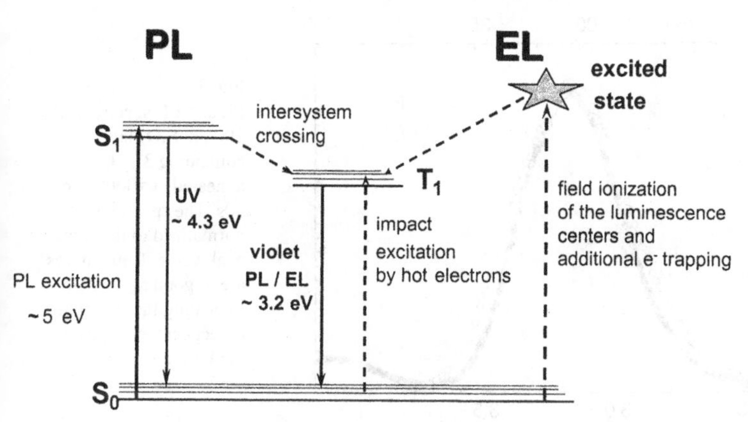

Fig. 2: Energy level scheme of an oxygen deficiency center consisting of a ground singlet state S_0, a first excited singlet state S_1 and a first excited triplet state T_1.

Based on the above results, the PL of Si-, Ge- and Sn-implanted SiO_2 layers is assumed to be caused by an oxygen deficiency center, whose generic energy level scheme is shown on the left hand side of Fig. 2. It consists of a ground singlet state S_0, a first excited singlet state S_1 and a first excited triplet state T_1. The PL can be understood as a radiative excitation from the ground state to S_1, followed by relaxation and intersystem crossing to T_1 and finally a radiative deexcitation back to the ground state. The radiative transition $T_1 \rightarrow S_0$ is optically forbidden by first order and occurs only because of the spin-orbit coupling. The increase of PL intensity in the order Si, Ge and Sn is explained by the heavy-atom effect which increases the spin-orbit coupling if one Si atom of the luminescence center is substituted by a heavier but isoelectronic Ge or Sn atom. Furthermore, intersystem crossing is a very efficient process leading to a much higher population of the T_1 state in comparison with the S_1 state, see also [2].

4. Electroluminescence

By application of a positive voltage to the top electrode (ITO) the MOS capacitors exhibit strong violet EL, which is well visible with the naked eye – of course, only in the case having selected an appropriate set of preparation parameters. Fig. 3 compares the normalized EL spectrum of Ge-implanted oxide films with the corresponding PL spectrum. The spectra show a main emission peak around 3.2 eV implying that this emission is caused by one and the same luminescence center. There is a nearly perfect congruence of the PL-and EL-spectra. The EL spectrum of Ge-implanted SiO_2 is composed of at least two modifications of the oxygen deficiency center having somewhat different excitation and emission energies [2]. The best devices based on Ge-implanted SiO_2 layers reach a power efficiency – defined as the ratio of optical output to electrical input power – of up to 0.5 %, which is very remarkable in the field of Si-based light emission. Furthermore, the absolute EL intensity of the Ge-implanted SiO_2 layer increases linearly with the injection current over more than three orders of

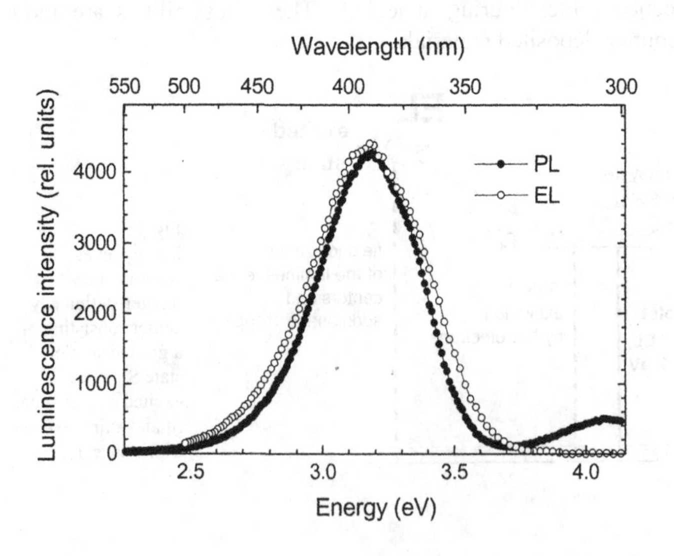

Fig. 3: PL and EL spectra of a 200 nm SiO_2 layer containing 3 % Ge, annealed at 1000°C for 30s. The spectra are normalized to the same peak value. Both curves are in good agreement, indicating that the same defect center is causing the luminescence.

magnitude. The shape of the EL spectrum of Ge-rich layers does not change at all with increasing injection currents.

As already mentioned, both the PL and EL are caused by one and the same luminescence center, but in the case of EL the excitation of the luminescence center is more complex and is strongly linked to the injection and transport mechanism of charge carriers. The relevant processes of charge injection and transport are schematically shown in Fig. 4. For the case of applying a high electric field (5-10 MV/cm) to a pure thermally grown SiO_2 layer electrons are injected from the cathode via Fowler-Nordheim tunneling (process 1). The implantation process creates defects in the oxide layer which appear as electron and hole traps in the SiO_2 band gap. Higher implantation doses will cause higher trap concentrations, and for very high doses Ge nanoclusters are formed. Traps and nanoclusters located close to the injecting interface can support the injection by trap assisted tunneling (process 2) or by tunneling from the conduction band of the Si substrate to the nanoclusters (process 3). Thus the electric field where tubnnel injection starts decreases with increasing trap concentration and the I-V characteristics shift to lower applied electric fields with increasing Ge concentration [3].

One mechanism to describe the charge carrier transport is the quasi-free movement of electrons within the conduction band of SiO_2 (process 4). In equilibrium the electrons can be characterized by an energy distribution depending on the position in the oxide layer. In the case of unimplanted SiO_2 layers the energy distribution of the electrons is mainly determined by phonon scattering and impact ionisation. It was shown that, under high field conditions (≥ 5 $MVcm^{-1}$), even the average electron energy can reach a value of 3...5 eV. Adapting these findings to the case of implanted oxide, a slight reduction of the electron energies due to the additional scattering at implantation-induced defects is expected. Electrons with a sufficiently high energy can be scattered at luminescence centers being in the ground state S_0, and the transferred energy is used to excite the luminescence centers. Another possibility is the charge carrier transport via traps by Poole-Frenkel conduction (process 5) or hopping conduction. In this scenario

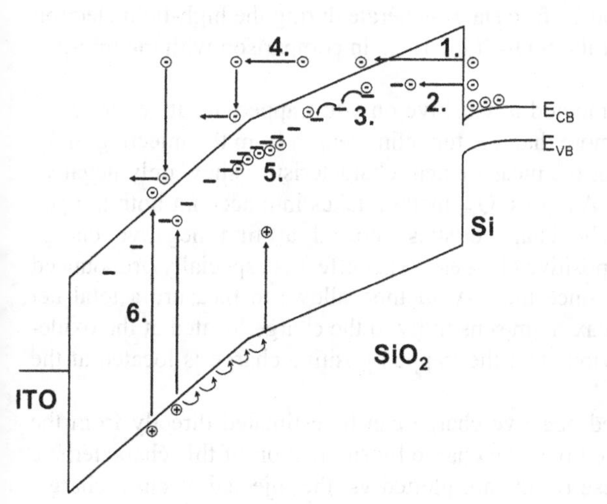

Fig. 4:
Band model of an ITO-SiO_2-Si structure demonstrating the injection and conduction mechanisms in electroluminescent SiO_2.
(1) FN-tunneling,
(2) Trap-assissted-tunneling (TAT),
(3) Hopping or PF conduction,
(4) free movement in the SiO_2 conduction band including scattering events,
(5) charge trapping,
(6) impact ionization or trap assisted impact ionization

luminescence centers will be ionized by the high applied electric field, which can be regarded as an excitation from the ground state S_0 to an „ionized state". The electrons move towards the metal electrode and leave the luminescence centers positively charged. If these centers trap an electron, they can relax – in most cases – back to the ground state. However, with a certain probability the centers relax to the T_1 state, too. Both excitation mechanisms are schematically demonstrated on the right hand side of Fig. 2. Recently it has been found [6] that the most probable way to excite luminescence centers is the impact excitation by hot electrons. Whereas the injection is explained by trap assisted tunneling of electrons from the substrate into the oxide, the electrons are transported via traps or by the quasi-free movement within the SiO_2 conduction band.

5. Charge Trapping

Because the EL devices are operated at high-electric fields (7–8 MV/cm), charge accumulation and stability of the oxide layers are the main problems for their application. Thus, we investigated negative and positive charge trapping phenomena in the Ge^+ ion implanted oxides which occur during EL device operation in the constant current regime [8]. For this end, thermally grown 80 nm thick SiO_2 films on (100) n-type Si wafers were implanted with Ge^+ ions at an energy of 50 keV to fluences of 6.5×10^{15} cm^{-2}. Thereafter RTA at 1000°C was performed for 6, 30 and 150 seconds. MOS capacitors for electrical measurements were fabricated using Al electrodes on front and back.

High-field electron injection from the silicon substrate into the oxide was performed at constant current ($J_{inj}=2\times10^{-5}$ A/cm^2) at room temperature. Charge trapping has been studied by combined measurements of the change of the voltage which was applied to the MOS structures at constant current regime (ΔV_{CC}), and the shift of flat-band voltage (ΔV_{FB}) of high-frequency (1 MHz) capacitance-voltage (C-V) characteristics performed at definite intervals. The use of the flat-band voltage shift was selected due to the minimum concentration of electron surface states generate during the high-field electron injection in this band gap depth at the SiO_2-Si interface in comparison with the mid-gap condition.

In our case the ΔV_{CC} vs. Q_{inj} method is sensitive only to trapped negative charge in the oxide volume at the distance more than the tunneling length from the injecting SiO_2-Si interface. It is worth noting that the measurement characteristics show only negative charge trapping in the oxide. The ΔV_{FB} vs. Q_{inj} method takes into account both trapped negative and positive charge. The characteristics showed at first negative charge trapping followed by trapping of positive charges. . This effect is especially pronounced for samples annealed for 150 s. Since the C-V method allows to measure a total net charge in the oxide and offers a maximum sensitivity to the charge located at the oxide-semiconductor interface, it is obvious that the trapped positive charge is located at the SiO_2-Si interface.

The parameters of the trapped negative charge can be estimated directly from the ΔV_{CC} characteristic and the trapped positive charge by subtraction of this characteristic from the ΔV_{FB} one. In Fig. 5 these results are plotted vs. the injected electron charge. Our trap analysis [8] demonstrated (see inset in Fig. 5), that three types of electron traps

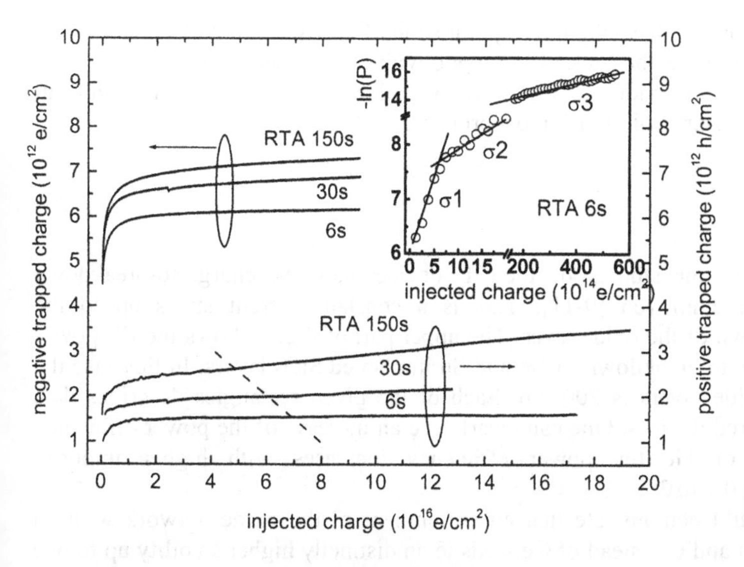

Fig. 5: Negative trapped charge (left scale) and positive trapped charge (right scale) as a function of electron fluence for different times of RTA treatment at 1000°C. The inset shows the trapping efficiency as a function of electron fluence for 6 s RTA treatment at 1000°C.

with average values of the capture cross-section of σ_1=2.6×10^{-15} cm^2, σ_2=6.3×10^{-16} cm^2 and σ_3=3.0×10^{-18} cm^2, and maximum concentrations (for 6 seconds RTA) of Q_{t1}^{max}=1.0×10^{12} cm^{-2}, Q_{t2}^{max}=1.0×10^{12} cm^{-2} and Q_{t3}^{max}=6.5×10^{12} cm^{-2} can be found.

Both, positive and negative charge trapping, increase with the annealing time of the Ge^+ ion implanted MOS devices. The enhancement of positive charge trapping at the SiO_2-Si interface can be correlated with the diffusion of Ge atoms and their embedding at the interface region during the high-temperature annealing. Experimental verification for that by Rutherford Backscattering Spectrometry (RBS) is demonstrated in Fig. 6. with the distribution of the implanted Ge as a function of the RTA time. The initial Ge peak shows a strong broadening during annealing and a second Ge peak is formed near the SiO_2-Si interface, as known from earlier investigations. Additionally, out-diffusion of the Ge through the surface of the SiO_2 layer was observed. For the longest RTA

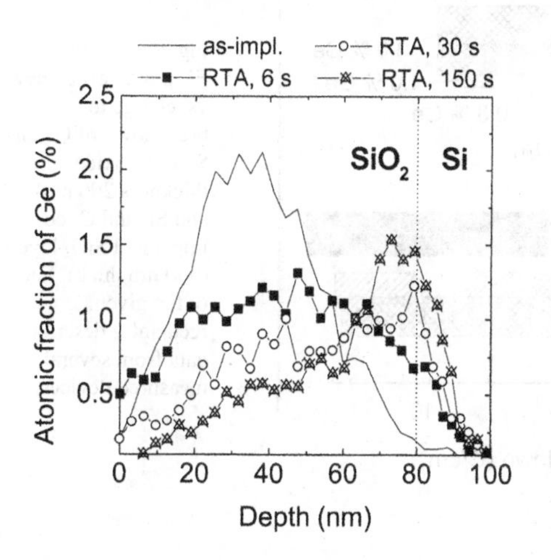

Fig. 6: Distribution of the implanted Ge (50 keV, 6.3x10^{15} cm^{-2}) in the SiO_2 layer, measured by RBS. The initially implanted Ge diffuses towards the SiO_2 / Si interface and forms an additional peak. After RTA with 150 s already 30 % of the implanted Ge diffused out through the surface.

treatment (150 s) about 30 % of the initially implanted Ge diffused out. The increase in negative charge trapping in the bulk of the oxide with increase in RTA time is not yet clear. A change of the oxide network structure due to the Ge redistribution in combination to Ge cluster evolution is most probable.

6. Device Stability

In the course of our studies of the EL device stability, charge-to-breakdown measurements were employed [9-11]. This is a constant current stress up to the destructive breakdown of the oxide layer. The upper part of Fig. 7 shows the EL power efficiency vs. charge to breakdown of Ge and Sn implanted SiO_2 layers. In this case the thickness of the oxide layers is 200 nm. Each of the given rectangles describes data from several measured devices. One can clearly see an increase of the power efficiency for Ge vs. Sn. For Ge the power efficiency increases with higher impurity concentrations from 0.1 to 3%.

Recently we could demonstrate that an enrichment of the oxide network with an double implant of Si and C instead of Ge leads to an distinctly higher stability up to one order of magnitude, see the lower part of Fig. 7 [11]. With decreasing Si/C concentration from 10 to 5% the EL rises. The reason might be the enhancement of non-radiative transitions with increasing Si/C concentration or the transformation of defects acting as luminescence centers into clusters. The efficiency of the Si/C rich devices is still smaller than that of the Ge and Sn rich devices. However, from the layers investigated here a further increase of the efficiency for even smaller concentrations is expected. This should be similar to the results from Ge or Sn implanted layers, where optimum conditions for the EL were found in a concentration range of 1…3%. Also, one has to mention that the maximum luminescence of such layers is in the range of 450-500 nm.

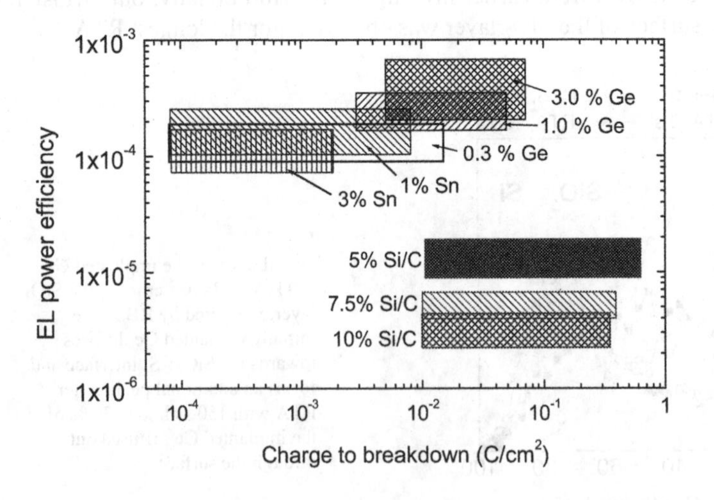

Fig. 7: EL power efficiency vs. charge to breakdown of Ge and Sn rich (oxide thickness 200 nm) and Si^+ and C^+ co-implanted SiO_2 layers (350 nm thick). Each of the given rectangles describes data from several measured devices.

7. Integrated Optocoupler

The suitability of Ge-implanted SiO_2 layers for optoelectronic applications was demonstrated by an all-silicon-technology based integrated optocoupler [4] schematically shown in Fig. 8. An optocoupler galvanically separates the circuit of a light emitter (transmitter) from the detection circuit (receiver), and the information is transmitted between them via light pulses. In this case the emitter circuit is almost identical with the device described before in chapter 4 and is isolated from the receiver circuit by a thick and transparent oxide layer. The receiver consists of a pin-diode made of amorphous silicon, which was deposited using a cluster tool for CVD processes. The sensitivity of the pin-diode is about 0.2 A/W at a wavelength of 400 nm. The transfer characteristics between the output and input current density showed linearity over three orders of magnitude. The integrated optocoupler can be used as a sensor in microsystems. Moreover, the relatively short emission wavelength could be advantageous for applications in biotechnology. If arranged in an array, the emitter devices could trigger a biochemical reaction in locally defined areas of the array.

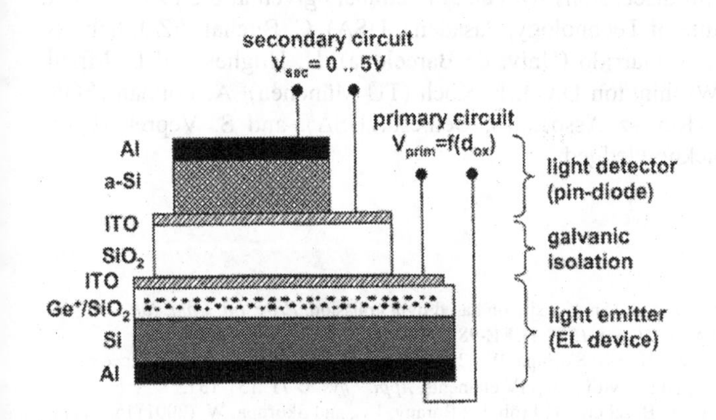

Fig. 8: Schematic structure of the integrated optocoupler consisting of a light emitter based on Ge-implanted SiO_2 layers and a light detector made of amorphous Si. Emitter and detector are galvanically isolated by a thick and transparent oxide layer.

8. Summary

A short review of our investigations devoted to the use of ion beam synthesized nanoclusters for silicon based light emission was presented. Blue-violet light emission was demonstrated based on Ge-implanted silicon dioxide layers thermally grown on silicon substrates. This version of silicon-based light emission relies on Ge-related defects in the amorphous ≡Si-O-Si≡network. The photoluminescence is excited by a singlet S_0-S_1 transition of a neutral oxygen vacancy. Whereas the PL excitation is a well-known mechanism, for the case of electroluminescence an interpretation was performed for the first time in the course of our studies. It was found that the most probable way to excite luminescence centers is the impact excitation by hot electrons injected from the silicon substrate into the silicon dioxide layer. Whereas this injection is explained by trap assisted tunneling of electrons, the electrons are transported via traps in the SiO_2 band gap and/or in the SiO_2 conduction band. The charge trapping in and the stability of these devices is investigated. Finally the application of the silicon-based light emitting devices for an integrated optocoupler arrangement is described.

9. Acknowledgements

The exciting work performed on ion beam synthesized Si- and Ge-rich silicon dioxide layers for the blue light emission would not have been possible without the support of and discussions with many colleagues at the Forschungszentrum Rossendorf: J. v. Borany, R. Groetzschel, K.-H. Heinig, M. Klimenkov, A. Markwitz, W. Möller, B. Schmidt, J. Sun, R.A. Yankov, J. Zhao; at the Technische Universität Dresden: H. Fröb, K. Leo; at the Institute of Semiconductor Physics in Novosibirsk: G.A. Kachurin and I.E. Tyschenko; at the Univ. di Catania: D. Pacifici, G. Franzò, F. Priolo; and at the Institute of Semiconductor Physics in Kiev: A.N. Nazarov, I.N. Osiyuk, and V.S. Lysenko. Moreover, helpful discussions with and/or seminars given at the FZR by H.A. Atwater (California Institute of Technology, Pasadena, USA), C. Buchal (FZ Jülich), S. Coffa (Univ. di Catania), B. Garrido (Univ. de Barcelona), H. Hughes and B. Mrstik (Naval Research Lab's, Washington D.C.), F. Koch (TU München), A. Polman (FOM Amsterdam), A. Revesz (Revesz Associates, Bethesda/USA), and S. Veprek (Univ. München) are gratefully acknowledged.

References

1. Skorupa, W., (1999) Ion beam processing for silicon based light emission, *Proc. Int. Conf. Ion Implantation Technology, Kyoto/Japan 1998* **IEEE-98EX144,** 827-832.
2. Rebohle, L., Borany, J.v., Fröb, H., and Skorupa, W. (2000) Blue photo-and electroluminescence of silicon dioxide layers ion implanted with group IV elements, *Appl. Phys.* **B 71**, 131-151.
3. Gebel, T., Rebohle, L. Zhao, J., Borchert, D., Fröb, H., Borany, J.v., and Skorupa, W. (2001) Ion beam synthesis based formation of Ge-rich thermally grown silicon dioxide layers : a promising approach for a silicon-based light emitter, *Mat.Res.Soc. Symp. Proc.* **638**, F18.1.1-F18.1.10.
4. Rebohle, L., Borany, J.v., Borchert, D., Gebel, T., Helm, M., Möller, W., and Skorupa, W. (2001) Efficient blue light emission from silicon, *J. Electrochem. Soc.: Electrochem. and Solid State Lett.* **7**, G57-G60
5. Rebohle, L., Borany, J.v., Borchert, D., Gebel, T., Helm, M., Möller, W., and Skorupa, W. (2001) Ion beam synthesized nanoclusters for silicon-based light emission, *Nucl. Instr. Meth.* **B188**, 28-34.
6. Rebohle, L., Gebel, T., Borany, J.v., Skorupa, W., Helm, M., Pacifici, D., Franzó, G., and Priolo, F. (2001) Transient behavior of the strong violet electroluminescence of Ge-implanted SiO_2 layers, *Appl. Phys.* **B 74**, 53-58
7. Skorupa, W., Rebohle, L., and Gebel, T. (2002) Group-IV nanocluster formation by ion beam synthesis, *Appl. Phys.* A (in print).
8. Gebel, T., Rebohle, L., Skorupa, W., Nazarov, A.N., Osiyuk, I.N., and Lysenko, V.S. (2002) Charge trapping in light emitting SiO_2 layers implanted with Ge^+ ions, *Appl. Phys. Lett.* (to appear on Sept. 30)
9. Gebel, T. (2002), Nanocluster-rich SiO_2 layers produced by ion beam synthesis: electrical and optoelectronic properties, Dissertation, Technische Universität Dresden, Germany
10. Gebel, T., Rebohle, L., Sun, J., Skorupa, W., Nazarov, A.N., and Osiyuk, I.N. (2002) Correlation of charge trapping and electroluminescence in highly efficient Si-based light emitters, *Physica E* (in print)
11. Gebel, T., Rebohle, L., Sun, J., Skorupa, W. (2002) Electroluminescence from thin SiO_2 layers after Si- and C-coimplantation, *Physica E* (in print)

Si/Ge NANOSTRUCTURES FOR LED

G.E. CIRLIN[1,2,3], V.G. TALALAEV[1,4], N.D. ZAKHAROV[1], P. WERNER[1]
[1] *Max-Planck-Institute for Microstructure Physics, Weinberg 2, D-06120 Halle/Saale, Germany*
[2] *Institute for Analytical Instrumentation RAS, Rizhsky 26, 198103 St.Petersburg, Russia*
[3] *A.F.Ioffe Physico-Technical Institute RAS, Politechicheskaya 26, 194021 St.Petersburg, Russia*
[4] *St.Petersburg State University, Petrodvorets, 198504 St.Petersburg, Russia*

1. Introduction

Si/Ge heterostructures are of great interest from the viewpoint of fundamental physics and the possibility of their integration into silicon-based technology for low-cost components in the fiber-optic communication wavelength [1]. Active optical Si/Ge components (photodetectors and light emitters) are expected to operate in the near-infrared spectral range (around 1.55 μm). High-speed near-infrared Ge photodetector integrated on Si chips was recently reported [2]. The most serious problem for application of near-infrared emission from Ge/Si heterostructures is the low luminescence efficiency ($\leq 10^{-4}$), especially at room temperature. The conservation momentum rule limits the luminescence efficiency because of the indirect band structure of SiGe systems. The desire to create Si-based light-emitters caused many attempts to overcome the low radiative efficiency in Si such as porous silicon [3], doping of Si with rare-earth impurities [4], inserting of direct band material (InAs) in a silicon matrix [5] etc. Another approach to overcome the problem of the indirect *k*-space transitions in Si/Ge system involves the spatial localization of the injected carriers in quasi-zero-dimensional Ge islands embedded into Si matrix (in the other words, the concept of the quantum dots, QDs [6]). In the last case, most useful is considered to be self-assembled quantum dot arrays, produced by the Stranski-Krastanow growth mechanism. It is well known that Ge/Si system is a prominent example of Stranski-Krastanow growth mode, where three-dimensional (3D) islands appear at the surface after exceeding of a certain *critical* thickness [7].

In the present work, we are concentrating on the fabrication of light-emitting structures during molecular beam epitaxy (MBE) in Si/Ge heteroepitaxial system. We report on i) the observation of a new kind of nanostructures formed by *sub-critical* Ge insertions in a Si matrix and their structural and optical properties, and ii) on the fabrication of defect-free multilayer structures containing Ge QDs layers in a Si matrix and exhibiting strong PL in the range of 1.55 μm at room temperature. In the first part, our approach is based on the following assumption: insertion of Ge submonalyers (SML) into

L. Pavesi et al. (eds.), Towards the First Silicon Laser, 79–88.

Si may lead to the formation of the ensemble of relatively small islands (with lateral sizes less or comparable with the holes Bohr radii). This can result in a partial lifting of the *k*-selection rule for radiative recombination and exciton formation (which may be stable up to room temperature) via electron and localised-hole interaction. This situation is possible if the Coulomb attraction energy is high enough to localise electrons near the potential barrier which is produced by Ge SML inclusions in the conduction band. It is known that for SML in other heteroepitaxial systems [8,9] the narrow photoluminescence (PL) line leads to the increase of the absorption (gain). The PL intensity will also increase if multiple layers are used with Ge SML separated by Si spacers. Due to relatively small strain accumulation in such a system, it is expected that there will be a low probability of dislocations and formation of structural defects. In fact, despite of the total amount of the Ge (sub) monolayers deposited is lower then the critical thickness we show that they exhibit (quasi) 3D properties when capped with a host (Si) material under certain growth conditions. In the second part, we report on the fabrication of defect-free multilayer structures containing Ge QD layers in a Si matrix and exhibiting strong PL in a range of 1.55 μm under room temperature. The result is reproduced on several sets of samples, which differed in the number of Ge QD layers, Ge layer thickness, intentional doping of Si spacer layers and cap composition. Great potential of this result for emitting device applications was the reason for our studies of the PL band origin. Such studies are carried out by combining PL, structural (transmission electron microscopy) and secondary ion mass-spectrometry (SIMS) measurements.

2. Experimental details

All structures are grown by molecular beam epitaxy (MBE) using a Riber SIVA 45 set-up on Si(100) semi-insulating or p-type substrates (conductivity 0.02 - 3 Ohm cm). The substrates are five inches in diameter, manufactured by OKMETIC. After chemical preparation by the method described in [10] the substrates are transferred into the MBE set-up loading chamber. The method of chemical treatment allows us to remove the oxide layer from the silicon surface at 840^{0}C in the growth chamber by direct radiating heating. During the growth process the rotation of the samples is used, and the temperature field inhomogeneity across the surface is about ≤5%. In order to grow Ge SMLs we use the SML epitaxy technique, which we have also used to grow the SML insertions in the A_3B_5 and A_2B_6 systems [11-13].

The structures consist of a Si 100 nm buffer layer, a Ge/Si superlattice (20 pairs) and a silicon 20 nm capping layer. The growth rates for Si and Ge are 0.5 and 0.02 – 0.3 Å/s, respectively. The substrate temperature (T_s) is varied within 600^{0}C –750^{0}C. Ge content is varied between 0.07 – 0.9 nm. Spacer thickness in between Ge-containing layers is 4 - 5 nm. In some cases, n-type doping of the spacers is used. If doping is applied, central part (2 nm) of the spacers is Ge:Sb grown at desired T_s with electron concentration ~ 10^{16} – 10^{18} cm^{-3}. The growth rates are controlled by two mass spectrometers with feedback which are set to 28 (Si) and 74 (Ge) masses. Total pressure during growth is better than 5x10^{-10} Torr. The surface is monitored *in situ* by reflection high-energy electron diffraction (RHEED). PL measurements are carried out in a standard lock-in configuration. Excitation is provided by the 488 nm line of an Ar^+ laser. For investigation of PL power dependency the laser beam is focused onto a sample area of 10^3 μm^2. For temperature PL measurements the samples are cooled in a He-flow cryostat. The PL signal is collected by a single grating monochromator coupled with a liquid-nitrogen-cooled Ge detector (Edinburgh Instruments Inc.) having the photoelectric threshold 1.7 μm (0.73 eV). PL spectra used for temperature and excitation dependencies are always normalised to the photodetector sensitivity. The samples are also investigated by different electron microscopic (TEM) and selected area electron diffraction (SAED) techniques using a JEM 4010 microscope with acceleration voltage 400 kV and a CM 20 (200 kV) microscope equipped with a field-emission gun.

3. Results and Discussion

3.1. SUB-CRITICAL Ge INCLUSIONS IN Si MATRIX

In this section we will discuss optical and structural properties of Si/Ge nanostructures formed by the deposition of Ge portions less than 0.5 nm which is the range where no 3D islands are formed directly during deposition, i.e. below the critical thickness.

3.1.1. *Undoped samples*

The surface morphology is monitored during the growth *in situ* by RHEED and shows no significant changes in comparison with the initial (2x2) surface reconstruction except of slight broadening of the main (0n) reflections. Thus, even on the upper layers of the superlattice the surface remains quite smooth and well pronounced 3D structural formation due to strain relaxation does not occur. Scanning tunnelling microscopy of *uncapped* samples shows quite smooth surfaces.

Typical sketch of the *capped* samples is presented in Fig. 1, a. The cross section images of grown structures taken in a diffraction contrast mode at relatively low resolution are shown in Fig.1b,c. The periodicity in growth [001] direction is 4.4 nm in both cases. It should be noticed that the images of both structures A and B look almost identical at these imaging conditions. However diffraction patterns show a distinct difference (Fig.2 a,b). In the diffraction pattern from superlattice formed by 0.5 ML Ge layers (specimen A) the number of Fourier harmonic **n** is twice less than in the case of 1 ML one (specimen B). It means that the thickness of incorporated Ge layers in the second case is practically mono atomic while in the first case it is up to 10 monolayers.

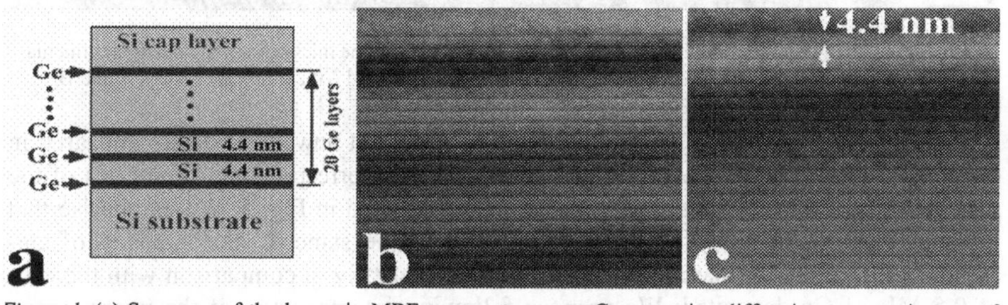

Figure 1. (a) Sequence of the layers in MBE grown structures. Cross section diffraction contrast images of specimen with 0.5 ML Ge (b) and 1 ML Ge (c) (samples A and B, respectively). The periodicity of grown structures in [001] direction is equal to 4.4 nm in both cases.

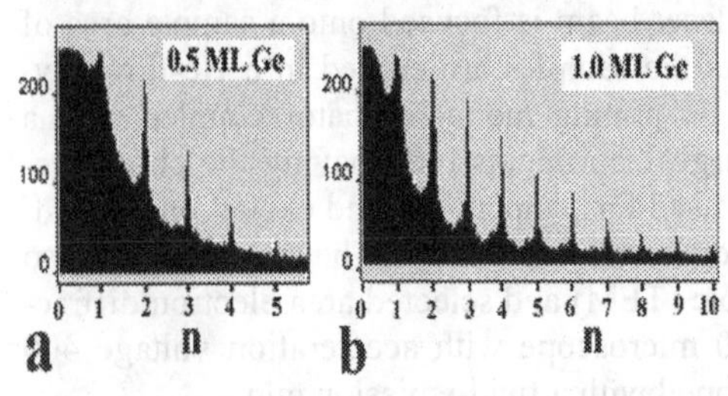

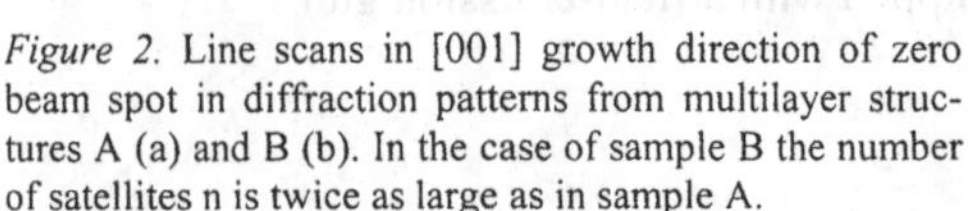

Figure 2. Line scans in [001] growth direction of zero beam spot in diffraction patterns from multilayer structures A (a) and B (b). In the case of sample B the number of satellites n is twice as large as in sample A.

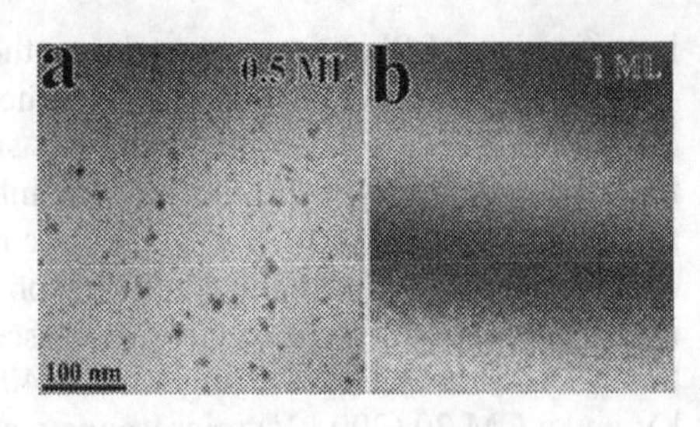

Figure 3. Plan view images of sample A (a) and B (b). The compositional fluctuations can be clearly seen in sample with 0.5 ML Ge, while the sample B (1 ML Ge) is very homogeneous.

The single monolayers of Ge are clearly seen in a high resolution cross-section image (Fig.4b) taken from very thin crystal region (t ≈30 nm) where the kinematical approximation still works well enough. According to this approximation $x = (k-1)/(f_{Ge}/f_{Si}-1)$, where f_{Ge}, f_{Si} - atomic scattering amplitudes for Ge and Si respectively, $k = \Delta I_x/\Delta I_{Si} = (I_{x,max} - I_{x,min})/(I_{si,max} - I_{si,min})$. It gives x=0.9±0.1 what indicates that these monolayers are practically pure Ge. In the case of sample A the compositional fluctuations are observed in Ge submonolayers (Fig.4a). The thickness of these fluctuations in growth direction [001] is about 0.8 nm what is in a quite good agreement with diffraction data in Fig. 2a. Besides, in the plane view images of sample A one can observe larger compositional inhomogenities ($N=1.7x10^{11}$ $1/cm^2$) which look like very small quantum dots. They are completely absent in sample B (Fig.3a,b).

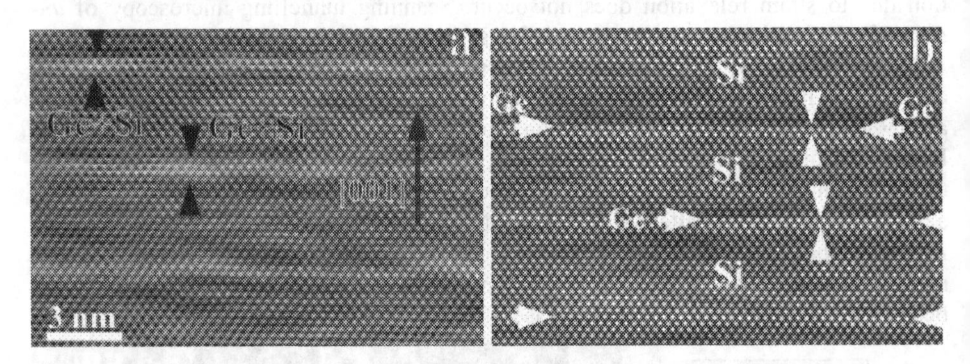

Figure 4. HRTEM cross-section images of samples A (a) and B (b). The thickness of compositional fluctuations measured along [001] growth direction are equal to ~1 nm and ~ 0.2 nm in specimens A and B.

For the same Si/Ge SML samples we have found that new PL lines are appeared in the spectra in comparison with Si substrate spectrum. Corresponding PL spectra of the same samples taken at helium temperatures are presented in Fig. 5. We emphasise that for an integer number of Ge (1 ML) deposited at the same T_s, the intensity of Ge-associated PL lines decrease by at least 20 times integrally in comparison with the case of 0.5 ML of Ge deposited. We propose following interpretation of the phenomenon [14].

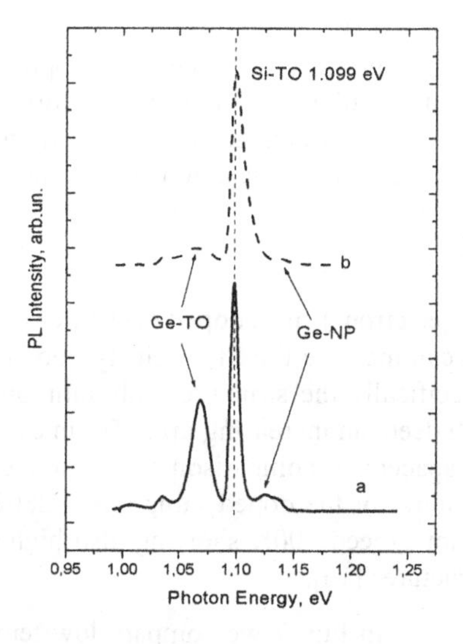

Figure 5. Low-temperature (15K) PL spectra for the Si/Ge multilayer structures containing 0.07 nm (a, solid line) and 0.14 nm (b, dashed line) Ge inclusions.

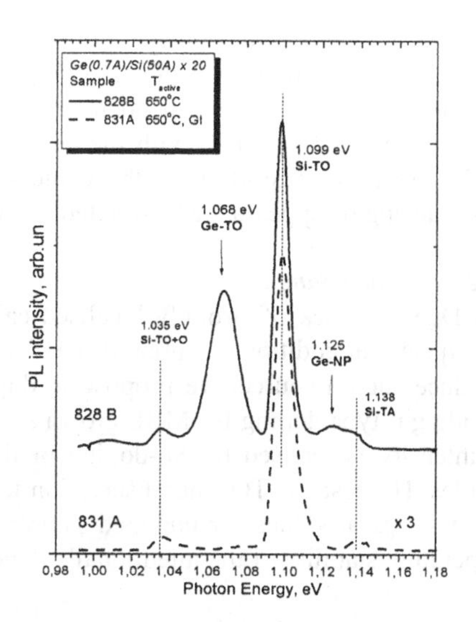

Figure 6. Low-temperature PL spectra taken at 15 K for the Si/Ge samples grown with (dashed line) and without (solid line) growth interruption.

It is well known from the theory of phase transformation in solids that above some size of (quasi) 3D inclusion they may transfer into platelets. This size is determined by balance of elastic strain and surface energies. To put it differently, below this critical size the 3D shape of the inclusion is more energetically preferable, whereas above the critical size the platelet is more stable. The deposition of a half ML of Ge firstly results in the formation of relatively small one-atomic layer thick islands on the Si surface because the quantity of material (Ge) is not enough to fill completely the whole surface. After covering by Si these platelets turned out to be imbedded in bulk Si. They will transform into 3D inclusions if their diameter is below the critical size or otherwise will keep their platelet shape. Such is indeed the case (Fig.3 a,b).

To prove this mechanism we have grown a set of the samples with and without growth interruption (GI) after Ge deposition. We expected that during the GI Ge adatoms efficiently migrate on the surface and form larger 2D islands than without GI. If so, the GI case should be similar to that we observe for full monolayer case. The size of these large islands is already far over the critical size that is why they will keep two-dimensional shape after Si overgrowth. In Fig. 6 we compare PL spectra from the samples containing 0.07 nm Ge deposited at 650^{0}C one of which (# 831A) is subjected to the 120 s GI and other (#828B) is grown without GI. In fact, we have observed well pronounced PL peaks labelled as Ge – TO and Ge - NP correlated with the formation of 3D islands for the sample with no GI but such features are absent for the GI case. TEM images taken for these samples show a very similar structure as presented in a Fig.3 a,b: morphological features corresponding to the formation of Ge nanoclusters for the sample without GI and otherwise smooth surface. It means that no 3D islands are formed during GI and consequent Si overgrowth due to the formation of the 2D nuclei with the lateral size exceeding critical one. These results are in agreement with the simple growth model proposed above.

High temperature growth ($T_s = 750^0C$) leads to the significant changes in the optical properties on Ge thickness dependencies. Deposition of 0.07 nm of Ge results in formation of modulated structure with the composition sinusoidal changes along [-113] and [1-12] directions, distortion of the cubic symmetry and in a relaxation of the misfit stress causing disappearance SL-related PL lines [15].

3.1.2. Doped Samples

Due to a lack of strongly localised carriers (electrons) in undoped structure, one can expect that additional source of the electrons can increase the PL intensity, too. To introduce such a source, we propose to dope specifically the structure with antimony providing n-type doping for MBE grown silicon. Indeed, an increasing up to 50 times of PL intensity is realised by Sb-doping of the Si spacers in comparison with undoped samples. Highest SL-TO line observation temperature for the doped sample is ~ 280K whereas typical values for undoped samples do not exceed 100K showing also higher temperature stability of the intentionally doped structures [16].

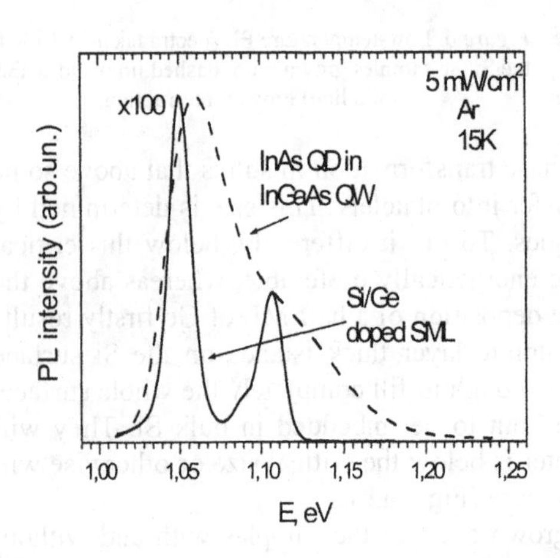

Figure 7. Comparison of PL spectra for InAs quantum dots in InGaAs quantum well and Sb-doped Si/Ge suprelattice samples.

In Fig. 7 we compare low temperature PL spectra for the samples containing Sb-doped superlattice and InAs QDs embedded in a 12% $In_xGa_{1-x}As$ quantum well which allows realising 1.3 μm emitting at room temperature. It was found that internal quantum efficiency for InGaAs/GaAs QDs structure in this particular case was close to 100% [17]. Note that the doped Si/Ge sample exhibits only two orders less in PL efficiency than direct band InAs QD in a GaAs matrix making promising their application in Si-based light-emitting devices.

3.2. CLOSE-TO-CRITICAL Ge INCLUSIONS IN Si MATRIX

Still at least two groups of authors [18-20] observed a relatively strong photoluminescence close to 1.55 μm at the room temperature originating from multilayer Ge/Si QD structures. H. Sunamura et al. [18] observed QD-related PL signal on three 7-period multiple structures having 4.4 and 5.9 monolayer Ge and K.Eberl, et al. [19,20] observed this phenomenon on the 5-times stacked 6.5 ML Ge/Si structure.

Concerning our results, in Fig. 8 we present typical cross-section high resolution TEM (a) and plan-view TEM (b) images for 20 layer Si/Ge with 0.7 nm of Ge in each layer, growth rate for Ge is 0.2 Å/s. The Ge layers are separated with 5 nm Si spacer. Despite of the small thickness of the spacer is used, no structural defects are observed trough the structure, not in Ge nanoislands neither in a Si matrix. The lateral size of the Ge islands is about 80 nm, the height differs layer-to-layer and is in a range 3-5 nm. The larger height is found to be in the middle of the structure whereas for both lower and upper QD sheets it becomes less in size. Average surface density of Ge QD extracted from the Fig. 8, b is about $1x10^{10}$ cm^{-2}. The islands are quite regularly arranged and exhibit well resolved square-based shape [21].

One of the goals during MBE runs is to influence on the electronic structure of the bands at the islands, especially a bending of bands at the Si/Ge^{QDtop} interface. Beside the lattice strain, we additionally incorporate (like in a previous section) a donor doping (antimony) of the spacers close to the top of Ge islands, which gives us an additional possibility to manipulate the bands bending. Indeed, PL measurements of the samples with different doping levels show that there is a significant effect of the Sb atom concentration on the optical properties. Secondary ion mass spectrometry (SIMS) is used for the doping atom concentration measurements using CAMECA setup. For a relatively small doping doze estimated from SIMS data ($n \sim 10^{17}$ cm^3) a maximum of PL intensity is achieved. Higher or lower Sb concentrations lead to a decrease of the PL intensity by several times. At the highest doping levels used ($n \sim (5-7)*10^{18}$ cm^3) the maximum of the PL peak even shifts towards higher energies.

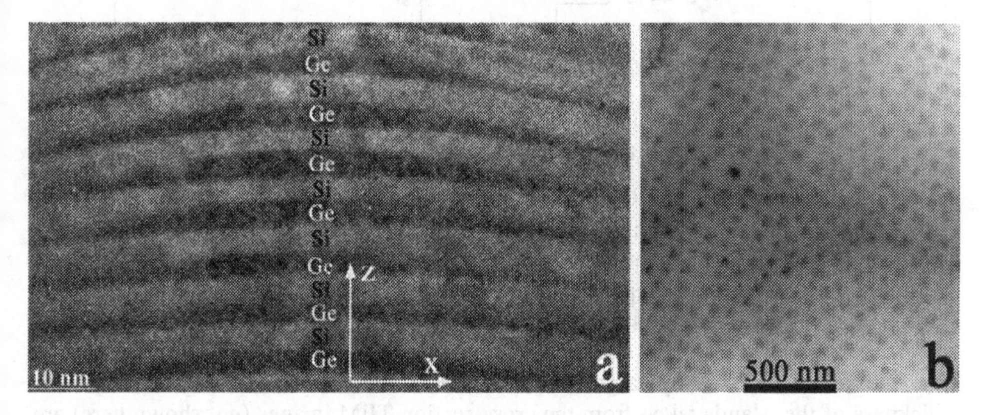

Figure 8. (a) HRTEM cross-sectional image and (b) plan-view TEM images for the Ge^{QD}/Si 20 layer sample. Each Ge layer contains 0.7 nm in average, spacer Si thickness is 5 nm, substrate temperature is 600°C.

Fig. 9,a presents PL spectra of the optimally doped multilayer Si/Ge structure (20 0.7 nm Ge layers, Si spacer is 5 nm, Sb doping is 10^{17} cm^3) taken at helium temperature and room temperature (RT). The signal related to the Ge islands, marked as QD, exhibits bright luminescence up to RT and the intensity is reduced only by a factor of 10 as compared measured at 8K. This result shows that the activation energies are relatively high to localize carriers in a vicinity of Si/Ge interface up to at least RT. Fig. 9,b displays the power dependence of the integrated PL intensity of the PL band labelled as QD in Figure 9,a, in double logarithmic scale. According to the formula $I = P^m$, where I correlates to the PL intensity and P to the power density, the factor m is found to be equal 1.6 within a wide excitation range of 3 to 6000 W/cm^2. A superlinear behavior of this dependence unambiguously manifests the increasing of the exciton oscillator strength with increasing of the non-equilibrium carriers at the Si/Ge interface. Such a behavior was never observed for type-II transition in a Si/Ge system (transition in real space). As an example, band alignment for the SiGe-QD structure [22] was reported with $m < 1$: $m = 0.78$ for the QD PL (type-II transition) and $m = 0.96$ for wetting layer PL (type-I transition). As opposite, our results suggest that the process of recombination in optimised Si/Ge QD acquire a character of type-I transition.

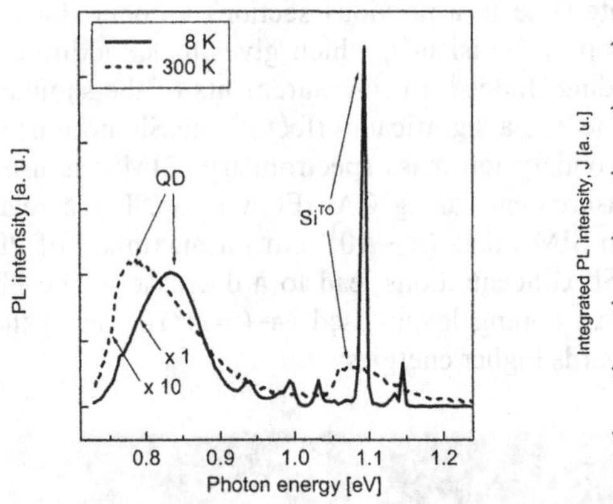

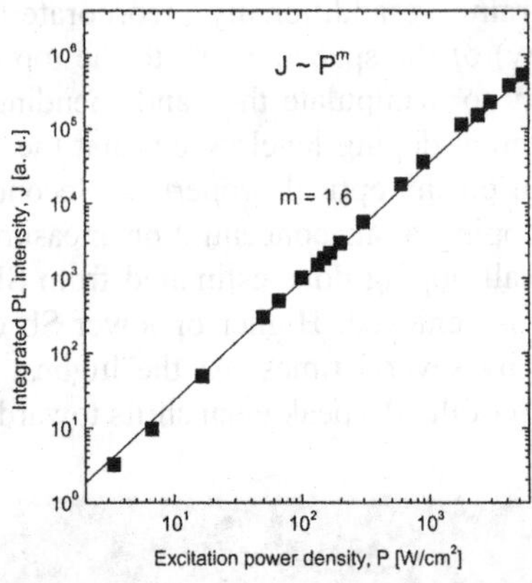

Figure 9. (a) PL spectra for the Ge^{QD}/Si optimised sample, excitation density is 1 W/cm^2; (b) power dependence taken at room temperature of the same sample

We have also found that Ge growth rate significantly influence on both sizes and shape of the islands and, hence, on the optical properties of the structures. In Fig.10 plan-view TEM image for the sample grown at the same growth conditions as the sample presented in Fig. 8,b except of the growth rate which is 0.02 Å/s in this case . The morphology of this sample is characterised by the presence of the two groups of the nanoislands having larger and smaller lateral sizes whereas for the case of higher growth rate the islands are mostly larger in size and distributed more homogeneously. The thickness of the islands taken from the cross-section TEM images (not shown here) are ~ 1-2 nm (platelet-like islands) for the sample shown in Fig. 10 (sample A) whereas ~ 3-5 nm (pyramid-like islands) for the previous sample (sample B).

Figure 10. TEM plan-view images for the Si/Ge multilayer sample containing 5.2 ML Ge grown with the growth rates of 0.02 A/s. The image size is 100 nm x 100 nm.

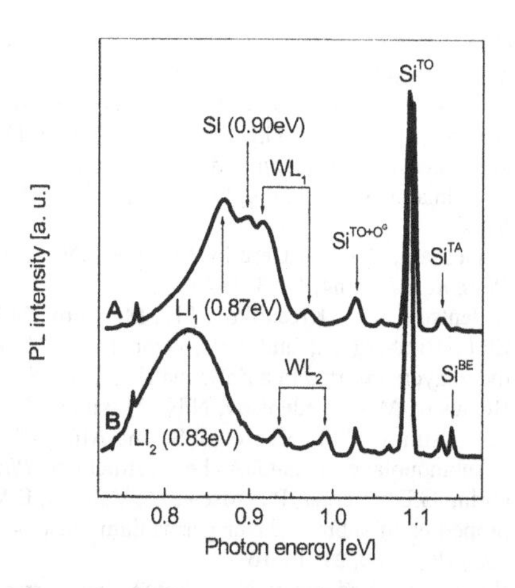

Figure11. Low temperature PL spectra (10 K) for the samples Aand B. WL_1,WL_2 and LI_1, LI_2 wetting layer and large islands related bands for the samples A and B, respectively, SI – small islands related band (sample A).

The sample B is characterised by the strong PL at room temperature, but PL intensity of the sample A is nearly vanishes at nitrogen temperature of observation. In Fig.11 we compare low-temperature PL spectra for both samples. As we expected from TEM investigation, two PL bands responsible for emitting from relatively small (band SI on the spectrum) and larger (LI_1 band) islands are observed for the sample A. Moreover, a signal from a wetting layer (WL_1) is resolved. At the same time, for the sample B only one island-related band (LI_2) is found. The different spectral position of the WL-related bands in samples A and B is correlated with the point that smoothening of the surface is energetically more favourable when low growth rate is used, which was shown previously for, e.g. InAs/GaAs system [23].

4. Acknowledgements

The authors thank Prof. N.N.Ledentsov, Prof. V.M.Ustinov and Dr. A.F.Tsatsul'nikov for fruitful discussions, Dr. V.A.Egorov, G.Gerth and A.Frommfeld for their participation during the growth experiments, Dr. B.V.Volovik and A.G.Makarov for some of PL measurements. G.E.Cirlin is grateful to the Alexander von Humboldt Foundation.

References

1. Hull, R and Bean, J.C. (1999) *Germanium Silicon: Physics and Materials: Optoelectronics in Silicon and Germanium Silicon*, Academie, San Diego.
2. Colace, L., Masini, G., Assanto, G., Luan, H.C., Wada, K., and Kimerling L.C. (2000) Efficient high-speed near-infrared Ge photodetectors integrated on Si substrates, *Appl. Phys. Lett.* **76**, 1231-1233.
3. Cullis, A.G., Canham, L.T., and Calcott, P.D.J. (1997) The structural and luminescence properties of porous silicon, *J. Appl. Phys.* **82,** 909–965.

4. Coffa, S., Franzò, G., and Priolo, F. (1996) High efficiency and fast modulation of Er-doped light emitting Si diodes, *Appl. Phys. Lett.* **69**, 2077-2079.
5. Heitz, R., Ledentsov, N.N., Bimberg, D.,.Egorov, A.Yu , Maximov, M.V., Ustinov, V.M., Zhukov, A.E., Alferov, Zh.I., Cirlin, G.E., Soshnikov, I.P., Zakharov, N.D., Werner, P., and Gösele, U. (1999): Optical properties of InAs quantum dots in a Si matrix, *Appl.Phys.Lett.* **74,** 1701-1703.
6. Grundmann, M., Bimberg, D., and Ledentsov, N.N. (1998) *Quantum Dots Heterostructures*, Wiley, New York.
7. Eaglesham, D.J. and Cerullo, M (1990) Dislocation-free Stranski-Krastanow growth of Ge on Si(100), *Phys. Rev. Lett.* **64**, 1943–1946.
8. Ledentsov, N.N., Krestnikov, I.L., Maximov, M.V., Ivanov, S.V., Sorokin, S.L., Kop'ev, P.S., Alferov, Zh.I., Bimberg D., and Sotomayor Torres, C.M. (1996) Ground state exciton lasing in CdSe submonolayers inserted in a ZnSe matrix, *Appl. Phys. Lett.* **69,** 1343-1345.
9. Belousov, M.V., Ledentsov, N.N., Maximov, M.V., Wang, P.D., Yassievich, I.N., Faleev, N.N., Kozin, I.A., Ustinov, V.M., Kop'ev, P.S., and Alferov Zh.I. (1995) Energy levels and exciton oscillator strength in submonolayer InAs-GaAs heterostructures, *Phys. Rev. B*, **51,** 14346-14351.
10. Cirlin, G.E., Werner, P., Gösele, U., Volovik, B.V., Ustinov, V.M., and Ledentsov, N.N. (2001): Optical properties of submonolayer germanium clusters formed by molecular-beam epitaxy in a silicon matrix. *Tech.Phys.Lett.* **27,** 14-16.
11. Wang, P.D., Ledentsov, N.N., Sotomayor Torres, C.M., Kop'ev P.S., and Ustinov, V.M. (1994) Optical characterization of submonolayer and monolayer InAs structures grown in a GaAs matrix on (100) and high-index surfaces, *Appl. Phys. Lett.* **64**, 1526-1528.
12. Cirlin, G.E., Petrov, V.N., Dubrovskii, V.G., Golubok, A.O., Tipisev, S.Y., Guryanov, G.M., Maximov, M.V., Ledentsov, N.N, and Bimberg, D. (1997) Direct formation of InGaAs/GaAs quantum dots during submonolayer epitaxies from molecular beams. *Czech.J.Phys.* **47,** 379-384.
13. Litvinov, D., Rosenauer, A., Gerthsen, D., and Ledentsov, N.N. (2000) Character of the Cd distribution in ultrathin CdSe layers in a ZnSe matrix, *Phys.Rev.B* **61**, 16819-16826.
14. Cirlin, G.E., Zakharov, N.D., Egorov, V.A., Werner, P., Ustinov, V.M., and Ledentsov, N.N. (2002) Mechanism of germanium nanoinclusions formation in a silicon matrix during submonolyer MBE, *Thin Solid Films* (in press).
15. Zakharov, N.D., Werner, P., Gösele, U., Gerth, G., Cirlin, G., Egorov, V.A., and Volovik, V.A. (2001) Structure and optical properties of Ge/Si superlattice grown at Si substrate by MBE at different temperatures, *Mat.Sci.Engeneering B* **87,** 92-95.
16. Cirlin, G.E., Zakharov, N.D., Werner, P., Makarov, A.G., Tsatsl'nikov, A.F., Ustinov, V.M., Ledentsov, N.N., Egorov, V.A., and Gösele, U. (2002) Structural and optical properties of the multilayer structures formed by Ge sub-critical insertions in a Si matrix, *Proc. MRS Spring 2002 Meetning, April 2002, SnFransisco, USA* (in press).
17. Mikhrin, S.S., Zhukov, A.E., Kovsh, A.R., Maleev, N.A., Ustinov, V.M., Shernyakov, Yu.M., Kayander, I.N., Kondrat'eva, E.Yu., Livshits, D.A., Tarasov, I.S., Maksimov, M.V., Tsatsul'nikov, A.F., Ledentsov, N.N., Kop'ev, P.S., Bimberg, D, and Alferov Zh.I. (2000) A spatially single-mode laser for a range of 1.25–1.28μm on the basis of InAs quantum qots on a GaAs substrate, *Semiconductors* **34,** 119-121.
18. Sunamura, H., Fukatsu, S., Usami, N.,and Shiraki, Y. (1995) Photoluminescence investigation on growth mode changeover of Ge on Si(100) *J. Cryst. Growth* **157**, 265-269.
19. Eberl, K., Schmidt, O.G., Duschl, R., Kienzle, O., Ernst, F., and Rau, Y. (2000) Self-assembling SiGe and SiGeC nanostructures for light emitters and tunneling diodes, *Thin Solid Films* **369**, 33-39.
20. Schmidt, O.G., Denker, U., Eberl, K., Kienzle, O., and Ernst, F. (2000) Effect of overgrowth temperature on the photoluminescence of Ge/Si islands, *Appl.Phys.Lett.* **77**, 2509-2511.
21. Cirlin, G.E., Talalaev, V.G., Zakharov, N.D., Egorov, V.A., and Werner, P. (2002) Room temperature superlinear power dependence of photoluminescence from defect-free Si/Ge quantum dot multilayer structures, *phys.stat.sol. (b),* **232,** R1-R3.
22. Wang, J., Jin, G.L., Jiang, Z.M., Luo, Y.H., Liu, J.L. and Wang, K.L. (2001) Band alignments and photon-induced carrier transfer from wetting layers to Ge islands grown on Si(001), *Appl.Phys.Lett.* **78,** 1763-1765 (2001).
23. Cirlin, G.E., Petrov, V.N., Golubok, A.O., Tipissev, S.Y., Dubrovskii, V.G., Guryanov, G.M., Ledentsov, N.N., and Bimberg, D. (1997): Effect of growth kinetics on the InAs/GaAs quantum dot arrays formation on vicinal surfaces. *Surf. Sci.,* **377-379,** 895-898.

OPTICAL SPECTROSCOPY OF SINGLE QUANTUM DOTS

JAN VALENTA[1], JAN LINNROS[2], ROBERT JUHASZ[2],
FRANK CICHOS[3], JÖRG MARTIN[3]

[1] *Department of Chemical Physics & Optics, Charles University, Ke Karlovu 3, CZ-121 16 Prague 2, Czech Republic*
[2] *Department of Microelectronics & Information Technology, Royal Institute of Technology, Electrum 229, S-164 21 Kista-Stockholm, Sweden*
[3] *Institute of Physics, Technical University Chemnitz, Weinholdbau, D-09107 Chemnitz, Germany*

1. Introduction

The technique of single quantum dot spectroscopy (SQDS) measurements of individual nanocrystals (NCs) is nowadays widely applied to study NCs of III-V and II-VI semiconductors [1]. It enables to observe effects that are hidden by significant inhomogeneous broadening inevitable in measurements on ensemble of NCs. The SQDS measurements revealed several intriguing phenomena such as photoluminescence (PL) intermittence, spectral diffusion and Stark effects [2,3]. Many of these phenomena seem to be explained by charging of the QD at high excitation power resulting in a local electric field [4]. (Accurate control of charging effects is, indeed, an essential issue for the development of future nanoelectronics.)

The application of the SQDS to study light emission from Si-NCs is complicated by two main reasons:

- *Very low emission rate*, which is a consequence of the indirect band-gap structure (conserved even in small Si-NCs [5]). Momentum-conserving phonons have to participate in optical transitions and the radiative lifetime is very long (typically about 0.1 ms at room temperature (RT) [6]).
- *Difficult fabrication of structures* in which Si-NCs are at same time *well defined, efficiently emitting and sufficiently diluted* in order to enable detection of PL from one NC.

The first published SQDS experiments on single Si nanostructures studied grains of porous Si (diluted into a colloid and deposited on a glass substrate [7]). In these samples an exceptionally high quantum efficiency of > 50 % was found [8] and later existence of one or more (up to 4) luminescing chromophores (small nanocrystals) in a single porous-Si grain was demonstrated [9]. Recently single colloidal Si-NCs coated by an organic monolayer were studied exhibiting fast PL decay (in the ns scale) and vibrational PL band corresponding possibly to Si-O-Si stretching vibrations [10].

In this paper we give a review of our single-dot photo- and electro-luminescence

L. Pavesi et al. (eds.), Towards the First Silicon Laser, 89–108.

spectroscopy experiments [11,12] performed on three types of Si-NCs. Future directions and application potential of Si-NCs are discussed in context of results published by other groups.

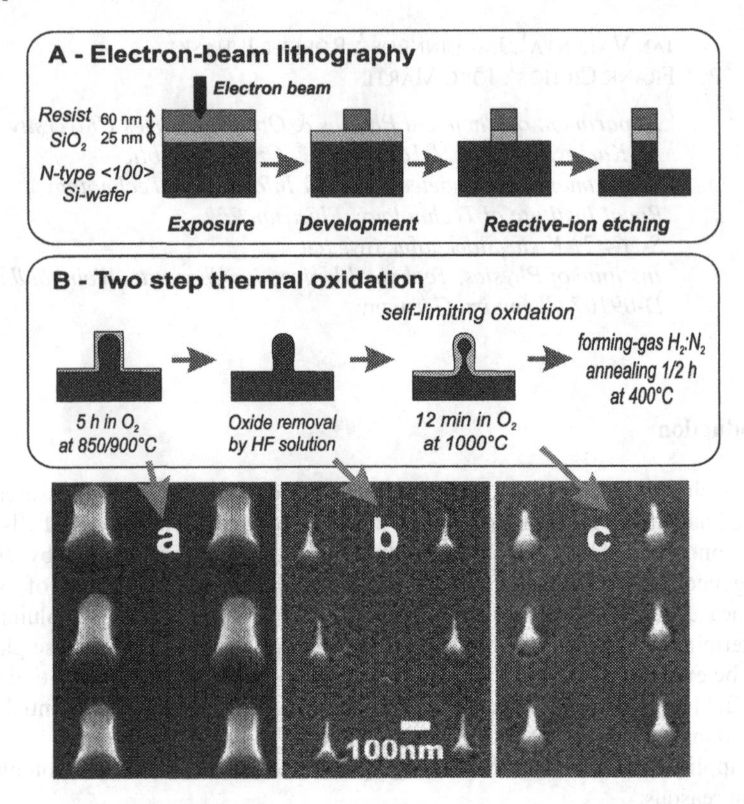

Figure 1. Procedure to prepare Si nanopillars by electron beam lithography, reactive ion etching and dry oxidation. The lower panel represents SEM images (45° tilt view) of the structure after initial patterning (a), after size-reduction by thermal oxidation and removal of oxide (b), and after the last oxidation step (c).

2. Sample preparation techniques

A choice of appropriate technique to prepare well-defined and diluted samples of NCs is the basis of a successful SQDS measurement. We fabricated and studied three substantially different silicon nanostructures.

2.1 ARRAYS OF SI NCS MADE BY ELECTRON BEAM LITHOGRAPHY

Electron beam lithography was used to form resist dots with diameters as small as 50 nm on an N-type (<100>, 20-40 Ω cm) Si wafer having a 25 nm thermal oxide layer. Reactive ion etching (RIE) using CHF_3/O_2-based chemistry was then performed to etch through the top SiO_2 layer followed by chlorine based RIE for Si etching. The resulting

200 nm tall pillars were subsequently thermally oxidized for 5h in O_2 gas at 850 °C or 900 °C. The different temperatures give slightly different consumption of Si, which combined with the range of different initial pillar diameters resulted in Si cores of different sizes ranging down to the few-nanometer regime. The oxide was then removed by buffered wet etching. A second oxidation followed at 1000 °C for 12 minutes. Finally, the samples were annealed for 30 minutes at 400 °C in a 1:9 mixture of H_2:N_2 gas to passivate surface states in order to enhance the PL. The preparation technology is schematically illustrated in Fig. 1 and also SEM images of the structure after several technological steps are shown.

The crucial point for achieving detectable PL is to find an optimal combination of the initial size of crystals (in our case it was 100 or 130 nm) and the oxidation parameters. Both wider and narrower pillars have no detectable PL as their Si core is either too large or it is completely consumed. Note that a phenomenon of self-limiting oxidation [13] plays probably a very important role in the formation of a Si-NC from a pillar. The rate of oxidation is significantly reduced on a few nanometer scale when the surface has a large curvature and stress builds up at the oxide-silicon interface. For pillar geometry, the largest curvature is at the top and one may therefore expect that an isolated SiO_2-embedded Si nanocrystal could remain here while other parts of the silicon core have been oxidized.

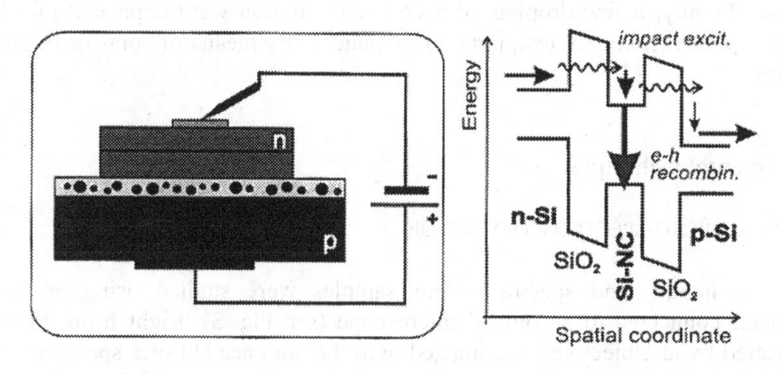

Figure 2. Sketch of the diode structure (left-hand side) and the energy band diagram with transport and recombination paths tentatively indicated (right-hand side)

2.2 ELECTROLUMINESCENT STRUCTURES WITH SI NCS MADE BY SI-ION IMPLANTATION

Electroluminescent structures were prepared by a procedure described in [14]. Thermal oxide layers (thickness of 12, 18, 50 or 100 nm) were grown on p-type (100) Si wafers (resistivity 20 Ωcm) and covered by a 210 nm thick amorphous Si layer. Then a dose of 0.3, 1 or 3 × 10^{17} cm^{-2} of $^{28}Si^+$ ions was implanted at 150 keV. The top amorphous layer served to centre the peak of the implantation profile in the oxide layer. Si-NCs were formed from the excess Si in the SiO_2 by annealing samples at 1100°C in N_2 atmosphere for 1 hour. In order to improve the electron injection into the nanocrystals a 160 nm thick poly-Si layer highly doped with phosphorus was deposited on top of samples. The final circular shape of diodes was defined by reactive ion etching. Fig. 2 illustrates the diode structure.

J-V characteristics of our diodes [14] shows a rectification factor of about 10 and the shape of curves indicates that the main transport mechanism is tunnel or field emission. The EL is excited most probably by impact excitation of electron-hole pairs in Si-NCs.

The fabricated structures with oxide thickness of 12 and 18 nm show EL under continuous forward bias at room temperature. The EL intensity is very stable and visible by a naked eye as a homogeneous glow for the best diodes.

2.3 COLLOIDS OF SI NCS MADE FROM POROUS SILICON

Porous Si (PSi) powder has been prepared by conventional electrochemical etching of a Si wafer (p-type, $\rho \sim 0.1\ \Omega\text{cm}$) in a HF-ethanol solution. Etching current density was kept relatively low (1.6 mA/cm^2) in order to obtain higher porosity and, consequently, resulting in a low mean size of Si-nc and a blue-shifting of the PL band (the peak is around 680 nm). The porous layer was scratched off from the substrate. A colloidal solution was obtained by mixing of the PSi powder with heptan or iso-butanol using both mechanical and ultrasound mixing. The solution was filtered by teflon membranes with pores of 200 nm (by filtering the colloid lost its yellow colour and became colourless). Finally, a few droplets of a colloidal solution were deposited on cleaned substrates – either Si wafers or quartz cover plates – by means of spin- or dip-coating techniques.

3. Experimental techniques

3.1 IMAGING MICRO-SPECTROSCOPY SET-UP

PL and EL images and spectra of our samples were studied using an imaging spectrometer connected to an optical microscope (see Fig. 3). Light from the sample was collected by an objective lens, imaged onto the entrance slit of a spectrometer and detected by an LN-cooled CCD camera (attached to the output of the spectrometer). Reflection images of studied structures were detected under illumination with the blue (442 nm) line of a cw He-Cd laser while PL was excited by the UV line (325 nm) of the same laser. The laser beam was directed towards the sample through the gap between the objective and the sample surface at grazing incidence.

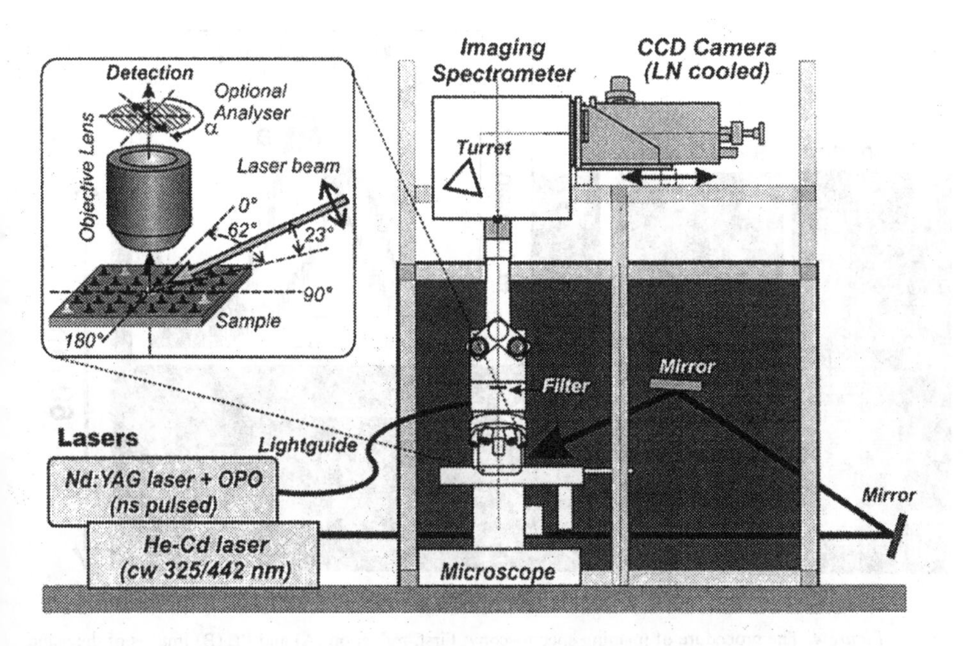

Figure 3. The compact and user-friendly imaging micro-spectroscopy set-up constructed for the single Si-NC spectroscopy. The top-left panel is a schematic drawing of the excitation-detection geometry.

The experimental procedure was as follows (see Fig.4): For each sample, first, the images of reflection and PL (or EL) were obtained using a mirror inside the spectrometer (entrance slit opened to a maximum). Then, an area of interest was placed in the center of the image, entrance slit closed to desired width (resolution) and the mirror was switched to a diffraction grating (mirror and two gratings are mounted on the same turret) in order to record a spectrum. PL was excited by the UV line (325 nm) of a He-Cd laser sent to the sample through the gap between the objective and the sample (grazing incidence). All spectra were corrected for spectral sensitivity of the detection system.

The spatial resolution d of our imaging system is limited by diffraction $d = 1.22\ \lambda\,(2{\cdot}NA)^{-1}$ where λ is the used wavelength and NA the numerical aperture. In our case ($NA = 0.7$) the resolution is about 500 nm in reflection and slightly worse in PL (see Fig. 6 A,B). The set-up allows us to detect a spectrum from any diffraction limited spot. However, we can often resolve the PL from individual NCs even in a case when the NC spacing is on the limit of resolution (0.5 μm - right-hand side of Fig. 2 A, B) because only a small fraction of dots is shining.

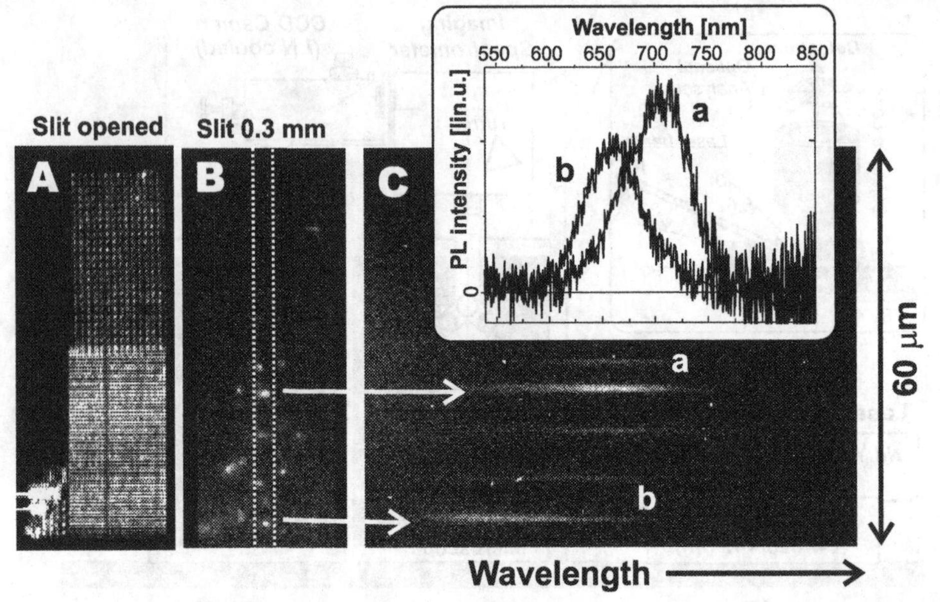

Figure 4. The procedure of imaging spectroscopy: First, reflection (A) and PL (B) images of the same part of a Si-NC sample are detected. Then a spot of interest is placed in the centre of image, the spectrometer entrance slit (dashed white rectangle) is partially closed, the mirror inside spectrometer is replaced by a diffraction grating and a spectral image is detected by the CCD camera. PL spectra of single shining dots appear as light traces on the image (panel C). The inset shows extracted PL spectra of the two emitting NCs (traces *a* and *b*). According to [12]

3.2 CONFOCAL MICRO-SPECTROSCOPY SET-UP

The home-build laser scanning (confocal) microscope (Fig. 5) is based on a computer-controlled 3D piezo-scanner which moves a sample. The microscope has an inverted epi-fluorescence configuration. Excitation beam (514 nm line of an cw Ar^+-ion laser) is expanded and focused on the sample by an objective lens (100x, NA=0.9). The size of an excitation spot (determining the resolution of the microscope) is about 350 nm (FWHM). The same objective collects PL emission which goes through a dichroic mirror to the detection system. The signal is detected by an avalanche photo-diode (APD) photon-counting module. We can also divide the signal 50:50 and use a pair of APDs to study correlation between incoming photons (using time-to-amplitude converter for short time intervals or a digital correlator). At the same time, a part of signal is used to detect spectrum with an imaging spectrometer and a LN-cooled CCD camera.

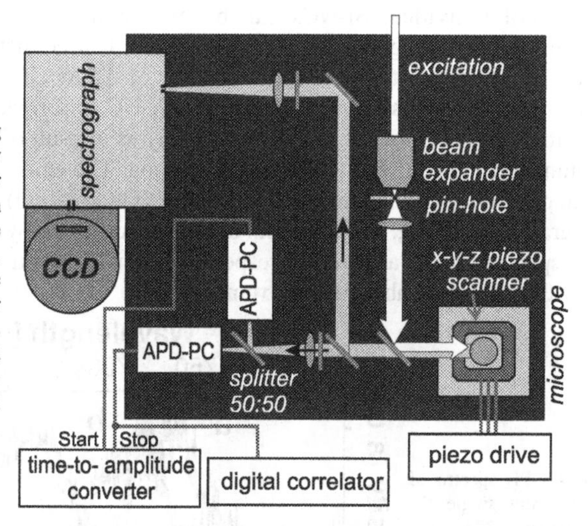

Figure 5. The laser scanning (confocal) micro-spectroscopy set-up: The exciting laser beam is focused by an objective lens onto a sample moved by a 3D piezo-scanner. The emitted light can be detected by a single or a pair of APD (detection of PL intermittency and autocorrelation) or by a LN-cooled CCD camera attached to a spectrometer (PL spectra).

4. Experimental results

4.1 PL SPECTRA OF SINGLE SI NANOPILLARS

Fig. 6A presents a reflection image of a matrix of Si-NCs made by oxidation of 100 nm wide pillars. The PL image of the same part of the sample (Fig. 6B) reveals a small number of emitting Si-NCs with various intensities. Intensity profiles of reflection and PL (Fig. 6 C) show a very good coincidence of peak positions, indeed, a strong indication that PL originates from individual pillars organized in a regular 2-D lattice.

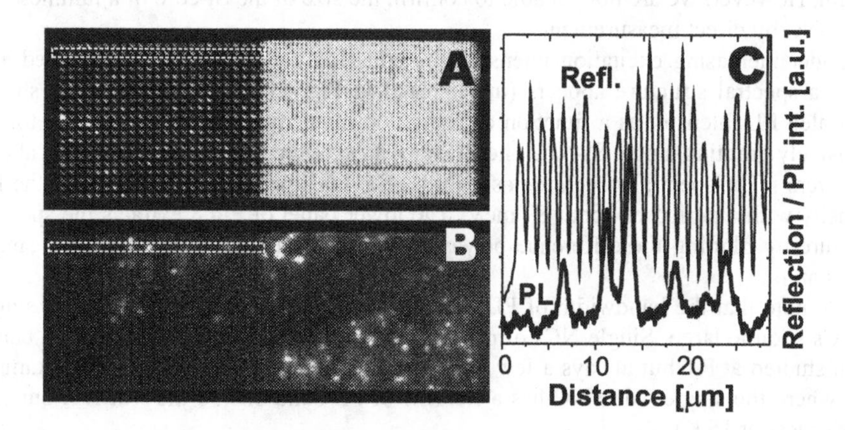

Figure 6.. Reflection (A) and PL (B) image of regular lattice of Si-NCs. Distance between neighbor NCs is 1 and 0.5 μm for the left and right-hand side of the sample, respectively. Panel (C) compares intensity profiles of reflection (a) and PL (b) signal taken from the same part of sample (indicated by dashed rectangles in the left upper corner of images A and B). According to [11].

PL spectra of individual Si-NCs can be measured at RT for the most intensively luminescing NCs. In Fig. 7, we plot PL spectra of three different dots. The detection time was 30 minutes at the excitation intensity of 0.5 W/cm^2 (the spectral resolution is about 10 nm). The PL spectrum of a single Si-NC is formed by a single band which peak position varies from dot to dot, most likely as a result of a variation in the amount of quantum confinement due to the size dispersion. The emission band can be fitted by a Gaussian peak lying in the range 1.58 - 1.88 eV (660-785 nm). Results of fits are plotted as bold gray lines in Fig. 7. Full-width at half-maximum (FWHM) is 122, 120, and 152 meV for spectra of dots a, b, and c, respectively (the individual spectra are at least twice narrower then the ensemble PL spectrum).

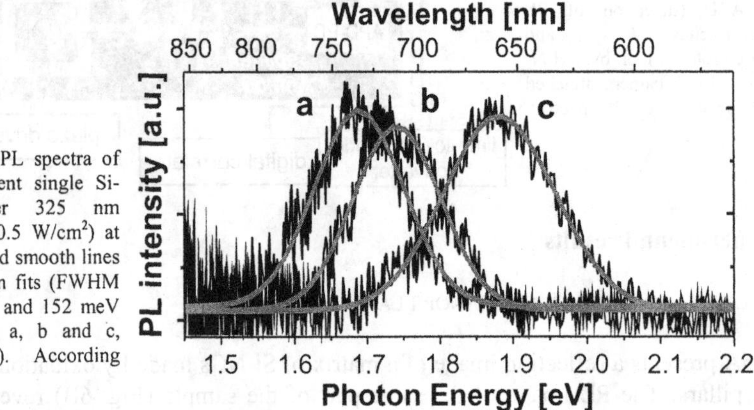

Figure 7. PL spectra of three different single Si-NCs under 325 nm excitation (0.5 W/cm^2) at RT. The bold smooth lines are Gaussian fits (FWHM is 122, 120 and 152 meV for spectra a, b and c, respectively). According to [11].

According to the experimental data of Wolkin et al. [15] and calculations of Reboredo et al. [16], a PL maximum at 1.7 eV corresponds to a NC diameter of about 3.6 nm. However, we are not yet able to confirm the size of the Si-core of a luminescing single-NC by direct measurement.

Under increasing excitation intensity, the emission band is slightly broadened and rarely a spectral structure appears (upper panel in Fig. 8). The inset in Fig. 8 shows integrated PL intensity as a function of excitation intensity (0.03 to 0.5 W/cm^2) for an intensively emitting single Si-NC. The PL starts to saturate at excitation intensity above 0.3 W/cm^2. The spectral structure has a form of a quasi-periodic modulation of the PL intensity with a period of about 80 meV. The lower panel of Fig.8 extracts the spectral structure by plotting the difference between the spectrum shown above and its single Gaussian fit.

We note that the bandwidth of PL spectra (120 - 210 meV) observed in our single Si-NCs is quite large. Single NCs of other semiconductors exhibit also broad PL bands when studied at RT but always a few times narrower than in our case (see for example [18] where the single CdSe NC has a FWHM of PL band of ~50 meV at RT, but few meV or less at 15 K).

There are mainly two possible reasons for the wide PL spectrum: dominating phonon-assisted transitions and spectral diffusion.

As we mentioned before Si-NCs conserve the indirect band-gap. Therefore, one or more phonons have to take part in optical transitions in order to conserve momentum

[5]. Recently, theoretical studies have shown that phonon-assisted transitions in Si-NC dominate over the full range of sizes and even at low-temperatures [17]. Participation of several phonons in radiative recombination of electron-hole pairs could also explain the quasi-periodic structure we observed for some dots (see Fig. 8). The period of about 80 meV does not correspond to any phonon energy in bulk Si (for optical transition LO-phonon is most significant - its energy in the Γ-point of bulk Si at RT is 64.4 meV) but it could be a surface phonon or local vibration energy. Further studies are under way.

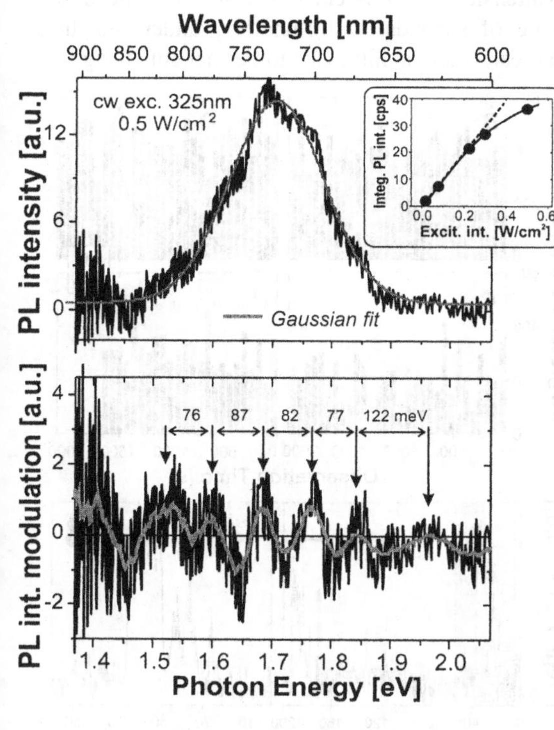

Figure 8. PL spectrum of efficiently emitting single Si-NC (cw excitation at 325 nm, 0.5 kW/cm^2) showing a spectral structure (upper panel). The quasi-periodic spectral structure is extracted from the spectrum above by subtracting its single Gaussian fit. The position of peaks and their distances are indicated. The intensity dependence of the integrated PL signal for this Si-NC is plotted in the inset.

We have to stress that the detection time for a PL spectrum from a single Si-NC is 30 min. Therefore, the inherent phonon-related structure of PL spectra can be smoothed out by spectral diffusion. Spectral diffusion is a shift of the emission line due to variations of the local field. It was observed in II-VI semiconductor quantum dots that the line-width is strongly dependent on the detection time as a consequence of important spectral diffusion even at low temperatures [2].

It would be desirable to measure PL of single Si-NC also at low temperatures. However, such experiment will be very difficult as the PL photon rate is further reduced due to the increasing lifetime of Si-NC excited states.

Fluctuations of the PL signal from single Si-NCs were studied first by repeated detection of PL images with exposure times of 1 min. This time interval was estimated as the shortest detection time necessary to obtain a reasonable signal to noise ratio. In Fig. 9, the integrated PL signal from four individual dots is plotted. Despite the quite long detection time one can see large fluctuations of the PL intensity for certain NCs (e.g. dots (b) and (c) in Fig. 9) whereas some NCs (e.g. dot (a) in Fig. 9) appear stable (i.e. PL variations are within detection noise, background signal being ~30 counts/min). We can speculate that the observed PL intensity fluctuations originate from PL intermittence (on-off switching) which is often observed in single NCs and molecules on the time scale of seconds and ms (see e.g. [18] for an intermittence study in CdSe NCs). In order to improve the time-resolution we performed additional experiments with an amplified gated CCD camera. It enables us to obtain a time-resolution of 0.2 s. One representative time-trace is shown in panel B of Fig. 10. The excitation was with the 514 nm line of an Ar-ion laser (intensity of 40 W/cm^2). We see that the on-off blinking takes place on the time scale of seconds. Unfortunately, under this high excitation shining nanocrystals tend to bleach (stop shining due to permanent changes).

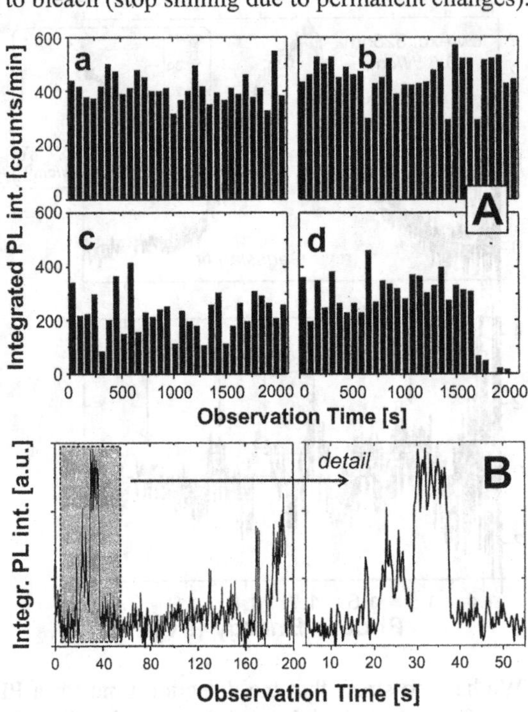

Figure 9. Temporal stability of single Si-NC emission: (A) Integrated PL intensity measured in 30 consecutive 1-min acquisitions is plotted against time (cw excitation 325 nm, 0.5 W/cm^2). Signal from four dots (a) to (d) is presented in the four panels. (B) PL blinking detected with an intensified CCD camera. The signal was detected in 0.2 s steps 1000 times (cw excitation, 514 nm, 40 W/cm^2). The right curve is a detail of one ON-interval (indicated by a rectangle on the left graph).

Finally, we performed polarization sensitive detection of single Si NC PL using a linear polarization filter (analyzer) inside a microscope (see inset in Fig. 3). This configuration allows checking the projection of an emitting dipole in the plane parallel to the sample surface. The results (published in [11]) indicate that PL from a majority of individual NCs has a high degree of linear polarization. On the other hand, there is almost no polarization memory - i.e. the orientation of PL polarization is independent of the exciting beam polarization. The high degree of linear polarization indicates that the orientation of an emitting dipole is quite stable. This suggests that NCs showing polarized PL are elongated in different directions [19] whereas NCs with non-polarized PL has either a random orientation of the emitting dipole (spherical shape) or the dipole is oriented perpendicularly to the sample surface. The absence of a polarization memory could be explained by the high energy of excitation - far from the emission wavelength. Thus, the absorbing and emitting states are very different.

4.2. PL QUANTUM EFFICIENCY IN SINGLE SI NANOCRYSTALS

The well-characterized experimental conditions allow us to estimate a PL quantum efficiency (QE) for the best emitting Si-NCs. We adopted two different approaches to calculate the QE.

The first method relies on the PL signal being well below saturation. Let us take an excitation intensity of P = 80 mW/cm^2. The corresponding photon flux is 1.3×10^{17} photons/s/cm^2 (i.e. 10 times lower than the saturation limit as defined before). The fraction absorbed in one NC is determined by the absorption cross section. This quantity was studied in detail by Kovalev and co-workers [5] in ensembles of Si-NCs. For excitation at 3.81 eV and detection at 1.65 eV they give $\sigma \sim 1\times10^{-14}$ cm^2. The signal count rate N is then equal to:

$$N = D\,\eta\,\sigma\,P \qquad (1),$$

where η is the PL QE and D is the overall detection efficiency of photons emitted by a NC (D = ~0.026 as estimated for our experimental set-up). The most intense NCs produce count rates of about 12 counts/s under P = 80 mW/cm^2. This gives us a high QE of 35 % (estimated accuracy is about ±50 %) for the best NCs. The majority of NCs seen in PL images on Fig. 2 and 5 have QE between 5 and 20 %.

The second estimation uses the saturated PL signal. This approach was used by Credo et al. to calculate a PL QE as high as 88 % for individual grains of porous Si [8]. The saturated count rate is calculated using the equation:

$$N_{sat} = D\eta / \tau \qquad (2),$$

where τ is the excited state lifetime. For the NC used in the above-described calculation, we found N_{sat} about 40 cps for the highest excitation intensity we used. Taking a lifetime τ = 100 μs (from measurements on ensembles of Si NCs [6]) Eq. 2 yields a QE of 15 % corresponding at least with a factor of two with the value obtained from Eq. 1. (However, the excited state lifetime may be quite different in the present structures.)

We want to stress the fact that only a few percent of Si-NCs in our structures show detectable PL, cf. Fig. 6B. This is in agreement with the report by Mason et al. that only about 2.8 % of porous Si particles are luminescing [7]. The reason could be non-radiative quenching by defects or possible escape of e-h pairs to the substrate. Indeed, the critical effect of defect passivation is clearly illustrated by the fact that the PL was totally quenched following the electron beam exposure during SEM imaging and could only be regained by a forming gas (hydrogen) anneal. Therefore, better isolation of NCs from the substrate and different oxidation conditions and forming-gas anneals are clearly worth to study. Whether it is possible to increase the fraction of luminescent dots and their QE or if the statistical nature of the occurrence of defects in NCs puts an upper limit to these numbers, remains unanswered.

4.3 ELECTROLUMINESCENCE OF SINGLE SI NANOCRYSTALS

The electroluminescence diode structures with oxide thickness of 12 and 18 nm show EL under a continuous forward bias at room temperature [14]. The EL intensity is very stable and looks quite homogeneous on a large scale (Fig. 10 A). However, the EL micro-images taken with high magnification (Fig.10 B) reveal that the emission is not fully homogeneous. It is composed from a quasi-homogeneous background emission on which a number of brighter spots is resolved (these spots can be better seen when increasing the contrast of the EL image - it was applied in Fig. 10B). Almost all spots are of the same size corresponding to the diffraction limit of the imaging optics. An EL intensity profile of one bright spot is plotted in Fig. 10 C.

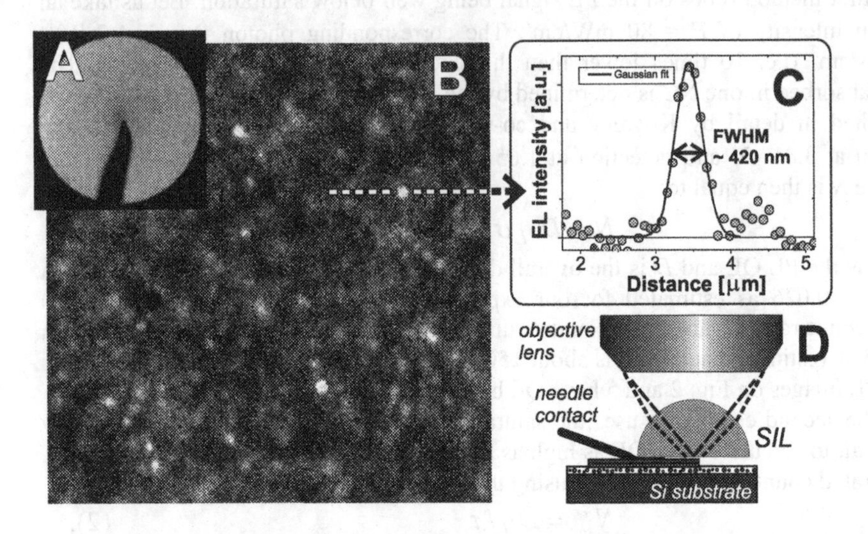

Figure 10. (A) The EL image of the whole LED structure (diameter of the top contact is 2 mm, the black object is a metal probe providing voltage bias). (B) The EL image (with increased image contrast) of a small part (area 23 × 23 μm^2) of a diode under forward bias of 10 V with application of a SIL lens. (C) EL intensity profile of one brightly emitting spot. (D) The sketch of the experimental arrangement with the SIL lens.

For some measurements we placed a solid-immersion lens (SIL) on the top contact of a diode (see Fig.10 D) in order to increase both the numerical aperture (NA) for signal collection and the spatial resolution of images. Our SIL is a hemisphere made of BK7 glass (n~1.517). The SIL allows imaging the small area below the centre of the hemisphere virtually without any aberration and increases the light collection efficiency [20]. Our tests confirmed an increase of magnification and of image resolution by a factor of 1.5 (almost equal to the SIL refraction index) in agreement with the classical relation for resolution (diffraction) limit of far field imaging $d = 1.22\,\lambda\,(2\,NA)^{-1}$.

The overall EL spectrum (collected from 1 mm^2 of the sample surface) of a typical diode with 12 nm thick oxide implanted with a dose of 1×10^{17} cm^{-2} is shown in Fig. 11 A. The PL spectrum (excited by the SHG green output of a Nd:YAG laser (532 nm, 40 Hz, 3 ns, 1 mJ/cm^2) through the top contact) is shown for comparison. The EL spectrum consists of two emission bands. The narrower short-wavelength peak is increasing with implantation dose and can be attributed to emission from oxide defects [21]. The main wide emission band around 800 nm is most probably due to e-h recombination in Si-NCs. In the PL spectrum, this band is somewhat narrower and comparable with the typical PL of similar samples made by Si^+-ion implantation of SiO_2 [6]. The decay of both PL and EL emission is long (tens of μs) and non-exponential (well described by a stretch-exponential curve) [14]. For both EL and PL the highest intensity is emitted from the diode implanted by the dose of 1×10^{17} cm^{-2}. We measured the external quantum efficiency of this diode and found the value of $\leq 3 \times 10^{-5}$ [22].

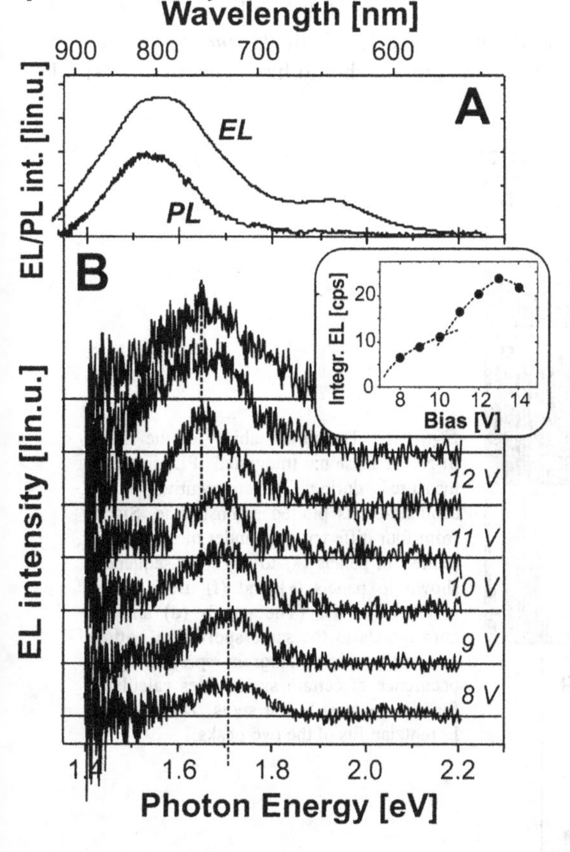

Figure 11. EL spectra for a sample with oxide thickness of 12 nm and implantation dose 1×10^{17} cm^{-2}.
(A) The overall EL spectrum from 1 mm^2 of the diode surface compared with a PL spectrum detected in the same place under excitation by the SHG output of a Nd:YAG laser..
(B) EL spectra of a single bright diffraction-limited spot under different biases of 8, 9, 10, 11, and 12 V (from bottom to top). The inset shows the integrated EL intensity vs bias.

The spectra of the brightest single EL spots can be measured by the imaging spectrometer. For each spot we observe spectra of various intensity, width and peak position. In Fig. 11 B we plot EL spectra of one EL spot taken under various biases from 8 to 14 V. The acquisition time of one spectrum is 30 min. For this spot, the EL signal saturates at about 11 cps (spectrally and spatially integrated signal from the whole diffraction limited spot) for a bias of 10 V but then it is again increasing up to 23 cps at a bias of 13 V. For biases above 11 V we see that the peak position jumps from 720 to 750 nm. Under low biases the EL spectrum can be quite narrow (down to about 120 meV, i.e. comparable to PL spectra of single Si-NC, see Fig. 7) but it widens considerably with increasing bias.

The fluctuations of the EL signal from single shining spots were studied by repeated acquisitions of EL images for 1 min (in accord to PL measurements - see Section 4.1). Fig. 12 A shows six sequences of 30 measurements of EL signal from different single bright spots under forward bias of 8 V (graphs *a* to *d*) and 14 V (graphs *e* and *f*). The intermittence of EL emission (on-off blinking) is observed for the majority of bright spots. The shortening of the on-off period with increasing bias and also the tendency of the EL signal to concentrate around a few distinct levels are clearly illustrated. This fact is visualised by constructing a histogram of integrated EL signals emitted by 10 single spots under 8 V (30 observations for each dot means total 300 values, Fig. 16 f). The histogram reveals two distinct maxima around signal levels of *25 and 50 cps*. (Let us remind that for a classical source of light the histogram should have a form of a (single) Poissonian distribution).

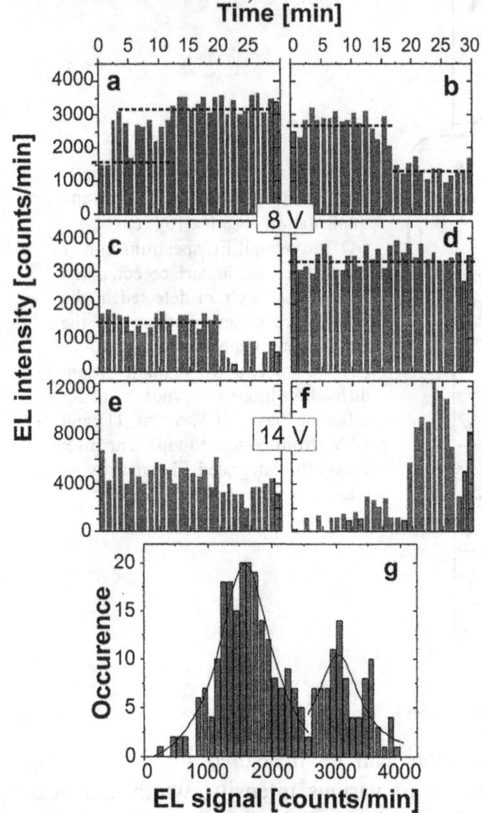

Figure 12. Temporal stability of the single Si-NC emission: Integrated PL intensity measured during 30 consecutive 1-min acquisitions is plotted against time. Signal from four different dots under bias of 8 V is plotted in panels (a) to (d). The sequences shown in panels (e) and (f) are detected under 14 V bias (The panels (d) and (e) corresponds to the same spot). The bottom part of figure is a histogram representing the occurence of certain signal rates calculated for the ten most bright spots. The lines are Lorentzian fits of the two peaks.

Polarization sensitive detection reveals a high degree of linear polarization of EL from the majority of single spots (Fig. 13). This indicates that the orientation of an emitting dipole is preferably along the oxide layer (perpendicular top the optical axis of imaging optics) [19].

Let us now roughly estimate the maximal EL signal, which could be detected, when a single Si-NC inside a diode is excited by passing current up to the level of saturation. We assume the maximum photon emission rate for EL of a single Si-NC to be the same as a saturated PL rate. (We found the saturated PL signal from the best Si-NCs to be about 40 cps – see Section 4.1). The experimental set-up was identical but the collection efficiency of EL photons from the diode is smaller compared to the detection of PL from Si-NCs placed on the surface of the Si substrate. This is because the NA aperture is reduced by the top poly-Si contact (n_{air}/n_{Si} = 1/3.65=0.274 - calculations for 750 nm) and there are also losses due to reflection from the poly-Si/SIL and SIL/air interfaces (reduction by 0.83 and 0.96, respectively) and, finally, due to absorption in the poly-Si contact (reduction by 0.965). The total reduction of the signal collection efficiency is then about 0.21 compared to the PL experiments. (The use of a SIL can increase the collection efficiency by a factor of ~ n_{SIL} =1.517.) From this we obtain the *expected saturated EL signal from a single Si-NC of about 13 cps*. This estimation is close to the value of 11-25 cps we observe (see Figs. 11 and 12).

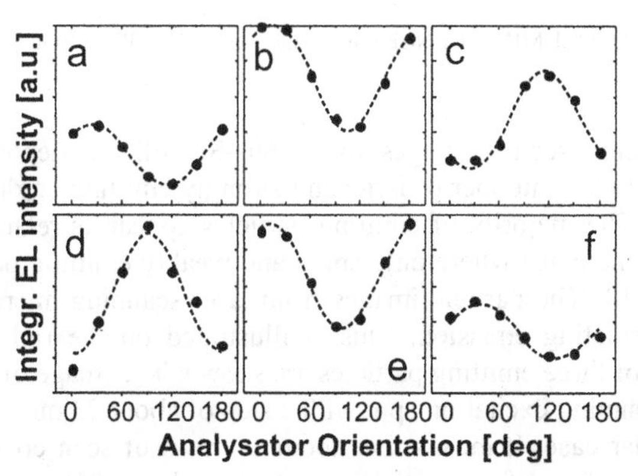

Figure 13. Polarization selective detection of EL from 6 different single spots (10 V, 0.68 mA). Spatially and spectrally integrated EL signal is plotted versus the analyser orientation. Experimental points are fitted with a squared sinusoid.

Therefore, we can conclude that the bright spots in EL images of our diodes are in fact single (eventually very few) Si nanocrystals which are efficiently excited by passing current. This statement is supported by the following observations:

(i) Spectral shape of the EL spectrum: At low bias (Fig.11) the EL band has a FWHM around 120 meV, the same as for PL spectra of single nanocrystals.

(ii) The EL intensity saturates around 10 cps which value corresponds well to the saturated PL signal from single Si-NC (as discussed above).

(iii) The EL signal from single spots is linearly polarized.

Now the question arises: how is it possible that we can observe EL of single Si-NC in this type of highly concentrated Si-NC structure?

Let us estimate the mean distance of Si-NCs in the oxide layer. According to Monte-Carlo simulations [22], using the TRIM code, an implantation dose of 1×10^{17} cm^{-2} creates a peak excess concentration of Si in SiO_2 of about 10 at. %, i.e. 5×10^{21} cm^{-3}. The TEM observations of similar implanted structures show that the mean diameter of NCs is about 3 nm [23]. This gives us a mean distance between nanocrystals of about 5 nm. But the TEM images revealed also that excess Si atoms precipitate on the Si/SiO_2 interfaces, which effect will cause significant losses of Si excess atoms in thin oxide layers (diffusion length of Si for a 1 h annealing at 1100°C is estimated to L ~ 6.8 nm). Therefore, the real inter-NC distance is probably significantly longer than 5 nm. Let us take an inter-dot distance of 7 nm, then there will be about 12 thousand NCs under a diffraction-limited spot of diameter of 620 nm.

However, the number of effectively emitting NCs is much reduced essentially due to two main reasons. The first one comes from the observation by Burrato's group [9] and by us that only a few percent, or less, of the Si-NCs in an ensemble shows observable PL signal. The majority of NCs is dark probably due to the presence of efficient non-radiative centers. The second reduction comes from the need of efficient impact excitation of e-h pairs in a NC. Electron flow through the oxide layer is most probably inhomogeneously distributed (NCs are to some extend "mapping" the current distribution in a thin oxide layer). There are probably many paths through which the current leaks without exciting EL (external quantum efficiency is only $\leq 3 \times 10^{-5}$). So, there could be a low probability that a Si-NC, free of non-radiative centers, will be in a position where impact excitation of EL is very efficient.

4.4 PHOTOLUMINESCENCE OF SINGLE GRAINS OF POROUS SI

Photoluminescence images of porous-Si colloids deposited on a substrate show a relatively low number of large and strongly emitting agglomerates of Si nanocrystalline grains. The majority of emitting particles appear as relatively weak diffraction limited spots. One place where only small and weakly emitting particles are observed is shown in Fig. 14. The particle images from laser-scanning microscope reveal features typical for fluctuating emission. This is illustrated on the right-hand side of Fig.14, where details of three emitting particles are shown. The image of an emitting particle is formed by a scanning excitation spot with a step of about 23 nm and dwell time of 10 ms in this particular case. It means that about 15 rows of scan covers a single particle spot (the apparent size of a particle is always about 350 nm (FWHM), which perfectly corresponds to the theoretical resolution of our microscope). When the particle is blinking on a time scale longer than the dwell time, we can observe dark rows inside the particle spot.

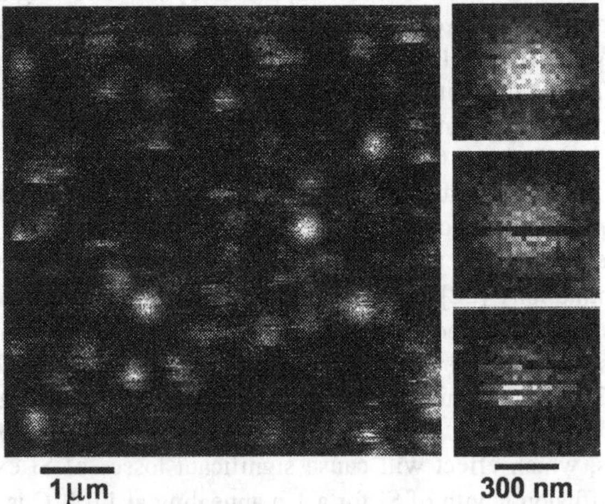

Figure 14. Laser-scanning microscope PL image of porous-Si particles deposited on a quartz substrate (excitation at 514 nm, 40 kW/cm^2). The left image was obtained by scanning an area of 6x6 μm in 256x256 steps with dwell time of 10 ms. The three small images are details of three emitting particles whose emission undergo on-off blinking.

PL intermittence is clearly observable when the excitation spot is stopped on the emitting particle and time changes of PL are detected. Fig. 15 gives an example of four PL time traces. The ON/OFF switching occurs on a time scale of seconds (ON and OFF interval statistics as a function of excitation intensity is currently investigated).

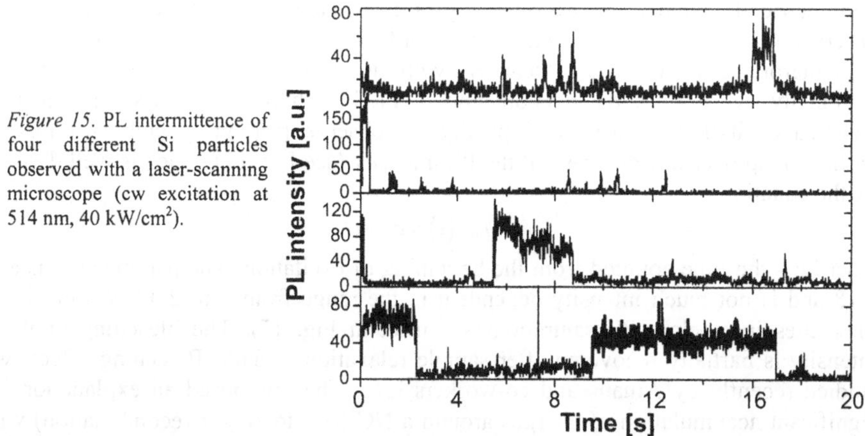

Figure 15. PL intermittence of four different Si particles observed with a laser-scanning microscope (cw excitation at 514 nm, 40 kW/cm²).

Measurement of luminescence intensity autocorrelation (normalized second order correlation function $g^{(2)}(\tau)$) is used to study dynamics of (single) molecules or nanocrystals [24]. For a single quantum emitter one can observe typical photon antibunching at very short time intervals ($g^{(2)}(\tau)\rightarrow 0$, for $\tau\rightarrow 0$) [25] and photon bunching due to emission intermittence at long time intervals. We measured PL correlation function $g^{(2)}(\tau)$ of single Si particles at various intensity of excitation. Unfortunately, there is not enough signal at short time intervals (< 0.01 ms), therefore, antibunching could not be observed. However, we see a high value (up to 3) of $g^{(2)}(\tau)$ on a time scale of seconds, which reflects photon bunching in ON-intervals. The value of $g^{(2)}(\tau)$ increases with increasing excitation (see Fig. 16 B).

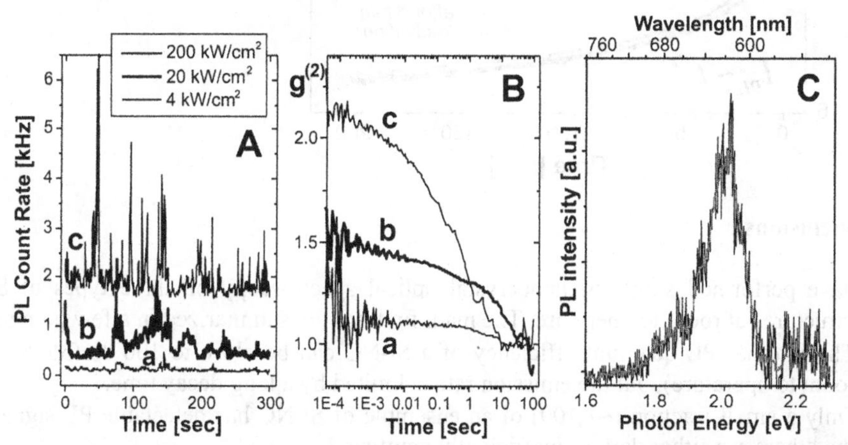

Figure 16. PL intermittence (A) and second order correlation function (B) measured for a single Si particle under excitation of 4, 20, and 200 kW/cm² (curves a, b, and c, respectively). The panel (C) shows a PL spectrum of a single Si particle.

The PL spectra of single Si particles are shifted to the blue compared to the above described Si-NCs. The spectrum is often quite narrow (FWHM ~140 meV) and asymmetrically extended to the low-energy side, where a second peak can be sometimes observed (see Fig. 16 C). This fact is in accord with observations by English et al. [10], who interpreted the low-energy peak as a vibronic band due to Si-O-Si stretching.

Finally, we discuss the phenomenon of PL bleaching which is significant under elevated excitation intensities. Typically we observe (for an ensemble of many Si particles) a power-law decrease of the PL intensity (see Fig. 17) which is well described by the equation

$$I_{PL}(t) = const \cdot t^{-\alpha}$$

where t is the time counted from the beginning of excitation. The parameter α is about 0.18 and is not much intensity dependent in the range from 1 to 2000 W/cm^2, i.e. for intensities just below PL saturation (see inset in Fig. 17). The bleaching of the PL intensity is partially recovered after sample relaxation in dark. Bleaching effects were studied recently by Amans and co-workers [26]. They proposed an explanation by a significant accumulation of charges around a NC (due to Auger recombination) which can break chemical bonds at the interface and create non-radiative centers.

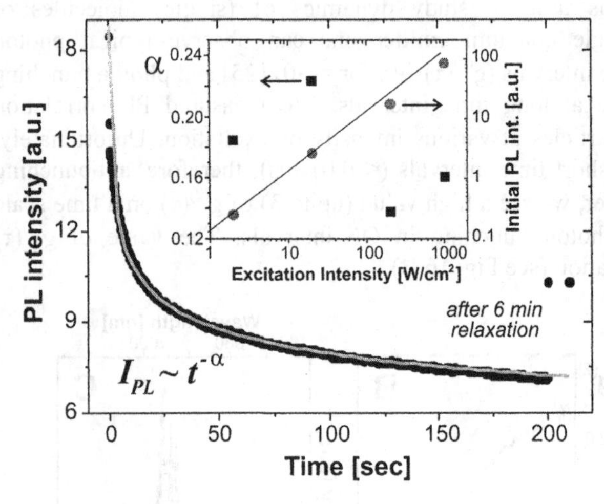

Figure 17. The PL bleaching in ensemble of Si particles under cw excitation (200 W/cm^2). The fit by a $t^{-\alpha}$ function is shown by a gray line. The inset shows intensity dependence of parameter α and of the initial PL intensity (log-log scale).

5. Conclusions

We have performed single Si nanocrystal optical spectroscopy on three types of Si nanostructures at room temperature. The main findings are summarized in a few points:

- The internal PL quantum efficiency of a Si-NC can be close to 100 % (35 % at room temperature), but the emission rate is limited by a long decay time.
- Only a small fraction (~1/100) of an ensemble of Si-NC has detectable PL signal, the others are either dark or inefficiently emitting.
- The PL spectrum can be as narrow as 120 meV (FWHM) and show spectral features that are probably due to vibrations on the surface of a Si-NC.
- The PL emission of a single Si-NC reveals intermittence - ON/OFF blinking - typically on the time scale of seconds.
- High excitation causes PL bleaching which is partially irreversible (or has a very long relaxation time).

- EL emission from single Si-NCs has characteristics very similar to single-dot PL, the level of saturated signal intensity is almost identical for PL and EL.

From the point of view of application of Si-NCs in light-emitting devices, there are both good and bad news. The high internal QE and characteristics of EL are promising, but the low emission rate, intermittence, and bleaching are disadvantages.

The future development should concentrate on smaller Si-NCs with blue-shifted emission [27] which have faster decay and higher emission rate (they are also better for single NC spectroscopy observation). The key issues of surface passivation [10,11] and photo-induced processes should be thoroughly investigated. An interesting new direction is combination of Si-NCs with photonic structures (microcavities, photonic crystals etc.) which enables to modify the spontaneous emission of Si-NC (frequency and direction of emitted photons) [28].

Acknowledgements

The authors would like to acknowledge discussions on the experimental parts with A. Galeckas and M. Vácha, and express their gratitude for being able to use e-beam lithography at Chalmers and at the KTH nanolab. The preparation of porous-Si colloids was successful grace to help of K. Dohnalová, K. Luterová, D. Nižňanský and J. Buršík. Partial funding was received from the KTH Faculty, Royal Swedish Academy of Sciences, DFG (Forschergruppe 388 "Laboratory Astrophysics"), and the European SBLED project. JV acknowledges support from the GAAV CR project B1112901.

REFERENCES

1. Gustafsson, A., Pistol, M-E., Montellius, L., Samuelson, L. (1998) Local probe techniques for luminescence studies of low-dimensional semiconductor structures, *J. Appl. Phys.* 84 (4), 1715-1775
2. Empedocles, S. A., Norris, D. J., Bawendi, M. G. (1996) Photoluminescence spectroscopy of single CdSe nanocrystallite quantum dots, *Phys. Rev. Lett.* 77, 3873-3876
3. Pistol, M-E., Castrillo, P., Hessman, D., Prieto, J.A., Samuelson, L. (1999) Random telegraph noise in photoluminescence from individual self-assembled quantum dots, *Phys. Rev. B* 59 (16), 10725-10729
4. Empedocles, S. A.and Bawendi, M. G. (1997) Quantum-Confined Stark-Effect in Single CdSe Nanocrystallite Quantum Dots, *Science* 278, 2114-2117
5. Kovalev, D., Heckler, K., Polisski, G., Koch, F. (1999) Optical properties of Si nanocrystals, *Phys. Status Solidi B* 215, 871-932
6. Linnros, J., Lalic, N., Galeckas, A., Grivickas, V. (1999) Analysis of the stretched exponential photoluminescence decay from nanometer-sized silicon crystals in SiO_2, *J. Appl. Phys.* 86 (11), 6128-6134
7. Mason, M.D., Credo, G.M., Weston, K.D., Buratto, S.K. (1998) Luminescence of Individual Porous Si Chromophores, *Phys. Rev. Lett.* 80 (24), 5405-5408
8. Credo, G.M., Mason, M.D., Buratto, S.K. (1999) External quantum efficiency of single porous silicon nanoparticles, *Appl. Phys. Lett.* 74(14), 1978-1980
9. Mason, M.D., Sirbuly, D.J., Carson, P.J., Burrato, S.K. (2001) Investigating individual chromophores within single porous silicon nanoparticles, *J. Chem. Phys.* 114 (18), 8119-8123
10. English, D.S., Pell, E.L., Yu, Z., Barbara, P.F., Korgel, B.A. (2002) Size tunable visible luminescence from individual organic monolayer stabilized silicon nanocrystal quantum dots, *Nano Letters* 2 (7), 681-685
11. Valenta, J., Juhasz, R., Linnros, J. (2002) Optical spectroscopy of single silicon quantum dots, *Appl. Phys. Lett.* 80 (6), 1070-1072
12. Valenta, J., Juhasz, R., Linnros, J. (2002) Photoluminescence from single silicon quantum dots at room temperature, *J. Luminescence* 98, 15-22
13. Liu, H.I., Biegelsen, D.K., Ponce, F.A., Johnson, N.M., Pease, R.F.W. (1994) Self-limiting oxidation for fabricating sub-5 nm silicon nanowires, *Appl. Phys. Lett.* 64 (11), 1383-1385

14. Lalic, N. and Linnros, J. (1998) Light emitting diode structure based on Si nanocrystals formed by implantation into thermal oxide, *J. Luminescence* 80 (1-4), 263-267
15. Wolkin M.V., Jorne J., Fauchet P.M., Allan G., Delerue C. (1999) Electronic states and luminescence in porous silicon quantum dots: The role of oxygen, *Phys. Rev. Lett.* 82 (1), 197-200
16. Reboredo, F.A., Franceschetti, A., Zunger, A. (2000) Dark excitons due to direct Coulomb interactions in silicon quantum dots, *Phys. Rev. B* 61 (19), 13073-13087
17. Delerue, C., Allan, G., Lannoo, M. (2001) Electron-phonon coupling and optical transitions for indirect-gap semiconductor nanocrystals, *Phys. Rev. B* 64, 193402
18. Banin, U., Bruchez, M., Alivisatos, A.P. ,Ha, T., Weiss, S.,Chemla, D.S. (1999) Evidence for a thermal contribution to emission intermittency in single CdSe/CdS core/shell nanocrystals, *J. Chem. Phys.* 110 (2), 1195-1201
19. Allan, G., Delerue, C., Niquet, Y.M. (2001) Luminescence polarization of silicon nanocrystals, *Phys. Rev. B* 63, 205301
20. Baba, M. Sasaki, T., Yoshita, M., Akiyama, H. (1999) Aberrations and allowances for errors in a hemisphere solid-immersion lens for submicron-resolution photoluminescence microscopy, *J. Appl. Phys.* 85 (9), 6923-6925
21. Yuan, J. and Haneman, D. (1999) Visible electroluminescence from native SiO_2 on n-type Si substrates, *J. Appl. Phys.* 86 (4), 2358-2360
22. Lalic, N. (2000) *Light emitting devices based on silicon nanostructures*, PhD. thesis, Stockholm
23. Linnros, J., Galeckas, A., Pereaud, A., Lalic, N., Grivickas, V., Hultman, L. (1998) , *Mat. Res. Symp. Proc.* 486, 249-254
24. Basché, T., Moerner, W.E., Orrit, M., Wild, U.P. (Ed.) (1997) Single-Molecule Optical detection, Imaging and Spectroscopy, VCH
25. Michler, P., Imamoglu, A., Mason, M.D., Carson, P.J., Strouse, G.F., Burrati, S.K. (2000) Quantum correlation among photons from a single quantum dot at room temperature, *Nature* 406, 968-970
26. Amans, D., Guillois, O., Ledoux, G., Porterat, D., Reynaud, C. (2002) Influence of light intensity on the photoluminescence of silicon nanostructures, *J. Appl. Phys.* 91 (8), 5334-5340
27. Belomoin, G., Therrien, J., Smith, A., Rao, S., Twesten, R., Chaieb, S., Nayfeh, M.H., Wagner, L., Mitas, L. (2002) Observation of a magic discrete family of ultrabright Si nanoparticles, *Appl. Phys. Lett.* 80 (5), 841-843
28. Valenta, J., Linnros, J., Juhasz, R., Rehspringer, J.-L., Huber, F., Hirlimann, Ch., Cheylan, S., Elliman, R.E. (2002) Photonic band-gap effects on photoluminescence of silicon nanocrystals imbedded in artificial opals, *J. Appl. Phys.* submitted

LUMINESCENCE FROM Si/SiO_2 NANOSTRUCTURES

Yoshihiko KANEMITSU
Graduate School of Materials Science,
Nara Institute of Science and Technology,
Ikoma, Nara 630-0101, Japan

1. Introduction

The discovery of the room-temperature luminescence from Si [1-3] and Ge nanocrystals [4,5] has stimulated considerable efforts in understanding optical properties of indirect-gap elemental semiconductor nanostructures. In particular, Si nanocrystals are receiving widespread interest because of their high quantum efficiency of light emission at room temperature. The photoluminescence (PL) and electroluminescence (EL) efficiency of crystalline Si (c-Si) nanoparticles have greatly increased in the last decade. Very recently, optical gain and stimulated emission have been reported in silicon nanoparticles [6,7]. In addition, there have been many different approaches towards useful Si light-emitting materials and devices compatible with current Si microelectronics [8,9]. The realization of bright Si light-emitting devices and silicon lasers will bring about a revolution in the semiconductor industry. Optical and electronic devices can be fabricated from the same material of silicon, and silicon will be the leading material for the future optoelectronics.

From the fundamental physics viewpoints, silicon is the key material for nanomaterials science, because silicon is an element semiconductor and there is no composition fluctuation even in the very small nanostructures. In addition, many different and unique Si based nanostructures and nanomaterials have been fabricated: Si clusters, Si polymers, Si nanoparticles, amorphous Si (a-Si), high-quality bulk Si crystals, Si based alloys and so on. The local structure of clusters and nanoparticles is controlled in silicon materials. Si nanostructures with SiO_2 surface layers have some advantages because Si/SiO_2 system is fully compatible with current Si technology. The well-characterized Si/SiO_2 nanostructures will be key materials for quantum device applications and molecular electronics. By changing the size and dimensionality of Si/SiO_2 nanostructures, we can control the delocalization volume of excitons and electrons and luminescence properties of Si nanostructures.

In this review paper, spectroscopic data on c-Si and a-Si nanostructures are summarized and the PL mechanism of Si/SiO_2 nanostructures is discussed for the development of efficient light-emitting devices and silicon lasers.

L. Pavesi et al. (eds.), Towards the First Silicon Laser, 109–122.

2. Size dependence of the luminescence spectrum

Many methods have been developed for the fabrication of light-emitting Si nanostructures. One of the simple methods is electrochemical etching of bulk materials. Light-emitting c-Si and a-Si nanoparticles have been prepared by electrochemical etching techniques [1,10]. The electrochemical anodization of c-Si wafers and a-Si:H films was carried out in HF-ethanol solution at a constant current density. Typical average sizes of c-Si and a-Si nanoparticles were 2-5 nm [11,12]. Bulk c-Si shows a sharp PL near 1.1 eV due to the TO-phonon-assisted optical transition. Our undoped a-Si:H film shows a broad PL near 1.35 eV [13]. The broad PL is due to the structural disorders and the strong exciton-phonon coupling in a-Si [14]. In both a-Si and c-Si bulk samples, luminescence is observed in the near-infrared spectral region only at low temperatures. However, the a-Si and c-Si nanoparticles show broad PL bands in the visible spectral region even at room temperature [1,10-13]. The formation of nanoparticles causes the blueshift of the PL spectra of a-Si and c-Si. The broad luminescence spectra are caused by contributions from all nanoparticles in the sample, because of the broad size distribution of nanoparticles.

In nanoparticle systems with a broad size distribution, resonant excitation spectroscopy is a powerful method to understand the PL mechanism [15,16]. By changing excitation laser wavelength, we can selectively excite individual nanocrystals with a certain size in the sample. Figure 1 shows PL spectra of H-passivated c-Si nanoparticles under different excitation energies at 7 K [17]. Under 1.785-eV excitation, a broad PL band (full PL) is observed in the red and infrared spectral region, and the peak energy is located near 1.5 eV. Resonant excitation at energies within the broad PL band results in fine structures in the spectrum. Steplike structures are observed under resonant excitation of the high energy side in the PL spectrum. With a decrease of excitation photon energy, the peak structure appears. These step and peak structures in the resonantly excited PL spectra can be explained by momentum-conserving phonon-assisted luminescence processes [15-17]. The energies of peak and steplike structures are almost consistent with those of one phonon [TA(Δ) or TO(Δ)] and two phonons [TA(Δ) + TO(Δ) or 2TO(Δ)], where TA(Δ) and TO(Δ) are transverse acoustic and transverse optical phonons, respectively, at the conduction-band-minimum Δ point, and the phonon energies of TA(Δ) and TO(Δ) are 18.5 and 57.5 meV, respectively [18]. This shows that Si nanocrystal is an indirect gap semiconductor. The absorption and emission of momentum-conserving phonons are needed during the light absorption and emission processes [19].

Under resonant excitation, the PL intensity in a-Si nanoparticles is much weaker than that in c-Si nanoparticles. In addition, the energy-dependence of the PL lifetime in the a-Si nanoparticle is different from that in c-Si nanoparticle [13]. These observations suggest that in a-Si nanoparticles the electron wavefunction is strongly localized within nanoparticles and the radiative recombination of carriers occurs at localized states rather than the extended state with a large density of states. However, in a-Si nanoparticles, the blueshift of the PL spectrum is observed. The PL mechanism in a-Si nanostructures will be discussed later.

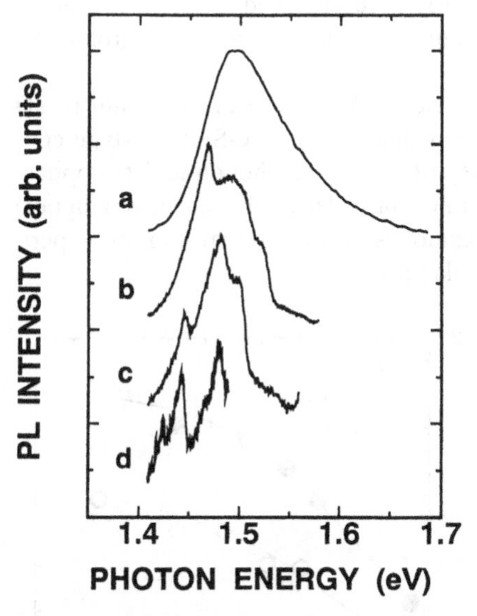

Figure 1. PL spectra of H-passivated c-Si nanoparticles under different excitation energies at 7 K: (a) 1.785, (b) 1.587, (c) 1.564, and (d) 1.503 eV. The step and peak structures appear in the PL spectrum under resonant excitation at energies with the PL band (From Ref. 17).

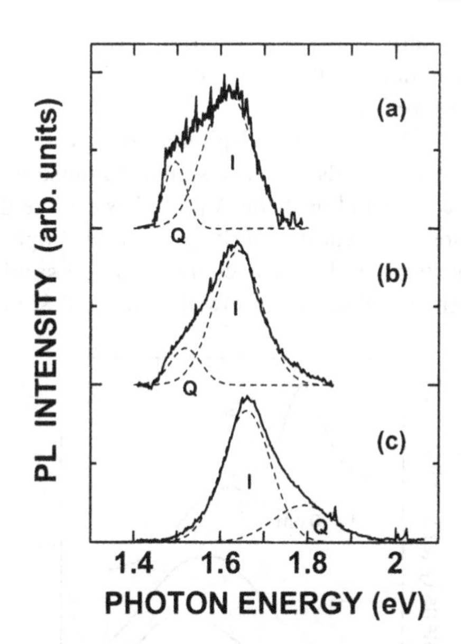

Figure 2. PL spectra of c-Si/SiO_2 single nanowells at 2 K: (a) 1.7, (b) 1.3, and (c) 0.6 nm thickness. The asymmetric PL spectra can be fitted by two Gaussian bands (the Q and I bands) (From Ref. 20).

Since it is difficult to control the size of nanoparticles exactly, the detailed size dependence of the optical properties of nanoparticles is not clear. Well-characterized Si nanostructures are desirable for understanding of the PL mechanism of nanoparticles. Si nanostructures with SiO_2 surface layers have some advantages because the SiO_2 layer is a stable surface material and a high barrier for electron confinement. One approach to control Si/SiO_2 nanostructures is the fabrication of two-dimensional (2D) nanowell structures. Si single nanowell structures were formed on SIMOX (separation by implanted oxygen) wafers [20,21]. The 2D thin Si layers were formed between thin surface SiO_2 (~30 nm) and thick buried SiO_2 layers (~400 nm). a-Si/SiO_2 multi-nanowell structures were formed by an electron-beam deposition technique [22,23]. The SiO_2 layer thickness was 3.0 nm and the a-Si layer thickness was varied from 0.8 to 2.0 nm.

Figure 2 shows thickness dependence of PL spectra in 2D c-Si/SiO_2 systems under 488-nm laser excitation at 2 K [20]. Efficient PL in the visible spectral region was observed in very thin well samples (less than 2 nm). The asymmetric PL spectra in the red and infrared spectral region can be fitted by two Gaussian bands, the weak PL band (denoted as Q) and the strong PL band (denoted as I), as shown in Fig. 2. The PL peak energy of the I band is almost independent of the well thickness and appears at ~1.65 eV. In contrast, the peak energy of the Q band depends on the thickness of the c-Si well and shifts to higher energy with a decrease of the Si well thickness [from Fig. 2 (a) to (c)]. This implies that the quantum confinement state in c-Si wells plays a key role in the Q band luminescence. Under selective excitation within the Q band, a TO-phonon structure in the PL spectrum and the PL anisotropy are observed under the resonant excitation of the Q band. The Q-band PL originates from optically anisotropic 2D nanowells [20].

The steplike TO-phonon structures in the selectively excited PL spectrum in 2D c-Si nanowells are very similar to those in 0D Si nanocrystals. In c-Si, the *p*-type conduction band near the Δ point has a large density of states, and phonon-assisted optical transitions cause overlapping of the optical transitions. Then, the quantized optical spectrum will appear continuous and similar featureless absorption spectra are experimentally observed in low-dimensional Si materials [20].

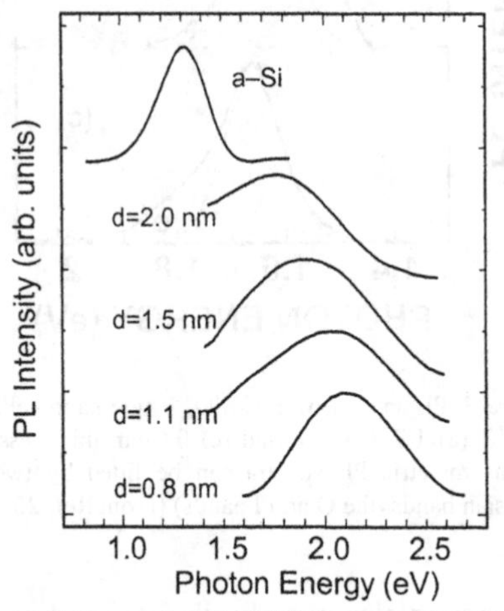

Figure 3. PL spectra of a-Si/SiO_2 nanowells at 10 K (From Ref. 22).

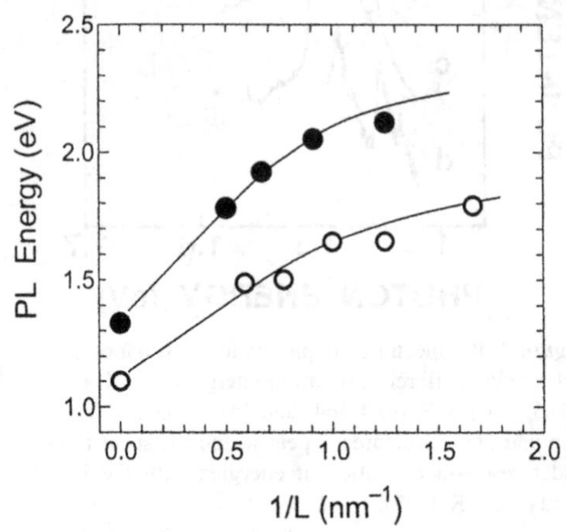

Figure 4. Well thickness dependence of the PL peak energy in c-Si/SiO_2 (open circles) and a-Si/SiO_2 nanowells (solid circles). The data are taken from Ref. 22 and 28. The solid lines are guides for the eye.

Figure 3 shows PL spectra of different a-Si/SiO_2 nanowell samples at 10 K under 325-nm laser excitation [22]. The PL spectra of these nanowell structures appear in the visible spectral region. The PL peak energies of a-Si/SiO_2 nanowell structures are blue-shifted from that of bulk a-Si (~1.35 eV). With a decrease of a-Si well thickness, the PL peak energy in a-Si/SiO_2 nanowell structures shifts to higher energy. In a-Si, there is a large Stokes shift between the absorption edge and the PL peak energy. With an increase of temperature, the PL intensity decreases and the Stokes shift increases. The PL spectrum at room temperature is probably attributed to radiative recombination in deep states rather than in the near band edge state of a-Si. In fact, the PL peak energy of a-Si nanowells is sensitive to the measurement temperature [24-26]. In a-Si/a-Si_3N_4 nanowell structures, the PL peak energy depends on the a-Si well thickness near room temperature, but the thickness-dependent PL spectrum is not clearly observed at low temperatures [25,26]. However, the pronounced PL blueshift is observed in a-Si/SiO_2 nanowell structures even at low temperatures, as shown in Fig. 3. The PL blueshift in a-Si/SiO_2 nanowell structures is much larger than that in a-Si/a-Si_3N_4 nanowell structures. The PL spectrum in a-Si/SiO_2 systems is sensitive to the well thickness [22,27]. In addition, the PL spectra of a-Si/SiO_2 nanowell structures become broad, compared to the case of bulk a-Si:H. The broad PL spectrum reflects the fluctuation of the well thickness and structural variations of the interface between Si and SiO_2.

The size dependence of the PL peak energy in c-Si/SiO_2 and a-Si/SiO_2 nanowells at low temperatures is summarized in Fig. 4 [22,28]. The PL peak energies in a-Si/SiO_2 nanowell structures are higher than those in c-Si/SiO_2 nanowell structures, because the PL energy in bulk a-Si is higher than that in bulk c-Si. With a decrease of the well thickness, the PL peak energy is blueshifted in both a-Si/SiO_2 and c-Si/SiO_2 nanowell structures. The well-thickness dependence of the PL peak energy implies that quantum confinement effects play an essential role in luminescence processes in both c-Si/SiO_2 and a-Si/SiO_2 nanowell structures [27-30]. However, the size-independent PL band (I band in Fig.2) is observed in c-Si/SiO_2 nanowell systems. The exciton localization at the interface between the Si well layer and the SiO_2 barrier layer is also important in the radiative and nonradiative recombination processes [20,28]. The polar bonds such as Si-OH and Si=O affect the electronic structure of the Si nanowell interior states. The interface effects will be discussed in section 4.

In a-Si/SiO_2 nanowell structures, the observed thickness- and temperature- dependence of the PL energy cannot be explained by simple quantum confinement models [24]. Here, we briefly discuss the mechanism of the blueshift of the PL spectrum in a-Si nanowells and nanoparticles. In a-Si, the localization length of carriers is estimated to be about 0.6 nm [31,32]. Therefore, no quantum confinement effects on the PL energy are expected for a-Si nanowells. However, it has been reported that the blueshift of the PL energy in a-Si nanowells and nanoparticles occurs with a decrease of the a-Si well thickness and the a-Si nanoparticle size [22,27,33]. In amorphous semiconductors, the mostly delocalized state (the extended state) and the weakly localized state (the band-tail state) are usually considered in the light absorption and emission processes. Radiative recombination usually occur after the carries are trapped into the band-tail states below the mobility edge [34,35]. Quantum confinement effects cause the blueshift of the extended state but no significant change of the deep localized states [36,37]. Then, quantum confinement effects might cause the spreading of the band-tail states. In addition, the recombination lifetime of carriers in the band-tail states becomes shorter because of the spatial confinement of carriers in a small volume [13,22]. The short lifetime of carriers reduces the energy relaxation of carriers into lower energy states and the recombination of carriers near the mobility edge contributes to the luminescence process. In addition, since the electrons are localized within the a-Si well, the optical properties are not sensitive to the surface structure. In nanostructure with well passivated surfaces, the thermally activated diffusion processes are inactive [38] and there is no long-range diffusion of carriers to nonradiative recombination centers. Therefore, it is concluded that the pronounced blueshift of the PL peak energy in a-Si nanostructures is caused by the quantum confinement and the spatial confinement of carriers.

3. Room-temperature luminescence in large–sized nanostructures

Unique optical properties of c-Si and a-Si nanoparticles and nanowells have been discussed briefly in the previous section. In particular, the temperature dependence of the

PL intensity of nanoparticles is interesting: The temperature quenching of the PL intensity in nanoparticles is much weaker than that of bulk crystals. This phenomenon is also observed even in impurity-doped semiconductor nanocrystals [39] and is very attractive from the standpoint of practical applications. Undoped and impurity-doped semiconductor nanocrystals show efficient luminescence at room temperature and will be used as nanophosphors in display applications. The thermal quenching of the PL intensity is drastically improved even in a-Si nanoparticles and a-Si/SiO_2 nanowells [24], although there exist nonradiative recombination centers related to structural disorder and dangling bonds in a-Si. The spatial confinement of electrons in a small volume reduces the nonradiative recombination rate and enhances the luminescence efficiency. From the viewpoints of device applications, it is very important and interesting to understand whether the spatial confinement of electrons modify the luminescence properties of c-Si nanostructures or not. In order to understand the importance of the spatial confinement of electrons, large-sized c-Si nanostructures have been fabricated, where the size of the nanostructure is much larger than the exciton Bohr radius in c-Si bulk form. The c-Si/SiO_2 nanowires were fabricated by anisotropic etching and thermal oxidation techniques [40]. The cross section of wires is the triangle of ~65 nm side length and the length of wire is ~1 mm or more.

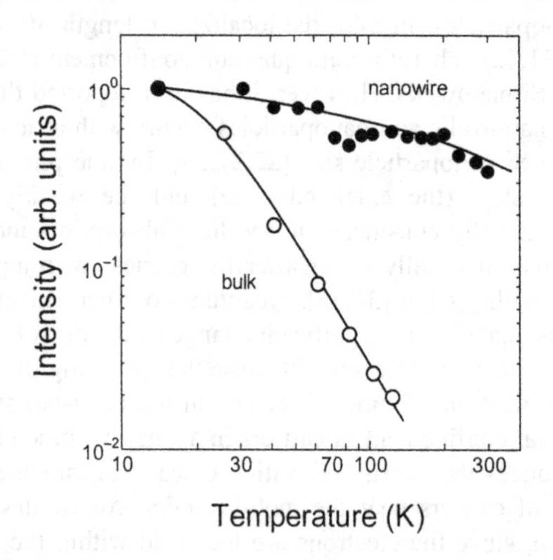

Figure 5. Temperature dependence of the spectrally integrated PL intensity in bulk c-Si and c-Si nanowire. The solid lines are guides for the eye.

Figure 5 shows the temperature dependence of the spectrally-integrated PL intensity in bulk c-Si, and the 65-nm c-Si nanowire. The spectrally integrated PL intensities (the TO-phonon assisted PL intensities) are plotted in the figure, because the PL bandwidth is sensitive to temperature. Bulk c-Si is an indirect gap semiconductor and its PL intensity abruptly decreases with an increase of temperature. The near-infrared PL in bulk c-Si completely disappears at temperatures above 150 K. The temperature dependence of the PL intensity in the 65-nm nanowires is completely different from that of bulk c-Si. The PL spectrum of the c-Si nanowire is quite similar to that of bulk c-Si [40]. TO-and TA-phonon assisted PL bands are clearly observed and their energy position in the

65-nm nanowire are exactly the same as that of bulk c-Si at low temperatures [41]. Since the size of the wire (~65 nm) is much larger than the bulk exciton Bohr radius (~4.5 nm), there is no difference in the PL spectrum between the nanowire and bulk c-Si samples at low temperatures. However, in c-Si nanowires, efficient near-infrared PL is observed even at room temperature. Our results support a picture that the enhancement of the PL intensity is also caused by the spatial confinement of electrons or excitons. The radiative recombination lifetime of carries and excitons is long, because the radiative recombination processes in bulk c-Si are phonon-assisted ones. Then, the long-range diffusion within the radiative lifetime occurs. However, the spatial confinement of excitons in nanowires prevents a long-distance diffusion to nonradiative recombination centers such as defects and impurities, because the size of the c-Si nanometer of 65 nm is smaller than the diffusion length of electrons (100 μm) [42]. In addition, the strains between the c-Si core and the surface SiO_2 layer enhance the radiative recombination rate of excitons in the nanowire, because the PL lifetime in the nanowire is shorter than that in bulk c-Si [43]. Therefore, it is believed that the room-temperature luminescence is due to the spatial confinement of excitons in c-Si nanowires and the modification of the near-band edge structure by strains.

In small-sized nanostructures, the PL enhancement is mainly caused by the quantum confinement effects. The PL energy can be controlled by changing the nanostructure size. However, the PL spectrum is sensitive to the fluctuation of the nanoparticle size and the nanowell thickness. The observed PL spectrum is broad. In large-sized nanostructures, on the other hand, optical properties are not sensitive to the surface structure and the size fluctuation. The well-controlled and large-sized Si nanostructures are one of the most useful materials for light-emitting devices showing a well-defined wavelength luminescence [44,45]. In other semiconductor systems, large-sized nanoparticles give an opportunity for the study of many body effects on luminescence processes, including highly dense excitons and electron-hole plasmas [46,47]. From the viewpoints of fundamental physics and device applications, large-sized nanostructures are unique materials in fundamental and applied physics.

4. Interface states in Si/SiO_2 nanostructures

When the size of semiconductor nanocrystals is smaller than the exciton Bohr radius in bulk form, the wavefunctions of excitons are delocalized over the nanoparticles. The optical responses of nanocrystals are sensitive to the nanocrystal size. In c-Si, the exciton Bohr radius is about 4.5 nm, and efficient visible PL is observed in 2-3nm sized c-Si nanoparticles. In these small c-Si nanoparticles, the band-gap energy is modified by the quantum confinement of electrons, and optical and excitonic responses are modified by the quantum confinement of excitons. In addition, because of a large ratio of surface to volume in c-Si nanoparticles, their optical properties are sensitive to the surface chemistry. In c-Si materials, good surface passivation is achieved by hydrogen termination and surface oxidation. Since c-Si is an elemental and nonpolar semiconductor and the deformation interaction between excitons and phonons is weak, it is considered that the optical and electronic properties of small c-Si nanostructures are influenced by the surface polar bonding such as Si-OH and Si=O [2,17,28,48-52].

In H-passivated c-Si nanoparticle experiments, as-prepared porous silicon has been used as samples, where porous silicon is prepared by electrochemical anodization of bulk c-Si wafer. The structure of porous silicon is complicated (wires and/or dots) [53,54] and the size distribution of nanostructures in porous silicon is broad [53-57]. It is not easy to determine the size of H-passivated c-Si nanoparticles in the porous silicon layers. On the other hand, isolated SiO_2-capped Si nanocrystals were fabricated by decomposition of SiH_4 gas and thermal oxidation techniques [2,3,51,55,58], sputtering techniques [59], thermal evaporation of Si [60], and so on. It is believed that the size determination of isolated SiO_2-capped Si nanocrystals is simple, compared to the case of porous silicon. The size dependence of the band-gap energy has been extensively studied in isolated surface-oxidized c-Si nanoparticles [55,60].

The estimated size of nanoparticles in porous silicon depends on the characterization method. The nanoparticle size in porous silicon samples has been extensively estimated by x-ray techniques (near-edge x-ray absorption and small angle x-ray scattering) and its values are smaller than the values determined by Raman scattering and transmission electron microscope analysis [55]. By comparison between the porous silicon data based on x-ray techniques [19,55,57] and the isolated SO_2-capped Si nanocrystals data [51,55,59], there is no significant difference in the PL energy between H-passivated and SiO_2-capped c-Si nanoparticles. Rather, it is difficult to determine the size and the PL energy of H-passivated c-Si nanoparticles in the porous layer without surface oxidation. The relationship between the PL energy and the size of H-passivated nanoparticles in porous silicon layers is under discussion [19].

In these inhomogeneous systems, resonant excitation spectroscopy is a powerful method to understand the PL mechanism. The resonantly excited PL spectrum is sensitive to the surface chemistry of c-Si nanoparticles, as shown in Fig. 6. In H-passivated c-Si nanoparticles (left in Fig. 6), the TO-phonon structures are clearly observed in resonantly excited PL spectra at low temperatures [51]. On the other hand, the TO-phonon structure becomes unclear in surface-oxidized c-Si nanoparticles. In particular, phonon structures are not observed in surface-oxidized Si nanocrystals under resonant excitation at energies above ~1.65 eV [51]. In addition, the luminescence Stokes shift of oxidized c-Si nanoparticles is larger than that of H-passivated c-Si nanoparticles [17], as shown in Fig. 7. It is considered that the localization processes of excitons and electrons at the Si/SiO_2 interface are important in the radiative recombination processes in surface-oxidized c-Si nanoparticles.

The importance of the exciton localization at the Si/SiO_2 interface is supported by the nanowell experiments [20]. In $c\text{-}Si/SiO_2$ single nanowell structures, two different PL band is observed in the red and infrared spectral region, as shown in Fig. 2. The one is the size-independent PL band (I band) and the peak energy appears at ~1.65 eV. The other is the size-dependent PL band (Q band) and the peak energy shifts to higher energy with a decrease of the c-Si well thickness, as shown in Fig.4. These results show that localized excitons at the interface and free excitons in the nanowell contribute to the luminescence process.

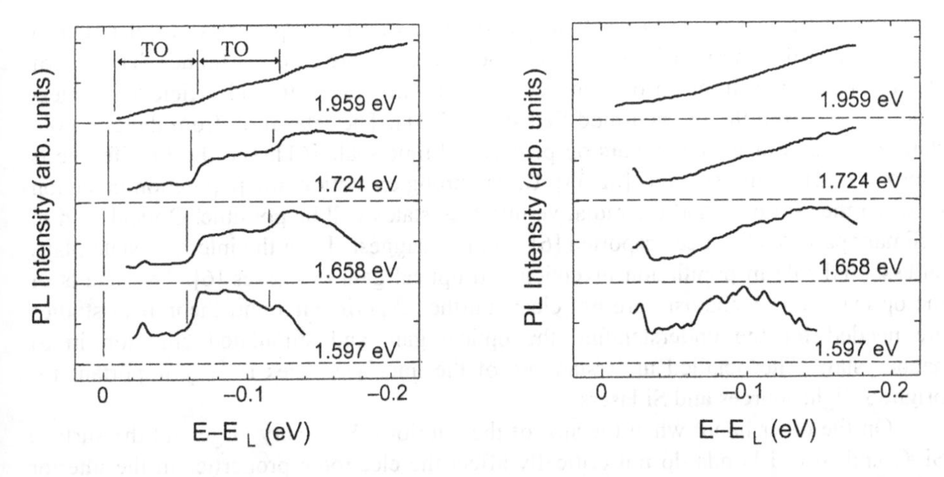

Figure 6. PL spectra of H-passivated (left) and SiO_2-capped c-Si nanoparticles (right) under resonant excitation at energies within the PL band. The excitation laser energies are shown in the figure (From Ref. 51).

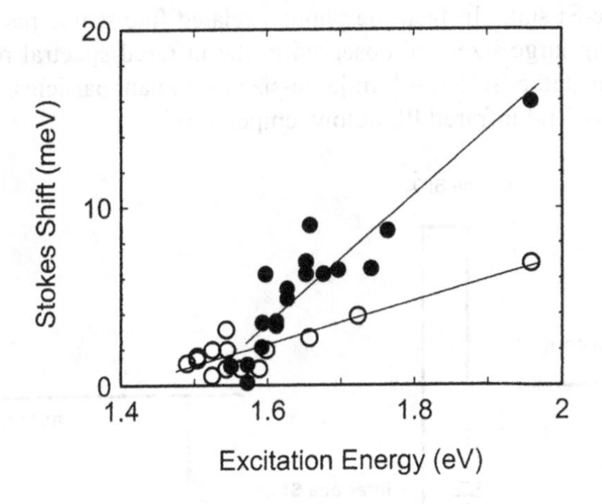

Figure 7. Luminescence Stokes shift of H-passivated (open circles) and surface-oxidized Si nanocrystals (solid circles) under resonant excitation at low 7 K (From Ref. 17).

The c-Si/SiO_2 nanoparticle and nanowell experiments suggest that efficient luminescence comes from the oxygen-modified states between c-Si and SiO_2 surface layer in small nanostructures. Oxide termination may have several effects including the formation of interface states between c-Si and SiO_2 layer and a modification of the electronic structure of the c-Si interior state. In c-Si/SiO_2 systems, the interface states are formed between c-Si and SiO_2, due to the lattice mismatch and the polar bonds. In particular, oxygen atoms at the interface induce the localization of excitons near the interface between c-Si and SiO_2, through the σ-n mixing and dipole interactions [49,50]. The electronic interactions at the interface states induce exciton localization and then cause broad PL with a large Stokes shift under resonant excitation in surface-oxidized Si nanocrystals.

The energy structure in small surface-oxidized c-Si nanoparticles is illustrated in Fig. 8. When the energy of the interior state (the c-Si core state) is higher than that of the interface state, light absorption occurs in the interior state and efficient luminescence comes from the oxygen-modified states [2,51]. The relaxation from the c-Si core state to the interface state occurs on picosecond time scale [61], and the PL lifetime is microseconds to milliseconds [62,63]. Under strong excitation, the population inversion between the c-Si core and the radiative interface states will be possible. Optical gain in c-Si nanoparticles has been reported [6], and it is suggested that the interface state plays an essential role in population inversion and optical gain processes [6]. The details of the optical gain mechanism are not clear. Further experimental and theoretical studies are needed for the understanding the optical gain and stimulated emission in Si nanocrystals. The detailed understanding of the interface states is very important for bright Si light sources and Si lasers.

On the other hand, when the size of the interior c-Si is very large and the surface Si=O and Si-OH bonds do not critically affect the electronic properties in the interior c-Si, the exciton energy of the interior c-Si is the lowest optical transition even in surface-oxidized Si nanocrystals. Then, the radiative recombination of excitons could occur in the interior c-Si state. In fact, the phonon-related fine structures in oxidized c-Si nanoparticles having large sizes are observed in the infrared spectral region, similar to the case of H-passivated c-Si [51,64]. In large-sized c-Si nanoparticles, the interior state contributes to the red and infrared PL at low temperatures.

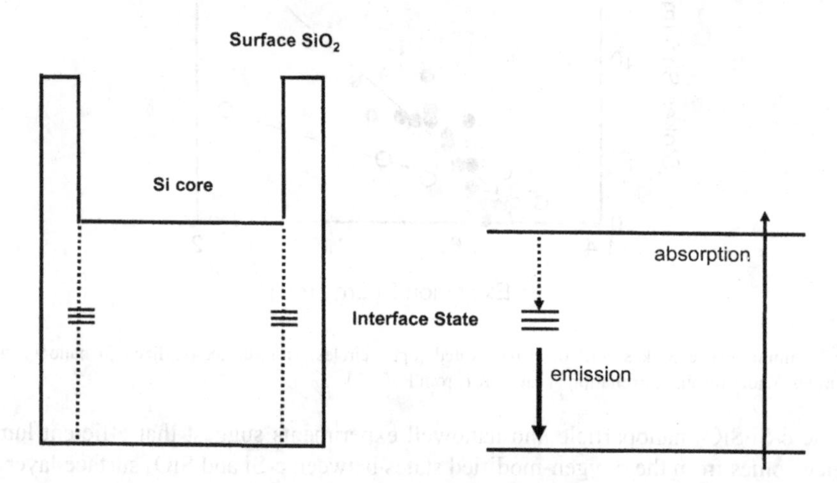

Figure 8. Electronic states of surface-oxidized Si nanocrystals. The interface state between the c-Si core and the surface SiO_2 layer plays an important role in efficient luminescence in small surface-oxidized Si nanocrystals.

There is an argument on the formation of the interface states in surface-oxidized c-Si nanoparticles. It has been pointed out that light absorption and light emission occur in the c-Si core state [53,55,58,59,64] in oxidized c-Si nanoparticles, similar to the case of H-passivated porous silicon. However, the PL processes in surface-oxidized c-Si nanoparticles depend on the nanocrystals size, as mentioned above. Since it is difficult to control the size of c-Si nanoparticles, the fabrication of well-controlled nanostructures and the characterization of the individual nanostructures are needed for the understanding of the electronic structures of individual Si/SiO_2 nanoparticles. Recently, single dot spectroscopy has been applied to c-Si nanoparticles systems [65-67]. However, the observed PL and EL spectra of individual c-Si nanoparticles are broad. The origin of the broad PL and EL bands is not clear. It can be pointed out that the exciton localization causes broad PL and enhances the exciton-phonon coupling. Si-based alloys [68], impurity-doped nanoparticles [69,70], and nanocrystalline Si multilayers [29,71] are fabricated, and a unique approach (defect engineering) to overcoming the intrinsic indirect nature of c-Si has been proposed and demonstrated [72]. These sample preparation and characterization studies will provide information on the efficient PL mechanism and the electronic properties of Si nanostructures.

5. Conclusion

Since the discovery of efficient visible PL from porous silicon and silicon nanocrystals in 1990, there are many studies on optical properties of c-Si nanoparticles. In addition, there have been many different approaches towards practical Si light-emitting devices. We have discussed here the PL mechanism of c-Si and a-Si nanostructures and the essential points for the achievement of efficient luminescence from Si materials. The luminescence properties of c-Si and a-Si nanostructures are drastically modified by quantum confinement and spatial confinement of electrons and excitons. However, the detailed electronic structures and optical properties in SiO_2-capped nanostructures are not clear and under discussion. The microscopic understanding of the Si/SiO_2 interface in nanostructures is important for the understanding of the optical gain mechanism and the development of silicon lasers.

The author would like to thank T. Matsumoto, S. Okamoto, Y. Hirai, S. Nihonyanagi, M. Iiboshi, Y. Fukunishi, S. Oda, K. Nishimoto, J.-N Chazalviel, Y. Takahashi, H. Kageshima, and K. Shiraishi for discussions and collaborations. The author also thanks many of the authors cited in this review for discussions.

References

1. Canham, L. T. (1990) Silicon quantum wire array fabrication by electrochemical and chemical dissolution of wafers, *Appl. Phys. Lett.* **57**, 1046.
2. Kanemitsu, Y., Ogawa, T., Shiraishi, K., and Takeda, K. (1993) Visible photoluminescence from oxidized Si nanometer-sized spheres: Exciton confinement on a spherical shell, *Phys. Rev. B* **48**, 4883.
3. Wilson, W. L., Szajowski, P. F., and Brus, L. E. (1993) Quantum confinement in size-selected, surface-oxidized silicon nanocrystals, *Science* **262**, 1242.
4. Maeda, Y., Tsukamoto, N., Yazawa, Y., Kanemitsu, Y., and Masumoto, Y., (1991) Visible photoluminescence of Ge microcrystals embedded in SiO_2 glassy matrices, *Appl. Phys. Lett.* **59**, 3168.
5. Kanemitsu, Y., Uto, H., Masumoto, Y., and Maeda, Y. (1992) On the origin of visible photoluminescence in nanometer-size Ge crystallites, *Appl. Phys. Lett.* **61**, 2187.
6. Pavesi, L., Dal Negre, L., Mazzoleni, C., Franzo, G., and Priolo, F. (2000) Optical gain in silicon nanocrystals, *Nature* **408**, 440.
7. Nayfeh, M. H., Barry, N., Therrien, J., Akcakir O., Gratton, E., and Belomoin, G. (2001) Stimulated blue

emission in reconstituted films of ultrasmall silicon nanoparticles, *Appl. Phys. Lett.* **78**, 1131.

8. Ball, P. (2001) Let there be light, *Nature* **409**, 974.
9. Iyer, S. S. and Xie, Y. -H (1993) Light emission from silicon, *Science* **260**, 40.
10. Wehrspohn, R. B., Chazalviel, J. N., Ozanam, F., and Solomon, I (1999) Spatial versus quantum confinement in porous amorphous silicon nanostructures, *Eur. Phys. J. B* **8**, 179.
11. Kanemitsu, Y. Uto, H., Masumoto, Y., Matsumoto, T., Futagi, T. and Mimura, H. (1993) Microstructure and optical properties of free-standing porous silicon films: Size dependence of absorption spectra in Si nanometer-sized crystallites, *Phys. Rev. B* **48**, 2827.
12. Bustarret, E., Sauvain, E., and Ligeon, M. (1997) High-resolution transmission electron microscopy study of luminescent anodized amorphous silicon, *Philos. Mag. Lett.* **75**, 35.
13. Kanemitsu, Y., Fukunishi, Y., and Kushida, T. (2000) Decay dynamics of visible luminescence in amorphous silicon nanoparticles, *Appl. Phys. Lett.* **77**, 211.
14. Street, R. A., *Hydrogenated Amorphous Silicon* (Cambridge Univ., Cambridge, 1991).
15. Calcott, P. D. J., Nash, K. J., Canham, L. T., Kane, M. J., and Brumhead, D. (1993) Spectroscopic identification of the luminescence mechanism of highly porous silicon, *J. Lumin.* **57**, 257.
16. Suemoto, T., Tanaka, K., Nakajima, A., and Itakura, T. (1993) T., Observation of phonon structures in porous Si luminescence, *Phys. Rev. Lett.* **70**, 3659.
17. Kanemitsu, Y., and Okamoto, S. (1998) Phonon structures and Stokes shift in resonantly excited luminescence of silicon nanocrystals, *Phys. Rev. B* **58**, 9652.
18. Shaklee, K. L., and Nahory, R. E. (1970) Valley-orbit splitting of free excitons? The absorption edge of Si, *Phys. Rev. Lett.* **24**, 942.
19. Matsumoto, T., Suzuki, J., Ohnumra, M., Kanemitsu, Y., and Masumoto, Y. (2001) Evidence of quantum size effect in nanocrystalline silicon by optical absorption, *Phys. Rev. B* **63**, 195322.
20. Kanemitsu, Y., and Okamoto, S. (1997) Photoluminescence from Si/SiO_2 single quantum wells by selective excitation, *Phys. Rev. B* **56**, R15561.
21. Takahashi, Y., Furuta, T., Ono, Y., Ishiyama, T., and Tabe, M. (1995) Photoluminescence from a silicon quantum well formed on separation by implanted oxygen substrates, *Jpn. J. Appl. Phys.* **34**, 950.
22. Kanemitsu, Y., Iiboshi, M., and Kushida, T. (2000) Photoluminescence dynamics of amorphous Si/SiO_2 quantum wells, *Appl. Phys. Lett.* **76**, 2200.
23. Nishimoto, K., Sotta, D., Durand, H. A., Etoh, K., and Ito, K. (1998) Visible photoluminescence from a-Si:H/SiO_2 superlattice fabricated by UHV evaporation, *J. Lumin.* **80**, 439.
24. Kanemitsu, Y., and Kushida, T. (2000) Size effects on the luminescence spectrum in amorphous Si/SiO_2 multilayer structures, *Appl. Phys. Lett.* **77**, 3550.
25. Miyazaki, S., Yamada, K., and Hirose, M (1991) Optical and electrical properties of a-Si_3N_4:H/a-Si:H superlattices prepared by plasma-enhanced nitridation technique, *J. Non-Cryst. Solids* **137/138**, 1119.
26. Miyazaki S., and Hirose, M. (1989) Amorphous silicon superlattices prepared by direct photochemical deposition, *Philos. Mag. B* **60**, 23.
27. Lockwood, D. J., Lu, Z. H., and Baribeau, J. -M. (1996) Quantum confined luminescence in Si/SiO_2 superlattices, *Phys. Rev. Lett.* **76**, 539.
28. Okamoto, S., and Kanemitsu, Y. (1997) Quantum confinement and interface effects on photoluminescence from silicon single quantum wells, *Solid State Commun.* **103**, 573.
29. Grom, G. F., Lockwood, D., J., McCaffrey, J. P., Labbe, H., J., Fauchet, P. M., White Jr, B., Diener, J., Kovalev, D., Koch, F., and Tsybeskov, L. (2000) Ordering and self-organization in nanocrystalline silicon, *Nature* **407**, 358.
30. Carrier, P., Lewis, L. J., and Dharma-wardana, M. W. C. (2002) Optical properties of structurally relaxed Si/SiO_2 superlattices: The role of bonding at interfaces. *Phys. Rev. B* **65**, 165339.
31. Tiedje, T., Abeles, B., and Brooks, B. G. (1985) Energy transport and size effects in the photoluminescence of amorphous-germanium/amorphous-silicon multilayer structures, *Phys. Rev. Lett.* **54**, 2545.
32. Nguyen, H. V., Lu, Y., Kim, S., Wakagi, M., and Collins, R. W. (1995) Optical properties of ultrathin crystalline and amorphous silicon films, *Phys. Rev. Lett.* **74**, 3880.
33. Park, N. M., Kim, T. S., and Park, S. J. (2001) Band gap engineering of amorphous silicon quantum dots for light-emitting diodes, *Appl. Phys. Lett.* **78**, 2575.
34. Tsang, C., and Street, R. A. (1979) Recombination in plasma-deposited amorphous Si:H: Luminescence decay, *Phys. Rev. B* **19**, 3027.
35. Wilson, B. A., Kerwin, T. P., and Harbison, J. P. (1985) Optical studies of thermalization mechanism in a-Si:H, *Phys. Rev. B* **31**, 7953.

36. Estes, M. J., and Moddel, G. (1996) Luminescence from amorphous silicon nanostructures, *Phys. Rev. B* **54**, 14633.
37. Allan, G., Delerue, C., and Lannoo, M. (1997) Electronic structure of amorphous silicon nanoclusters, *Phys. Rev. Lett.* **78**, 3161.
38. Collins, R. W., Paesler, M. A., and Paul, W. (1980) The temperature dependence of photoluminescence in a-Si:H alloys, *Solid State Commun.* **34,** 833.
39. Tanaka, M., and Masumoto, Y. (2000) Very weak temperature quenching in orange luminescence of ZnS:Mn^{2+} nanocrystals in polymer, *Chem. Phys. Lett.* **324**, 249.
40. Kanemitsu, Y., Nihonyangaki, S., Sato, H., and Hirai, Y. (2002) Efficient radiative recombination of indirect excitons in silicon nanowires, *Phys. Stat. Sol. (a)* **190**, 755.
41. Dean, P. J., Haynes, J. R., and Flood, W. F. (1967) New radiative recombination processes involving neutral donors and acceptors in silicon and germanium, *Phys. Rev.* **161**, 711.
42. Davis, G. (1989) The optical properties of luminescence centres in silicon, *Phys. Rep.* **176**, 83.
43. Nihonyanagi, S., and Kanemitsu, Y. (2002) Mechanism of room temperature luminescence in silicon nanowires, submitted for publication.
44. Tsybeskov, L., Moore, K. L., Duttagupta, S. P., Hirschman, K. D., Hall, D. G., and Fauchet, P. M. (1996) A Si-based light-emitting diode with room-temperature electroluminescence at 1.1 eV, *Appl. Phys. Lett.* **69**, 3411
45. Tsybeskov, L., Hirschman, K. D., Duttagupta, S. P., Zacharias, B. M., Fauchet, P. M., McCaffrey, J. P., and Lockwood, D. J (1998) Nanocrystalline-silicon superlattice produced by controlled recrystallization, *Appl. Phys. Lett.* **72**, 43.
46. Kanemitsu, Y., Inagaki, T. J., Ando, M., Matsuda, K., Saiki, T., and White, C. W. (2002) Photoluminescence spectrum of highly excited single CdS nanocrystals studied by a scanning near-field optical microscopy, *Appl. Phys. Lett.* **81**, 141.
47. Bacher, G., Weigand, R., Seufert, J., Kulakovskii, D., Gippius, N. A., Forchel, A., Leonardi, K., and Hommel, D. (1999) Biexciton versus exciton lifetime in a single semiconductor quantum dot, *Phys. Rev. Lett.* **83**, 4417.
48. Kanemitsu Y., and Okamoto, S. (1997) Resonantly excited photoluminescence from porous silicon: Effects of surface oxidation on resonant luminescence spectra, *Phys. Rev. B* **56**, R1696.
49. Kageshima, H., and Shiraishi, K. (1997) Microscopic mechanism for SiO_2/Si interface passivation: Si=O double bond formation, *Surf. Sci.* **380**, 61.
50. Kageshima, H., and Shiraishi, K. (1997) First-principles study of photoluminescence from silicon/silicon-oxide interfaces, *Mater. Res. Soc. Proc.* **486**, 337.
51. Kanemitsu, Y., Okamoto, S., Otobe, M., and Oda, S. (1997) Photoluminescence mechanism in surface-oxidized silicon nanocrystals, *Phys. Rev. B* **55**, R7375.
52. Wolkin, M. V., Jorne, J., Fauchet, P. M., Allan, G., and Delerue, C. (1999) Electronic sates and luminescence in porous silicon quantum dots: The role of oxygen. *Phys. Rev. Lett.* **82**, 197.
53. Cullis, A. G., Canham, L. T., and Calcott, P. D. J. (1997) The structural and luminescence properties of porous silicon, *J. Appl. Phys.* **82**, 909.
54. Cullis, A. G., and Canham, L., T. (1991) Visible light emission due to quantum size effects in highly porous crystalline silicon, *Nature* **353**, 335.
55. Schuppler, S., Friedman, S. L., Marcus, M. A., Aldler, D. L., Xie, Y.- H., Ross, F. M., Chabal, Y. J. Harris, T. D., Brus, L. E., Brown, W. L., Chaban, E. E., Szajowski, P. F., Christman, S. B., and Citrin, P. H. (1995) Size, shape, and composition of luminescent species in oxidized Si nanocrystals and H-passivated porous Si, *Phys. Rev. B* **52**, 4910.
56. Zhang., Q., and Bayliss, S. C. (1996) The correlation of dimensionality with emitted wavelength and ordering of freshly produced porous silicon, *J. Appl. Phys.* **79,** 1351.
57. Binder, M., Edelmann, T., Metzger, T. H., Mauckner, G., Goerigk., G., and Peisl, J. (1996) Bimodal size distribution in p-porous silicon studied by small angle x-ray scattering, *Thin Solid Films* **276**, 65.
58. Brus, L. E., Szajowski, P. F., Wilson, W. L. Harris, T. D., Schuppler, S., and Citrin, P. H. (1995) Electronic spectroscopy and photophysics of Si nanocrystals: Relationship to bulk c-Si and porous Si, *J. Amer. Chem. Soc.* **117**, 2915.
59. Takeoka, S., Fujii, M., and Hayashi, S. (2000) Size-dependent photoluminescence from surface-oxidized Si nanocrystals in a weak confinement regime, *Phys. Rev. B* **62**, 16820.
60. van Buuren, T., Dinh, L. N., Chase, L. L., Siekhus, W. J., and Terminello, L. J. (1998) Changes in the electronic properties of Si nanocrystals as a function of particle size. *Phys. Rev. Lett.* **80**, 3803.

61. Matsumoto, T., Futagi, T., Mimura, H., and Kanemitsu, Y. (1993) Ultrafast decay dynamics of luminescence in porous silicon, *Phys. Rev. B* **47,** 13876.
62. Kanemitsu, Y. (1993) Slow decay dynamics of visible luminescence in porous silicon: Hopping of carries confined on a shell region in nanometer-size Si crystallites, *Phys. Rev B* **48**, 12357.
63. Kanemitsu, Y. (1994) Luminescence properties of nanometer-sized Si crystallites: Core and surface states, *Phys. Rev. B* **49**, 16845.
64. Kovalev, D., Heckler, H., Ben-Chorin, M, Polisski, G., Schwartzkopff, M., and Koch, F. (1998) Breakdown of the k-conservation rule in Si nanocrystals, *Phys. Rev. Lett.***81**, 2803.
65. Ito, K., Ohyama, S., Uehara, Y., and Ushioda, S. (1995) Visible light emission spectra of individual microstructures of porous Si, *Appl. Phys. Lett.* **67**, 2536.
66. Mason, M. D., Credo, G. M., Weston, K. D. and Buratto, S. K. (1998) Luminescence of individual porous Si chromophores, *Phys. Rev. Lett.* **80**, 5405.
67. Valenta, J., Juhasz, R., and Linnros, J. (2002) Photoluminescence spectroscopy of single silicon quantum dots, *Appl. Phys. Lett.* **80,** 1070.
68. Leong, D., Harry, M., Reeson, K. J., and Homewood, K. P. (1997) A silicon/iron-disilicide light-emitting diode operating at a wavelength of 1. 5 μm, *Nature* **387**, 686.
69. Mimura, A., Fujii, M., Hayashi, S., Kovalev, D., and Koch, F. (2000) Photoluminescence and free-electron absorption in heavily phosphorus-doped Si nanocrystals, *Phys. Rev. B* **62,** 12625.
70. Han, H. S., Seo, S. Y., and Shin, J. H. (2001) Optical gain at 1.54 μm in erbium-doped silicon nanoclusters sensitized waveguide, *Appl. Phys. Lett.* **79**, 4568.
71. Lu, Z. H., and Grozea, D. (2002) Crystalline Si/SiO_2 quantum wells, *Appl. Phys. Lett.* **80**, 255.
72. Ng., W. L., Lourenco, M. A., Gwilliam, R. M., Ledain, S., Shao, G., and Homewood, K. P. (2001) An efficient room-temperature silicon-based light-emitting diode, *Nature* **410**, 192.

ELECTRONIC AND DIELECTRIC PROPERTIES OF POROUS SILICON

D. KOVALEV, J. DIENER
Technische Universität Muenchen, Physik-Department E16
85747, Garching, Germany

1. Introduction

Bulk Si is, due to its indirect band-gap electronic structure, a very inefficient emitter, even at liquid He temperatures. In recent years several approaches were developed towards improving the efficiency of light emission from Si-based structures. All of them were based on the lifting of the lattice periodicity inducing an uncertainty in the k-space and therefore altering the indirect nature of this material. Additionally, a better overlap of wavefunctions of electrons and holes results in a larger oscillator strength of optical transitions. Some examples are: SiGe [1] or Si-SiO_2 superlattices [2] or Si nanocrystal assemblies [3]. The largest quantum yield that has been achieved under optical excitation of Si nanocrystals is of the order of 10% and is already comparable with that of direct band-gap quantum dots assemblies. The purpose of the paper is to survey the experimental work which has been carried out towards detailed understanding of the basic optical properties of silicon nanocrystal assemblies.

Si nanocrystals can be prepared in different ways. The most widely discussed in the literature is a system known as porous silicon [4]. It has attracted much interest due to the simplicity of the preparation procedure and the high emission efficiency under optical excitation (up to 10%). Structural investigations have confirmed that it consists of Si nanocrystals of different size (typically a few nm) and shape, that retain the diamond lattice structure of bulk Si [5]. Si nanocrystals prepared by aerosol procedure or by thermal precipitation of Si atoms implanted in SiO_2 layers show strong near infrared and visible PL with characteristics very similar to that of porous silicon [3].

2. Photoluminescence properties of silicon nanocrystals

Silicon nanocrystal assemblies consist of a nearly infinite number of individual absorbers having different band-gaps. The clear evidence that emitting states are driven to higher energies by confinement is coming from the PL measured under high energy of excitation, when all crystallites in the distribution are excited (see Fig. 1). The PL, depending on morphology of the crystallite ensemble, can be continuously tuned with small increments over a very wide spectral range from the Si band-gap to the green

L. Pavesi et al. (eds.), Towards the First Silicon Laser, 123–130.

region. Unfortunately the large inhomogeneous emission line width (of the order of 500 meV) arising from the residual size and shape distribution obscures the spectroscopic information it contains.

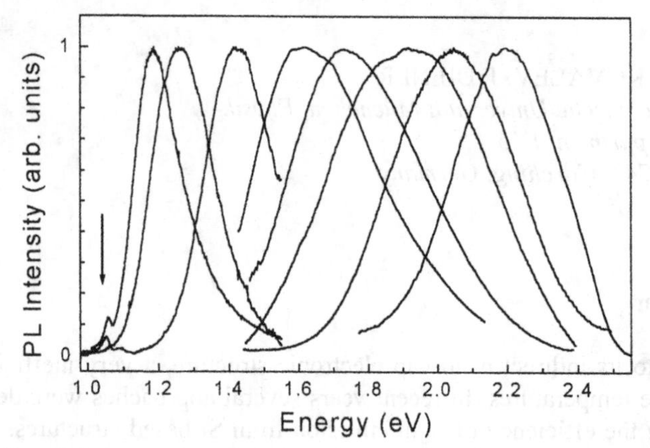

Figure 1. Tunability of the porous silicon PL band. Various etching parameters have been used. The arrow shows the spectral position of the lowest energy free exciton transition in bulk silicon.

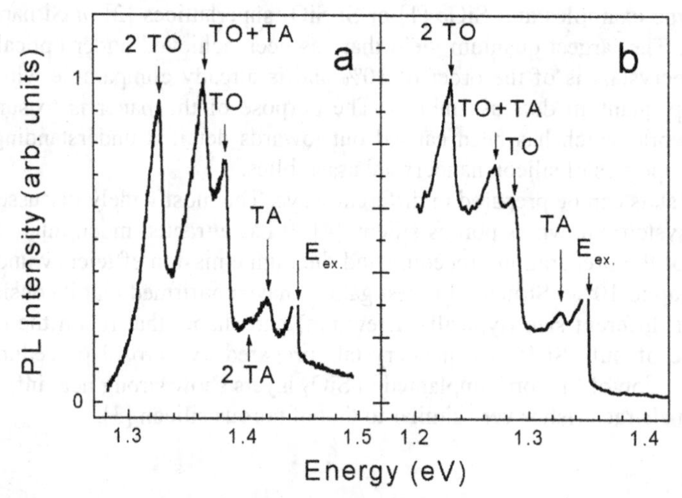

Figure 2. Resonant PL spectra of naturally (a) and heavily oxidized (b) porous silicon. The arrows show the energy position of Si TA and TO momentum-conserving phonons with respect to the exciton ground state.

One of the general predictions of the theory is that in smaller Si nanocrystals the probability of NP transitions should increase with respect to phonon-assisted (PA) processes [6]. We used energy-selective optical spectroscopy to estimate the relative strengths of the NP transitions and the various momentum conserving phonon assisted processes [7]. The resonant PL spectra excited at the edge of the luminescence band show distinctive emission peaks from quantum confined exciton states recombining via NP, TA- and TO-phonon assisted processes (see Fig.2). Since both NP and PA processes are possible (in both emission and absorption) peaks are replicated at 1 and 2 phonon energies below the NP peak. The observation of a peak structure in the resonant spectrum confirms inhomogeneous nature of broadening of nonresonantly excited PL.

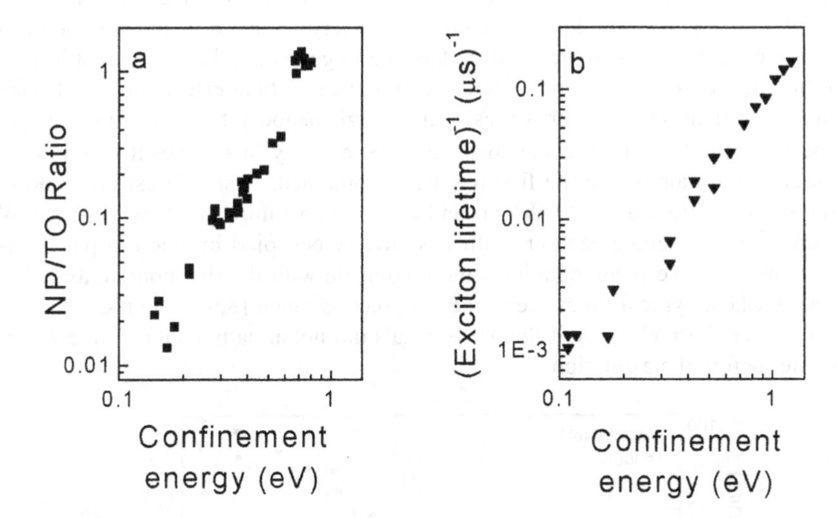

Figure 3. Influence of quantum confinement on the basic properties of excitons localized in Si nanocrystals. Confinement energy dependence: a). Relative strength of no-phonon and TO-phonon assisted exciton recombination channels. b). Inverse lifetime of the optically active exciton state.

In Fig.3a we plot the relative probabilities of NP to TO-phonon assisted transitions, estimated from amplitude of spectral peaks, as a function of confinement energy in the spectral range of 1.3-1.9 eV. One can see that in the vicinity of the crystalline Si band-gap, in the weak confinement regime, TO-phonon assisted processes dominate. At 1.8 eV NP processes begin to take over. Si crystallites having SiO_2 shells behave in general in a similar way. However the strength of NP processes is significantly enhanced at the same emission energies. This extremely large variation of the relative NP processes oscillator strength can be explained only on the basis of quantum confinement model.

On Fig. 3b we plot the inverse radiative recombination time of excitons in the same range of confinement energies. This value is proportional to the oscillator strength of the allowed exciton transition. Both dependencies are very similar. One has to mention that the appearance of NP transitions does not imply their direct nature: the exciton has a final probability to recombine either with or without phonon assistance. Because even at highest confinement energies the exciton lifetime is on the order of μs the optical transitions are still indirect.

3. Nonlinear optical properties of silicon nanocrystals

One of the most important properties of the nanocrystal assemblies is their strong nonlinear optical response. At low excitation levels when no more than one *e-h* pair is present per nanocrystal, the Auger rate is zero in contrast to bulk semiconductors, where the Auger rate is not zero at any pumping power. In that sense, at low temperature and weak excitation power, the absence of transport makes the Auger-process less efficient in a nanocrystal system compared with a bulk semiconductor. On the other hand, at intense excitation, when a double occupation of a nanocrystal can be achieved, the local concentration of 2 *e-h* pairs in the nanocrystals due to geometrical confinement is very high: on the order of 10^{18}-10^{20} cm^{-3}. Therefore, nonlinear optical effects in the PL take place when more than one *e-h*-pair is present in a single nanocrystal. The long radiative decay time $\tau_{rad.}$~10^{-6}-10^{-2} s of excitons localized in Si nanocrystals makes it very easy to excite a second *e-h* pair before the first one has recombined. A simple estimate shows that at room temperature at ~100 W/cm^2 and at He temperature at ~1 W/cm^2 of CW optical excitation in average each crystallite is already occupied by one *e-h* pair. Furthermore, slow radiative recombination cannot compete with the fast nonradiative Auger processes whose typical times are in the nanosecond range [8,9] and an excited *e-h* pair recombines radiatively only if the nanocrystals did not already contain an *e-h* pair or free carrier before the excitation.

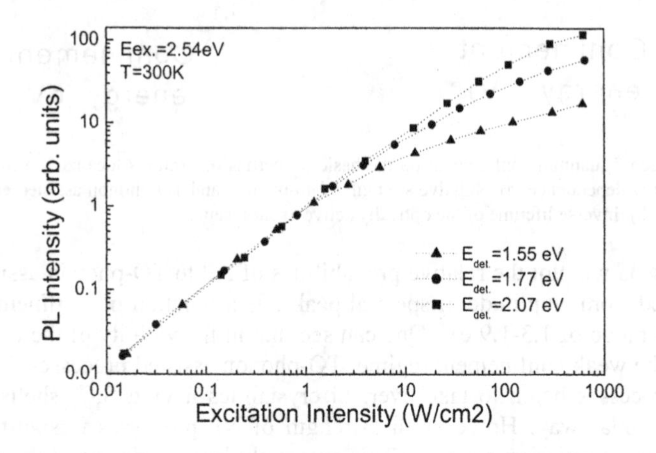

Figure 4. PL intensity versus CW excitation intensity for three different detection energies. $E_{ex.}$=2.54 eV.

This behavior can be readily seen in Fig. 4 where we plot the dependence of the PL intensity on the CW excitation density. The PL intensity increases linearly with the excitation intensity up to 2 W/cm^2, above this value saturation effects are observed. The density of electronic states and, therefore, nanocrystal absorption cross section are larger for smaller values of emission energies and $\tau_{rad.}$ is longer for smaller nanocrystal bandgaps. That is why the deviation from the linear behavior is stronger, and appears at lower excitation intensities for lower detection energy. One can see that at weak optical excitation the PL intensity has a linear dependence and at an infinitely high one the quantum yield is limited by the emission of one photon per nanocrystal per exciton lifetime. This is a strong principle limitation of the emissivity of silicon nanocrystals which seems to be not easy to overcome.

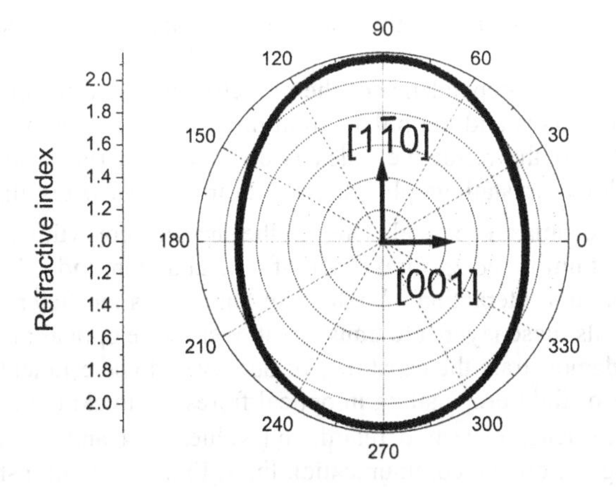

Figure 5. Polar plot of the measured refractive index value versus the polarization direction of the incident light. Crystallographic directions are shown by arrows. λ=633 nm. Optical axis coincides with $[1\bar{1}0]$ crystallographic direction.

4. Dielectric properties of porous silicon

Over recent years, nanostructuring of semiconductors has been considered as an alternative way to the search for new materials. A key idea is an introduction of optical anisotropy due to reduction of the symmetry of the bulk crystals via ordered nanostructuring. We investigate properties of optical devices based on macroscopically nanostructured porous silicon layers: optical in-plane anisotropy of porous silicon layers (retarders) and polarisation-dependent response of Bragg reflectors. For an etched (100) Si surface the preferential alignment of nanocrystals in the [100] growth direction results in a larger dielectric constant for the light polarized along [100] direction [10,11]. However, for most of practical applications an in-plane birefringence is required. To achieve in-plane uniaxial symmetry we employed the electrochemical etching of lower symmetry (110) Si wafers. In Ref. 12 the selective crystallographic pore propagation in equivalent [010] and [100] directions tilted to the (110) surface plane has been demonstrated. The pores having directions along the [100] and [010] crystallographic directions are tilted to the (110) surface and their projection on the (110) surface is aligned along the $[1\bar{1}0]$ direction. The application of an electric field to a spatially anisotropic dielectric structure (e.g. to a dielectric wire) results in a different dielectric constant (ε) dependent on the polarization direction of the electric field due to the difference in the screening of the applied electric field by induced surface polarization charges [13]. Therefore the etching of (110) Si wafers results in a reduction of the symmetry of the optical properties from an isotropic cubic to an in-plane uniaxial one and gives different refractive indices $n_{[1\bar{1}0]}$ and $n_{[001]}$ for light polarized along the $[1\bar{1}0]$ and [001] crystallographic directions, respectively. The sign of the difference depends on the surface plane morphology of layers which is not known due to extreme technical difficulties with correct imaging of nanometer sized pores. Bulk Si is no anisotropic crystal but its porous modification becomes intrinsically uniaxial for this plane due to anisotropic dielectric nanostructuring.

Fig.5 shows the dependence of the refractive index value on the angle between the crystallographic axes and the polarization direction of normally incident light. The uniaxial symmetry of the refractive index is evident, the largest value is for the light polarized in the $[1\bar{1}0]$ crystallographic direction (direction of wires alignment). Thus the layer is a negative uniaxial crystal. One of the unique properties of birefringent PSi layers is an almost unlimited spectral range of optical action. Indeed, because bulk Si is not-polar material and interaction of light with phonons is inefficient, contrary to other birefringent crystals these layers are transparent over entire optical range from visible to far-infrared. To demonstrate the capability of our layers to be efficient controllers of the polarisation state of light propagating in optical fibres the thickness of layers was intentionally chosen to achieve desired retardation values ($\lambda/4$ and $\lambda/2$ conditions) in the transparency range of optical communication lines. Fig. 6 (left side) shows how linearly polarised incident light having wavelength 1.53 μm transforms into circularly polarized light. Respectively the right side shows how a $\lambda/2$ retardation condition can be achieved at λ=1.3 μm. The analysis of the angular polarization pattern of the transmitted light shows a high retardation accuracy: the deviation from the ideal circularly or linearly polarized light is on the order of 0.1 % for $\lambda/4$ and $\lambda/2$ plates conditions what is in fact an experimental accuracy (see Fig. 6).

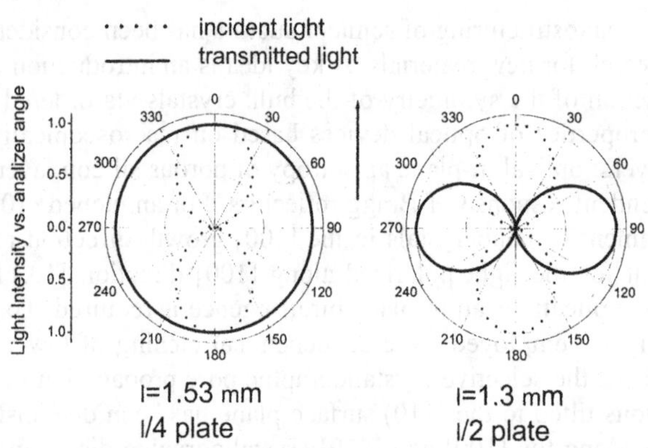

Figure 6. Polar plots of the intensity of incident (dashed lines) and transmitted (solid lines) light as a function of the angle of the analysing Polaroid. Measurements are performed at $\lambda/4$ (left side) and $\lambda/2$ (right side) retardation values. Birefringence direction (45 ° to [001] axis) is shown by the arrow.

At this point we want to emphasise the fundamental specifics of our system. The layers are quasi-bulk: except of a crystalline Si sponge-like network they contain a large fraction of empty space inside. The most remarkable property of our system is that incorporation of an additional dielectric substance in the pores reduces the anisotropy and thus decreases the phase shift of the electric field of the transmitted light. The phase shift is defined by the dielectric constant of the substance and the filling factor of the pores. Therefore, an analysis of the polarization of the transmitted light allows controlling the amount of condensed fluid, i.e. the filling factor of the pores and dynamics of the phase transition. Since these layers are highly sensitive to dielectric surrounding they seem to be a good candidates for sensors and for studying physics of the phase transitions in a nanometer geometrical scale.

Previous approach is based on in-plane nanowires alignment. The key idea of the second one is alternative variation of the dielectric constant in-depth of the (110) PSi layers. The combination of the variation of the etching rate in depth with an in-plane optical anisotropy results in the variation of the refractive index in three dimensions. Each layer additionally has in-plane optical anisotropy and, therefore, a stack of layers, acting as a distributed Bragg reflector (DBR), should have two spectrally distinct reflection bands, depending on the polarization direction of the incident linearly polarized light. These structures can yield optical effects which are difficult to achieve with conventional Bragg reflectors. Fig. 7 shows the reflectivity of such a DBR at almost normal incidence. Contrary to standard porous silicon DBRs the spectral position of λ_{Bragg} depends on the polarization direction of the incident light. For E parallel to [001] (short dashed line) the first order λ_{Bragg} appears around 1060nm but is significantly shifted towards longer wavelength (~ 1170 nm) for E parallel to $[1\bar{1}0]$ crystallographic direction (solid line). A similar polarisation-dependent optical properties exhibit microcavities based on (110) birefringent porous silicon layers.

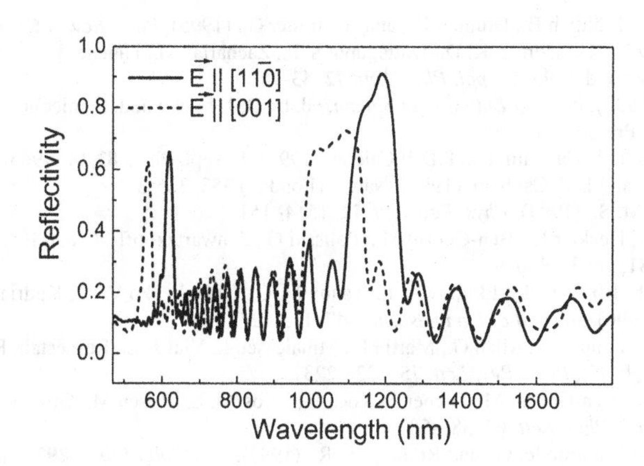

Figure 7. Spectrally and polarization-resolved reflection of a (110) distributed Bragg reflector.

Conclusions

To conclude, even nanometre-size silicon crystallites do not become a direct semiconductor. A consequence of the more direct nature of the recombination is an increase of the oscillator strength, physically seen as a shortening of lifetime of the optically active exciton state by two orders of magnitude over the whole spectral range. The second most important consequence of the carrier confinement is a drastic enhancement of nonradiative Auger-recombination due to high effective concentration of carriers localized inside the crystallites. For technical applications it is obviously a disadvantage: the luminosity of a device based on silicon nanocrystals is limited to one photon per nanocrystallite per exciton lifetime, which is still in a microsecond range even at room temperature. For any light emitting device application an electrical excitation is desirable. But the efficiency of the PL is particularly a result of the "disabling" of nonradiative recombination channels due to strongly reduced exciton transport, so that there is a contradiction which seems not to be easy to overcome.

However, porous silicon layers seem to be a good candidate for passive optical devices. For example, we presented here an artificially created highly in-plane birefringent material produced by anodic etching of (110) oriented Si wafers. Our structures exhibit values of the anisotropy parameter Δn larger than any birefringent crystal found in nature. The average refractive index can be easily adjusted to desired values by varying the porosity. The spectral range of efficient birefringence extends from near to far infrared. Additionally, multilayered birefringent porous silicon structures can yield optical effects which are difficult to achieve with conventional Bragg reflectors and microcavities i.e. they can be used for sensing of the polarization direction of light.

We acknowledge the support of this work by Deutsche Forschungsgemeinschaft.

References

1. Zrenner A., Fröhlich B., Brunner K., and Abstreiter G., (1995), *Phys. Rev. B* **52**, 16608-16611.
2. Tsybeskov L. , Hirschman K. D., Duttagupta S. P., Zacharias M., Fauchet P. M., McCaffrey J. P., and D.J. Lockwood, (1998), *Appl. Phys. Lett.* **72**, 43-45.
3 Brus L., (1996), in *Light Emission in Silicon*, ed. by D. J. Lockwood, Semiconductors and Semimetals (Academic Press, NY).
4. A.G. Cullis, L.T. Canham, and P.D.J. Calcott, (1997), J. Appl. Phys. **82**, 909-965.
5 A.G. Cullis and L.T. Canham, (1991), Nature (London) **353**, 335-336.
6. Hybertsen M. S., (1994), *Phys. Rev. Lett.* **72**, 1514- 1517.
7. Kovalev D., Heckler H., Ben-Chorin M., Polisski G., Schwartzkopff M., and Koch F., (1998*), Phys. Rev. Lett.* **81**, 2803-2806.
8. Chepic D. I., Efros Al. L., Ekimov A. I., Ivanov M. G., Kharchenko V. A., Kudriavtsev I. A., and T.V. Yaseva, (1990), *Journal of Luminescence* **47**, 113-127.
9. Delerue C., Lanoo M., Allan G., Martin E., Mihalcescu I., Vial J. C., Romestain R., Muller F., and Bsiesy A, (1995), *Phys. Rev. Lett.* **75**, 2228-2231.
10. Kovalev D., Ben-Chorin M., Diener J., Koch F., Efros Al. L., Rosen M., Gippius A., Tikhodeev S. G., (1995), *Appl. Phys. Lett.* **67**, 585-587.
11. Mihalcescu I., Lerondel G., and Romerstain R., (1997), *Thin Solid Films*, **297**, 245-249.
12. Chuang S. F., Collins S. D., and Smith R. L., *Appl. Phys. Lett.* **55**, 675-677.
13. Bruggeman D. A. G., (1935), *Annalen der Physik* (Paris) **24**, 636-648.

SILICON TECHNOLOGY USED FOR SIZE-CONTROLLED SILICON NANOCRYSTALS

M. ZACHARIAS, J. HEITMANN, L.X. YI, E. WILDANGER*, AND R. SCHOLZ
Max Planck Institute of Microstructure Physics
Weinberg 2, 06120 Halle, Germany.
**Caltech, Pasadena, CA, USA.*

1. Introduction

Nanoscience represents a rapidly expanding area of research that includes aspects of many areas such as physical science, materials science and engineering. The growth and investigation of low dimensional semiconductor structures offer new possibilities for light emitting devices. In a bulk semiconductor electrons and holes can move freely through the crystal. Reducing the semiconductors dimensions may drastically change their recombination properties. Size dependent optical properties, greater electron/hole overlap for enhanced photoluminescence efficiency, and discrete single electron-hole charging are some of the consequences of carrier confinement.

Nanocrystals, more often called quantum dots in case of direct semiconductors, of different materials have been successfully prepared over the last decade as free particles, capped with various ligands or in a host matrix. Because of their enormous surface-to-volume ratio, nanocrystals may be highly affected by surface chemistry. Depending on the chemical reactivity of the semiconductor and the surrounding material a strong degradation of optical properties caused by changes in the ratio of radiative to non-radiative recombination channels could occur. Carriers might be transferred from the nanocrystal to surface or defect states which effects the actual recombination path.

There are now relatively well-developed methods for the synthesis of zero-dimensional (0D) Si nanoclusters by using laser pyrolysis of silane which leads to a molecular beam of freely propagating nanoparticles [1]. Size selection has been realized by using a simple molecular beam chopper. However, the deposition of nanoclusters prepared in this way for device applications and mass production is rather unlikely. After a period of intense research on the preparation and stabilization of porous silicon with deep insight into the physics and chemistry of porous and nanocrystalline materials, today's Si nanocrystal research is concentrated on the preparation of nanocrystals in an oxide host matrix. Methods applied for Si nanocrystal preparation are ion implantation into amorphous SiO_2 [2], co-sputtering of Si and SiO_2 [3], or thermal evaporation of SiO powder in a reactive atmosphere [4]. Annealing of such films at 1100°C under protecting atmosphere, normally nitrogen, leads to the growth of Si nanocrystals in the films either by diffusion and Oswald ripening or due to phase separation. With these

L. Pavesi et al. (eds.), Towards the First Silicon Laser, 131–138.

methods, size control is realized by control of the Si content. Smaller sizes require less Si in the oxide matrix. A lower Si content does result in smaller crystals but also in a reduced number of Si crystals. Because the PL intensity linearly scales with the number of crystals an undesired reduction in PL intensity is normally observed. Consequently, the question still remains for Si nanocrystals research: How can Si nanocrystals be fabricated with independent control of size and density?

In this paper, we will demonstrate the fabrication of ordered arranged Si nanocrystals realizing size and density control as well as establishing a cheap way for a fabrication technique on whole wafers. The fabrication technique as well as the resulting properties of the nanocrystals will be discussed. A combined infrared and luminescence study of intermediate stages of phase separation offers new insight into the luminescence properties of oxides containing Si clusters and crystals and is discussed, too.

2. Experimental details

The basic idea of our method is to force an ultra thin SiO layer, embedded between two SiO_2 layers, to undergo a phase separation and crystallization process. For this goal, amorphous SiO/SiO_2 superlattices were prepared on 4 inch wafers by evaporation of SiO powder in vacuum or with additional oxygen pressure as first demonstrated in [5,6]. A simple vacuum chamber evacuated to a pressure of around 10^{-7} mbar was used. The substrate temperature was held at 100°C during the processing. The amorphous SiO/SiO_2 superlattices were prepared on a Si substrate by alternating reactive evaporation of SiO powder in vacuum (below 10^{-7} mbar) or in oxygen atmosphere with an oxygen pressure of 10^{-4} mbar. This changes the stoichiometry x of the deposited SiO_x film alternatively between 1 and 2 realizing the desired amorphous superlattice with nanometer thicknesses. After film deposition the samples were annealed at 1100°C for 1h under N_2 atmosphere to enforce phase separation and crystallization. Selected samples were annealed at intermediate temperatures under nitrogen atmosphere to study the phase separation process [7]. Selected samples were investigated by transmission electron microscopy using bright field and dark field images for monitoring the as-prepared amorphous and the crystallized state of our films. The Fresnel defocus method was applied to get a better contrast. In addition, dark field images were used for estimation of the size distribution.

Optical properties were investigated using photoluminescence (PL) spectroscopy, infrared (IR) spectroscopy and resonant luminescence excitation spectroscopy. For luminescence the samples were excited by the 325 nm line of a HeCd laser. The PL signal was focused to the entrance of a 500 mm monochromator and detected by a nitrogen cooled CCD matrix camera. All spectra are corrected for spectral response of the measurement system. For IR investigation a Fourier IR spectrometer was used in the range of 500 to 1500 cm^{-1}.

3. Results and discussion

Our method was first presented in [5, 6] showing a schematic model of the preparation approach, bright field TEM images of crystallized SiO/SiO_2 superlattice films, and the luminescence blue shift of a whole set of samples. We will concentrate here on more recent and detailed investigations of the superlattices. Prior to the deposition of superlattices, the growth rate of the SiO and SiO_2 layers was estimated by using samples with a thickness of 100 nm. The actually deposition time to be used for thin layers was then estimated resulting in an accuracy of 0.5 nm. The real thickness can only be determined from selected samples by TEM. Due to the low element contrast of a SiO layer compared to a SiO_2 layer, we have to apply the Fresnel defocus contrast method which enables to monitor the layered structure but might virtually change the thickness observed. The best way to monitor the real thickness is by just using the luminescence peak position after crystallization for control.To understand the phase separation and crystallization process of the amorphous SiO/SiO_2 superlattices we studied the annealing behavior of thin SiO layers by a combination of IR spectroscopy, PL measurement, and TEM investigations [7, 8]. It was found that the phase separation process and crystallization can be divided into three stages:

- Stage 1 for annealing temperatures up to 600 °C is characterized by a PL signal at 560 nm and a corresponding 880 cm^{-1} IR absorption.
- Stage 2 for annealing temperatures between 600°C and 900°C still represents an amorphous state of the superlattice. Stage 2 is characterized by a more pronounced IR shift of the Si-O-Si bridging mode above 1000 cm^{-1}, the vanishing of the PL at 560 nm and the development of a new PL band which gradually shifts its position to 900 nm with increasing temperature.
- Stage 3 is observed for annealing temperatures above 900°C. It is associated with a very strong red or near infrared luminescence with the peak position depending on the SiO layer thickness, no further changes in the Si-O related absorption band are observed. Si nanocrystals are detected in the former SiO layer by TEM and XRD.

For more details and discussion of the intermediate states of phase separation please see [7,8].

Fig. 1 (a) presents the dark field images of a film with rather thin SiO sublayers capped by around 3 nm SiO_2 barriers which results in crystals having a size of (2.8 ± 0.2) nm with a Gaussian shaped size distribution shown in Fig. 1 (b). A superlattice with slightly thicker SiO layers is presented in Fig. 2 (a) with nanocrystals of a size of (3.3 ± 0.35) nm and a corresponding size distribution as shown in Fig. 2 (b). In contrast, the bulk like crystallized SiO film in Fig. 3 (a) shows much larger sizes and a log normal size distribution (Fig. 3 (b)). The obviously narrower size distribution, and the size control independent of Si content clearly proves the advantages of the described system.

The same films are now investigated under the same conditions by PL measurements. As can be seen in Fig. 4, the PL of the superlattices shows a well pronounced blue shift which depends on the SiO layer thickness, i.e. crystal size. Also, the PL of the crystallized bulk SiO film is much broader and its intensity is at least a factor of 10 less compared to the superlattices. However, in case of the superlattices, the PL signal remains strong even for small sizes. This is an additional proof of the remaining high density of the nanocrystals because the PL intensity scales with the number of crystals.

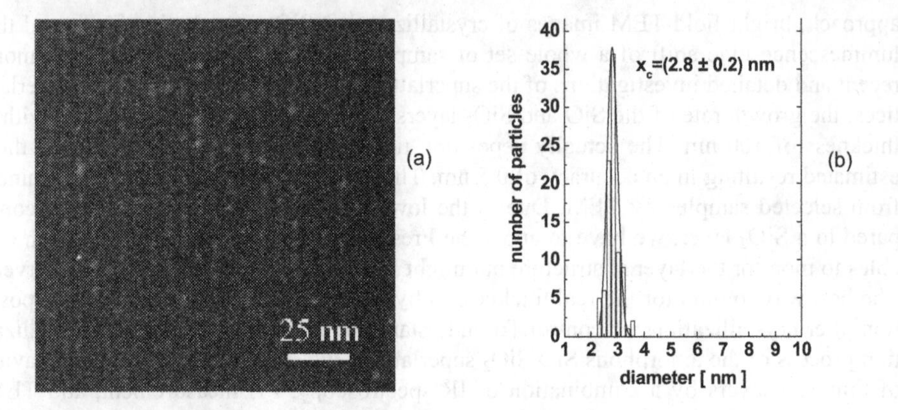

Figure 1. Dark field image (a) of a crystallized SiO/SiO_2 superlattice with 2-3 nm SiO layers and the corresponding size distribution (b).

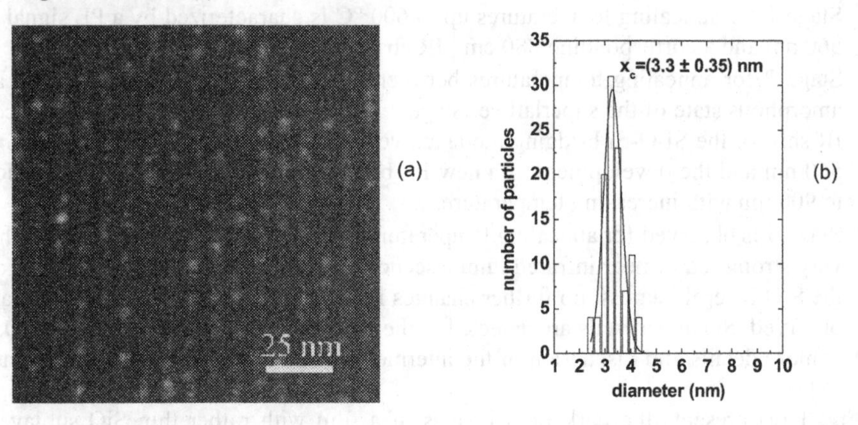

Figure 2. (a) Dark field image of a crystallized SiO/SiO_2 superlattice with slightly thicker SiO layers than in Fig. 1, (b) the correspondent size distribution.

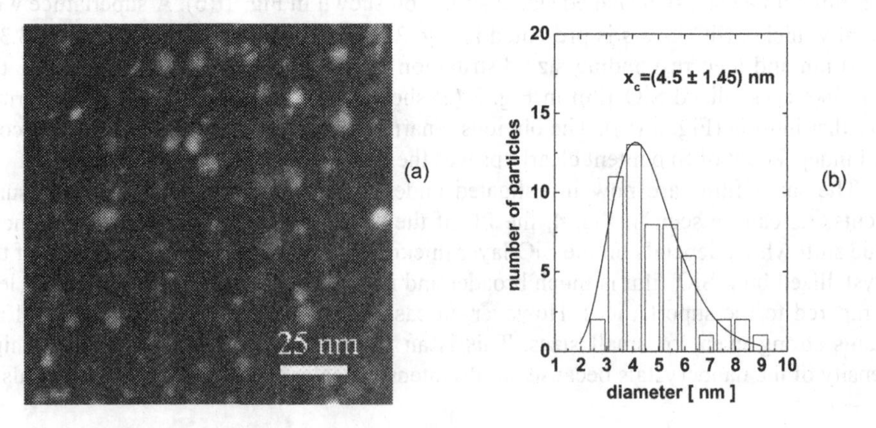

Figure 3. Dark field image of a crystallized bulk SiO film (a) showing a large number of nanocrystals with a log-normal size distribution (b).

To demonstrate the controlled shift and the Gaussian profile for the luminescence in Fig. 5 four different samples are presented with PL maximum at 777 nm (a), 811 nm (b), 853 nm (c), and 878 nm (d). Please note, that the power density used for excitation was of the order of only 1.9 mW/cm^2 to avoid any power dependent effect on the PL shape and position. The full width at half maximum (FWHM) of the samples were 116 nm (a), 127 nm (b), 127 nm (c), and 125 nm (d), respectively. For comparison, the FWHM of the bulk SiO film is 178nm. However, it still means that the line width is very large and in the range between 250 meV and 330 meV. The indirect nature of the radiative transition is causing the broadening of the luminescence line if compared to results reported for direct semiconductors. Even at low temperatures a very broad line is still observed. In addition, there might be some carrier tunneling involved from smaller nanocrystals to larger ones, i.e. from larger band gaps to smaller band gaps. Especially for the bulk SiO film an influence of the broader and asymmetric size distribution is expected.

The still indirect nature of the radiative recombination for the Si nanocrystals was proven by resonant excitation of the photoluminescence at low temperatures [9]. We observed well pronounced phonon steps with values in agreement to TO and TA phonons or a combination of both. An increasing ratio of the NO-phonon to the TO and TA phonon assisted processes with decreasing crystal size was observed which is discussed in more details in [9]. The breakdown of k-conversation for smaller crystals combined with an increasing recombination probability proves that the signal is caused by quantum confinement.

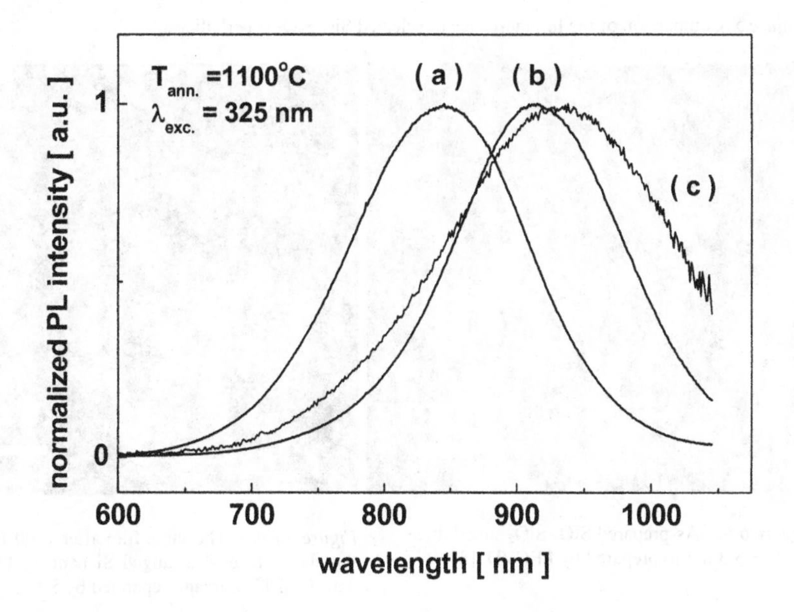

Figure 4. Comparison of the luminescence of crystallized SiO/SiO_2 superlattices having nanocrystals of a size of (a) 2.8 nm, and (b) 3.3 nm with a crystallized bulk-like SiO film (c) having a size of 4.5 nm.

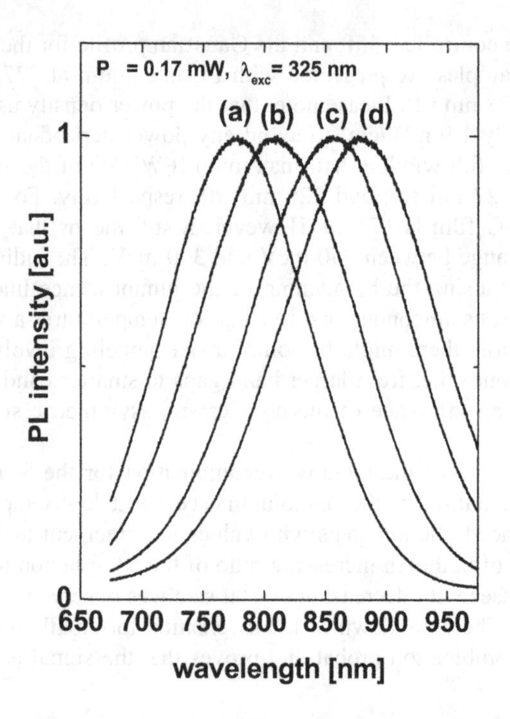

Figure 5. Comparison of the luminescence of selected SiO/SiO_2 superlattices.

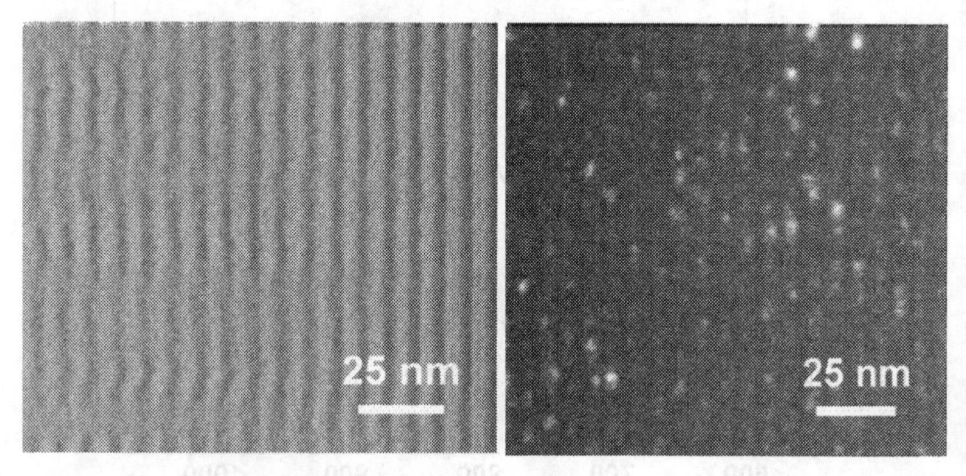

Figure 6 (a). As-prepared SiO_x/SiO_2 superlattice (3.5/3.8 nm) prepared by PECVD. [8]

Figure 6 (b).): The same film after 1100°C annealing shows layered arranged Si nanocrystals in the dark field TEM image separated by SiO_2. [8]

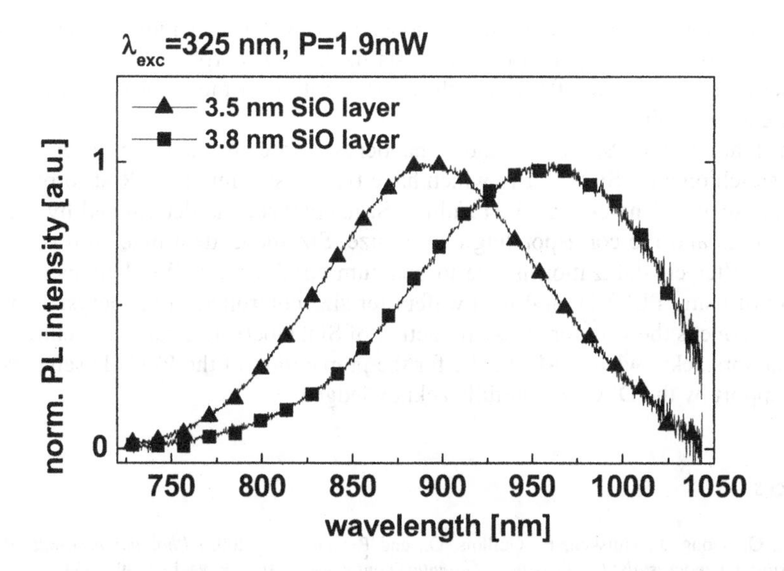

Figure 7: PL spectra of two superlattices prepared by PECVD with SiO layers of 3.5 and 3.8 nm. [8]

In Fig. 6 we demonstrate the TEM images of an as-prepared (a) and annealed (b) sample prepared by plasma enhanced chemical vapor deposition (PECVD). A phase separation similar to the evaporated superlattices is observed. The average size of the nanocrystals in Fig. 6 (b) was estimated as (3.25 ± 0.45) nm for an initial SiO layer thickness of 3.5 nm. This clearly proves that Si nanocrystals can be prepared by a mass production technique like PECVD. The slightly wavy structure is an artifact of a not optimized process. Fig. 7 shows the PL spectra of the annealed samples with 3.5 and 3.8 nm layer thicknesses. The two PECVD prepared samples show a blue shift (from 958 to 894 nm). The blue shift for smaller SiO layer thickness is caused by quantum confinement as shown already for the evaporated superlattices. The PL signal of the PECVD samples is around 20 times weaker compared to samples prepared by the evaporation technique. This might be due to the difference in the growth technique. No hydrogen is involved in case of the evaporated samples. However, for the PECVD samples the used silane results in hydrogen in the films. Hydrogen will influence the phase separation and hence the crystallization process. Further work is needed to study samples prepared by PECVD and to optimize the involved processes.

4. Summary

Si nanocrystals can be fabricated by using a process which is fully compatible to Si technology. The crystals can be arranged in a specific depth and for a specific number of layers and with a specific density. The approach is based on the preparation of an amorphous SiO/SiO_2 superlattice followed by thermal annealing for phase separation and crystallization. Using a three stage model we developed a more comprehensive understanding of phase separation and crystallization. Amorphous Si clusters grow mainly between 600°C and 900°C and the crystallization of the amorphous Si clusters takes place above 900°C.

Si rings, amorphous Si clusters and Si nanocrystals are correlated to different states of a non-stoichiometric SiO_x matrix which have typical signatures in IR absorption and photoluminescence. The size of the resulting Si nanocrystals is determined by the SiO layer thickness and the corresponding cluster size. Size-dependent blue shift of the luminescence after crystallization is due to quantum confinement. We demonstrated the possibility of using PECVD on 4 inch wafers for size controlled Si nanocrystals preparation which opens the way for mass production of Si nanocrystals based structures.

The authors acknowledge M. Reiche for the preparation of the PECVD samples. Financial support by the DFG is gratefully acknowledged.

References

[1] Ledoux, G., Gong, J., Huisken, F., Guillois, O., and Reynaud, C. (2002) *Photoluminescence of size-separated silicon nanocrystals: Confirmation of quantum confinement*, Appl. Phys. Lett. **80**, 4834.

[2] Shimizu-Iwayama,T., Fujita, K., Nakao, S., Saitoh, K., Fujita, T., and Itoh, N. (1994) *Visible photoluminescence in Si+ -implanted silica glass,* J. Appl. Phys. **75**, 7779.

[3] Hayashi, S., and Yamamoto, K. (1996) *Optical properties of Si-rich SiO2 films in relation with embedded Si mesoscopic particles*, J. Lumin. **70**, 352.

[4] Kahler, U., Hoffmeister, H. (2001) *Visible light emission from Si nanocrystalline composites via reactive evaporation of SiO*, Opt. Materials **17**, 83.

[5] Zacharias, M., Heitmann, J., and Gösele, U., *Superlattice Process Controls size of Si nanocrystals on Si wafers* (2001) MRS Bulletin **26**, 975.

[6] Zacharias, M., Heitmann, J., Scholz, R., Kahler, U., Schmidt, M., and Bläsing, J. (2002) *Size-controlled highly luminescent silicon nanocrystals: A SiO/SiO_2 superlattice approach*, Appl. Phys. Lett. **80**, 661.

[7] Yi, L.X., Heitmann, J., Scholz, R., and Zacharias, M. (2002) *Si rings, Si cluster, and Si nanocrystals-different states of ultra thin SiO layers*, Appl. Phys. Lett., in press.

[8] Zacharias, M., Yi, L.X., Heitmann, J., Scholz, R., Reiche, M., and Gösele, U., *Size-controlled Si nanocrystals for photonic and electronic applications*, Solid State Phenomena, in press.

[9] Heitmann, J., Kovalev, D., and Zacharias, M., submitted.

STRUCTURAL AND OPTICAL PROPERTIES OF SILICON NANOCRYSTALS EMBEDDED IN SILICON OXIDE FILMS

M. MIU, A. ANGELESCU, I. KLEPS, M. SIMION, A. BRAGARU
National Institute for Research and Development in Microtechnologies (IMT-Bucharest) P.O.Box 38-160, 72225 Bucharest Romania
tel. 4021.4908412, fax:4021.4908238, e-mail: mihaelam@imt.ro

1. Introduction

The photoluminescence properties of silicon nanostructured materials are strongly influenced by thermal treatments. To develop silicon based room temperature light emitting materials in addition to porous silicon fabricated by anodization, other kinds of films fabricated by sputtering, gas vaporization, and chemical vapor deposition were investigated [1.2]. Thermal treatments have different influences upon these structures: porous silicon is stabilized and its photoluminescence (PL) intensity is increased, while in substoichiometric SiO_x films, annealing causes Si nanocrystal formation which leads to PL emission. The influence of thermal treatments on photoluminescence emission of silicon nanocrystal embedded in oxide films is studied in this paper. The samples were investigated by IR and Raman measurements to determine the structure and chemical bonds. Their optical properties were studied by PL measurements.

2. Experiments

2.1. FABRICATION OF SILICON OXIDE STRUCTURES

Three silicon oxide sandwich structures were prepared, in order to observe the changes induced by thermal treatments:

SiO / Si structures; the SiO films with thickness about 1μm were deposited onto p-type Si(100) substrates. Silicon monoxide was deposited from SiO powder source, by vacuum evaporation method; deposition was performed in a Balzers equipment BA510, at a pressure of $5x10^{-6}$ torr, using a target with a perfored cap.

SiO_2 / SiO / Si structures; the SiO/Si samples were covered with a SiO_2 film with a thickness of about 0.7μm, in order to increase the film stress and to enhance the Si crystallites formation.

SiO_2 / SiO / SiO_2 / Si structures; the started material is SiO_2/Si-p. The SiO films with 1μm thickness were deposited onto these substrates, followed by the LPCVD deposition of SiO_2 films with 0.7μm thickness.

L. Pavesi et al. (eds.), Towards the First Silicon Laser, 139–144.

2.2. FORMATION OF SI ISLANDS INTO SILICON OXIDE THIN FILMS UPON THERMAL ANNEALING

Nanocrystal size of few nanometers is required to obtain an efficient visible light emission [3]. In order to Si nanocrystals in silicon oxide, the samples were annealed in N_2 atmosphere for 60 min at high temperature. For the study of the nanocrystals formation process we have used two annealing temperatures: 1000°C and 1100°C. To improve the PL emission two additional thermal processes were applied on some samples: an annealing process in a forming gas (F.G.), 10% H_2 + 90% N_2, for 120 min., at 400°C; a rapid thermal annealing (R.T.A.) process [4]. A schematic view of the experiments is presented in table 1:

TABLE 1: A schematic view of the experiments

EXPERIMENTAL STRUCTURES	THERMAL TREATMENTS
SIO/SI	thermal annealing at 1000°C for 60 min (N_2 atmosphere)
	thermal annealing at 1100°C for 60 min. (N_2 atmosphere)
	1. thermal annealing at 1000°C for 60 min. (N_2 atmosphere) + 2. thermal annealing at 400°C for 120 min. (F.G.)
SIO_2 / SIO / SI	thermal annealing at 1000°C for 60 min (N_2 atmosphere)
	thermal annealing at 1100°C for 60 min. (N_2 atmosphere)
	1. thermal annealing at 1000°C for 60 min. (N_2 atmosphere) + 2. thermal annealing at 400°C for 120 min. (F.G.)
	1. thermal annealing at 1000°C for 60 min. (N_2 atmosphere) + 2. rapid thermal annealing (N_2 atmosphere)
SIO_2 / SIO / SIO_2 / SI	thermal annealing at 1000°C for 60 min (N_2 atmosphere)
	thermal annealing at 1100°C for 60 min. (N_2 atmosphere)
	1. thermal annealing at 1000°C for 60 min. (N_2 atmosphere) + 2. thermal annealing at 400°C for 120 min. (F.G.)
	1. thermal annealing at 1000°C for 60 min. (N_2 atmosphere) + 2. rapid thermal annealing (N_2 atmosphere)

3. Results and discussion

3.1 STRUCTURAL CHARACTERISATION

The structural characterisation of the experimental films, Si nanocrystallite parameters and chemical bonds were made by IR and Raman spectroscopy. Both the influences of structure samples and of the modifications induced by thermal annealing on Raman spectra are investigated. The evolution of Raman spectra obtained for the same type of structures, subjected to different thermal treatments is presented in the fig. 1.

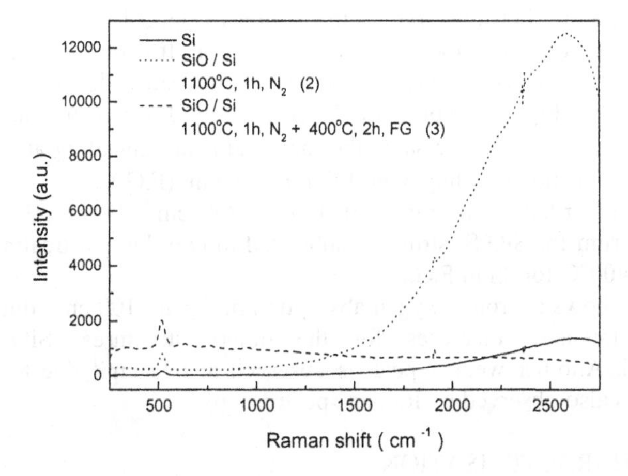

Figure 1: Raman spectra obtained for SiO/Si structure, subjected to different thermal treatments

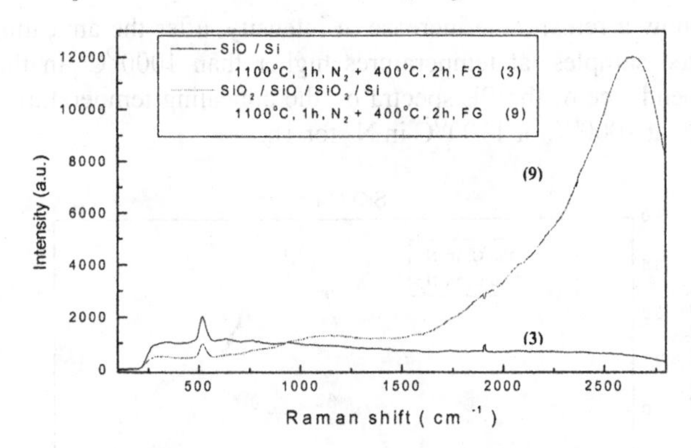

Figure 2:: Raman spectra obtained for 2 types of structures, subjected to the same thermal treatment.

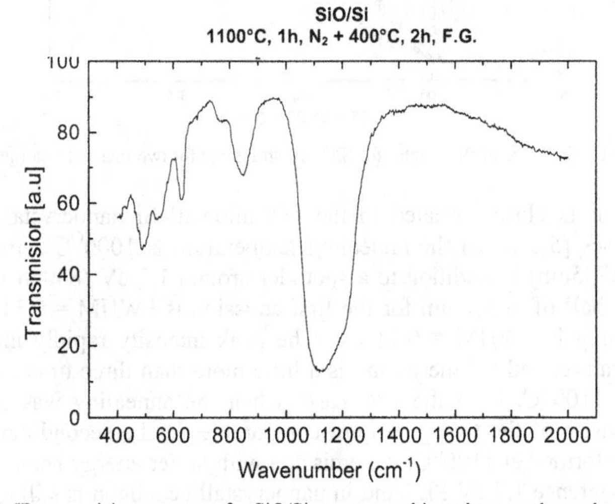

Figure 3: IR transmision spectrum for SiO/Si structure subjected to two thermal treatments

The analysis of the 300-700 cm^{-1} range, the region of optical phonons for bulk c-Si, allowed the evaluation of Si nanocrystal size distribution. It is observed a narrow peak, near to 520 cm^{-1}, and its increase in intensity indicates the nucleation of the nanocrystalline phase. Figure 2 reports the Raman spectra for two structures SiO/Si and SiO_2/SiO/SiO_2/Si, subjected to the same thermal treatment: annealing at 1000°C for 60 min. (N_2 atmosphere) and annealing at 400°C for 120 min. (F.G.).
IR spectra were recorded in the range of 400 – 1800 cm^{-1}. Figure. 3 shows the IR transmision spectrum for SiO/Si structure subjected to two thermal treatments, 1100°C for 1h in N_2 and 400°C for 2h in F.G..
The IR spectrum shows a strong oxygen absorption peaks at 1100 cm^{-1} due to the Si-O-Si bonds. Its intensity increases for the others structures, SiO_2/SiO/Si and SiO_2/SiO/SiO_2/Si. Another weaker peak is observed at 625 cm^{-1} due to multiphonon processes and it is also observed by Raman spectroscopy.

3.2 OPTICAL CHARACTERISATION

The PL spectra show a remarkable increase in intensity, after the annealing process on all the investigated samples, at temperatures higher than 1000°C. In the fig. 4 it is presented the dependence of the PL spectra on the annealing temperature, for a simple thermal annealing, at 1000°C or 1100°C in N_2 for 1h:

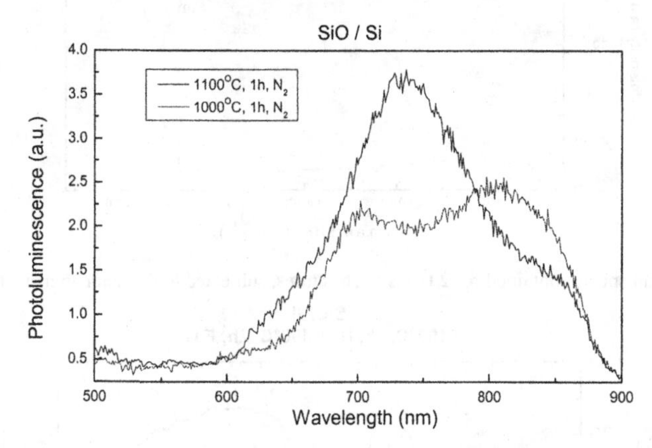

Figure 4: Dependence of PL spectra for SiO / Si structures for two thermal treatments;

This PL behaviour is closely related to the formation of Si nanocrystals due to the annealing processes [5]. When the annealing temperature is 1000°C a main PL peak around 1,65 eV (735nm) in addition to a shoulder around 1.5 eV (840nm) is obtained. The full width at half of maximum for the first emission is FWHM = 0.31 eV, and for the second emission is FWHM = 0.11 eV. The peak intensity rapidly increases with annealing temperature, and the increasing is a little more than three times, compared to the first one for 1100°C. If in the first case, when the annealing was performed at 1000°C, the intensity of the two peaks are comparable, in the second case, when the annealing was performed at 1100°C, the emission with larger energy become dominant. The origin of the intense 1.7 eV PL band in nanocrystalline silicon is still controversial.

For silicon based light emission at 740 nm, the PL intensity is sensitive to the crystallite size according to the quantum size effect (QSE) [6].

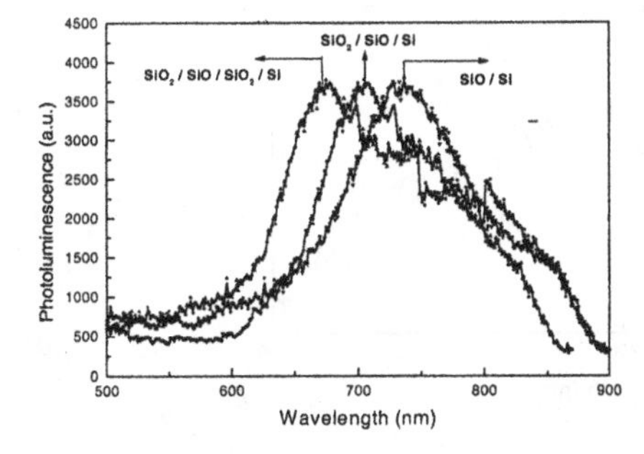

Figure. 5: Dependence of PL spectra for SiO/Si, SiO_2/SiO/Si and SiO_2/SiO/SiO_2/Si structures subjected to the same thermal treatment

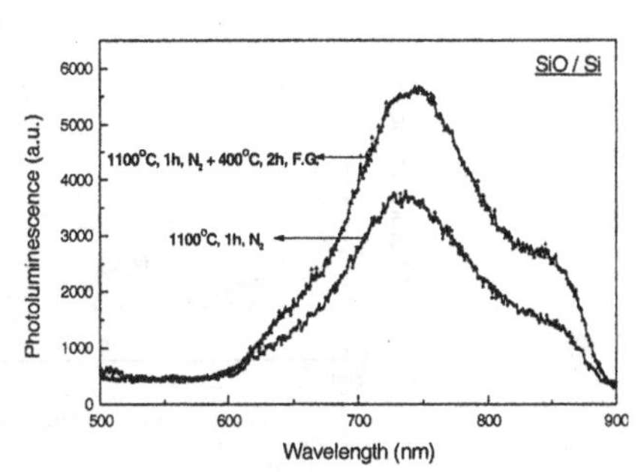

Figure 6: Photoluminescence spectra for SiO/Si sandwich structure before and after additional thermal treatment in F.G.;

In the fig. 5, the PL peaks energy for the structures annealed at 1100^{0}C, slightly shifts to higher energy, and the value of FWHM decreases with the increase of the structure complexity by using of a SiO_2 additional layer. Two phenomena could be taken in consideration for this behaviour:
- a decrease in the average size of the native silicon clusters with increasing of oxygen content;
- an increase of the stress in the silicon nanocrystals induced by SiO_2.
An interesting behaviour is observed in SiO_2/SiO/SiO_2/Si structures: an additional peak around 2.1 eV (580 nm) appears. The intensity of this third peak is much smaller, about 3 times.
In order to observe the modifications induced by extended annealing processes, we performed two supplementary treatments: an annealing process in forming gas at 400^{0}C for 2 hours, and a rapid thermal annealing process. In the fig. 6, PL spectra before and after annealing in F.G. are shown. All the investigated samples, with one or more silicon oxide films, show the same behaviour.

The peak positions are at 1,47 eV and 1,67 eV with a FWHM of 0.31 eV and 0.15 eV, respectively. During the extended annealing treatment in F.G., the intensity increases of about 1.5 with respect to that after a N_2 annealing process, but the peaks position is unchanged. The intensity modifications are due to the additional hydrogen which appears in the samples. In the fig. 7, we compare the PL spectra before and after a rapid thermal annealing treatment. It is observed an increase of the PL peak intensity and a shift to higher energy.

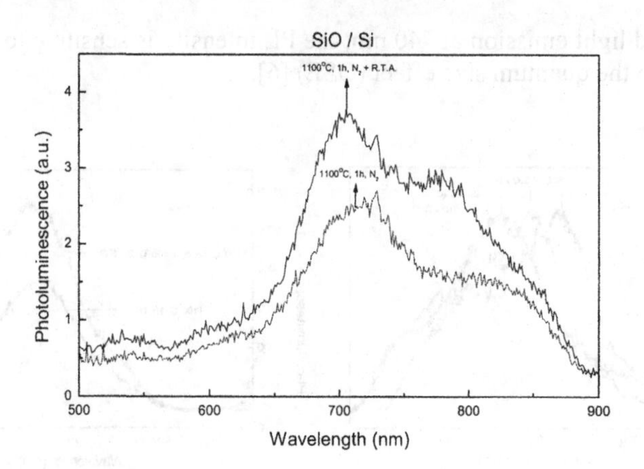

Figure 7: Photoluminescence spectra for SiO / Si sandwich structure before and after rapid thermal annealing;

4. Conclusions

Three types of silicon oxide sandwich structures: SiO/Si, SiO2/SiO/Si and SiO_2/SiO/SiO_2/Si were studied. Structural and chemical bond modifications induced by different thermal treatments were determined by IR and Raman measurements. The PL spectra of various structures were also compared. After different thermal treatments the presence of PL peaks is an evidence of the formation of Si nanocrystals. Two main tendencies were observed:

- a PL shift to higher energy by introducing of SiO_2 layer adjacent to SiO layer;
- an increase of PL peak intensity with temperature, from 1000°C to 1100°C, and by supplementary annealing treatments.

The authors acknowledge the support of the European Community through the INCO-COPERNICUS program, contract SBLED No. 7037 / 1998.

References

1. Tong, S., Liu, X., Gao,T., Bao, X., (1997) Intense violet-blue photoluminescence in as-deposited amorphous Si:H:O films, Appl. Phys. Lett. 71 (5), 698 – 700;
2. Kanemitsu, Y., (1996) Mechanism of visible photoluminescence from oxidized silicon and germanium nanocrystallites, Thin Solid Films 276, , 44-46;
3. Makimura, T., Kunii, Y., Ono, N., Murakani, K., (1998) Silicon nanoparticles embedded in SiO_2 films with visible photoluminescence, Appl. Surface Science 127-129, 388-392;
4. Komoda, T., Weber, J., Homewood, K.P., Hemment, P.L.F., Sealy, B.J., (1996), Effect of forming gas on the photoluminescence from nanocrystalline silicon formed by Si^+ implantation into SiO_2 matrix, Nuclear Instruments and Method in Physics Research B 120, 93-96.
5. Inokuma, T., Wakayama, Y., Muramoto, T., Aoki, R., Kurata, Y., Hasegawa, S., (1998) Optical properties of Si cluster and Si nanocrystallites in high-temperature annealed SiO_x films, J. Appl. Phys. 83 (4), 2228-2234;
6. Umezu, I., Yoshida, K., Sugimura, A., Inokuma, T., Hasegawa, S., Wakayama, Y., Yamada, Y., Yoshida, T., (2000) A comparative study of the photoluminescence properties of a-SiO_x film and silicion nanocrystallites, J. Non-Crystalline Solids 266-269, 1029-1032;

STIMULATED EMISSION IN SILICON NANOCRYSTALS

Gain measurement and rate equation modelling

L. DAL NEGRO, M. CAZZANELLI, Z. GABURRO, P. BETTOTTI, L. PAVESI
INFM and Dipartimento di Fisica, Università di Trento
Via Sommarive 14, 38050 Povo Italy
F. PRIOLO, G. FRANZÒ, D. PACIFICI
INFM and Dipartimento di Fisica e Astronomia,
Università di Catania, Corso Italia 57, I-95129 Catania, Italy
F. IACONA
CNR-IMM, Sezione di Catania,
Stradale Primosole 50, I-95121 Catania, Italy

1. Optical gain in silicon nanocrystals: Introductory remarks

Silicon is the electronic material *per excellence*. Integration and economy of scale are the two keys ingredients for the silicon technological success. Silicon has a band-gap of 1.12 eV, which is ideal for room temperature operation, and an oxide (SiO_2), which allows the processing flexibility to place today more than 10^8 devices on a single chip. The continuous improvements of silicon technology have made possible to grow routinely 200 mm single silicon crystals at low cost, and even larger crystals are now under development. The high integration levels reached by the silicon microelectronic industry have permitted high-speed device performances and unprecedented interconnection levels [1]. However, today the required interconnections between devices are sufficient to cause critical propagation delays, over heating and information latency. To overcome this *interconnection bottleneck* is together the main motivation and opportunity for silicon Microphotonics, where attempts to combine photonics and electronic components on a single Si chip or wafer are strongly pursued. In addition, photonics aims to combine the power of silicon microelectronics with the advantages of photonics. In this way it is expected that the continuous increase of chip performances predicted by Moore's law can be ultimately faced.

Silicon Microphotonics has boomed in the recent years [2-5]. Several silicon photonics devices have been demonstrated, *e.g.*, silicon based optical waveguides with extremely low losses and small curvature radii [3], tunable optical filters, fast switches (ns) [6] and fast optical modulators (GHz) [7], fast CMOS photodetectors [8], integrated Ge photodectors for 1.55 μm radiation [5,9]. Micromechanical systems or photonic crystals have been demonstrated, and switching systems are already commercial. On the other hand, the main limitation of silicon photonics is the lack of any practical Si-based light sources, either efficient light emitting diodes (LED) or Si lasers.

L. Pavesi et al. (eds.), Towards the First Silicon Laser, 145–164.

Silicon is an indirect band-gap material, thus the dominant light emission mechanism is a phonon-mediated process with low probability. Typical spontaneous recombination lifetimes are in the ms range. In standard bulk silicon, competitive non-radiative recombination rates are much higher than the radiative ones, and most of the excited electron-hole pairs recombine non-radiatively. This yields very low internal quantum efficiency ($\eta_i \approx 10^{-6}$) for bulk silicon luminescence. In addition, fast non-radiative processes such as *Auger* or *free carrier* absorption severely prevent population inversion for silicon optical transitions at the high pumping rates needed to achieve optical amplification. Despite of all, during the nineties many different strategies have been employed to overcome these materials limitations [4]. For example, low dimensional silicon consists in nanostructures where the material electronic properties are modified by quantum size effects. A steady improvement in silicon LED performances has been achieved and silicon LEDs are now only a factor of ten out of the severe market requirements [10,11]. In addition at the end of 2000 and during 2001 many breakthroughs have been demonstrated showing that this field is very active and still promising. The main future challenge for silicon Microphotonics still remains the demonstration of a silicon-based laser action to hopefully engineer a silicon laser.

In 2000 it was demonstrated that highly packed silicon nanocrystals (Si-nc) exhibit positive optical gain with broad gain spectra around 800nm [12]. These results have been confirmed in [13], where it was also shown the very fast dynamics of stimulated emission. Even laser oscillations in the visible have been very recently claimed for Si-nc and spatially coherent radiation with speckle patterns have been reported [14]. Exploiting very small Si-nc, stimulated emission in the blue has been observed by two different groups [15,16]. Other groups are still working towards a better understanding of the physics involved and trying to confirm these observations. In this paper we report on our recent work.

2. Critical review of the VSL technique

Some doubts have been raised about the reliability of the VSL technique to measure optical gain in silicon nanocrystals. The aim of this section is to make clear that, despite that the difficulties of the VSL have to be carefully considered, accurate optical gain values can be extracted from VSL measurements. In the VSL method, the sample (either transparent or absorbing) is optically excited by an intense laser, which is focused by a cylindrical lens to form a narrow stripe on the sample surface. The length ℓ of the stripe can be varied through a variable slit. An amplified spontaneous emission (ASE) signal $I_{A.S.E.}$ is collected from the edge of the sample as a function of ℓ (Fig. 1). As a result of population inversion achieved at high pumping rates, spontaneously emitted light is amplified and an intense, and partially coherent ASE signal grows up exponentially with the excitation length. Distinctive characteristics in $I_{A.S.E.}$ appear, since the ASE bandwidth is appreciably narrower than that of spontaneous emission and exhibits a soft threshold behaviour [17,18]. In addition, the presence of ASE shortens significantly the luminescence decay time [13,16,19,20]. It is customary to interpret the VSL data within a one dimensional amplifier model [20]

$$I_{A.S.E.}(\ell) = \frac{J_{sp}(\Omega)}{g_{\text{mod}}}\left(e^{g_{\text{mod}}\ell} - 1\right) \qquad (1)$$

where J_{sp} is the spontaneous emission intensity emitted in an appropriate emission solid angle Ω and g_{mod} is the net modal gain of the material. The net modal gain is $g_{mod}=\Gamma g_m-\alpha$, where Γ is the optical confinement factor of the waveguide (Fig. 2), g_m the material gain and α the propagation losses. For simplicity of notation, we put $g_{mod}=g$ hereafter. From a fit of the experimental data with equation (1), g can be deduced for every wavelength within the spontaneous emission spectrum.

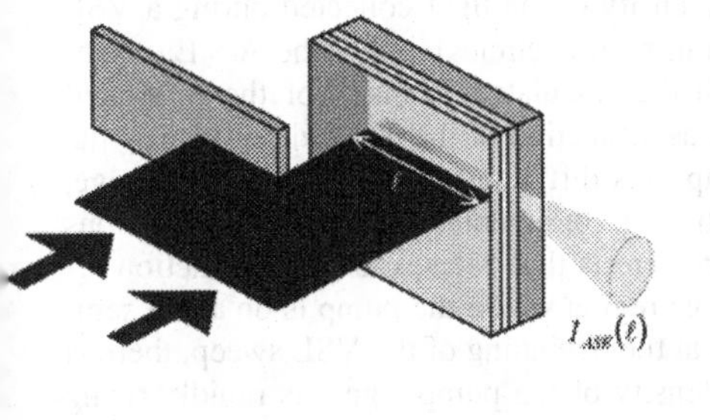

Figure 1. Sketch of the variable stripe length configuration. The amplified spontaneous luminescence intensity $I_{A.S.E.}$ is collected from the edge of the sample as a function of the excitation length ℓ. The laser beam is focused on a thin stripe by a cylindrical lens

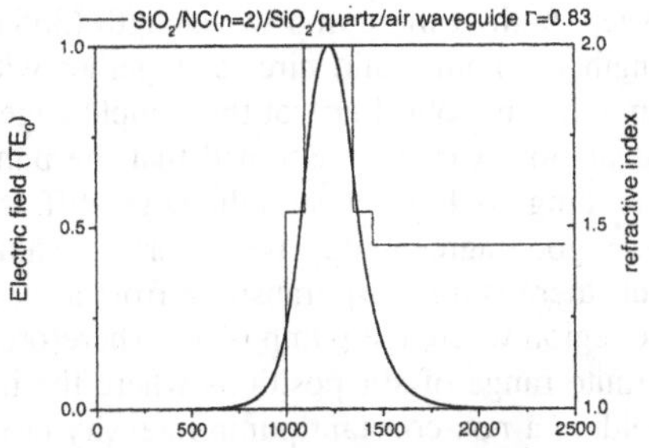

Figure 2. Spatial mode profile of the waveguide structure for one representative NS sample 1A. Calculation has been performed by assuming a NS refractive index of 2 and an effective 4 layers planar waveguide.

Equation (1) assumes that g is independent of ℓ. This condition is verified only if the gain-length product $g\ell$ is small enough to avoid gain saturation. A quantitative estimation of $g\ell$ can be derived by the following relation [20]

$$g\ell = \log\left(\frac{\lambda^5 I_{sat}}{\pi h c^2 \Omega \Delta\lambda}\right) \qquad (2)$$

where λ is the emission wavelength, $\Delta\lambda$ is the emission linewidth, I_{sat} the saturation intensity, and Ω the ASE solid angle. The maximum gain-length product yielding unsaturated gain follows directly from the condition $I_{A.S.E.}=I_{sat}$, within a four level one dimensional amplifier model. For our experimental conditions we deduce a working constraint $g\ell < 10$. Similar estimations can also be deduced by using the Lindford formula for inhomogeneous broadened lasers [17,18].

$$I_{A.S.E.} = \eta I_s \left(\frac{\Omega}{4\pi}\right) \frac{(G-1)^{\frac{3}{2}}}{(G \ln G)^{\frac{1}{2}}} \qquad (3)$$

where G is the single pass gain $G = \exp(g\ell)$, and η the luminescence quantum yield. For our experimental conditions we deduce $g\ell < 15$. For higher values of $g\ell$, the application of Eq. (1) is not justified.

A second important requirement for the validity of the Eq. (1) is that the excitation intensity must be constant over the whole ℓ-range. In addition to monitor the laser intensity on the sample, one has to consider also diffraction effects caused by the edge of the variable slit. The irregular edge of the sample also complicates the interpretation of the results. Fresnel diffraction of the pump beam at the slit edge yields a non-constant pump intensity profile on the sample surface [21]. A way to characterise this diffraction profile is to look at the diffused pump light. This measurement does not require additional experimental set-up, as it can be performed like a VSL sweep collecting, in the 90 degree direction, the light of the pump itself, instead of the A.S.E signal, as a function of the pumping length ℓ. Figure 3 shows the intensity of the light collected during a VSL sweep, both at the pump wavelength (365 nm, hollow circles) and at the A.S.E. wavelength (750 nm, solid circles) together with the calculated intensity of the diffracted pump beam (solid line), at the sample edge, as a function of the slit width [19]. For the calculation, we have assumed that the pump gets diffracted by the variable slit edge, according to Fresnel straight edge diffraction. Experimental points and calculations show good agreement. The important feature here is that because of the diffraction effects there is no sharp transition from a sample region where the pump is on and a sample region where the pump is off. Therefore, at the beginning of the VSL sweep, there is a finite range of slit positions where the intensity of the pump signal is rapidly rising yielding a non-constant pump intensity profile on the sample surface. We must discard the VSL A.S.E. data that fall into this range of slit positions because there will be necessarily a rapid increase of A.S.E. signal (almost exponential) associated to the rapid increase of the pump intensity. Blind application of Eq. (1) within that diffraction region would lead to fake gain artefacts.

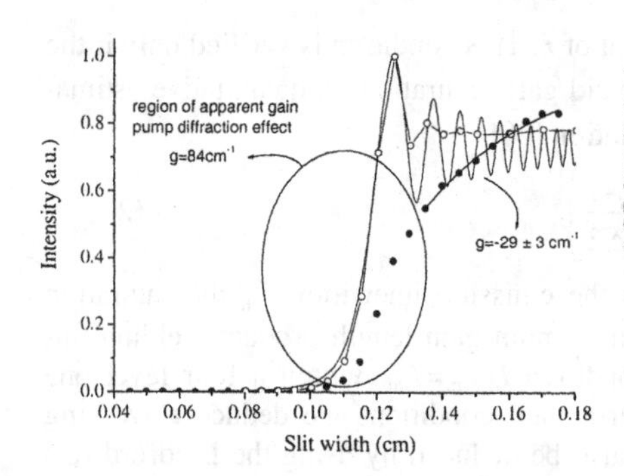

Figure 3. Edge emission (solid dots) and diffracted pump intensity profile (hollow dots) as a function of the slit width. The sample was the 5A. Laser wavelength: 365 nm. Laser intensity: 3 kW/cm^2. Variable slit was 7 cm from the sample surface. Data were fitted with Eq. (1) and gain was extracted in two different regions.

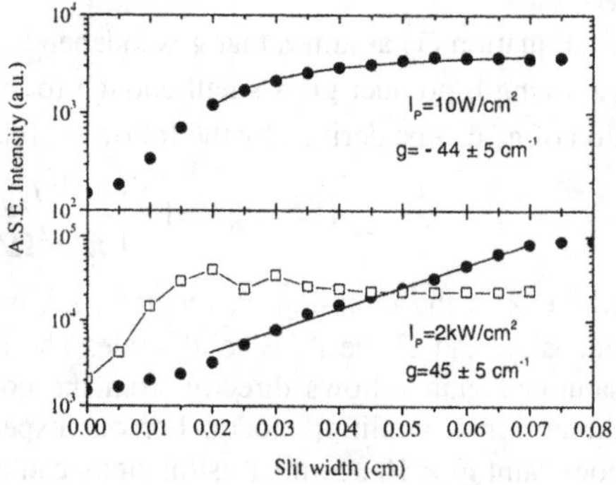

Figure 4. Amplified spontaneous emission A.S.E. (black dots) and scattered pump laser (hollow squares) as a function of the slit width. The sample was a SiO2/Si-nc/SiO2 waveguide with the same structure as the sample used for Fig. 3, but temperature of thermal treatment was 1200°C. The A.S.E. data were collected at the fixed detection wavelength of 800nm. Data were fitted with Eq. (1).

In Figure 3, the pumping intensity can be considered constant on the sample only for $\ell > 0.13$ cm, where the zero of the ℓ scale is arbitrary. A fit of the experimental data with Eq. (1), in the region where diffraction effects are dominant, yields an artificial positive gain coefficient of 84 cm^{-1}. However, this artefact has a very peculiar signature, which is its independence on the pump intensity on a wide pumping range. This is clearly a non-physical characteristic since population inversion, and hence optical gain, strongly depends on the pumping conditions. On the contrary, if the data are fitted for $\ell > 0.13$ cm negative gain values (*i. e.* optical losses) of – 29 cm^{-1} can be correctly deduced. Outside the critical range of diffraction, VSL yields very accurate measurements. Another example is shown in Fig. 4, where losses at low pumping power and gain at high pumping power were measured [22]. It is worth noticing that the gain and loss values extracted from the fit with Eq.1 are stable with respect to the fitting procedures, *i.e.*, they do not depend on the subset of ℓ used in the fitting, as long as they lay outside the range affected by diffraction. Other approaches can be followed to minimise the diffraction problems in VSL measurements, such as by using lenses to image the slit width on the sample or by positioning the variable slit directly on the sample surface [23]. However, the proposed *pump diffraction analysis* is easily applied in all the situations, and helps in taking into a proper consideration many diffraction effects.

A third important issue concerning VSL measurements is that of the constant coupling of the guided emission mode with the detection optics. The coupling efficiency must be independent of the pumping length ℓ in order to apply Eq. (1). In Ref. [23], it was pointed out that, when the numerical aperture (NA) of the collecting lens does not match the one of the waveguide, artefacts in VSL measurement can easily arise. To check the constancy of the coupling the "shifting excitation spot" (SES) technique has been proposed. This method was originally introduced to measure the optical losses in laser waveguide structures [24,25]. We applied this technique to our measuring conditions and noticed a further effect, which accounts for some observations of Ref. [23]. In our measurement, a silicon nanocrystal waveguide with a core thickness of w_1=250 nm was excited by a laser spot of 80 μm diameter and of 365 nm wavelength (UV extended Argon laser). The laser spot was formed by passing the beam through a movable aperture just after the cylindrical lens used in the VLS. Collection of the luminescence guided by the sample was performed via a 40X microscope objective with NA=0.65, the same aperture that we used for the VSL measurements. The numerical aperture of the waveguide was estimated to be $NA \approx \lambda/\pi w_1$ =0.44, and hence all the light emitted from the sample edge can be collected in our experiments. Figure 5 shows SES and VSL measurements performed on the same sample. In SES measurement (central panel) the luminescence signal decreases exponentially when sweeping the laser spot from the sample edge. An effective loss parameter α can be extracted which is related to both absorption and propagation losses in the waveguide. Moreover, this α value coincides within the experimental errors with the optical loss value measured from VSL when using a low pump power (bottom panel in Fig. 5). This is a direct prove that in our experimental condition the coupling efficiency both in VSL and SES is constant. The inset of Fig. 5 shows the wavelength dependence of the losses: at short wavelengths almost constant losses are measured while for large wavelengths the losses decrease.

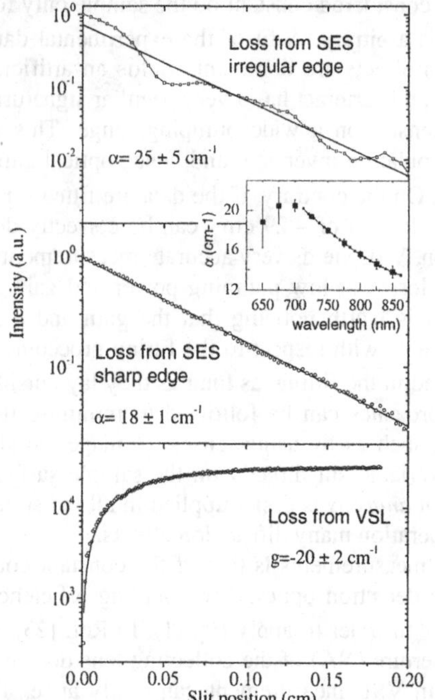

Figure 5. Edge emission intensity versus slit position/width when performing shifting exciting spot (SES) measurements (top and middle panel) and VSL measurements. Data are relative to sample of Fig. 3. The top panel shows a SES measurement on an irregular edge of the sample; the centre panel shows a loss measurement obtained on a sharp edge; the bottom panel shows the result of a low power VSL measurement on a similar quality edge. Fit of the experimental data with the Beer's law for SES data or with Eq. (1) for VSL data are reported on the graph. The inset shows the wavelength dependence of the losses measured by SES on a sharp edge of the sample.

Figure 6.Top view of the waveguide facets obtained by a scanning electron microscope (SEM). The top photograph shows an example of very irregular edge shapes yielding SES oscillations (see Fig.5 top). The bottom panel shows a good example where the smoothness of the sample edge has been improved by a cleaning procedure. In this last case very reproducible SES results can be obtained (see Fig. 5 middle).

The quality of the sample edge significantly affects the shape of the measured SES curve: complex oscillatory behaviour is possible for damaged edge (see top panel in Fig. 5). Fit to the average decrease yields larger losses than those measured from the same waveguide but with a better edge quality.

Ref. [23] reported that the SES signal can even increase, when sweeping the excitation spot away from the sample edge. We observed this effect only either when the optical axis of the waveguide formed an angle different from zero with respect to the optical axis of the collecting lens (geometrical misalignment effects), or when the focal plane of the objective was well within the sample (Fig. 7). The latter situation deserves some careful discussion. An optical objective is characterised by the numerical aperture NA and by the depth of field *D*. In the visible, $D \approx \lambda / NA^2$ where λ is the wavelength. Objectives with low NA can have significant D. Non-constant coupling and artefacts in SES or VSL measurements can arise due to the use of objectives with small NA when the focal plane of the objective is placed within the sample. Sharp increase of the luminescence when sweeping the exciting laser spot away from the sample edge could be due to the increasing overlap between the focal plane of the objective and the exciting spot itself. This is demonstrated in Fig. 7, where SES measurements are performed for different locations of the objective focal plane. When the focal plane is placed on the sample edge, an exponential decrease of the signal is measured. When the focal plane is moved well into the sample, the luminescence initially increases due to the increased overlap between the laser spot and the focal plane of the objective and then decreases again. Note the almost symmetric shape observed when the focal plane is well within the sample. Similar increases in the signal are also observed in VSL measurements when no careful alignment is performed. A clear check on the validity of VSL data is their power dependence (Fig. 4).

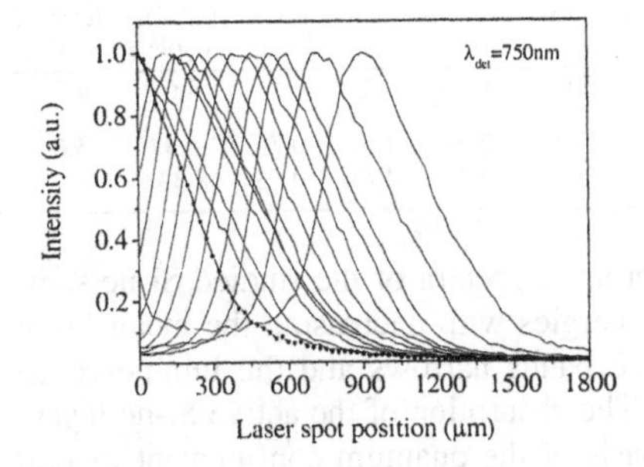

Figure 7. Edge emission versus slit position for different place of the collecting objective focal plane. The focal plane was moved from the sample edge into the sample with a constant step of 50 μm. Emission was collected at a detection wavelength of 750nm. The dotted line refers to the SES measurement performed with the focal plane on the sample edge. The sample was the same as in Fig. 4.

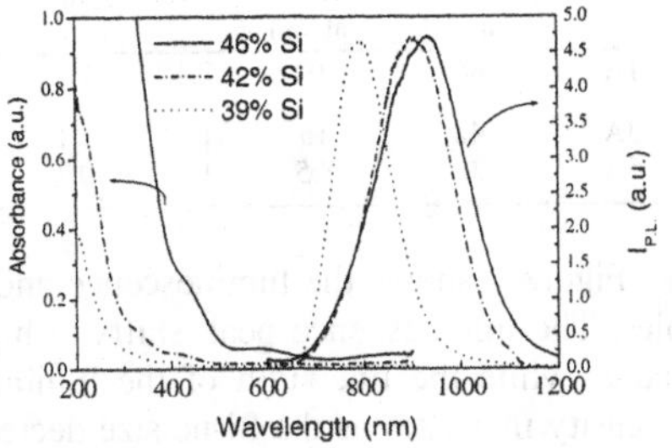

Figure 8. Room temperature absorbance and luminescence spectra of the NS samples annealed at 1250°C. The luminescence spectra were excited by the 488 nm Argon line at a pumping power of 5mW and recorded by a germanium detector. Absorbance was measured by a standard UV-visible double beam spectrophotometer.

3. Optical gain in PE-CVD silicon nanocrystals

We studied Si-nc samples produced by high temperature annealing of substoichiometric silicon oxide (SiO_x) thin films grown by plasma enhanced chemical vapor deposition (PECVD). The structural and luminescence properties of such systems have been fully discussed in [26]. Here we focus on a set of three different samples produced with an annealing temperature at 1250 °C, for one hour and with different total Si content in the deposited oxide: 46 at. % (named 1A), 42 at. % (named 3A) and 39 at. % (named 5A). Some of their characteristics are reported in Table 1. The annealing temperature at 1250°C maximizes the PL intensity for a fixed annealing time of 1h. The oxide layer containing Si-nc was 250 nm thick and was embedded between two 100 nm thick stoichiometric SiO_2 layers to form a waveguide in order to perform VSL measurements accurately. The waveguide was formed on a transparent quartz substrate. An example of the optical mode profile and of the refractive index profile is shown in Fig. 2. By such simulation the optical mode confinement factor Γ can be computed. The mean radii of the investigated Si-nc sample, the FWHM of the size distribution, the linear refractive indexes, the waveguide Γ factors, the Si-nc volume density and the measured gain cross sections are all summarized in Table 1.

TABLE 1. Principal structural and optical parameters of the samples. Total Si in Si-nc refers to the total atomic percentage of Si in silicon nanocrystals measured by X-ray absorption measurements [27]. $R_{Si\text{-}nc}$ and $\Delta R_{Si\text{-}nc}$ are the mean value and the width of the radii distribution of the Si-nc measured by TEM, λ^{PL}_{max} and $\Delta\lambda^{PL}$ the wavelength of the maximum and width of the luminescence emission band, I^{PL} the luminescence intensity in arbitrary units for a fixed low pumping power, n the refractive index of the samples measured by m-line measurements at 633 nm, Γ the optical confinement factor, ρ the Si-nc density in the sample estimated by X-ray measurements and σ_g the Si-nc gain cross section at 750 nm.

Name	Total Si in the film (at. %)	Si in Si-nc (at. %)	$R_{Si\text{-}nc}$ (nm)	$\Delta R_{Si\text{-}nc}$ (nm)	λ^{PL}_{max} (nm)	$\Delta\lambda^{PL}$ (nm)	I^{PL}	n	Γ	ρ (cm^{-3}) $x10^{18}$	σ_g (cm^2) $x10^{-17}$
1A	46	19	2.1	1.2	938	215	0.86	2	0.83	4.6	4.3
3A	42	13	1.7	1.1	906	188	3.39	1.82	0.76	6.3	3.6
5A	39	8.5	1.5	0.7	795	136	4.68	1.66	0.63	8.33	-

Figure 8 shows the luminescence and absorbance spectra of the studied Si-nc samples. The luminescence peak shifts to higher energies with decreasing the mean Si-nc radius while the linewidth of the luminescence bands narrows and the luminescence intensity increases as the Si-nc size decreases. The absorption of the active Si-nc layers increases at high energy as expected on the basis of the quantum confinement model. Remarkable is the big Stokes-shift between absorption edges and emission peaks. No appreciable absorption has been measured in the wavelength region where the luminescence maxima occur. A large debate on the origin of the luminescence is still present. Many theoretical [28-32] as well as experimental [33-35] works attribute the luminescence to radiative recombinations of excitons trapped in localized states formed at the Si=O bonds (silanone) present at the interface between the Si-nc and the SiO_2 matrix.

We have performed CW excitation VSL measurements by using an UV extended Argon laser (λ_{exc}=365 nm, maximum power up to 600 mW). The laser beam has been focused through a cylindrical lens on a straight line about 60 μm wide and 1.5 mm long. A movable slit has been used to vary the length of the excited sample region which coincided with the waist of the gaussian laser beam [36]. This yields a reduced accessible length range of roughly 0.8 mm. The amplified spontaneous emission spectra have been collected through a double monochromator and detected by a photomultiplier operating in photon counting mode.

Fig. 9 and 10 show different VSL curves obtained on samples 1A and 3A at different detection wavelengths and at different pumping intensities. The maximum measured peak net modal gain (in sample 3A) was g=52 ± 5 cm^{-1} at 760 nm for 3kW/cm^2 (see Fig.10). Note that the peak gain wavelength is blue-shifted by 100 nm with respect to the luminescence peak wavelength (Table 1). The modal gain spectrum for this sample is wide and flat: an appreciable gain coefficient of 30 cm^{-1} has been measured at 860 nm, too. The excitation intensity dependence of the modal gain allows estimating a transparency threshold of 0.5 kW/cm^2. From Eq. (1), the modal gain spectrum can be evaluated by using

$$g = \frac{1}{\ell}\left[\ln\left(\frac{I_{ASE}(2\ell)}{I_{ASE}(\ell)} - 1\right)\right] \tag{4}$$

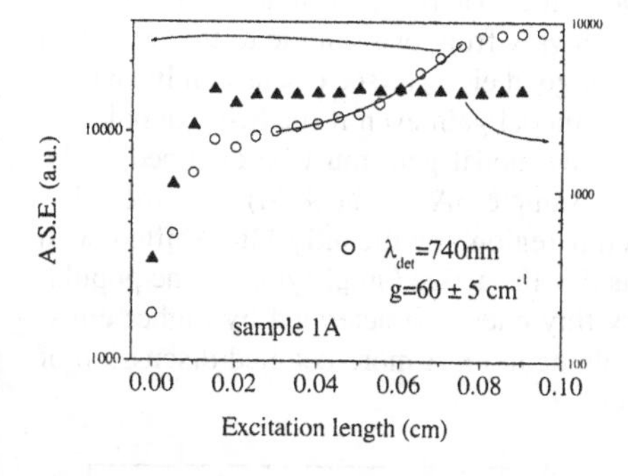

Figure 9. Room temperature VSL curves. Symbols refer to data, lines are a fit. Sample 1A. VSL at 740 nm (open circles) at pumping intensities of 3kW/cm^2. The gain coefficient extracted from the fit is 60 ± 5 cm^{-1}. The black triangles are the laser beam diffraction curve measured at 365 nm.

Figure 10. Room temperature VSL curves on sample 3A as a function of the pumping intensities. The detection wavelength was 760nm. Switching from 0.05kW/cm^2 pumping intensity to 1kW/cm^2 the optical losses turn into optical gain. The values of optical gain became saturated at an intensity of 3kW/cm^2.

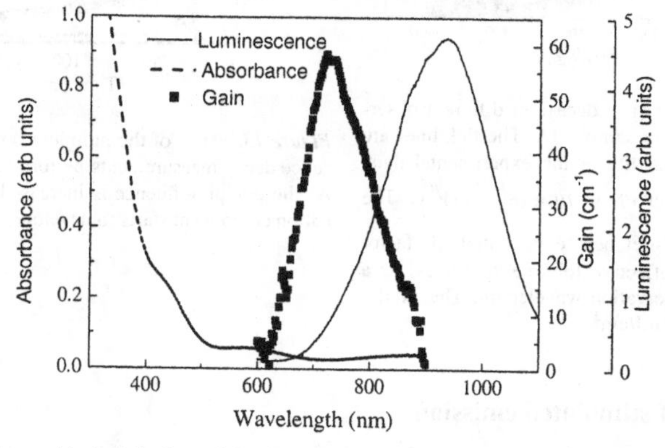

Figure 11. Absorbance (dashed line), modal gain spectrum (solid squares) and luminescence spectrum (solid line) for sample 1A.

Figure 11 compares the modal gain spectrum of sample 1A with the luminescence spectrum of the same sample. Modal gain values as high as 60 cm^{-1} have been measured at 730 nm for an excitation intensity of 3kW/cm^2. A blue-shift of the gain with respect to the luminescence is observed also here. The modal gain linewidth is 120 nm wide while the luminescence linewidth is 215 nm. Table 1 summarizes the VSL measurements on the three samples. The optical gain is given as the nanocrystal gain cross-section, σ_g, which is determined from the net modal gain by using

$$\sigma_g = \frac{g}{\Gamma \rho} \tag{5}$$

where ρ is the Si-nc density and Γ the optical confinement factor. Different wavelengths show different σ_g. Note that VSL measurements are sensitive only to the net modal gain of the material, namely the difference between the material gain and the waveguide propagation losses. Since propagation losses can vary from one sample to the other and even within the same sample if the edges are not good enough (see discussion in chapter 2) it is not always possible to measure net VSL modal gain even if the Si-nc population is inverted. In fact in a steady state condition the modal gain must be clamped at the losses value. We think that this is the case of sample 5A (39 at.% Si). For the other samples we measured optical gain in a spectral region significantly blue-shifted with respect to the luminescence one. This suggests that the active (amplifying) Si-nc population can be just the small fraction of the very tiny ones, characterized by higher emission cross sections and increased surface to volume ratio. A more detailed discussion of likely emission models will be given in chapter 5.

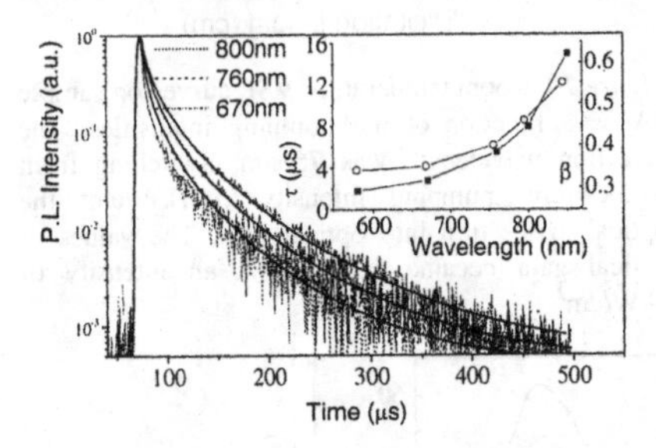

Figure 12. Luminescence decays at different observation wavelengths of sample 3A. The tick lines are stretched exponential fits of the experimental data, by using the formula $I(t) = I(0)\exp\left(-(t/\tau)^{\beta}\right)$. The inset shows the stretched exponential β factor (squares) and the effective lifetime τ (circles) as a function of the observation wavelength. The excitation fluence was 0.5mJ/cm²

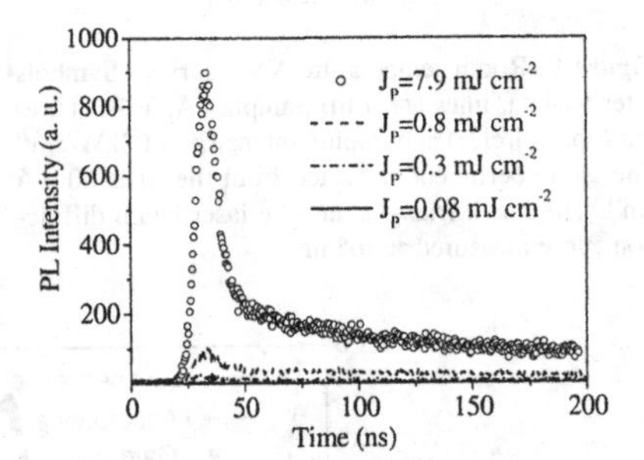

Figure 13. Effect of the pumping power on luminescence decay measurements performed on sample 3A. As the pumping fluence is increased a fast recombination component starts to develop.

4. Dynamics of stimulated emission

To investigate the stimulated emission dynamics, time-resolved experiments were performed both in a standard 45 degree photoluminescence configuration and in the usual 90 degree VSL configuration (TR-VSL). The former was used to measure stimulated emission lifetimes directly when the modal volume V_a is high to allow the observation of sizeable effects (see section 6). The latter is used to measure the stimulated emission build-up time as the signal photons are amplified along the waveguide axis. Propagation losses should be considered in this case to compare results obtained with the two geometries. We used high fluencies short optical pulses (6 ns, 10 Hz, 430 nm) produced by an optical parametric oscillator pumped by the third harmonic or the third harmonic

(4 ns, 355 nm) of a Nd-YAG pulsed laser to excite the samples. The estimate of the fluence is affected by an error of about a factor 3. We detected the radiation through a single grating spectrometer and a Hamamatsu Streak camera with 10 ps resolution.

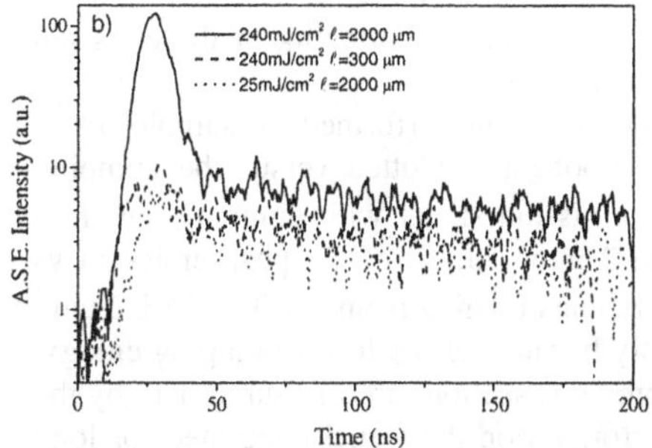

Figure 14. Pumping length ℓ and pumping fluence dependence of A.S.E. decay for sample 3A. The excitation length ℓ and the excitation fluency are reported in the inset. Note that the intensity of the slow decay tails is almost the same in all the measurements, since the slow recombination component is strongly saturated in the high pumping condition yielding population inversion. Optical losses of about 20-30cm^{-1} can in fact be extracted from the slow component behaviour versus the pumping length.

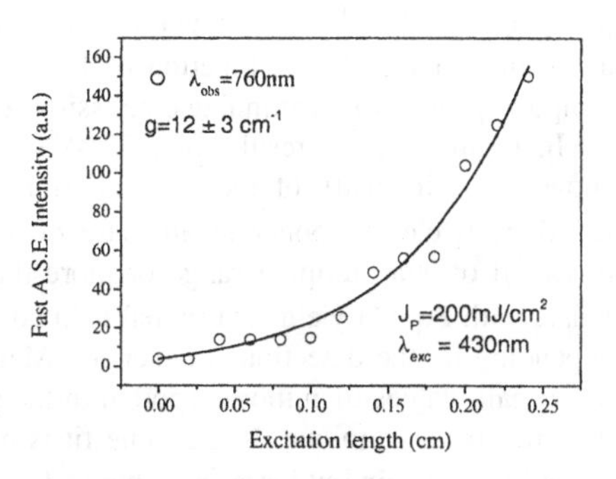

Figure 15. Points: Amplified Spontaneous Emission (A.S.E.) peak intensity at 760nm versus the excitation length and for a pump fluence of 200mJ/cm^2. Full line: fit of the experimental data with Eq. (1) which yields a net modal gain values of 12 ± 3 cm^{-1}. Modal gain values ranging between 8 cm^{-1} and 20 cm^{-1} are measured depending on the detection wavelength in a wavelength region extending from 700nm to 850nm.

Figure 12 shows the luminescence decays at low fluence (J_p) obtained on sample 3A (42 at.% Si). The decay curves are stretched exponentials, with observation energy dependent lifetimes in the range of several μs, as already reported [37].

However at high J_p, time resolved measurements in VSL configuration show a new fast component (full width at half maximum, FWHM, of about 10ns) which is superimposed to the usual *slow* component (Fig.13). Some very peculiar characteristics of the fast component are: i) it disappears when either the excitation length ℓ is decreased at a fixed J_p or when J_p is decreased for a fixed ℓ (see Fig. 14); ii) it shows a super linear increase vs ℓ for high J_p which can be fitted with the usual one-dimensional amplifier equation yielding a net optical gain of 12 ± 3 cm^{-1} at 760nm (Fig. 15); iii) it shows a threshold behavior in the emitted intensity (Fig. 16): at low J_p the emission is sublinear to a power 0.5 in a double logarithmic scale suggesting a strong Auger limited regime, while for higher J_p population inversion is achieved and a superlinear increase to a power ≈3 is measured, suggesting the onset of the stimulated regime; iv) the FWHM of the fast component signal decreases with J_P and a further sharp decrease is observed when the stimulated emission regime is entered, as shown in Fig. 17 ; v) the FWHM decreases at high J_P as a function of the the excitation length ℓ (Fig. 18); vi) it is strongly blue-shifted with respect to the standard CW and to the slow luminescence of Si-nc (Fig. 19).

Figure 14 shows that, in the VSL configuration, a stronger fast component (15 ns) appears for long pumping slit (2000 μm) *and* high fluence (240 mJ/cm^2). The intensity of such fast component decreases if either the slit width (Figure 14) or the fluence (Figure 13) are decreased. Fast component in the luminescence decay of heavily photo excited porous silicon has been attributed to Auger recombinations. Auger recombinations are indeed able to explain the appearance of a fast non radiative recombination at high pumping rates but they can explain neither the superlinear increase of the VSL signal as a function of ℓ and as a function of J_p, nor the significant shortening of the emission temporal profile when stimulated emission starts to set in.

In Figure 15, the results of a TR-VSL measurement performed on sample 3A is shown. The intensity of the fast emission component is plotted versus the pumping length ℓ. A clear exponential increase of the emission signal with pumping length is observed over a pumping range of more than 2 mm. The fit of the peak emission vs ℓ data with Eq. (1) yields net modal gain coefficients ranging from 8 ± 3 to 20 ± 2 cm^{-1} depending on the detection wavelength. Mainly because of the lower pumping energy, lower gain values than those reported in the previous sections are measured here by the time resolved technique. If the same fit is performed on the signal integrated for long times (μs), no gain but losses are obtained.

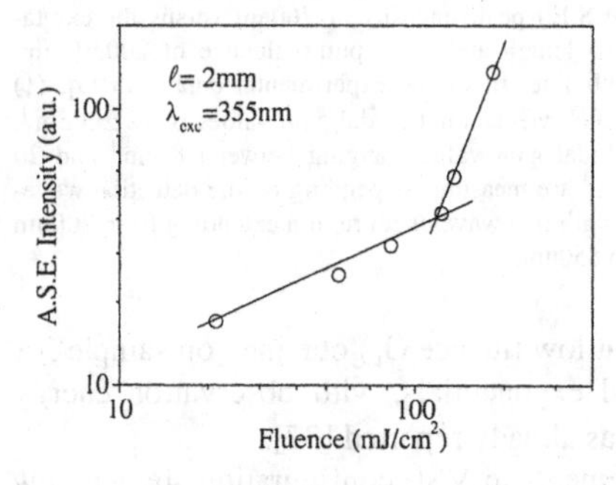

Figure 16. Amplified spontaneous emission peak intensity versus the pumping fluence for the sample 3A. The sample has been excited at a fixed excitation length of 2 mm. A clear threshold appears separating two different recombination behaviors. A sub-linear emission regime typical of Auger recombinations is present under the threshold while a superlinear emission regime is characteristic of stimulated emission above threshold. Here the threshold fluence and superlinear slope are different than in [19] because of different sample, pumping wavelength and focusing conditions.

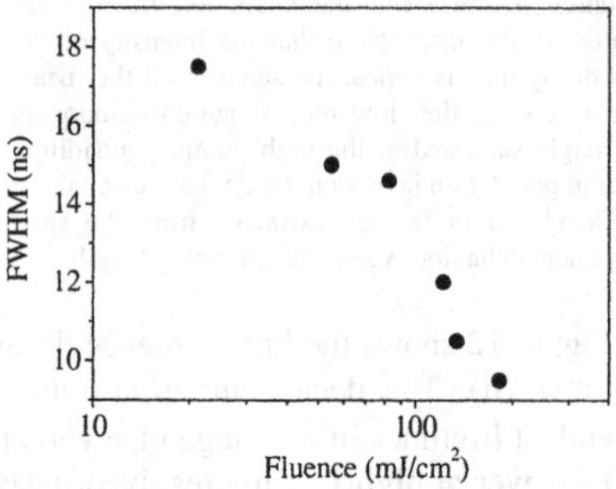

Figure 17 FWHM of the ASE fast component as a function of the pumping fluence. The FWHM values are extracted from a gaussian fit. As soon as the stimulated emission threshold is reached, a sharp decrease of the emission time width sets in. Same sample and excitation condition as in Fig. 16.

In Fig. 18 we show the full width at half maximum (FWHM) of the fast ASE decay component as a function of the pumping length. The FWHM has been extracted by a gaussian fit of the fast ASE peak. As expected on the basis of a more refined VSL theory, where all the recombination dynamics also depend on the spatial coordinate ℓ, the stimulated emission rate must be dependent on the pumping length. The fast component becomes faster as ℓ is increased (see Fig. 18), because of the propagation of the photon flux in the waveguide structure that consistently speed up the emission rate. Anyway, a detailed and simple theory of space-time VSL problems is still lacking and we will not be lured here to enter this subtle realm.

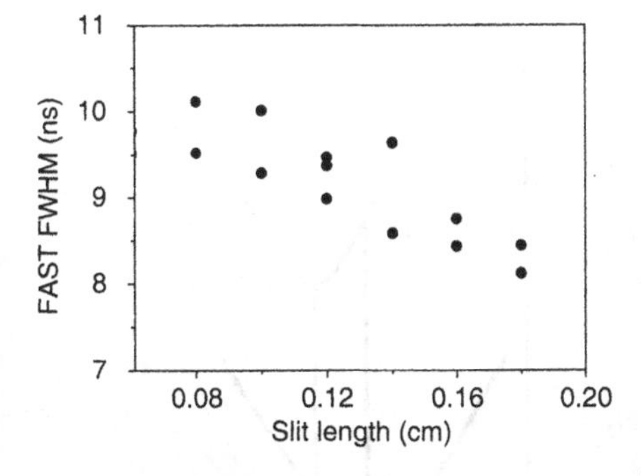

Figure 18. Full width at half maximum of the fast component of the A.S.E. decay measured for various excitation lengths on sample 3A. Pumping wavelength was 430 nm. The scatter of the points gives an indication of the errors in the measurements. The excitation fluency was 110 mJ/cm^2. Minor differences with the data reported in [19] are due to a different definition of FWHM.

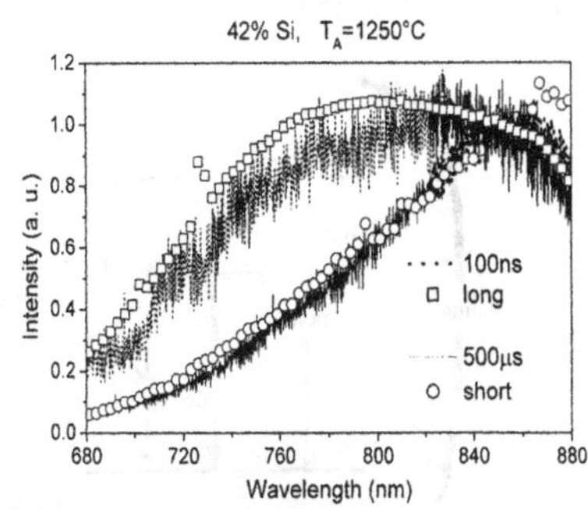

Figure 19. The solid lines are the lineshape of the time resolved luminescence at a fluence of 15 mJ/cm^2 measured and time integrated for a time interval of 500 μs or of 100 ns after the arrival of the exciting pulse. The symbols refer to the amplified spontaneous emission lineshape for an excitation length of 100 μm or of 600 μm measured in the CW VSL experiment for the 3A sample.

In Figure 19 we show the measured spectral dependence of the fast and slow components in time resolved luminescence. Because of the reduced propagation losses, in a time resolved luminescence experiment stimulated emission lifetimes can be measured at considerably lower fluencies than in TR-VSL. On the other hand, since the effective photon propagation length is much smaller the rate of increase above threshold is weak. A strong blue-shift of the Si-nc light emission is observable in the early nanoseconds after the pulse excitation (fast component). The comparison with the CW-VSL measurements, also reported in Fig. 19, shows that the fast component has a spectral shape similar to the A.S.E. band measured for long ℓ (where strong amplification is expected) whereas the slow component spectrally overlaps the ASE signal measured for short ℓ (where negligible amplification is expected). The spectral matching between the PL fast component and the long-slit A.S.E. band also suggests that the fast PL component measured with high fluence optical pulses is originated by stimulated emission in Si-nc and that stimulated emission lifetime can be measured directly.

5. The microscopic gain picture: towards a theoretical understanding

The observation of optical gain in both ion implanted [12] and PE-CVD formed Si-nc demonstrates that this effect is not exclusively related to the Si-nc preparation processes but it is more likely an intrinsic property of the Si-nc and their interfaces. Unfortunately, the key material parameters that enter directly in the optical gain physics are not so simplistic to be fully mastered yet, and additional researches must be oriented to investigate this intriguing subject. Despite of a clear understanding of the optical gain mechanism is still lacking, the field is under rapid improvement and other groups reported evidences of light amplification in Si-nc related systems (see other contributions in this book and [13-16]).

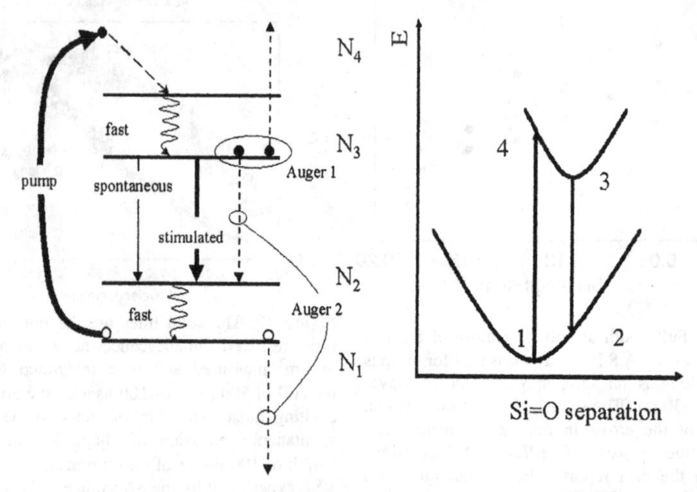

Figure 20. Effective four level system that has been introduced to model qualitatively the recombination dynamics under gain conditions. Two different kinds of Auger recombinations are considered and can be studied on the basis of the rate equations of the relaxation dynamics.

Figure 21. Schematic of the energy configuration diagram of the silicon nanocrystals in an oxygen rich matrix. Localised radiative states are formed inside the nanocrystal band-gap by the interface oxygen atoms. The excited nanocrystal state can occur at a different lattice coordinate with respect to the ground state. Level labelling refers to transitions in Fig. 20.

Here, we propose an effective four-level model (Figure 20) to treat qualitatively the strong competition among losses, Auger recombination and stimulated emission. This is still a phenomenological model, which does not refer to a developed theory of the optical properties of Si-nc in SiO_2 and of their interfaces. However, we can suggest a possible nature for this four-level model.

From x-ray absorption studies and ab-initio DFT calculations [27], strong evidences emerge that there is a transition region between Si-nc and the oxide matrix, which constitutes a shell of stressed silica all around Si-nc. The thickness of this region is estimated about 1 nm, in very good agreement with X-ray experiments [27]. This stressed SiO_2 could enhance the formation of oxygen-related states, like silanone bonds [28-32] at the interface between Si-nc and SiO_2 or at small Si inclusion in SiO_2. The energetic of silanone-like bond as a function of the Si=O interatomic distances is typical of a four level scheme, with different local surface atoms rearrangements for the Si-nc ground and excited state configurations, originating both a significant Stokes shift between absorption and luminescence and a wide energy dispersion of the related states [28,31,32,]. Within this scheme, photon excitation induces a strong structural relaxation either of the small H-saturated and O-saturated nanocrystals leading to new transitions involving surface localized states or of the localized Si=O bonds. A sketch of such energy-configuration diagram is shown in Fig. 21. In this picture, levels 1 and 4 are associated with absorption transitions in the Si-nc ground state configuration while levels 2 and 3 are associated to the localized states in the excited state configuration. Within this model, the broad and blue-shifted gain spectrum of Si-nc could be understood in terms of molecular-like inhomogeneous broadening mechanisms, such as different atomic surface configurations or strain fields, acting on small interface localized atoms or small silicon inclusions efficiently pumped by Si-nc through energy transfer. It remains clear however that accurate theories must still be developed for a detailed explanation of the relevant Si-nc gain physics.

6. Rate equation treatment including ASE and Auger

We solve now the system of rate equations associated with the four-level scheme. In this way we can take into account the severe competition of stimulated emission with Auger recombinations. In fact, as in other quantum dot based systems [55,56] a severe competition with efficient non-radiative processes, such as Auger and Excited State Absorption, is present in Si-nc which causes very fast dynamics in the optical gain. We consider the following major features of our experiments: i) an almost complete absence of optical absorption at the peak gain emission for all the amplifying Si-nc samples, Fig. 8; ii) a moderate low pumping threshold of about 0.5 kW/cm^2 for population inversion, under steady state conditions; iii) time resolved VSL with fast recombination component and power threshold behavior, which depends on the excitation slit length, iv) strong competition of the stimulated emission rate against the strong Auger dynamics that set in when two excited electron-hole pairs are present within the same Si-ns. We must stress again that such a nanosecond recombination dynamics is typical of Auger non-radiative processes in porous Silicon [57,58] and hence the fast dynamics of its own is not enough to claim for stimulated emission. Anyway, a strong competition between Auger fast processes and stimulated emission can still be present in Si-nc, and can be properly accounted for only within a four level recombination scheme. Indeed, in some of our samples, Auger recombinations prevail over stimulated emission. As an example, sample 5A shows a fast recombination dynamics in the ns range, but, unlike the amplifying samples 1A and 3A, no exponential increase of the VSL signal, or intensity threshold, have been measured. On the contrary, negative gain of the order of -5cm^{-1} has been found even at the maximum available pumping fluence. The loss behavior of sample 5A can be explained by the low refractive index of its light guiding layer (Tab.1). Because of the resulting low modal confinement, losses overcome the modal gain. It is necessary to probe different samples and the influence of all the different parameters before claiming that gain or no gain is measured

Items i) and ii) and iv) can be explained within the effective four-level model reported in Fig. 20. We assume that levels 2-4 are empty before the excitation occurs. Once the pumping starts, levels begin to be populated and two different Auger non-radiative recombination processes can in principle take place. The first mechanism consists in an electron relaxation from energy level 3 to level 2 with the energy given to a second electron which is also present in the same level 3 and which is promoted to higher lying levels in the conduction band, from which a very fast relaxation to level 4 occurs. This Auger process involves two electrons in the same level, and therefore the rate of such a process depends quadratically through a coefficient C_{A1} on the level population N_3. Another Auger mechanism involving an electron in the third level and a free hole in the valence band edge (level 1) can in principle occur, where the hole is sent deep in the valence band and then very rapidly relaxes again to the band edge. This process, has to be proportional, through a coefficient C_{A2}, to the product of the population of the emitting level 3 (N_3) and the hole concentration in the valence band edge (N_h). N_h equals the total concentration of electrons in the various excited levels, that is $N_h = N_2+N_3+N_4$. Order of magnitude estimate of the Auger coefficient for Si-nc can be obtained from the Si bulk value ($\approx 10^{-30}$ cm^{6}s^{-1}) divided by the Si-nc density ($\approx 10^{19}$cm^{-3}) which yields $C_A \approx 10^{-11}$ cm^{3}s^{-1}. As expected in low dimensional systems, the data are simulated with a larger effective $C_A \approx 10^{-10}$ cm^{3}s^{-1}. Within the four level scheme we are proposing, the relaxation times of electrons from levels 4 and level 2 are so fast that N_4 and N_2 are always almost empty. Therefore $N_h \approx N_3$. Hence, the following set of coupled rate equation has to be integrated:

$$
\begin{aligned}
\frac{dN_1}{dt} &= -\sigma_P \phi_P(t) N_1 + \Gamma_{21} N_2 \\
\frac{dN_2}{dt} &= \frac{N_3}{\tau} - \Gamma_{21} N_2 + B n_{ph}(N_3 - N_2) + (C_{A1} + C_{A2}) N_3^2 \\
\frac{dN_3}{dt} &= -\frac{N_3}{\tau} - B n_{ph}(N_3 - N_2) + \Gamma_{43} N_4 - C_A N_3^2 \\
\frac{dN_4}{dt} &= C_{A1} N_3^2 + \sigma_P \phi_P(t) N_1 - \Gamma_{43} N_4 \\
\frac{dn_{ph}}{dt} &= V_a B n_{ph}(N_3 - N_2) - \frac{n_{ph}}{\tau_{ph}} + \beta \frac{N_3}{\tau_R}
\end{aligned}
\tag{6}
$$

where N_i represent the level population densities (i=1,..,4), σ_P is the absorption cross section at the wavelength of the pump, ϕ_P is the time dependent pumping photon flux, Γ_{ij} are the relaxation rates from the i to the j energy levels, τ is the total lifetime of the emitting level N_3, B is the stimulated transition rate which implicitly contains the gain cross section σ, n_{ph} is the emitted photons number, V_a is the optical mode volume, τ_{ph} is the photon lifetime, β is the spontaneous emission factor and τ_R is the radiative lifetime of N_3. C_A is an effective Auger coefficient equal to $2C_{A1} + C_{A2}$, taking into account both of the two particles Auger processes (*e-e* or *e-h*) and affecting the dynamics of the emitted photons.

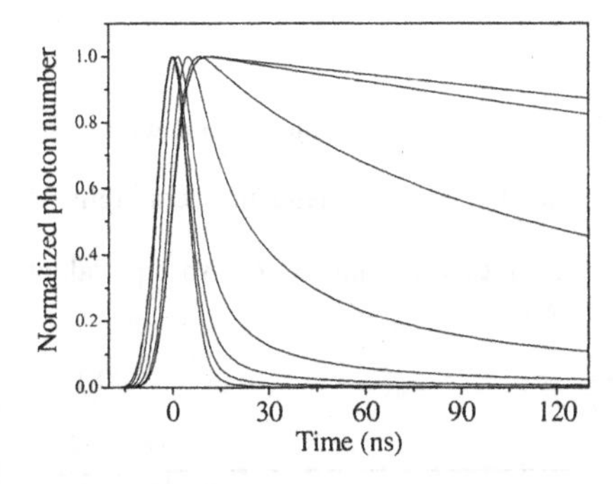

Figure 22. Simulations of the normalized PL intensity as a function of the incident photon flux ϕ_P. The peak of the incident photon flux ϕ_P was varied between 10^{19} and 10^{25} photons $s^{-1}cm^{-2}$. The main parameters used in the simulation were the pump absorption cross section $\sigma_P=10^{-15}cm^2$, the emission cross section $\sigma=10^{-17}cm^2$, the active center concentration $N=6x10^{17}$ cm^{-3}, the spontaneous emission factor $\beta=4.5x10^{-4}$, the optical losses $\alpha=3cm^{-1}$. These values were mainly extracted from experimental data or from the literature. No Auger recombination has been considered here. Other model parameters used in the simulations are the recombination rate $1/\tau=10^6$ s^{-1}, the intermediate state ultra fast recombination rate Γ_{43} and Γ_{21} both fixed at 10^{15} s^{-1}, the radiative recombination rate $1/\tau_R=10^{-3}s^{-1}$, and finally the sample physical thickness $d=10^{-4}cm$.

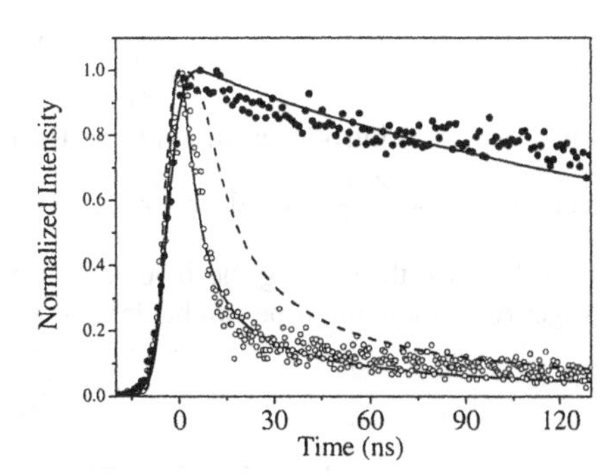

Figure 23. Time decay of the photoluminescence signal measured at 0.4 mJ/cm^2 (black circles) and at 1 mJ/cm^2 (open circles). Pumping wavelength 430 nm. The solid lines are the simulations of the emitted photon obtained by solving the set of rate equations. In the simulations we used an effective Auger coefficient $C_A=50x10^{-11}$ cm^3s^{-1} (yielding an Auger lifetime of 6 ns at the peak fluence), $\phi_P=3x10^{20}$ photons $s^{-1}cm^{-2}$ and $\phi_P=3x10^{23}$ photons $s^{-1}cm^{-2}$ while all the other parameters are the same as in Figure 22. The dashed line shows the effect of the Auger lifetime shortening when the stimulated emission is neglected. The effective Auger coefficient has been set to $C_A=30x10^{-11}$ cm^3s^{-1} (yielding an Auger lifetime of 6.46 ns at peak fluence) to have the same Auger lifetime as in the previous simulation. The peak incident photon flux is again $\phi_P=3x10^{23}$ photons $s^{-1}cm^{-2}$.

The effect of pumping on the recombination dynamics is illustrated in Fig. 22 where the normalized number of emitted photons is simulated by solving the set of rate equations as a function of the pumping rate, and by neglecting for simplicity the Auger coefficients. The set of parameters used is a reasonable one, while no systematic attempt has been made to fit existing data to derive the parameters. A fast recombination component builds up as the pumping rate becomes high enough to create population inversion. A more realistic comparison among the measured luminescence decays and the numerical simulations is reported in Fig. 23 and 24, where also the contribution of the Auger recombination has been included in the simulations. At this stage of the development of our model no comparison with TR-VSL data is possible. Good qualitative agreements are observed for a reasonable set of simulation parameters both at low and high pumping rates. Note that such agreement cannot be reached when stimulated emission is turned off, unless the Auger contribution is set extremely large ($C_A=900x10^{-11}$ cm^3s^{-1} with an Auger lifetime of 70 ps at peak fluence). Within the four level recombination model it turned out that an emission (gain) cross section of the order of 10^{-17} cm^2 is large enough to compensate for Auger processes with typical lifetime as fast as 2-10 ns.

It is possible to observe optical gain whenever the stimulated emission rate is greater than the Auger recombination rate. From Eq. (6), we define a stimulated emission lifetime as:

$$\tau_{se} = \frac{1}{Bn_{ph}} = \frac{4}{3}\pi R_{nc}{}^{3}\frac{1}{\xi\sigma c n_{ph}} \tag{7}$$

where we have introduced the volume fraction of the emitting center ξ and where the relation, $B = \frac{\sigma c}{V}$, valid under the assumption of monochromatic incident light, is used. It is worth noticing the inverse dependence of τ_{se} on ξ and on σ. An equivalent Auger recombination time can be defined as follows:

$$\tau_A = \frac{1}{2C_A N_3} \tag{8}$$

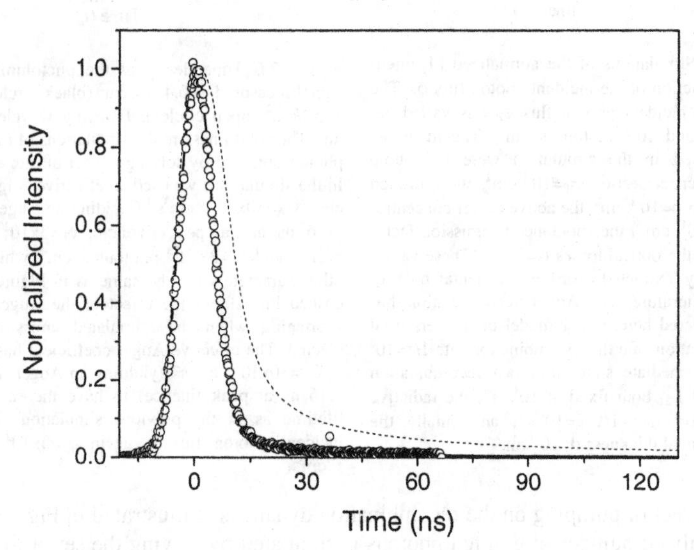

Figure 24. Experimental luminescence decay (open circles) at a pumping fluence of 15 mJ/cm^2. The pumping wavelength was 430 nm. The lines are simulation with (full line) or without (dashed line) positive optical gain. In the simulation we considered ϕ_P=3x10^{24} photons s^{-1}cm^{-2} and C_A=30x10^{-11} cm^{3}s^{-1} yielding an Auger lifetime of 2.3 ns at peak fluence. The physical sample thickness is d=0.3x10^{-4}cm both in the experiment and in the simulation. The density of emitting centers and the emission cross section that yield the best agreement with experimental data are respectively N=6x10^{18}cm^{-3} and σ=1.5x10^{-17}cm^2. All the other simulation parameters are the same as in Figure 22 and 23.

It is clear from the discussion that to observe optical gain $1/\tau_{se} \geq 1/\tau_A$; or equivalently, if we define a competition factor $C = \tau_A/\tau_{se}$, $C \geq 1$. This sets a precise requirement on the volume fraction, enlightening the role of Si-nc density on optical gain. The proposed rate equation model fits qualitatively the experimental data. Fig. 24 shows an example. It turns out that, in our best samples, gain cross section of the order of 10^{-17} cm^2 are large enough to compensate for Auger processes even when they are as fast as 1 ns.

A problem still unsolved in these simulations is the large photon flux increase needed to reproduce the experimental trend in Fig. 23 and 24.

7. Conclusions

Gain has been measured on a set of PECVD Si-nc samples produced by high temperature annealing at 1250°C of SiO_x thin films with different silicon contents. Continuous-wave UV excitation VSL measurements show a low threshold (0.5 kW/cm^2) positive optical gain of the order of 50 cm^{-1} at pumping intensities of 3 kW/cm^2. Large Stokes-shift between luminescence and gain has been found. High power time-resolved luminescence and VSL have been performed, showing the onset of stimulated emission. Superlinear light emission and stimulated emission lifetime shortening have been demonstrated. A four-level model which includes amplified spontaneous emission and Auger processes has been introduced and a simple criterion for the onset of optical gain in Si-nc samples has been proposed. Good qualitative agreement with measured time-resolved decays has been found.

Despite the large quantity of data here presented and the refined model we show, many points have still to be clarified on the optical gain. Main aspects that require further work are: i) the optimized material parameters and waveguide geometry, ii) pump and probe experiments in waveguide configuration, iii) the role of nanocrystal interaction, iv) the role and composition of the SiO_2 embedding matrix, v) the modeling of TR-VSL, vi) the precise nature of the four levels in the model, in particular the location and role of Si-O bonds, vii) a theoretical prediction for all the main parameters of the four level model.

This work has been supported by the National Institute for the Physics of Matter (INFM) through the advanced research project RAMSES. We acknowledge fruitful discussion with S. Ossicini and his theoretical group.

References

1 International Technology Roadmap for Semiconductors, Interconnect (http://public.itrs.net.), 2000 Update.
2 Soref R. A., Proc. of IEEE vol. 81, p. 1687 (1993).
3 Kimerling L. C., Appl. Surf. Science, 159–160, 8–13 (2000).
4 Bisi O., Campisano S. U., Pavesi L., and Priolo F., Silicon based microphotonics: from basics to applications (Amsterdam: IOS press) (1999).
5 Masini G., Colace L., and Assanto G., Mat. Science Eng. B89, 2–9 (2002).
6 Irace A., Coppola G., Breglio G., and Cutulo A., IEEE J. Sel. Top. Quantum Elect. 6, 14 (2000).
7 Li B., Jiang Z., Zhang X., Wang X., Wan J., Li G., and Liu E., Appl. Phys. Lett. 74, 2108 (1999).
8 Csutak S. M., Schaub J. D., Wu W. E., and Campbell J. C., IEEE Photon. Technol. Lett. 14, 516 (2002).
9 Winnerl S., Buca D., Lenk S., Buchal Ch., Mantl S., and Xu D.-X., Mat. Science Eng. B89, 73–76 (2002).
10 Gelloz B., and Koshida N., J. Appl. Phys. 88, 4319 (2000).
11 Green M. A., Zhao J., Wang A., Reece P. J., and Gal M., Nature 412, 805 (2001).
12 Pavesi L., Dal Negro L., Mazzoleni C., Franzò G., and Priolo F., Nature 408, 440 (2000).
13 Khriachtchev L., Rasanen M., Novikov S., and Sinkkonen J., Appl. Phys.Lett. 79, 1249 (2001).
14 Nayfeh M. H., Rao S., and Barry N., Appl.Phys.Lett. 80, 121 (2002).
15 Nayfeh M. H., Barry N., Therrien J., Akcakir O., Gratton E., and Belomoin G., Appl. Phys. Lett. 78, 1131 (2001).
16 Luterová K., Pelant I., Mikulskas I., Tomasiunas R., Muller D., Grob J., Rehspringer J. L., and Hönerlage B., Appl. Phys. Lett. 91, 2896 (2002).
17 Svelto O., and Hanna D. C., "Principles of Lasers", Plenum Press (1998).
18 Svelto O., Taccheo S., and Svelto C., Optics Communications 149, 277 (1998).
19 Dal Negro L., Cazzanelli M., Daldosso N., Gaburro Z., Pavesi L., Priolo F., Pacifici D., Franzò G., and Iacona F., Physica E (2003).
20 Milloni P. W., and Eberly J. H., "Lasers", John Wiley & sons, New York (1988).
21 Shaklee K. L., Nahaory R. E., and Leheny R. F., J. Lumin. 7, 284 (1973).

22 Dal Negro L., Pacifici D., Bettotti P., Gaburro Z., Cazzanelli M., and Pavesi L., submitted to Appl. Phys. Lett. (2002).
23 Valenta J., Pelant I., Linnros J., Appl. Phys. Lett. 81, 1396 (2002).
24 Mogensen P. C., Smowton P. M., and Blood P., Appl.Phys.Lett. 71, 1975, (1997).
25 Smowton P. M., Herrmann E., Ning Y. et al., Appl.Phys.Lett. 78, 2629, (2001).
26 Priolo F., Franzò G., and Spinella C., J. Appl. Phys. 87, 1295 (2000).
27 Daldosso N. et al., Physica E (2003) in press, M. Luppi, S. Ossicini, phys. stat. sol. (a) (2003) in press.
28 Filonov A. B., Ossicini S., Bassani F., and Arnaud d'Avitaya F., Phys. Rev. B 65, 195717 (2002).
29 Wolkin M. V., Jorne J., Fauchet P. M., Allan G., and Delerue C., Phys. Rev. Lett. 82, 197 (1999).
30 Zhou F., and Head J. D., J. Phys. Chem. B 104, 9981 (2000).
31 Baierle R. J., Caldas M. J., Molinari E., and Ossicini S., Solid State Communications 102, 545 (1997).
32 Puzder A., Williamson A. J., Grossman J. C., and Galli G., Phys. Rev. Lett. 88, 97401 (2002).
33 Kobitski A. Yu, Zhuravlev K. S., Wagner H. P., and Zahn D. R. T., Phys. Rev. B. 63, 115423 (2001).
34 Kanemitsu Y., Ogawa T., Shiraishi K., Takeda K., Phys. Rev. B. 48, 4884 (1993).
35 Klimov V. I., Schwarz Ch. J., McBranch D. W., White C. W., Appl. Phys. Lett. 73, 18, 2603 (1998).
36 Pavesi L. Dal Negro L., Cazzanelli M., Pucker G., Gaburro Z., Prakash G., Franzò G., Priolo F., Proceedings of SPIE, vol 4293, 162 (2001).
37 Linros J., Galeckas A., Lalic N., Grivickas V., Thin Solid Films 297, 167 (1997).
38 Dumke W P, Phys. Rev. 127, 1559 (1962).
39 Ng W.L., Lourenço M. A., Gwilliam R. M., Ledain S., Shao G., and Homewood KP, Nature 410, 192 (2001).
40 Soref R. A., Friedman L., and Sun G., Superlattices and Microelectronics 23, 427 (1998).
41 Dehlinger G., Diehl L., Gennser U., Sigg H., Faist J., Ensslin K., Grützmacher D., and Müller E. Science 290, 2277 (2000).
42 Dehlinger I., Brunner K., Hackenbuchner S., Zandler G., Abstreiter G., Schmult S., and Wegscheider W., Appl Phys Lett 80, 2260 (2002).
43 Lynch S. A., Dhillon S. S., Bates R., Paul D. J., Arnone D. D., Robbins D. J., Ikonic Z., Kelsall R. W., Harrison P., Norris D. J., Cullis A. G., Pidgeon C. R., Murzyn P., and Loudon A., Mat. Science Engin. B89, 10-12 (2002).
44 Pavlov S. G., Hübers H. W., Rummeli M. H., Zhukavin R. Kh., Orlava E. E., Shastin V. N., and Riemann H., Appl Phys Lett 80, 4717 (2002).
45 Franzò G., Iacona F., Vinciguerra V., and Priolo F., Mat. Science Eng. B 69/70, 338 (1999).
46 Coffa S., Libertino S., Coppola G., and Cutolo A., IEEE J. Quant. Electr. 36, 1206 (2000).
47 Han H.-S., Seo S.-Y., and Shin Y.-H., Appl. Phys. Lett. 79, 4568 (2001).
48 Daldosso N., PHD Thesis, Université J. Fourier, Grenoble, France (2001).
49 Pavesi L., La Riv. Nuovo Cimento 20, 1-76 (1997).
50 Iacona F., Franzò G., Moreira E. C., Priolo F., J. Appl. Phys. 89, 8354 (2001).
51 Franzò G., Irrera A., Moreira E. C., Miritello M., Icona F., Sanfilippo D., Di Stefano G. F., Fallica F., Priolo F., Appl. Phys. A74, 1 (2002).
52 Lin C.-F., Chung P.-F., and Miin-Jang Chen Wei-Fang Su, Optics Lett. 27, 713 (2002).
53 Heikkilä L., Küüsela T., and Hedman H. P., Superl. Microstructures 26, 157 (1999).
54 Borkar S. Y., and Paniccia M., Intel Developer UPDATE Magazine 31, April 11 (2002).
55 V.I.Klimov et al., Science 290, 314, (2000)
56 A.V.Malko et al., Appl. Phys. Lett. 81, 1303, (2002)
57 Delerue C., Lanoo M., Allan G., Martin E., Mihalcescu I., Vial J. C., Romenstain R., Muller F., Bsiesy A., Phys. Rev. Lett. 75, 2228 (1995).
58 M'ghaïeth R., Maâref H., Mihalcescu I., Vial J. C., Phys. Rev. B. 60, 4450 (1999).

LASING EFFECTS IN ULTRASMALL SILICON NANOPARTICLES

MUNIR H. NAYFEH
Department of Physics, University of Illinois at Urbana-Champaign
1110 W. Green Street, Urbana, Illinois 61801 USA [m-nayfeh@uiuc.edu]

1. Synthesis of silicon nanoparticles

We pulverize crystalline silicon using an electrochemical treatment that involves gradually immersing the wafer into a bath of HF and H_2O_2 while arranging for an electrical current to skim the top skin of the wafer [1-4]. H_2O_2 catalyzes the etching producing ultra small structures and cleans impurities and produces a higher electronic and chemical quality with an ideal hydrogen termination and no oxygen [4]. This process erodes the surface layer, producing weakly interconnected nanostructures. The wafer is then immersed in an ultrasound bath [1-2], causing the fragile nanostructured network to crumble into particles. The procedure produces a family of discrete size Si_nH_x particles that are 1.0 ($Si_{29}H_{24}$), 1.67 (Si_{123}), 2.15, 2.9, and 3.7 nm in diameter which can be consequently separated [5]. This is unlike the abundance spectrum of uncapped clusters Si_n that has been known to exhibit no discrete magic numbers for n >20. The smallest four are ultra bright blue, green, yellow, and red luminescent particles respectively. A thin graphite grid is immersed in the colloid of the 1 nm particles and imaged by high-resolution TEM [3, 6] as shown in Fig. 1. Electron photo spectroscopy shown in Figure 2 shows that the particles are composed of silicon with less than 10 percent oxygen.

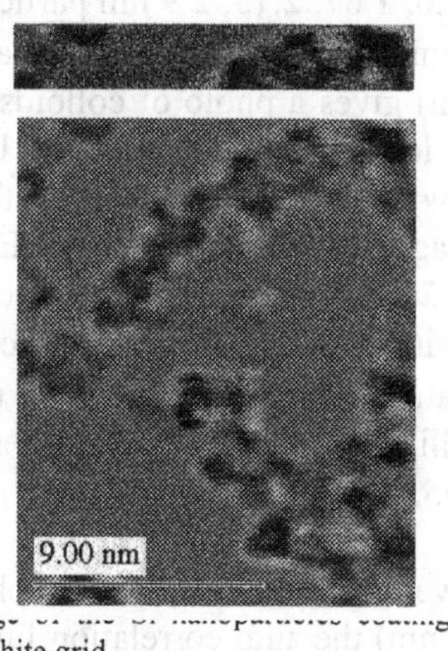

graphite grid

Counts / Second
C
O
Si

electron photo spectroscopy

L. Pavesi et al. (eds.), Towards the First Silicon Laser, 165–180.

2. Emission/detection of single nanoparticles

We prepare a colloid of the 1 nm particles. The colloid is excited by 355 nm pulsed radiation [3]. The blue emission is observable with the naked eye, in room light, as shown in Figure 3. The excitation, i.e., the absorption monitored at a specific emission wavelength (product of absorption and emission) was recorded on a photon counting spectro-fluorometer with a Xe arc lamp light source and 4 nm bandpass excitation and emission monochromator. Figure 4 gives the spectrum for excitation wavelength at 330, 350, 365, and 400 nm, showing a strong blue band that maximizes for 350 nm.

Figure 3 A photo of a Si colloid excited by 355 nm radiation.

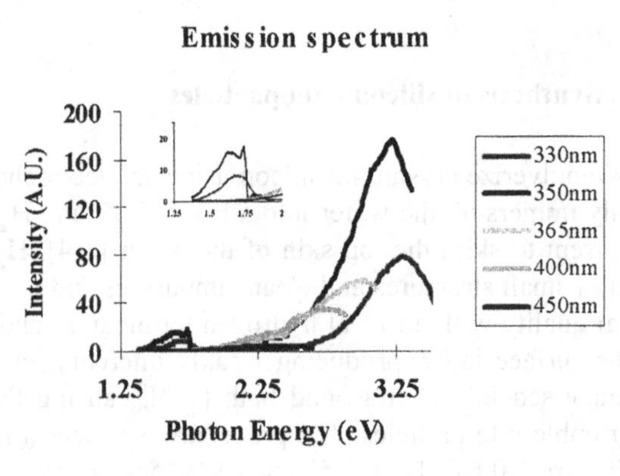

Figure 4 Emission and excitation spectra (inset) of the particles

We mapped out the excitation of the members of the family sizes in the range 250 nm and 800 nm, while monitoring the emission in the range 400-700 nm [5]. We identified the resonance excitation structure (Figure 5) of the 1.0, 1.67, 2.15, 2.9 nm particles at 3.44 ± 0.1, 2.64, and 2.39, and 2.11 eV, producing emission bands with maxima at 410, 540, 570, and 610nm respectively. Figure 6 (bottom) gives a photo of colloids of the magic sizes under the irradiation from an incoherent low intensity commercial UV lamp at 365 nm, showing the characteristic red, yellow, green, blue colors. A thin graphite grid is immersed in the colloid mixture and imaged by high-resolution TEM. Fig. 7 shows the atomic planes in close ups of the 1.67, 2.15, 2.9 and 3.7 nm particles [5]. Organization or self-assembly requires size uniformity. Under certain conditions, the particles segregate according to size upon crystallization. Over time, 5 to 100 μm crystals have formed in a water colloid. Colloidal crystallites were placed on glass and illuminated with light from a mercury lamp (at 360, 395, 450, 520, 560, 610 nm) as shown in Figure 6 (top)

Because UV is not friendly to biological molecules, we used near-infrared two photon excitation (700 to 900 nm). Figures 8-9 give (at 780 nm) the auto correlation function of the fluctuating time-series of the luminescence of the 1 nm particles with progressive dilution to demonstrate the sensitivity of the detection: 5.4 Si particles; a fluorescein standard with 1.5 molecule; 2.75 Si particles; and a single Si particle (0.75) in the focal volume. The measurements yield a particle size of ~ 1nm, consistent with direct imaging by TEM. They also yield a brightness fourfold larger than that of fluorescein [1].

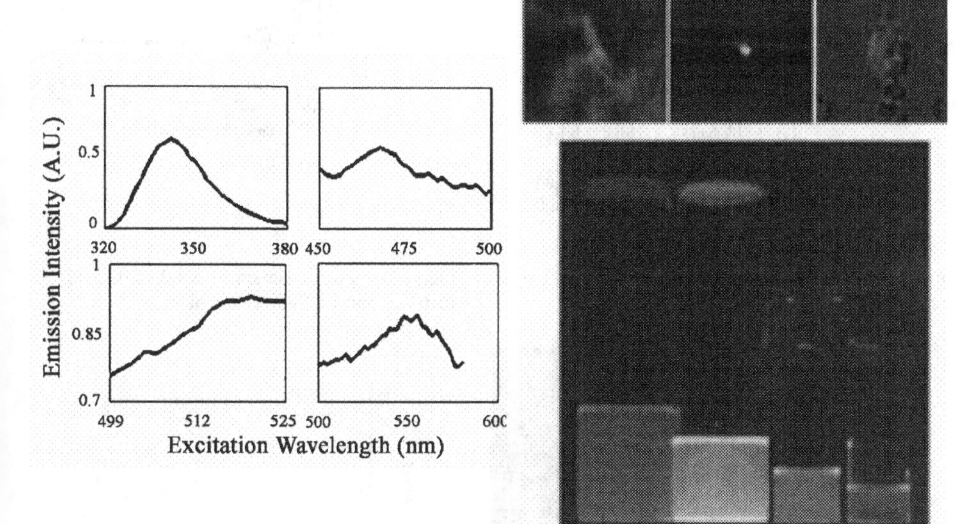

Figure 5 (top) Excitation spectra of the particles monitored at maximum emission

Figure 6 (bottom) Si colloids (top) Si crystallites excited by 355 nm radiation.

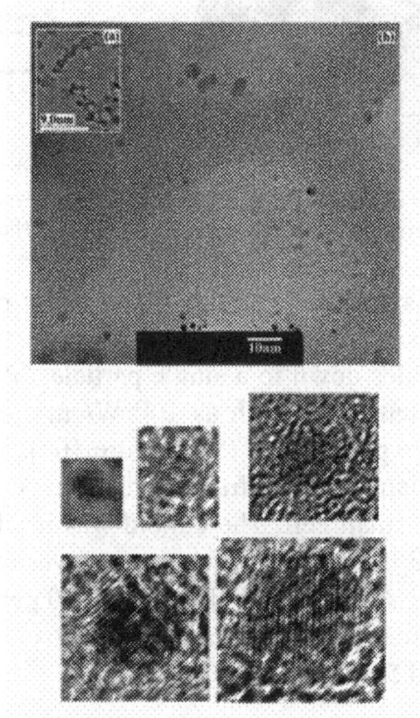

Figure 7 Transmission electron microscopy images of the Si nanoparticles

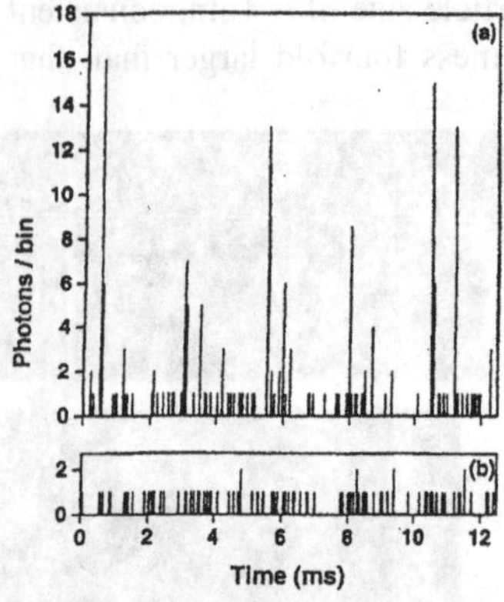

Figure 8a-b Raw traces of (a) Si particles a fluorescein standard and (b) a fluorescein standard

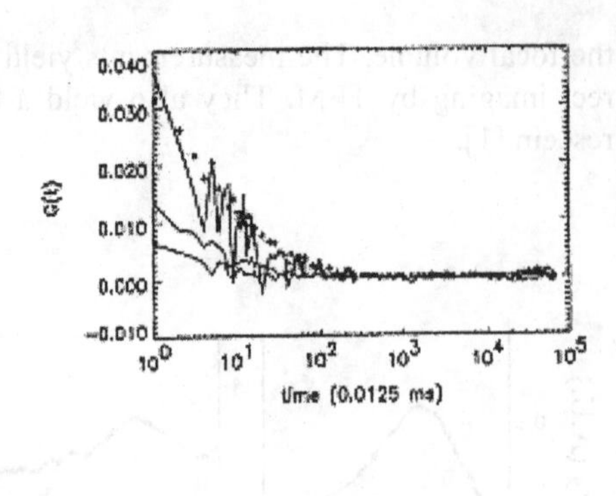

Figure 9 Auto correlation of 5.4,2.75,0.75 Si particles, and 1.5 fluorecein molecules

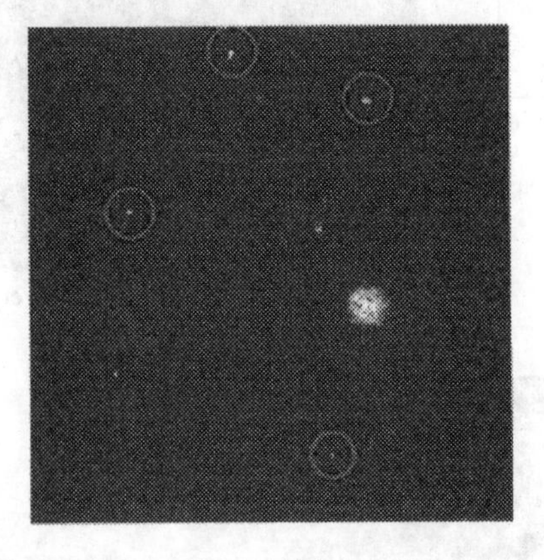

Figure 10a Luminescent images of frozen Si particles in a demonstrating the ability to observe / image single ones.

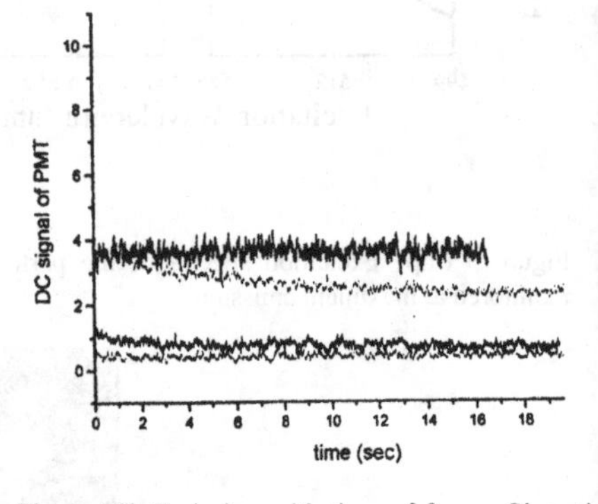

Figure 10b Emission with time of frozen Si parti in a gel while "parking" the excitation on statioı particles shows the photostablity

The photostability was tested by targeting stationery particles frozen in a gel [2]. Figure 10a gives examples of luminescent images, demonstrating the ability to observe / image of clusters of particles down to a single particle. "Parking" the excitation beam, focused to an average intensity as high as 10^6 W/cm^2, on stationery particle cluster shows that the particles are photostable (See Figure 10b). We expect the particles to be less stable under UV radiation than infrared radiation. A 1-cm^3 colloid of particles in propanol was placed in a small cell. The colloid was irradiated by 30 ns, 30 pulses/s, pulsed laser radiation at 355 nm with an average power of 20 mW and an interaction volume of 1-cm^3. The emission intensity dropped by 50 percent after 3 hours of irradiation. Under the same irradiation condition and geometry, blue dyes such as coumarine and stibline decay at 8 and 50 fold faster.

3. Solubility and Aggregation

The procedure used here to disperse silicon produces hydrogen passivated particles. The colloid dispersions of H-terminated Si-np in several organic solvents (Cl-benzene, xylenes, heptane, THF) were not stable and rapidly formed amorphous precipitates [7, 8]. A resonication restored their transparency only for short time. This poses problems for applications requiring a high particle density in solution, such as using particle solutions as optical gain media. Terminating particles with organic tails enhances their solubility and prevent aggregation. Alkylation and aminization of the particles significantly increases their solubility in organic solvents and the stability of colloid dispersions [8, 9]. Aminized and alkylated Si nanocrystals in organic solvents (heptane, THF, CH_2Cl_2) were found to remain stable for months, indicating an increase in solubility and reduction of bulk aggregation [7]. In contrast to particles alkylated with butyl amine [7] and 1-pentene [8], carboxyl acid-functionalized particles were not dissolvable in heptane. However, they completely dissolved in CH_2Cl_2 and THF and more polar methanol, forming colloids that are stable for months.

The aggregation of *the particles* was characterized by gel-permission chromatography (GPC) in THF at 365 nm. Prior to the modification GPC showed a single monodispersion but there is observed aggregation which may be explained by interconnection of the particles through H-bonding of Si-OH species formed because of partial oxidation of the surface. Contrary to the H-terminated particles, the GPC plot of particles functionalized with butyl amine (Si-NH-C_4H_9) showed a much narrower peak. This narrowing points to the absence of bulk aggregates (even after vacuum drying). The GPC of 1 pentane treated particles showed a weak broad shoulder to a sharp peak. Low intensity of this shoulder is evidence of low possibility of cross-linking for the hydrosilylation of 1-alkenes. The low polydispersity of Si-np after modification [7, 8] may have resulted from an increased resistance of alkylated Si surface of nanoparticles to oxidation, which likely is the main reason of aggregation of H-terminated Si-np. The GPC of the acid-functionalized particles dispersed in THF showed monodispersion but contained some cross-linked particle's aggregates.

N-alkyl-capping with alkylamine (Si-NH-C linkage) was achieved as follows. A chlorobenzene colloid of particles is saturated with Cl_2 gas. The dissolved chlorine gas chlorinates the particles by electrophilic replacement of hydrogen with chlorine. Distilled butylamine ($C_4H_9NH_2$) was then added to the dried particles and was heated at the end of which the dried particles were redispersed in heptane by ultrasonication and filtered. Hydrosilylation of 1-alkene (Si-C linkage) was accomplished as follows. Dispersion of ultra small Si-np in xylenes was reacted with 1-pentene; the residue was dried under vacuum and ultra sonicated in heptane, giving a clear solution. The dispersion of alkylated particles in organic solvents (heptane, THF, CH_2Cl_2) remained stable for month [7]. Nano-probe H NMR indicates that about 25% of Si-H groups were involved into hydrosilylation. Carboxyl (Si-C linkage) attachment was carried out also. A bi-functional molecule containing terminal double bond and terminal ester-protected acid group was used in a hydrosilylating process. After completion of the reaction, the ester groups were converted to -COOH groups by hydrolysis in acidic (basic) media.

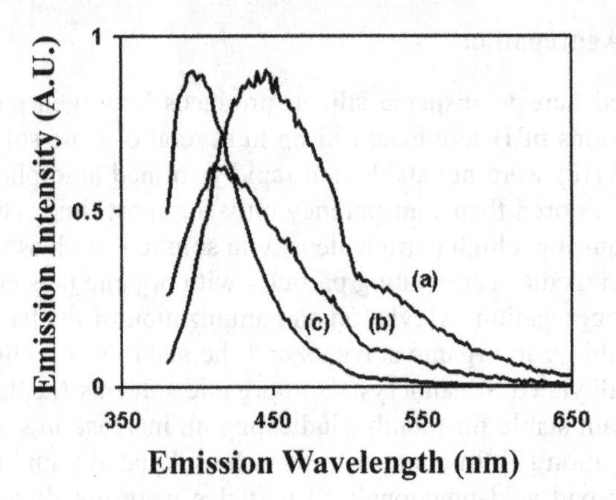

Figure 11 The unnormalized emission spectra of the dispersed particles at 365 nm, (a) Si-N linkage, (b) Si-C linkage, and (c) Si-H termination.

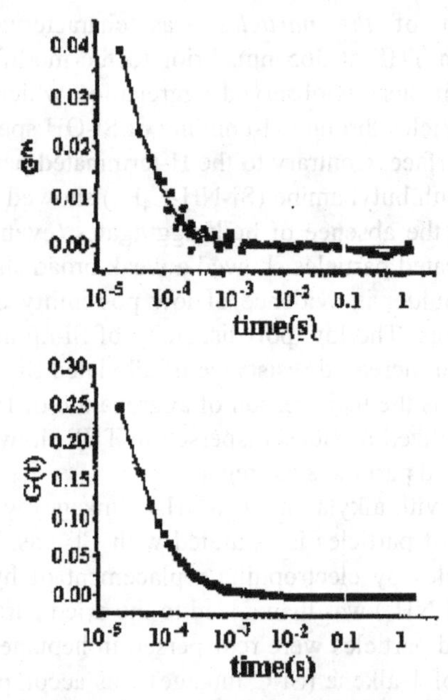

Figure 12 The auto correlation function G as a function of time derived from the emission of (a) the particles in a colloid along with that of (b) a coumarine standard under two-photon excitation using mode locked near infrared. The functions belong to one particle and 0.2 molecules in the focal volume.

Fig 11 gives the emission spectrum at 365 nm of the dispersed Si-H, Si-N, and Si-C passivated particles [9-10]. In the H-terminated case, the spectrum is dominated by a strong blue band with a tail band that extends into the visible, diminishing at ~ 600 nm. Apart from a shift in the blue band, the spectra have similar shape for the aminized particles [7]. For instance, the peak of the blue band emission of the hydrogenated particle shifts from 410 nm to 450 nm upon aminization. On the other hand, in the case of Si-C, the figure shows that the process narrows the emission band by decreasing the red wing, causing a minor blue shift of less than 5-10 nm. The HUMO-LUMO absorption edge is calculated to be 3.5 ± 0.3, 3.25 ± 0.3, and 3.55 ± 0.3 eV for H-, N-, and C-termination respectively, indicating that measured emission correlates with the band gaps [9-10]. Auto correlation fluctuation spectroscopy (FCS) (Fig 12) was used to determine the brightness after functionalization. From the photon counting histograms, we find the brightness to be two-fold smaller than coumarine.

4. Particle films

In addition to the novel properties that the individual particles provide, there is a potential to engineer additional properties by synthesizing two and three dimensional arrays of the particles. The ability to produce monosize particles gives us the opportunity to manipulate interspacing with atomic precision to tailor new elemental silicon-based material with unique optical and electrical properties. We used precipitation from a volatile solvent to reconstitute the particles into thin films [2, 11]. By gentle evaporation, micocrystallites are demonstrated on high quality Si or silicon oxide, or in free standing mode. The structures are optically clear. Optical imaging in initial experiments shows colloidal crystals of 5 to 50 μm across (Fig. 13). We experimented with slower growth rates using slower evaporation rates at reduced temperatures to produce larger, flatter, clearer, and more uniform films. Careful regulation of the temperature allows some adjustment of the destabilization.

Figure 13 Image of colloidal crystals of 5 to 20 μm across.

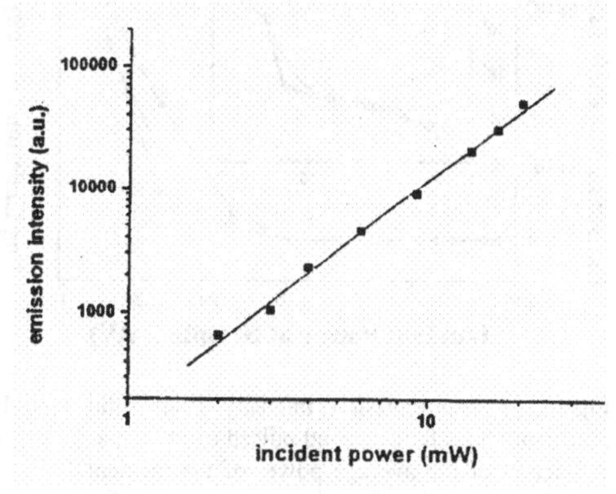

Figure 14 Power dependence of emission of dispersed particles

More recently, we developed a procedure for delivery of nanoparticles from a particle solution, using electrochemical deposition processes. Thin coatings of the particles on metals, foils, or silicon substrates are demonstrated. Modulation of the conductivity of the Si substrate using oxide masking allowed selective area deposition [12]. Addition of metal ions to the particle solution allows us to produce fluorescent composites of films of nanoparticles and metal {oxide). Scanning electron and fluorescence microscopy of the substrate may show tree-like network assembly of particles [13]. Avoidance of closed loops and preference for angle of branching of 90°-120° are observed. The building block of the tree-network is not individual particles but spherical particle aggregate of ~ 150 nm in diameter.

5. Lasing effects

We examined stimulated and lasing effects in the silicon nanoparticles under femtosecond infrared excitation. We first examined dispersed particles in a colloid [2]. We performed power dependence studies of the emission under a single-particle condition. Figure 14 shows quadratic dependence as in two-photon processes. Next we examined the dependence from aggregates of particles. Aggregates are found on the solid precursor before sonification and dispersion into individual particles. The intensity of the emission from such aggregates has a sharp threshold, with highly nonlinear emission, rising by several orders. Figure 15 shows the emission intensity under femtosecond pulsed excitation as a function of the average power of the incident radiation focused to a spot of 0.75 μm in diameter. The inset is the log-log of the data. At the threshold there is a dramatic increase in the slope of the power pumping curve from ~ 1.6 to ~ 11-12, similar to stimulated emission in inverted systems. The contrast is more dramatic in the case of CW excitation (not shown). The fluorescence is extremely weak, basically immeasurable for low powers. However, at an average comparable to the threshold intensity in the pulsed case, the CW exhibits somewhat similar threshold [14].

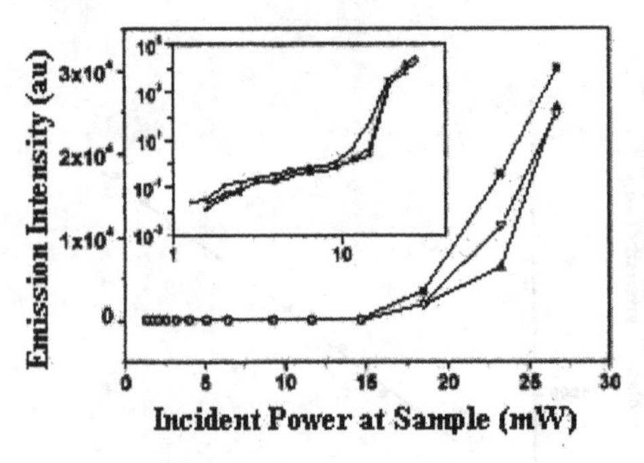

Figure 15 The emission intensity from solid precursor under femtosecond pulsed excitation as a function of the average power of the incident radiation.

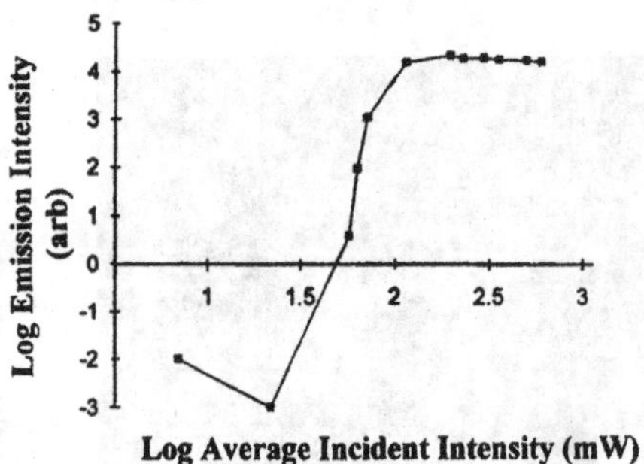

Figure 16 Emission as a function of the average incident intensity for a thin film of Si nano particles close to one side, provides a reflector of 2π solid angle, i.e., high feedback

The region that exhibits the threshold behavior on the solid precursor [14] is abundant, but it is spotty, shallow, and uncontrollable. We created large, thick, uniform and under control layers of the ultra small material by reconstituting the dispersed particles as described above. Figure 16 gives the emission intensity from such films as a function of the average incident intensity. It is typical of the response from any part of the film [15]. For low intensity, the emission is finite, but at an average intensity of ~ 10^6W/cm^2 (~ 20-25 mW), the emission exhibits a sharp threshold, rising by many orders of magnitude. Beyond the threshold, there sets in a low order power dependence that may saturate at the highest intensities.

Figure 17 displays the interaction of the incident beam with the micro crystallites [15]. It is normally incident (normal to the plane of the figure). The top four examples show that directed blue beams emerge, propagating in the interior of the crystallite, in the plane of the sample, locally normal to the closest side. Figure 17(bottom) shows two cases when the opposite faces of a crystallite are close. The beam, in this case, strikes the opposite face forming a weaker bright spot. The blue beams are characterized by a threshold. When the incident intensity is reduced, the beam fades away, and disappears, while the interaction spot remains bright (frames in Figure 18). Geometrically, the blue beam is favored when the incident beam strikes close to a face of micro crystallites.

More recently [16], we reported similar behavior at ~ 610 nm in aggregates of ultra small elemental 2.9 nm silicon nanoparticles. The aggregates are excited by radiation at 560 nm from a mercury lamp. Intense directed beams, with a threshold, manifest the emission (Fig 19-20). We observe line narrowing, Gaussian beams, and speckle patterns, indicating spatial coherence. Figure 21 gives the intensity of the red beam with the pumping intensity. Above threshold, the emission slows to a linear or sub linear response. The insets of Figure 21 give the line profile of the separated beam spot, fitted to Gaussian functions with a good result. Gaussian beams are characteristic of the output of lasers, and those that exist inside spherical mirror optical resonators.

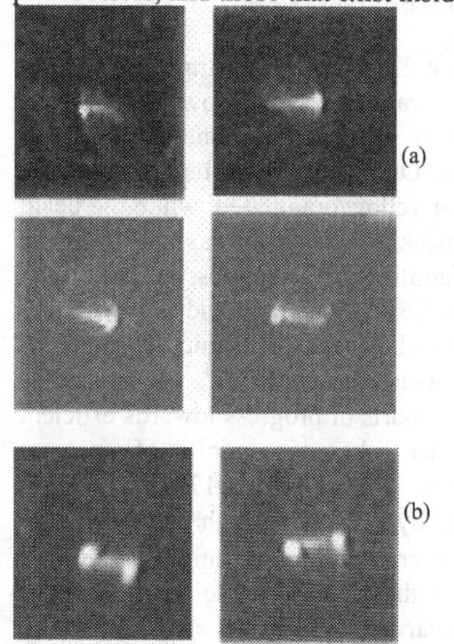

Figure 17 Photoluminescence images from the micro crystallite region. The interaction spot appears as "white" spot. (a) shows four examples of a blue beam. (b) shows two examples of a blue beam between opposite faces.

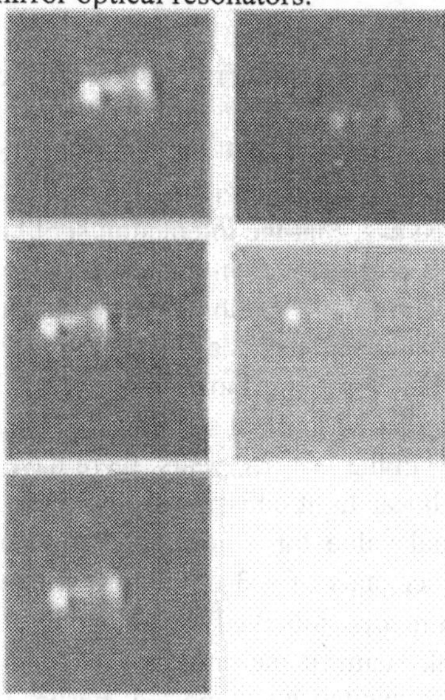

Figure 18 Photoluminescence images from the micro crystallites showing five frames of the blue beam taken under decreasing incident intensity.

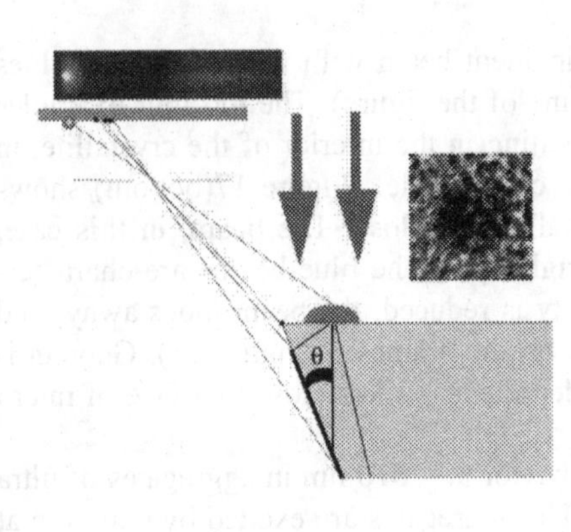

Fig. 19 A cluster 8 μm in diameter is on a glass plate imaged in the horizontal plane (left inset) An actual image of the cluster under 560 nm Hg lamp illuminations. (right inset) A TEM image of 2.9 nm Si nanoparticles, a constituent of the cluster.

Fig 20 Images of the interaction of the incident beam with the active Si nanoparticles aggregate (as in Figure 19) with decreasing pumping intensity (left to right). The lower spot is the interaction. The upper is a red beam spot that emerges

We analyze in terms of prism/filter RGB. Figure 22 gives the spectral analysis of a line profile along the separated spot, through the interaction spot. It shows the domination of the red for all points on the beam, while in the interaction region, we see strong mixing of red and green. There is no blue in both. One indeed can observe, with the naked eye, that the interaction spot has a mixture of yellow/orange/red, while the beam has a defined red. In some images, we can see speckle patterns as was shown in the close up; those exhibit, under standard tests, the familiar characteristics, indicating the spatial coherence. We checked the response with white and blue light and then with 560 nm below the threshold of red beam, but found no such speckle. This micro lasing constitutes an important step towards the realization of a laser on a chip.

In the last two years, there have been several reports of progress towards efficient nonlinear light emission from silicon, suggesting that Si-based material is of great potential value for a new generation of light sources. In one effort [17], luminescent nanocrystals of ~3 nm across embedded in quartz by implanting high energy Si ions into quartz, followed by annealing at 1100 C° were created. Optical amplification of a weak beam at the emission wavelength, traversing the medium while the medium is being excited was reported. Earlier [18], super linear emission in the range 600-1400 nm from porous Si, oxidized at elevated temperatures was reported, but it was long-lived (ms), unstable, and was attributed to heat. Several studies of gain related measurements have been made including waveguiding effects in the measurement of optical gain in a layer of Si nanocrystals, optical gain in Si/SiO2 lattice, and stimulated emission in blue-emitting Si+-implanted SiO2 films [19].

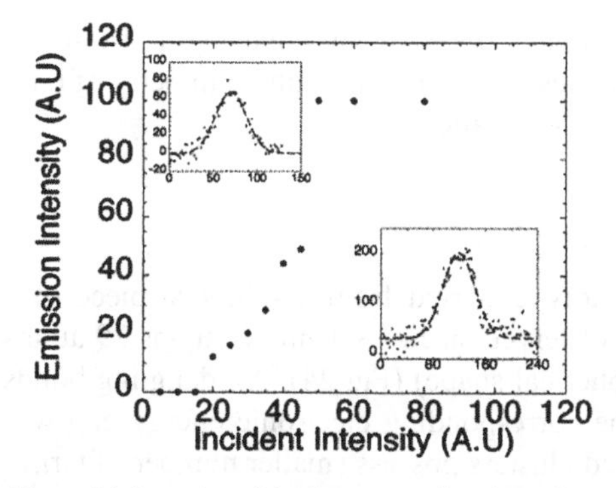

Figure 21 Intensity of the beam as a function of the incident pumping intensity . (Insets) line profiles across the beam at low(up) and high incident intensity (bottom) with Gaussian fits.

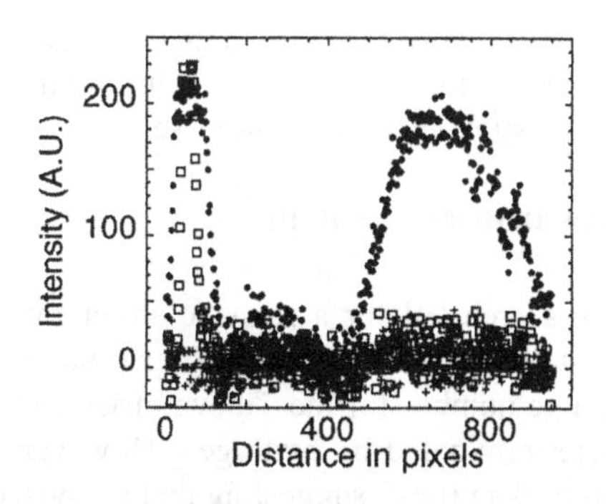

Figure 22 RGB spectral analysis of a line profile along the beam (right) and going through the interaction spot (left) (solid circle) red, (open square) green, and (cross) blue

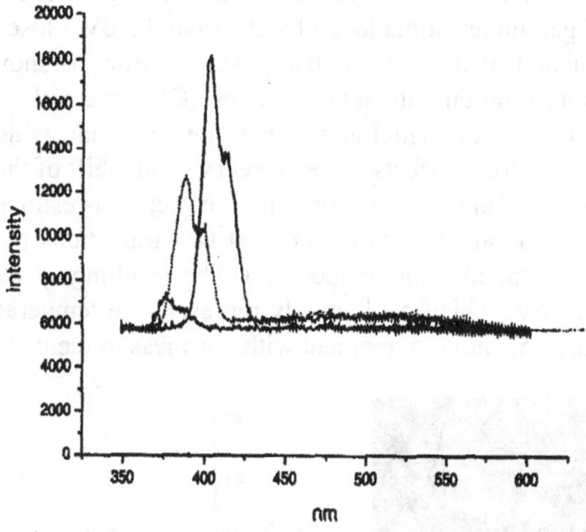

Figure 23 The second harmonic emission spectra for the three excitation wavelengths 780, 800, and 832 nm from left to right (at different incident intensities) respectively

6. Second harmonic generation

Bulk Si is known to have negligible nonlinearity, being zero at the second order level (not allowed because of centrosymmetry), and very small at the third-order level. We recently reported the first observation of second harmonic generation in films of ultra small silicon nanoparticles. Figure 23 gives the emission spectra for the three excitation wavelengths 780, 800, and 832 nm (not of the same intensity) [11]. Each shows a peak with a shoulder on the red wing. The shoulders are at 390, 400, 416 nm, half the wavelengths of the incident beam. The peaks in the spectra are at 380, 390, and 406 nm, i.e., blue shifted by 10 nm from the shoulder in each of the spectra. Those and other measurements show that the emitted wavelength tracks the incident wavelength. These results point to a mechanism in ultra small particles that break the centrosymmetry of bulk Si, the symmetry that inhibits second harmonic generation.

7. Structural prototype

The search [9] for a realistic structural prototype started from a spherical piece of a crystalline Si which, for the experimentally observed size of ~ 1 nm, contains 29 atoms (magic number for the Td symmetry and spherical shape) (Fig 24). All dangling bonds were terminated by hydrogen. However, the corresponding electronic energy gap was larger than 6 eV, suggesting that the observed clusters possess smaller number of terminating hydrogen/oxygen, with a part of the dangling bonds saturated by a nanocrystal surface reconstruction. By eliminating 12 H atoms we arrived at a structure of $Si_{29}H_{24}$ with 6 reconstructed surface Si-Si dimers similar to Si(001) surface 2x1 reconstruction. The resulting $Si_{29}H_{24}$ system was then relaxed using the Density Functional Theory (DFT) with the PW91 exchange-correlation functional, giving, after correcting for the well-known DFT gap underestimation, a band gap of 3.5 eV, close to the one observed.

We next calculated of the optical absorption spectrum as shown in Figure 25. We employed the Configuration Interaction singles (CIS) method which constructs the wavefunctions with correct spatial and spin symmetries and which accommodates, to the first order, the excitonic effects. The correct overall shift of the spectrum due to the incomplete treatment of the electron correlation effects was estimated from preliminary quantum Monte Carlo calculations of the first two transitions (T1->T2) which correspond to the edge of the absorption spectrum. The resulting spectrum with a Gaussian broadening of 0.14 eV, which qualitatively represent the temperature and size averaging, given in Figure 25, shows agreement with our measurement.

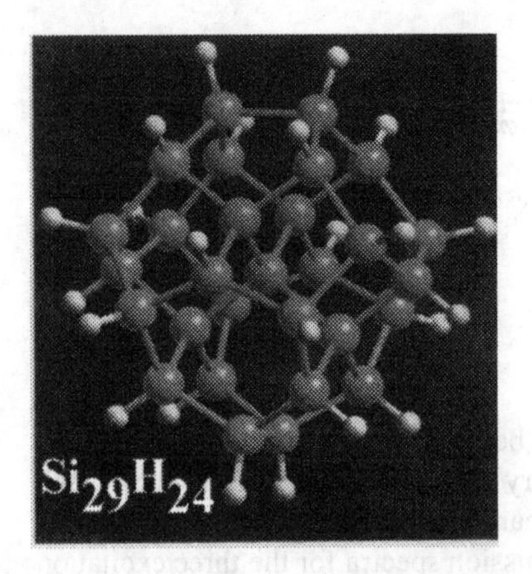

Figure 24 A configuration structure of a particle that contains 29 silicon atoms, with 5 core atoms and 24 on the surface. Hydrogen terminating atoms are in white.

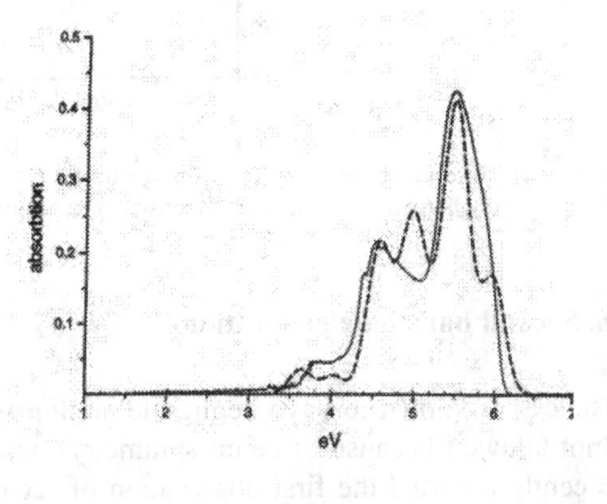

Figure 25 The calculated (dot) and experimental (solid) absorption (normalized) of the 29 atom particle.

We evaluated the cluster polarizability, a uniquely defined property, using GGA/PW91 and 6-31G* basis set, giving ~ 793 (a.u.). A general definition of a dielectric constant in semiconductor nanoparticles is not possible since the energy levels are discrete. To arrive at an effective "dielectric constant" one has to define an appropriate "cluster interior volume". For an effective interior radius (~ 4.2 A) we get a dielectric constant of ~ 5.7. This value is close to estimations done before by Allan at al. and Zunger et al, but the Penn's model-based analysis gives, for a diameter of 0.8 nm, a more reduced static dielectric constant of 2 [20].

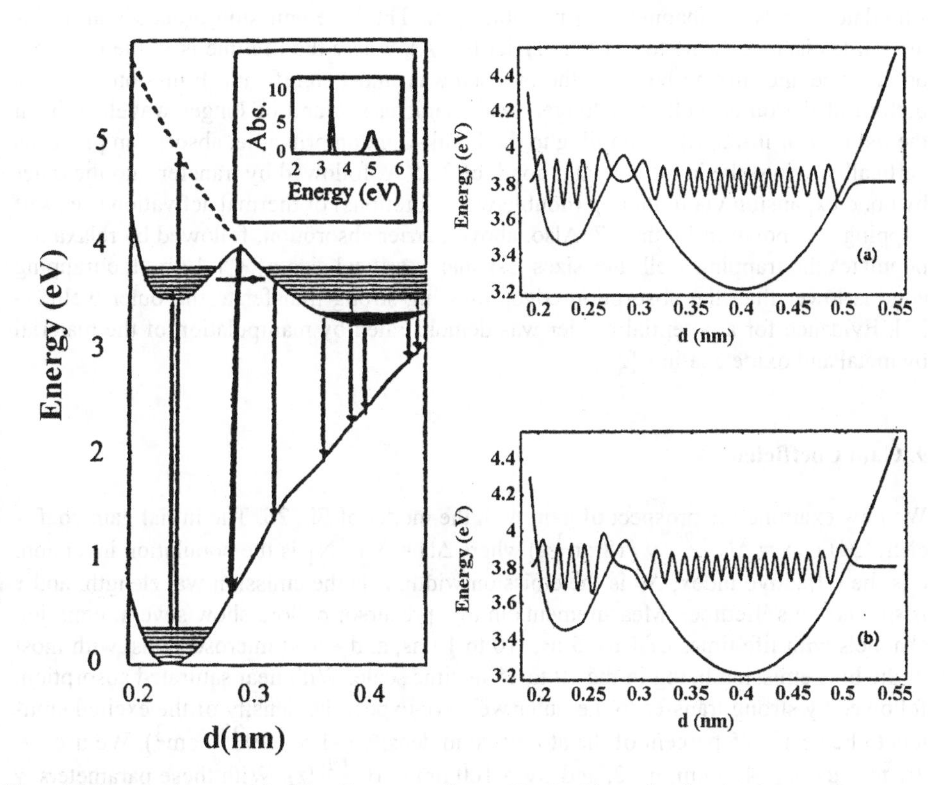

Figure 26. Interatomic potential of the dimmers in 1.03 nm crystallites showing the ground and the first excited electronic states, along with the pathways for excitation and emission. The inset is the transition probability in absorption.

Figure 27. Double well vibrational wave function (8,33) in a 1.03 nm crystallite. (a) bonding and (b) antibonding.

8. Structure of Si-Si bonds

Figure 26 gives a schematic of the interatomic potential of the Si-Si reconstructed bond[21] calculated by Allan et al using LDF in a 1.03 nm diameter particle, and the various pathways for absorption and emission calculated by Nayfeh et al[22]. The excited state is a double well with a potential barrier. The inner well (at 2.35A) is associated with the tetrahedral configuration and radiates on a long time scale of milliseconds. The outer well, a new state found only in ultra small nanoparticles (< 1.75 nm across), is a trap well (at 3.85 A) that radiates with life times of 5 ns to 100 μs. The calculations were performed with the silicon atoms terminated by hydrogen. Only minor changes such as a shift of the bottom of the outer well inward are expected for oxygen terminated dimers since they are less amenable to expansion. The state to which the outer well radiate vertically down is the ground electronic state. But, at extended bond lengths, the ground state is high lying and unpopulated, hence this system constitutes a stimulated emission channel, and possible gain. The blue emission proceeds at an interatomic distance at the top of the barrier (~3 A), where the lifetime is in the nanosecond regime and mixing between the two states is most significant. Emission from the bottom of the outer well is of longer time characteristic and of longer wavelength (in the red or near infrared). According to the Frank-Condon principle, absorption proceeds vertically up into the inner well at a bond of 2.35 A, followed by transfer into the outer by bond expansion via tunneling (double well vibrations) or thermal activation (i.e., self trapping) as shown in Figure 27. Also, above-barrier absorption, followed by relaxation populates the trapping well. For sizes less than a critical size of ~ 1.4 nm, the trapping edge is lower than the absorption edge, allowing strong transfer to the outer well[21-22]. Evidence for a potential barrier was demonstrated by manipulation of the material by metal and oxide coatings[23].

9. Gain Coefficient

We now examine the prospect of gain with the model of Fig 26. The initial gain coefficient [24] is $\gamma = \Delta N \lambda^2 \Delta\nu / (8\pi n^2 \tau)$ where $\Delta N = N_2 - N_1$ is the population inversion, n is the refractive index, $\Delta\nu$ is the emission width, λ is the emission wavelength, and τ is spontaneous lifetime. Measurements in the precursor colloid show several emission channels with life times of 1 ns, 5 ns, 10 to 15 ns, and ~ 100 microseconds, with most of the blue emission being in the 10 to 15 ns time scale. With near saturated absorption, followed by strong transfer to the outer well, we expect the density of the excited emitters to be nearly 25 percent of the atomic solid density (~1.5×10^{22} / cm^3). We use τ = .01 to 1 μs, λ = 400 nm, n= 2, and $\Delta\nu$ = 100 nm (~10^{14}Hz). With these parameters, $\gamma \sim 1.5 \times 10^3$ to 1.5×10^5 cm^{-1}. This gain allows considerable growth over microscopic distances. The number of spontaneous emission modes is given by $p = 8\pi\Delta\nu n^3 V / \lambda^2 c$ where V is the active volume. Using an active cross section of (2 μm)^2 and an active sample thickness of 0.2 μm, we get p~10^3, sizable even for microscopic volume.

10. Conclusion

We developed a process for converting bulk silicon into ultra small nanoparticles. The particles - which are 1 (29 Si atoms), 1.67 (123 Si atoms), 2.15, 2.9 nm in diameter - can be formed into colloids, crystals, films, and collimated beams. Unlike bulk silicon, a dull material, ultra small silicon nanoparticles are spectacularly efficient at emitting light in RGB colors. In addition to being ultra bright, reconstituted films of particles exhibit stimulated emission. Light-emitting Si devices could eventually result in a laser on a chip, new generation of Si chips, and extend the functionality of Si technology from microelectronics into optoelectronics and biophotonics. There are hundreds of potential applications. Among the most promising are single-electron transistors for consumer electronics, charge-based nanomemory devices, nano-transistors, -switches, -displays, -sensors and -assays, fluorescent labeling, laser on a chip, and optical interconnects.

Applications of silicon nanoparticles transcend their applications in laser development. Work in Si nanomaterial finds also impetus in the biomedicine and microelectronics as silicon is the backbone of the industry. This new class of fluorescent semiconductor particles has physical characteristics superior to all presently available fluorescing substances allowing the development of a new generation of significantly improved biomedical markers, having superior detection sensitivity, improved photo and chemical stability, and of nanoparticles size to provide good spatial resolution. Intricate chemical and biological attachments can be applied to the particles to make them smart and compatible to human body conditions such that they selectively home in or attach to any specific sub cellar component of molecule.

Research in silicon nanoparticles may impact our understanding of a wide range of astrophysical environment including reflection nebula, planetary nebula, HII regions, halos of galaxies, diffuse interstellar medium (DISM), and the solar corona. Recently, red luminescent Si nanoparticles have been proposed as carriers for the broad red dust luminescence band that has been observed in these environments [25]. Spectroscopic properties of the band point to a luminescence phenomena, most probably photoluminescence, originating from some dust component. However, the real nature of the carrier is still under debate, and no blue extended emission has been observed. Examining the formation and stability of our discrete family of particles under astrophysical environment conditions would be very valuable in this regard [26]. Searching for bright infrared emitting particles is under way in our laboratory.

The authors acknowledge the State of Illinois Grant IDCCA No. 00-49106, US NSF grant No. BES-0118053, the US DOE Grant DEFG02-ER9645439, NIH Grant RR03155, Motorola, and the University of Illinois at Urbana-Champaign.

References

1 O. Akcakir, J. Therrien, G. Belomoin, N. Barry, E. Gratton, and M. Nayfeh Appl. Phys. Lett. 76, 1857 (2000)

2 M. H. Nayfeh, E. Rogozhina, and L. Mitas, in Synthesis, Functionalization, and Surface Treatment of Nanoparticles, M.-Isabelle Baratron, ed., American Scientific Publishers, (2002); M. H. Nayfeh, J. Therrien, G. Belomoin, O. Akcakir, N. Barry, and E. Gratton, Proc. Mat. Soc. 638, F9.5.1-12 (2000); J. Therrien, G. Belomoin, and M. H. Nayfeh, Proc. Mat. Soc. 582, 11.4.1-6 (1999).

3 G. Belomoin, J. Therrien, and M. Nayfeh, Appl. Phys. Lett 77, 779, (2000)

4 Zain Yamani, Howard Thompson, Laila AbuHassan, and Munir H. Nayfeh, Appl. Phys. Lett. 70, 3404 (1997)

5 G. Belomoin, J. Therrien, A. Smith, S. Rao, R. Twesten, S. Chaieb, L. Wagner, L. Mitas, and M. Nayfeh, Appl. Phys. Lett 80, 841, (2002)

6 J. Therrien, G. Belomoin, and M. Nayfeh, Appl. Phys. Lett 77, 1668 (2000)

7 E. Rogozhina, G. Belomoin, J. Therrien, P. Braun, and M. H. Nayfeh, Appl. Phys. Lett., 78, 3711 (2001)

8 Elena V. Rogozhina, Gennadiy A. Belomoin, Munir H. Nayfeh and Paul V. Braun, MRS Conference, Boston, MA, Nov. 26 – 30, 2001, Book of Abstracts, W 6.10.; E. V. Rogozhina, G. A. Belomoin, J. M. Therrien, M. H. Nayfeh, P. V. Braun (under preparation for JACS).

9 L. Mitas, J. Therrien, G. Belomoin, and M. H. Nayfeh, Appl. Phys. Lett. 78,1918 (2001)

10 G. Belomoin, E. Rogozhina, J. Therrien, P. V. Braun, L. Abuhassan, M. H. Nayfeh, L. Wagner, L. Mitas, Phys. Rev. B **65**, 193406 (2002)

11 M. H. Nayfeh, O. Akcakir, G. Belomoin, N. Barry, J. Therrien, and E. Gratton, Appl. Phys. Lett. 77, 4086 (2000)

12 A. Smith, G. Belomoin, , M. H. Nayfeh, T. Nayfeh, (to be published)

13 A. Smith, M. H. Nayfeh, M. Alsalhi, A. Alaql, J. of Nanosciece and Nanotechnology 2 (5), (2002) (In press).

14 M. Nayfeh, O. Akcakir, J. Therrien, Z. Yamani, N. Barry, W. Yu, and E. Gratton, Appl. Phys. Lett.75, 4112 (1999)

15 M. H. Nayfeh, N. Barry, J. Therrien, O. Akcakir, E. Gratton, and G. Belomoin, Appl. Phys. Lett. 78, 1131 (2001)

16 M. H. Nayfeh, S. Rao, N. Barry, J. Therrien, G. Belomoin, A. Smith, and S. Chaieb, Appl. Phys. Lett. 80, 121 (2002)

17 L. Pavesi, L. Del Negro, C. Mazzoleni, G. Franzo, and F. Priolo, Nature 408, 440 (2000); A. Borghesi, A. Sasella, B. Pivas, L. Pavesi, Solid. State Commun. 87, 1 (1993).

18 H. Koyama, and p. Fauchet, Appl. Phys. Lett. 73, 3259 (1998); H. Koyama, L. Tsybeskov, and P. Fauchet, J Luminescence 80, 99 (1999); H. Koyama, and P. Fauchet, J. Appl. Phys. 87, Feb 15, 2000; H. Koyama, and P. Fauchet, MRS 536, 9 (1999)

19 J. Valenta et al., Appl. Phys. Lett. 81, 1396 (2002); K. Luterová et al., J. Appl. Phys. 91, 2896 (2002); Leonid Khriachtchev et al., Appl. Phys. Lett. 79, 1249 (2001)

20 L-W. Wang, and A. Zunger, Phys. Rev. Lett. **73**, 1039 (1994); R. Tsu, L. Ioriatti, J. Harvey, H. Shen, aand R. Lux, Mater. Res. Soc. symp. Proc. **283**, 437 (1993); G. Allan, C. Delerue, M. Lannoo, E. Martin, Phys. Rev. B 52, 11982 (1995); D. R. Penn, Phys. Rev. **128**, 2093 (1962)

21 G. Allan, C. Delerue, and M. Lannoo, Phys. Rev. Lett. 76, 2961 (1996)

22 M. Nayfeh, N. Rigakis, and Z. Yamani, Phys. Rev. B 56, 2079 (1997); MRS 486, 243 (1998)

23 Z. Yamani, N. Rigakis, and M. H. Nayfeh, Appl. Phys Lett. 72, 2556-2558 (1998); Z. Yamani, O. Gurdal, A. Alaql, and M. Nayfeh, J. Appl. Phys. 85, 8050-8053 (1999); M. H. Nayfeh, Z. Yamani, O. Gurdal, and A. Aql, Proc Mat. Res. Soc. 536, 191 (1998); Z. Yamani, A. Alaql, J. Therrien, O. Nayfeh, and M. Nayfeh, Appl. Phys. Lett. 74, 3483 (1999)

24 A. Yariv , *Quantum Electronics*, John Wiley & Sons (1975)

25 A. Witt, R. Schild, and J. Kraiman, APL **281**, 708 (1994); K. Gordon, A. N. Witt, and B. C. Friedmann, APJ **498**, 552 (1998); M. Cohen, C. M. Anderson, and A. Cowley, APJ 196, 179 (1975); J. M. Perrin, and J. P. Sivan A&A **255**, 271 (1992)

26 M.H. Nayfeh, S. R. Habbal, and L. Mitas, to be published

ON FAST OPTICAL GAIN IN SILICON NANOSTRUCTURES

L. KHRIACHTCHEV and M. RÄSÄNEN
Laboratory of Physical Chemistry, P. O. Box 55, FIN-00014 University of Helsinki, Finland

1. Introduction

During long time application of the well-developed silicon technology to optoelectronics had been limited by the extremely poor generation of light by bulk silicon. However, the properties of silicon were found to depend on its structure at the nanometer scale, and the bright photoluminescence (PL) from porous silicon (p-Si) was discovered by Canham in 1990 [1]. The band gap of silicon nanocrystals can be up to 3 eV depending on their size, and they can emit light with efficiency as high as 1%. Amorphous and crystalline Si nanostructures embedded in SiO_2 constitute another emitting material of considerable interest [2]. The Si/SiO_2 nanoscale materials can be prepared with a number of procedures, such as repeated growth of Si and SiO_2 ultra-thin layers [so-called Si/SiO_2 superlattice (SL)], co-deposition of Si and SiO_2 (silicon rich silica SiO_x, $x < 2$), implantation of Si into SiO_2, *etc.* Thermal annealing can be used to promote the formation of Si nanocrystals. Unfortunately, the light emission decays in most cases relatively slowly, in microseconds, limiting the achievable repetition rate to 1 MHz, which is not sufficient for modern communication applications. An elegant way to proceed to short light pulses in silicon nanostructures is to speed up the emission kinetics by employing laser philosophy based on stimulated rather than spontaneous emission. In this situation, the silicon-laser pulse duration would be controlled by the existence of population inversion, and its lifetime can be efficiently shortened in the process of light amplification. The first experimental indications of light amplification (optical gain) in Si nanocrystals embedded in a SiO_2 matrix in the 1.4 – 2.0 eV energy region were reported by Pavesi *et al.* [3]. They suggested that the population inversion occurred between states of the Si/SiO_2 interface, and the gain occurred via a three-level scheme. Recently we have found that this optical gain can be very *fast* (~ 1 ns), which is of essential importance for applications [4]. In this work, we describe the experimental data supporting the existence of fast optical gain in Si nanocrystals embedded in a SiO_2 matrix. The possible sources of PL and optical gain are discussed.

2. Experiment

The optically studied SiO_x films (x = 1.7–1.9) had typical thickness of 2 μm. The Si/SiO_2 superlattices (up to 600 Si/SiO_2 periods) were constituted with 1 and 1.5 nm Si

L. Pavesi et al. (eds.), Towards the First Silicon Laser, 181–190.

layers and 2 nm SiO_2 layers. The substrates were Si or fused quartz plates. The samples were manufactured in the Electron Physics Laboratory (Helsinki University of Technology), and the details of the deposition procedures can be found elsewhere [4-8]. The as-grown samples show amorphous Si inclusions, and the PL is seen at ~1.8 eV. The samples described in this article were typically annealed at 1150 °C in nitrogen atmosphere for 1 hour.

The Raman and CW PL measurements were performed using an Ar^+ laser (Omnichrome 543-AP, ~ 50 mW at the sample, 488 nm) and a single-stage spectrometer (Acton SpectraPro 500I) equipped with a CCD camera (Andor InstaSpec IV) providing resolution down to ~ 2 cm^{-1}. Raman scattering and PL light was transmitted without polarization analysis through collecting optics, a holographic filter (Kaiser Super-Notch-Plus), and an optical fiber. In order to measure PL with time resolution, the samples were exposed to ~ 5-ns pulses of a tunable optical parametric oscillator with a frequency doubler (Continuum), and the emission was analyzed with a single-stage spectrometer (Spex 270M) and a gated intensified CCD camera (Princeton Instruments). In the "transverse" detection geometry, the radiation was directed at ~45° to the sample surface, and the detected light was collected in the orthogonal direction to the sample surface. In the *waveguiding* geometry, the excitation was orthogonal to the sample surface, and the detection was done from the film edge. The PL spectra are given without correction to optical sensitivity of the equipment. The transmission and reflection spectra were recorded with a fiber-optics spectrometer (SD2000, Ocean Optics) and a broadband light source (DH-2000, Top Sensor System).

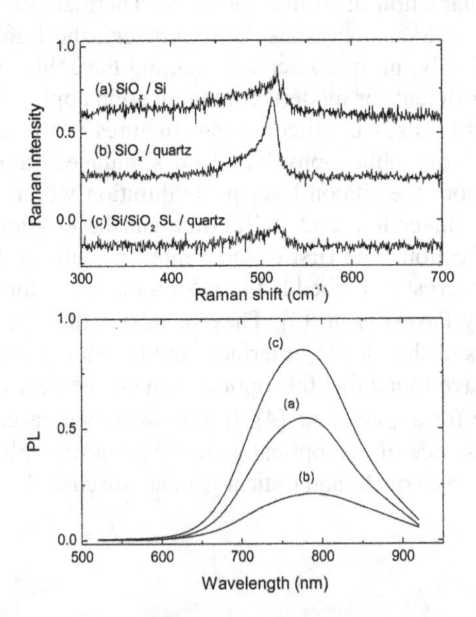

Figure 1. Raman and PL spectra of three annealed (1150 °C) samples: (a) SiO_x, $x = 1.7$ on Si substrate, (b) SiO_x, $x = 1.7$ on quartz substrate, (c) Si/SiO_2 SL with 1-nm Si layers on quartz substrate. CW excitation at 488 nm is used.

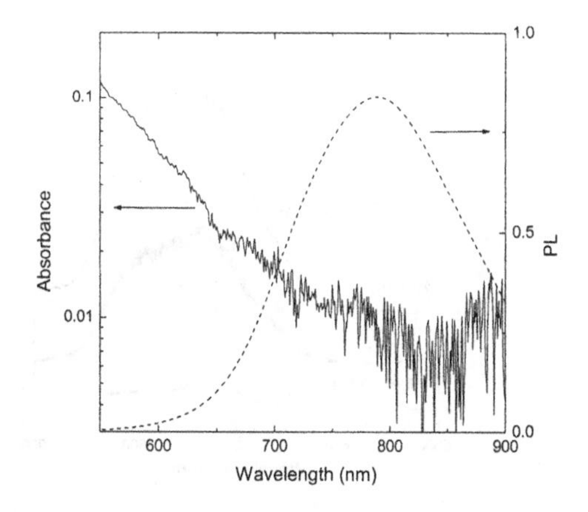

Figure 2. Absorption and PL spectra of a SiO_x film ($x = 1.7$) on quartz substrate.

3. Experimental results

Figure 1 presents Raman and PL spectra of three samples annealed at 1150 °C. The PL positions (~ 1.6 eV) are similar for these samples. The Raman scattering bands at 515-517 cm^{-1} evidences the annealing-induced formation of Si nanocrystals, and the characteristic crystalline size is estimated to be 3-4 nm using the phonon confinement model [9]. Annealing at 1150 °C leads to high transparency of the material at energies below 2 eV, and no absorption band in the PL spectral range is observed (see Fig. 2). The absorption curve in Fig. 2 was obtained by summarizing the transmission and reflection spectra as describe elsewhere [10], which gave the result to the accuracy of scattering. For the annealed samples, the PL spectra obtained with pulsed excitation at various wavelengths (266 to 488 nm) were very similar to each other and to the corresponding spectra obtained with CW excitation at 488 nm (see Fig. 3). Two exponents with similar amplitudes and with lifetimes of ~ 1 and 10 μs fit quite successfully the decay of the "slow" PL at 720 nm (SiO_x, $x = 1.7$). The decay rate at this scale is practically independent of the excitation wavelength and intensity, and it is very similar for various samples studied [8]. Importantly, the PL lifetime remarkably decreases for shorter wavelengths showing the inhomogeneous distribution of the emitters [4].

The experiments on optical gain were performed with stripe-like excitation in the waveguiding geometry (detection from the layer edge), which allowed long propagation length for photons in the pumped volume [4]. If the refractive index of the film material is high enough, the ~ 2-μm layer on quartz transmits suitable

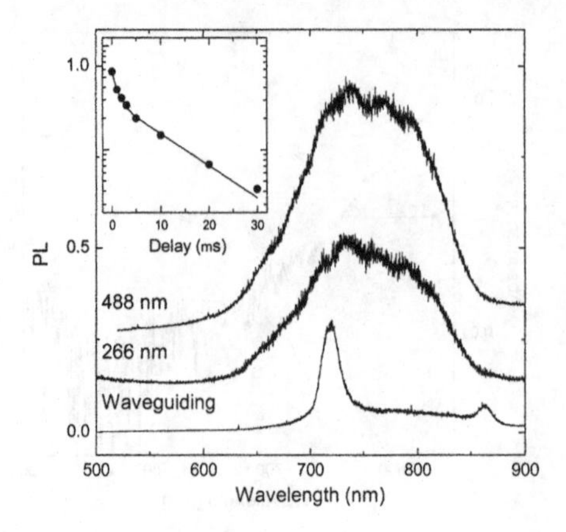

Figure 3. PL spectra measured with pulsed excitation. The two upper curves were measured in the transverse geometry with excitation at 488 and 266 nm, and the lower curve was obtained in the waveguiding geometry with excitation at 380 nm. The insert presents the PL decay fitted by two exponents.

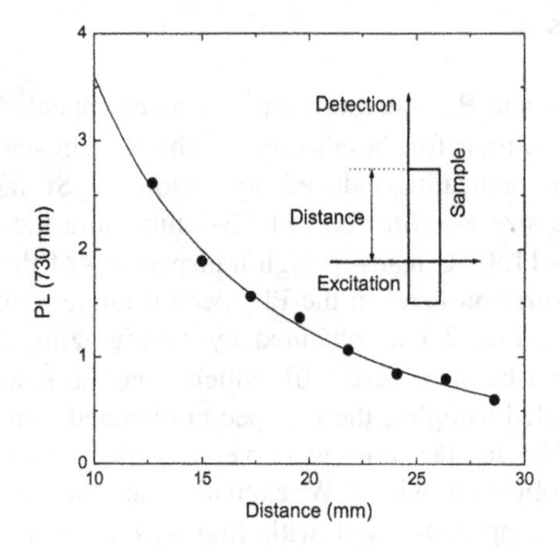

Figure 4. PL intensity as a function of the distance between the excitation spot (CW, 488 nm) and the SiO_x film edge (x = 1.7) measured in the waveguiding geometry as shown in the insert.

wavelengths only (see Fig. 3). The waveguiding in these samples is very efficient, and we detected good PL signals when the excitation is even several centimeters from the edge. In fact, the estimated waveguiding loss coefficient was as low as 0.4 cm^{-1}. To perform this estimate, the optical detection system was focused to the film edge, and the excitation spot (~ 0.5 mm in diameter) was at some distance, $L >> 1$ mm, from the film edge (see Fig. 4). In this scheme, the detected signal is proportional to $\exp(-k_{WG}L)/L$ where k_{WG} is the waveguiding loss constituted mainly by the absorption loss and the interface loss. The excitation was performed at the long distance from the edge, which prevented us from artifacts in the detected signal being possible for the excitation near the edge [11].

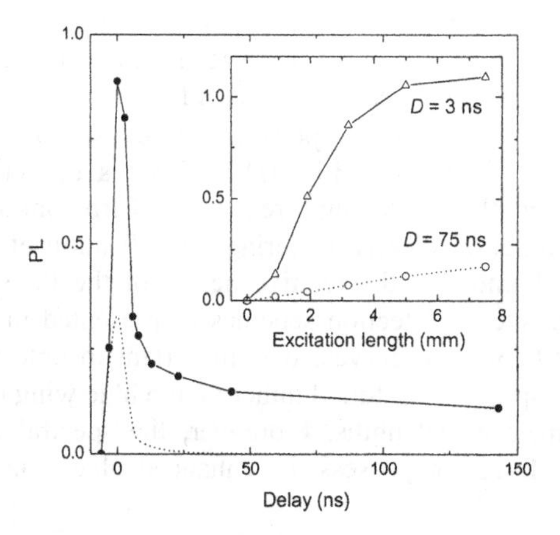

Figure. 5. Fast 720-nm emission measured in the waveguiding geometry from the SiO_x (x = 1.7) film with excitation at 380 nm. The pumping pulse is shown by dots. The insert presents emission with delays 3 and 75 ns as a function of the excitation length. The gate width is 5 ns.

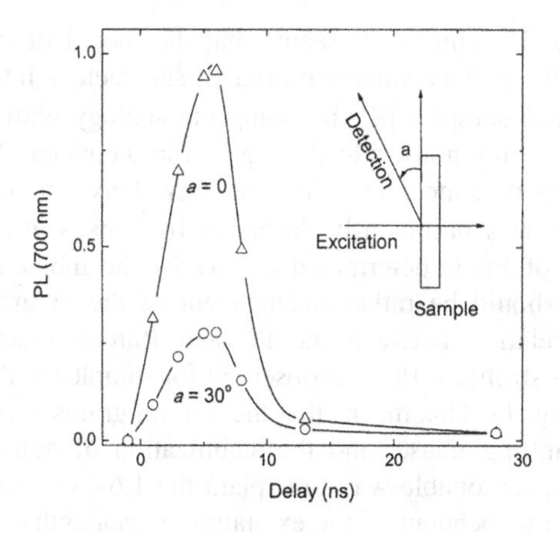

Figure 6. Angular dependence of the fast emission at 700 nm measured with 380 nm excitation (SiO_x, x = 1.7). The longer-lived emission components are equalized for the two dependencies. The experimental geometry is presented in the insert. The excitation distance is ~ 5 mm and the excitation pulse energy density is ~ 10 mJ/cm^2.

When certain samples were pumped with 5-ns pulses at 380 nm in the waveguiding geometry, we found short (a few ns) PL spikes at the early stages of the decay (see Fig. 5). The effect was found for the annealed SiO_x film with x = 1.7 [4] and for the annealed Si/SiO_2 SL with 1.5-nm Si layers, i.e. for materials with the larger amount of Si when the film works as a good waveguide. The excitation was performed at a distance of > 10 mm from the film edge. The fast PL spikes relatively decreased at lower pumping intensity and shorter excitation lengths. The relative height of the PL spikes strongly depends on a balance between various excitation and detection parameters (excitation length and energy density, width of the excitation stripe, detection aperture and direction, *etc.*). Ratios as large as 10 were obtained between PL signals at delays $D = 3$ and 23 ns in the best tries for the annealed SiO_x sample ($x = 1.7$) at 720 nm. For this sample, we measured PL as a function of the excitation stripe length, and the result obtained for a pumping pulse energy density of ~10 mJ/cm^2 is presented in the insert of Fig. 5. The long-lived PL ($D = 75$ ns) is a practically linear function of the excitation stripe length in this length region. In sharp contrast, the PL spike ($D = 3$ ns) is a non-linear function clearly featuring amplification at intermediate excitation lengths. It was also found that the relative height of the PL spike is different in the waveguiding and transverse detection schemes as presented in Fig. 6 where the long-lived PL is equalized for both curves. It is important to note that these "anomalous" effects with short PL spikes were found rather on the blue wing of the PL spectra (< 750 nm) than at the longer wavelengths. Moreover, the spectral contour of the fast PL around 720 nm was found to possess the enhanced blue wing as compared with the long-lived PL.

4. Discussion

First, we discuss the PL emitter. It seems that the model of molecular-like emitting centers localized at the Si/SiO_2 interface offers a satisfactory interpretation for the 1.6-eV PL of the annealed samples [8]. By using the analogy with oxidized p-Si [12,13], emitting centers containing non-bridged oxygen (Si=O covalent bonds) can grow in the Si/SiO_2 network upon annealing. The analogy between oxidized p-Si and Si nanocrystals in SiO_2 was particularly discussed by Prokes *et al.* [14]. Based on this model, the position of PL is determined mainly by the molecular structure and local morphology, i.e. it should be rather independent of the Si grain size. In accord, no straightforward correlation between the PL and Raman spectra was found in our experiments, and the strongest PL was observed for samples with very weak crystalline Raman bands (see Fig. 1). This means that the 3-4 nm grains seen in our Raman spectra do not form the emitting phase, and the stabilization of non-bridged oxygen at the Si/SiO_2 interface is a reasonable way to explain the 1.6-eV emission. Furthermore, the PL position is quite independent of the excitation wavelengths (see Fig. 3). Gole *et al.* reported the same experimental fact for oxidized p-Si and used it as an argument to support an interface-based source of the PL [13]. Indeed, this observation is clearly consistent with the model of excitation trapped at a lower-energy interface state emitting at ~1.6 eV. In this model, the quantum confinement in the Si nanocrystal can participate in the stage of photon absorption and the exciton decays to a radiative oxygen-related surface defect. This image is also consistent with the relatively short lifetime of the emission in our experiments (1 - 10 μs) as compared with 50 – 100 μs usually reported for isolated Si nanocrystals in p-Si [1]. It looks probable that the lifetime measured in our samples is determined by a localized molecule-like center trapping quickly the excitation. One should also keep in mind that the PL spectra can be influenced by nonradiative recombination at the Si/SiO_2 interface [15].

The direct absorption of the Si=O covalent bonds do not appear in the PL spectral region as seen in the absorption spectrum presented in Fig. 2, and there are evident reasons for this. First, the known simulations of the related structures with Si=O covalent bonds show that the equilibrium bond length essentially elongates in the excited state, which necessarily leads to the blue-shifted absorption with respect to the relaxed emission [16]. Furthermore, the Stokes-shifted emission can be in general caused by interaction of the emitting molecule with the local surrounding. This means that one should search for the Si=O absorption band in the UV spectral region where it overlaps with the absorption of Si nanocrystals.

Second, it is worth noting that the absorption with extensive inhomogeneous broadening ($\Delta\omega / \omega_0 \sim 1/4$) associated directly with the long-lived PL ($\tau \sim 10^{-5}$ s) should be very weak even if it is not shifted [10]. The absorption cross section is connected with the spontaneous lifetime τ via the relation [17]:

$$\sigma = \left(\lambda_0 / 2n\right)^2 g(\omega) / \tau \tag{1}$$

where λ_0 is the central wavelength in vacuum, n is the refractive index, and $g(\omega)$ describes broadening of the line ($\int g(\omega)d\omega = 1$). Using realistic values $\lambda_0/2n$ = 250 nm and $g(\omega_0)/\tau \sim 10^{-9}$, we obtain for the cross section an order of 10^{-18} cm^2. Absorbance of 0.01, which should be easily detected in our experiments, corresponds in the 2-μm layer to an absorption coefficient of $\sim 10^2$ cm^{-1} requiring $\sim 10^{20}$ cm^{-3} concentration of the absorbers. For 3-nm Si crystallites, this concentration would mean a filling factor of ~ 50 %, which is definitely too much for the present rather transparent material. This estimate suggests that the absorption via this mechanism (Si=O bonds) should be very weak in our samples, which is actually in agreement with the experiment.

We have recently suggested that the origin of the short PL spikes at the early stages of PL evolution is the amplified spontaneous emission in the presence of population inversion [4]. The PL light propagating in the Si/SiO_2 layer is amplified if the short pumping pulses invert the population. Indeed, population inversion should show thresholds with respect to excitation energy density and stripe length, and this was demonstrated experimentally. By using the amplitude of the short spikes as a function of the excitation length (see Fig. 5), we extracted gain of ~ 6 cm^{-1} at 720 nm, and this value should be considered as a lower and quite rough estimate. More detailed analysis of the time dependencies is limited by the time resolution of our detection system.

The presence of amplification is strongly supported by the angular dependence of the PL decay (see Fig. 6). In the waveguiding geometry, the spontaneous emission propagates a long distance in the pumped area, and the fast amplification is efficient. For large detection angles, the collected light propagates a shorter distance in the pumped area, and the fast amplified PL relatively decreases. However, the extraction of numerical information from the angular dependencies of emission is difficult due to the unknown angular dependence of the propagation length. The situation is further complicated, for instance, by the probable interference of the known fast spontaneous PL reported for the relevant samples [18]. Based on the data presented in Fig. 5 (insert), we emphasize that the fast PL spikes are not fully from the short-lived spontaneous emission. As an additional argument, the fast spontaneous emission decreases for higher annealing temperatures but the fingerprints of optical gain is not seen for low annealing temperatures (< 1100 °C). One should also keep in mind the inhomogeneous properties of the gaining phase, which decreases the reliability of numerical estimates. In any case, the observed short lifetime of the population inversion is very important for applications because it allows short light pulses to be generated with a high repetition rate.

Similarly to the 1.6-eV PL, it is plausible to assign the optical gain to oxygen-related defects in the Si/SiO_2 network, for instance, to the Si=O bonds stabilized at the Si/SiO_2 interface. This structure might be spectrally similar to the molecular species with Si=O bonds theoretically analyzed by Zhou and Head [16]. In fact, these species clearly offer a four-level amplification system. The excitation of the species prepares it at the repulsive part of the excited-state potential curve. The relaxation of the excited species towards the energy minimum occurs via elongation of the Si-O distance. The energy minimum of the excited state allows transition to the attractive part of the ground-state potential curve whose initial population is negligible, thus, producing the conditions for population inversion. The remaining question is how the excited-state species is prepared upon optical pumping. The direct absorption of light at < 3 eV somewhat contradicts with the *ab initio* calculations indicating a larger excitation energy [16]. However, the calculations are performed for the isolated species whereas the real species are embedded in a SiO_2 matrix, which can change the energetic. As the second and possibly more realistic way, one can consider absorption by Si grains with energy transfer to the interface species as it was discussed earlier for the PL mechanism.

It is not obvious at the present time if the active centers must be localized on the Si *crystallite* surface. In our opinion, the gaining structures can in principle be embedded by the amorphous Si inclusions, for instance, in amorphous Si/SiO_2 SLs so that the claim concerning "necessarily" crystalline structure of the gaining Si-based material is very conditional. Nevertheless, the practical detection of the amplification is complicated for amorphous material by the relatively strong absorption probably extending the gain coefficient [7].

It should be emphasized that our consideration does not support the three-level amplification scheme suggested in Ref. 3. In the three-level scheme, the gain coefficient at a given wavelength is always smaller than the absorption coefficient. As estimated in the waveguiding geometry, the absorption coefficient in the annealed SiO_x ($x = 1.7$) material is reliably smaller than 1 cm^{-1} at 720 nm. This small absorption does not allow us to assign the 6 cm^{-1} gain to the simple three-level system. Otherwise, the attenuation factor lonely due to absorption would be larger than 400 for a distance of 1 cm, which is in clear contradiction with the result presented in Fig. 4. The proposed four-level scheme is free of such complications because the absorber in question does not exist here without pumping.

The shortening of amplified spontaneous emission lifetime in the four-level system is a well-known phenomenon. While describing a four-level laser, very fast relaxation processes are usually assumed, the population inversion occurs between the "middle" levels, and its dynamics is given by the equations [17]:

$$\begin{aligned} \dot{N} &= W_p N_g - BqN - N/\tau \\ \dot{q} &= V_a BqN - q/\tau_c \end{aligned} \qquad (2)$$

where W_pN_g describes the pumping from the ground state, BqN corresponds to the stimulated emission, B is the Einstein coefficient, q is the number of photons, V_a is the volume, and q/τ_c describes losses. When the pumping W_p is suddenly switched off, the population inversion and photon density decay with a lifetime controlled by stimulated emission rate ($\sim Bq$) rather than by spontaneous emission rate ($1/\tau$).

The experimental results indicate that optical gain does not follow qualitatively the shape of the PL spectrum but it rather exists in our samples at wavelengths shorter than 750 nm. In our opinion, this fact agrees with the proposed model for the gaining phase and corresponds well to the shortening of the emission lifetime towards the blue. On one hand, the gain should be proportional to the corresponding Einstein coefficient, i.e. to $1/\tau$. On the other hand, the population inversion upon intense pulsed pumping is rather insensitive to the lifetime until the lifetime is much longer than the time needed for pumping and trapping of the exciton. This means that the lifetime of the populated state is allowed to be tens of nanoseconds in the pulsed regime. This qualitative consideration leads to the conclusion that the gain profile should not reproduce the PL profile, but it should be rather shifted to the blue.

5. Conclusions

We investigated structural, emitting and gaining properties of annealed SiO_x films and Si/SiO_2 SLs with a variety of Si contents. Clear parallelism in annealing-induced modifications of their microstructure and emission was found. In particular, this similarity suggests a uniform light emitting mechanism in these two materials. The as-grown samples are amorphous, disordered Si inclusions are evident in Raman spectra for the samples with higher Si contents, and the PL is seen at ~ 1.8 eV [8]. Annealing at 1150 °C in nitrogen atmosphere leads to growth and ordering of Si inclusions, and the typical crystalline size is estimated to be 3 - 4 nm. For all samples (both SiO_x films and Si/SiO_2 SLs) an increase of PL at ~ 1.6 eV is observed upon annealing and its resulting maximum is quite independent of the initial sample architecture. The 1.6-eV PL of the annealed samples is practically identical for pulsed excitation at various wavelengths (from 266 nm to 488 nm), and two exponents with lifetimes of ~ 1 and 10 μs fit the slow PL. No clear correlation between the Raman scattering and PL spectra for the annealed samples is found. All these observations support an excitation transfer from optically generated excitons to an oxygen-related emission center, similarly to the model of oxidized p-Si, rather than the direct radiative recombination of excitons confined in Si crystallites.

The experiments show the presence of ultra-short PL spikes under pumping with 5-ns light pulses in the waveguiding geometry, which is explained in terms of fast optical gain in Si nanocrystallites embedded in a SiO_2 matrix. The observed short lifetime of population inversion potentially allows generation of short (~ 1 ns) amplified light pulses in the Si/SiO_2 material. As measured at 720 nm, the optical gain in the present samples is estimated to be 6 cm^{-1}. The gain seems to increase at shorter wavelengths, which reasonably correlates with the decrease of the PL lifetime. The presented facts correspond to the model of amplification in a four-level scheme involving the ground and excited states of oxygen-related defects stabilized at the Si/SiO_2 interface, i.e. the same centers as suggested for spontaneous PL.

The Academy of Finland supported this work. Sergei Novikov is thanked for sample preparation.

References

[1] For a review, see Cullis A. G., Canham L. T., and Calcot P. D. J. (1997) The structural and luminescence properties of porous silicon, J. Appl. Phys. **82**, 909-965.

[2] For a review, see Bettotti P., Cazzanelli M., Dal Negro L., Danese B., Gaburro Z., Oton C. J., Vijaya Prakash G., and Pavesi L. (2002) Silicon nanostructures for photonics, J. Phys.: Condens. Matter. **14**, 8253- 8281.

[3] Pavesi L., Dal Negro L., Mazzoleni C., Franzo G., and Friolo F. (2000) Optical gain in silicon nanocrystals, Nature (London) **408**, 440-444.

[4] Khriachtchev L., Räsänen M., Novikov S., and Sinkkonen J. (2001) Optical gain in Si/SiO2 lattice: Experimental evidence with nanosecond pulses, Appl. Phys. Lett. **79**, 1249-1251.

[5] Khriachtchev L., Räsänen M., Novikov S., Kilpelä O., and Sinkkonen J. (1999) Raman scattering from very thin Si layers of Si/SiO_2 superlattices: Experimental evidence of structural modification in the 0.8 - 3.5 nm thickness region, J. Appl. Phys. **86**, 5601-5608.

[6] Khriachtchev L., Kilpelä O., Karirinne S., Keränen J., and Lepistö T. (2001) Substrate-dependent crystallization and enhancement of visible photoluminescence in thermal annealing of Si/SiO_2 superlattices, Appl. Phys. Lett. **78**, 323-325.

[7] Khriachtchev L., Novikov S., and Kilpelä O. (2000) Optics of Si/SiO_2 superlattices: Application to Raman scattering and photoluminescence measurements, J. Appl. Phys. **87**, 7805-7813.

[8] Khriachtchev L., Novikov S., and Lahtinen J. Thermal annealing of Si/SiO_2 materials: Modification of structural and emitting properties. J. Appl. Phys. (accepted).

[9] Campbell I. H. and Fauchet P. M. (1986) The effects of microcrystal size and shape of the one phonon Raman spectra of crystalline semiconductors, Solid State Commun. **58**, 739-741.

[10] Khriachtchev L. (2002) Comment on "Optical absorption measurements of silica containing Si nanocrystals produced by ion implantation and thermal annealing" [Appl. Phys. Lett. **80**, 1325 (2002)], Appl. Phys. Lett. **81**, 1357-1358.

[11] Valenta J., Pelant I., and Linnros J. (2002) Waveguiding effects in the measurements of optical gain in a layer of Si nanocrystals, Appl. Phys. Lett. 81, 1396-1398.

[12] Wolkin M. V., Jorne J., Fauchet P. M., Allan G., and Delerue C. (1999) Electronic states and luminescence in porous silicon quantum dots: The role of oxygen, Phys. Rev. Lett. **82**, 197-200.

[13] Gole J. L., Dudel F. P., Grantier D., Dixon D. A. (1997) Origin of porous silicon luminescence: Evidence for a surface-bound oxyhydride-like emitter, Phys. Rev. B **56**, 2137-2153.

[14] Prokes S. M., Carlos W. E., Veprek S., and Ossadnik Ch. (1998) Defect studies in as-deposited and processed nanocrystalline Si/SiO_2 structures, Phys. Rev. B **58**, 15632-15635.

[15] Kanemitsu Y., Iiboshi M., and Kushida T. (2000) Photoluminescence dynamics of amorphous Si/SiO_2 quantum wells, Appl. Phys. Lett. **76**, 2200-2202.

[16] Zhou F. and Head J. D. (2000) Role of Si=O in the photoluminescence of phorous silicon, J. Phys. Chem. B **104**, 9981-9986.

[17] Svelto O. (1982) *Principles of lasers*, Plenum Press, New York, Ed. 2.

[18] Tsybeskov L., Vandyshev Ju. V., and Fauchet P. M. (1994) Blue emission in porous silicon: Oxygen-related photoluminescence, Phys. Rev. B **49**, 7821-7824.

EXPERIMENTAL OBSERVATION OF OPTICAL AMPLIFICATION IN SILICON NANOCRYSTALS

M. IVANDA[1,2], U.V. DESNICA[1], C. W. WHITE[3] AND W. KIEFER[2]
[1]*Ruđer Bošković Institute, P.O.Box 180, 10002 Zagreb, Croatia,*
E-Mail: ivanda@rudjer.irb.hr
[2]*Institut für Physikalische Chemie, Universität Würzburg,*
Am Hubland, D-97074 Würzburg, Germany,
[3]*Oak Ridge National Laboratory, P.O.Box 2008, Oak Ridge,*
T 37831-6057, USA

1. Introduction

Silicon, as an indirect band gap semiconductor, is a poor light emitter and, therefore, has been considered as unsuitable for optoelectronic application. The situation has been changed with discovery of light emission from porous silicon [1-3], but due to instability and relatively poor light emission, this material has not yet found functional application. Recent discovery of optical gain in silicon nanocrystals embedded in various matrices promises soon fabrication of silicon laser [4]. The light amplification was demonstrated by pump and probe technique using picosecond light pumping pulses and variable strip method (VSM). The similar observation of amplified spontaneous emission was found by an experiment with nanosecond light pumping pulses [5]. In this paper we provide further evidences for the optical gain in nanocrystaline silicon in fused silica matrix by using VSM and continuous wave laser excitation. Moreover, due to observed strong spatial directionality and narrow line width, these findings are direct evidences for the lasing emission in nanocrystalline silicon.

2. Experimental

Silicon nanocrystals were produced by implantation of 400 keV Si ions of doses $1x10^{17}$ cm^{-2} and $6x10^{17}$ cm^{-2} into fused silica substrate (1 mm thick Corning 7940) to generate a nearly uniform concentration of approximately $0.5x10^{22}$ cm^{-3} and $2x10^{22}$ cm^{-3} of excess silicon atoms (the samples S1 and S2, respectively). At this energy, the projected range of excess Si is ~600 nm and the profile has a full width at half maximum of ~300 nm. After implantation, annealing was carried out at 1100°C for 1 h in flowing Ar+4%H_2, resulting in the formation of nearly spherical and randomly oriented Si nanoparticles with an average diameter of 3.5 nm in S1 and 5.5 nm in S2. More details about the samples are presented in the refs. [6,7]. The samples were irradiated with the krypton and argon ion lasers using excitation lines at 647 nm, 514.5 nm and 488nm. A cylindri-

L. Pavesi et al. (eds.), Towards the First Silicon Laser, 191–196.

cal lens was used to focus laser beam on the surface sample forming a strip of 15 μm in wide and of variable length. Only the central part of the laterally focused spot was used to excite the sample ensuring the constant laser power density on the sample. Measurements show that in this case the laser power density is independent on the strip length. An objective of focal length of 75 mm, f/1.4, was used to focus the light from the edge of the sample onto the first slit of the spectrometer. The spectrometer slit width was set to be 50 μm. The spectra were recorded with a Spex double-monochromator equipped with a CCD detector in red spectral region, and Dilor Z24 triple-monochromator in the green and blue spectral regions.

3. Results and discussion

In the variable stripe length (VSL) method, part of the spontaneous luminescent radiation with the spontaneous emission in an appropriate emission solid angle Ω, J_{sp} propagates along the stripe axis, and is amplified in the excited region by stimulated emission. The amplified spontaneous emission intensity $I_{A.S.E.}$, that is emitted from the sample edge (observation angle $\phi=0^o$) is linked with the excited stripe region length l by the relation [6]

$$I_{A.S.E.}(\lambda,l)=\frac{J_{spont}(\lambda,\Omega)}{g(\lambda)-\alpha(\lambda)}\left[e^{[g(\lambda)-\alpha(\lambda)]l}-1\right], \quad (1)$$

where g is the modal gain coefficient and α is the passive propagation losses in the waveguide.

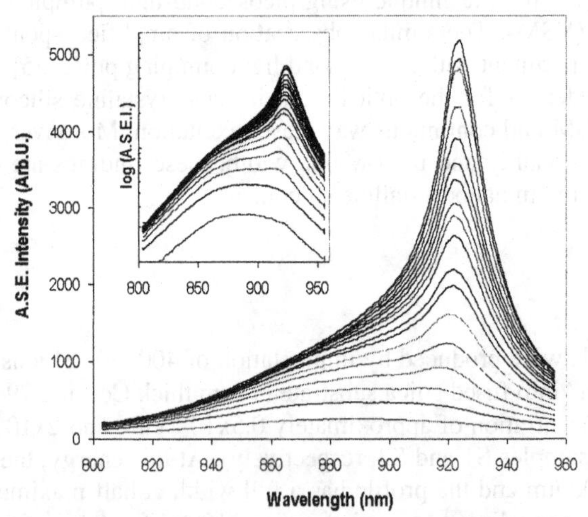

Figure 1. Room temperature amplified spontaneous emission (A.S.E.) spectra of the sample S2 in dependence on the strip length starting with 50 μm for the low intensity spectrum with a constant increase by 50 μm up to 700 μm for the high intensity spectrum. The stimulated emission appeared at 922 nm. The inset shows the same, but using logarithmic value for the A.S.E. intensity.

On the basis of Eq. (1), by dividing the emission spectrum at stripe length $2l$, $I_{ASE}(\lambda,2l)$, with the emission spectrum at length l, $I_{ASE}(\lambda,l)$, it is possible to derive the relation for the spectral dependence of the net modal gain

$$g(\lambda)-\alpha(\lambda)=\frac{1}{l}\left[\ln\left(\frac{I_{ASE}(\lambda,2l)}{I_{ASE}(\lambda,l)}-1\right)\right] \quad (2)$$

Figure 1 shows the A.S.E. spectra recorded from the sample edge as a function of the stripe length using the VSL technique. At length of 150 μm an emission at 922 nm appeared. Contrary to the background signal, this emission sharply increased with excitation length indicating the presence of stimulated emission. The appearance of this new line is more evident by using a log-scale as shown on the inset of Fig. 1.

Figure 2 a) shows the A.S.E. intensity at 922 nm as a function of stripe length. The fit of Eq. (1) on the data gives the net modal gain of 33 cm^{-1}. Figure 2b) shows the A.S.E. intensity at 880 nm. In this case the fit gives a negative value, g-a=-41 cm^{-1}, i.e. optical losses. Figure 2 c) shows the laser power intensity on the sample measured by a power meter as a function of stripe length. It is evident that linear dependence was provided for the range of stripe length used for the gain measurement.

Figure 3 shows the spectral dependence of the net modal gain for different stripe lengths. It is positive in the spectral interval 900-946 nm, i.e. in the range of the observed peak at 922 nm.

Figure 4a) shows the dependence of the A.S.E. spectrum on the observation angle ϕ. Here, the angle ϕ is defined in the plain of deposited layer with respect to the centre of the excitation strip and the optical axes of the one-dimensional amplifier. When the stripe length and the excitation power are fixed, and the observation angle ϕ is increased an exponential decrease in the intensity of the A.S.E. line at 922 nm occurs for the angles $\phi>1^\circ$ as shown on the inset of Fig. 4 a). Here it is also evident that the emission is concentrated on a very narrow angle of 1-2 degrees. In Figure 4 b), for the angles $\phi>1^\circ$, it is evident that each spectrum exhibits exponential tail at high-energy wing which is typical for thermalised distribution of excitons. This could be related to the emission from the band-edge excitons in silicon nano-particles, or in SiO_x interface layer.

The distribution of excitons is described by the Boltzmann statistics ~$\exp(-E/kT_{eff})$, where T_{eff} is the effective temperature of the excitons. The fits on the spectra on Fig. 4 b) gave the effective temperatures T_{eff} = 835 K for the spectrum taken at ϕ=5.18°, and T_{eff} = 719 K for ϕ=1.73°. The small discrepancy could be due to contribution of the broad peak that appears for the observation angles below 1.73°.

Figure 5 shows the difference spectrum of the spectra taken at ϕ=0° and ϕ=1.73°, i.e. with and without the broad peak at 880 nm. The difference spectrum consists of narrow emission line at 922 nm (of 8 nm FWHM) and of the broad peak with maximum at 880 nm. The energy difference between these peaks is 56 meV. The same value for the energy difference was observed in the photo-luminescence spectra of porous silicon, where it is explained with momentum conserving TO-phonon replicas [8].

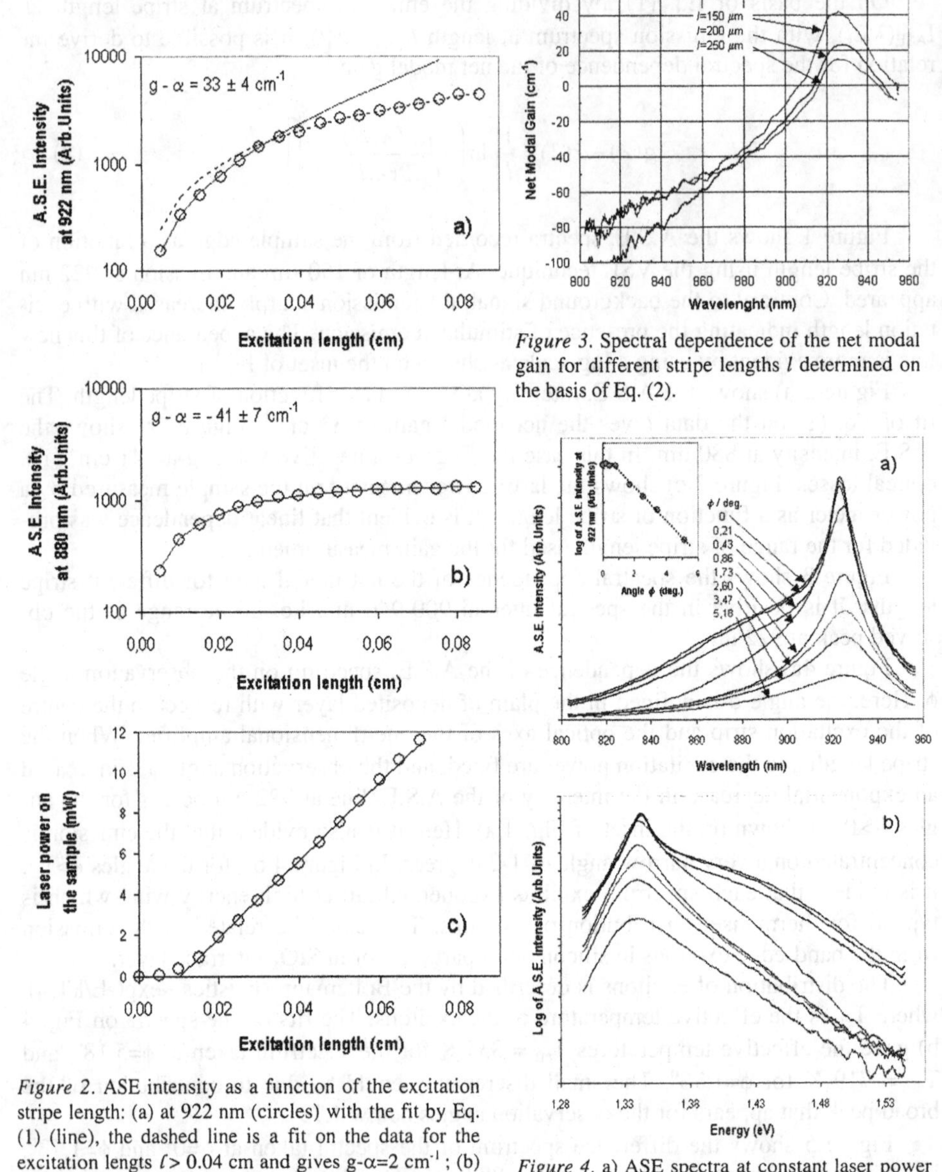

Figure 3. Spectral dependence of the net modal gain for different stripe lengths l determined on the basis of Eq. (2).

Figure 2. ASE intensity as a function of the excitation stripe length: (a) at 922 nm (circles) with the fit by Eq. (1) (line), the dashed line is a fit on the data for the excitation lengts $l > 0.04$ cm and gives g-α=2 cm^{-1} ; (b) at 880 nm (circles) with the fit by Eq. (1) (line); (c) the excitation laser power on the sample in dependence on the excitation length (circles); the line is drawn for visual aid.

Figure 4. a) ASE spectra at constant laser power excitation P=60 mW/cm^2 and l=0.26 cm as a function of the observation angle ϕ. The inset shows the exponential decrease of the intensity at 922 nm for $\phi>1^o$; the line is drawn for visual aid. b) The same as in a), but the intensity is in log-scale showing an exponential decrease with energy for $\phi \geq 1.73^o$.

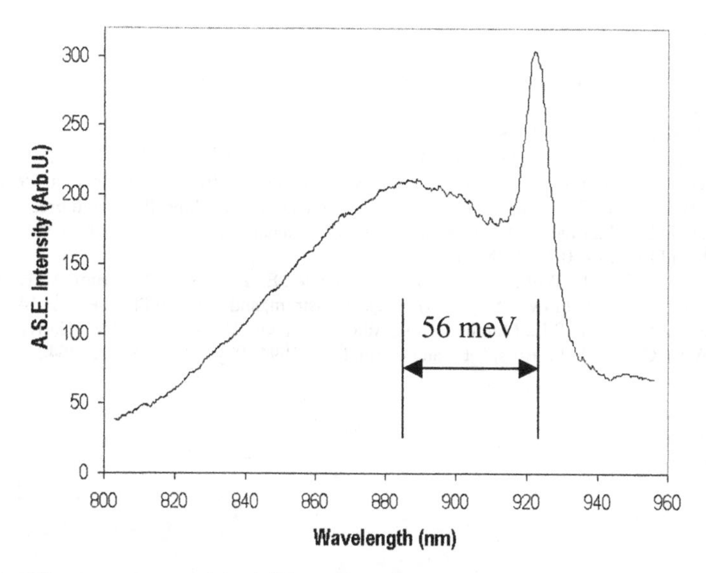

Figure 5. The difference spectrum of the A.S.E. spectra taken at the angles $\phi=0$ and $\phi=1.73$ degree, i.e. the spectra with and without broad luminescence peaked at 880 nm.

Due to quantum confined effects the optical transition in nc-silicon can be of two kinds: first, allowed by the confinement with no-phonon transition, and second, momentum conserving phonon-assisted transition involving absorption (or emission) of TO phonon along with the photon. The peak at 880 nm exists even at small excitation length as shown on Fig. 1. With further increase of the excitation length, the peak at 922 nm starts to appear. For this reason the origin of stimulated emission at 922 nm could be in 880 nm emission mediated by the absorption of TO phonons.

4. Conclusion

We have measured light amplified spontaneous emission (A.S.E.) from silicon nanocrystals in fused silica. The net modal gain (g-α) was measured by variable stripe method using continuous wave laser excitation. For the A.S.E. peak at 922 nm, the net modal gain of 33 cm^{-1} was observed. The peak is sharp, of 8 nm in full width at half maximum, and has strong directionality properties. The corresponding similar A.S.E. peaks were also observed with green and blue laser excitations. The evidences of stimulated emission from the band edge excitons mediated by the absorption of TO phonons are also presented.

This work was supported by the Ministry of Science and Technology of the Republic of Croatia under contracts TP-01/0098-23 and 00980303, and from the Deutsche Forschungs-gemeinschaft (Sonderforschungsbereich 410, Teilprojekt C3). Oak Ridge National Laboratory is managed by UT-Battelle, LLC, for the U.S. Dept. of Energy under contract DE-AC05-00OR22725.

References:

1. Canham, L.T. (1990) Appl. Phys. Lett. **57**, 1045.
2. Cullis, A.G. and Canham, L.T. (1991) Nature **353**, 335.
3. Fauchet, P.M. (1999) J. Lumin. **80**, 53.
4. Pavesi, L., Dal Negro, L., Mazzoleni, C., Franzo, G., and Priolo, F., (2000) Nature (London) **408,** 440.
5. Khriachtchev, L., Räsänen, M., Novikov, S., and Sinkkonen, J., (2001) Appl. Phys. Lett. **79**, 1249.
6. White, C.W., Budai, J.D.,Withrow, S.P., Zhu, J.G., Pennycook, S.J., Zuhr, R.A., Hembree, Jr, D.M., Henderson, D.O., Magruder, R.H., Yacaman, M.J., Mondragan, G., Prawer, S., (1997) Nuclear. Instrum. and Meth. in Phys. Res. B **127/128**, 545.
7. White, C.W., Budai, J.D.,Withrow, S.P., Zhu, J.G., Sonder, E., Zuhr, R.A., Meldrum, A., Hembree Jr, D.M., Henderson, D.O., Prawer, S., (1998) Nuclear. Instrum. and Meth. in Phys. Res. B **141**, 228.
8. Schuppler, S., Friedman, S.L., Marcus, M. A., Adler, D.A., Xie, Y.–H., Ross, F.M., Harris, T.D., Brown, W.L., Chabal, Y.J., Brus, L.E., and Citrin, P.H. (1994) Phys. Rev. Lett. **72**, 2648.

OPTICAL AMPLIFICATION IN NANOCRYSTALLINE SILICON SUPERLATTICES

PHILIPPE M. FAUCHET
Department of Electrical and Computer Engineering
University of Rochester, Rochester NY, USA
JINHAO RUAN
Department of Physics and Astronomy
University of Rochester, Rochester NY, USA

1. Introduction

The ever-increasing integration density and operating frequency of electronic chips is causing an electrical interconnect bottleneck to emerge [1]. Today, the overall performance of a multichip computing system is dominated by the limitations of the interconnections between chips and it is predicted that this problem will migrate to the single chip level after one decade. Interconnects are now the major bottleneck not only in terms of limiting performance, such as speed and signal integrity, but also in terms of power use and heat dissipation. Material innovations and traditional scaling will stop satisfying performance requirements when the feature size approaches 50 nm. Without a solution to the "interconnect problem," the growth of the semiconductor industry may come to a halt. Finding a solution to this problem is at least as crucial as any breakthrough in individual device performance (e.g., silicon quantum dot transistors) or any revolutionary advance in computing architecture (e.g., quantum computing).

Figure 1 illustrates the enormous heat dissipation problem that is facing the semiconductor industry. It shows how the power density required to operate a chip increases as the feature size on a chip decreases. The power density found in four other situations is shown for comparison.

Figure 2 shows how the maximum total interconnect length on a chip increases over the years. This surprising number results from the increasing use of parallelism in microelectronic chips and clearly identifies interconnects as a major limiting factor. Recently, a top-down system analysis has shown that optical interconnects have a significant advantage over copper interconnects even for intra-chip communications in almost all situations [2].

Using optical means to provide the interconnect is an attractive solution because it offers many benefits [3]. For example, there is essentially no distance-dependent loss or distortion of the signal, there are no deleterious fringing-field effects, there is no heat dissipation in the interconnection itself, and finally the interconnect does not have to be

L. Pavesi et al. (eds.), Towards the First Silicon Laser, 197–208.

redesigned as the clock speed increases. As a result, optical interconnects to electronic chips have been identified as a critical new technology by the International Technology Roadmap for Semiconductors. However, the practical implementation of optical interconnects is not without major challenges. One key problem is the lack of a suitable laser that is fully compatible with silicon microelectronic manufacturing.

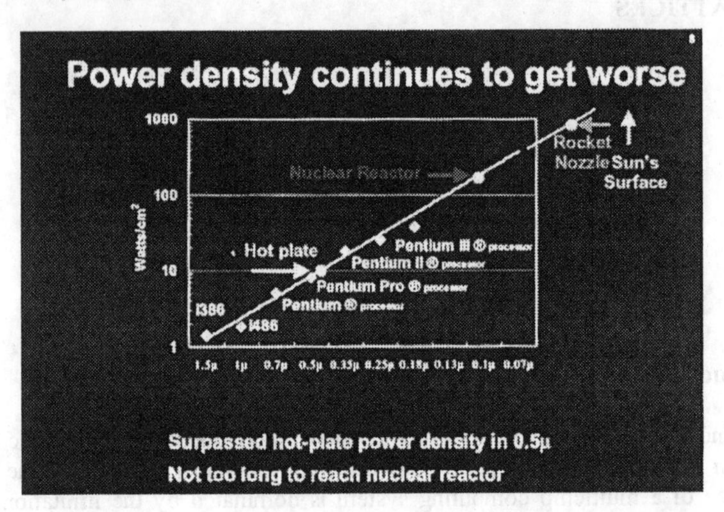

Figure 1: After [4]. Power density required to operate a chip plotted vs feature size.

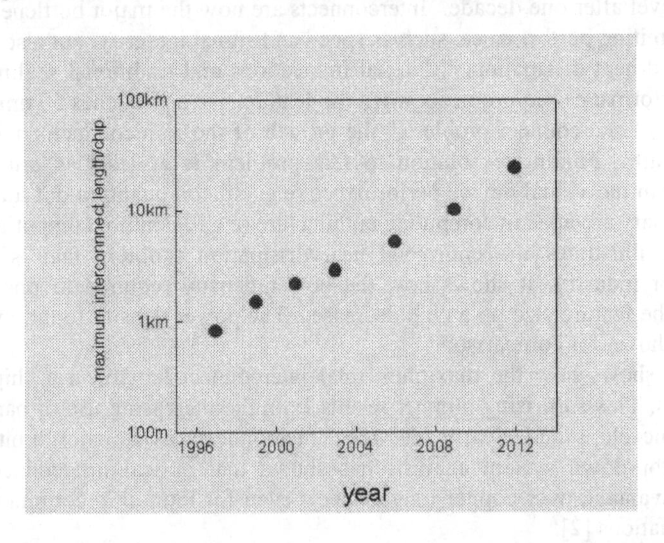

Figure 2: After [1]. Increase of the maximum total interconnect length on a chip in the next decade.

Silicon can already be used for most of the elements of a complete interconnect system (a desirable situation for integration with silicon chips): waveguides, photodetectors, and modulators. For example, we have very recently shown that one-dimensional photonic bandgap structures made of a porous silicon microcavity impregnated with liquid crystal molecules can act as an electrically-addressable mirror [5]. Such an element can be useful as an optical switch. However, one critical missing link is an appropriate silicon light source. Silicon is not an efficient light emitter because its indirect bandgap makes it necessary for a phonon to participate in the radiative recombination of an electron and a hole. In 1990, it was shown that under specific conditions silicon can become an efficient light emitter [6]. Since then, we and others have demonstrated that efficient (>10%) photoluminescence, which is due to the presence of silicon quantum dots or wires (nanocrystals) of sizes below 5 nm [7], can occur in samples prepared by a wide variety of techniques [8]. We achieved success in manufacturing stable porous silicon light emitting devices (LEDs) and integrating them with silicon circuitry on the same silicon chip [9]. Recently, the power efficiency of silicon LEDs has been increased from 0.1% to 1 % [10,11]. However, no practical implementation of these discoveries has followed yet because of the low brightness and low modulation frequency (limited to ~ 10 MHz [12]) of the devices and the fact they emit light over a wide spectral and angular distribution.

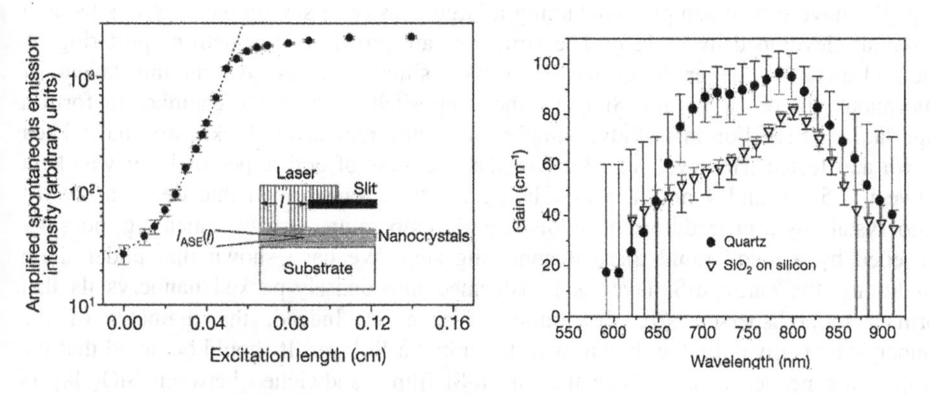

Figure 3: After[14]. (left) Variable stripe length (VSL) measurement on a SiO_2 sample containg Si nanocrystals. (right) Gain spectrum measured on two different samples.

A silicon laser would circumvent the speed limitation because its maximum modulation frequency is not limited by the recombination lifetime. Its output is also highly directional, unlike that of an LED. In addition, a silicon laser would avoid the need for integration of III-V semiconductors or other light-emitting materials onto silicon. A silicon laser made using "conventional" silicon manufacturing technologies would be a disruptive new technology that may have a large impact on the future of the semiconductor industry. A laser consists of an amplifying medium inserted in an optical cavity that provides feedback and creates a directional, coherent output beam. In our group, we have made highly resonant cavities in silicon (distributed Bragg reflector mirrors and optical microcavities [13]) but until recently nobody had been able to achieve amplification (gain) in silicon. The situation dramatically changed in 2000 when evidence for optical gain was presented in silicon nanocrystals that were produced by implanting silicon in silicon dioxide layers grown on silicon wafers [14] (Figure 3). In this work, ~3 nm silicon nanocrystals luminescent in the near infrared around 1.5 eV were inserted into a crude waveguide. The light emitted by the material was amplified as it traveled along the waveguide. Net modal gains between 10 and 100 cm^{-1} were measured using the variable stripe length (VSL) method. This unexpected result has now been independently confirmed [15,16,17], although a recent report pointed to possible problems with the VSL method [18]. To explain gain, Pavesi proposed a three-level system based on our earlier work [19], where pumping occurs between the valence and conduction bands inside the quantum dot and gain involves Si=O double bonds at the interface between the silicon nanocrystals and the oxide matrix. A very recent report confirms the energetic position of the Si=O levels [20], whereas a variation on this model has very recently been proposed [21] to explain gain. Very recently, Chabal confirmed the presence of Si=O bonds during the initial stage of oxidation of clean Si(100) surfaces. [22]

2. Sample preparation and characterization

We have grown samples containing a large density of silicon nanocrystals using a technique developed by us [23]. The structures are grown by magnetron sputtering. A thick (1μm) SiO_2 layer is grown on n-type silicon wafers. Alternating layers of amorphous silicon (a-Si) and SiO_2 are then deposited in the same chamber to form a superlattice (SL). For this study, samples with different layer thicknesses have been grown and tested. However, in all cases, the thickness of both types of layer was kept between 2.5 nm and 4 nm. The a-Si layers are then transformed into dense arrays of nanocrystals by a procedure that involves high temperature furnace annealing, possibly preceded by a quick rapid thermal annealing step. We have shown that under these conditions, the entire a-Si layer is transformed into densely-packed nanocrystals that form a monodispersed size distribution (Figure 4). Indeed, the diameter of the nanocrystals is given by the thickness of the initial a-Si layer. It should be noted that the temperature needed to crystallize the thin a-Si films sandwiched between SiO_2 layers increases steeply with decreasing a-Si layer thickness [23]. This behavior has been shown to exist in both Si and Ge [24] and a theoretical model has been proposed to explain this behavior [25].

Figure 5 shows TEM pictures of two crystallized superlattices. Note that when the a-Si layer thickness increases above 10 nm, the annealing treatment may produce rectangular bricks instead of spherical dots [26]. The origin of these bricks, which seem to be characterized by a ratio of the long to short axes equal to an integer, is at present not well understood. In the samples used to demonstrate gain, the nanocrystals are spherical. Folded acoustic mode Raman spectroscopy and small angle X-ray scattering have shown that even after crystallization the interface between the layers remains very smooth, on the order of one monolayer [26]. These observations, coupled with our TEM measurements, lead us to conclude that the nanocrystal size distribution is very narrow, on the order of one monolayer. To our knowledge, our technique leads to the narrowest nanocrystal size distribution in Si. Very recently, the size distribution in crystallized a-SiO/SiO_2 has been quantified and it was found that in these samples, which are identical to ours except that the nanocrystal separation in the layers is larger than in our samples, the size fluctuation is on the order of one monolayer [28]. Performing studies on monodisperse distributions promise to minimize potentially complicating factors such as energy transfer from smaller to larger crystallites, to simplify modeling, and to optimize the percentage of active nanocrystals.

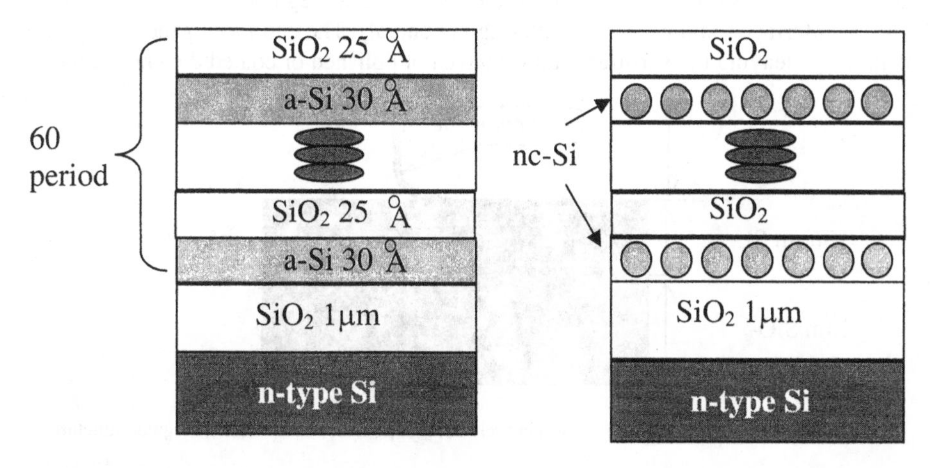

Figure 4: The as-grown superlattice structure is showed on the left. After annealing, the entire a-Si is transformed into densely-packed nanocrystals with identical size as shown on the right.

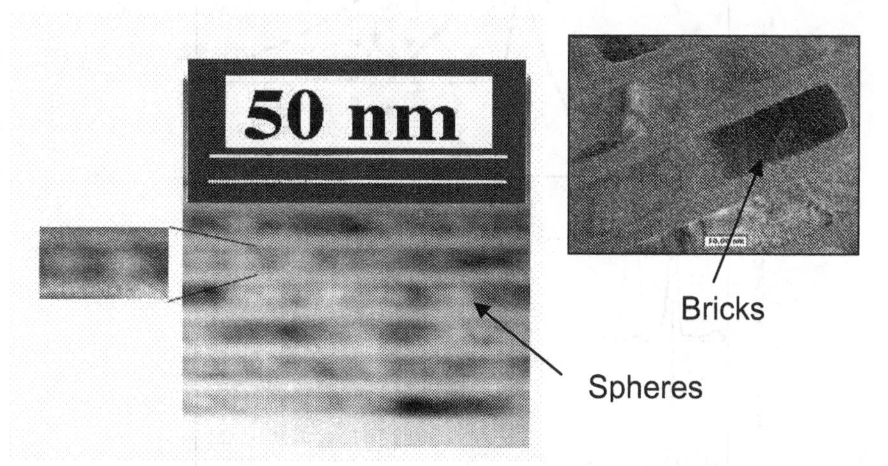

Figure 5: After Grom (2000). TEM pictures showing two crystallized nc-Si superlattices. When the thickness of the a-Si layer is small, the nanocrystals are spherical as shown on the left. As the thickness increases the nanocrystal shape is rectanglar, as shown on the right.

In all the samples tested so far, the number of Si/SiO_2 periods has been kept constant at 60, which translates into a thickness of approximately 0.3 to 0.4 μm. The SL forms the active core of the waveguide. The waveguide structure, shown in Figure 6, is designed to be single mode. Since the calculated mode confinement factor Γ is close to unity and the volume fraction of nanocrystals is ~ 50%, the modal gain, defined as the product of the material gain, Γ and the volume fraction of nanocrystals, is close to the material gain, making the conditions more favorable for observing gain than in the other published experiments. The possible disadvantages of our structures are the very close proximity of the nanocrystals (which raises questions about the quality of the interfaces) and the lack of a SiO_2 cladding layer at the top of the waveguide (which could increase losses by interface roughness and scattering in general). The end face of the waveguide is obtained by cleaving the substrate and we have not polished or coated the end facets.

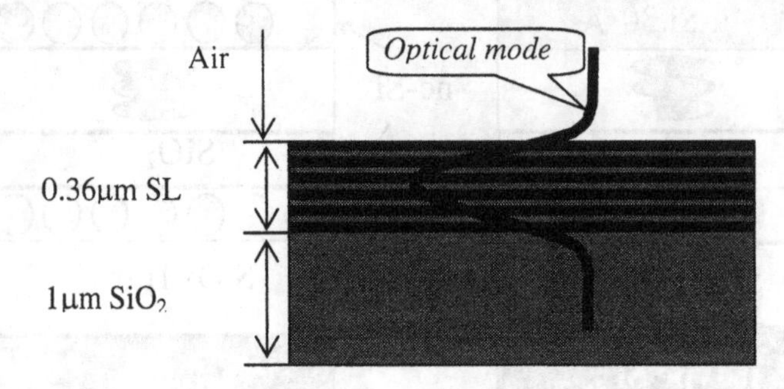

Figure 6: The waveguide mode is almost entirely in the superlattice core of our waveguide structure.

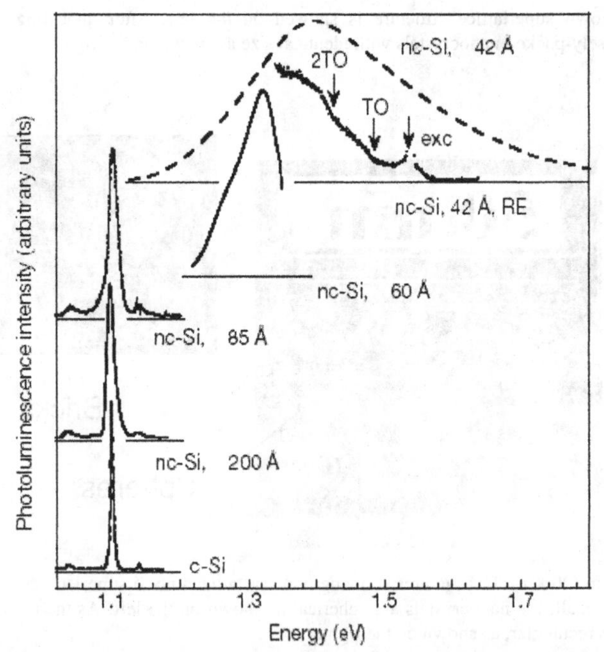

Figure 7: After [26]. Non-resonant photoluminescence spectra in c-Si and nc-Si superlattices with different nc-Si layer thicknesses. The resonant excitation energy and the energies of optical phonons in c-Si are shown by arrows. The spectra baselines are indicated and shifted for clarity.

Despite the very narrow nanocrystal size distribution, the measured normal-incidence PL linewidth is very large for the sizes of interest here. Figure 7 shows measured PL spectra for a series of samples in which the thickness of the a-Si layers was changed from 20 nm to 4.2 nm [26]. For nanocrystals above ~ 6nm, the linewidth is close to that of bulk silicon, whereas when the size decreases below ~ 6 nm, the PL linewidth rapidly increases to 200-300 meV and even beyond. The magnitude of this broadening appears to be inconsistent with our narrow size distribution, unless the homogeneous linewidth for single nanocrystals is itself very broad. Recent experiments have shown that the linewidth of one single small Si nanocrystal is at least 120 meV [27]. Recent calculations also suggest that the homogeneous linewidth of small nanocrystals should be much broader than that measured and calculated in II-VI quantum dots [29]. Thus the results of Fig. 7 can be explained at least in part by homogeneous broadening. In addition, they suggest a critical size (~ 6 nm) below which this broadening may become important.

The normal incidence PL has been measured in the samples tested for gain. The PL intensity increases linearly with pump power in the range of interest and the spectrum remains unchanged. These results indicate that Auger recombination does not play an important role and confirm that our size distribution is very narrow.

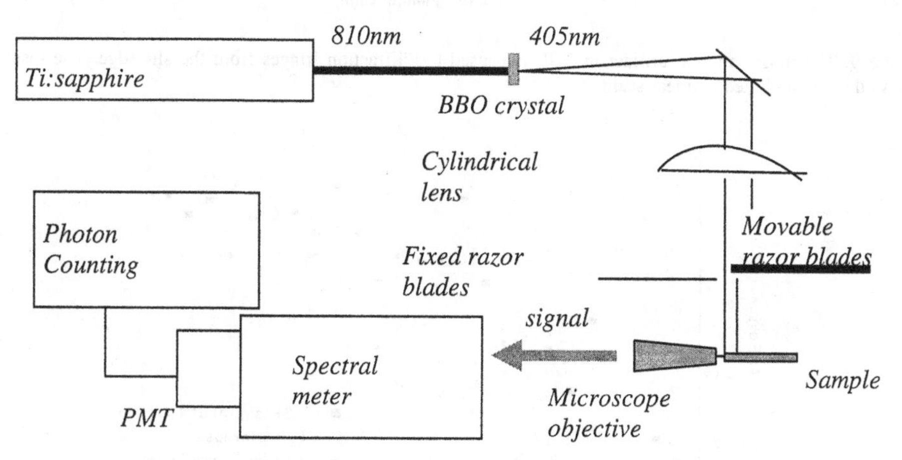

Figure 8: Experiment setup for the variable stripe length (VSL) measurements

3. Experimental arrangement

Our experimental set up is shown in Figure 8. The measurement is performed in the standard variable stripe length configuration. The pump intensity decreases by less than 20% over the maximum stripe length of 1 mm. Because the waveguide core is relatively thick, a 40 X microscope objective with a N.A. of 0.5 can be used while keeping any artifacts related to the amount of light that is collected to a minimum. The pump source is a frequency-doubled femtosecond Ti:sapphire laser. The pump wavelength is 405 nm, which efficiently injects carriers into the nanocrystals while keeping a homogeneous excitation as a function of depth. The intensity at the sample has been varied from ~ $1 W/cm^2$ to more than $1 kW/cm^2$. Artifacts such as diffraction fringes due to the slit edge have been identified, as shown in Figure 9. We avoided this artifact by starting the fit to the data for distances beyond the first diffraction peak.

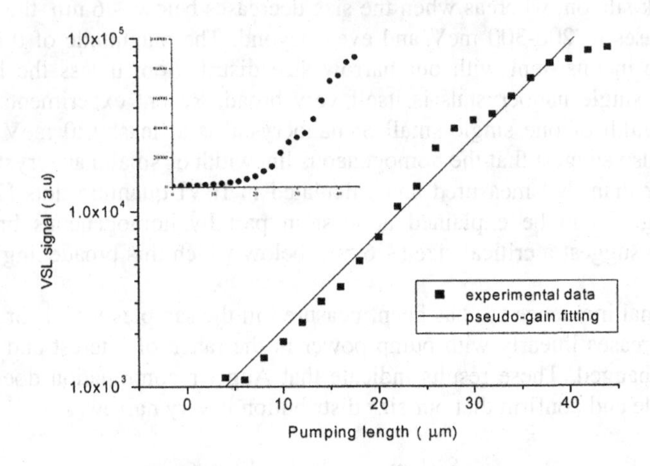

Figure 9: The most common artifact in VSL is caused by diffraction fringes from the slit edge. The inset shows the signal plotted in linear scale

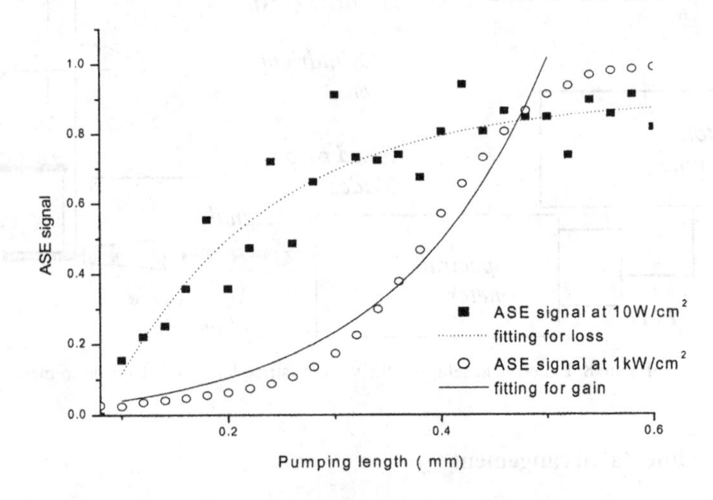

Figure 10: Amplified spontaneous emission(ASE) measured on a sample that exhibits gain at two different pump intensities. The solid squares show loss for low pump intensity and the open circles show gain at high pump intensity. The signal was measured at 800nm.

4. Results

We have grown and tested approximately 20 wafers, in an attempt to perform a parametric study of the gain involving parameters such as nanocrystal size and SL processing conditions. At this time, only one wafer yielded indication of gain. Indeed, it is our experience that VSL measurements are very sensitive to many experimental parameters such as output facet imperfections or pump laser conditions. Presently, we cannot state whether the conditions of this sample are the only ones that produce gain.

Figure 10 show results obtained on a sample prepared from the one wafer that did exhibit gain. When the pump intensity is raised from 10 W/cm^2 to 1 kW/cm^2, we observe a switch from a loss of $\alpha \sim 20\ cm^{-1}$ to a gain of $g \sim 100\ cm^{-1}$. These results strongly suggest that under intense pumping, it is possible to produce enough gain to overcome the waveguide losses. The magnitude of the gain is a factor of ~ 5 greater than in recent results obtained by Dal Negro [16]. We attribute this enhancement to the greater density of nanocrystals in our sample.

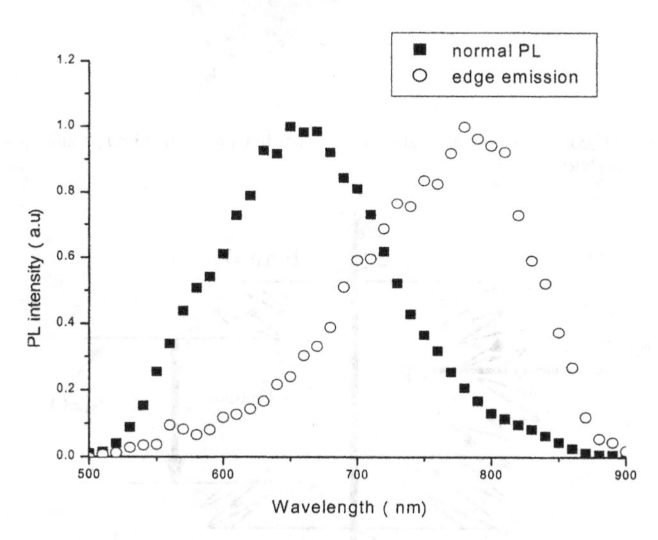

Figure 11: A pronounced and repeatable red shift was observed between the normal incidence PL and edge emission.

Figure 11 compares the spectrum of the normal-incident PL with that of the edge emission. We observe a pronounced and repeatable red shift of the gain spectrum compared to the normal PL. This is in contrast with Dal Negro's recent data that show a pronounced blue shift of the gain spectrum compared to the normal incidence PL [16]. It should be noted however that both our gain spectra are similar. It is in fact our normal incidence PL that is at much shorter wavelengths than in their samples, presumably because our nanocrystals are smaller. The near coincidence of our respective gain spectra while our respective nanocrystal sizes are different may be an important fact for the development of a comprehensive theory.

5. Discussion

We need to develop first a qualitative microscopic model and then a comprehensive model that includes both the details of the electronic properties of this complex material and the key parameters of the device structures. Figures 12 and 13 provide a first attempt at providing a simplified model.

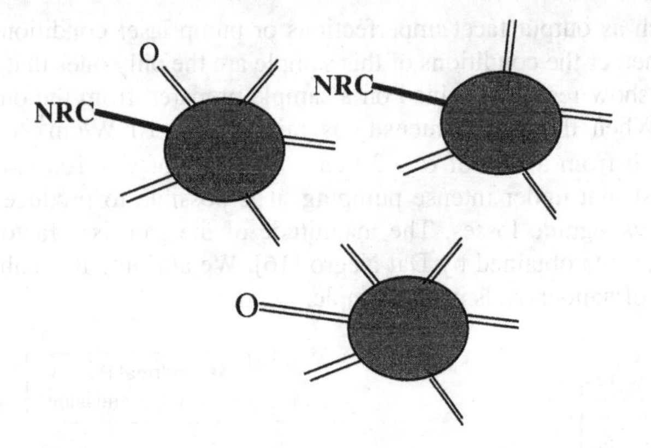

Figure 12: Typical collection of several nanocrystals having both Si=O bonds and non-radiative recombination centers (NRC).

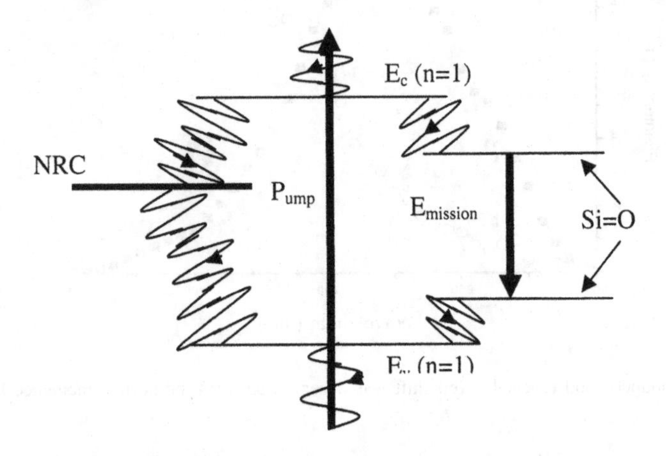

Figure 13: Schematic energy level diagram for a small (<3nm) nanocrystal . The free LUMO and HOMO states are shown in the center. The NRC is shown on the left and the Si=O state is shown on the right.

Let us consider a collection of nanocrystals as shown in Figure 12. On the surface of each nanocrystals, there are several Si=O bonds and either one or no non-radiative recombination center (NRC). If there were no Si=O bonds and luminescence resulted from radiative recombination of free excitons, those nanocrystals that contain a non-radiative center would be "dark", meaning that they would not luminesce. The presence of Si=O bonds opens a new radiative recombination pathway that is energetically favored if the nanocrystal is small enough [19]. Each nanocrystal could contains several Si=O bonds, and as it has very recently been shown [21], their energy levels do not depend on how many other such bonds exist on any given nanocrystal. The fundamental reason is that once carriers are captured on these bonds, their electronic wavefunction is deeply localized on the bonds itself [19]. Competition between capture by the NRC and the Si=O bonds makes luminescence possible even in the presence of a defect.

Figure 13 shows a possible energy level diagram for a small (< 3nm) nanocrystal. Optical pumping occurs well above the bandgap that is increased by quantum confinement. Relaxation of photoinjected carriers to the LUMO and HOMO states takes place on a time scale ~ 1 psec [30]. Capture by both type of surface bonds also takes place on a ~ 1 psec time scale, as is typical of defects. If the charge is captured by the NRC (which has arbitrarily been described as a single trap state), recombination out of that single trapped state typically occurs on a much longer time scale. Calculations have shown that recombination involving dangling bonds in Si nanocrystals take ~ 1 μsec [31]. We have shown previously that radiative recombination involving the Si=O bond occurs on a μsec time scale [19]. The final step involving recombination of carriers captured by Si=O bonds can be seen as hole capture, a fast process (~ psec). Thus, we see that the Si=O/nanocrystal system behaves as a four-level system, and that spontaneous radiative recombination involving the Si=O bond competes efficiently with recombination involving the NRC.

If one performs a simple numerical estimate for the parameters of our samples, including a gain of 100 cm^{-1} and assuming that 90% of the nanocrystals are dark (i.e., do not emit light), we find a spontaneous emission cross section of ~ 2×10^{-18} cm^2 and about ten excited electron-hole pairs per nanocrystal. This cross-section implies a radiative lifetime of ~ 10 μsec, which is consistent with theory and experiments involving Si=O bonds [19]. However, the presence of 10 free excitons in the nanocrystals does not seem realistic, as very efficient Auger recombination, involving 2 or more pairs, would rapidly deplete the population and thus preclude the possibility of finding the required number of emitters per nanocrystal. We propose a possible way out of this dilemma. Suppose that about 5 electron-hole pairs are injected in each nanocrystal. Within 1 psec of formation, they would all be localized at different Si=O bonds on the surface. If Auger recombination involving these highly localized, spatially-separated carriers is not as efficient as it is inside the bulk of a nanocrystal, then it will be possible to excite a large number of such surface bonds without triggering fast recombination via Auger processes. It would then be possible to populate a large number of light-emitting surface bonds, and the possibility of having multiple radiative sites per nanocrystal becomes a strong possibility. The validity of this simple model depends on the expected Auger rate for carriers that are highly localized and spatially separated. Theorists need to produce model calculations that would validate (or not) the concept.

It should be noted that the original proposal and demonstration of the involvement of Si=O bonds in the radiative recombination in silicon nanocrystals [19] was made for isolated nanocrystals passivated by Si-H bonds, not for nanocrystals embedded in an oxide matrix. Whether the actual localized radiative state is Si=O [19] or the situation is more complex [21] remains to be seen, but the exact nature of the light-emitting bond is not critical for this discussion.

6. Conclusions

We have designed efficient waveguide structures containing a large number of silicon nanocrystals of identical size. These samples have been characterized by a variety of techniques, including photoluminescence, Raman scattering, and TEM. The variable stripe length method was used to measure the presence of gain upon transverse pumping with a frequency-doubled CW modelocked Ti:sapphire laser. In one wafer, we have observed switching from attenuation to gain when the pump intensity was increased. A model was proposed and briefly discussed, which may explain the observations.

This research was supported in part by the Semiconductor Research Corporation and the National Science Foundation. The authors would like to thank Rishikesh Krishnan, Jianliang Li and Chris Striemer from the University of Rochester for technical support and Luca Dal Negro from the University of Trento for useful discussions regarding the experimental setup. PMF also acknowledges useful discussions with Dr. Christophe Delerue regarding theory and with Dr. Polman regarding modeling.

REFERENCES

1. Semiconductor industry association (SIA) roadmap (1997), http://public.itrs.net
2. Kapur, P. and Saraswat, K.C., *"Optical interconnects for future high performance integrated circuits,"* presented at the European Materials Research Society, Strasbourg, June 2002; Kapur, P., Ph.D. Dissertation, Stanford University, 2002
3. Miller, D.A.B., Proc. IEEE **88**, 728 (2000)
4. Pollack. F, *"New microarchitecture challenges in the coming generations of CMOS process technologies"* plenary talk at the MICRO-32 ACM/IEEE International Symposium on Microarchitecture, Haifa, Israel, (1999)
5. Weiss, S.M. and Fauchet, P.M., *"Electrically tunable porous silicon active mirrors,"* to appear in Physica Status Solidi (2002)
6. Canham, L.T., Appl. Phys. Lett. **57**, 1046 (1990)
7. Fauchet, P.M. and von Behren, J., Phys. Stat. Sol. (b) **204**, R7 (1997)
8. Fauchet, P.M., IEEE J. Select. Topics in Quantum Electron. **4**, 1020 (1998)
9. Hirschman, K.D. *et al.*, Nature **384**, 338 (1996)
10. Gelloz, B. and Koshida, N., J. Appl. Phys. **88**, 4319 (1999)
11. Green, M.A. *et al.*, Nature **412**, 805 (2001)
12. Peng, C. and Fauchet, P.M., Appl. Phys. Lett. **67**, 2515 (1995)
13. Chan, S. and Fauchet, P.M., Appl. Phys. Lett. **75**, 274 (1999); Chan, S. *et al.*, J. Am. Chem. Soc. **123**, 11797 (2001)
14. Pavesi, L. *et al.*, Nature **408**, 440 (2000)
15. Khriachtchev, L. *et al.*, Appl. Phys. Lett. **79**, 1249 (2001)
16. Dal Negro, L. *et al.*, *"Optical gain in PECVD grown silicon nanocrystals,"* presented at the SPIE meeting, Seattle, July 2002 and to appear in the conference proceedings
17. Luterová, K. *et al.*, Appl.Phys.Lett., **91**, 2896 (2002)
18. Valenta, J. *et al.*, Appl. Phys. Lett. **81**, 1396 (2002)
19. Wolkin, M.V. *et al.*, Phys. Rev. Lett. **82**, 197 (1999)
20. Puzder, A. *et al.*, Phys. Rev. Lett. **88**, 97401 (2002)
21. Ossicini, S. *et al.*, *"Si nanostructures embedded in SiO_2: Electronic and optical properties,"* presented at the SPIE meeting, Seattle, July 2002 and to appear in the conference proceedings. (2002); Ossicini, S. this volume (2003)
22. Chabal, Y.J., *et.al.*, Phys Rev. B **66**, 161315 (2002)
23. Tsybeskov, L. *et al.*, Appl. Phys. Lett. **72**, 43 (1998)
24. Zacharias, M. *et al.*, Appl. Phys. Lett. **74**, 2614 (1999)
25. Zacharias, M. *et al.*, Phys. Rev. B **62**, 8391 (2000)
26. Grom, G.T. *et al.* Nature **407**, 358 (2000)
27. Valenta, J. *et al.*, Appl. Phys. Lett. **80**, 1070 (2002)
28. Zacharias, M. *et al.*, Appl. Phys. Lett. **80**, 661 (2002)
29. Delerue. C *et al.*, Phys.Rev. B **64**, 193402 (2001)
30. Shank C. V. *et al.*, Phys. Rev. Lett. **50**, 454 (1983)
31. Delerue, C *et al.*, Phys.Rev. B 48, 11024 (1993)

OPTICAL GAIN FROM SILICON NANOCRYSTALS

A critical perspectives

A. POLMAN
FOM-Institute AMOLF
Kruislaan 407, 1098 SJ Amsterdam, The Netherlands
R.G. ELLIMAN
Department of Electronic Materials Engineering
Research School of Physical Sciences and Engineering
Australian National University, Canberra, ACT 0200, Australia

1. Introduction

It has generally been considered impossible to fabricate a silicon laser. The reason being that –due to its indirect bandgap– silicon has a small cross-section for stimulated emission. As a result, optical losses due to free carrier absorption are dominant. It has been proposed that Si nanocrystals offer a solution to this problem. The optical properties of silicon nanocrystals are quite well understood. This is the result of extensive research over the past ten years on porous Si as well as on Si nanocrystals embedded in an SiO_2 matrix. It is generally found that well-prepared and passivated Si nanocrystals exhibit photoluminescence in the wavelength range between 500 and 1100 nm. The luminescence is attributed to the recombination of quantum-confined excitons, the emission energy thus being strongly dependent on the nanocrystal size. SiO_2 is the ideal matrix for Si nanocrystals as it can passivate dangling bonds that may cause non-radiative quenching. Indeed, many of the optical characteristics of Si nanocrystals in SiO_2 prepared by different methods, as well as oxidized porous Si, are very similar. The radiative recombination process can be entirely understood assuming a "classical" model of recombination of excitonic singlet and triplet states of the excited Si nanocrystals. In this model, the optical transition is indirect in nature, and has a relatively small cross-section. In recent years, several applications of Si nanocrystals have been explored, and include light-emitting diodes,[1,2] non-volatile memories,[3,4] and sensitized optical amplifiers.[5,6] The fabrication of an optical amplifier or laser based on interband transitions has been considered impossible because –by analogy with bulk Si– the cross-section for free carrier absorption was thought to be higher than that for stimulated emission. Yet, in an article published in 2000, Pavesi *et al.* [7] claimed that optical gain could be achieved using Si nanocrystals, contrary to earlier predictions. Central in this claim is the presumption that the observed light emission from silicon nanocrystals is not due to the recombination of "free" excitons but rather to the recombination of electron-hole pairs trapped at an interface state. This could, according to the authors, reduce the deleterious effect of free carrier absorption. A three-level model was introduced to explain the ob-

L. Pavesi et al. (eds.), Towards the First Silicon Laser, 209–222.

served optical gain, with the intermediate level attributed to a Si=O double bond at the interface between the nanocrystal and the surrounding silicon oxide matrix.

In this paper we will first summarize recent work on the optical properties of Si nanocrystals made by ion implantation into SiO_2 thin films. These data provide extensive evidence for the "classical" excitonic model, in contrast to the surface defect model proposed by Pavesi *et al.*[7] We focus on four key issues that are important in engineering optically active Si nanocrystal assemblies: 1) passivation, 2) the homogenous and inhomogeneous line width, 3) interaction between nanocrystals, and 4) the recombination kinetics and internal quantum efficiency.

Next, we will provide experimental and theoretical arguments that the three-level Si=O defect model cannot be correct in the wavelength range at which optical gain was reported. We speculate that that optical gain could possibly be achieved in a four-level model in which the Si nanocrystals act as a sensitizer for defects or impurities in the SiO_2 host (such as e.g. Er). However, such a model does not involve quantum-confined excitons in the gain transition. We conclude that optical gain in Si nanocrystals remains an open question.

2. Optical properties of Si nanocrystals

2.1 PASSIVATION

Figure 1, taken from Min *et al.*[8] shows a photoluminescence (PL) spectrum from Si nanocrystals made using implantation of 50 keV Si ions into a thermally grown SiO_2 film on Si, at a fluence of 5×10^{16} cm^{-2}. The sample was annealed in vacuum at 1100 °C for 10 min. to induce the nucleation and growth of the nanocrystals. High-resolution transmission electron microscopy (reported in Ref. 8) shows the presence of nanocrystals with diameters ranging from 1 to 3 nm. The PL spectrum shows two broad features, peaking at 600 nm (2.1 eV) and 750 nm (1.6 eV), respectively. PL lifetime measurements at 15 K show a fast decay (τ<100 ns) for the band around 600 nm, and a slow decay (τ=0.63 ms) at 750 nm. The fast component for shorter wavelengths is attributed to luminescence from defect states in the glass that result from the ion implantation process. The defect luminescence originates from a very thin surface layer, as it was found to disappear after removing a 15 nm surface layer from the Si implanted film by wet etching.[9] The luminescence around 750 nm is attributed to luminescence from Si nanocrystals.

Passivation with H or D can be used to remove the defect band.[8] Figure 1(a) shows spectra after 600 eV D implantation at fluences of (0.9, 1.8, and 3.3)$\times10^{15}$ cm^{-2}. As can be seen, the defect band completely disappears after D implantation. Figure 1(b) shows that after passivation, thermal annealing at 400 °C leads to an eight-fold increase in the PL intensity, presumably due to the in-diffusion of the D that then passivates the deeper lying Si nanocrystals. Annealing at temperatures of 500 °C and higher reduces the passivation effect, presumably due to the out-diffusion of D. The PL lifetime remains unchanged upon the 400 °C passivation anneal, indicating that the increase in PL is due to an increase in the density of active nanocrystals in the film. Vice versa, it indicates that samples that are not well passivated contain a large density of optically inactive nanocrystals. In such films it is impossible to achieve optical gain, as the large un-bleacheable fraction of nanocrystals will cause loss that cannot be overcome with gain.

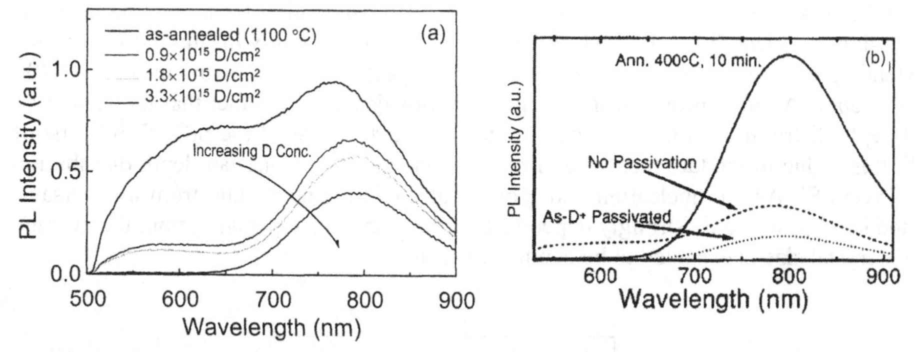

Figure 1 (a) Room temperature PL spectra of Si nanocrystal-doped SiO_2 made by 50 keV Si implantation (5×10^{16} Si/cm^2), annealed at 1100 °C for 10 min. and passivated by 600 eV deuterium implantation at (0.9, 1.8, and 3.3) $\times10^{15}$ D/cm^2. An Ar laser at 457.9 nm was used as an excitation source. (b) PL spectra before and after D passivation (same as in (a)), and after thermal annealing at 400 °C for 10 min. (From Min *et al.*, Ref. 8)

2.2 HOMOGENOUS AND INHOMOGENEOUS LINEWIDTH

The luminescence feature peaking at 750 nm (1.65 eV) has a spectral width of 300 meV. The width can be partly ascribed to inhomogeneous broadening due to the large size distribution of Si nanocrystals in the film. In addition, homogeneous broadening plays a role. Recent measurements by Valenta *et al.*[10] on single nanocrystals show a homogeneous line width of 120-160 meV. A similar value is derived from phonon features in PL excitation spectra by Diener *et al.*[11] We thus conclude that the 300 meV line width of the PL spectrum for the passivated sample in Fig. 1 is due to the broad size distribution, convoluted with a ~120 meV homogeneous line width. The large homogeneous broadening can also explain the fact that sometimes PL spectra from Si nanocrystals have tails that extend well beyond the wavelength of the bandgap of bulk Si.[12] It can also partly explain the multi-exponential decay in PL lifetime measurements, as will be discussed in section 2.3.

The fact that inhomogeneous broadening plays a role firstly appears from the fact that the lifetime measured on a particular sample varies with emission wavelength. For the passivated sample in Fig. 1 it ranges from 10 μs at 650 nm to 50 μs at 850 nm.[13] The luminescence lifetime measured at 750 nm is consistent with a theoretical value for nanocrystals with a diameter of 2.5 nm.[14] The long lifetimes are characteristic for an indirect-bandgap semiconductor. The fact that the lifetimes remain long even for wavelengths as small as 500 nm indicates that even particles as small as 1 nm possess an indirect bandgap. Given the inhomogeneous component in the spectral width it is possible to tailor the PL spectrum by tailoring the nanocrystal size distribution. For example, annealing in an oxygen ambient leads to oxidation of the Si nanocrystals, and hence a reduction of the size. This is shown in Fig. 2, which shows PL spectra (on a logarithmic scale) of samples that were annealed at 1000 °C in 1 Atm. O_2 at times ranging from 3 to 29 min. (data from Brongersma *et al.*, Ref. 15). The inset shows the expected change in size distribution: upon annealing the density of large nanocrystals decreases, while the density of small nanocrystals first increases. This is also seen in the PL spectra: the component at 900 nm continuously decreases with increasing oxidation time, while the signal at 600 nm first gradually increases. Using this oxidation technique the peak PL emission wavelength can be blue-shifted from 900 to 700 nm, with luminescence tails extending to 500 nm. Note that in all cases the spectral width remains in the 300–400 meV range. Another origin of inhomogeneous broadening is the fact that the nanocrystal depth distribution is not constant throughout the thickness of the SiO_2 film.[9] First of all, this is due to the fact that Si ion implantation leads to a Gaussian depth distribution of excess Si. As the nucleation and growth rates of Si nanocrystals from a supersaturated solid solution are strongly dependent on the degree of supersaturation, the average nanocrystal size is expected to be depth dependent.

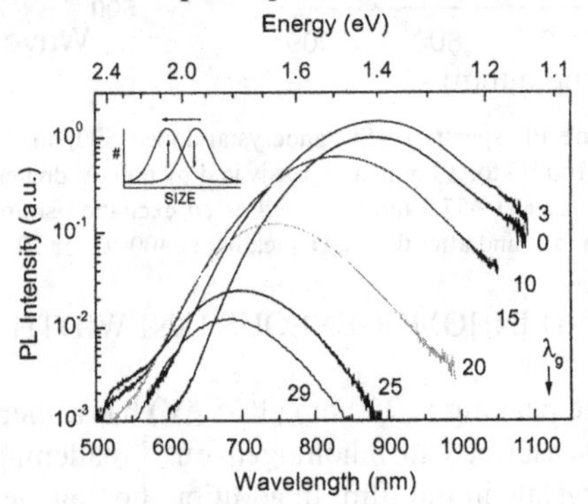

Figure 2 Room temperature PL spectra (plotted on logarithmic scale) of Si nanocrystals embedded in SiO_2 after thermal oxidation at 1000 °C in O_2 at 1 Atm., for times ranging from 0-29 min. (indicated in the figure). After oxidation, passivation was carried out using 600 eV D implantation followed by a 400 °C , 10 min. anneal. (From Brongersma *et al.*, Ref. 15)

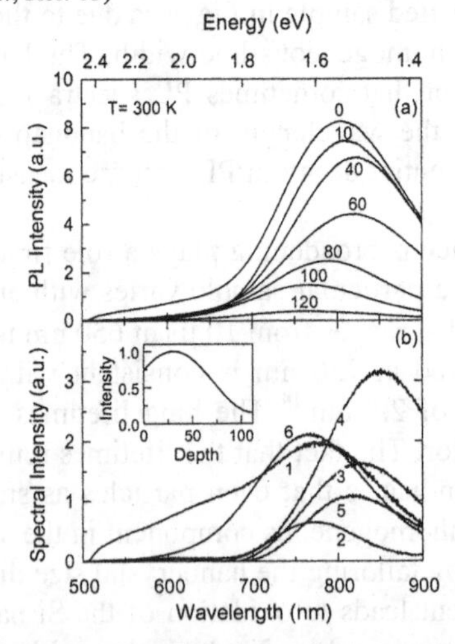

Figure 3 (a) Room temperature PL spectra obtained from a SiO_2 film containing Si nanocrystals, after etching in buffered HF for times ranging from 0 to 120 s. (b) Difference spectra obtained by subtracting spectra in (a) for subsequent etch steps, and corrected in such a way that the spectral intensity at a fixed wavelength is proportional to the average concentration of nanocrystals emitting at that wavelength. The inset shows the pump intensity profile as a function of depth in the SiO_2 film. (From Brongersma *et al.*, Ref. 9)

Figure 3 shows PL spectra taken after subsequent etching of a Si nanocrystal doped film in buffered HF for times ranging from 0–120 s (data from Brongersma *et al.*).[9] By subtracting the data taken after each subsequent etch and correcting for the fact that the pump beam creates a standing wave pattern leading to an inhomogeneous pump rate throughout each film, the normalized PL spectrum for each layer can be derived. It follows that small nanocrystals with emission peaking at ~750 nm are found predominantly near the surface and substrate interface, while larger nanocrystals, with emission peaking at ~ 850 nm, are found in the center of the film. The latter may be due to increased particle coarsening at high concentration, or increased interaction between nanocrystals at high concentration (see section 2.3), which leads to a spectral shift to larger wavelength. The relatively high density of small nanocrystals near the Si substrate is attributed to the effect of the substrate as a sink during precipitation. It is important to realize that the large inhomogeneous component to the emission spectrum of Si nanocrystals (due to the size variations, both locally and a as a function of depth) will cause undesirable self-absorption and optical losses.

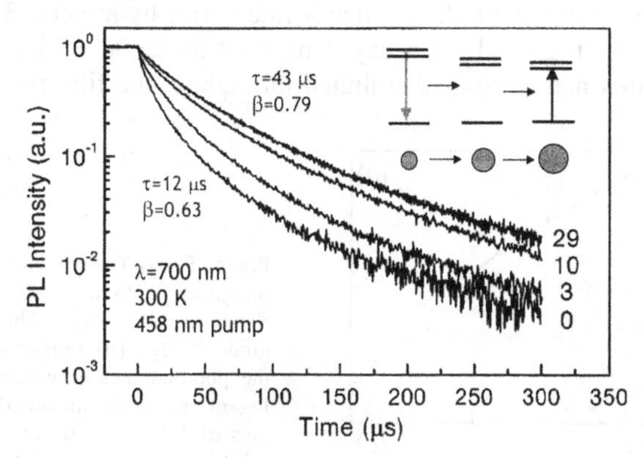

Figure 4 Normalized room temperature PL decay traces taken at 700 nm from samples oxidized at 1000 °C for 0, 3, 10, and 29 min. The inset shows a schematic of energy transfer between nanocrystals (From Brongersma *et al.*, Ref. 15)

2.3 MULTI-EXPONENTIAL DECAY

Figure 4 shows PL decay measurement of several oxidized samples, all taken at room temperature at the same wavelength of 700 nm (from Brongersma *et al.*, Ref. 15). The decay for the unoxidized sample is strongly non single-exponential. It can be fitted with a stretched exponential ($I(t)=\exp(-t/\tau)^{\beta}$) with a characteristic decay time of τ=12 μs and β=0.63.[15] One explanation for this non-exponential behavior is that interaction between Si nanocrystals takes place in the highly concentrated nanocrystal-doped film, in which energy is transferred from small nanocrystals (large bandgap) to large nanocrystals (small bandgap) as indicated schematically in the inset of Fig. 4. In this way, due to the inhomogeneous distribution of nanocrystals in the film, different nanocrystals, even with the same size, can decay at different rates. Figure 4 also shows decay rates after oxidation. As can be seen, the decay rate becomes more uniform, with τ=43 μs, and β=0.79. This is consistent with the fact that upon oxidation, the nanocrystal density and

the interparticle distance increase, thus leading to reduced interaction. The nanocrystal interaction mechanism was first demonstrated to occur by Kagan *et al.*[16] for highly concentrated ensembles of CdSe quantum dots. Detailed PL decay measurements by Linnros *et al.*[17] on differently prepared Si nanocrystal ensembles in SiO_2 made by ion implantation strongly indicate that nanocrystal interaction also takes place in that material. The interaction model is further supported by the data by Vinciguerra *et al.*[18] who found single-exponential decay in nanocrystals ensembles at low concentration.
We note that the large homogenous broadening observed for Si nanocrystals will also lead to multi-exponential decay. This is because a measurement at a fixed wavelength collects luminescence from nanocrystals with different size (within the homogeneous linewidth), and hence different decay time.

As an interesting corollary, we also note that in a thin-film system, purely single-exponential decay can never be achieved because the local optical density of states[19] is a strong function of position in the film. Calculations by De Dood *et al.*[20] show that in a 60 nm thick SiO_2 film on Si the local density of states varies by a factor 3 throughout the film. This implies that the radiative decay rate varies by a factor 3, and thus, if the quantum efficiency is high and the decay dominated by radiative decay, the integrated decay rate of excited nanocrystals distributed throughout the film thickness is not single-exponential.

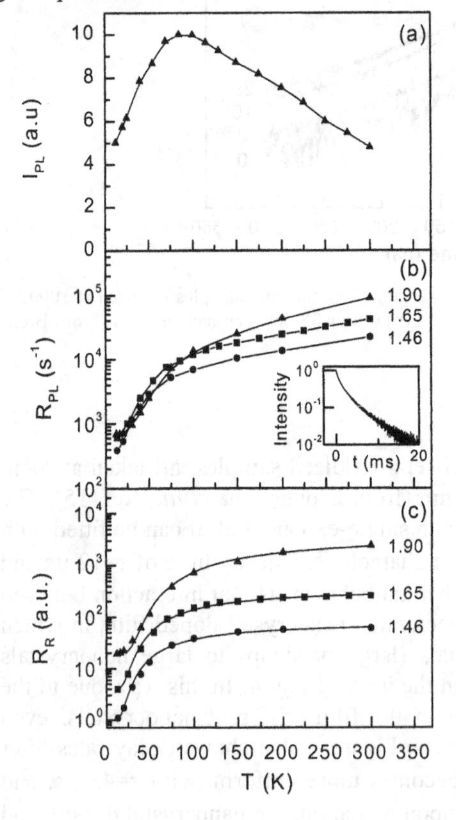

Figure 5 (a) Temperature dependence of the integrated photo luminescence intensity, I_{PL}, of Si nanocrystals in SiO_2. The solid line serves to guide the eye. (b) Temperature dependence of the photoluminescence decay rate, R_{PL}, on a logarithmic scale, measured at emission energies of 1.46 eV, 1.65 eV, and 1.90 eV. The solid lines through the data serve to guide the eye. The inset shows a typical decay trace taken at 1.65 eV and 15 K on a logarithmic intensity scale. (c) Temperature dependence of the radiative rate at emission energies of 1.46 eV, 1.65 eV, and 1.90 eV, obtained from a multiplication of the temperature dependent I_{PL} data in (a) and R_{PL} data in (b). Each data set is expressed in arbitrary units and was multiplied by a different constant factor to facilitate comparison. The solid curves are best fits of a model that takes into account the exchange splitting of the energy levels of quantum-confined excitons in the Si nanocrystals. (From Brongersma *et al.*, Ref. 13

2.4 RECOMBINATION MODEL AND QUANTUM EFFICIENCY

The nanocrystal luminescence properties depend strongly on the temperature.[13] Figure 5(a) shows PL spectra taken at 12, 100, and 300 K, taken from Brongersma *et al.*[13] When increasing the temperature from 12 to 300 K, a 60 meV shift in the peak emission energy is seen (from 1.60 eV to 1.54 eV). This is attributed to the change in bandgap energy with temperature. For comparison: the band gap variation with temperature for bulk Si is 50 meV over the temperature range from 10 to 300 K.[21] The luminescence intensity is strongly temperature dependent, as can be seen in Fig. 5(a). As the measurements were performed in the low-pump power regime, this implies that the quantum efficiency is temperature dependent: at room temperature it is less than 50 %. It should be noted that the internal quantum efficiency is a strong function of preparation and thermal annealing conditions. For example, in samples prepared by Fujii *et al.*[22] and Kovalev *et al.*[23] it was found that the PL intensity was constant with temperature. It was noted that the cool-down time after annealing is a critical parameter determining the quantum efficiency.[24]

The luminescence lifetime is strongly temperature dependent,[13,22,25] as can be seen in Fig. 5(b). For example, at 1.65 eV it decreases from 2 ms at 10 K to 50 μs at 300 K. By multiplying the data in Fig. 5(a) and Fig. 5(b) a relative measure of the radiative decay rate can be derived at each wavelength.[13] These data are presented in Fig. 5(c) for emission energies of 1.4 eV, 1.65 eV, and 1.90 eV. At each energy, the temperature dependence of the radiative lifetime is in perfect agreement with a model for the recombination of singlet and triplet excitons, with the singlet and triplet energies split by an exchange energy, Δ, as described in Ref. 13. The ratio of singlet-to-triplet decay rates increases from 300 to 800 with increasing emission energy (decreasing nanocrystal size), as expected. It is found that Δ ranges from 8 meV at low energy to 17 meV at higher energy, in agreement with values of the exchange energy calculated using effective mass theory.[26] The value for Δ is much smaller than the 71 meV that was reported earlier for oxidized nanocrystals by Kanemitsu *et al.*[27] whose luminescence is thought to originate from states at the Si-SiO_2 interface.

2.5 SUMMARY

In summary, a detailed study of the optical properties of silicon nanocrystals shows a coherent and comprehensive picture, in which all data can be described by a model in which quantum-confined excitons recombine in Si nanocrystals with a size-dependent indirect bandgap. Important key factors to take into account when designing nanocrystal assemblies with optimized properties are:

- Luminescence from SiO_2 films containing Si nanocrystals is not always due to nanocrystals,
- Passivation may be required to activate "optically dead" nanocrystals that are unbleacheable under optical pumping,
- Si nanocrystals in SiO_2 show large homogeneous and inhomogeneous broadening,
- The large inhomogeneous component (due to the size distribution) causes self-absorption and optical losses,
- Multi-exponential decay is a result of: 1) interaction between nanocrystals, 2) homo-

geneous broadening, and 3) variation in the local optical density of states,

- The internal quantum efficiency of Si nanocrystals can be well below 100 %, depending on the annealing treatment,
- The nanocrystal recombination kinetics are very well described by the single-triplet model of quantum confined excitons,
- Si nanocrystals in SiO_2 thus behave as indirect-bandgap semiconductors, with small optical cross-sections.

None of the above analysis points at a model in which the luminescence is due to the recombination of Si=O surface states, as proposed in Ref. 7. While these important factors must be considered to optimize the optical properties of Si nanocrystals, the problem of free carrier absorption in Si nanocrystals still remains. Unless the cross-section of this effect is several orders of magnitude smaller in Si nanocrystals than in bulk Si, it will be impossible to fabricate a Si nanocrystal laser.

3. Optical gain in Si nanocrystals

In the report[7] that claimed the observation of net optical gain in Si nanocrystals an entirely new mechanism for exciton recombination in Si nanocrystals was suggested. This new claim has triggered great interest and excitement. The reported net optical gain was based on several observations:

1) Optical transmission data that showed an absorption band near 800 nm. This band was attributed to a Si surface state (Si=O double bond) that was then assigned as the initial state in the gain transition,
2) Determination of the gain cross-section from these data by comparing the measured absorbance at 800 nm with that of the nanocrystal absorbance near 500 nm, for which the cross-section was measured,
3) The proposal of a three-level model for population inversion, including the Si=O surface state,
4) A comparison of this model with theoretical results reported in the literature describing the size-dependent energy levels of Si nanocrystals with Si=O surface states,
5) Variable-stripe length measurements of the optical emission around 800 nm from the output facet of the waveguide that were fitted with an amplified spontaneous emission model,
6) Normal-incidence pump-probe measurements that showed a ~15 % change in probe intensity upon pumping a 100 nm-thick nanocrystal doped film. This value was then converted to a gain coefficient of $\sim 10^4$ cm^{-1}.

In the remainder of this paper we report a critical assessment of each of these 6 observations, either by comparing with existing literature or with experiments performed in our own labs.

3.1 Si=O ABSORPTION BAND

Central in the interpretation of the data in Ref. 7 is the presumption that the observed light emission from silicon nanocrystals is not due to the recombination of "free" excitons but rather due to the recombination of electron-hole pairs trapped at an interface state. This could, according to the authors, reduce the deleterious effect of free carrier absorption, which usually prevents optical gain from indirect band gap semiconductors. The trap state was attributed to a Si=O double bond at the interface between the nanocrystal and the surrounding silicon oxide matrix.

Figure 6 shows the optical absorption data for a sample made using similar irradiation conditions as in Ref. 7 (80 keV, 1×10^{17} Si/cm^2).[28] Optical density is derived from transmission measurements assuming that changes in transmission are entirely due to absorption. This graph reproduces the data in Ref. 7. Figure 6 also shows is a measurement on a sample implanted with 400 keV Si at 3×10^{17} Si/cm^2. Interestingly, a completely different feature is now observed in the transmission measurement. Detailed analysis, reported in Ref. 28, reveals that the features observed in transmission experiments are not due to absorption, but rather to an interference effect, related to the refractive index modification created by the Si implant (see inset in Fig. 6). All transmission data can be fitted with a simple Fresnel model for a multilayer system with multiple reflection. Thus, the existence of a Si=O state cannot be deduced from the optical absorption data in Ref. 7.

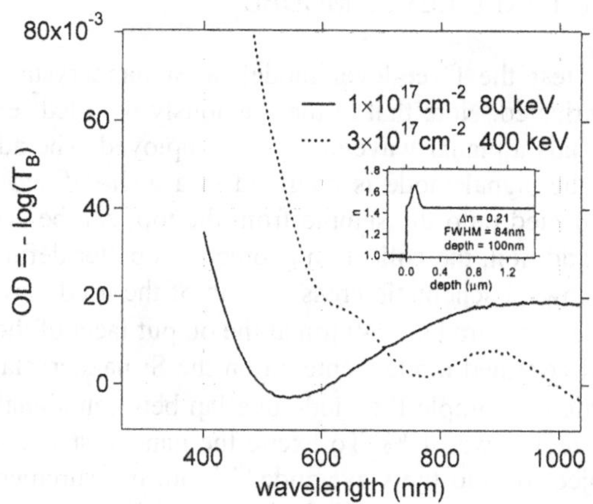

Figure 6 Optical density as a function of wavelength extracted from balanced transmission measurements for samples implanted with 80 keV (solid) and 400 keV (dashed) Si ions. The inset shows the refractive index profile for the 80 keV implant. (From Elliman et al., Ref. 28)

3.2 Si=O ABSORPTION CROSS-SECTION

In Ref. 7, a cross-section for stimulated emission was also derived (3×10^{-16} cm^2) from the apparent "absorption band". The finding described in section 3.1 demonstrates that this derivation is incorrect. More detailed measurements of the absorption spectrum can be performed using photo-thermal deflection spectroscopy. Figure 7 shows such spectra for the two samples in Fig. 6, taken from Ref. 28. For both samples an absorption edge is observed for smaller wavelengths, but no signature of an absorption band around 800 nm is found. From the data, an upper limit of the absorption cross-section of any transitions hidden in the noise of the measurements can be derived: 10^{-18} cm^2. This value is much too small to explain the reported optical gain.

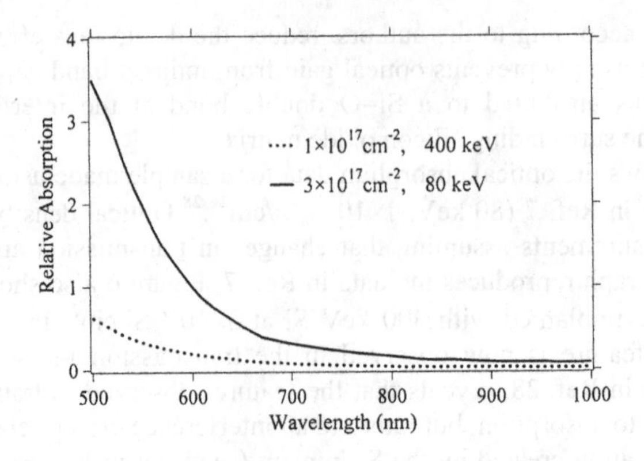

Figure 7 Relative absorbance as a function of wavelength measured by photothermal deflection spectroscopy for samples implanted with 80 keV (solid) and 400 keV (dashed) Si ions. (From Elliman *et al.*, Ref. 28)

3.3 TESTING THE THREE-LEVEL MODEL

To experimentally test the three-level model, a Si nanocrystal doped SiO_2 channel waveguide was fabricated. Note that in the previously reported[7] experiments on optical gain in Si nanocrystals a planar waveguide was employed. The advantage of a channel waveguide is that the signal mode is confined in a well-defined cross-section. In this way the pump, projected onto the sample from the top, can be accurately aligned with the waveguide. In addition, the collection geometry is better defined. Figure 8(a), taken from Kik *et al.*,[5] shows a schematic cross-section of the used waveguide geometry. An optical mode profile, measured at 1.49 μm at the output facet of the waveguide is shown in Fig. 8(b); a well-confined mode, centered on the Si nanocrystal-doped region is observed. In this particular sample the mode overlap between signal (at 1.49 μm) and Si nanocrystal doped region was 1 %. To excite the nanocrystals a continuous-wave 458 nm pump was projected onto the waveguide.[29] From measurements of the nanocrystal PL and rise time, as the pump was switched on and off, the pump rate was found to be 4×10^4 s^{-1} at 250 mW, well above the decay rate of 1.4×10^4 s^{-1} at 800 nm. Thus, assuming the simple three-level model proposed in Ref. 7, inversion should be created.

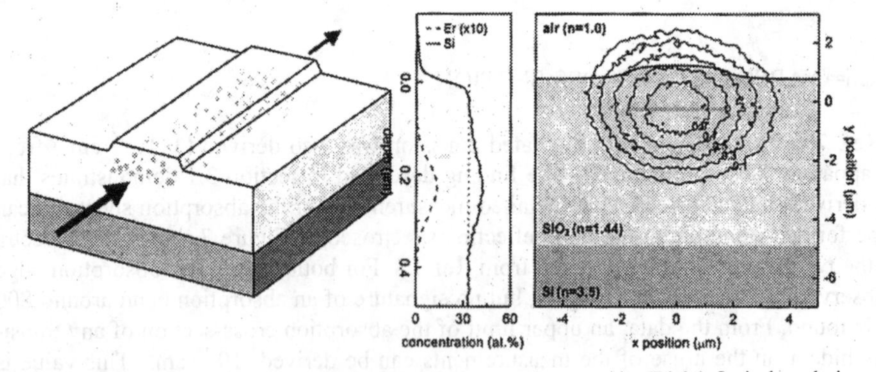

Figure 8 (left) Schematic of Si nanocrystal doped SiO_2 channel waveguide. (Right) Optical mode image at λ=1.49 μm measured at the output facet of the waveguide. (From Kik *et al.*, Ref. 5)

Figure 9 shows the normalized emission intensity at 850 nm coupled from the output facet of the waveguide, as a function of the length of the illuminated stripe (same technique as in Ref. 7) at two pump powers, 25 and 250 mW. [29] Except for a scaling factor of 4.6 the two curves are nearly identical. In the variable-stripe-length method, the emission intensity $I(l)$ as function of illumination length l is then given by:

$$I(l) = (I_{spont} / (g-\alpha)) \times (e^{(g-\alpha)l} - 1) \quad (1)$$

with I_{spont} the spontaneous emission intensity, g the gain, and α the overall waveguide loss, all defined per unit length.[30] Eqn. (1) fits the data very well, and we find $(g-\alpha) \approx -10$ cm^{-1} for both pump powers.[31] This demonstrates that g=0 and that the behavior in Fig. 9 is entirely due to waveguide loss: $\alpha \approx 10$ cm^{-1}. As no optical gain is observed we thus demonstrate that the simple three-level model is incorrect and that it should be modified either to include non-linear processes under high-intensity pumping, or should be replaced by a four-level model.[32] For example, optical gain could possibly be achieved in a four-level model in which the Si nanocrystals act as a sensitizer for defects or impurities in the SiO_2 host.

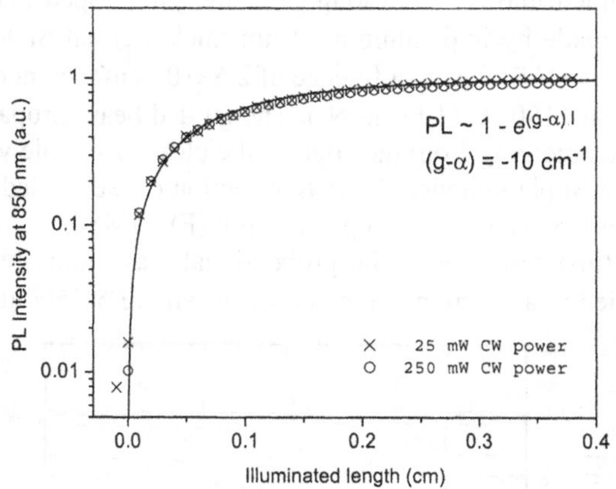

Figure 9 Normalized optical emission intensity at 850 nm from a Si nanocrystal doped waveguide (1×10^{17} Si/cm^2), as a function of illuminated length (pump wavelength 458 nm). The pump stripe geometry was the same as in Ref. 7. We attribute the data point for the shortest length to an artifact, e.g. the improper definition of length (for short length) and/or scattering of pump light. (From Kik *et al.*, Ref. 29)

3.4 COMPARING THE Si=O THREE-LEVEL MODEL WITH LITERATURE

The proposal for a Si surface state related model was carefully compared with the theory by Wolkin *et al.*[33] This model predicts that for nanocrystals emitting at 800 nm (the wavelength at which optical gain was reported), the emission is not due to a trap state, but rather due to recombination of quantum-confined excitons centered in the nanocrystals. In this model, a three-level model was only found for wavelengths < ~640 nm; a four level model was found for wavelengths < ~500 nm. These wavelengths are much different to those for which gain was reported.

3.5 VARIABLE-STRIPE LENGTH MEASUREMENTS AND CORRECTIONS

Recent literature[34] demonstrates that the use of the variable-stripe method is sensitive to several artifacts. First of all, diffraction of the pump beam from the slit will create an inhomogeneous intensity profile on the illuminated section of the waveguide. Second, a confocal effect takes place, in which the collection efficiency of light emitted from the waveguide is not constant over the length of the waveguide, but depends on the position of the focus of the collecting microscope objective. The challenge in performing variable-stripe measurements now lies in clear partitioning of 1) raw measured data, 2) precise determination of correction factors and their error bars, including at least diffraction and confocal effects, 3) correction of the raw data, and determination of their error bars, 4) fitting Eqn. (1) to the corrected data.

3.6 PUMP-PROBE EXPERIMENTS IN WAVEGUIDE GEOMETRY

Improved pump-probe experiments were undertaken to verify the results reported in Ref. 7. Instead of using a transmission geometry, which restricts the interaction length to the thickness of the implant distribution (~100 nm in the case of Ref. 1), an 800-nm probe beam was coupled into a ~2 cm long Si nanocrystal-doped planar waveguide.[35] The waveguide was made by implanting a 12 μm thick thermal SiO_2 layer (thermally grown on Si) with 600 keV Si ions to a fluence of 2.6×10^{17} cm^{-2}. Nanocrystals were then formed by annealing at 1100 °C (1 hr. in N_2). The guided beam propagated ~15 mm in the waveguide before emerging from the edge of the cleaved sample where its intensity was measured using a Si photodiode. A 2.5 mm section of the guided beam was irradiated with a 355 nm pump beam having a pulse width (FWHM) of 25 ns and a repetition rate of 30 Hz. The time response of the probe signal was monitored following each pump pulse and studied as a function pump energy density (3.5-3500 μJ/cm^2).

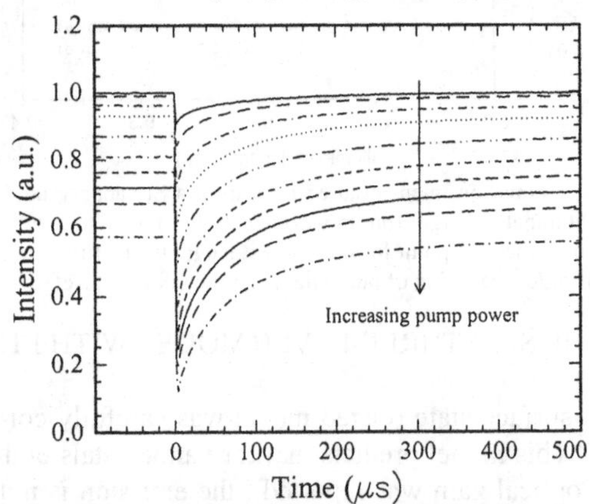

Figure 10 Time dependence of the normalized 800 nm probe intensity measured on exit from the waveguide, measured after excitation with a 25 ns pulse at 355 nm (at t=0). Curves from top-to-bottom correspond to pump energy densities of 3.5, 7.0, 15.5, 35, 70, 175, 350, 700, 1750 and 3500 μJ/cm^2, respectively. From Elliman *et al.*, Ref. 35)

Figure 10, taken from Ref. 35, shows time-dependent measurements of the probe output following the pump pulse. The data clearly show that the probe beam is attenuated by the pump and that the attenuation increases with increasing pump power – no evidence for optical gain was observed. The recovery of the probe signal is also evident, occurring on a time scale of 45-70 μs (depending on the pump power); a value similar to the PL lifetime of nanocrystals at 300 K. This suggests that the observed transient absorption is related to an excited-state absorption process such as free-carrier absorption. Whilst these are preliminary measurements they suggest that –similarly to bulk Si– free-carrier absorption could be a major limiting factor that prohibits the achievement of optical gain in Si nanocrystals.

Finally we mention measurements by Mimura et al.,[36] who have measured the infrared optical absorption spectra of P-doped Si nanocrystals. They attributed the measured absorption tails to intervalley transitions of free electrons supplied by the P dopants. The (extrapolated) absorption coefficient due to free carrier absorption around 800 nm was found to be $\sim 10^2$ cm^{-1} for highly doped nanocrystals.

4. Conclusions

The optical properties of Si nanocrystals made using Si ion implantation into SiO_2, and subsequent annealing, are quite well understood. The broad range of data presented in the first section of this paper was shown to be consistent with a model in which quantum-confined excitons recombine in Si nanocrystals with an indirect bandgap that varies with nanocrystal size. From these data it is evident that key factors of importance for the design of nanocrystal assemblies with optimized luminescence properties are: 1) passivation, 2) homogeneous and inhomogeneous broadening, 3) interaction between nanocrystals, and 4) the luminescence quantum efficiency. Si nanocrystals in SiO_2 behave as indirect-bandgap semiconductors, with small optical cross-sections.

The report by Pavesi *et al.*[7] that claimed the observation of net optical gain in Si nanocrystals is very intriguing. A critical assessment of each of the six observations made in the paper was presented. We conclude that the proposed simple three-level mode is incorrect. We speculate that optical gain could possibly be achieved in a material in which the Si nanocrystals act as a sensitizer for defects or impurities in the SiO_2 host (such as e.g. Er). However, such a model does not involve quantum-confined excitons in the gain transition. More detailed measurements of the variable-stripe technique are required to evaluate data taken using this technique. Pump-probe measurements on planar waveguides show no gain, but instead a transient enhanced absorption, possibly due to free carrier absorption. We conclude that optical gain from silicon nanocrystals remains an open question.

This work is part of the research program of the Foundation for Fundamental Research on Matter (FOM) and is financially supported by the Dutch Organization for the Advancement of Science (NWO). We gratefully acknowledge the contributions of our co-workers. They are listed in the references quoted in the captions.

References

[1] P. Photopoulos and A. G. Nassiopoulou, Appl. Phys. Lett. **77**, 1816 (2000).
[2] A. Irrera, D. Pacifici, M. Miritello, G. Franzò, F. Priolo, F. Iacona, D. Sanfilippo, G. Di Stefano, and P.G. Fallica, Appl. Phys. Lett. **81**, 1866 (2002).

[3] S. Tiwari, F. Rana, H. Hanafi, A. Hartstein, and E.F. Crabbé, Appl. Phys. Lett. **68,** 1377 (1996).
[4] M.L. Ostraat, J.W. De Blauwe, M.L. Green, L.D. Bell, M.L. Brongersma, J. Casperson, R.C. Flagan, and H. A. Atwater, Appl. Phys. Lett. **79,** 433 (2001).
[5] P.G. Kik and A. Polman, J. Appl. Phys. **91,** 534 (2002).
[6] H.-S. Han, S.-Y. Seo, and J.H. Shin, Appl. Phys. Lett. **79,** 4568 (2002).
[7] L. Pavesi, L. Dal Negro, C. Mazzoleni, G. Franzò, and F. Priolo, Nature **408,** 440 (2000).
[8] K.S. Min, K.V. Shcheglov, C.M. Yang, H.A. Atwater, M.L. Brongersma, and A. Polman, Appl. Phys. Lett. **69,** 2033 (1996).
[9] M.L. Brongersma, A. Polman, K.S. Min, and H.A. Atwater, J. Appl. Phys. **86,** 759 (1999).
[10] J. Valenta, I. Pelant, and J. Linnros, Appl. Phys. Lett. **81,** 1398 (2002).
[11] J. Diener, D. Kovalev, H. Heckler, G. Polisski, and F. Koch, Phys. Rev. B. **63,** 73302 (2001).
[12] F. Priolo, G. Franzò, D. Pacifici, V. Vinciguerra, F. Iacona, and A. Irrera, J. Appl. Phys. **89,** 264 (2001).
[13] M.L. Brongersma, P.G. Kik, A. Polman, K.S. Min, and H.A. Atwater, Appl. Phys. Lett. **76,** 351 (2000).
[14] M. Hybertsen, Phys. Rev. Lett. **72,** 1514 (1994).
[15] M.L. Brongersma, A. Polman, K.S. Min, E. Boer, T. Tambo, and H.A. Atwater, Appl. Phys. Lett. **72,** 2577 (1998).
[16] C. R. Kagan, C. B. Murray, M. Nirmal, and M. G. Bawendi, Phys. Rev. Lett. **76,** 1517 (1996).
[17] J. Linnros, N. Lalic, A. Galeckas, and V. Grivickas, J. Appl. Phys. **86,** 6128 (1999).
[18] V. Vinciguerra, G. Franzò, F. Priolo, F. Iacona, and C. Spinella, J. Appl. Phys. **87,** 8165 (2000).
[19] E. Snoeks, A. Lagendijk, and A. Polman, Phys. Rev. Lett. **74,** 2459 (1995).
[20] M.J.A. de Dood, L.H. Slooff, A. Moroz, A. van Blaaderen, and A. Polman, Phys. Rev. A. **64,** 33807 (2001), P.G. Kik, "Energy transfer in erbium doped optical waveguides based on silicon", Ph.D. Thesis, FOM-Institute AMOLF (2000), p. 102.
[21] S.M. Sze, *Physics of semiconductor devices* (John Wiley and Sons, New York, 1981).
[22] S. Takeoka, M. Fujii, and S. Hayashi, Phys. Rev. B **62** 16820 (2000).
[23] D. Kovalev, private communication (2002).
[24] M. Fujii, private communication (2002).
[25] J. Diener, D.I. Kovalev, G. Polliski, and F. Koch, Appl. Phys. Lett. **74,** 3350 (1999).
[26] P.D.J. Calcott *et al.*, J. Phys. Cond. Matt. **5,** L91 (1993).
[27] Y. Kanemitsu, Phys. Rev. B **53,** 13515 (1996).
[28] R.G. Elliman, M.J. Lederer, and B. Luther-Davies, Appl. Phys. Lett. **80,** 197 (2002).
[29] P.G. Kik, M.J.A. de Dood, and A. Polman, to be published.
[30] Note that Eqn. (1) deviates from Eqn. (1) in Ref. 7, in that it does not include the linear term l in the prefactor, see: K.L. Shaklee, R.E. Nahaory, and R.F. Leheney, J. Lumin. **7,** 284 (1973).
[31] We note that, as discussed in section 3.5, variable-stripe length methods are sensitive to artifacts due to diffraction and confocal effects. The data by Kik *et al.* presented here are not corrected for such effects. However, the fact that similar curves are found for low and high pump power is a strong indication that no optical gain is observed.
[32] A. Polman, presented at the MRS Fall Meeting, Boston, November, 2001.
[33] M.V. Wolkin, J. Jorne, P.M. Fauchet, G. Allan, and C. Delerue, Phys. Rev. Lett. **82,** 197 (1999).
[34] J. Valenta, I. Pelant, and J. Linnros, Appl. Phys. Lett. **81,** 1398 (2002).
[35] R.G. Elliman, M.J. Lederer, N. Smith, and B. Luther-Davies, presented at 13th International Conference on Ion Beam Modification of Materials, Kobe, Japan, September 1-6, 2002, to be published in Nucl. Instr. and Meth. B (2003).
[36] A. Mimura, M. Fujii, S. Hayashi, D. Kovalev, and F. Koch, Phys. Rev. B. **62,** 12625 (2000).

OPTICAL GAIN MEASUREMENTS WITH VARIABLE STRIPE LENGTH TECHNIQUE

J. VALENTA[1], K. LUTEROVÁ[2], R. TOMASIUNAS[3,4],
K. DOHNALOVÁ[1,2], B. HÖNERLAGE[3] and I. PELANT[2]
[1] *Charles University, Faculty of Mathematics and Physics, Ke Karlovu 3, 121 16 Praha 2, Czech Republic*
[2] *Institute of Physics, Academy of Sciences of the Czech Republic, Cukrovarnická 10, 162 53 Praha 6, Czech Republic*
[3] *IPCMS, GONLO, UMR7504, CNRS-ULP, 23 Rue du Loess, 67037 Strasbourg Cedex, France*
[4] *Institute of Materials Science and Applied Research, Vilnius University, Sauletekio 10, 2040 Vilnius, Lithuania*

1. Introduction

Optical gain in semiconductors plays a decisive role when evaluating potential of a given semiconducting material in photonic applications: The occurrence of positive optical gain is a necessary (albeit insufficient) condition for realisation of a semiconductor laser diode. It is therefore highly desirable to have a simple and reliable experimental method enabling us to get information about the presence/absence of gain and, in affirmative case, about its magnitude. Despite the fact that semiconductor lasers are pumped by electric current, the method must be purely optical so as to circumvent possible complications arising from the necessity of fabrication of suitable electric contacts and p-n junctions, detrimental Joule heating etc. In other words, the method should be based on optical rather than electrical pumping.

A method meeting all these requirements was proposed by Kerry L. Shaklee and coworkers from Bell Telephone Laboratories in 1971 [1,2]. The method is based on exciting a specimen of semiconductor with a laser beam focused, using a cylindrical lens, into a narrow stripe. The length of this pumped "line" is let to vary and light output (amplified spontaneous emission) is measured as a function of the stripe length l. This experimental technique, later on called "variable stripe length (VSL) method", became rapidly widespread and was contributing significantly to the development of a family of semiconductor lasing compounds (for instance already in 1971 (!) Shaklee et al. demonstrated high optical gain in GaN [2]). Besides such a practise-related exploitation, the VSL method was applied successfully also for studying the laws of new excitonic recombination channels in highly excited semiconductors (e.g. biexciton radiative decay and stimulated emission in CuCl [3] and exciton-exciton collisions in GaSe [4]). Later in eighties and nineties the method proved its potential also in studying low-dimensional semiconductor structures [5,6,7] and continues to be widely used to this day – not only

L. Pavesi et al. (eds.), Towards the First Silicon Laser, 223–242.

in semiconductor optoelectronics but also for investigation of luminescence and lasing potential of e.g. polymer substances [8,9].

In the course of thirty years, however, some of the basic principles, conditions and limitations of the method seem to be lost to a considerable extent [10]. The aim of this contribution is to recall and review the fundamentals of the VSL method, to discuss its week points and application limits. In particular, we wish to turn reader's attention to possible drawbacks of the method when measuring relatively low gain values, as expected in silicon nanostructures. The article is organised as follows: In Sec. 2 the principle, experimental realisation and mathematical description of the method in terms of kinetic equations are presented. Sec. 3 reviews application of the VSL technique to the measurement of high gain in direct-gap semiconductors and displays some typical examples. Sec. 4 is devoted to potential difficulties when the method is used in low-gain materials and Sec. 5 compares briefly the VSL method with other experimental ways to measure optical gain. Sec. 6 is devoted to summary and conclusions.

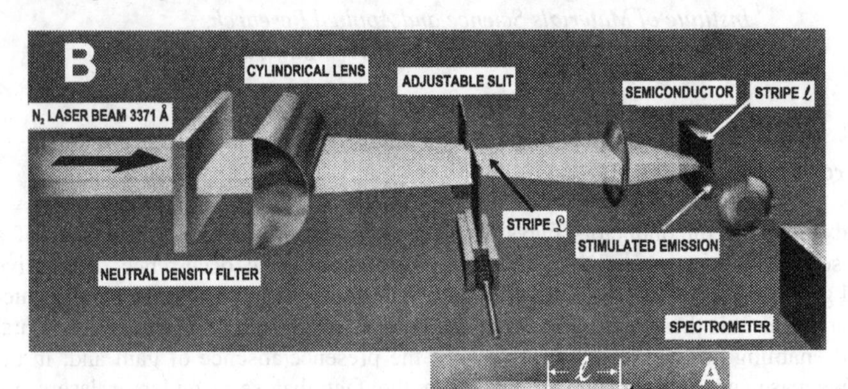

Figure 1.
(A) Schematics of the VSL method principle.
(B) View of a typical experimental set-up. According to [11].

2. The basis of the VSL method

Let us suppose a luminescent semiconductor sample in the form of a parallelepiped, cut or cleaved from a standard polished wafer (Fig. 1A). Photoluminescence is excited with a laser beam focused into a narrow line. In order to assure a homogeneous excitation density distribution over the whole stripe length l, the use of laser sources possessing a rectangular beam cross section (not a Gaussian beam) is required. An appropriate laser system from this point of view is either an excimer or N_2-laser. Fortunately, also excitation wavelengths of these lasers (308 nm in a XeCl laser or 337 nm in a N_2-laser) provide suitable high-energy photons to excite most semiconductors across the band gap. Thus these lasers appear to be almost ideal excitation sources in most cases. The use of other laser systems is, of course, not excluded (frequency doubled or tripled Nd:YAG laser etc.) but attention must be paid to guarantee a constant excitation profile along the stripe, especially when long stripes are in use.

To focus the laser beam into a stripe, the use of a cylindrical lens is obviously indispensable. Fig. 1B displays the relevant set-up in more detail [11]. As a rule, it is not the first stripe, located at the focal plane of the cylindrical lens that is projected onto the sample; this primary stripe is usually imaged by a second (spherical) lens and only this "secondary" image represents the excited region on the surface of the sample. In the focal plane of the cylindrical lens is located an adjustable slit by means of which the length of the primary image ℓ can be varied. Such a location of the slit is dictated by the necessity of having clean cuts in the both ends of the stripe. Disadvantage may be a laser-induced damage of the slit jaws due to high excitation density but shifting the slit slightly out of the focus can prevent it.

The double projection follows two goals: First, the magnification ratio M of the spherical lens image is selected to be less than unity which increases the accuracy of determination of the stripe length $l = M\,\ell$. This is of importance especially when the investigated range of l is small, $l \leq 100$ μm. Second, the condition $M < 1$ also minimises the *unwanted effect of the diffraction pattern* that originates inevitably on the adjustable slit jaws. The stripe width t after two-lens projection usually varies between 10 μm and 20 μm. Owing to the high absorption coefficient α of semiconductors in the UV region, the penetration depth of the excitation stripe is very low, $\alpha^{-1} \cong 0.1 - 1$ μm.

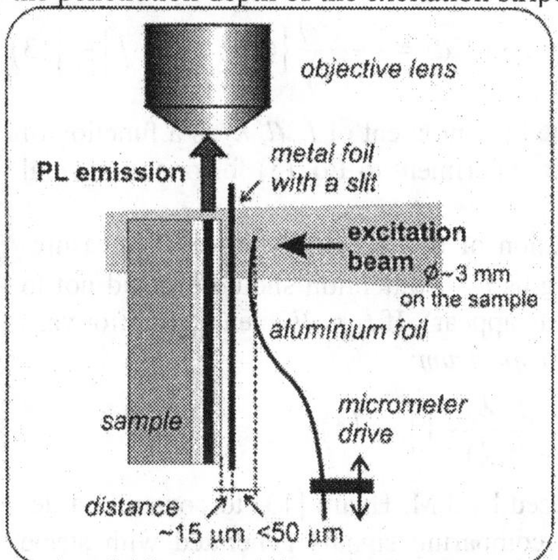

Figure 2. The set-up with a metallic slit and moving jaw close to the sample. The detection part consists of a microscope objective lens and imaging spectroscope with a CCD camera (not shown).

Recently, we used successfully an alternative set-up (see Fig. 2): The sample was excited through a fixed metal slit (width of 10 μm or larger) placed close to the sample surface (20-50 μm). The resulting stripe length was varied by moving a metal foil over the slit with the aid of a micro-stepping motor. The signal was collected by microscope objectives (with various numerical apertures - NA), focused on the input slit of an imaging spectrograph and detected by a CCD camera [12]. This arrangement ensured the constant position of the excitation spot on the sample surface, negligible influence of light diffraction (diffraction effect could be restricted to the first 10 μm of the stripe), and perfect control of the experiment geometry.

The recombination radiation originating in the excited stripe (luminescence or spontaneous emission) is emitted randomly to all directions. Spontaneous photons can induce acts of stimulated emission and this effect is monitored along the axis of the stripe by measuring the total light output intensity I_{tot} leaving the sample facet (Fig. 1A). Considering the stripe as a one-dimensional object ($l \gg t$) along the x-axis and provided population inversion is reached, the excited part of sample acts as a single-pass amplifier for photons coming in the direction of the stripe. The net gain G is defined as a relative change of light intensity passing an infinitesimal distance dx

$$G = \left[g(\lambda) - K \right] = \frac{dI(x,\lambda)}{dx} \frac{1}{I(x,\lambda)} \qquad (1)$$

where K stands for constant losses and $g(\lambda)$ is a gain coefficient (negative absorption coefficient)[#]. The total change of detected intensity when increasing stripe length by dx is a sum of a gain magnification of the incoming light and spontaneous emission $I_{sp}(\lambda)$ from the stripe of length dx:

$$dI_{tot}(x,\lambda) = \left[g(\lambda) - K \right] \cdot I_{tot}(x,\lambda) \cdot dx + I_{sp}(\lambda) \cdot dx. \qquad (2)$$

Here, it is generally accepted that $I_{sp}(\lambda)$ and $g(\lambda)$ are independent of x and, more importantly, the coupling of emission to detector is constant, independent of x.

Solving this linear inhomogeneous differential equation we retrieve the classical VSL equation [1,11]

$$I_{tot}(l,\lambda) = \exp\left[G(\lambda) \cdot l \right] \cdot I_{sp}(\lambda) \cdot \int_0^l \exp\left[-G(\lambda) \cdot x \right] \cdot dx = \frac{I_{sp}(\lambda)}{G(\lambda)} \left\{ \exp\left[G(\lambda) \cdot l \right] - 1 \right\} (3)$$

that enables us to evaluate net gain G from the measurement of $I_{tot}(l, \lambda)$ as a function of l. The simplest approach how to do it is to fit experiment to Eq. (3) for one or several selected wavelengths.

More complex analysis involves combination of two spectrally resolved measurements $I_{tot}(l_1, \lambda)$ and $I_{tot}(l_2, \lambda)$ at two stripe lengths l_1, l_2 (attention should be paid not to go to the stripe lengths where saturation effects appear). If $l_1 = 2l_2$ then straightforward algebra leads to an analytic expression for *gain spectrum*:

$$G(\lambda) = \left(1/l_2 \right) \ln \left(\frac{I_{tot}(l_1, \lambda)}{I_{tot}(l_2, \lambda)} - 1 \right). \qquad (4)$$

Let us mention that the relation (4) was used by J.M. Hvam [13] to construct a device that records directly gain spectrum by comparing signals generated with stripe lengths periodically changing between l_1 and l_2.

3. Application of the VSL technique to high-gain semiconductors

We shall demonstrate in short how the VSL technique can be applied to high gain (basically direct-gap) semiconductors. In this most simple case we may assume $G.l \gg 1$ and

---

[#] Note: In the language of (semiconductor) laser physics the net gain G is called cavity gain $g_C = g_{mod} - \alpha_i - \alpha_m$, where g_{mod} is a modal gain, α_i and α_m, mean internal losses (inside the cavity) and mirror losses, respectively.

Eq. (3) reduces to $I_{tot}(l, \lambda) \propto \exp(G.l)$. Therefore, an exponential increase of I_{tot} with l is a direct indication of gain and the presence of stimulated emission. Fig. 3A displays one of early examples [2]: Data obtained on GaN needles at T = 2 K exhibits a linear increase of I_{tot} with l in logarithmic scale. Relevant values of G (10^3 – 10^4 cm^{-1}) can be read from the slope. At the same time this figure reveals an important feature: For longer excitation lengths l the data (full lines) deviate from the straight lines. This is due to *saturation* of the stimulated emission, a quite natural effect occurring when photon density is increased [14] because the population inversion becomes progressively exhausted and I_{tot} ultimately varies linearly with l only (gain appearing in Eq. (3) can be referred also as a small signal gain). A simple four-level model [15] predicts that a critical stripe length l_c for deviation of the experimental data from the small signal approximation (Eq. (3)) is inversely proportional to the pump intensity, $l_c \propto 1/I_{pump}$, provided G grows linearly with I_{pump}. Such behavior can be indeed found in Fig. 3A. Therefore, the appearance of a saturated region in VSL plots (like in Fig. 3A) is being generally accepted as additional evidence for stimulated emission.

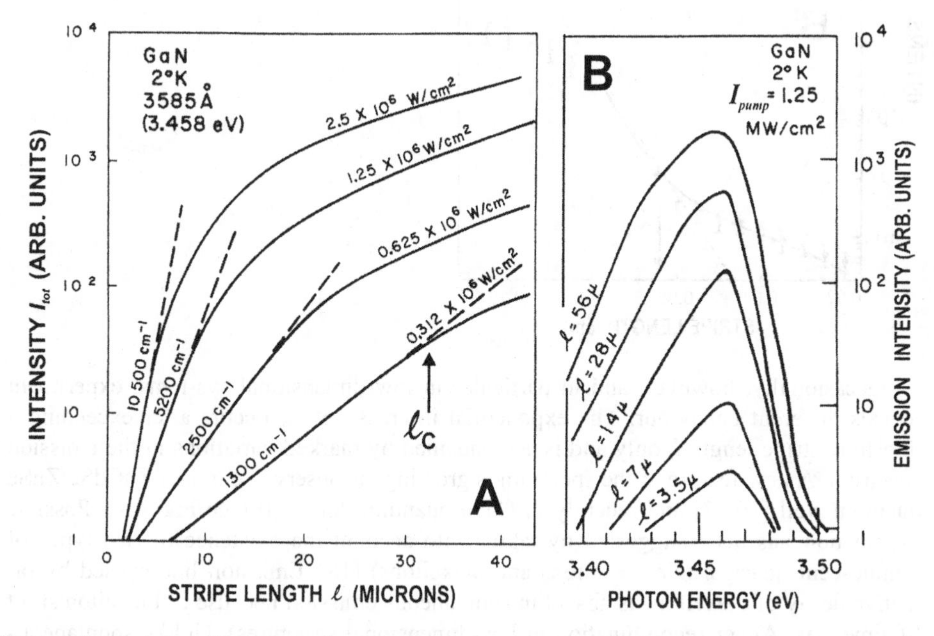

Figure 3. (A) Variation of light output from a GaN sample as a function of stripe length. Pump intensities (N_2-laser) and G values are given at each curve. (B) Evaluation of emission spectrum with increasing stripe length for pump intensity I_{pump} = 1.25 MW/cm^2. According to [2].

In the GaN example under debate G is high, up to 10^4 cm^{-1}, i.e. its magnitude is close to the value of the optical absorption coefficient for photons with above band gap energy. One can thus interpret this particular gain as being due to band-to-band recombination of electron-hole plasma and conceive that both spontaneous and stimulated emission originate basically from identical types of excitation (homogeneous line broadening). The last claim is supported by virtually invariable emission line shape with increasing stripe length (Fig. 3B). This or similar observation (possibly a small red shift and line narrowing) justifies the use of Eq. (3) for fitting the data.

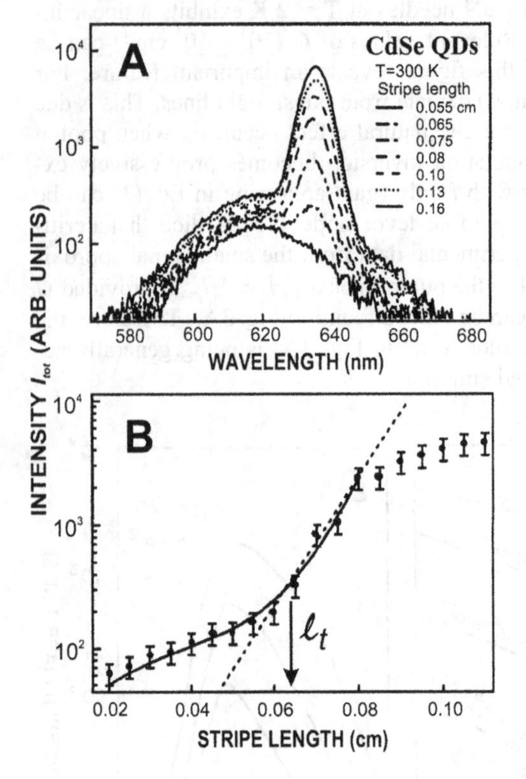

Figure 4. (A) Emission spectra of a CdSe QD film recorded using the VSL technique. (B) Data from (A) plotted as a function of the stripe length. The solid line is a fit to Eq.(5) G_{bx}=155 cm^{-1}. I_{pump} = 10 GW/cm^2, pulse duration 100 fs. A threshold stripe length l_t is indicated. According to [18].

Occasionally, however, and in particular in low-dimensional sys-tems, experiment reveals different behaviour. The exponential in-crease of I_{tot} occurs after exceeding a threshold stripe length l_t only and is accompanied by marked variations in the emission spectrum. A new feature in the spectrum is growing, as observed e.g. in a ZnCdSe/ZnSe quantum well [16,17] and recently in CdSe quantum dots [18] (see Fig. 4A). Possible explanation has been suggested by taking into account the existence of two types of luminescent quasiparticles (excitons and biexcitons) [18]. Emission line caused by radiative decay of biexcitons is absent in spontaneous emission because of biexciton short lifetime (fast Auger recombination in low-dimensional structures). Unlike spontaneous emission, biexcitons dominate the regime of stimulated emission and the VSL formula Eq. (3) can be rewritten as

$$I_{tot} = A_x \cdot l + \frac{A_{bx}}{G_{bx}}\left[\exp(G_{bx} \cdot l) - 1\right] \quad (5)$$

where A_x and A_{bx} are constants proportional to spontaneous emission intensities of excitons and biexcitons, respectively, and G_{bx} is the biexciton net gain. Application of Eq. (5) enables to fit the experimental points both below and above $l_t \approx 6$ mm, see solid line in Fig. 4B, and to extract G_{bx}= 155 cm^{-1}.

Finally notice a trivial mathematical consequence of Eq. (3): If $(|G|.l) << 1$ then $\exp(Gl) \cong (1 + Gl)$ and thus $I_{tot}(l, \lambda) = I_{sp}(\lambda).l$. It means that in this case the monitored light emission increases *linearly* with l independently of whether $G > 0$ (true light amplification) or $G < 0$ (light attenuation along the stripe); the method is therefore hardly capable to distinguish stimulated emission from optical absorption. This indicates some problems we can encounter when studying low gain values using the VSL technique.

4. Pitfalls of the VSL method in investigation of low-gain materials

In the simple theory of the VSL method described in Section 2 there are several, more or less evident, conditions which should be fulfilled. The conditions are namely:

- *Constant excitation intensity over the whole excited stripe* (as well as homogeneity of the studied sample)
- *Negligible width of the stripe* (the theory does not account for the stimulated emission which can occur by photons going in direction other than parallel to the stripe)
- Geometrically *perfect facet of a sample*
- *Constant coupling efficiency of emission from any part of the stripe to the detector*

Some of these conditions cannot be easily met for typical samples of Si nanocrystals (Si-NCs) and similar materials. Possible problems are mainly due to:

- *Low absorption of exciting light* by Si-NCs (the concentration of Si-NCs is usually relatively low in the sample) – an important fraction of light can be reflected and refracted in the sample especially when the substrate is transparent.
- *Emitted light is guided by the sample* - the Si-NCs are often fabricated in the form of a thin layer which acts as a waveguide because of the difference in refractive indexes. Waveguiding influences significantly the coupling of emitted light to the detector optics (numerical aperture (NA) matching above all).

Now we will describe in detail several factors affecting the precision of the VSL measurement in samples with low gain.

4.1. Effect of the transparent substrate

Samples containing semiconductor quantum dots are often prepared on a transparent substrate (silica or glass slides). In this case a perfect perpendicular adjustment of the sample with respect to the pump beam must be assured; otherwise an artifact resembling to gain manifestation can be observed (Fig. 5). By opening the stripe from zero to l_l, we start to excite the luminescent film "from the back" at first (Fig. 5A), due to total internal reflection in the substrate. In this way several luminescent stripes l'_l may appear and the collecting optical system records I_{tot} growing fast with increasing l ($<l_l$). This fast growth is moreover enhanced by total internal reflection of luminescence originating in the stripes l'_l. Once the pump beam has reached the sample corner ($l > l_l$), the film is being excited from the face and further growth of I_{tot} is considerably slower since the efficient rear excitation is stopped, as well as the total internal reflection of luminescence does not apply any more. The result is shown in Fig. 5B: an abrupt break on the

$I_{tot} = I_{tot}(l)$ plot, very much alike a "saturation" (however, the breaking point does not shift with varying pump intensity).

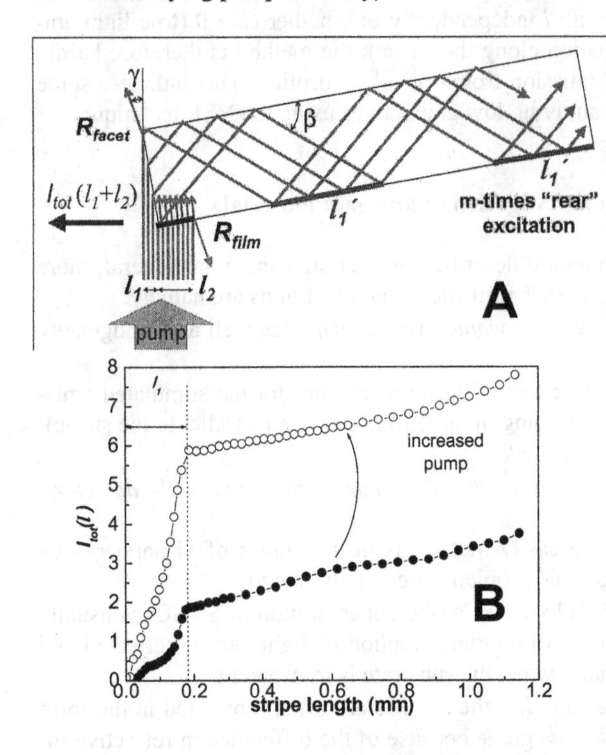

Figure 5. (A) Ray optics scheme of the VSL technique applied to a sample on a transparent substrate, whose facet is tilted by an angle γ to the pump beam. (B) Experimental demonstrations of the effect of a tilted silica substrate bearing a luminescent film without gain (γ = 10°, substrate thickness 1 mm and length 10 mm).

4.2. EFFECTS OF THE FINITE STRIPE WIDTH

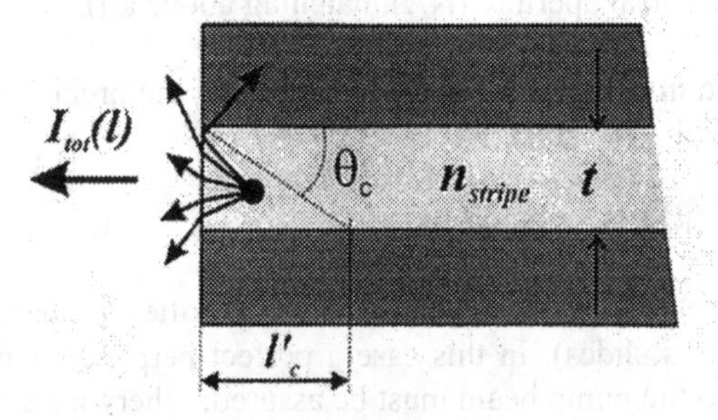

Figure 6. Limited light extraction efficiency due to total internal reflection.

In case of a low luminescence signal it might appear advantageous to increase the stripe width t. This can, however, entail an unwanted effect, similar to the limitation of light extraction efficiency from a planar face of a light-emitting diode (Fig. 6). Total internal reflection leads to a critical angle θ_c and a corresponding critical length $l'_c = t/\tan(\arcsin(1/n_{stripe}))$. For $l \leq l'_c$ only part of the emission I_{tot} can leave the sample, accordingly the plot $I_{tot} = I_{tot}(l)$ is expected to exhibit a noticeable change at l'_c, likewise that shown in Fig. 5B. For $n_{stripe} = 1.7$ and $t = 0.5$ mm we get $l'_c = 0{,}68$ mm. Similar effect can be caused by a slightly oblique sample edge [19].

4.3. COUPLING OF THE EMITTED LIGHT TO A DETECTION SYSTEM

The issue of the output coupling of light emitted inside a solid-state material is important especially for fabrication of efficient light-emitting devices (see e.g. Ref. 20 and references therein) and for light-guide coupling.

In the theory of VSL technique (Section 2, Eq. (2)) there is an implicit assumption that $I_{sp}(\lambda)$ and $g(\lambda)$ are independent of x (homogeneity of the sample) and, more importantly, the *coupling of emission to detector is constant*, independent of x.

4.3.1. The shifting-excitation-spot measurement

In order to check validity of the assumption that signal from any infinitesimal part dx of the excited stripe is equally coupled to the detection system, we propose a simple modification of the VSL arrangement which we call "shifting-excitation-spot" (SES) [12] measurement. The moving shield is replaced by a moving slit perpendicular to the fixed slit (see right-hand part of Fig. 7). Consequently, only a tiny rectangular area of the sample is excited. This excited segment is moved by changing the distance x from the edge of the sample and relevant changes of detected signal are observed. If the above mentioned assumption is correct, we should observe only a slow exponential decrease of the signal (due to losses, because light amplification is precluded owing to dividing the excited stripe into small segments). With increasing distance x of the segment from the sample edge, we therefore expect:

$$dI_{SES}(x,\lambda) = I_{sp}(\lambda)\cdot\exp\left[-\left(\alpha(\lambda)+K\right)\ x\right]\cdot dx \quad (6)$$

where $\alpha(\lambda)$ is a residual attenuation that is in Eqs. (1), (2) overwhelmed by gain $g(\lambda)$.

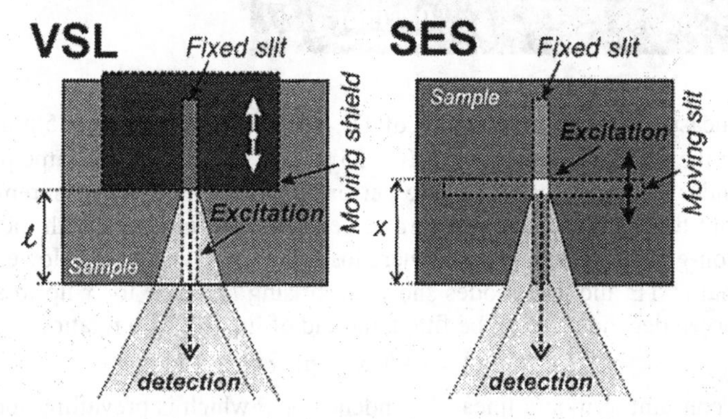

Figure 7. The modification of the VSL experimental set-up (left-hand side) for the shifted excitation spot (SES) measurements (right-hand side) [12].

For illustration we present here results of the SES measurement obtained on a sample prepared by Si-ion implantation (implantation energy of 400 keV and total dose of 4×10^{17} cm^{-2}) into a synthetic silica substrate. Annealing at 1100°C in N_2 atmosphere for

1 hour lead to formation of Si-NCs in a thin layer parallel to the silica surface [21]. Room temperature photoluminescence (PL) excited by the 325 nm line of a He-Cd laser shows typical wide band in the red spectral region 650-900 nm [22, 23] (see Fig. 8A, curve a). This ordinary PL measurement is performed by exciting and observing the front face of the implanted plane. However, when detecting PL from the edge of the implanted layer we observe huge changes of spectral shape (Fig. 8A, curve b). The main features are the TE and TM modes guided by the nanocrystalline layer forming a planar waveguide (due to the relatively high implantation energy, the resulting layer composed of Si-NCs is buried in the SiO_2 matrix providing the refractive index contrast indispensable for waveguiding - details will be published elsewhere [24]). The modes can be distinguished using a polarizer (See Fig. 8A, curves c, d). The edges of samples were polished to a good optical quality (Fig. 8B).

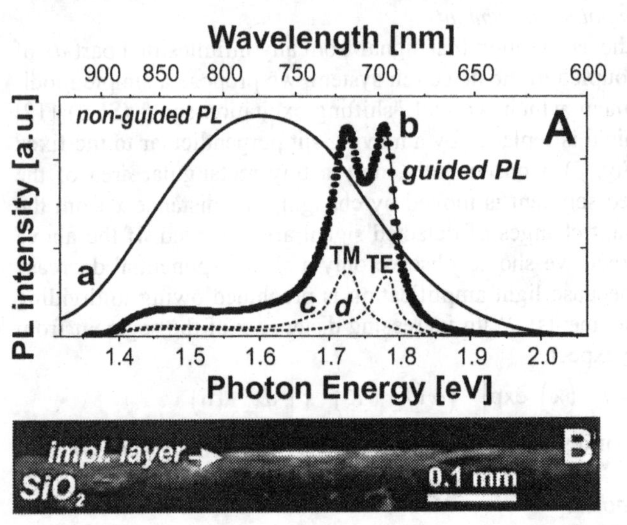

Figure 8. (A) PL spectra of a silica sample implanted with dose of 4×10^{17} cm^{-2} (λ_{exc}=325 nm and I_{exc} = 0.26 W/cm^2). Curves a and b were detected in directions perpendicular (non-guided) and parallel (guided) to the implanted plane, respectively. Dashed lines (c and d) are the polarization resolved TM and TE modes. (B) Microscope image (0.65 × 0.01 mm) of the PL emerging (in direction perpendicular to the plane of the figure) from the implanted layer. The facet of sample is slightly illuminated by a halogen lamp. According to [12].

The SES measurement at wavelengths of TE, TM modes and PL at 825 nm are plotted in Fig. 9B; the excited segment was (dx = 25 μm) × 100 μm. At the same part of the sample and under identical conditions we performed also the VSL measurement (Fig. 9A). It turns out that Eq. (6) — a *decrease* in detected light intensity with increasing x — holds for non-guided PL signal only where loss of 29 cm^{-1} was found (lower curve in Fig. 9B). Signal at TE and TM modes shows surprising *increase* for x up to about 0.5 mm. The observed dependence can be fitted, instead of Eq. (6), by a relation

$$I_{SES}(x,\lambda) = a + b \cdot x \cdot \exp[-cx]. \quad (7)$$

There is a constant term and linear dependence on x which is prevailing for small x. For larger x the exponential decay dominates.

If there is no amplification of PL by stimulated emission, the output from the VSL measurement must give the same information as integration of the SES plot from x = 0 to l. The numerical integral of the SES measurement presented in Fig. 9B is plotted in Fig. 9C. One can see that the initial parts of curves for TE and TM modes have a superlinear shape, which can be well fitted with the VSL Eq. (3). Gain values of 29 cm^{-1} and

22 cm^{-1} were found almost identical for TE and TM modes, respectively, with the results shown in Fig. 9A. This result demonstrates that the gain-like behaviour in Fig. 9A represents rather a consequence of a non-constant coupling of guided PL to the detection system than a true gain.

4.3.2 Interplay between output angle of guided emission and detection NA

The main reason for the observed peculiar SES dependence is the interplay between the output direction of the guided modes and the NA of detection optics (see inset in Fig. 9D). To illustrate this fact we performed the SES measurement using objectives with different numerical apertures of 0.075, 0.13, and 0.3. Results are plotted in Fig. 9D. A gradual increase of the SES signal for x up to 0.5 mm is observed with NA 0.075 and 0.13, but NA = 0.3 gives a constant signal already at $x \sim 0.1$ mm.

Therefore, our experiments demonstrate that the VSL method cannot be used in its classical form when a waveguiding effect takes place in the studied sample. In such a case, we have to consider Eq. (2) in a more general form, namely

$$dI_{tot}(x,\lambda) = \beta(x)\cdot\left[g(\lambda) - K\right]\cdot I_{tot}(x,\lambda)\cdot dx + \beta(x)\cdot I_{sp}(\lambda)\cdot dx. \quad (8)$$

Here we introduce a coupling function $\beta(x)$. The general integral of this differential equation is:

$$I_{tot}(x) = \exp\left[G(\lambda)\cdot\int\beta(x)\cdot dx\right]\cdot I_{sp}\int\beta(x)\cdot\exp\left[-G(\lambda)\cdot\int\beta(x)\cdot dx\right]\cdot dx. \quad (9)$$

When $\beta(x)$ is constant we retrieve the classical VSL Eq. (3). The coupling function $\beta(x)$ can be determined from the SES measurement that obviously yields

$$I_{SES}(x,\lambda) = I_{sp}(\lambda)\cdot\exp\left[-\left(\alpha(\lambda)+K\right)\ x\right]\cdot\beta(x)\cdot dx. \quad (10)$$

By performing the VSL and the SES measurement under otherwise identical condition we can check if the SES signal $I_{SES}(x,\lambda)$ increases with x and whether a complicated coupling of PL signal to the detection takes place. Then either the experimental condition should be modified (with aim to have $\beta(x)$ constant) or the function $\beta(x)$ has to be determined from the SES measurement (by comparing Eqs. (7) and (10)) and solving Eq. (9) (numerically) to find $G(\lambda)$. For our samples the real net optical gain $G(\lambda)$ turns to be close to zero under excitation by a cw laser, λ_{exc}=325 nm and a pump intensity up to 1 W/cm^2.

Another example of combining the VSL and SES measurements in silicon quantum dots is shown in Fig. 10. The investigated sample is schematically displayed in Fig. 10(a). It consists of a layer of highly concentrated Si-NCs embedded in a SiO_2 matrix. The sample was prepared in the following way. At first, p-type c-Si wafer $\rho \sim 0.1$ Ωcm was etched in a solution composed of 6 volumes of 50 wt. % HF and 15 volumes of added ethanol (14.3 % HF concentration). The etching current density was kept relatively low (1.6 mA/cm^2) so as to achieve prolonged etch time leading to higher porosity and, consequently, low mean size of Si-NCs and blue shifting of the photoluminescence band [25]. Spontaneous luminescence spectrum of resulting porous silicon (por-Si) was peaked at ~680 nm. (This particular approach was chosen to maximize the contribution of phononless radiative transitions inside Si-NCs [26] and thus to increase the chance of stimulated emission to overcome losses due to the free carrier absorption [27].)

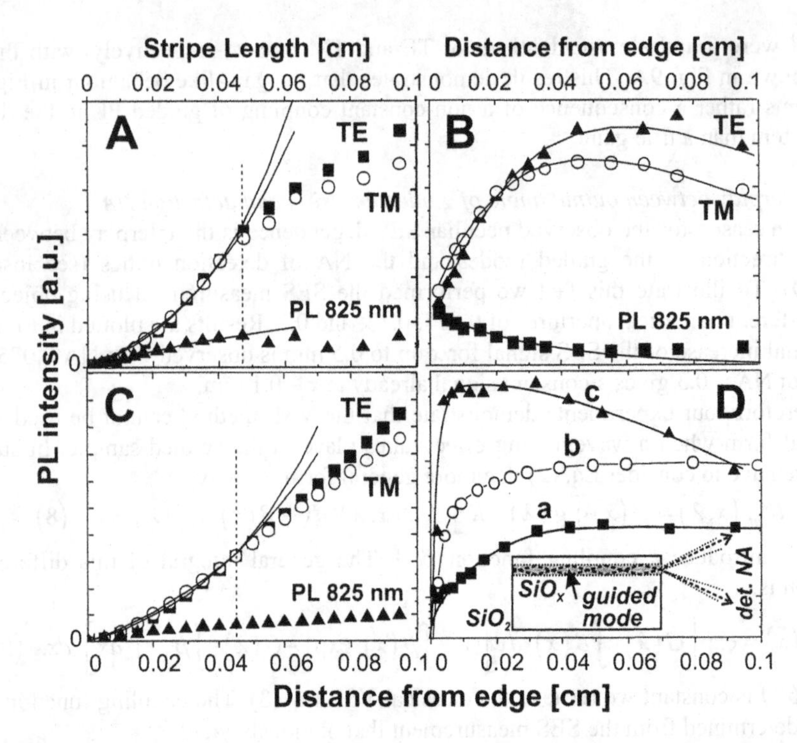

Figure 9. (A) Results of the VSL measurement at the peak of TE and TM mode and for non-guided PL around 825 nm. The fits (lines) with Eq. (3) give values of $G = 35$ cm^{-1} and 28 cm^{-1} for TE and TM modes, respectively, and losses of 11 cm^{-1} for non-guided PL. (B) Results of the SES measurement performed under identical conditions as the VSL. The lines are fits using Eqs. (7) and (6) for TE (TM) and PL, respectively. (C) Integration of data from panel (B). The gain fits (lines), when using formally again Eq. (3), give values of $G =$ 29 cm^{-1} and 22 cm^{-1} for TE and TM modes, respectively. (D) The SES measurements using the detection NA of 0.075 (*a* - black squares), 0.13 (*b* - white dots), and 0.3 (*c* - black triangles). Lines are guides to the eye. The inset shows interplay between the exit angle of the guided mode and the detection NA. According to [12].

The por-Si layer was then mechanically scratched and dispersed in spin-on-glass P507 sol with the initial concentration of 1mg/200 μl. After 15 min treatment in an ultrasonic bath the suspension was let to solidify in a glass cuvette 1x1 cm^2 at 40 °C for 6 days. During solidification the por-Si powder was sedimenting to the bottom of the cuvette which resulted into creation of a brightly luminescent film of close-packed Si-NCs. Thickness of the film (measured by using fluorescence microscope image) was 62 μm which yields mean nanocrystal concentration of about 1.10^{18} NCs/cm^3.

The described transformation of a por-Si layer into a film of close-packed Si-NCs in a transparent SiO_2 matrix had three aims in view: (i) Achieving of a higher concentration of Si-NCs, (ii) avoiding excess heating of the nanocrystals since the non-absorbed excitation power propagates basically unattenuated through the matrix and (iii) an additional blue shifting and emission intensity enhancement due to phosphorus [28,29] contained in P507 spin-on-glass. In the outlook for the future, such a close-packed system of nanocrystals should also assure better transport properties than native por-Si or layers prepared by standard Si^+-implantation.

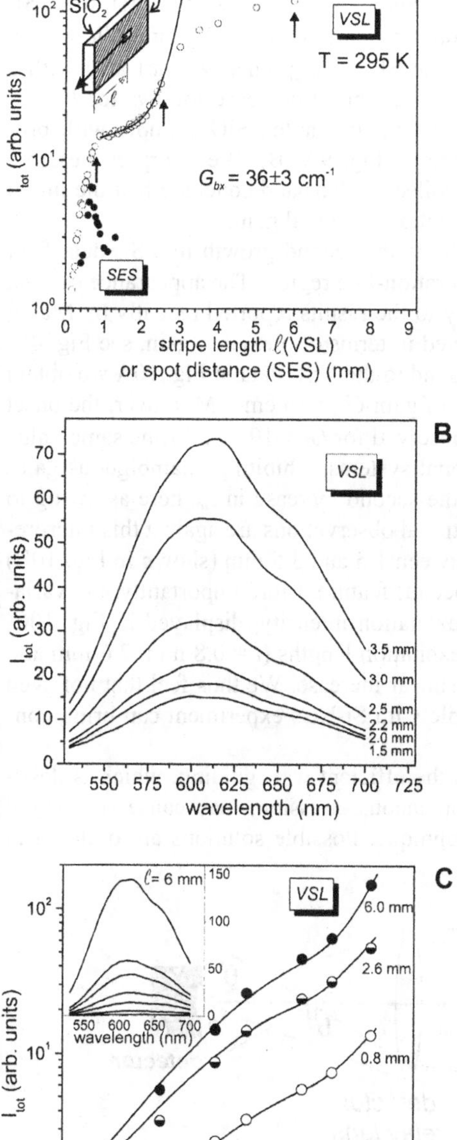

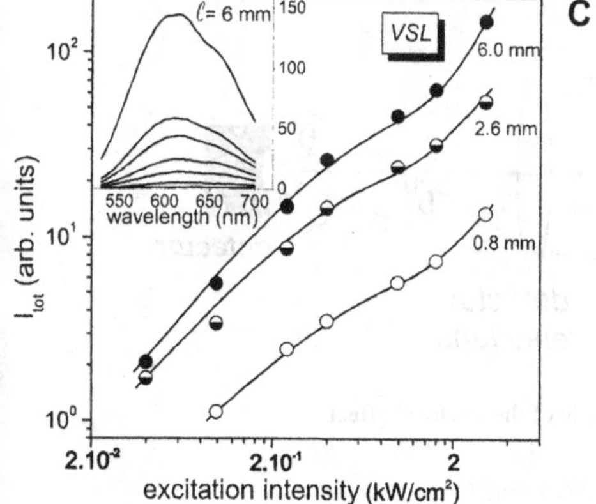

Figure 10. (A) Room-temperature VSL (○) and SES (●) data taken on a long sample of Si-NCs prepared by etching of crystalline Si (see text). The solid line is a fit to Eq. (5) with $G_{hx} = 36$ cm^{-1}. Pump intensity ~3 kW/cm^2, 308 nm, 18 ns. (B) Emission spectra corresponding to (a), measured at several different stripe lengths. (C) Light output I_{tot} as a function of pump intensity at three different stripe lengths (indicated by arrows in A). The lines are guides for the eye. Inset: Evolution of the emission spectrum with pump intensity for $l = 6$ mm.

Once solidified, the sample was turned out of the cuvette and used as a free-standing layer for VSL and SES measurements. Photoexcitation was provided with a XeCl excimer laser (308 nm, 18 ns) at room temperature.

Room temperature experimental data obtained by using the both methods is displayed in Fig. 10A. There are two interesting features worth discussing. First, the VSL measurement (o) starts up with an initial exponential part up to $l_o \cong 0.8$ mm followed by a linear section resembling to "saturation". The SES data points (•) start also with a nonlinear growth terminating at the same l_o but an abrupt decrease follows for $l > l_o$. Such a behavior is similar to that found in the Si^+-implanted SiO_2 sample with pronounced waveguiding effects as discussed above (Fig. 9A, B). We interpret therefore this observation as the aperture effect of the collecting lens and conclude that the initial exponential part in the VSL measurements is not due to a real gain.

The second interesting feature in Fig. 10A is the second growth in VSL data for l above approx. 2 mm, followed again by a saturation-like region. The appearance of such a "threshold" stripe length resembles strongly to the results reported recently by Malko et al. in CdSe nanoparticles [18] and interpreted in terms of biexciton gain, see Fig. 4B. Applying Eq. (5) with an appropriate background to the VSL data in Fig. 10A we obtain a good fit (solid line) and a reasonable value of gain G_{bx} =36 cm^{-1}. Moreover, the onset of the saturation-like region ($l \geq 2.8$ mm) is observed for $Gl \approx 10$, i.e. for the same value of the Gl product like in other low-dimensional systems exhibiting unambiguous gain! [16,18] It is thus very tempting to interpret the second increase in I_{tot} here as owing to the presence of real gain. However, two additional observations are against this interpretation: Emission spectra taken at several l between 1.5 and 3.5 mm (shown in Fig. 10B) exhibit no indication of a new developing spectral feature. More importantly, the variation of the light output I_{tot} as a function of excitation intensity, displayed in Fig. 10C, shows very similar behavior for three fixed excitation lengths (l = 0.8 mm, 2.6 mm and 6 mm) without any significant break to superlinear increase. We thus feel that not even in case of this especially prepared long sample with Si-NCs experiment can bring conclusive evidence for optical gain.

These experiments thus demonstrate that the efficient waveguiding, which is desirable for more efficient amplification of spontaneous emission, may cause substantial problems in application of classical VSL technique. Possible solutions are outlined in Section 6.

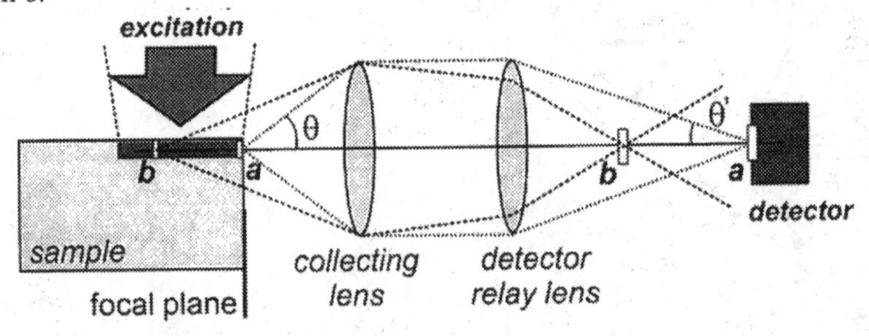

Figure 11. The basic principle of the confocal effect.

4.3.3 Confocal effect in the detection of the VSL signal

Another effect, which could contribute to the non-constant coupling efficiency of light coming from different points of a sample, can be called the confocal effect.

The common arrangement for collection and detection of emitted light in luminescence spectroscopy is an imaging of the emitting spot located on the sample facet (point *a* in Fig. 11) to the detector (or the input slit of the spectrometer). High NA and short focus collecting lens is preferred in order to have a good efficiency of light collection, while the relay lens is usually of longer focus (in order to match the spectrometer NA). Therefore the image is usually magnified in the detection plane. In the measurement of a weak gain with the VSL method a long excited stripe is sometimes used. In such a case the emitting points of the sample which are far from the output facet (point *b* in Fig. 11) will be imaged in a plane in front of the detector. Then, if the detector size is small (or the input spectrometer slit is narrow) some light will be lost around the detector. The effect will be stronger when the collecting lens NA is higher (i.e. the depth of focus is smaller), magnification of the optical system is higher and the detector size smaller.

In fact this is the same effect as that used in confocal microscopes to achieve depth resolution so we can use theory developed for confocal microscopes [30]. The VSL experiment corresponds to the observation of fluorescence plane reflectors with pinhole or slit detector in confocal microscope. For emitters far from the focus the detected intensity should decrease as $1/z^2$ or $1/z$ for pinhole and slit detector, respectively [30].

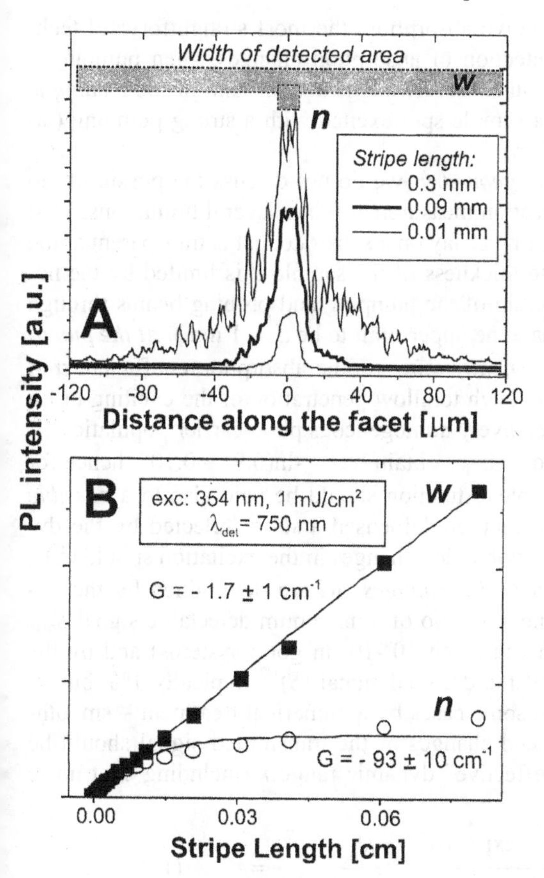

Figure 12. Illustration of the confocal effect: VSL measurement with a high NA_{det}=0.45, when either all signal coming from the sample facet (240 µm wide area) or just a signal from the area corresponding to the image of the excited stripe (10 µm) is detected. (A) PL intensity profiles of the output facet (illustration of defocusing of emission from points distant from the edge), (B) the VSL stripe dependence for the two widths of detection area.

To illustrate the confocal effect we show in Fig. 12 VSL measurements on Si-ion implanted silica layers using an objective lens with NA = 0.45. The CCD images show that for long stripes the emitting spot becomes significantly larger. Restriction of the detection area in the image plane from 240 μm (which covers all the signal, area *w* in Fig. 12A) to a narrow area *n* of about 10 μm (corresponding to the width of excited stripe) causes significant loss of signal, which is observed as increased losses (93 instead of 1.7 cm^{-1}).

It should be quite easy to avoid the confocal effect using low NA optics (contrary to call for high NA dictated by efficient coupling of guided light – Section 4.3.2) and a large-area detector (spectrometer slit). In addition the effect can be identified and corrected using the SES measurement.

5. Comparison of the VSL technique with other methods of gain measurement

5.1. DETECTION OF ABSORPTION CHANGES BY THE PUMP-AND-PROBE TECHNIQUE

Because optical gain is defined as negative absorption, the most straightforward technique to measure it consists in the detection of absorption changes when pumping a sample to the population inversion. In other words, this is a *pump-and-probe technique* (P&P) in which optical absorption in a sample spot excited with a strong pumping (laser) beam is tested by a weak probe beam.

The *P&P technique in transmission geometry* (we do not discuss the possibility to calculate absorption changes from reflection measurement) has several limitations. First of all, the sample should be self-supporting or lay on a substrate that is transparent at the investigated range of wavelengths. The thickness of the sample *d* is limited by the necessity to achieve a perfect spatial overlap of the pumping and probing beams through the whole excited volume (let us estimate the upper limit to be $d = 1$ mm). *At the pumping wavelength the sample has to be optically thin*, i.e. the absorption coefficient α of the material under study must be low enough to allow penetration of the exciting beam through the sample and creation of a relatively homogeneous photocarrier population (if a 10% decrease of excited light is allowed, we obtain $\alpha d = -\ln(0.9) = 0.105$, hence for $\alpha = 10^3$ cm^{-1} *d* is limited to 1 μm). (Note: Attention should be paid also to a *possible lensing effect* – the test beam may be focused/defocused and/or deflected by the dynamic lens created by the induced refraction index changes in the excitation spot [31].)

The smallest detectable absorption (gain) changes $\Delta\alpha$ are determined by the dynamic range of the detection system (i.e. the ratio of a maximum detectable signal S_{max} to the detector noise level N_{det}, which can reach 10^4-10^5 in good systems) and by the shot noise (it is equal to square root of the detected signal $(S)^{1/2}$, typically 1%, but its influence can be effectively reduced in some cases by a numerical treatment – smoothing). In other words, the relative induced changes of the transmitted signal should be equal or greater than the reciprocal "effective" dynamic range *r* (including shot noise and signal treatment):

$$\frac{I_0 \exp\left[-(\alpha+\Delta\alpha)d\right]-I_0\exp(-\alpha d)}{I_0\exp(-\alpha d)} \geq \frac{N_{det}+\sqrt{S}}{S_{max}} = r^{-1}. \quad (11)$$

Here I_0 is input test beam intensity and α is the sample absorption coefficient without pumping. After simple operations we found (for $r >> 1$):

$$\Delta\alpha d \geq \ln(1 + 1/r) \cong 1/r. \quad (12)$$

For a system with the effective dynamic range of 10^3 we get $\Delta\alpha.d \geq 10^{-3}$. For a typical sample of Si-NCs in a SiO_2 matrix with $d \sim 0.5$ μm we need $\Delta\alpha = G \geq 20$ cm^{-1}, therefore the P&P technique seems not to be suitable for such low-gain samples.

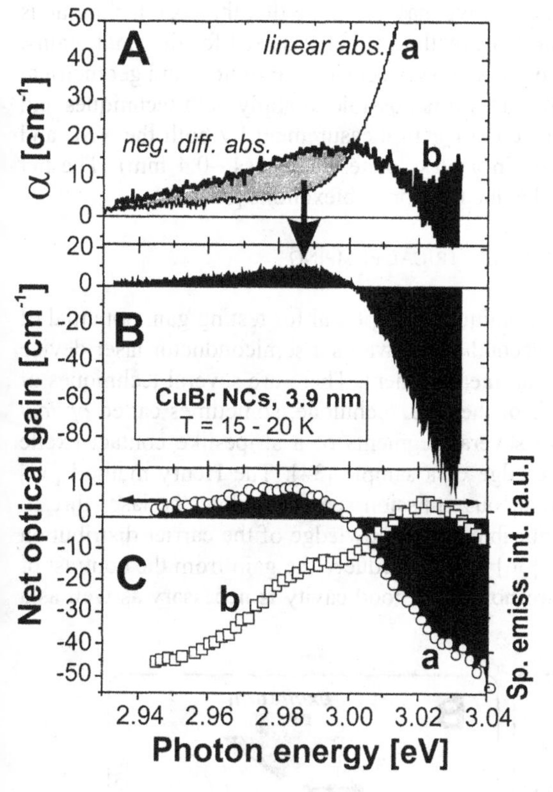

Figure 13. Comparison of the P&P and VSL techniques for measurement of gain in CuBr-NCs (mean radius 3.9 nm, T=15-20 K).
(A) Linear absorption at the edge of excitonic band (a) compared with the negative differential absorption (b) produced by band-to-band excitation at 308 nm (150 kW/cm^2).
(B) (B) Net optical gain is a difference between curves *a* and *b* in (A).
(C) Spectra of net gain (a) and spontaneous emission (b) calculated from VSL measurements (excitation at 308 nm, 110 kW/cm^2).
According to [33].

The VSL method is in principle also a kind of P&P technique where spontaneous emission itself plays a role of the test beam. Therefore, the method is applicable for those wavelengths only where spontaneous emission is strong enough. The advantage of the VSL technique is a simpler experimental arrangement with just one laser beam (and, moreover, principal acceptance of non-transparent substrates). The lower limit for the $G.l$ product should be basically the same as given by Eq. (12) but the specific features of the VSL technique (Section 3) set this limit to $G.l \cong 1$. There is also an upper limit that stems from two facts. First, exponential VSL section saturates generally [32] for $G.l_C \sim 5$ (see e.g. Fig. 3A). Second, inevitable pump light diffraction prevents exciting homogeneously stripes shorter than about 10 μm. Thus high gains which produce saturation of the VSL signal within l~10 μm, i.e. $G \geq 5 \times 10^3$ cm^{-1}, turn to be measurable with difficulty. Such high gains, however, occur only exceptionally (e.g. in GaN). All these considerations are summarised in Tab. I.

TABLE I. Estimation of limits for application of the VSL and P&P techniques

Method	lower limit	upper limit	notes
P&P	$Gd \geq 10^{-3}$	—	$\alpha(\lambda_{pump})d \leq 0.1$ $d \leq 1$ mm
VSL	$Gl \geq 1$	$G \leq 5 \times 10^3$ cm^{-1}	

Taking into account that the stripe length l can be as much as by four orders of magnitude larger that a typical film thickness d, we may conclude that the P&P technique is good for high-gain materials while the VSL method is better suited for the small gains. The applicability of both methods is, of course, very sensitive to optical and geometrical properties of studied samples. In some cases it is possible to apply both techniques and compare them. In Fig. 13 we show results of gain measurement by both the VSL and P&P methods in CuBr-NCs embedded in a glass plate (thickness ~0.4 mm). The net gain of about 10 cm^{-1} is interpreted to be due to trapped biexcitons [33].

5.2. Optical gain measurement with electrical pumping

The above-mentioned P&P and VSL techniques are optimal for testing gain potential of a given semiconductor material. The second step towards a semiconductor laser device is, however, an electrically pumped gain measurement. There are several techniques of this type. One of them is the analogue of the VSL technique, sometimes called *optical stripe length method* (OSLM), where several segments of a stripe-like contact excite electroluminescence starting from the edge of a sample [34]. The Henry method [35] calculates gain from the device spontaneous emission spectra (below the laser threshold). The method is not generally applicable since knowledge of the carrier distribution function is required. The Hakki-Paoli [36] method deduces the gain from the contrast of Fabry-Perot oscillations below the threshold, so a good cavity is necessary as well as a good spectral resolution.

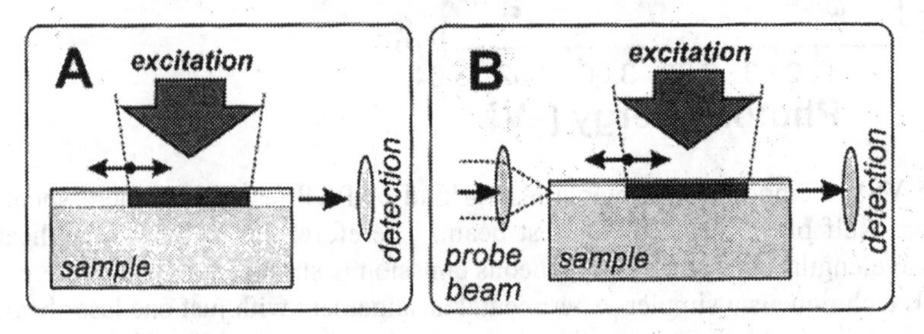

Figure 14. (A) Alternative VSL measurement inside the sample, (B) P&P experiment under the same conditions with a probe beam coupled to the waveguide

6. Conclusions

The VSL method is suitable for gain measurements from $G.l \cong 1$ to $G \leq 5 \times 10^3$ cm^{-1}. The technique looks very simple and is tempting to underestimate systematical experimental errors. However, there is quite a lot of "snags" that one has to be aware of. Attempting to emphasize the best laboratory practice for the VSL method, we can say: *Check exact adjustment of the experiment, make the excited stripe as narrow as possible, combine the VSL and SES measurements under identical conditions, try to modify the experimental conditions until the SES results follow Eq. (6).* There seems to be one simpler alternative how to avoid the "coupling" and edge-related problems: To perform the VSL measurements inside the sample, i.e. to start the excited stripe not at the facet but at a certain distance from the edge (Fig. 14A) [37]. The very best experiment would be to couple a weak spectrally broad light beam into the waveguide and look for absorption/gain changes during optical pumping (pump-and-probe experiment under the same conditions as the VSL, Fig. 14B).

Finally, it should be kept in mind that the exponential shape of the VSL curve, followed by a saturation region, is not sufficient to prove the presence of gain. Investigating emission spectral changes should complete this observation, along with variation of the emission intensity as a function of pump intensity (break in the slope) and time resolved studies (pulse shortening).

This work was supported by the projects A1010809 and AVOZ 1010914 of GAAVCR, the project 202/01/D030 of GAČR, by the NATO (PST.CLG.978100) and by the Royal Swedish Academy of Sciences. One of us (R.T.) appreciates the financial support from the Lithuanian State Foundation for Science and Studies. We are grateful to Prof. J. Linnros for stimulated discussions, to Dr. S. Cheylan for fabrication of high quality samples and to Dr. A. Galeckas, Dr. P. Gilliot and Dr. O. Crégut for experimental assistance.

References

1. Shaklee, K.L. and Leheny, R.F. (1971) Direct observation of optical gain in semiconductors, *Appl. Phys. Letters* **18**, 475-477.
2. Dingle, R., Shaklee, K.L., Leheny, R.F. and Zetterstrom, R.B. (1971) Stimulated emission and laser action in Gallium Nitride, *Appl. Phys. Letters* **19**, 5-7.
3. Shaklee, K.L., Leheny, R.F. and Nahory, R.E. (1971) Stimulated emission from the excitonic molecules in CuCl, *Phys. Rev. Letters* **26**, 888-891.
4. Capozzi, V. and Staehli, J.L. (1983) Spontaneous and optically amplified luminescence from exciton-exciton collisions in GaSe at liquid-He temperature, *Phys. Rev.* **B28**, 4461-4467.
5. Bylsma, R.B., Becker, W.M. Bonsett, T.C., Kolodziejski, L.A., Gunshor, R.L., Yamanashi, M. and Datta, S. (1985) Stimulated emission and laser oscillations in $ZnSe-Zn_{1-x}Mn_xSe$ multiple quantum wells at ~453 nm, *Appl. Phys. Letters* **47**, 1039-1041.
6. Gutowski, J., Diesel, A., Neukirch, U., Weckendrup, D., Behr, T., Jobst, B. and Hommel, D. (1995) Exciton dynamics and gain mechanismsms in optically pumped ZnSe-based laser structures, *phys. stat. sol. (b)* **187**, 423-434.
7. Butty, J., Peyghambarian, N., Kao, Y.A. and Mackenzie, J.D. (1996) Room temperature optical gain in sol-gel derived CdS quantum dots, *Appl. Phys. Letters* **69**, 3224-3226.
8. McGehee, M.D., Gupta, R., Veenstra, S., Miller, E.K., Díaz-García, M.A. and Heeger, A.J. (1998-I) Amplified spontaneous emission from photopumped films of a conjugated polymer, *Phys. Rev.* **B58**, 7035-7039.
9. Díaz-García, M.A., De Ávilla, S.F. and Kuzyk, M.G. (2002) Dye-doped polymers for blue organic diode lasers, *Appl. Phys. Letters* **80**, 4486-4488.
10. It is interesting to note that the references 1, 3 and 11 have been cited according to Science Citation Index (since 1980) almost 400 times but in many articles nowadays the VSL technique is applied even without citing the original papers.

11. Shaklee, K.L., Nahory, R.E. and Leheny, R.F. (1973) Optical gain in semiconductors, *J. Luminescence* **7**, 284-309.
12. Valenta, J., Pelant, I. and Linnros, J. (2002) Waveguiding effects in the measurement of optical gain in a layer Si nanocrystals, *Appl. Phys. Letters*, **81**, 1396-1398.
13. Hvam, J.M. (1978) Direct recording of optical-gain spectra from ZnO, *J. Appl. Phys*. **49**, 3124-3126.
14. Saleh, B.A.E. and Teich, M.A. (1991) *Fundamentals of Photonics*, J. Wiley & Sons, New York.
15. Shaklee, K. L., Nahory, R.E. and Leheny, R.F. (1972) Stimulated recombination highly excited semiconductors, in *Proc. 11th Int. Conf. Phys. Semicond., Warszava 1972, Vol. 2*, Polish Scientific Publishers, Warszawa, pp. 853-862.
16. Tomasiunas, R., Pelant, I., Hönerlage, B., Lévy, R., Cloitre, T. and Aulombard, R.L. (1998-II) Stimulated emission and optical gain in a single MOVPE-grown ZnCdSe-ZnSe quantum well, *Phys. Rev.* **B57**, 13077-13085.
17. Mikulskas, I., Luterová, K., Tomasiunas, R., Hönerlage, B., Cloitre, T. and Aulomard, R.L. (1998) Light amplification due to free and localized exciton states in ZnCd Se GRINSCH structures, *Appl. Phys.* **A67**, 121-124.
18. Malko, A.V., Mikhailovsky, A.A., Petruska, M.A., Hollingsworth, J.A., Htoon, H., Bawendi, M.G. and Klimov, V.I. (2002) From amplified spontaneous emission to microring lasing using nanocrystal quantum dot solids, *Appl. Phys. Letters* **81**, 1303-1305.
19. Mohs, G., Aoki, T., Shimano, R., Kuwata-Gonokami, M. and Nakamura, S. (1998) On the gain mechanism in GaN based diodes, *Solid State Com.* **108**, 105-109.
20. Benisty, H. (1999) Physics of light extraction efficiency in planar microcavity light-emitting diodes, in H. Benisty, J.-M. Gérard, R. Houdré, J. Rarity and C. Weisbuch (eds.), *Confined Photon Systems*, Springer-Verlag, Berlin, pp. 393-405.
21. Elliman, R.G., Lederer, M.J. and Luther-Davis, B. (2002) Optical absorption measurements of silica containing Si nanocrystals produced by ion implantation and thermal annealing, *Appl. Phys. Letters*. **80**, 1325-1327.
22. Guha, S. (1998) Characterization of Si^+ ion-implanted SiO_2 films and silica glasses, *J. Appl. Phys.* **84**, 5210-5217.
23. Linnros, J., Lalic, N., Galeckas, A. and Grivickas, V. (1999) Analysis of the stretched exponential photoluminescence decay from nanometer-sized silicon crystals in SiO_2, *J. Appl. Phys.* **86,** 6128-6134.
24. Valenta, J., Pelant, I., Luterová, K., Tomasiunas, R., Cheylan, S., Elliman, R.G., Linnros, J. and Hönerlage, B., An active planar optical waveguide made from luminescent silicon nanocrystals, submitted to *Appl. Phys. Letters*.
25. Canham, L.T. (1990) Silicon quantum wire array fabrication by electrochemical and chemical dissolution of wafers, *Appl. Phys. Letters* **57**, 1046-1048.
26. Hybertsen, M.S. (1994) Absorption and emission of light in nanoscale silicon structures, *Phys. Rev. Letters*. **72,** 1514-1517.
27. Dumke, W.P. (1962) Interband transitions and laser action, *Phys. Rev.* **127,** 1559-1563.
28. Fujii, M., Mimura, A., Hayashi, S. and Yamamoto, K. (1999) Photoluminescence from Si nanocrystals dispersed in phosphosilicate glass thin films: Improvement of photoluminescence intensity, *Appl. Phys. Letters*. **75,** 184-186.
29. Mimura, A., Fujii, M., Hayashi, S., Kovalev, D. and Koch, F. (2000) Photoluminescence and free-electron absorption in heavily phosphorus-doped Si nanocrystals, *Phys. Rev.* **B62** 12625-12627.
30. Corle, T.R. and Kino, G.S. (1996) *Confocal Scanning Optical Microscopy and Related Imaging Systems*, Academic Press, New York.
31. Klingshirn, C. (1995) *Semiconductor Optics*, Springer-Verlag, Berlin.
32. Klingshirn, C. and Haug, H. (1981) Optical properties of highly excited direct gap semiconductors, *Phys. Reports* **70**, 315-398.
33 Valenta, J., Dian, J., Gilliot, P. and Hönerlage, B. (2001) Photoluminescence and optical gain in CuBr semiconductor nanocrystals, *phys. stat. sol.* (b) **224**, 313-317.
34. Bognár, S., Grundmann, M., Stier, O., Ouyang, D., Ribbat, C., Heitz, R., Sellin, R., and Bimberg, D. (2001) Local modal gain of InAs/GaAs quantum dot lasers, *phys. stat. sol. (b)* **224,** 823-826.
35. Henry, C.H., Logan, R.A. and Merritt, F.R. (1980) Measurement of gain and absorption spectra in AlGaA buried heterostructure lasers, *J. Appl. Phys*. **51,** 3042-3050.
36. Hakki, B.W. and Paoli, T.L. (1975) Gain spectra in GaAs double-heterostructure injection lasers, *J. Appl. Phys.* **46,** 1299-1306.
37. We thank to Dr. L. Khriachtchev for suggesting us this idea.

THEORY OF SILICON NANOCRYSTALS

C. DELERUE AND G. ALLAN
Institut d'Electronique et de Microélectronique du Nord, Dept. ISEN, 41 boulevard Vauban, 59046 Lille Cedex France
M. LANNOO
Laboratoire Matériaux et Microélectronique de Provence, ISEM Place G. Pompidou, 83000 Toulon, France

1. Introduction

The intense luminescence observed for porous silicon [1] has raised extremely interesting problems related to the possibility of using silicon in optoelectronics. One likely explanation is quantum confinement, induced by the formation of nanocrystallites. However, the experimental data reveal a complex situation characteristic of several radiative as well as non radiative channels. Our main aim in this chapter will thus be to review the relevant theoretical information concerning the optical properties of Si nanocrystals in order to identify these channels.

In parts 2 and 3 we summarize theoretical results concerning the quantum confinement effect and the different channels for the recombination: i) direct transitions across the quantum confined gap, ii) the recombination of self-trapped excitons, iii) recombination at surface Si=O bonds and iv) non radiative recombination via dangling bonds and the Auger effect. In the next parts, we present with more details recent results on phonon-assisted transitions, on the polarization of the light and on intraband transitions in *n*-type nanocrystals.

2. Accurate calculations of the quantum confinement effect

Si nanostructures are particularly interesting not only for their optical properties but also because recent developments [2,3,4] have made possible Si devices with feature size below 10 nm. These devices have shown exciting low temperature transport properties [5] with promising applications in microelectronics (single electron transistors and memories). An accurate description of the electronic properties of Si nanostructures with arbitrary geometries in the range 1-10 nm is thus needed. *Ab initio* selfconsistent methods such as the local density approximation (LDA) can only be used for small clusters (< 1000 atoms) with high symmetry [6]. They are not suitable for realistic situations. Semi-empirical, non self-consistent methods (such as pseudopotentials [7,8] (PP), tight-binding [9,10,11,12] (TB), or $\mathbf{k} \cdot \mathbf{p}$ [13]) can solve much larger problems. Semi-empirical methods are designed to make the best possible approximation to the

L. Pavesi et al. (eds.), Towards the First Silicon Laser, 243–260.

self-consistent one-particle Hamiltonian H_0 in bulk material, either in the whole first Brillouin zone (PP, TB) or around specific k-points ($\mathbf{k}\cdot\mathbf{p}$). They involve adjustable parameters (e.g. effective masses, TB interaction parameters...) that are fitted to experimental data or *ab initio* bulk band structures. These parameters are then transferred to the nanostructures (i.e. $H=H_0$ inside the nanostructure) with appropriate boundary conditions. The better the bulk description and boundary conditions, the better the electronic structure we expect in nanostructures.

Recently we have presented an improved TB description designed to give accurate results over the whole range of sizes [14]. For this, we fit the TB parameters not only on bulk band energies in the first Brillouin zone but also on effective masses. This was not the case of previous TB treatments but is essential if one wants to get accurate results for large size nanostructures where the effective mass approximation (EMA) or $\mathbf{k}\cdot\mathbf{p}$ models become exact. We also include spin-orbit coupling. From improved computer codes, we are able to apply TB to nanostructures with feature size in the range 1-12 nm. This extended range allows us to study precisely the overlap region where both TB and $\mathbf{k}\cdot\mathbf{p}$ or EMA are accurate. The energy of the highest valence state and of the lowest conduction state is fitted in the whole range of size with the following expressions:

$$E_v(d) = \frac{K_v}{d^2 + a_v d + b_v} \tag{1}$$

$$E_c(d) = \frac{K_c}{d^2 + a_c d + b_c} + E_g(\text{bulk}) \tag{2}$$

where d is the characteristic dimension of the nanostructure ; K, a and b are adjustable constants ; $E_g(\text{bulk}) = 1.143$ eV is the bulk bandgap energy. This expression is more accurate than the widely used fit K/d^α when a large range of dimensions is considered. It correctly behaves like $1/d^2$ in large structures, so that it can be considered as valid over the whole range of sizes. The fits are reported in Table 0 for all nanostructures. The bandgap energy mainly depends on the transverse section of the wires or volume of the dots as long as their shape is not too prolate [8]. Notice that this result cannot however be extended to higher excited states.

TABLE 1. Parameters for the fits to the energy of the highest valence state and lowest conduction state for various Si nanostructures. The energies are given in eV and the lengths in nm. d is the diameter for spheres, the width for wires and cubes, and the thickness for films.

	K_v	a_v	b_v	K_c	a_c	b_c
(100) Films	-1326.2	1.418	0.296	394.5	0.939	0.324
(110) Films	-1019.3	10.371	-3.713	1123.2	0.535	1.481
[100] Cylinders	-3448.4	2.194	1.386	2811.6	1.027	0.396
[100]Wires (001)x(001)	-2817.9	1.988	0.708	2378.6	0.883	0.400
[100] Wires (011)x($0\bar{1}1$) faces	-2981.7	2.386	1.087	2162.0	1.129	0.138
[100] Cylinders	-2551.8	2.970	0.813	2860.8	1.330	2.650
[100] Wires ($0\bar{1}1$)x(001) faces	-2217.5	3.177	0.130	2684.3	2.237	1.408
Spherical Dots	-6234.0	3.391	1.412	5844.5	1.274	0.905
Cubic Dots (100) faces	-3967.0	2.418	0.522	4401.0	1.138	0.889

We now proceed to test our TB model against various other descriptions such as

PP, LDA, or other TB models in the case of small spherical Si dots. Our results are first compared with those of Ref. [8] calculated with pseudopotentials in the range 1-4 nm. As shown in Fig. 1, they are in good agreement. Comparison with the *ab initio* LDA calculations of Ref. [6] corrected for the bulk bandgap error of 0.65 eV shows that our results are extremely good down to the smallest clusters, except for some oscillations in the LDA that do not exist in TB. Finally, the agreement with other more complex TB models with overall band structure of similar quality, such as the $sp^3d^5s^*$ model of Ref. [15] or the nonorthogonal TB model of Ref. [16] is also excellent for small ($d_0 < 4$ nm) crystallites. In consequence, the results previously published by our group [10,11] for small spherical Si dots remain valid.

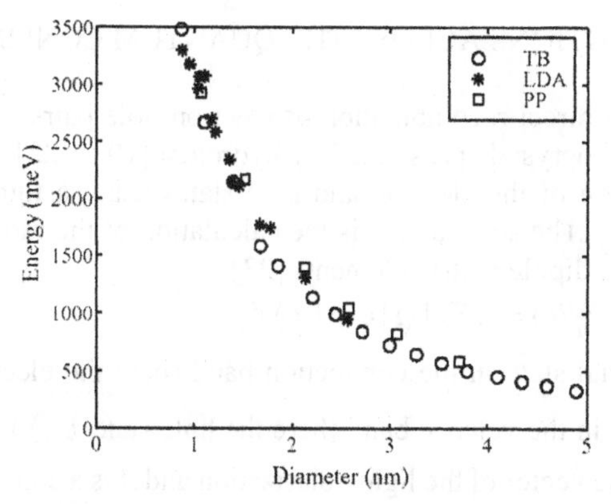

Figure 1. Comparison of confinement energy in spherical Si dots, between our tight binding (TB) model and pseudopotentials (PP) or LDA.

The gap defined above (called hereafter E_g^0) is the independent particle gap defined as the difference of eigenvalues. To compare with experiments, we need to calculate the excitonic gap. A common practice has been to subtract from the eigenvalue-gap the screened direct electron-hole attraction. However the whole procedure is not clearly justified and conflicting points of view [17-19] have been expressed concerning its validity. To clarify this problem, we have recently performed GW calculations followed by the resolution of the Bethe-Salpeter equation for Si nanocrystals [20]. We express the excitonic gap E_g^{exc} as the difference between the quasi-particle gap E_g^{qp} (the difference between the separate electron and hole quasi-particle energies) and E_{coul}, the attractive interaction between these two quasi-particles,

$$E_g^{exc} = E_g^{qp} - E_{coul} = E_g^0 + \delta\Sigma - E_{coul} \quad (3)$$

where E_g^{qp} is written as the sum of the independent particle value E_g^0 and a self-energy correction $\delta\Sigma$. The main result that we obtained [20] is that there is strong cancellation between the two large quantities $\delta\Sigma - \delta\Sigma_b$ (where $\delta\Sigma_b$ is the bulk value equal to 0 in TB and to 0.65 eV in LDA) and E_{coul}, such that $E_g^{exc} \approx E_g^0 + \delta\Sigma_b$. This means that the single particle calculations should yield accurate results for the excitonic gap. These results have been recently confirmed by diffusion quantum Monte Carlo calculations [21].

3. Possible channels for the recombination

In this section we consider first the radiative recombination across the quantum confined gap of nanocrystallites and show that this cannot explain the full body of experimental data. We then demonstrate the existence of intrinsic surface self-trapped excitons which luminesce at smaller energies in very small clusters. In nanocrystals exposed to oxygen, we show a theoretical model in which electronic states appear in the bandgap when a Si=O bond is formed, in good agreement with experiments. Finally, we discuss two mechanisms for the non radiative recombination, on Si dangling bonds or via the Auger effect.

3.1. DIRECT TRANSITIONS ACROSS THE QUANTUM CONFINED GAP

We consider here the direct recombination of electron-hole pairs, i.e., no-phonon processes, in spherical nanocrystals passivated by hydrogen [10,11,22]. The first step of our work is the calculation of the electron and hole states using a tight binding technique described in Ref. [22]. The second step is the calculation of the radiative recombination rates defined from the dipole matrix elements [23]

$$W_{if}(\mathbf{e}) = AE_{if} \mid \langle i \mid \mathbf{e}\cdot\mathbf{p} \mid f \rangle \mid^2, \tag{4}$$

where $|i\rangle$ is the initial state in the conduction band (here the electron state $|\psi_e\rangle$) and $|f\rangle$ is the final state in the valence band (here the hole state $|\psi_h\rangle$). E_{if} is the energy of the transition, $\mathbf{e}$ is the vector of the light polarization and A is a constant. The momentum matrix element $\langle \psi_e | \mathbf{e}\cdot\mathbf{p} | \psi_h \rangle$ is developed in the tight binding basis where the atomic orbitals are replaced by Gaussians [24]. To include the effect of the temperature, we define a thermal average of the recombination rates which is justified by the fact that the thermalization of the electron and the hole after excitation in the bands is more efficient than the radiative recombination (the radiative recombination rate is smaller than $10^{7}\ \mathrm{s}^{-1}$ as already shown in Ref. [11]).

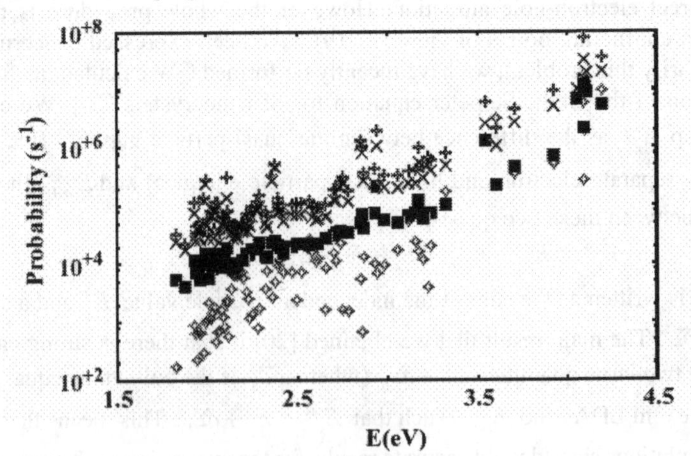

Figure 2. Recombination rate as a function of the nanocrystal bandgap: sum of no-phonon and phonon-assisted processes (+); no-phonon (diamonds); transitions assisted by optical phonons (X); by TA phonons (squares).

The results are given in Fig. 2. The radiative recombination is in average quite slow. The mixing of different *k*-states gives optical properties intermediate between an indirect gap and direct gap material. The strong decrease of the radiative recombination rate for lower photon energy is obviously due to the indirect nature of the silicon band gap which gives a radiative recombination rate equal to zero in the limit of bulk silicon (in a first order dipole theory). Another feature of Fig. 2 is the important scattering of the recombination rate.

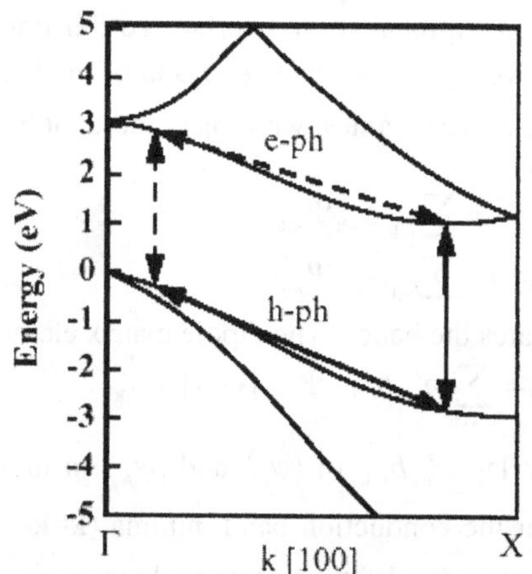

Figure 3 The Si band structure near the gap region along the (100) direction. The processes for phonon-assisted transitions are illustrated.

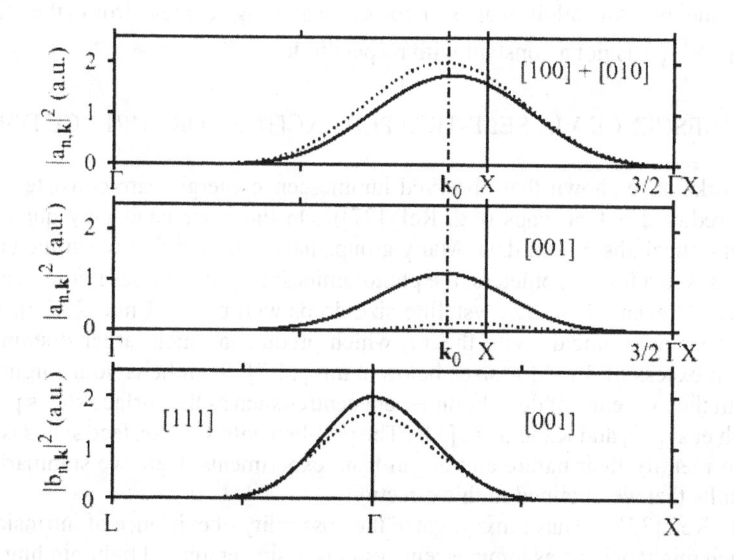

Figure 4 . Top: projection ($|a_{n,\mathbf{k}}|^2$) of the lowest electron state in an ellipsoid on the bulk states $\Psi_{n,\mathbf{k}}$ for **k** along [100] and [010] (sum of the two). Middle: same for **k** along [001]. Bottom: projection ($|b_{n,bfk}|^2$) of the highest hole state for **k** along [001] (right) and [111] (left). Solid lines: ellipsoid long axis of 1.90 nm, short axis of 1.36 nm. Dashed line: long axis of 2.17 nm, short axis of 1.36 nm.

To understand this peculiar behaviour, we analyse in detail the nature of the no-phonon transitions. We come back to important points which have been discussed in detail in Ref. [11,25]. Because of the indirect gap, band edge transitions in bulk Si are only possible with the assistance of phonons to supply the momentum in a second order process (Fig. 3). In nanocrystals, the strong confinement of the electron and hole wave functions in real space leads to a spread of the wave functions in momentum space. Thus radiative recombination can proceed by direct no-phonon transitions and the oscillator strength is directly proportional to the reciprocal space overlap. To illustrate this effect, we plot in Fig. 4 the weight of the lowest electron state $|\psi_e\rangle$ and of the highest hole state $|\psi_h\rangle$ in momentum space. To obtain these values, we project the tight binding eigenfunctions in the basis of the bulk states

$$|\psi_e\rangle = \sum_{n,\mathbf{k}} a_{n,\mathbf{k}} \Psi_{n,\mathbf{k}},$$
$$|\psi_h\rangle = \sum_{n,\mathbf{k}} b_{n,\mathbf{k}} \Psi_{n,\mathbf{k}}, \tag{5}$$

where the index n enumerates the bands. The dipole matrix element is given by

$$\langle \psi_e | \mathbf{e}\cdot\mathbf{p} | \psi_h \rangle = \sum_{n,n',\mathbf{k}} a^*_{n,\mathbf{k}} b_{n',\mathbf{k}} \langle \Psi_{n,\mathbf{k}} | \mathbf{e}\cdot\mathbf{p} | \Psi_{n',\mathbf{k}} \rangle. \tag{6}$$

Fig. 4 shows that the overlap $a^*_{n,\mathbf{k}} b_{n',\mathbf{k}}$ of $|\psi_e\rangle$ and $|\psi_h\rangle$ in momentum space is small because $|\psi_e\rangle$ is centred at the conduction band minima ($\mathbf{k}=\mathbf{k}_0$) and $|\psi_h\rangle$ is centred at $\mathbf{k}=0$. It explains why the radiative lifetime remains long in silicon crystallites [11,26]. Since $a_{n,\mathbf{k}}$ and $b_{n',\mathbf{k}}$ are oscillatory functions of $\mathbf{k}$, oscillatory factors enter the optical matrix elements. An additional source of scattering comes from the fact that $\langle \Psi_{n,\mathbf{k}} | \mathbf{e}\cdot\mathbf{p} | \Psi_{n',\mathbf{k}} \rangle$ is not a constant with respect to $\mathbf{k}$.

3.2. LUMINESCENCE VIA SELF-TRAPPED EXCITONS OR SURFACE DEFECTS

Several works have shown that observed luminescence energies are consistently lower than the predicted optical gaps (e.g. Ref. [27]). On the other hand they quantitatively agree with optical absorption data. Many groups have reported that when the crystallite size decreases to a few nanometers, the photoluminescence in air does not increase much beyond 2.1 eV even when the crystallite size drops well below 3 nm [28,29]. This observation does not coincide with theory, which predicts a much larger opening of the bandgap, in excess of 3 eV for sizes below 2 nm [11,7]. Such behaviour might be consistent with the existence of deep luminescent centres such as the surface states postulated by F. Koch et al [30] and Kanemitsu [31]. The problem with such surface states is that it is difficult to identify their nature and origin from experiments. Here we summarize theoretical results that we obtained in this direction.

In Ref. [32], we have investigated the possibility of existence of intrinsic surface states which might behave as luminescent systems. Using empirical tight binding and first principle local density calculations, we have shown that such states indeed exist under the form of self-trapped excitons. In small clusters containing less than $\lessapprox$100 Si atoms, we obtain that there is a very strong atomic relaxation in the excited state, and the exciton tends to localize itself on a single Si-Si bond. This situation is likely to occur near the surface where the relaxation requires less energy. A favourable system is obtained for dimer bonds passivated by hydrogen atoms or by silicon oxide [32]. Light emission from these trapped excitons is predicted in the infrared or the near visible. We are thus led to the interpretation that part of the luminescence could be due to such surface states while optical absorption is characteristic of quantum confinement effects. Note that in larger nanocrystals (>100 Si atoms) the self-trapped state is likely to be metastable. On the experimental side, self-trapped excitons have been invoked in Si/SiO_2 multilayers [33] and in small Si particles [34,35].

The surface passivation plays an important role in the luminescence. In a recent work [36], oxygen-free porous Si samples with different porosities and emitting throughout the visible spectrum (red to blue, following Ref. [37]) were examined. After exposure to air, a red shift of the photoluminescence was observed, which can be as large as 1 eV for blue luminescent samples that contain crystallites smaller than 2 nm. In contrast, no red shift at all was detected when the samples were kept in pure hydrogen or in vacuum. All these results suggest that the electron-hole recombination in oxidized samples occurs via carriers trapped in oxygen-related localized states that are stabilized by the widening of the gap induced by quantum confinement. Thus we have applied electronic structure calculations to various situations involving oxygen atoms at the surface of Si clusters. As expected because of the large offset between bulk SiO_2 and Si (4 eV), for normal Si-O-Si bonds, we do not find any localized gap state. Similar results are obtained for Si-O-H bonds. But, when nanocrystalline Si is oxidized and a Si-O-Si layer is formed on the surface, the Si-Si or Si-O-Si bonds are likely to weaken or break in many places because of the large lattice mismatch at the Si/SiO_2 interface [38]. Some mechanisms can act to passivate the dangling bonds [39]. A Si=O double bond is more likely to be formed and stabilize the interface, since it requires neither a large deformation energy nor an excess elements. Such bonds have been suggested at the Si/SiO_2 interface [39].

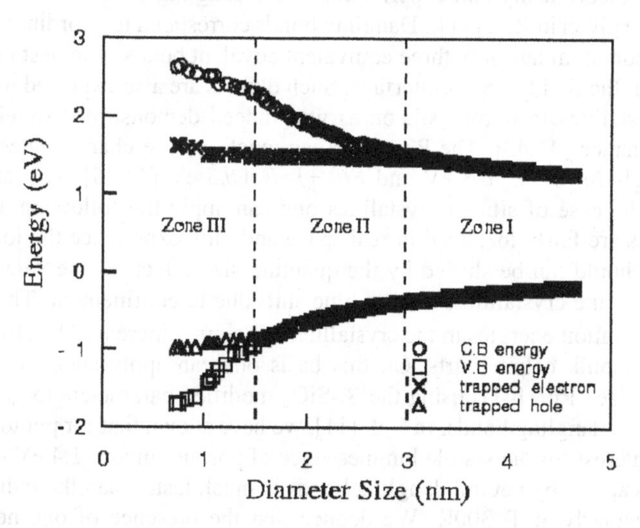

Figure 5. Electronic states in Si nanocrystals as a function of cluster size and surface passivation. The trapped electron state is a *p*-state localized on the Si atom of the Si=O bond and the hole trapped state is a *p* -state on the oxygen atom.

The electronic structure of Si clusters with one Si=O bond (the other dangling bonds being saturated by hydrogen atoms) has been calculated as a function of the cluster size (Fig. 5). Three different recombination mechanisms are suggested, depending on the size of the cluster. Each zone in Fig. 5 corresponds to a different mechanism. In zone I, recombination is via free excitons. As the cluster size decreases, the photoluminescence energy increases, exactly as predicted by quantum confinement. There is no red shift whether the surface termination is hydrogen or oxygen, since the bandgap is not wide enough to stabilize the Si=O surface state. In zone II recombination involves a trapped electron and a free hole. As the size decreases, the photoluminescence emission energy still increases, but not as fast as predicted by quantum confinement, since the trapped electron state energy is size-independent. In zone III, recombination is via trapped excitons. As the size decreases, the photoluminescence energy stays constant, and there is a large photoluminescence red shift when the nanocrystallite surface is oxidized. Comparison between experimental and theoretical photoluminescence energy is good (Fig. 4 of Ref. [36]), in spite of the simplicity of the model. Note finally that our calculations have been basically confirmed by density functional and quantum Monte Carlo calculations [40].

3.3. NON-RADIATIVE RECOMBINATION

In this part we summarize results concerning two efficient channels for nonradiative recombination: surface dangling bonds and the Auger effect. Both are important for an understanding of the properties of porous silicon.

From previous work on the Si-SiO_2 interface the electron-hole recombination on silicon dangling bonds is known to be due to the multiphonon capture of the electron and hole [41]. We discuss here a calculation of the corresponding capture rate based on what is known both theoretically and experimentally for dangling bonds at the Si-SiO_2 interface (details are given in Ref. [11]). Dangling bonds correspond to coordination defects in which the silicon atom has only three equivalent covalent bonds. The best known case is the Pb centre at the Si(111)-SiO_2 interface. Such defects are also expected to occur at the surface of crystallites in porous silicon as was indeed demonstrated by electron paramagnetic resonance [42,43]. The Pb centre can exist in three charge states +, 0, - with ionisation levels $E(-/0)=E_c-0.3$ eV and $E(0/+)=E_v+0.3$ eV [44,45]. To transpose these properties to the case of silicon crystallites one can apply the following arguments: i) deep gap states are fairly localized in real space and only experience the local situation. Their energy should not be shifted by the quantum size effect; ii) the valence and conduction states of the crystallite exhibit a blue shift due to confinement. This means that the thermal ionisation energies in the crystallites are simply increased by ΔE_c or ΔE_v with respect to their bulk counterparts. On this basis one can apply standard multiphonon capture theory (see Ref. [41]) using the Si-SiO_2 modified parameters to get the capture properties due to dangling bonds. In Ref. [11], we have shown that for photon energies in the range of interest for the visible luminescence of porous silicon (1.4 eV - 2.2 eV), the non radiative capture by neutral dangling bonds is much faster than the radiative recombination, particularly at T=300K. We deduce that the presence of one neutral silicon dangling bond at the surface of a crystallite in porous silicon kills its luminescence above 1.1 eV, in agreement with experiments [46].

The Auger effect is one of the most efficient channels for electron-hole (eh) recombination in bulk silicon. We summarize here results of calculations showing that it remains as efficient even in very confined systems. A detailed account of this work is given in Refs. [47,48]. In the Auger process, an eh pair recombines non radiatively transferring its energy to a third carrier. In bulk silicon, the Auger probability per unit time and volume is equal to An^2p+Bp^2n where p and n are respectively the hole and electron concentrations. A and B are not known accurately, values being reported between 10^{-30} and 10^{-32} cm^6/s [49]. If we extrapolate these data to nanocrystallites taking the concentrations n and p to correspond to one carrier confined in a spherical volume $4\pi R^3/3$. Such extrapolated Auger lifetimes lie between 0.1 and 100 nanoseconds for crystallite radius $R<2.5$ nm which is several orders of magnitude faster than the radiative lifetime. A detailed calculation of the Auger recombination probability with the Fermi rule [49] in which the matrix elements are those of the screened electron-electron interaction between initial and final states consisting of Slater determinants give values close to the simple estimation [47,48].

The Auger recombination gives a coherent interpretation of several experiments on porous silicon: i) the saturation of the photoluminescence at high excitation [47,48] ; ii) the voltage quenching of the photoluminescence [50] is observed on n-type porous silicon samples cathodically polarized in an aqueous solution of sulfuric acid; iii) the peculiar spectral width (≈ 0.25 eV) of the electroluminescence of n-type porous silicon cathodically polarized in a persulfate aqueous solution [51]; the quenching of the photoluminescence due to P doping of porous Si [52]; persistent hole burning experiments [53].

4. Phonon-assisted transitions

Semiconductor quantum dots can be considered as artificial atoms since their electronic spectrum consists of discrete atomic-like energy levels. When the quantum dots are built from direct gap semiconductors, their low temperature photoluminescence is mainly characterized by sharp spikes with linewidth lower than a few hundreds of meV, confirming the atomic-like nature of the emitting state [54]. The situation becomes more complex for nanocrystals made of indirect gap semiconductors like silicon because no-phonon transitions are only weakly allowed and there is a size-dependent competition between no-phonon and phonon-assisted transitions. This has been confirmed for porous silicon where several experimental studies [55,56] show that phonon-assisted transitions remain efficient in spite of the confinement effect. On the theoretical side the importance of phonon-assisted transitions has been supported by the work of M. Hybertsen [57]. However this calculation is based on simple effective-mass theory that is too crude for small nanocrystals. Here we present a more quantitative approach to these electron-phonon related phenomena. Details can be found in Ref. [22].

To calculate the recombination rates for phonon-assisted transitions, we start again from Eq. (4). But now, the states $|i\rangle$ and $|f\rangle$ include the coordinates of the electron and of the nuclei. Working within the adiabatic approximation, the matrix element of the momentum becomes

$$\langle i|\mathbf{e}\cdot\mathbf{p}|f\rangle = \langle \chi_i|\langle\psi_i|\mathbf{e}\cdot\mathbf{p}|\psi_f\rangle|\chi_f\rangle \tag{7}$$

where $|\chi_i\rangle$ and $|\chi_f\rangle$ are the vibrational states of the system that are functions of the nuclear coordinates. They are built in the harmonic approximation from independent harmonic oscillators corresponding to *3N* normal modes Q_j (*N* is the number of atoms in the system). $|\psi_i\rangle$ and $|\psi_f\rangle$ in Eq. (7) are the electronic wave functions which can be defined for any set of nuclear coordinates. We expand the momentum matrix element to first order in the normal modes

$$\langle\psi_i|\mathbf{e}\cdot\mathbf{p}|\psi_f\rangle = \langle\psi_i|\mathbf{e}\cdot\mathbf{p}|\psi_f\rangle_0 + \sum_{j=1}^{3N}\mathbf{A}_j\langle\chi_i|Q_j|\chi_f\rangle \tag{8}$$

where $\mathbf{A}_j = \left(\partial\langle\psi_i|\mathbf{e}\cdot\mathbf{p}|\psi_f\rangle/\partial Q_j\right)$. The index 0 means that the quantities are calculated at the equilibrium positions. Using the transformation $Q_j = \sqrt{\hbar/2\omega_j}\,(a_j^+ + a_j)$ as function of the creation and annihilation operators and averaging over n_j, the number of phonons in mode *j*, we obtain the following expressions for the recombination rates of the processes involving one phonon of energy $\hbar\omega_j$ [22]

$$\text{one-phonon emission:}\quad \mathrm{W_{if}}(\mathbf{e}) \approx A\left|\sqrt{\frac{\hbar}{2\omega_j}}\mathbf{A}_j\cdot\mathbf{e}\right|^2\left\{\bar{n}_j+1\right\}$$

$$\text{one-phonon absorption:}\quad \mathrm{W_{if}}(\mathbf{e}) \approx A\left|\sqrt{\frac{\hbar}{2\omega_j}}\mathbf{A}_j\cdot\mathbf{e}\right|^2\left\{\bar{n}_j\right\} \tag{9}$$

where $\bar{n}_j = \left[\exp\left(\hbar\omega_j/kT\right)-1\right]^{-1}$. The heavy part of the work is the evaluation of the coupling coefficients which are obtained numerically for all the modes *j* of the nanocrystal. For each mode *j*, it requires to calculate the wavefunctions - therefore the Hamiltonian - and the optical matrix elements when the nuclei are displaced from their equilibrium sites according to Q_j. The vibrational modes of the nanocrystals are calculated using a valence force field model [59]. In Ref. [22] we discuss the fact that direct use of Eq. (9) is not possible in nanocrystals with bandgap larger than $\gtrsim 2$ eV because multiphonon processes become important for low-frequency acoustic modes. An appropriate treatment described in this work leads to a broadening of the peaks but the total amplitude of the peaks remains given by Eq. (9).

Fig. 6 shows the recombination rates at 4K calculated for a 2.85 nm nanocrystal. At this temperature, the phonon absorption process is negligible and the recombination proceeds by phonon emission. We also plot the luminescence spectrum below the no-phonon line calculated assuming that the intensity is directly proportional to the recombination rate. It shows, in agreement with the EMA results of Ref. [57], that optical modes dominate and that the contributions from transverse acoustic (TA) modes are smaller. On Fig. 6, we also plot the luminescence spectrum for a single quantum dot taking into account the multiphonon broadening with acoustic modes. The broadening of the peaks is large, of the order of 10s of meV due not only to the breakdown of the selection rules but also to multiphonon effects. Thus the intrinsic luminescence of small silicon nanocrystals (below ≈4 nm) should not be dominated by sharp spikes like for quantum dots made of direct semiconductors.

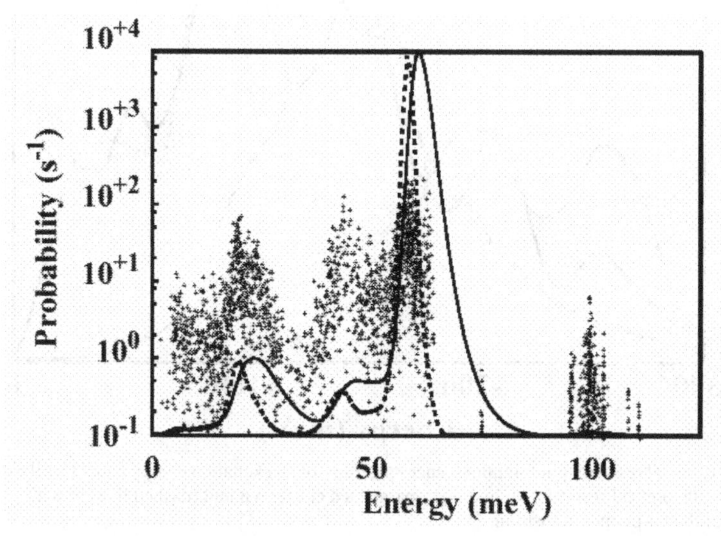

Figure 6. Radiative recombination rate at 4K (+) with respect to the energy of the phonon involved in the transition for a 2.85 nm hydrogen passivated nanocrystal. The dotted line shows in arbitrary units the luminescence intensity of a single cluster assuming it is directly proportional to the recombination rate. The full line shows the luminescence intensity broadened by the multiphonon effect with low-frequency acoustic modes.

Figure 2 compares the strength of no-phonon (W_{no}), one-phonon acoustic (W_{ac}) and one-phonon optical (W_{op}) processes. W_{ac} and W_{op} are obtained by summing over the individual phonon contributions in the corresponding energy range, respectively 15-30 meV and 45-70 meV. The comparison with the resonant photoluminescence experiments of Kovalev *et al.* [58] performed on oxidized porous silicon is delicate since our calculations correspond to hydrogen terminated crystals. However, the comparison leads to two conclusions: i) the calculation gives $W_{op}/ W_{ac} \approx 10$ in full agreement with experiment; ii) on the contrary, for gaps in the 1.8-2 eV range, our predicted values for W_{no}/ W_{op} are about 150 times smaller. The origin of such a difference must then be ascribed to the existence of strained interface regions in the oxidized samples. However it cannot be due to the deep defects mentioned earlier (self-trapped excitons, Si=O defects) since their Stokes shift is by far too large to be observable in resonant photoluminescence. One must then look for states with weaker localization due to more or less distorted bonds in the vicinity of the interface region. This is quite similar to what happens in theoretical descriptions of amorphous silicon. We have shown recently [60] that such distorted bonds are responsible for the band tails of localized states of *a*-Si that can extend over regions containing 10-100 atoms. Furthermore we have calculated the corresponding radiative recombination rates that turn out to be in the 10^5-10^6/sec range [61], i.e. ≈100 times larger than our values for ideal clusters.

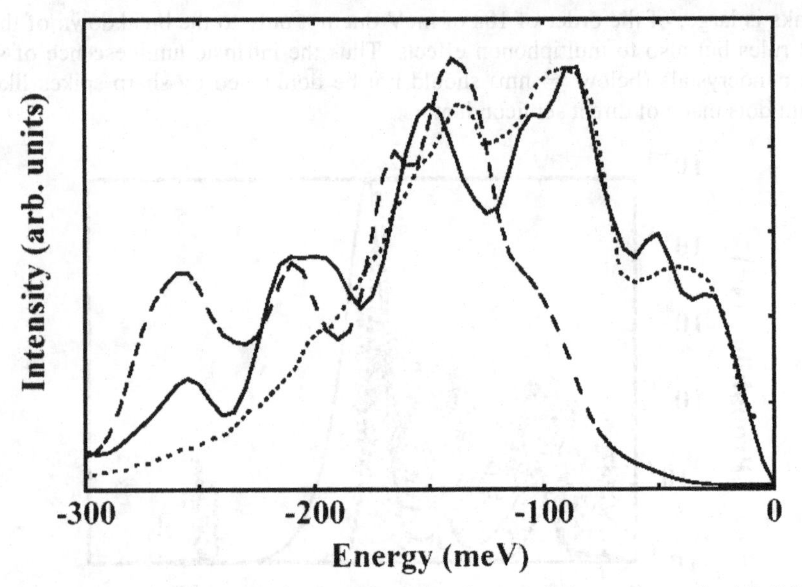

Figure 7. Experimental resonant photoluminescence spectrum for an excitation energy equal to 1.9 eV (full line) compared to the theoretical result for an hydrogen-passivated cluster without (dashed line) or with (dotted line) enhancement of the no-phonon transition.

Thus our view of the resonant photoluminescence of oxidized nanocrystals is that it is dominated by such transitions between localized states due to strained bonds. To confirm this point of view, we have calculated a resonant photoluminescence spectrum and compared it to experimental results of the authors of Ref. [58]. Details of the calculation are given in Ref. [22]. The result is plotted in Fig. 7. The predicted curve with W_{no} calculated for the ideal clusters does not compare well with experiment. On the other hand, if the relative value of W_{no} is increased by a factor ≈ 150 then the agreement becomes strikingly good.

5. Luminescence polarization of silicon nanocrystals

The luminescence of silicon nanocrystals is sometimes characterized by a huge degree of linear polarization (up to 30 %) [62-65]. An interpretation of these experiments has been proposed on the basis of a dielectric model [63,64] where porous silicon is described by ellipsoidal Si crystals embedded in an effective dielectric medium with a lower dielectric constant. The local electric field in the nanocrystal is anisotropic which makes the constant A in Eq. (4) dependent on the polarization vector $\mathbf{e}$. However the polarization anisotropy which results from the dipole moments $\langle i | \mathbf{e} \cdot \mathbf{p} | f \rangle$ is not considered in these models. In Ref. [66], we have considered the polarization of the luminescence that arises from the anisotropy of the optical moments. We summarize here some results.

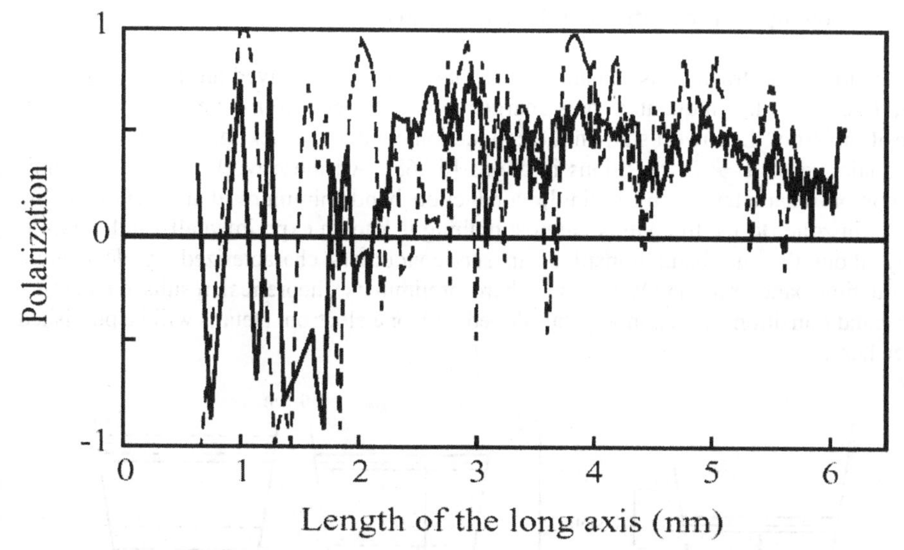

Figure 8 The polarization degree for Si ellipsoids oriented along a [100] direction (straight line) or a [110] direction (dashed line) calculated as a function of the length of the long axis, which is equal to $\sqrt{2}$ times the length of small ones (T = 300 K).

We consider Si ellipsoids defined three principal axes of respective length *a, b, c*, with *a* oriented along a [100] or [110] axis with $b=c=a/\sqrt{2}$. The degree of linear polarization of the emitted light is defined by

$$\sigma = \frac{W(\mathbf{e}_{//}) - W(\mathbf{e}_{\perp})}{W(\mathbf{e}_{//}) + W(\mathbf{e}_{\perp})} \qquad (10)$$

where $\mathbf{e}_{//}$ is along the long axis of the ellipsoid and $\mathbf{e}_{\perp}$ is perpendicular to $\mathbf{e}_{//}$. The degree σ of polarization for no-phonon transitions is plotted in Fig. 8 with respect to length *a*. We see that for $a<2$ nm, σ is scattered and gets positive or negative values, but for larger sizes σ tends to become positive in average. The results obtained for ellipsoids with a long axis oriented along [100] and [110] directions do not coincide. Similar curves are obtained for phonon-assisted transitions [66]. All these results are rather surprising at first glance and they have no equivalence in the case of direct gap semiconductors. A detailed analysis made in Ref. [66] shows that the oscillations are due to the indirect nature of the Si band gap which leads to a dependence of the optical matrix elements on the oscillatory overlaps between electron and hole states in momentum space (in agreement with the discussion of section 3.1). However, in a statistical ensemble of crystallites elongated in a given direction and with size larger than 2-3 nm, we obtain that the light is in average polarized along this direction. We have checked that this conclusion holds true for other types of asymmetric nanocrystals like cylinders or boxes.

Recent photoluminescence experiments on porous silicon have shown that the positive polarization degree is larger under resonant excitation than under non resonant conditions [53]. This is interpreted by the authors by a polarization of the dipole moment in anisotropic nanocrystals, which is totally supported by our calculations.

6. Intraband optical transitions in Si nanocrystals

Intraband optical transitions in nanocrystals have recently received an increasing attention. Recent works on quantum dots charged with one electron (n-type) reported optical transitions from the lowest state in the conduction band to the next higher state which is the analogue of $S \rightarrow P$ transitions in atoms [67,68]. All these works concerned semiconductors characterized by a single conduction band minimum at the centre of the Brillouin zone (**k**=0). In contrast, almost nothing is known experimentally and theoretically about the intraband transitions in semiconductors characterized by degenerate conduction band minima. We present here preliminary theoretical results concerning intraband transitions in Si nanocrystals doped with one electron. Details will be published elsewhere.

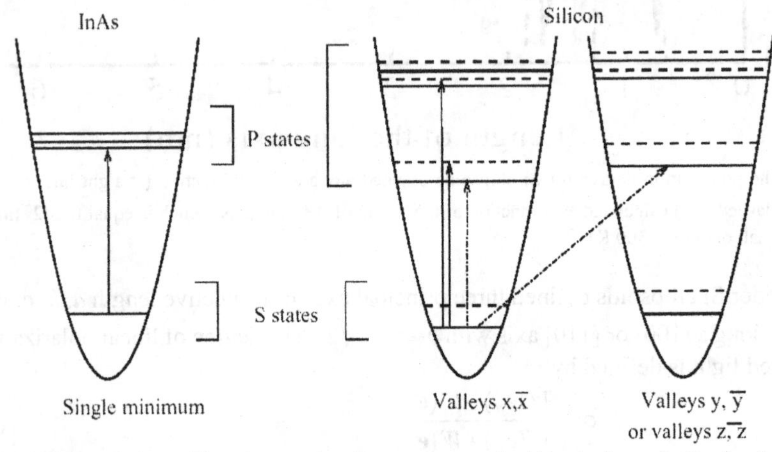

Figure 9. S → P optical transitions in a semiconductor nanocrystal with a single conduction band minimum such as InAs or in a Si nanocrystal with six valleys. In Si, the P levels are split due to the anisotropy of the effective masses. The inter-valley coupling lifts the degeneracies between x and $\bar{x}$ (resp. y and $\bar{y}$, z and $\bar{z}$) valleys (dashed lines). Optical transitions occur between states in the same valley or in different valleys (arrows with straight line: no-phonon transitions, arrows with a dashed-dotted line: one-phonon transitions).

We consider crystallites bounded by (100) equivalent planes of dimensions $L_xL_yL_z$ but similar results are obtained for spherical shapes. In a quantum dot made of a semiconductor such as InAs with a single conduction band minimum (Fig. 9), the lowest confined state is non degenerate (S-like) and the next higher state is threefold degenerate (P-like) if L_x=L_y=L_z. The infrared optical absorption is characterized by a sharp peak corresponding to the direct $S \rightarrow P$ transition. In silicon, the conduction band is characterized by six equivalent (100) valleys at wave vectors $\mathbf{k}_{0l}$ ($l \in \{x,\bar{x},y,\bar{y},z,\bar{z}\}$ and $|\mathbf{k}_{0l}| = k_0 \approx 0.85(2\pi/a)$ where a is the bulk lattice parameter, x stands for the (100) valley and $\bar{x}$ for the (-100) valley) and by anisotropic effective masses (transverse mass m_t=0.19 m_0, longitudinal one m_l =0.92m_0). The quantum confinement gives S, P states in each valley [14] (Fig. 9). However, the P levels are split in two groups due to the strong anisotropy in the effective masses. If the shape is slightly anisotropic ($L_x \neq L_y \neq L_z$) which is

likely in real quantum dots, the degeneracy between *x*, *y* and *z* valleys is lifted. The degeneracy between valleys *x* and $\bar{x}$ (resp. *y* and $\bar{y}$, *z* and $\bar{z}$) is also lifted by inter-valley couplings [66]. This leads to a rich electronic structure (Fig. 9).
The calculation of the optical properties requires including phonon-assisted (one-phonon) processes since no-phonon transitions between different minima in bulk Si are strictly forbidden due to **k** selection rule. However, we have seen that in systems where the wave functions are confined in real space and thus spread in reciprocal space, the **k** selection rule is broken and no-phonon transitions become allowed [66,22]. Thus we use the same approach as in Sect. 4 to calculate no-phonon and phonon-assisted transitions.

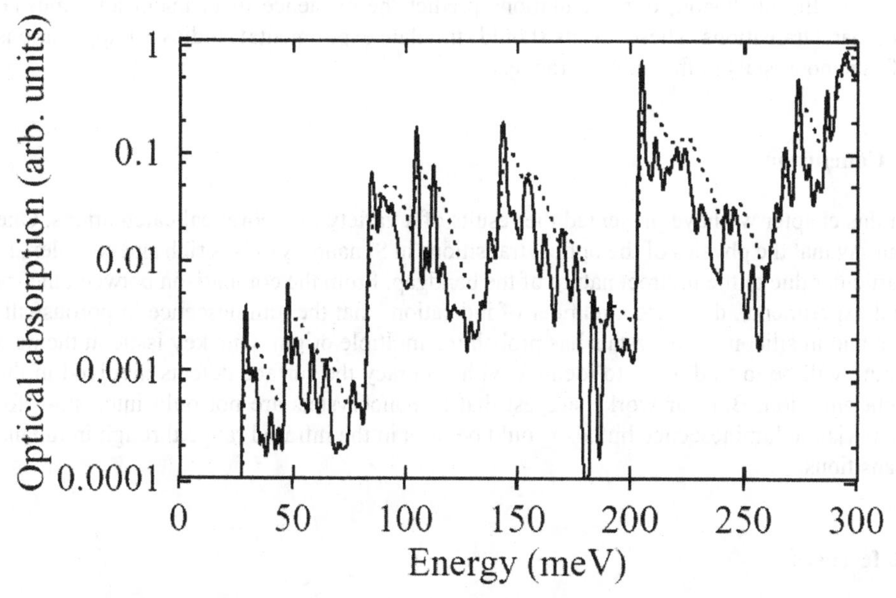

Figure 10. Straight line: optical absorption spectrum calculated for a Si crystallite of dimensions L_x=*5a*, L_y=*6a*=*3.3* nm and L_z=*7a* where *a*=5.42Å (*T* = 4K). The calculation includes 25 electron states in the conduction band and 8475 vibrational modes. The dashed line includes the broadening due to multiphonon couplings to acoustic phonons. Some no-phonon transitions are indicated (NP).

The optical absorption spectrum at 4K for a crystallite containing 1909 Si atoms is shown in Fig. 10. The only possible initial state is the ground state derived from the valleys *x* and $\bar{x}$ and phonon-assisted transitions only proceed by emission of phonons. The spectrum consists in series of peaks corresponding to all possible excited states and all phonons *j*. Interestingly, and in contrast to direct gap semiconductors, the spectrum for a single crystallite is broad because there are many excited states derived from the six conduction band minima of Si and because a large number of vibrational modes are involved due to the confinement which leads to a spread of the wave functions in the reciprocal space. A deeper analysis [69] shows that there are three types of transitions: i) no-phonon transitions within the same valley which are the analogues of $S \rightarrow P$ transitions in direct gap semiconductor nanocrystals, with a similar efficiency (total oscillator strengths in the 0.2-1.1 range); ii) transitions within the same valley but with the assistance of phonons, mainly optical ones at high energy (60-63 meV) and acoustic ones at very low energy

(1-10) meV. These modes are mainly derived from phonons at $\mathbf{k} \approx 0$ and $\mathbf{k} \approx (\pm 2k_0,0,0) \equiv (\pm 0.3(2\pi/a),0,0)$, as required by the **k** conservation rule ($x \rightarrow x$ and $x \rightarrow \bar{x}$). The efficiency of these transitions is high (oscillator strength of the order of 0.5), comparable to direct ones;iii) this third category corresponds to excited states in valleys y and z. Because initial and final states are derived from different valleys, the transitions are mainly assisted by phonons, with a main contribution from optical phonons at energy close to 56-60 meV. Once again, these results can be understood in terms of **k** conservation rules since the transition from valleys x and $\bar{x}$ at $(\pm k_0,0,0)$ to valleys y and $\bar{y}$ at $(0,\pm k_0, 0)$ requires phonons with wave vectors close to $(\pm k_0, \pm k_0,0)$.

In conclusion, our calculations predict the existence of efficient and unusual intraband transitions. These results should stimulate experimental studies for applications of Si nanocrystals in the infrared range.

7. Conclusion

In this chapter we have presented the results of a variety of theoretical calculations. One can say that the physics of the optical transitions in Si nanocrystals is rich and complex, in particular due to the indirect nature of the bandgap. From the comparison between theory and experiments, there are a number of indications that the luminescence in porous silicon and in silicon nanocrystals has probably a multiple origin. One key issue in the next future will be to find ways to identify with accuracy the surface defects involved in the light emission. But our works suggest that Si nanocrystals are not only interesting for their visible luminescence but also could be used in the infrared range through intraband transitions.

References

[1] L.T. Canham, Appl. Phys. Lett. 57, 1046 (1990).
[2] A.C. Seabaugh, A.H. Taddiken, E.A. Beam, J.N. Randall, Y.C. Kao, Electron. Lett. 29, 1802 (1993).
[3] E. Leobandung, L. Guo, Y. Wang, S.Y. Chou, Appl. Phys. Lett 67, 938 (1995).
[4] A. Fujiwara, Y. Takahashi, K. Murase, M. Tabbe, Appl. Phys. Lett. 67, 2957 (1995).
[5] *Single charge tunneling : Coulomb blockade phenomena in nanostructures*, ed. by H. Grabert and M. Devoret, Proceedings of a NATO Advanced Study Institute on Single Charge Tunneling held in Les Houches, France (1991), Plenum Press,New York, 1992.
[6] B. Delley, F. Steigmeier, Appl. Phys. Lett. 67, 2370 (1995).
[7] Lin-Wang Wang, A. Zunger, J. Chem. Phys. 100, 2394 (1994).
[8] A. Zunger, Lin-Wang Wang, Appl. Surf. Science 102, 350 (1996).
[9] J.C. Slater, G.F. Koster, Phys. Rev. 94, 1498 (1954).
[10] J.P. Proot, C. Delerue, G. Allan, Appl. Phys. Lett. 61, 1948 (1992).
[11] C. Delerue, G. Allan, M. Lannoo, Phys. Rev. B 48, 11024 (1993).
[12] Shang Yuan Ren, Phys. Rev. B 55, 4665 (1997).
[13] T. Takagahara, K. Takeda, Phys. Rev. B 46, 15578 (1992).
[14] Y.M. Niquet, C. Delerue, G. Allan, and M. Lannoo, Phys. Rev. B 62, 5109 (2000).
[15] J.M. Jancu, R. Scholz, F. Beltram, F. Bassani, Phys. Rev. B 57, 6493 (1998).
[16] P.B. Allen, J.Q. Broughton, A.K. MacMahan, Phys. Rev. B 34, 859 (1986).
[17] S. Ögüt, J.R. Chelikowsky and S.G. Louie, Phys. Rev. Lett. 79, 1770 (1997).
[18] R.W. Godby and I.D. White, Phys. Rev. Lett. 80, 3161 (1998).
[19] A. Franceschetti, L.W. Wang, and A. Zunger, Phys. Rev. Lett. 83, 1269 (1999).

[20] C. Delerue, G. Allan, and M. Lannoo, Phys. Rev. Lett. 84, 2457 (2000).
[21] A.R. Porter, M.D. Towler, and R.J. Needs, Phys. Rev. B 64, 035320 (2001).
[22] C. Delerue, G. Allan, and M. Lannoo, Phys. Rev. B 64, 193402 (2001).
[23] D.L. Dexter, *Solid State Physics, Advances in Research and Applications*, edited by F. Seitz and D. Turnbull (Academic, New York, 1958), Vol. 6, p. 360.
[24] J. Petit, G. Allan, and M. Lannoo, Phys. Rev. B 33, 8595 (1986).
[25] M.S. Hybertsen, *Light Emission from Silicon*, edited by S.S. Iyer, L.T. Canham, and R.T. Collins (Materials Research Society, Pittsburgh, 1992), p. 179.
[26] J.C. Vial, A. Bsiesy, F. Gaspard, R. Hérino, M. Ligeon, F. Muller, R. Romestain, and R.M. Macfarlane, Phys. Rev. B 45, 14171 (1992).
[27] D.J. Lockwood, Solid State Communications 92, 101 (1994).
[28] J. Von Behren, T. Van Buuren, M. Zacharias, E.H. Chimowitz and P.M. Fauchet, Solid State Comm. 105, 317 (1998).
[29] S. Schuppler et al, Phys. Rev. B 52, 4910 (1995).
[30] F.Koch, V. Petrova-Koch, T. Muschick, in *Light Emission from Silicon*, ed. by J.C. Vial, L.T. Canham and W. Lang, p. 271, J. Lum. 57, (Elsevier-North-Holland, 1993).
[31] Y. Kanemitsu, Phys. Rev. B 49, 16845 (1994).
[32] G. Allan, C. Delerue, and M. Lannoo, Phys. Rev. Lett. 76, 2961 (1996).
[33] B.V. Kamenev and A.G. Nassiopoulou, J. Appl. Phys. 90, 5735 (2001).
[34] M.H. Nayfeh, N. Rigakis, and Z. Yamani, Phys. Rev. B 56, 2079 (1997).
[35] M.H. Nayfeh, N. Barry, J. Therrien, O. Akcadir, E. Gratton, and G. Belomoin, Appl. Phys. Lett. 78, 1131 (2001).
[36] M.V. Wolkin, J. Jorne, P.M. Fauchet, G. Allan, and C. Delerue, Phys. Rev. Lett. 82, 197 (1999).
[37] H. Mizuno, H. Koyama and N. Koshida, Appl. Phys. Lett. 69, 1 (1996).
[38] A. Ourmazd, D.W. Taylor, J.A. Rentschler and J. Bevk, Phy. Rev. Lett 59, 213 (1987).
[39] F. Herman and R.V. Kasowski, J. Vac. Sci. Thechnol. 19, 395 (1981).
[40] A. Puzder, A.J. Williamson, J.C. Grossman, and G. Galli, Phys. Rev. Lett. 88, 097401 (2002).
[41] D. Goguenheim, and M. Lannoo, J. Appl. Phys. 68, 1059 (1990).
[42] M.S. Brandt, and M. Stutzmann, Appl. Phys. Lett. 61, 2569 (1992).
[43] H. J. Von Bardeleben, D. Stievenard, A. Grosman, C. Ortega and J. Siejka, Phys. Rev. B 47, 10899 (1993).
[44] I.H. Poindexter, and P.J. Caplan, Prog. Surf. Sci. 14, 201 (1983).
[45] D. Goguenheim, and M. Lannoo, Phys. Rev. B 44, 1724 (1991).
[46] B.K. Meyer, D.M. Hofmann, W. Stadler, V. Petrova-Koch, F. Koch, P. Omling and P. Emanuelsson, Appl. Phys. Lett. 63, 2120 (1993).
[47] C. Delerue, M. Lannoo, G. Allan, E. Martin, I. Mihalescu, J.C. Vial, R. Romestain, F. Müller, and A. Bsiesy, Phys. Rev. Lett. 75, 2228 (1995).
[48] I. Mihalescu, J.C. Vial, A. Bsiesy, F. Müller, R. Romestain, E. Martin, C. Delerue, M. Lannoo, and G. Allan, Phys. Rev. B 51, 17605 (1995).
[49] P.T. Landsberg, *Recombination in Semiconductors* (Cambridge University Press, 1991).
[50] A. Bsiesy, F. Muller, I. Mihalcescu, M. Ligeon, F. Gaspard, R. Hérino, R. Romestain and J.C. Vial, in *Light Emission from Silicon*, J.C. Vial, L.T. Canham and W. Lang, Editors, p. 29, J. Lum. 57, (Elsevier-North-Holland, 1993).
[51] A. Bsiesy, F. Muller, M. Ligeon, F. Gaspard, R. Hérino, R. Romestain and J.C. Vial, Phys. Rev. Lett. 71, 637 (1993).
[52] A. Mimura, M. Fujii, S. Hayashi, D. Kovalev, and F. Koch, Phys. Rev. B 62, 12625 (2000).
[53] D. Kovalev, H. Heckler, G. Polisski, and F. Koch, Phys. status solidi, B 215, 871 (1999).
[54] K. Brunner, G. Abstreiter, G. Bohm, G. Trankle, and G. Weimann, Phys. Rev. Lett. 73, 1138 (1994); S.A. Empedocles, D.J. Norris, and M.G. Bawendi, Phys. Rev. Lett. 27, 3873 (1996); D. Gammon, E. S. Snow, B. V. Shanabrook, D.S. Katzer, and D. Park, Phys. Rev. Lett. 76, 3005 (1996).
[55] P. D. J. Calcott, K. J. Nash, L. T. Canham, M. J. Kane and D. Brumhead, J. Phys. Condens. Matter 5, L91 (1993). T. Suemoto, K. Tanaka, A. Nakajima, and T. Itakura, Phys. Rev. Lett. 70, 3659 (1993).
[56] L. Brus, in *Light Emission in Silicon. From Physics to Devices*, edited by D. Lockwood, series Semiconductors and Semimetals, Vol. 49 (Academic Press, 1998), p. 303.
[57] M.S. Hybertsen, Phys. Rev. Lett. 72, 1514 (1994).
[58] D. Kovalev, H. Heckler, M. Ben-Chorin, G. Polisski, M. Schwartzkopff, and F. Koch, Phys. Rev. Lett. 81, 2803 (1998).
[59] R. Tubino, L. Piseri, G. Zerbi, J. Chem. Phys. 56, 1022 (1972) ; D. Vanderbilt, S.H. Taole, S. Narasimham, Phys. Rev. B 40, 5657 (1989).

[60] G. Allan, C. Delerue, and M. Lannoo, Phys. Rev. B 57, 6933 (1998).
[61] G. Allan, C. Delerue, and M. Lannoo, Phys. Rev. Lett. 78, 3161 (1997).
[62] A.V. Andrianov, D.I. Kovalev, N.N. Zinov'ev, and I.D. Yaroshetskii, JETP Lett. 58, 427 (1993).
[63] D. Kovalev, M. Ben Chorin, J. Diener, F. Koch, Al.L. Efros, M. Rosen, N.A. Gippius, and S.G. Tikhodeev, Appl. Phys. Lett. 67, 1585 (1995).
[64] P. Lavallard and P.A. Suris, Solid State Commun. 25, 267 (1995).
[65] W.H. Zheng, J.-B. Xia, and K. W. Cheah, J. Phys., Condens. Matter. 9, 5105 (1997).
[66] G. Allan, C. Delerue, and Y.M. Niquet, Phys. Rev. B 63, 205301 (2001).
[67] M. Shim and P. Guyot-Sionnest, Nature 407, 981 (2000).
[68] C. Wang, M. Shim, and P. Guyot-Sionnest, Science 291, 2390 (2001).
[69] G. Allan and C. Delerue, unpublished.

GAIN THEORY AND MODELS IN SILICON NANOSTRUCTURES

Paper dedicated to the memory of Claudio Bertarini

STEFANO OSSICINI[a], C. ARCANGELI[b], O. BISI[a], ELENA DEGOLI[a], MARCELLO LUPPI[b], RITA MAGRI[b], L. DAL NEGRO[c], L. PAVESI[c]
[a] *INFM-S3 and Dipartimento di Scienze e Metodi dell'Ingegneria, Università di Modena e Reggio Emilia;* [b] *INFM-S3 and Dipartimento di Fisica, Università di Modena e Reggio Emilia;* [c] *INFM and Dipartimento di Fisica, Università di Trento*

1. Introduction

The main goal in the information technology is to have the possibility of integrating low-dimensional structures showing appropriate optoelectronic properties with the well established and highly advanced silicon microelectronics present technology [1,2,3,4]. Therefore, after the initial impulse given by the work of Canham on visible luminescence from porous Si [5], nanostructured Si has received extensive attention both from experimental and theoretical point of view during the last ten years (for review see Refs. [6,7,8,9,10,11,12]). This activity is mainly centered on the possibility of getting relevant optoelectronic properties from nanocrystalline Si, which in the bulk crystalline form is an indirect band gap semiconductor, with very inefficient light emission in the infrared. Although some controversial interpretations of the visible light emission from low-dimensional Si structures still exist, it is generally accepted that the quantum confinement, caused by the restricted size, and the surface passivation are essential for this phenomenon [12].

Here we will review our activity in the field of the theoretical determination of the structural, electronic and optical properties of Si nanocrystals (Si-nc). The present work aims at answer a very important question related to the origin of the enhanced photoluminescence in Si-nc embedded in SiO_2. In fact, optical gain has been recently observed in ion implanted Si-nc [13] and in Si-nc formed by plasma enhanced chemical vapor deposition and annealing treatments [14]. We propose, here, an analysis of the experimental findings based on an effective rate equation model for a four level system; moreover looking at our theoretical results for the optical properties of Si-nc we search for structural model that can be linked to the four level scheme. As final outcome, due to the results for the optoelectronic properties of Si-nc in different interface bond configurations, we demonstrate that in order to account for the striking photoluminescence properties of Si-nc it is necessary to take carefully into account not only the role of quantum confinement, but also the role of the interface region surrounding the Si-nc. We'll firstly give a description of the rate equation model we have used and of the obtained results (section 2). After an introduction of the different theoretical methods (section 3) used in our calculations, we'll describe our results on the electronic and optical

L. Pavesi et al. (eds.), Towards the First Silicon Laser, 261–280.

properties of Si-nc in their ground and excited states and we will link these results to the model used for the rate equations. In section 5 we'll be concerned, instead, with the determination of the optimized geometry for Si nanodots and clusters; in particular, we will investigate the role of oxidation processes at the surface of the Si-nc. Section 6 will be devoted to the description, including preliminary results, of our *ab*-initio gain calculation for Si-nc. Finally section 7 will be devoted to the development of a model for the calculation of the optoelectronic properties of Si-nc embedded in a SiO_2 matrix.

2. The four level scheme and the rate equation model

Although a full theoretical calculation of the stimulated emission processes in Si-nc is still lacking due to the formidable difficulties, a rather simplified phenomenological approach is possible. On the basis of the experimental findings we propose a rate equation model to explain the major features of the gain experiments [13,14]: i) the almost complete absence of optical absorption at the peak gain emission for all the amplifying Si-nc samples; ii) a moderate low pumping threshold of about 0.5 kW/cm^2 for population inversion. Items i) and ii) can be explained within an effective four-level model. The corresponding set of rate equations for photons and carriers includes both amplified stimulated emission and Auger recombination. In particular, we assume that levels 2-4 are empty before the excitation occurs. Once the pumping starts, levels begin to be populated and two different Auger non-radiative recombination processes can in principle take place. The first consists in an electron relaxation from energy level 3 to level 2 with the energy given to a second electron which is also present in the same level 3 and which is promoted to higher lying levels in the conduction band, from which a very fast relaxation to level 4 occurs. This Auger process involves two electrons in the same level, and therefore the rate of the process depends quadratically through a coefficient C_{A1} on the level population N_3. Another Auger mechanism involving an electron in the level 3 and a free hole in the valence band edge (level 1) can in principle occur, where the hole is sent deep in the valence band and then very rapidly relaxes again to the band edge. This process has to be proportional, through a coefficient C_{A2}, to the product of the population of the emitting level 3 (N_3) and the hole concentration in the valence band edge (N_h). N_h equals the total concentration of electrons in the various excited levels, that is $N_h = N_2+N_3+N_4$. Within the four level scheme we are proposing, the relaxation times of electrons from levels 4 and level 2 are so fast that N_4 and N_2 are always almost empty. Therefore $N_h \approx N_3$. Hence, the following set of coupled rate equation has to be integrated, where N_i represent the level population densities (i=1,..,4), σ_P is the absorption cross section at the wavelength of the pump, ϕ_P is the time dependent pumping photon flux, Γ_{ij} are the relaxation rates from the i to the j energy levels, τ is the total lifetime of the emitting level N_3, B is the stimulated transition rate which implicitly contains the gain cross section σ, n_{ph} is the emitted photons number, V_a is the optical mode volume, τ_{ph} is the photon lifetime, β is the spontaneous emission factor and τ_R is the radiative lifetime of N_3. C_A is an effective Auger coefficient equal to $2C_{A1} + C_{A2}$, taking into account both of the two particles Auger processes (*e-e* or *e-h*):

$$
\begin{aligned}
\frac{dN_1}{dt} &= -\sigma_P \phi_P(t) N_1 + \Gamma_{21} N_2 \\
\frac{dN_2}{dt} &= \frac{N_3}{\tau} - \Gamma_{21} N_2 + B n_{ph}(N_3 - N_2) + (C_{A1} + C_{A2}) N_3^{\,2} \\
\frac{dN_3}{dt} &= -\frac{N_3}{\tau} - B n_{ph}(N_3 - N_2) + \Gamma_{43} N_4 - C_A N_3^{\,2} \\
\frac{dN_4}{dt} &= C_{A1} N_3^{\,2} + \sigma_P \phi_P(t) N_1 - \Gamma_{43} N_4 \\
\frac{dn_{ph}}{dt} &= V_a B n_{ph}(N_3 - N_2) - \frac{n_{ph}}{\tau_{ph}} + \beta \frac{N_3}{\tau_R}
\end{aligned}
\tag{1}
$$

It is possible to observe optical gain whenever the stimulated emission rate is greater than the Auger recombination rate. From Eq. (1) it is possible to define a SE lifetime as:

$$\tau_{se} = \frac{1}{B n_{ph}} = \frac{4}{3} \pi R_{nc}^{\ 3} \frac{1}{\xi \sigma c n_{ph}} \tag{2}$$

where ξ isthe Si-nc volume fraction and $B = \frac{\sigma c}{V}$, valid under the assumption of monochromatic incident light, is used.

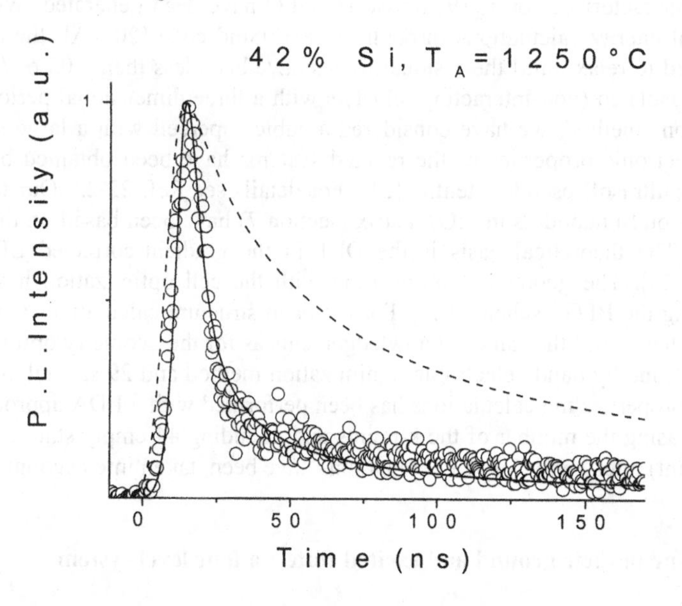

Figure 1. (symbols) RT time resolved decay curve [14]. Detection wavelength was 750 nm. (full line) Emitted photon numbers obtained by solving the four levels model with an Auger lifetime of 20 ns, a gain cross section of 1.2×10^{-17} cm^2, a nc density of 6×10^{18} cm^{-3} and a pumping photon flux of 10^{18} photons/(s cm^2). (dashed line) Emitted photon numbers obtained by solving the four levels model with the same parameters and neglecting stimulated emission. All the curves are normalized to unity.

An equivalent Auger recombination time can be defined as follows:

$$\tau_A = \frac{1}{2C_A N_3} \tag{3}$$

It is clear that to observe gain $1/\tau_{se} \geq 1/\tau_A$ or equivalently, if we define a competition factor $C = \tau_A/\tau_{se}$, $C \geq 1$. This poses a condition on the volume fraction. The proposed rate equation model fits qualitatively the experimental data. Fig. 1 shows an example.

3. Methods of calculations of the structural, electronic and optical properties

For the calculations of the structural and optoelectronic properties of Si-nc both in their *ground* and *excited states* (section 4) we adopt the Hartree-Fock scheme within the Neglect of Differential Overlap (NDO) approximation, using the MNDO-AM1 scheme for geometries and INDO for optical spectra [15]. In the excited state one electron is taken from the highest occupied molecular orbital (HOMO) (hole in the valence) and placed in the lowest unoccupied molecular orbital (LUMO) (electron in the conduction), also here the geometry is optimised and the optical properties are calculated fully taking into account the excitonic coupling [16,17]. For what concerns the calculations on silicon nanoclusters (section 5) the theoretical basis has been the Density Functional Theory in the Local Density Approximation (DFT-LDA) [18] with a norm-conserving pseudopotentials approach to treat the electron-ion interaction. Pseudopotentials in the Kleinman and Bylander factorized form [19] for Si, H and O have been generated. We have performed total energy calculations through the FHI98md code [20]. All the atoms have been allowed to relax, until the residual forces have been less than 0.05 eV/Å. In order to simulate isolated (non-interacting) clusters with a three dimensional periodic boundary conditions method, we have considered a cubic supercell with a large side length. The optoelectronic properties of the relaxed systems have been obtained by a similar codes using ultrasoft pseudopotential [21]. For details see Ref. 22-23. Our total energy calculations on Si nanodots in SiO_2 matrix (section 7) have been based on the CASTEP code [24]. The theoretical basis is the DFT in the gradient-corrected LDA version (GGA) [25,26]. The geometry optimization with the cell optimization has been performed using the BFGS scheme [27]. For the band structure calculations at relaxed geometry, we have used the same acknowledgements as for the geometry optimization run but with a "band by band" electronic minimization method and 29 special k-points. For the optical properties the calculations has been performed with a LDA approach and not GGA, increasing the number of the k-points and including 44 empty states. Only direct (same k-point) inter-band (VB-CB) transitions have been taken into account [22,23]

4. Si-nc in their ground and excited states: a four level system

Here we describe the investigation of the structural and electronic properties of small hydrogenated nanoparticles, including the effect of incorporation of a Si=O silanone unit: the investigation was conducted through Hartree-Fock techniques [16,17].

For Si-H nc we found that the absorption gap, in the stable relaxed *ground state* geometry, shows a strong blue shift with respect to the bulk Si that is due to both confinement and relaxations effects. Fig 2 (a) shows the case of the $Si_{44}H_{42}$ cluster [16].

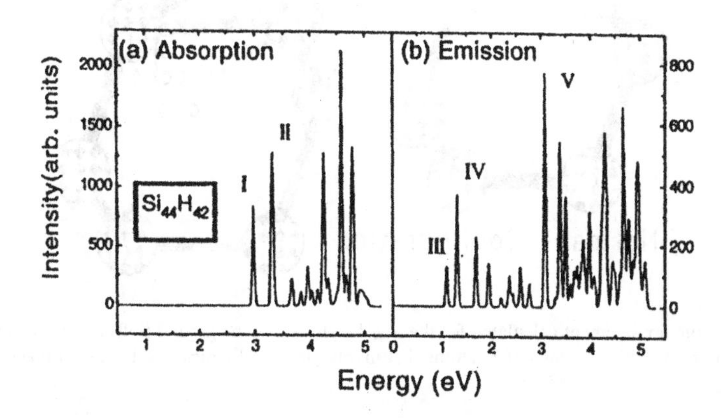

Figure 2. Optical spectrum calculated for the $Si_{44}H_{42}$ cluster in its (a) ground and (b) excited state configuraton, corresponding to absorption and emission spectra respectively.

We then examine the optical properties of the same cluster in their *excited state*. With the geometry optimized for this electronic *excited state*, we again study the optical properties. Fig. 2 (b) shows the result. The difference is striking: the first transition is drastically red shifted. This spectrum is more properly associated with photoluminescence, whereas Fig. 2 (a) should correspond to absorption only. The origin of the red shift is related to the relaxation of the particle to a locally distorted equilibrium configuration.

What we find in fact is that the particle in the excited state undergoes a strong symmetry-lowering spontaneous distortion, extremely localized on two neighboring surface Si atoms, each bonded just to one H atom. In the distortion the Si atoms bend and stretch their bond, keeping the distance to their internal neighbors almost unchanged. One of the H atom moves to bridge the bond, so that a surface configurational defect is created, with no complete rupture of any bonds in the particle (see Fig. 3).

A similar effect is found for the case of the insertion of a oxygen double bonded (DoB) to a Si atom at the surface substituting two H atoms. Incorporation of such a Si=O unit perturbs both absorption and emission properties. The absorption is red shifted with respect to the Si-H corresponding case and the emission is another time strongly lowered with respect to the absorption (see Fig. 4). These results are due to the fact that the presence of the Si=O bond promotes in the ground state a slight distortion of the Si-Si backbonds, distortion that in the excited states is enhanced, together with the shortening of the Si=O bond length [17].

Similar results have been obtained by Filonov and coworkers [28] also for pyramidal Si nanocluster as witnessed by Fig. 5, where the optimized cluster structures for both ground and excited state configurations for a $Si_{30}OH_{38}$ cluster are shown.

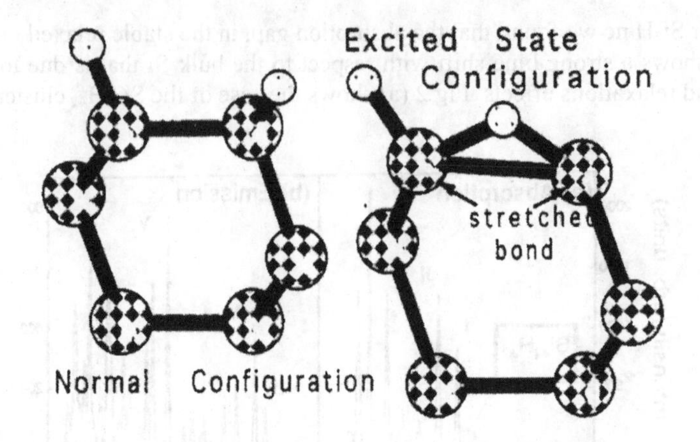

Figure 3. Atomic arrangement calculated for the $Si_{44}H_{42}$ cluster, in the ground (left) and in the excited (right) state. Shown are the six fold rings of Si atoms that involve the locally distorted bond and reach the center of the cluster.

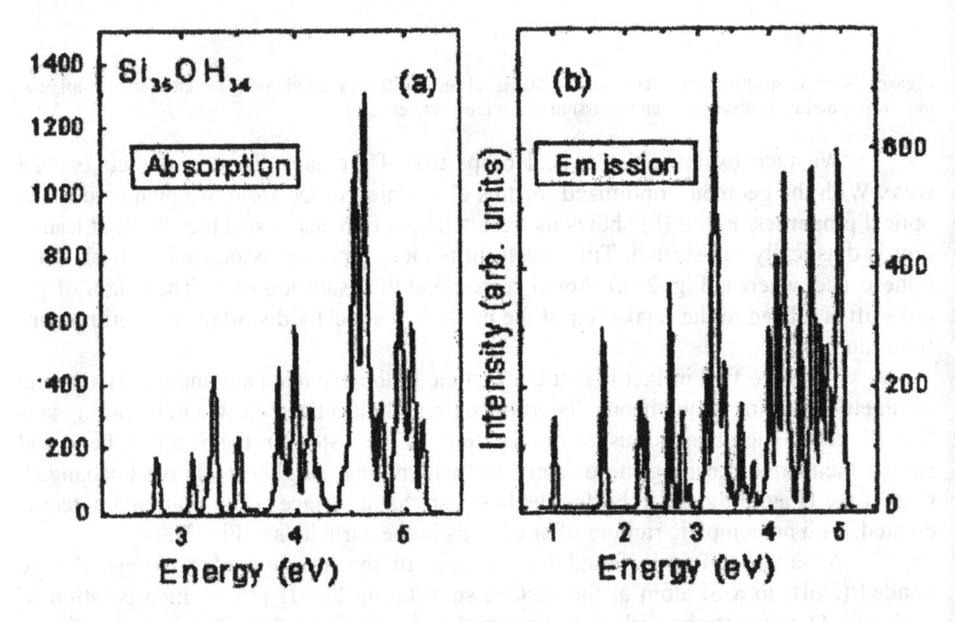

Figure 4. Optical spectra for the $Si_{35}OH_{34}$ particles obtained by the substitution of a SiH_2 unit with a Si=O double bond. (a) absorption spectrum, (b) emission spectrum, both calculated in their relaxed geometry.

Also in this case the calculation scheme used was a Hartee-Fock one. Once more we see that the Si atoms that underwent the maximum shift from their ground state positions are the Si atoms in the vicinity of the double bonded O atom.

It is important to underline that the difference between the absorption and emission spectra are , in all cases, due to the very different spatial localization of the

HOMO and LUMO states. As showed in Fig. 6 for the case of the $Si_{44}H_{42}$ cluster, on going from absorption to emission, the HOMO state, previously localized on the internal shell of Si atoms, is now strongly localized on the external Si atoms involved in the distortion. The same happens for the LUMO+1 state, this state is also partially involved in the lowest transition [16]. The structural distortion originates strongly localized states

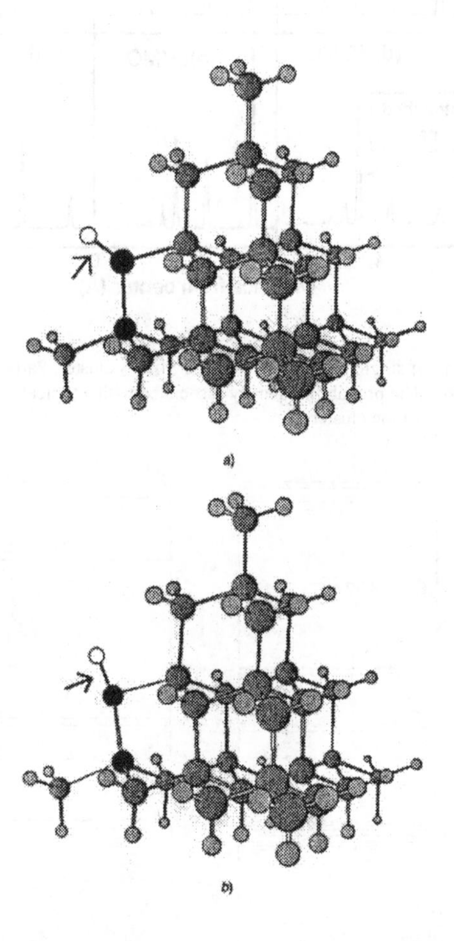

Figure 5. $Si_{30}OH_{38}$ cluster structure for the ground (a) and excited (b) states. The Si atoms underwent the maximum shift from their equilibrium ground state positions are marked in black. Arrows indicate the Si=O bonds.

The picture that emerges is that of a four level scheme (previously proposed in section 2), where level 1 and 4 are the HOMO amd LUMO levels of the ground state and the levels 2 and 3 are , instead the HOMO and LUMO levels of the excited states.

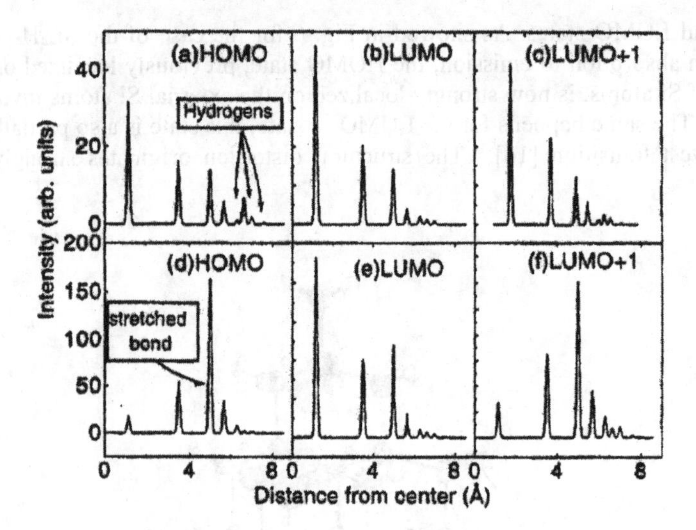

Figure 6. Spatial localization of single particle states in the $Si_{44}H_{42}$ cluster. Parts (a) to (c) refer to the absorption and (d) to (f) to emission. The probability density associated with a typical atom in a given shell is plotted vs. its distance from the center of the cluster.

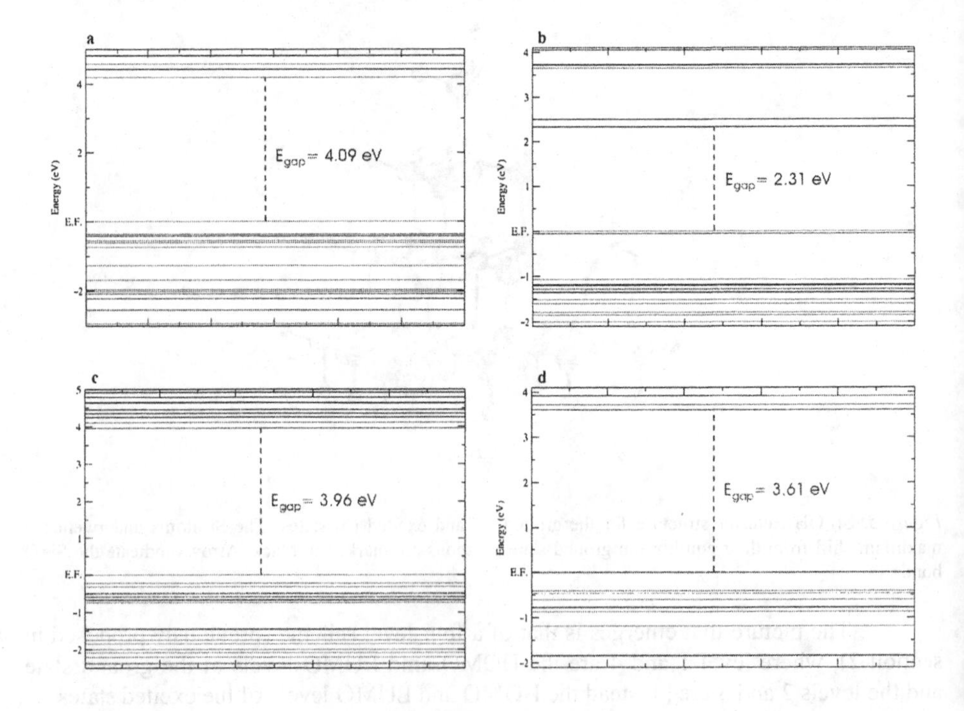

Figure 7. Energy levels (k-point Γ) for all the considered clusters at relaxed geometry: a) $Si_{14}H_{20}$, b) $Si_{14}H_{16}O_2$, c) $Si_{14}H_{20}O_2$, d) $Si_{12}H_{16}O_2$.

5. Electronic and optical properties of Si nanoclusters: role of surface termination and of multiple oxidation

To better understand the role of surface termination and of multiple oxidation on the structural and optoelectronic properties of small silicon crystalline clusters, we have looked at the differences between similar systems with different types of Si-H and Si-O bonds [22,23]. We have built up a small cluster of 14 silicon atoms (3.7 x 5.7 x 6.7 Å) with a crystalline Td-interstitial structure and we have recovered it with hydrogen atoms in order to passivate all the dangling bonds ($Si_{14}H_{20}$ cluster). Then we have substituted some Si-H or Si-Si bonds with Si-O bonds. We have mainly focused our attention on three Si-O bond configurations: in $Si_{14}H_{16}O_2$ cluster we have obtained two Si=O double bonds (DB), deleting four Si-H bonds on two Si atoms of the structure; in $Si_{14}H_{20}O_2$ we have formed two Si-O-Si bridges breaking two Si-Si bonds in the crystalline skeleton; and in $Si_{12}H_{16}O_2$ we obtained two Si-O-Si bridges substituting two Si atoms of the silicon structure with O. All the clusters structures have been relaxed through geometry optimization runs.

As expected, the H passivation has left the crystalline Si skeleton unchanged. In the case of Si=O DoBs ($Si_{14}H_{16}O_2$), the changes are limited practically only to the Si-Si-Si bond angle under the Si=O bond (that is aligned along the (001) direction), and to the correspondent Si-Si bond lengths that are only a little strained. Concerning the $Si_{14}H_{20}O_2$ and $Si_{12}H_{16}O_2$ clusters there is an appreciable modification of the Si structure near the Si-O-Si bridges, with huge angles variations, but still with a small effect on the bond length.

In Fig. 7 the energy levels at k-point Γ for all the clusters are plotted. The comparison between the hydrogenated case and the Si=O DoB shows a huge reduction of the energy gap (E_g) which goes respectively from 4.09 eV to 2.31 eV.

This reduction is due to the birth of new interface-related states which strongly modify the band edges. Looking at the energy levels for the two Si-O-Si bridges cases, it's evident how the Si-O interface affects less strongly their behaviour. For the $Si_{14}H_{20}O_2$ we register a very small E_g reduction, only 0.13 eV, while for $Si_{12}H_{16}O_2$ the variation is more evident: E_g=3.61 eV, but less than for the Si=O DoB. It is important to remember that all these values are affected by the underestimation due to the LDA approach (the scissor operator value for these confined systems can be evaluated around 0.8-1 eV). The analysis of the distribution of HOMO (highest occupied Kohn-Sham orbital) and LUMO (lowest unoccupied Kohn-Sham orbital) helps to clarify the situation. In Fig. 8 are reported the isosurfaces at fixed value for the HOMO and LUMO of all the clusters. Only in the Si=O DoB case both HOMO and LUMO are localized around the oxygen of the Si-O bonds, while for the two cases with Si-O-Si bridges only part of the LUMO is around it; the HOMO actually maintains mainly the character of the hydrogenated structure, located along peculiar Si-Si bonds of the silicon skeleton.

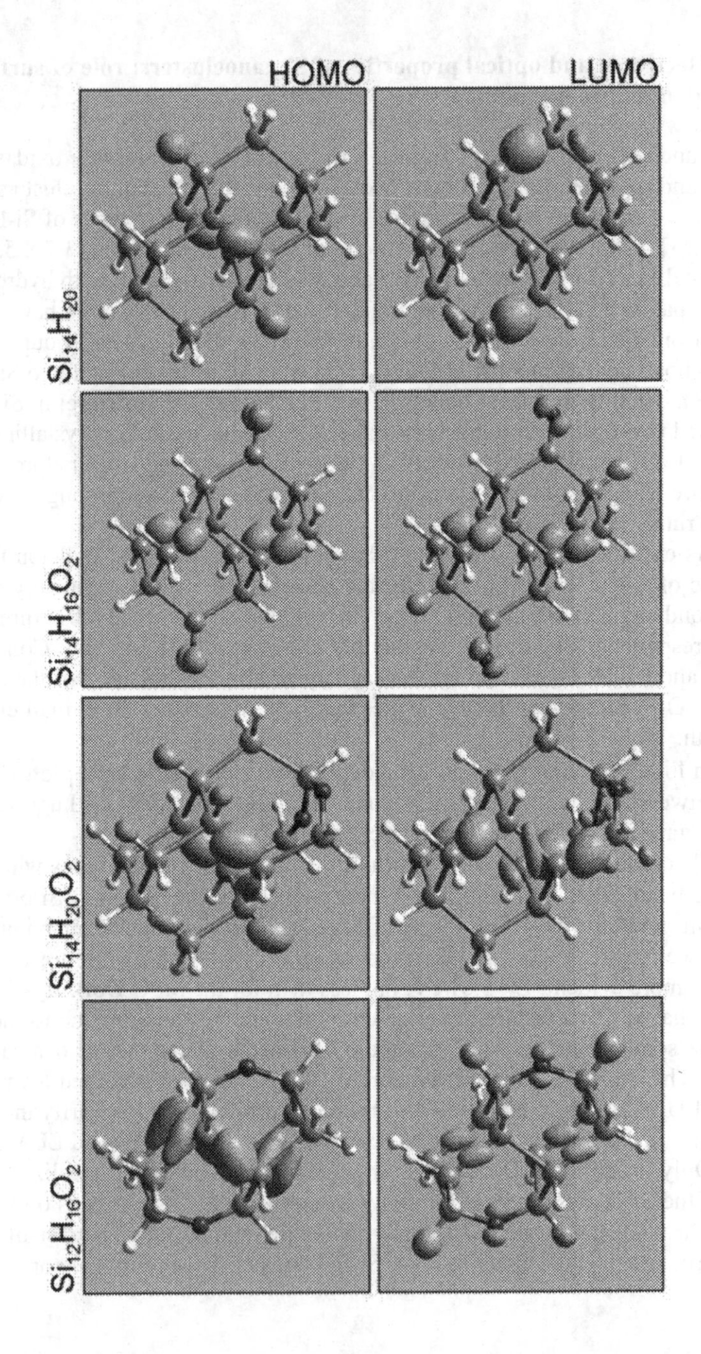

Figure 8. Isosurfaces of the square modulus of the highest occupied (HOMO) and lowest unoccupied (LUMO) Kohn-Sham orbitals, for all the considered clusters at optimised geometry. The isosurfaces are plotted at 50% of their maximum amplitude. In the stick and ball representation of the final geometry, white spheres stand for H, light grey for Si and dark grey for O atoms.

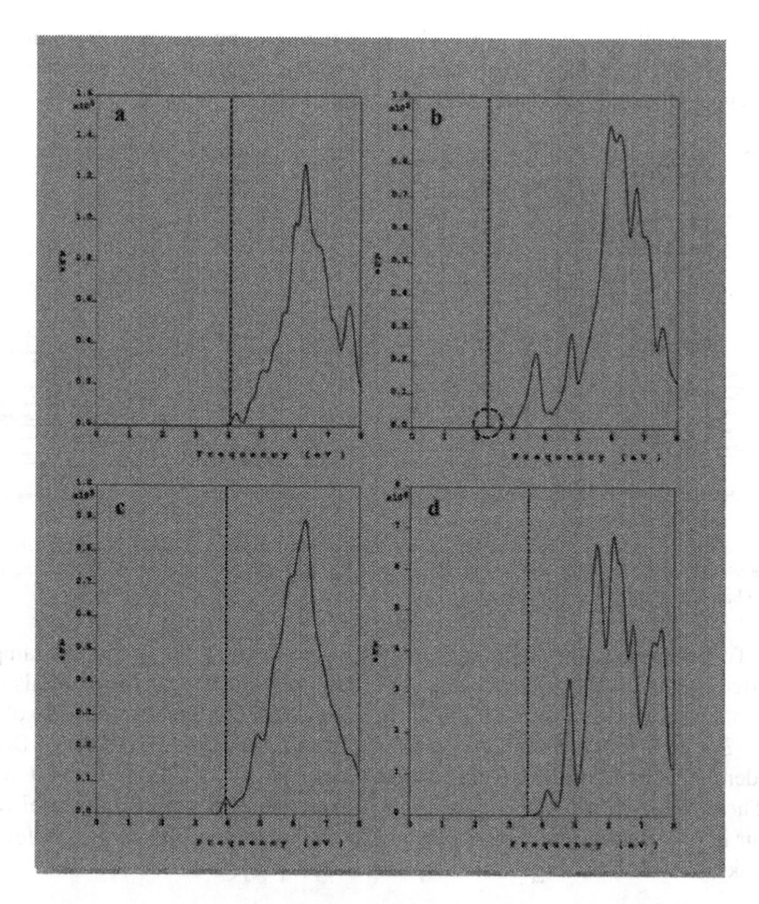

Figure 9. Absorption spectra (direct inter-band transitions) for all the considered clusters (relaxed geometry): a) $Si_{14}H_{20}$, b) $Si_{14}H_{16}O_2$, C) $Si_{14}H_{20}O_2$, d) $Si_{12}H_{16}O_2$. The vertical lines indicate the calculated E_g.

This suggest that, in presence of Si=O DoB, the band edges are dominated by the surface states related to Si-O bonds and so the first transitions in an absorption process involves only the Si-O interface states; while for the Si-O-Si bridges the occupied states are still mainly related to the internal states along Si-Si bond directions, so that the character of the first optical transitions is still influenced by the crystalline silicon structure.

The optical properties reflect what we have observed for the electronic ones. In Fig. 9 we have reported absorption spectra for all the clusters. New features at the gap onsets characterize the absorption line-shapes. In the Si=O DoB case the first transitions have a very low probability: only a very tiny peak (highlighted by circle in Fig. 9b)) is present at energies around E_g. Also the $Si_{12}H_{16}O_2$ presents a similar behaviour. These results go in the same direction of what reported by Vasiliev *et al.* [29] concerning the prohibition of the direct dipole transitions between HOMO and LUMO. At the same time the red shift in the absorption onset of these last two cases (respect to hydrogenated one) compares well with the red shift observed in photoluminescence from high porosity porous Si samples after O exposition [30] and the experimental results for the band gap behaviour of heavily oxidized Si nanoparticles [31].

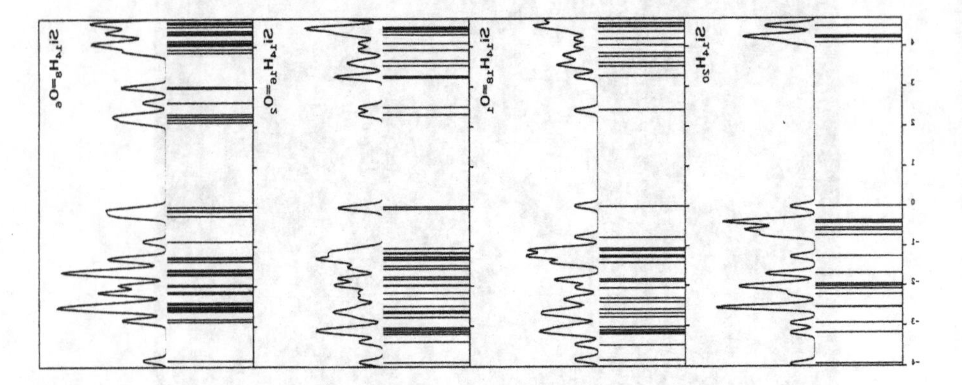

Figure 10. Calculated energy levels at the Γ point and total density of states (T-DOS) for the $Si_{14}H_n{=}O_m$ clusters. In the y-axis are the energies calculated in eV. The alignment has been done simply equating the top of the valence bands in the different calculations.

To better simulate the progressive exposure to air of the porous Si samples we have varied systematically the number of DoBs at the clusters surface and also the dimension of the cluster itself ($Si_{10}H_n{=}O_m$, $Si_{14}H_n{=}O_m$). A non linear reduction of the energy gap (E_g) with the DoBs number is shown, for all the cluster sizes. Each DoBs acts independently from the others causing the formation of localized states at the band edges. These findings provide a consistent interpretation of the experimental red shift behaviour in oxidized porous Si samples [30] and of recent results on the photoluminescence peak bandwidth of single silicon quantum dots [32].

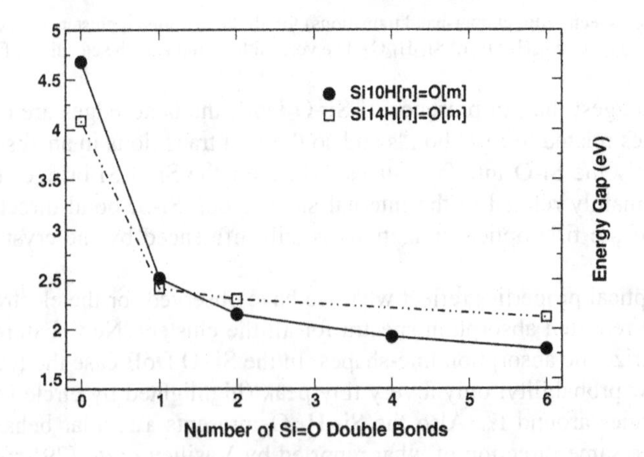

Figure 11. Electronic energy gap as a function of the number of Si=O double bonds at the cluster surface. Circle: Si10H*n*=O*m*, square: Si14H*n*=O*m*. The zero values correspond to the fully-H case for each cluser types.

The energy levels and total density of states (T-DOS) for the $Si_{14}H_n$=O_m clusters are reported in Fig. 10. The results confirm that in presence of DoBs the HOMO-LUMO transitions are allowed, even if not always very probable. The insertion of a single DoB at the surface produces a strong gap red shift which diminishes by increasing the NDs size (Si atoms number). New localized states related to the DoB appear within the gap at both bands. Their relative positions seem not to be strongly affected by size differences.
The gap values are in fact held within 0.2 eV. Increasing the oxygen coverage level, i.e. adding new DoBs, the gap tends to decrease. The reduction however is not linear with the number of DoBs.

For all the cluster sizes, in fact, the strongest red shift is achieved with the first DoB; a second DoB produces a further reduction but weaker than the first and the more we add DoBs the more their contributions is smaller. Thus a sort of saturation limit seems to be reached. The plot of the electronic energy gap vs. the DoBs number (Fig. 11) shows clearly this behavior. A bigger number of DoBs at the surface can be seen as a longer exposure to air, i.e. oxidation, of the porous Si samples. Assuming that the excited state doesn't change radically the ground state band edges situation, our results reproduce perfectly the observed porous Si photoluminescence red shift behaviour for intentionally oxidized samples [30].

6. "Ab-initio" calculation of gain

For the wave intensity inside a medium we have $I(z) = I_0 e^{\gamma(\omega) z}$. Therefore $\gamma(\omega)<0$ means absorption, $\gamma(\omega)>0$ gain. $\gamma(\omega)$ is directly proportional to the imaginary part of the dielectric function through $\gamma(\omega) = -[\omega/n(\omega)c]\epsilon_2(\omega)$, where n is the refraction index, thus it is possible to calculate the coefficient $\gamma(\omega)$ through $\epsilon_2(\omega)$. In the single particle approach and in the dipole approximation one has:

$$\varepsilon_2(\omega) = \frac{8e^2\pi}{m^2 V \omega^2} \sum_{m,n} \sum_{k} \left| \vec{P}_{m,n}(k) \right|^2 \delta(\varepsilon_{m,n}(k) - \hbar\omega) [f_m - f_n]$$

Where f_m and f_n are the quasi-Fermi level occupation probabilities. In the ground state is $f_m = 1$ and $f_n = 0$, thus we have always $\gamma(\omega)<0$ and only absorption. In order to have gain some higher lying level must have an higher f_n than some lower lying level (population inversion). Thus in order to calculate gain we must have a model for population inversion and we must calculate $\epsilon_2(\omega)$.

This has been done in the case of GaAs bulk. We started from our calculated eingenvalues/eigenstates through pseudopotential method. The positions of the quasi Fermi levels are determined from the calculated density of states

$$N = \int \rho(E) f_n(E, F_n)\, dE$$

Where N is the number of electron/holes created in the systems.

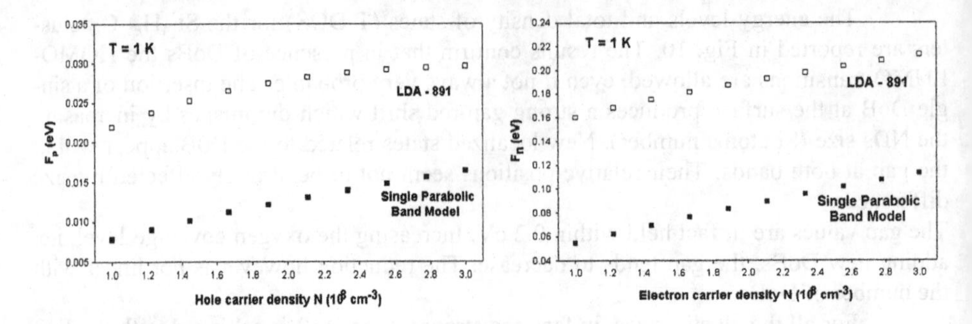

Figure 12. Quasi-Fermi level energy vs electron and hole concentrations in GaAs bulk.

Using the calculated quasi-Fermi levels one fill the states above the Fermi level and deplete a corresponding number of states below the Fermi level. Figure 12 shows our calculated quasi-Fermi population for electron and holes as function of the carrier density. The result for a LDA calculation using 891 k-points is compared with that of a single parabolic model. Choosing a specific electron and hole populations we have calculated the new $\epsilon_2(\omega)$ for this population inversion system. As the following figure shows, this enables to calculate gain (negative absorption), near the gap value.

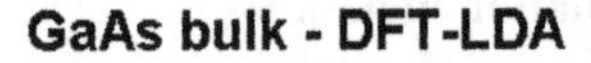

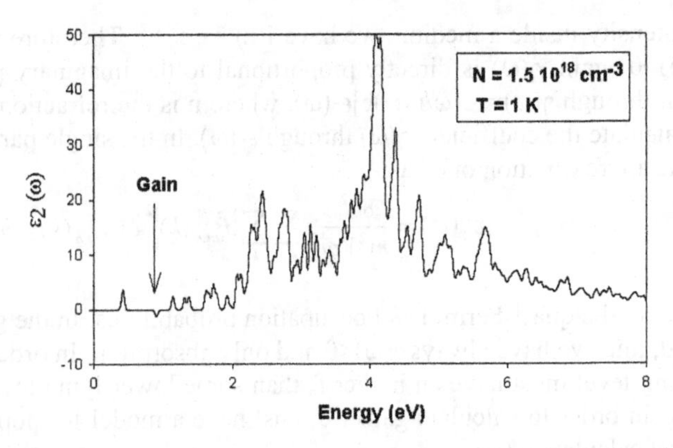

Figure 13. The calculated imaginary part of the dielectric function taking into account the population inversion.

Ab-initio gain calculations have been, also, performed on the hydrogenated silicon cluster ($Si_{14}H_{20}$) described in section 5. The geometry of the cluster has been relaxed through a total energy calculation based on DFT-LDA using a pseudopotential approach. In detail we have worked with norm-conserving semi-local non-relativistic and non-spin polarised Bachelet-Hamann-Schülter (BHS) pseudopotentials with an energy cutoff of 16 Ha.

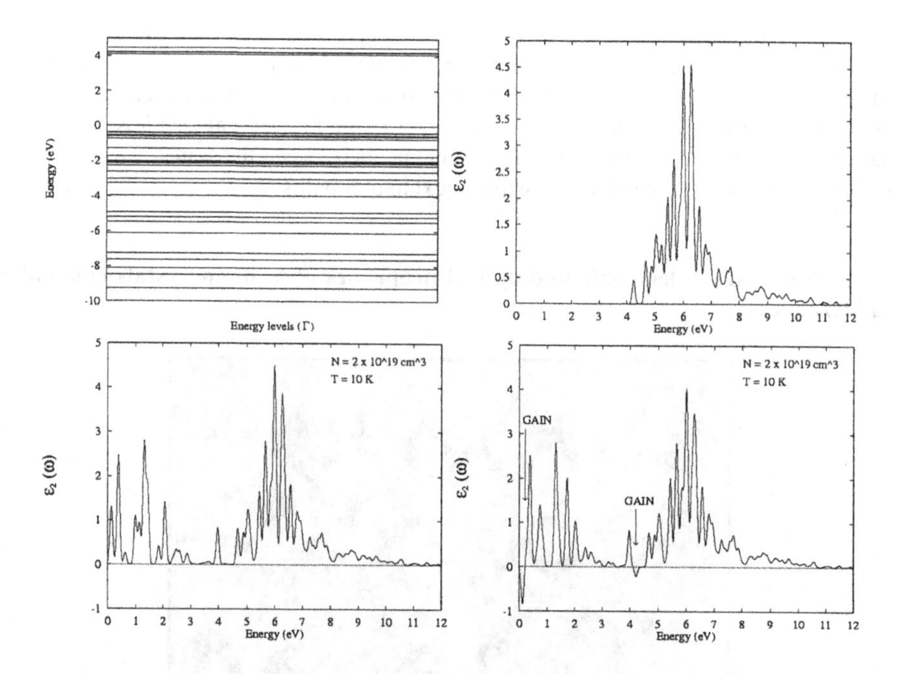

Figure 14. Top left) Calculated enegy levels at the Γ point in the brillouin zone for the $Si_{14}H_{20}$ cluster. The energies are refered to the higher valence level. Top right) Imaginary part of the dielectric function for the $Si_{14}H_{20}$ cluster. Bottom left) Calculated absorption spectra for the $Si_{14}H_{20}$ cluster after having depleted the highest valence level and filled the lowest conduction level with a $2x10^{19}$ cm^{-3} electron-hole pair density. Bottom right) Calculated absorption spectra for the $Si_{14}H_{20}$ cluster after having depleted the highest valence level and filled the second conduction level with a $2x10^{19}$ cm^{-3} electron-hole pair density.

The electronic and optical properties have been computed: figure 14 (Top left) shows the energy levels at the Γ point, actually, in the case of clusters, where the number of atoms is really small, we haven't a real band structure but simply discrete energy levels. The computed energy gap is of 4.1 eV. Starting from the electronic results and through the theory of inter-band transitions we have computed the imaginary part of the dielectric function ϵ_2 (see Fig. 14 Top right) in the optical limit. The gaussian brodening of 0.05 used to produce this figure hide the real onset of the function that correspond to a transition energy of 4.22 eV, bigger that the gap value. This is due to the fact that the transition probability between the highest valence level and the lowest conduction level is null: this transition is forbidden. To perform gain calculations we have considered two different models of population inversion: in the first case (see Fig. 14 Bottom left) we have depleted the highest valence level and filled the lowest conduction level with a $2x10^{19}$ cm^{-3} electron-hole pair density at a temperature of 10 K, while in the second one (see Fig. 14 Bottom right) we have depleted the highest valence level and filled the second conduction level in the same conditions of carrier density and temperature.

As can be clearly seen there are in both cases a lot of new peaks in the low energy range with respect to the ground state ε_2; these are due to the new inter- and intra-band transitions induced by the considered population inversion. In the first case no gain is present; actually, as previously said, the transition between the involved states is forbidden. In the second case, instead, we can observe two negative absorption peaks; the gain observed at 0.12 eV is due to the intra-band transition from the second to the first conduction levels, while the 4.22 eV negative peak is a consequence of the inter-band transition from the second conduction level to the highest valence level. This calculation, simply based on the optical matrix elements between the states involved in the transition, can give an idea of the possible deexcitation paths.

7. Structural, electronic and optical properties of Si nanocrystals embedded in a SiO_2 matrix

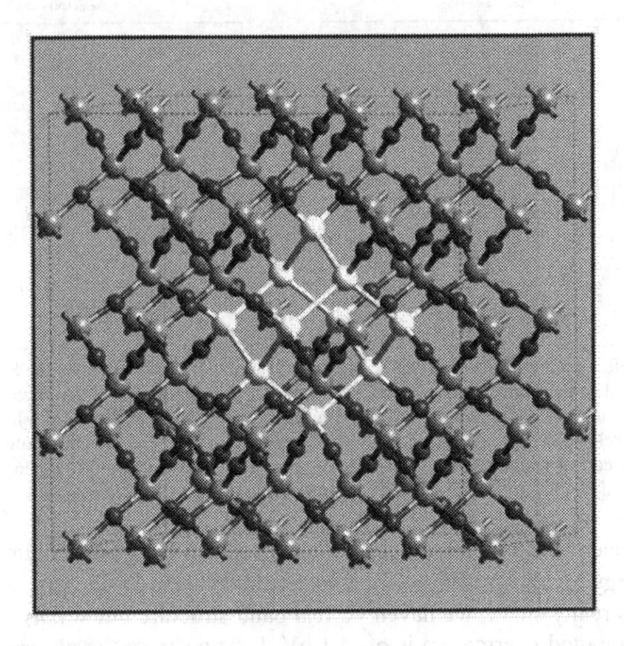

Figure 15. Stick and ball pictures of the final optimized structure of Si_{10} in SiO_2. The dark gray spheres represent O atoms, light gray Si and white the Si atoms of the nanocrystal.

In this section our goal has been to build up a simple model to study, for the first time, the properties of Si nanocrystals embedded in SiO_2 matrix from a theoretical point of view [22,23]. We wanted at same time two fundamental qualities: a silicon skeleton with a crystalline behavior for simulating the Si-nc and the simplest Si-SiO_2 interface, with the minimum number of dangling bonds or defects.

For these reasons we have started with a cubic cell (l=14.32 Å) of SiO_2 betacristobalite (BC) which is well known to have one of the simplest Si/SiO_2 interface because of its diamond-like structure [33]. We get the cell repeating two times along each cartesian axe the unit cell of SiO_2 BC. Then we obtained a small dot simply deleting some oxygen atoms of the SiO_2 matrix and linking together the silicon atoms left with dangling bonds, as shown in Fig. 15. In this way we have built an initial supercell of 64 Si and 116 O atoms with 10 Si bonded together to form a small crystalline skeleton (Td interstitial symmetry) with a very highly strained bond length with respect to the bulk case: 3.1 Å. On this system we have performed geometry optimization runs leaving free to relax all the atoms and all the cell parameters.

Fig. 15 shows the final relaxed supercell structure after the geometry optimization for the Si-nc (Si_{10}) in SiO_2. Looking at the dot behavior we find that the skeleton is still crystalline-like (diamond) with a Si-Si bond length of 2.67 Å, that means a strain of 14% respect to the bulk case. This rearrangement causes a complex deformation of the SiO_2 matrix around the dot both in bond lengths and angles.

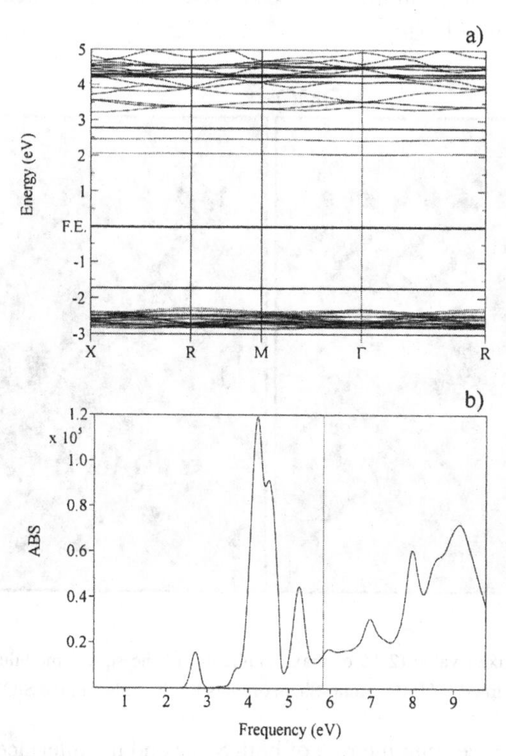

Figure 16. a) Band structure along high symmetry points of the BZ for the Si NC in SiO_2 at relaxed geometry. b) Absorption spectrum of the Si NC in SiO_2. The vertical line indicates the calculated E_g for SiO_2 BC (bulk).

It seems as if the lack of oxygen atoms and the compression of the dot volume determine an empty region that the SiO_2 structure tends to fill up stressing bonds and angles in various ways. Nevertheless the deformation doesn't affect the entire SiO_2 matrix. It is actually possible to find still a good BC crystalline structure, in terms of angles and bond-length, at a distance from the dot's atoms of 0.8-0.9 nm. This means that the dot is surrounded by a cap-shell of stressed SiO_2 BC with a thickness of about 1 nm which progressively goes towards a pure crystalline BC. This view is also supported by the analysis of the electronic properties as described below.

Fig. 16a shows the results for the electronic properties of the relaxed structure. The calculated E_g is 2.07 eV, that must be compared with the value of 5.84 eV for the E_g

of BC SiO_2 (bulk) (calculated with the same technique). The strong reduction is originated by the presence, at the valence and conduction band edges, of confined, flat, states completely related to the Si-nc, whereas deep inside the valence and conduction bands the more k-dispersed states related to the SiO_2 matrix are still present.

In Fig. 17 the HOMO and LUMO isosurfaces at fixed value are reported; we clearly see that the distribution is totally confined in the Si-nc region with some weight on the interface O atoms. These dot-related states originate strong absorption features in the optical region as witnessed from Fig. 16b. These features are entirely new and can be at the origin of the photoluminescence observed in the red optical region for Si-nc immersed in a SiO_2 matrix [34].

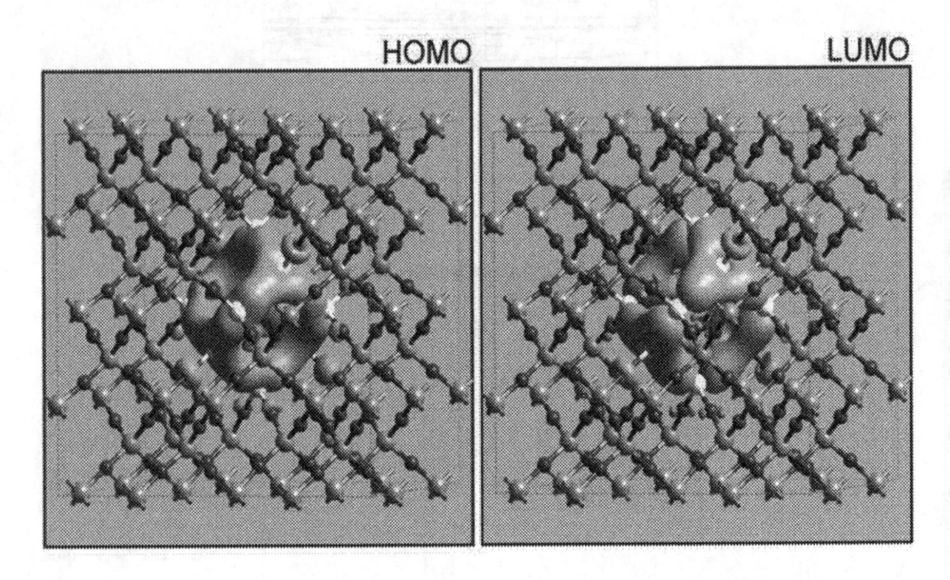

Figure 17. Isosurfaces at fixed value (25% of max. amplitude) of the square modulus of highest occupied (HOMO) and lowest unoccupied (LUMO) Kohn-Sham orbitals for the Si NC in the SiO2 matrix.

Moreover our result concerning the role of both Si-nc and the interface Si-O region with respect to the absorption process is in close agreement with the X-ray absorption fine structure measurements [35] that indicate the presence of an intermediate region between Si-nc and the SiO_2 matrix that participates in the light emission process [36].

8. Summary

The present work aims at answer a very important question related to the origin of the striking optical properties of Si nanoclusters. We have demonstrated, by first principle calculations that the structural, electronic and optical properties of Si-nc strongly depend not only on the dimensions of the nanocrystals, i.e. on the quantum confinement, but also on the different passivation regimes, i.e. on the interface region. Starting from

H-covered Si-nc we have shown that single bonded O atoms originate small variations in the electronic properties and huge changes in the structural properties. On the contrary double bonded O atoms make a small contribution to geometry variations, but a strong reduction in the energy gap. Moreover a non-linear reduction of the energy gap with the double bonds number has been found. In the case of Si-nc embedded in SiO_2 matrix, our results show that the SiO_2 cage is only slightly deformed by the presence of the Si-nc, that new electronic states are originated within the SiO_2 band gap and that both the Si-nc region and the interface region play a role in the optical properties. These results help to clarify the experimental outcomes regarding the electronic and optical properties of heavily oxidized Si-nc and the structural, electronic and optical properties of Si-nc dispersed in SiO_2. Hartree-Fock calculations for the ground and excited states of the Si-nc have been interpreted within a four level model system, which has also been used to construct a rate equation system that fits qualitatively the experimental data on gain in Si-nc embedded in a SiO_2 matrix. Finally preliminary results of ab-initio calculation of gain in Si nanostructures have been presented.

This paper is dedicated to the memory of Claudio Bertarini. Claudio, former member of our research group, perished September 2001 in a car crash. We all remember his active presence, his scientific and computational skill, his unforgettable silent smile. Financial support by INFM-PRA Ramses and CNR PF MADESSII is acknowledged.

References

1. Sze, S. M. (1998) *Modern Semiconductor Device Physics*, John Wiley & Sons.
2. Davies, J. H. (1998) *The Physics of Low-Dimensional Semiconductors: An Introduction* , Cambridge University Press, Cambridge.
3. Borisenko, V. E. Filonov, A. B. Gaponenko, S. V. Gurin, V. S. (2001) *Physics, Chemistry, and Application of Nanostructures*, World Scientific, Singapore.
4. Tsu, R. (2000) Appl. Phys. A 71, 391.
5. Canham, L. T. (1990) Silicon quantum wire array fabrication by electrochemical and chemical dissolution of wafers, *Appl. Phys. Lett.* **57**, 1046.
6. Bensahel, D. Canham, L. T. Ossicini, S. (1993) *Optical Properties of Low Dimensional Silicon Structures*, Kluwer, Dordrecht, Amsterdam.
7. Hamilton, B. (1995) Porous silicon, *Semicond. Sci. Technol.* **10**, 1187.
8. Kanemitsu, Y. (1995) Light emission from porous silicon and related materials, *Phys. Rep.* **263**, 1.
9. John, G. S. and Singh, V. A. (1995) Porous silicon: theoretical studies, *Phys. Rep.* **263,** 93.
10. Canham, L. (1997) *Properties of Porous Silicon*, INSPEC, The Institution of Electrical Engineers, London.
11. Cullis, A.G. Canham, L. T. and Calcott, P.D. J. (1997) The structural and luminescence properties of porous silicon, *J. Appl. Phys.* **82,** 909.
12. Bisi, O. Ossicini, S. Pavesi, L. (2000) Porous silicon: a quantum sponge structure for silicon based optoelectronic, *Surf. Sci. Reports* **38**, 5.
13. Pavesi, L. Dal Negro, L. Mazzoleni, C. Franzò G. and Priolo, F. (2000) Optical gain in silicon nanocrystals, *Nature* **408,** 440.
14. Dal Negro, L. Cazzanelli M. Daldosso, N. Gaburro, Z. Pavesi, L. Priolo, F. Pacifici, D. Franzò, G and Iacona, F. (2002) in press Stimulated emission in plasma enhanced chemical vapour deposited silicon nanocrystals, *Physica E.*
15. Dewar, M. J. S. *et al.* (1985) , *J. Am. Chem. Soc.* **107**, 3902.
16. Baierle, R. J. Caldas, M. J. Molinari, E. and Ossicini, S (1997) Optical emission from small Si particles, *Solid State Comm.* **102**, 545.
17. Caldas, M. J. (2000) Si nanoparticles as a model for porous Si, *phys. stat. sol. (b)*, **217**, 641.

18. Dreizler, R.M. and Gross, E.K.U. (1990) *Density Functional Theory. An Approach to the Quantum Many-Body Problem,* Springer-Verlag, Berlin.
19. Kleinman, L. and Bylander, D.M. (1982) Efficacious Form for Model Pseudopotentials, *Phys. Rev. Lett.* **48**, 1425.
20. Bockstedte, M. Kley, A. Neugebauer, J. and Scheffler, M. (1997) Comput. Phys. Commun. **107**, 187.
21. Vanderbilt, D. (1990) Soft self-consistent pseudopotentials in a generalized eigenvalue formalism, *Phys. Rev. B* **41**. 7892.
22. Luppi, M. and Ossicini, S. (2002) in press Oxygen role on the structural and optoelectronic properties of silicon nanodots, *phys. stat. sol. (a).*
23. Luppi, M. and Ossicini, S. (2002) in press Oxygen role on the optoelectronic properties of silicon nanodots, *Material Science and Engineering B.*
24. Milman, V. Winkler, B. White, J. A. Pickard, C. J. Payne, M. C. Akhmatskaya, E. V. Nobes, R. H. (2000) Int. J. Quant. Chem. **77**, 895.
25. Perdew, J. P. (1986) Density-functional approximation for the correlation energy of the inhomogeneous electron gas, *Phys. Rev* ***B*** **33**, 8822.
26. Perdew, J. P. Burke, K. and Ernzerhof, M. (1996) Generalized Gradient Approximation Made Simple, *Phys. Rev. Lett.* **77** 3865; (1997) **78**, 1396.
27. Fisher T.H., Almlöf, J. (1992) J. Phys. Chem. **96**, 9768.
28. Filonov, A. B., Ossicini, S., Bassani, F. Arnaud D'Avitaya, F. (2002) Effect of oxygen on the optical properties of small silicon pyramidal clusters, *Phys. Rev* ***B*** **65**, 195317.
29. Vasiliev, I. Chelikowsky, J. R. and Martin, R. M. (2002) Surface oxidation effects on the optical properties of silicon nanocrystals, *Phys. Rev. B* **65**, 121302(R).
30. Wolkin, M. V. Jorne, J. Fauchet, P. M. Allan, G. and Delerue, C. (1999) Electronic States and Luminescence in Porous Silicon Quantum Dots: The Role of Oxygen, *Phys. Rev. Lett.* **82**, 197.
31. van Buuren, T. Dinh, L. N. Chase, L. L. Sekhaus, W. J. Terminello, L. J. (1998) Changes in the Electronic Properties of Si Nanocrystals as a Function of Particle Size, *Phys. Rev. Lett.* **80**, 3803.
32. Valenta, J. Juhasz, R and Linnros, J. (2002) Photoluminescence spectroscopy of single quantum dots, *Appl. Phys. Lett.* **80**, 1070.
33. Kageshima, H. Shiraishi, K. (1996) Double bond formation mechanism for passivating Si/SiO_2 interface states, Scheffler, M. Zimmermann R., *Proc. 23rd Int. Conf. Phys. Semicon.*, World Scientific, Singapore, p. 903.
34. Prakash, G. V. Daldosso, N. Degoli, E. Iacona, F. Cazanelli, M. Gaburro, Z. Puker, G. Dalba, G. Rocca, F. Ceretta Moreira, E. Franzò, G. Pacifici, D. Priolo, F. Arcangeli, C. Filonov, A. B. Ossicini, S. and Pavesi, L., (2001) Structural and Optical Properties of Silicon Nanocrystals Grown by Plasma-Enhanced Chemical Vapor Deposition, *J. Nanosci. Nanotech.* **1**, 159.
35. Daldosso, N. Dalba, G. Grisenti, R. Dal Negro, L. Pavesi, L. Rocca, F. Priolo, F. Franzò, G. Pacifici, D. and Iacona, F. (2002) in press X-ray absorption study of light emitting silicon nanocrystals, *Physica E.*
36. Daldosso, N., Luppi, M. Ossicini, S. Degoli, E., Magri, R. Dalba, G. Fornasini, P. Grisenti, R. Rocca, F. Pavesi, L. Priolo, F. and Iacona, F (2002) submitted Role of the interface region on the optoelectronic properties of silicon nanocrystals embedded in SiO_2, *Phys. Rev. B*

Si-Ge QUANTUM DOT LASER: WHAT CAN WE LEARN FROM III-V EXPERIENCE?

N.N. LEDENTSOV
Abraham Ioffe Institute of Physics and Technology of the RAS
Polytechnicheskaya 26, 194021, St.Petersburg, Russia, and
Institut für Festkörperphysik, Technische Universität Berlin, Hardenbergstr. 36, D-10623 Berlin, Germany

1. Introduction

Si-Ge system offers a significant extension to traditional Si-based microelectronics. However, applications of the system would be further greatly expanded if it can be used for high-speed optical transmitters and interconnects. A straightforward idea to achieve this goal is to use the device designs, which are already successfully applied for direct-gap semiconductor materials, particularly, to III-V materials. These are diode lasers based on double heterostructures [1, 2] and on heterostructures with reduced dimensionality [3], so called, quantum wells (QWs), quantum wires (QWWs) and quantum dots (QDs). An idea to achieve lasing in indirect gap materials by using double hetrostructure concept was first mentioned by Kroemer in 1963 [2]. In his paper H. Kroemer proposed to use the double heterostructures (DHS) for carrier confinement in the active region of the diode laser and wrote that "laser action should be obtainable in many of the indirect gap semiconductors and improved in the direct gap ones, if it is possible to supply them with a pair of heterojunction injectors". Attempts to achieve lasing SiGe-Si DHSs and QWs did not result is significant success, however, as also in the case of other types of indirect-gap materials, for example, AlGaAs DHSs with high Al content (x>0.5), or in type-II GaAs-AlAs quantum QWs. A different approach to achieve lasing in semiconductors was first proposed by Basov, Vul and Popov in 1959 [4], who considered unipolar carrier injection. Population inversion between ionised impurities and free carriers was thought as a gain mechanism through impurity ionisation upon application of pulsed electric field. Boundaries of the sample providing the reflection of light were proposed for a laser feedback mechanism. For Si-based optoelectronics such an opportunity is particularly important, because optical transitions in the latter case are linked only to one band and the problem of indirect crystal band structure in silicon is lifted. In 1971 an extension of the unipolar laser approach was proposed by Kazarinov and Suris [5]. The authors proposed to use population inversion between different electron subbands in a specially designed QW superlattice. The laser based on such approach (cascade laser) was realised in 1985 by Faist *et al.* [6]. The success of the cascade laser is linked, however, to direct-gap III-V materials and not to Si-based systems, in spite of the fact that the hystory of intraband lasing in indirect gap materials (e.g. in p-doped Ge) is quite long [7]. The interest towards Si-based unipolar

L. Pavesi et al. (eds.), Towards the First Silicon Laser, 281–292.

p-doped Ge) is quite long [7]. The interest towards Si-based unipolar laser is recently revived (see, for example, [8, 9]) and the progress in this field is also summarised in this book. A wave of new interest to potential realization of Si-based laser was initiated by self-organized Stranski-Krastanow (S-K) Si-Ge quantum dots (QDs). This activity is motivated by the fast progress of QD lasers in direct gap III-V materials systems. However, a way towards Si-Ge QD lasers appeared to be much longer, than initially expected. In my contribution I will concentrate on several promising approaches to achieve lasing in Si-based structures, which are all related to epitaxial nanostructures [10, 11] and address also key laser design issues.

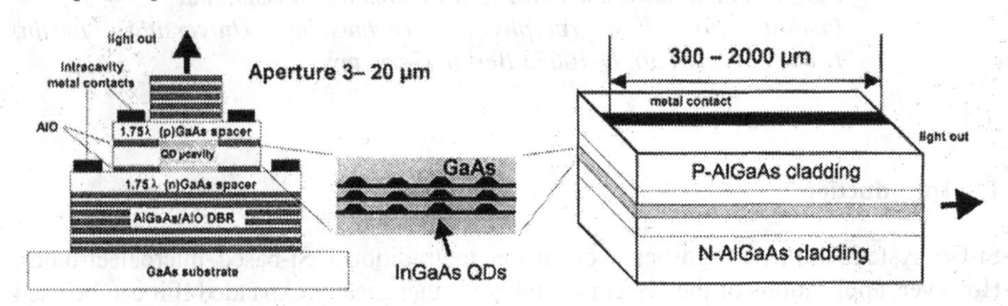

Figure 1. Schematic representation of edge and surface emitting lasers. In vertical-cavity surface-emitting lasers (VCSELs) all-semiconductor (e.g. AlAs-GaAs) or selectively-oxidized AlO-GaAs distributed Bragg reflectors (DBRs) are typically used to achieve high reflectivity (to the left). Active region typically represents ultrathin layers or, more recently, stacks of coherent nanoinclusions (quantum dots) as it is shown in the figure. Geometrical dimensions of the structures are arbitrary. For single-mode operation, the thickness of the waveguide region in edge-emitting structure should not exceed typically 0.6-0.8 µm, and for vertical-cavity surface-emitting structure the lateral dimensions of microcavity should stay at 3-5 µm, depending on particular design

2. Laser Designs

Prior to going to issues of applicability of the QD media for stimulated emission in Si, let us briefly discuss the key other requirements for lasing, which are related to optimal laser designs. Examples of the two basic laser structures are shown in Fig. 1. These are edge-and surface emitting lasers. The particular examples are based on III-V materials and use QDs as active media of the devices. In edge-emitting devices, the active medium, for example a thin layer, is placed in a waveguide region having a larger refractive index than the surrounding cladding layers. The laser light is confined in this narrow waveguide. It is diffracted at the facet exit at typically large angles of 30°-60°. The advantage of edge-emitting laser is a compact output aperture realised simultaneously with high light output power. High reflection and antireflection dielectric coatings are deposited usually on the rear and front facets, respectively. In a vertical cavity surface-emitting laser (VCSEL) the photons are bounced in a high finesse cavity in vertical direction. The cavity is very short and the optical gain per cycle is very low. Thus, it is of key importance to ensure very low losses at each reflection, otherwise, lasing will either not be possible, or will require too high current densities, not suitable for continuous wave operation. VCSELs were proposed as early as 1961 [4]. The authors foresaw a possibility of lasing in the direction perpendicular to the p-n junction plane. They wrote that the boundaries of the current injection region may serve as reflectors, thus, "a "resonant cavity" is formed". The practical realization of surface lasing took place in 1979, when Kenichi Iga used metallic mirrors on the top and bottom surfaces of the wafer [12]. The same year Burnham, Scifres and Streifer filed a US patent [13] describing Fabry-Perot cavities transverse to the plane of the wafer. One realization described the use of alternating semiconductor layers to form distributed Bragg reflector (DBR) structures which act as high reflectivity mirrors. VCSELs can be made very small, may operate at threshold currents as low as few microamperes, and are produced in a very production-friendly planar technology. Circular output aperture VCSELs do not suffer from the astigmatism of edge-emitting lasers. A temperature increase does not result in a significant wavelength shift, which is characteristic of conventional edge-emitting lasers, caused by the bandgap narrowing of semiconductors with increasing temperature. However, VCSELs can be easily realised only if suitable heterocouple is existing to fabricate DBRs. The requirements are: perfect lattice matching between the materials and a significant difference in the refractive indices. This is the case of the GaAs-AlAs materials, but does not hold, for example, for the InGaAsP-based materials lattice matched to InP. VCSELs are also less suitable for high-power operation. As the light output aperture of the VCSEL is equal to the current injection region, as opposite to the case of edge-emitter, fabrication of high power VCSELs having a compact aperture necessary for single lateral mode operation is challenging. Currently, single mode operation in conventional VCSEL design provides only few milliwatt output powers.

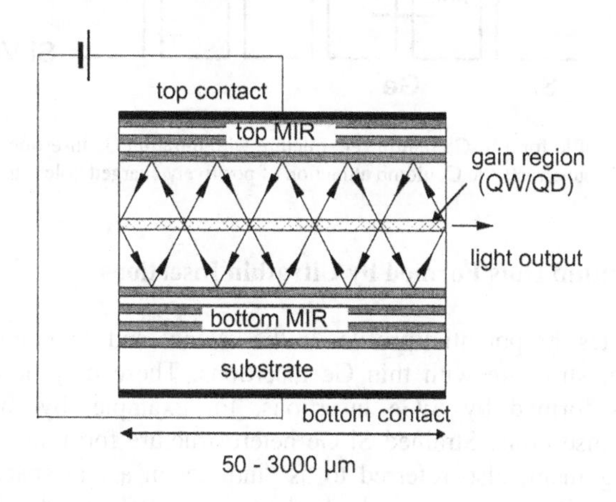

Figure 2. Tilted cavity laser (TCL) design for in-plane light output. Top and bottom multilayer interference reflectors (MIR) are calculated on the basis of the material system chosen (e.g. $In_{0.5}Ga_{0.5}As$ - $In_{0.48}Al_{0.52}As$ on InP) and the angle of incidence chosen. The cavity width is calculated on the basis of the wavelength to be selected and the angle of incidence defined by the DBRs. The number of periods is chosen to ensure high MIR reflectivity for the selected wavelength at which the device should operate. The length of the device is chosen to ensure no lasing for conventional in-plane mode, by selecting cavity length of front facet reflectivity for normal incidence small. For true single-mode operation the length of the device should be about 100 μm.

To overcome this problem an advanced laser design was proposed recently, called *tilted cavity mode laser* (TCL) [14]. It is shown schematically in Fig. 2. TCL may be realized as a surface or an edge emitter. For surface emission direct light outcoupling may be used for properly adjusted tilt angles. However, in other scenarios, the light output may be released via surface diffraction pattern, or via near field zone coupling to optical fibre [14]. Wavelength stabilisation in a TCL is achieved by adjustment of the optical cavity thickness and composition and the MIR design. In case of Si-based applications using of conventional DHS approach is complicated by the problem of light confinement. Lack of suitable heterocouple to silicon with exactly the same lattice parameter, and significant difference in the refractive indices and the forbidden gaps makes current-injection edge-emitting lasers challenging, even if the appropriate active medium is developed. As opposite, realization of VCSELs or TCLs seems to be much easier, as high-reflectivity multilayer dielectric DBRs or MIRs can be easily deposited. Another approach is using of photonic bandgap crystals [15] for light confinement. In this approach a special care is to be taken to avoid nonradiative recombination of non-equilibrium carriers at the etched surfaces.

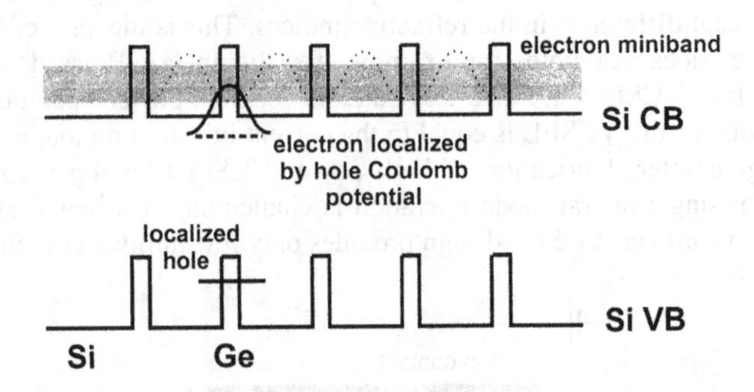

Figure 3. Potential profile for a Si-Ge multilayer structure with ultrathin Ge insertions. Note that the spike in the conduction band counteracts the Coulomb attraction of positively charged hole trapped in Ge inclusion.

3. Si-Ge Quantum Dots Formed by Ultrathin Insertions

Figure 3 illustrates the potential profiles in the valence and the conduction band of the Si-Ge multilayer structure with thin Ge insertions. These may be thin Ge (or SiGe) layers, or QDs formed by SiGe insertions, for example, by monolayer, or sub-monolayer [16] insertions. Strained Si-Ge heterostructure form so-called type-II (staggered) band alignment, also referred to as "indirect in a real space" (see *Figure 3*). Type-II QWs and QDs may have a high electron and hole overlap integral in the real space only if the spike in the conduction band does not prohibit formation of a stable s-like confined exciton ground state. This question was considered in detail for type-II GaSb QWs in a GaAs matrix [17]. Thus, to keep good electron and hole wavefunction overlap the thickness of the Ge insertion is to be below some certain value, at which the binding energy of the *s*-like exciton state formed by the localized hole in the Ge insertion and an electron in the Si matrix is larger than the binding energy of the *p*-like exciton state. The dependence of the electron binding energy in units of the exciton Rydberg in Si as a function of the power of the spike in the conduction band is shown in Fig. 4, where $\boldsymbol{a_B}$ is the Bohr radii, and :

$$\kappa = m_e \ V_e \ L/\hbar^2, \qquad (1)$$

where m_e is the elctron effective mass, V_e is the height of the potential spike in the conduction band, and L is the thickness of Ge insertion. Assuming comparable electron effective masses in Si and Ge and a height of the potential spike of 0.1 eV one may conclude that the exciton ground state has the lowest binding energy to effective Ge layer thickness of 0.6-0.7 nm.

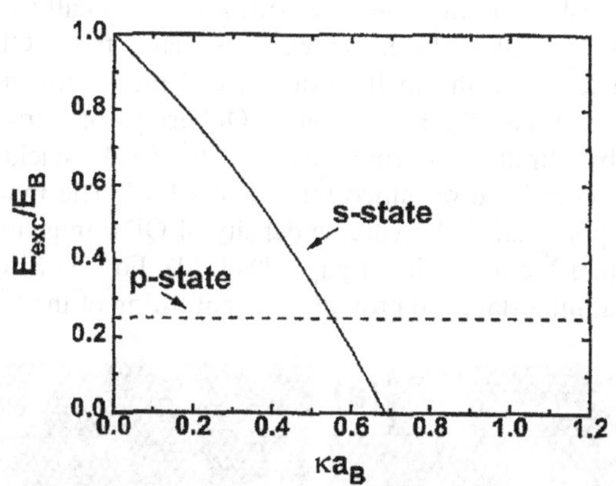

Figure 4. Reduction of the ground state exciton bindinge energy with increase in the power of the potential spike in the conduction band.

Indeed, photoluminescence studies [18, 19] performed in a wide range of excitation densities demonstrated that ultrathin Ge insertions do not demonstrate a characteristic high energy shift of the luminescence with excitation density, which is characteristic for type-II structures. This is quite the opposite to the situation with thicker Ge insertions [19], or Stranski-Krstanow SiGe QDs [20] and in agreement with a relatively small potential spike in the conduction band, which can be estimated as ~0.1 eV for strained Ge inclusions in Si [21]. At the same time the localisation energy of electrons remain fairly small and the thermal excitation of electrons into the electron miniband in Si occurs at fairly low (20-30 K) temperatures. This process is accompanied by reduction of the energy separation between the Ge-related photoluminescence (PL) and the corresponding Si-related PL lines and quenching of the Ge-related emission [18]. The effect of a strong decrease of the Ge-related PL is mostly linked to the trapping of thermally escaped electrons by nonradiative surface or Si substrate states. Doping of the active region with Sb creates significant equilibrium concentration of electrons preventing electron depletion. This strongly reduces temperature dependence of the PL emission from ultrathin Ge insertions in Si and allows it's observation up to room temperature [18]. The key condition to achieve lasing is to ensure sufficient modal gain to overcome both internal and external losses. In QD structures the modal gain is proportional to the volume density of the QDs. Thus, small islands, which can provide higher volume density of QDs are advantageous for high-gain applications. In indirect gap materials it is of particular importance also to have efficient lateral confinement of the excitons on a scale comparable or smaller to it's Bohr radii. This is necessary to provide efficient lifting of the k-selection rule. In this sense it is of topmost importance to optimize growth conditions to fabricate ultradense arrays of ultrasmall (<3-4 nm) quantum dots.

Taking into account that the modal gain is proportional to the optical confinement factor [11], and that no intentional waveguide can be normally used, the total thickness of the stacked Si-Ge insertions should to be large enough, for example, 1-5 μm.

What are the possible ways to fabricate ultradense arrays of Si-Ge QDs? It was previously demonstrated in III-V and II-VI materials systems that submonolayer or slightly above-monolayer lattice-mismatched deposits might, in appropriate growth conditions, spontaneously form high-density arrays of very small (3-5 nm) uniform and flat QDs [16] on the crystal surface. As was just said, these QDs are particularly advantageous for materials with small exciton Bohr radii allowing efficient exciton confinement on one side, and the realisation of QD arrays of ultrahigh density on the other. QDs formed by ultrathin insertions can be also easily stacked with thin spacer layers in a vertically correlated or anticorrelated way [16]. The modal gain of the QD media is roughly proportional to the volume density of QDs; thus, ultrahigh modal gain can be realised (up to 10^5 cm^{-1} in direct-gap QDs [16]). For the Si-Ge system using of such QDs is of crucial importance to provide efficient lifting of the k-selection rule.

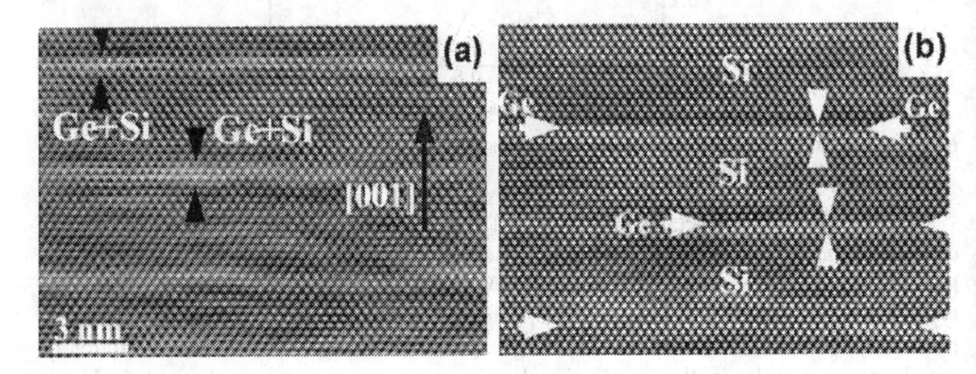

Figure 5. HRTEM images of Ge insertions in Si with nominal thickness of 0.07 nm (a) and 0.136 nm (b).

In *Figure 5* we show cross-section high-resolution transmission electron microscopy (HRTEM) image of Ge-Si multilayer structures with different nominal thickness of Ge insertions: 0.5 ML (a) and 1ML (b). It follows from Figure 5 that flat Ge islands are formed in both cases. At the same time submonolayer insertions demonstrate smaller lateral size and much higher density of the islands [18] and seem to be more advantageous for potential laser applications. Thicker Ge insertions have an advantage of more efficient localisation of holes, and the compromise is to be reached on the basis of detailed growth studies. In *Figure 6* we show photoluminescence spectra of Si-doped Ge insertions in a Si matrix recorded in surface geometry. As it follows from *Figure 6*, increase in the excitation density, or decrease in the observation density causes a strong superlinear growth and some narrowing of the linewidth of the TO-phonon related luminescence from Ge QDs. We note that stimulated emission mechanism with participation of phonons is characteristic for excitons, which are indirect in the *k*-space [22]. The nature of this effect [18] is not completely clear. In edge geometry superlinear growth of the PL intensity at some threshold excitation density was also observed, but the intensity of both zero phonon QD line and that of the TO-phonon replica were growing proportionally. More systematic studies are needed to understand the effect.

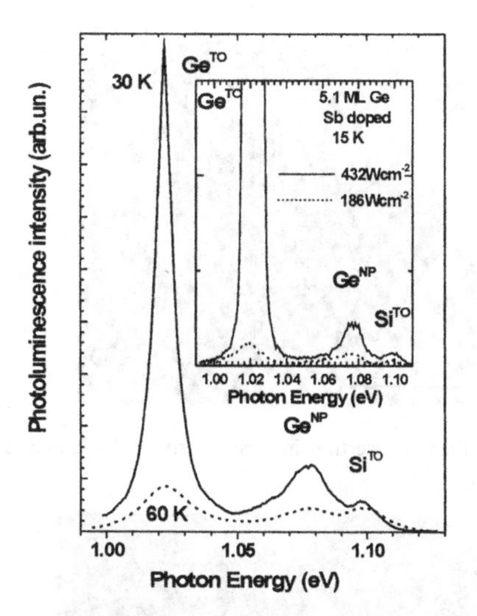

Figure 6. Photoluminescence spectra of Sb-doped Ge insertions in a Si matrix at different temperatures and excitation densities recorded in surface geometry.

4. Stranski-Krastanow Si-Ge QDs

Low density of QDs and low electron and hole wavefunction overlap represent significant problems to achieve reasonable gain using relatively large (~10 nm) three-dimensional S-K QDs. Very long luminescence decay times of about 4 μs are reported evidencing low wavefunction overlap in a real space and only weak lifting of the *k*-selection rule [23]. However, using of Sb-doped Si-Ge S-K QDs providing sufficient concentration of electrons in the vicinity of localized holes, a photoluminescence in the important 1.5 μm wavelength range was reported. The structure represented closely stacked laterally large flat QDs [24] formed by 0.7 nm Ge insertions. Effective localization of holes, which results in suppression of the hole diffusion to defects and free surfaces together with a high concentration of equilibrium electrons may ensure efficiency of radiative recombination sufficient for particular device applications. SK QDs can be, probably, used for light-emitting diodes, where high gain is not needed.

5. InAs QDs in a Si(SiGe) matrix

Direct incorporation of InAs QDs into a Si matrix was proposed as a possible solution to create efficient Si-based light-emitters [25]. Under optimised growth conditions coherent InAs inclusions can be grown, as it is confirmed by HRTEM studies and HRTEM modelling [26]. An example of a HRTEM image characteristic to InAs QDs in a Si matrix is shown in *Figure 7.* The dots have a lateral size of about 2-6 nm and a height of 2-3 nm. HRTEM modelling indicated that the images underestimate the size of the QDs, and the InAs dots having a diameter of 1.2 nm can't be resolved.

Figure 7. HRTEM image of InAs coherent insertion in a Si matrix. InAs lattice is revealed by elongated streaks. Arrows indicate InAs inclusions.

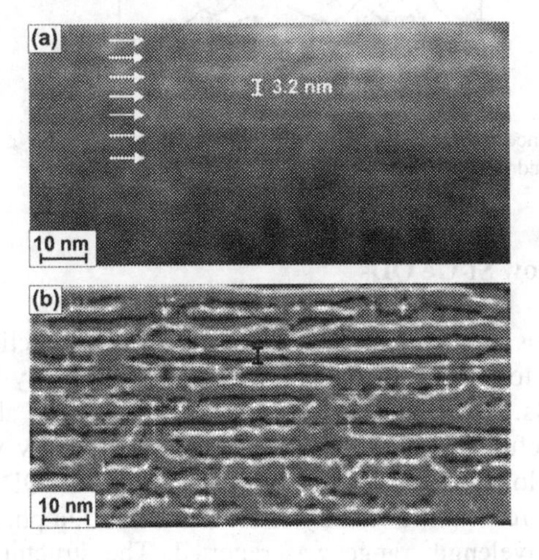

Figure 8. Diffused high-resolution plan-view TEM image (a) of the (311)A-grown GaAs-AlAs SL. Note the stripe pattern with a 3.2 nm periodicity. The processed (level equalized) image (b) evidences the 3.2 nm periodicity more clearly. Semiconductor industry will reach 3.2 nm lateral resolution in 25 years from now, assuming that the Moore's law continues. The factory cost will be $ 100 bln by that time.

Luminescence studies of InAs-Si QD structures have been performed [26]. Intense PL in the 1.3 μm range was found for the structures with 3D InAs nanoislands. As opposite, no PL in this range was found for the structures containing only InAs wetting layer, capped with Si prior to the QD formation. PL emission was composed of several broad TO-phonon replicas of the zero-phonon line and demonstrated a strong blue shift with excitation density increase [26], characteristic to type-II QDs indirect in a real space. This points to localisation of only one type of carriers, most probably, holes in QDs. The PL decay time was close to 400 ns, which is about two orders of magnitude longer than in conventional InAs-GaAs QDs, but an order of magnitude shorter than in the case of the Si-Ge Stranski-Krastanow QDs. Electron maintain their indirect band nature reducing the probability of the radiative recombination. Very small InAs QDs may overcome this disadvantage. However, growth of such QDs is a serious task, which is still to be carefully adjusted. Once the injected electrons will be trapped in the InAs nanoinclusions by a Colomb attraction of the localised holes, efficient radiative recombination and lasing may become probable. Some flexibility exists also in combining InAs and Ge QD growth using approaches of QD seeding and vertical coupling. SiGe matrix can be used for cladding of InAs QDs to reduce the height of the potential spike in the conduction band. Plastically-relaxed Si-Ge layers give an opportunity to reduce strain in the InAs QD and suppress the spike in the conduction band further.

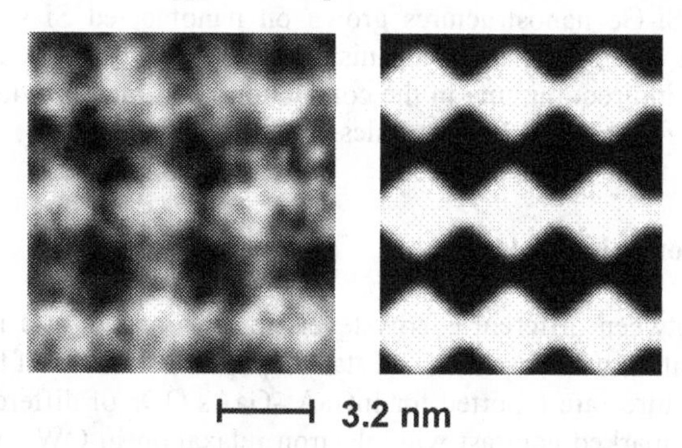

Figure 9. Level equalised diffused cross-section HRTEM image of the corrugated (311)A GaAs-AlAs SL (left) and the idealised cross-section image (right).

6. Quantum wires and quantum dots on corrugated surfaces

Si surfaces are known to exhibit spontaneous quasi-periodic nanofaceting (see e.g. Refs. [27,28]). This opens an opportunity to fabricate in situ vertically correlated arrays of QWWs and QDs [29]. This approach was claimed initially for GaAs-AlAs corrugated superlattices (SL) with 3.2 nm lateral periodicity and 2 nm thickness modulation of the GaAs and AlAs layers grown on spontaneously nanofaceted (311)A substrates [30]. Some authors, however, doubted the growth model initially-proposed in 1991. For example, in Ref. [31] the authors proposed that normal (AlAs-on-GaAs) interface is intermixed and the inverted interface has a corrugation height of only 0.34 nm. More recently it was demonstrated [32-34] that the original model [30] is valid and high-quality lateral superlattices with ultrafine lateral periodicity and strong lateral confinement can be fabricated. *Figure 8* demonstrates plan-view processed HRTEM images of the GaAs-AlAs corrugated SL with 1.7 nm-thick GaAs and AlAs layers. 10 corrugated GaAs and AlAs layers are imaged simultaneously evidencing good vertical correlation of QWWs. Compositional modulation can be also resolved in cross-section HRTEM geometry; even the contrast is rather weak (see *Figure 9*) due to the lack of chemically-sensitive reflection along the [-233] direction, parallel to the surface corrugation. A perfect 3.2 nm lateral periodicity can be most clearly revealed in Fourier transform images. It is evidenced by the spots related to the lateral periodicity and by the checkerboard arrangement of the spots of the reciprocal lattice [33. 34]. The latter effect is caused by the anticorrelated character of GaAs and AlAs layer corrugations (see *Figure*

9, right). The most important result for Si-Ge structures is the possibility to realise efficient luminescence in type-II structures using growth on nanofaceted surfaces. It was shown that type-II GaAs-AlAs (311)A SLs demonstrate high efficiency of radiative recombination in the short-wavelength spectral range at room temperature, in a marked contrast to SLs on (100) and (311)B surfaces grown side by side, but having no, or much weaker interface corrugation, respectively [33, 34]. Most recently, luminescence in the green and yellow spectral range at room temperature was reported [34] from GaAs-AlAs (311)A short-period SLs. This observation is certainly of very high potential importance for Si-Ge nanostructures grown on nanofaceted Si surfaces. Having interface corrugation as an efficient mechanism for lifting of the k-selection rule and mixing of direct and indirect minima in the conduction band may provide an important solution to the field of Si-based optoelectronics.

7. Si-Ge quantum dot unipolar laser

Carrier relaxation between different energy levels in QDs occurs at a reduced rate as compared to the quantum well case. Excited state depopulation times of the order of 10-30 ps at low temperatures are reported for InGaAs-GaAs QDs of different design [35]. These times are in a marked contrast with electron relaxation in QWs, where the phonon-related depopulation of the excited subband occurs at times typically below 1 ps. Very fast phonon-assisted depopulation of the excited subband acts as a parasitic leakage mechanism preventing realisation of population inversion for stimulated photon emission. This causes threshold current densities to be high making robust continuous wave operation of the device at room temperature difficult. Once using of QDs instead of QWs reduces the importance of the parasitic channel, the threshold current density to achieve population inversion can be also dramatically reduced. One should note, however, that the size and the shape uniformity of QDs in this case are to be well controlled and the density of QDs is to be very high to ensure sufficient optical gain to overcome losses. Si-Ge growth on nanofaceted surfaces or growth of well-ordered vertically coupled Si-Ge QDs may represent a possible solution. Type-II band alignment in Si-Ge structures requests using of p-type Ge-Si QD-based cascade lasers. The size of the QD is not necessarily must be small enough, as lifting of the k-selection rule for interband recombination is not needed. The lateral size of the QDs can be large enough to adjust the appropriate energy spacing between the hole sublevels due to size quantization in vertical direction involved in population inversion and laser gain.

8. Conclusions

Nano-engineering provides powerful tools for dramatic modifications of optical properties of the material. However, the goals for the Si-Ge laser nanotechnology must be first clearly understood and strictly defined. Not every type of epitaxial nanoobjects is advantageous for laser applications. As opposite, some of the “solutions” may lead to nowhere and even worsen the situation. As opposite, a number of direct epitaxial approaches can be used for realisation of stimulated emission in Si. The most promising approaches for interband lasing are:

- using of dense arrays of closely stacked small Ge QDs, formed, for example, by sub-monolayer deposition,
- using of small InAs QDs coherently inserted in a Si or SiGe matrix, or using of composite InAs-Ge QDs to improve uniformity and achieve better flexibility in the device design,
- using of Ge (SiGe) growth on corrugated Si surfaces having a fine lateral periodicity and pronounced corrugation height.

Different types of QD structures are needed for cascade lasers, where uniformity and the phonon relaxation rates, but not the size, are of the major concern.

In addition to appropriate design of the active media, a careful check of the optimum device geometry is needed. Lack of an ideal heterocouple to Si with perfect lattice matching, large difference in refractive index, similar crystal structure and defect-free interfaces makes device design issues of crucial importance. Using VCSEL and TCL concepts adjusted to dielectric DBRs may be advantageous both for interband and unipolar Si-based devices. We acknowledge helpful discussions with Zh.I. Alferov, D. Bimberg, R. Hetiz, I.P. Ipatova, D. Litvinov, V.A. Shchukin, I.P. Soshnikov, V.M. Ustinov, P. Werner, and N.D. Zakharov are acknowledged. The author is grateful to the Volkswagen-Stiftung and the Peregrinus-Stiftung for support.

References

1. Alferov, Zh.I. and Kazarinov, R.F. (1963) Double Heterostructure Laser, Authors Certificate No 27448, Application No 950840 with a priority from March, 30, 1963.
2. Kroemer, H. (1963) A proposed class of heterojunction injection lasers, *Proc. IEE* **51**, 1782-1783
3. Dingle, R and Henry C.H. (1976) Quantum effects in heterostructure lasers, U.S. Patent No. 3982207.
4. Basov, N.G., Vul B.M., and Popov Yu.M. (1959) Quantum mechanical semiconductor generators and amplifiers of electromagnetic oscillations, *JETP* **37**, 416.
5. Kazarinov, R.F. and Suris, R.A. (1971) Possibility of the amplification of electromagnetic waves in a semiconductor with superlattice, *Fizika i Tekhnika Poluprovodn.* **5**, 797-800 - *Sov. Phys. - Semicond.* **5**, 707-709.
6. Faist, J, Capasso, F., Sivco, D. L., Hutchinson, A.L., Sirtori, C., Chu, S. N. G. and Cho, A.Y. (1994) Quantum cascade laser: Temperature dependence of the performance characteristics and high T_0 operation, *Appl. Phys. Lett.* **65**, 2901-2903.
7. see for a review (1991) *Optical and Quantum Electron.* **23** (2), Special Issue on Far-infrared Semiconductor Lasers.
8. Bormann, I., Brunner, K., Hackenbuchner, S., Zandler, G., Abstreiter, Schmult, G.S. and Wegscheider, W. (2002) Midinfrared intersubband electroluminescence of Si/SiGe quantum cascade structures *Appl. Phys. Lett.* 80, 2260-2262.
9. Diehl, L., Sigg, H., Dehlinger, G., Grützmacher, D., Möller E., Gennser, U., Sagnes, I., Fromherz T., Campidelli, Y., Kermarrec, O., Bensahel, D., and Faist, J. (2002) Intersubband absorption performed on p-type modulation-doped $Si_{0.2}Ge_{0.8}$/Si quantum wells grown on $Si_{0.5}Ge_{0.5}$ pseudosubstrate *Appl. Phys. Lett.* **80**, 3274-3276
10. Shchukin, V.A. , Ledentsov, N.N., and Bimberg, D. (2003) *Epitaxial Nanostructures*, Springer, Berlin.
11. Bimberg, D., Grundmann, M., and Ledentsov N.N. (1999) *Quantum Dot Heterostructures*, Wiley, Chichester.
12. Soda, H., Iga, K., Kitahara, C., and Suematsu, Y. (1979) GaInAsP-InP surface-emitting injection lasers, *Jpn. J. Appl. Phys.* **18**, 2329-2330.
13. Burnham R.D.; Scifres D.R.; Streifer W. (1982) Transverse light emitting electroluminescent devices, US patent No. 4309670, filed: September 13, 1979.
14. N.N. Ledentsov and V.A. Shchukin (2002) Long wavelength lasers using GaAs-based quantum dots,

Proceedings of SPIE Vol. #4732, Photonics and Quantum Computing Technologies for Aerospace Applications IV (Eds. Donkor, E. , Hayduk, M.J., Pirich, A.R., Taylor, E.W.) SPIE, 2002, pp. 15-26.

15. Galli, M., Agio, M., Andreani, L. C., Belotti, M., Guizzetti, G., Marabelli, F., Patrini, M., Bettotti, P., Dal Negro, L., Gaburro, Z., Pavesi, L., Lui, A. and Bellutti, P. (2002) Spectroscopy of photonic bands in macroporous silicon photonic crystals, *Phys. Rev.* B **65**, 113111.
16. Krestnikov, I.L., Ledentsov, N.N., Hoffmann, A., Bimberg, D. (2001) Arrays of Two-Dimensional Islands Formed by Submonolayer Insertions: Growth, Properties, Devices (review), *phys. stat. sol.* (a) **183**, 207-233.
17. Ledentsov, N.N., Böhrer, J., Beer, M., Heinrichsdorff, F., Grundmann, M., Bimberg, D., Ivanov, S.V., Meltser, B.Ya., Yassievich, I.N., Faleev, N.N., Kop'ev, P.S., and Alferov, Zh.I. (1995) Radiative states in type-II GaSb/GaAs quantum wells, *Phys.Rev.* **B 52**, 14058-14066.
18. Makarov, A., Ledentsov, N.N., Tsatsul'nikov, A.F., Cirlin, G.E., Egorov, V.A., Ustinov, V.M. (2003) Studies of optical properties of ultradense arrays of Ge quantum dots in a Si matrix, *Semiconductors* **37**, *in print.*
19. Lenchyshyn, L.C. , Thewalt, M.L.W., Houghton, D.C., Noel, J.-P., Rowell, N.L., Sturn, J.C. and Xiao, X. (1993) Photoluminescence mechanisms in thin $Si_{1-x}Ge_x$ quantum wells, *Phys. Rev.* B **47**, 16655 -16658.
20. Dashiel, M.W. , Denker, U., Müller, C., Costantini, G., Manzano, C. , Kern K., and Schmidt, O.G. (2002) Photoluminescence of ultrasmall Ge quantum dots grown by molecular beam epitaxy at low temperature, *Appl. Phys. Lett.* **80**, 1279-1281 (2002).
21. Baier, T., Mantz, U., Thonke, K., Sauer, R., Schäffler, F., and Herzog, H.–J. (1994) Type-II band alignment in $Si/Si_{1-x}Ge_x$ quantum wells from photoluminescence line shifts due to optically induced band-bending effects: Experiment and theory, *Phys. Rev.* B **50**, 15191-15196.
22. à la Guillaume, C. B., Debever, J.-M., and Salvan, F. (1969) Radiative Recombination in Highly Excited CdS, *Phys. Rev.* **177**, 567-580
23. Fukatsu, S., Sunamura, H., Shiraki, Y., Komiyama, S. (1997) Phononless radiative recombination of indirect excitons in a Si/Ge type-II quantum dot, *Appl. Phys. Lett.* **71**, 258-260
24. Cirlin, G.E., Talalaev, V.G., Zakharov, N.D., Egorov, V.A., Werner, P. (2002) Room Temperature Superlinear Power Dependence of Photoluminescence from Defect-Free Si/Ge Quantum Dot Multilayer Structures, *phys. stat. sol.* (b) **232**, R1-R3.
25. Ledentsov, N.N. (1996) Ordered arrays of quantum dots, Proceedings of the 23rd International Conference on the Physics of Semiconductors, Berlin, Germany, July 21-26, 1996, Eds.: Scheffler, M. and Zimmermann, R., World Scientific, Singapoure, v. 1, pp. 19-26.
26. Heitz, R., Ledentsov, N.N., Bimberg, D., Egorov, A.Yu., Maximov, M.V., Ustinov, V.M., Zhukov, A.E., Alferov, Zh.I., Cirlin, G.E., Soshnikov, I.P., Zakharov, N.D., Werner, P., and Gösele, U. (1999) Optical properties of InAs quantum dots in a Si matrix, *Appl. Phys. Lett.* **74**, 1701-1703.
27. Phaneuf, R. J., and Williams, E. D. (1987) Surface phase separation of vicinal Si(111), *Phys. Rev. Lett.* **58**, 2563–2566
28. Phaneuf, R. J., Bartelt, N. C., Williams, Ellen D., Swiech ,W., and Bauer, E. (1993) Crossover from metastable to unstable facet growth on Si(111) *Phys. Rev. Lett.* **71**, 2284–2287.
29. Shchukin, V.A., Borovkov, A.I., Ledentsov, N.N. and Kop'ev, P.S. (1995) Theory of quantum wire formation on corrugated surfaces, *Phys. Rev.* B **51**, 17767-17779.
30. Nötzel, R., Ledentsov, N.N., Däweritz, L., Hohenstein, M. and Ploog, K. (1991) Direct synthesis of corrugated superlattices on non-(100)-oriented surfaces, *Phys.Rev.Lett.* **67**, 3812-3815.
31. Lüerßen, D., Dinger, A., Kalt, H., Braun, W., Nötzel, R., Ploog, K., Tümmler, J. and Geurts, J. (1998) Interface structure of (001) and (113)A GaAs/AlAs superlattices, *Phys. Rev.* B **57**, 1631–1636.
32. Ledentsov, N.N., Litvinov, D., Rosenauer, A., Gerthsen, D., Soshnikov, I. P., Shchukin, V.A., Ustinov, V.M., Egorov, A.Yu., Zukov, A.E., Volodin, V.A., Efremov, M.D., Preobrazhenskii, V.V., Semyagin, B.P., Bimberg, D and Alferov, Zh.I. (2001) Interface structure and growth mode of quantum wire and quantum dot GaAs-AlAs structures on corrugated (311)A surface *J. Electr. Mat.* **30**, 463-470.
33. Ledentsov, N. N., Litvinov, D., Gerthsen, D., Ljubas, G.A., Bolotov, V.V., Semyagin, B.R., Shchukin, V.A., Soshnikov, I.P., Ustinov, V.M. and Bimberg, D. (2002) Quantum wires and quantum dots on corrugated (311) surfaces: potential applications in optoelectronics, in: Proceedings of SPIE "Quantum Dot Devices and Computing", Eds.: Lott, J.A., Ledentsov, N.N., Malloy, K.J., Kane, B.E., Sigmon, T.W., 21 January, 2002 San Jose, USA, Vol. 4656, SPIE, pp. 33-42.
34. Litvinov, et al. . (2002) „Ordered arrays of vertically-correlated GaAs and AlAs quantum wires grown on a GaAs (311)A surface" *Appl. Phys. Lett.* **81**, 1080-1082.
35. Ledentsov, et al. (1996) Direct formation of vertically coupled quantum dots in Stranski-Krastanow growth, *Phys. Rev. B* **54**, 8743-8750.

PROMISING SiGe SUPERLATTICE AND QUANTUM WELL LASER CANDIDATES

GREG SUN
Department of Physics, University of Massachusetts at Boston
Boston, MA 02125, USA
RICHARD SOREF
Sensors Directorate, Air Force Research Laboratory, AFRL/SNHC,
Hanscom AFB, MA 0173, USA
ZORAN IKONIC
School of Electronic and Electrical Engineering,
University of Leeds, Leeds LS2 9JT, United Kingdom

1. Introduction

Strained-layer SiGe/Si heterostructures with 1D quantum confinement offer a pathway to lasing at long, very long, and far-infrared wavelengths (also known as TeraHertz). The purpose of this paper is to describe promising SiGe/Si superlattice (SL) and multi-quantum-well (MQW) laser candidates that would be pumped *electrically or optically or by phonons only*. The designs are IV-IV strained-balanced QW/barrier heterostructures. We shall exclude from our discussion the Ge-in-Si quantum dot lasers described elsewhere at this conference. The lasers proposed here use radiative transitions between valence subbands or minibands. The valence-band offsets in SiGe/Si MQWs are generally larger than the conduction band offsets, which is one reason to focus upon the valence subbands. In the future, we expect that SiGe/Si conduction intersubband devices will become viable; however, the shallowness of the conduction QWs will probably limit those devices to THz operation. Several research groups at this conference have reported the resonant interaction of SiGe/Si valence states with boron acceptor states, thus we shall not discuss impurity-related lasers.Along with wavefunction engineering, "strain engineering" has become vitally important for SiGe/Si photonic devices within the past four years. Today, we use the in-plane strains of QW and barrier as "variables" in our design space, strains selected to minimize the buildup of net strain across a "thick" MQW stack. Generally, we believe that the growth of the IV-IV strained-layer SL upon a relaxed $Si_{1-z}Ge_z$ buffer-on-Si has become a critical enabling technology for practical $Si_{1-x}Ge_x/Si_{1-y}Ge_y$ emitters and detectors. The SiGe buffer on Si acts as a "virtual substrate" (VS) for the device, and the VS cubic lattice parameter is chosen to be anywhere from 5.431Å (Si) to 5.656Å (Ge). Epitaxial growth of the coherently strained IV-IV MQW upon the VS allows balancing of the compressive QW strain by the tensile barrier strain in the MQW stack, giving zero cumulative strain, thereby

L. Pavesi et al. (eds.), Towards the First Silicon Laser, 293–306.

permitting the growth of ~ 10μm-high stacks whose thickness is not constrained by a critical-thickness limit (the stack height has been "severely limited" in prior MQWs grown directly upon Si). Here, we shall assume the use of a VS-on-Si in all cases.

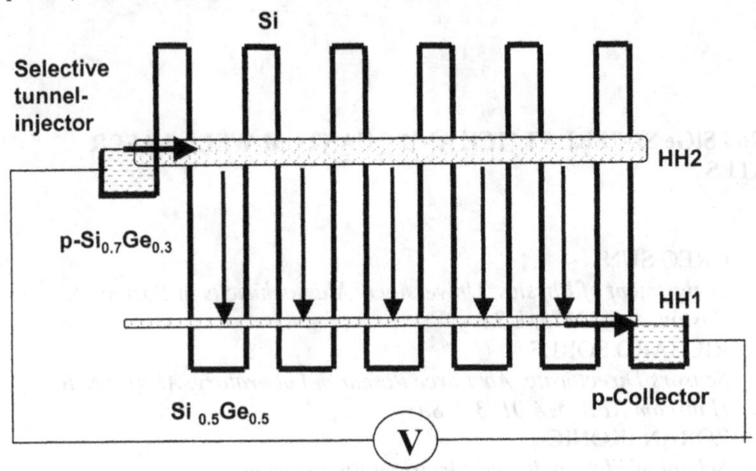

Figure 1. The QPL: p+-i-p diode showing lasing between two valence minibands of SiGe/Si SL.

2. Electrically Pumped Lasers

The tempo of research on SiGe/Si MQW lasers has quickened dramatically in the last three years and clearly the quantum cascade laser (QCL) approach is the most practical at present. The value and practical importance of the quantum cascade approach are demonstrated by the SiGe/Si experimental results presented at this conference by Gruetzmacher *et al*, Diehl *et al*, and Kelsall *et al*, who represent research consortia in Austria, Switzerland and the United Kingdom. Their electroluminescent SiGe cascade "emitters" demonstrate a wide spectral range (tunable by design). They report emission at wavelengths ranging from 8 to 120μm, so the design space is quite wide. At shorter wavelengths such as 8μm, the epitaxy is more demanding than at 120μm because the 8μm devices require more Ge content in the wells (more highly strained layers) and the barriers are quite thin, just a few monolayers of Si. For those seeking a low-cost silicon-based laser technology, the hope is that once the SiGe QCLs are demonstrated they can be optimized quickly to give performance comparable to the III-V QCLs already on the market. The 30 to 120μm wavelength (THz) range of operation, appears to be a good niche for IV-IV QCLs. In this regime, the QCL temperature of operation and output power may both be higher than that of the III-V QCL competition. This SiGe superiority is expected because, on the laser transition, the unwanted nonradiative optical phonon scattering is weaker in Group IV-IV than in Group III-V (non-polar vs polar scattering). In addition, there is evidence that the unwanted THz free-carrier absorption within the IV-IV laser cavity will be smaller than that in the III-V device.

Various electrically pumped IV-IV SL and MQW laser approaches have been reviewed in the book chapter by Kelsall and Soref [1] where 52 references are cited. Generally, the electrical approaches fall into the QCL and QPL (quantum parallel laser) categories. The QPL [2,3,4,5] has never been tested but is promising. The SiGe/Si QPL is an intrinsic valence interminiband SL laser with p contacts, and this p-i-p device would be biased into a nearly flat band condition with selective injection of holes into the upper miniband. (Fig.1) The QWs here are pumped in parallel, not series, but, for high-gain designs, the injection current falls off noticeably across the QPL SL. Hence, in practice, the QPL would be limited to a relatively small ~ 20 period SL, or to a series of QPLs that are "stitched together electrically" in a p-i-p-i-p-i-p cascade arrangement—a series of parallel-pumped SLs.

At this workshop, Gruetzmacher, Diehl, and Kelsall cover several of the excellent SiGe/Si QCL designs that are available, thus we shall focus on an alternative, the HH2 inverted mass QCL not touched upon by them. Generally, the QCL will utilize some combination of the HH1, LH/SO1, and HH2 subbands. We at Hanscom pioneered some of the LH1 to HH1 designs that are now being transitioned to practice [6,7,8].

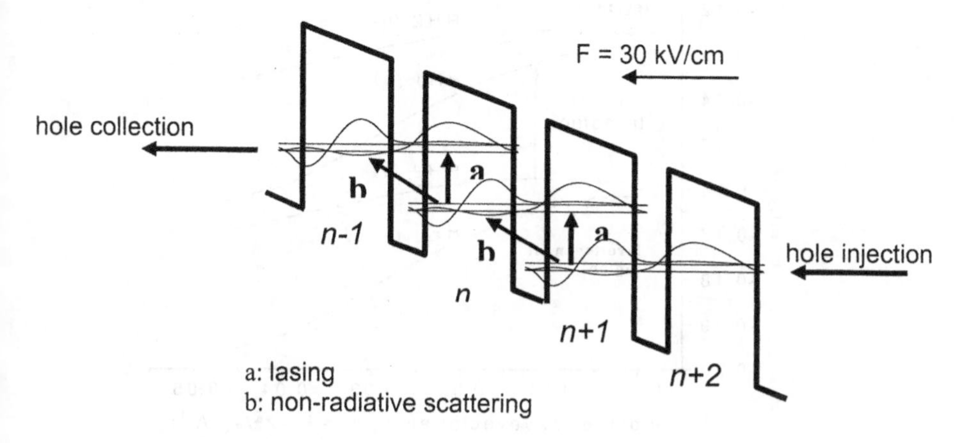

Figure 2. Band diagram of SiGe/Si HH2/HH1 laser at a bias of F=30kV/cm.

A "quantum staircase laser" (QSL) is a simple cascade without SL-injector sections. It can be engineered to place the HH2 subband adjacent in energy to HH1 (unlike the traditional LH1 HH1 pairing) [9,10]. In this special case, we have designed a strain-balanced SiGe/Si SL THz laser that employs the HH2 to HH1 transition in which an inverted effective mass is engineered for the HH2 subband near zone center [9-11]. An anti-crossing occurs between the ground state (HH1) and first excited state (HH2) subbands with their envelope functions localized separately in two neighboring QWs (nth and n+1th wells) at a critical field. The operating field is chosen to be slightly larger than the anti-crossing field so as to produce a small energy spacing between the two anti-crossing subbands that form a doublet. Figure 2 illustrates the case of four QWs of 90Å $Si_{0.8}Ge_{0.2}$ and 35Å Si with two active doublets per QW which form a four-level system under a bias field of 30kV/cm, which is just above the critical field of 26.4kV/cm. Two p^+ contacts are needed (not shown): the first p+ contact injects holes selectively into the doublets of the first QW, while the second collects holes from the doublets of the last QW. We calculate the well-and-barrier parameters that yield HH2 inverted effective mass, which exists over several structural parameter regions in alloy composition, well and barrier widths under biases. Figure 3 shows the calculated in-plane dispersion of the two involved subband doublets of coupled QWs with the same structure parameters (that represent the behavior of an N-well QSL) under the same operating field (30kV/cm). The energy spacing between the two subbands within a doublet is 3meV. The population distribution between the HH doublets follows the Boltzmann function because of the small energy separation and strong envelope-function overlap, which leads to population inversion between HH2 and HH1 in the two neighboring doublets.

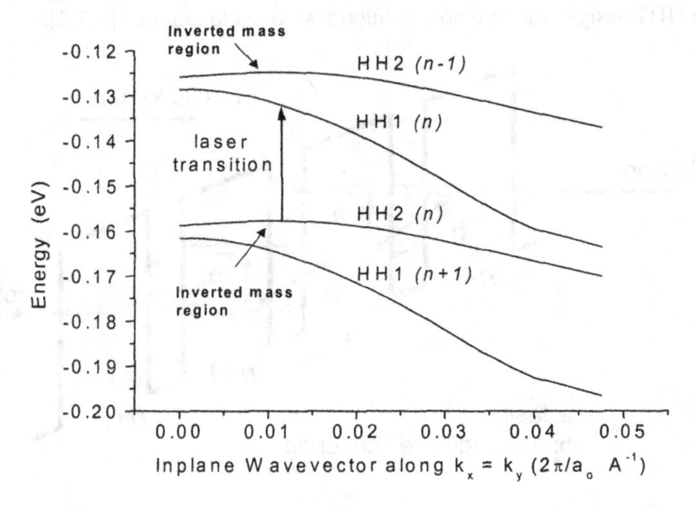

Figure 3 Dispersion of subbands in the two involved doublets under bias of 30kV/cm

We want to eliminate optical phonon emission from the upper laser state and make the radiative-to-nonradiative branching ratio as large as possible. Since the SiGe structure contains Si-Si, Si-Ge, and Ge-Ge lattice vibration modes (with optical phonon energy 64.5, 50.8, and 37.2meV, respectively), we engineer the SL so that the laser photon energy is less than the lowest-energy Ge-Ge phonons. Hole-acoustic-phonon intersubband scattering rates are calculated along with the THz spontaneous emission rates to establish a relationship between the total population and injected current density. The result of optical gain as a function of injected current density is shown in Fig.4. The peak gain of 450cm^{-1} at 7.3THz is expected to be larger than the QSL cavity losses such as free carrier absorption.

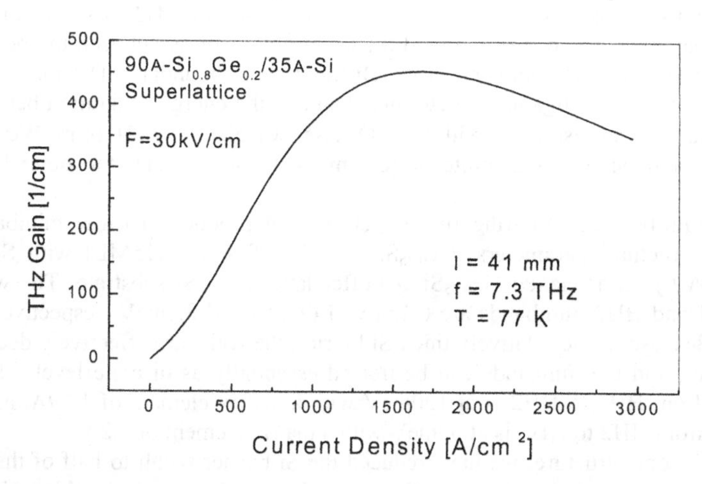

Figure 4 Calculated gain as a function of injected hole current density.

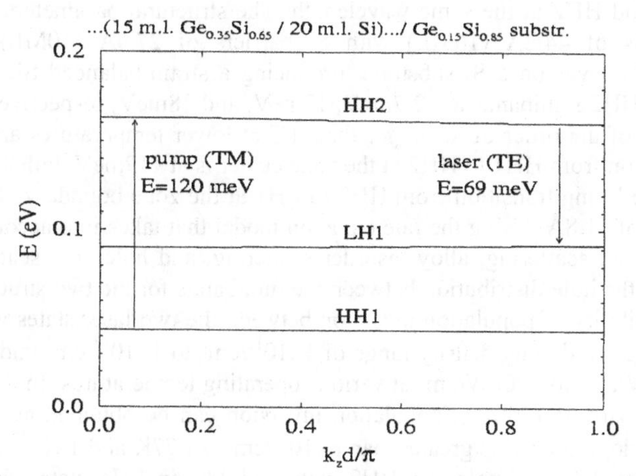

Figure 5.. Miniband diagram of 15ML $Si_{0.35}Ge_{0.65}$/20 ML Si.

3. Optically Pumped Lasers

We have explored the possibility of constructing optically pumped lasers in simple SiGe/Si SLs with each period consisting of a single QW. The SL is doped p-type in order to populate the lowest-energy miniband with holes. The pumping, relaxation, and lasing processes all occur within the lowest three minibands, namely HH1, LH1 and HH2. Specifically, the pumping transition is from the bottom of HH1 to the top of HH2 as shown in Fig.5 where each miniband has a k_z dispersion, and the lasing transition is between the bottom of HH2 and the top of the LH1 miniband at the Brillioun zone boundary. The range of structural parameters under investigation is determined so that the energy separation between the two pumping states is in resonance with the CO_2 laser wavelength of 10.6μm. We have studied two SL structures: one with quite narrow miniband widths and the other having broad minibands.

The strain-balanced SL (the first structure) that produced narrow minibands has the following structural parameters: $Ge_{0.35}Si_{0.65}$ wells of 41.6Å (15ML) with Si barriers of 54.0Å (20ML) on a relaxed $Ge_{0.15}Si_{0.85}$ buffer layer on a Si substrate. The widths of the HH1, LH1 and HH2 minibands are 0.1meV, 1.6meV, and 2.8meV, respectively, as shown in Fig.5. Because of the relatively thick Si barrier, the wells are effectively decoupled from one another, and the minbands can be treated essentially as discrete levels. The pumping transition from HH1 to HH2 is at 120meV with a matrix element of 10.9Å, and the lasing transition from HH2 to HH1 is at 69meV with a matrix element of 7.2Å.

In the second structure, we have reduced the Si barrier width to half of that in the first one to enhance the coupling between the wells, thus producing much wider minibands. The wells had a very small increase in width by 1ML in order to maintain the pumping transition between HH1 and HH2 at the same wavelength. The structural parameters are as follows: $Ge_{0.35}Si_{0.65}$ wells of 44.2Å (16ML) with Si barriers of 27.0Å (10ML) on a relaxed $Ge_{0.22}Si_{0.78}$ buffer layer on a Si substrate, producing a strain-balanced SL. The widths of HH1, LH1 and HH2 minibands are 2.7meV, 13meV, and 18meV, respectively, as shown in Fig.6. They are of the order of, or larger than, kT at lower temperatures around 77K. The pumping transition from HH1 to HH2 at the zone center is at 119meV with a matrix element of 9.6Å, and the lasing transition from HH2 to LH1 at the zone boundary at 57meV with a matrix element of 11.8Å. Using the rate equation model that takes into account the acoustic and optical phonon scattering, alloy disorder scattering, and hole-hole scattering [12], we have calculated the hole distribution between the minibands for the two structures. We have studied the possibility of population inversion between the two laser states within HH2 and LH1 minibands for a doping density range of $1\times10^{11}/cm^2$ to $1\times10^{12}/cm^2$ under the pumping power of $100kW/cm^2$ to $300kW/cm^2$ at various operating temperatures. In the first structure with thicker Si barriers (20ML), population inversion can be obtained at 77K, but not at 150K when the doping level is greater than $6\times10^{11}/cm^2$. At 77K and $1\times10^{12}/cm^2$ doping, the first structure has 1.6% of holes on HH2 state and 1% on LH1 state when pumped by $100kW/cm^2$, while at $300kW/cm^2$ it has 4.5% of all holes on HH2 and 2.5% on LH1. The population inversion depends on both the optical pumping power and the hole density (the latter due to hole-hole scattering), as shown in Fig.7.

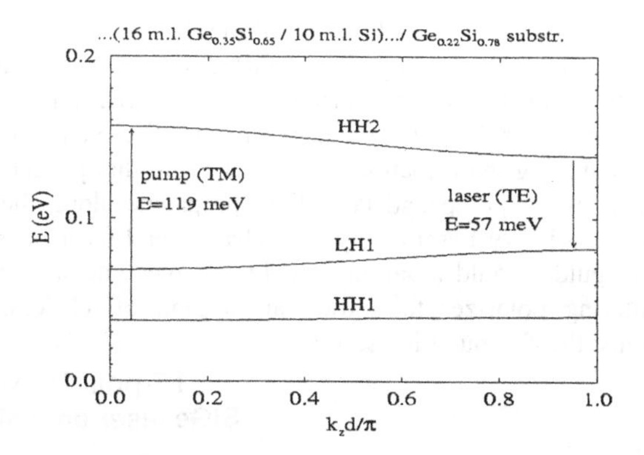

Figure 6 Miniband diagram of 16ML $Si_{0.35}Ge_{0.65}$/10 ML Si.

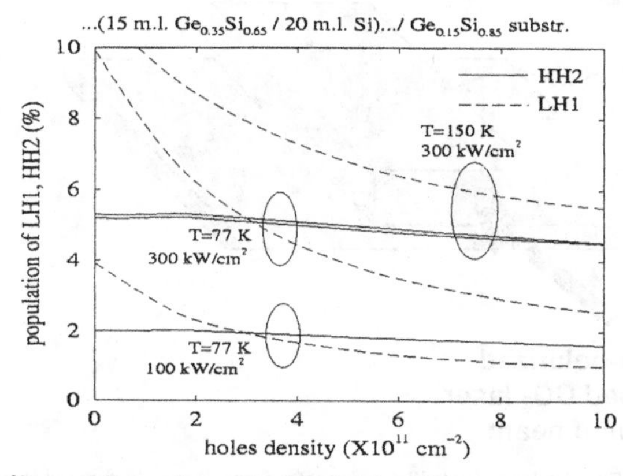

Figure 7 The population of lower and upper laser states as a function of pump power, temperature, and hole density.

It is interesting to point out that the second structure with much broader minibands showed no population inversion between the HH2 and LH1 minibands in the above parameter range. The result is somewhat surprising in the sense that intuitively one may think the much broader minibands should work in favor of creating population inversion between them as holes in HH2 will likely occupy states near the zone boundary while holes in the LH1 miniband will likely stay away from there. We believe, among the various scattering processes, the hole-hole scattering is the dominant process at the doping levels of interest, and the broader minibands and closer energy spacing between the laser states further enhance this scattering mechanism which limits the possibility of population inversion in structures of broader minibands. We would also like to point out that this is a preliminary study, further investigation is needed to optimize the structure, and possibly including multiple coupled QWs in each period of the SL. The pumping source needs to be *z*-polarized (TM mode in a ridge waveguide resonator) to facilitate transition between HH1 and HH2 minibands, while the lasing transition between HH2 and LH1 will be in-plane polarized (TE mode in the ridge waveguide resonator). We envision a waveguided laser resonator with end-pumping via a focused CO_2 laser beam that produces end-fire emission from the SiGe/Si SL (both pump and laser light propagate along the long axis x of the guide). This would be a Fabry-Perot cavity with cleaved end facets as shown in Fig.8. The semiconductor waveguide would be an etched ridge waveguide whose emitting facet was coated with a filtering polarizer to allow transmission of TE emitted photons while discriminating against the TM pumping source.

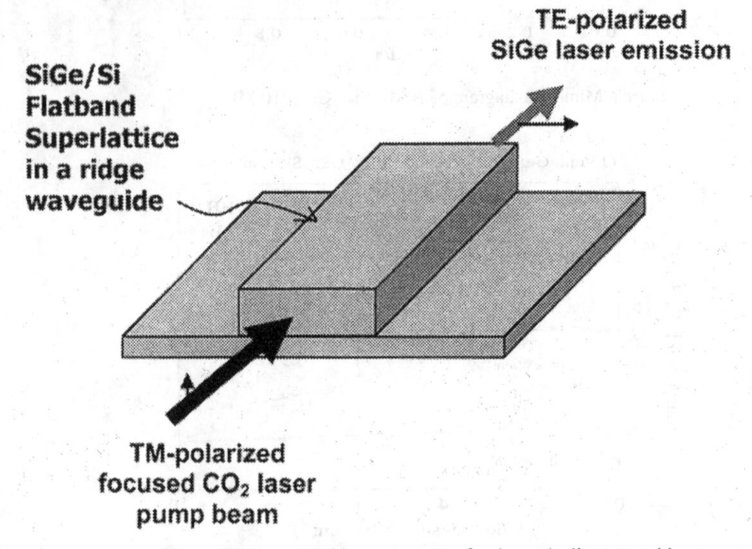

Figure 8 Waveguided laser resonator for the optically pumped laser.

4. Phonon Pumped Lasers

When a temperature gradient such as 77 to 300K is set up along the growth axis z of a properly designed p-doped SiGe/Si SL resonator, then far-infrared laser emission from the resonator appears feasible, simply by virtue of the absorption of optical phonons from the HH1 miniband into a higher miniband such as HH2 [13,14]. For such absorption, we engineer the zone-center HH2-HH1 spacing to be approximately the same as the LO phonon energy, about 50meV. From the dispersion diagram discussed below, it follows that the laser photon energy will be less than $\hbar\omega_o$ (LO), typically $\hbar\omega_L$(laser) ~ 25meV. Our proposed design (Fig.9) consists of a p-type doped $Si_{1-x}Ge_x$/Si SL active region sandwiched between two, thick, undoped $Si_{1-y}Ge_y$ layers that act as upper and lower waveguide claddings for the higher index SL, and as SL buffer layers whose lattice constant is about midway between the QW and barrier lattice parameters in order to obtain a strain-balanced SL. A Si substrate layer is present and a Ge heat buffer layer (HBL) is deposited atop the upper $Si_{1-y}Ge_y$ buffer. The metal coating deposited on the top and bottom of the semiconductor stack completes the far-infrared strip waveguide, and each heat sink is in contact with that plasmon film. The laser resonator would be therefore a waveguided edge-emission Fabry-Perot cavity with cleaved end facets, and the semiconductor waveguide would be either a rib or a strip guide with surface-plasmon coatings as shown in Fig.10.

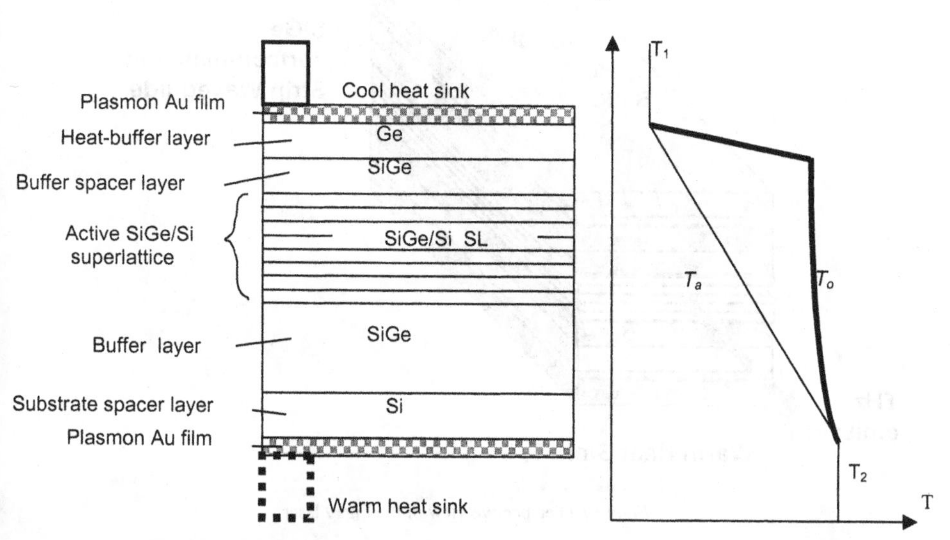

Figure 9 Cross-section end view of proposed phonon-pumped laser. The acoustic- and optical phonon temperature gradient across the structure is shown.

Phonon pumping in Fig.9 is achieved by keeping one heat sink at the cool temperature T_1 while the other sink is maintained at $T_2 > T_1$. The "warm" T_2 silicon layer generates optical and acoustic phonons that are transmitted to the active SL layers. A requirement in Fig.9 is that the energy spectra of the optical phonons in the substrate, in the Si-rich buffers, and in the active regions, overlap each other – and that the spectra are dramatically different from the HBL spectra; or ideally, we need to have a gap in the optical phonon spectrum of the HBL material at the optical-phonon frequencies of the active SL. When these requirements are met, optical phonons generated in the Si substrate will be able to travel freely into the active region, but because of the phonon frequency gap in the HBL as well as their small group velocity, optical phonons will be reflected at that interface and remain inside the active region. At the same time, acoustic phonons traversing the waveguide are free to penetrate through the HBL into the cold heat sink. Consequently, one can establish in the active region a substantial distance range (up to the heat diffusion length, i.e., a few μm) over which the optical phonons (T_o) are ``warmer" than the acoustic phonons (T_a). A schematic of these two different effective temperatures ($T_o > T_a$) is shown in the Fig.-9 temperature-gradient plots. Conveniently, a pure Ge layer can be used in Fig.9 for the HBL since Ge has a very different optical phonon spectrum from that of Si and is known to provide confinement of Si optical phonons. The Ge layer, with a 4% lattice mismatch to Si, can be bonded to the $Si_{1-y}Ge_y$ upper cladding on top of the SL.

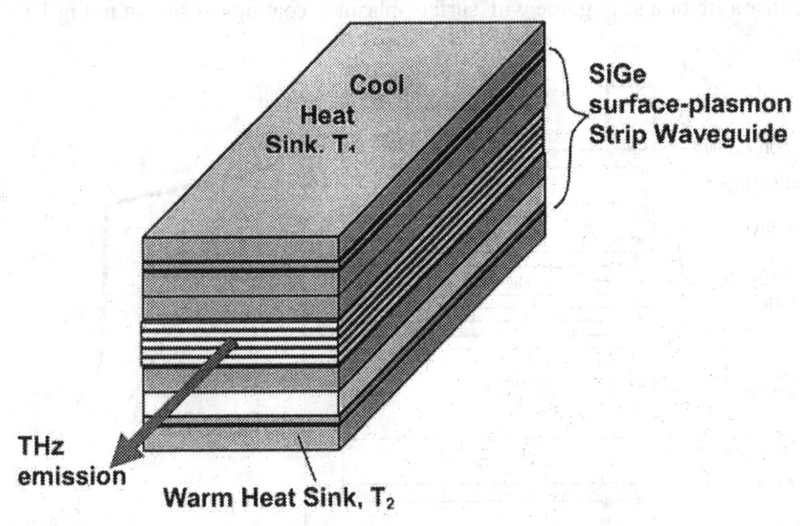

Figure 10 Perspective view of the Fig. 9 laser.

We selected an optimum active region for the structure as a p-doped strain-balanced $Si_{0.94}Ge_{0.06}/Si$ SL with 68Å wells and 35Å barriers, yielding $\hbar\omega_L < \hbar\omega_o$ along with the 4-level-like dispersion discussed below. This is a flat band SL, and the relaxed $Si_{0.97}Ge_{0.03}$ claddings serve as free-standing SL buffers. For the three lowest-energy valence states HH1, LH1, HH2, we have calculated the miniband dispersion as a function of SL wavevector k_z with the result given in Fig.11. The LH1 level that intervenes between HH2 and HH1 does not influence laser operation because the low-energy TE-polarized HH2-LH1 radiative transition is suppressed by the Fig.-10 waveguide resonator. We see that the energy minimum of the HH1 miniband lies at the Brillouin zone center, and that the minimum of the excited HH2 miniband occurs at the SL Brillouin zone boundary. The energy relaxation inside each miniband occurs primarily via interaction with acoustic phonons (with some participation of the hole-hole scattering at higher densities). The inter-miniband HH2-HH1 relaxation, on the other hand, is due mainly to the interaction with optical phonons because the energy separation between the two minibands is comparable to the energy of the Si optical phonon. If the phonons are non-polar, as is the case in SiGe, the intra-miniband scattering is expected to be faster than the inter-miniband process. As long as this is true, the distribution of carriers within each miniband can be characterized by the effective temperature of the acoustic phonons, T_a, while the distribution between different minibands is characterized by the effective temperature of the optical phonons, T_o.

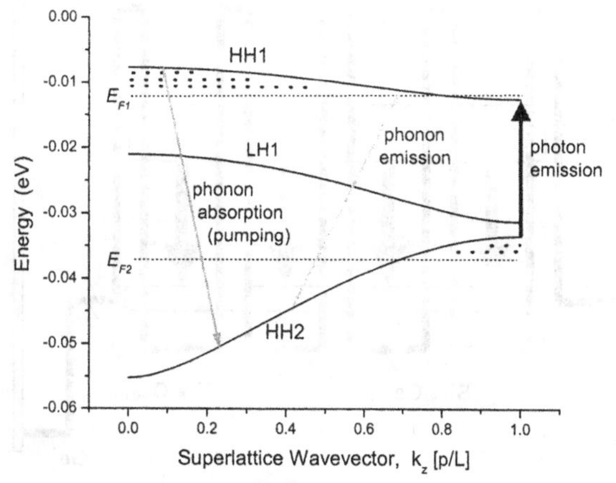

Figure 11 Calculated k_z dispersion of a strain-balanced $Si_{0.94}Ge_{0.06}$/Si SL illustrating phonon-pumping, relaxation, quasi-Fermi-levels, hole populations, and four-level THz lasing.

It can be seen from Fig.11 that due to the opposite energy-dispersion curvature of the two minibands, it is possible to achieve local-in-k_z population inversion near the Brillouin zone boundary, even though the total population of HH2 is always less than that of HH1. It is interesting to note that the operating principle of this phonon-pumped laser is quite similar to that of a conventional band-to-band laser. Holes that are pumped by optical phonons into the HH2 miniband quickly relax toward (via acoustic phonon interaction) and populate the energy minimum within HH2 (the upper laser state) near the SL Brillouin zone boundary. Those zone-edge holes undergo a vertical lasing transition to the maximum of HH1 (the lower laser state), and they finally relax toward the bottom of the HH1 miniband at the center of SL Brillouin zone – again via a fast acoustic-phonon process. Figure 12 illustrates the band-edge diagram of the Fig.-9 structure under the flat-band condition, along with the zone-edge THz vertical emission within each QW (the wells are pumped and emit in parallel; there is no cascade). The energy of the TM-polarized lasing transition at the end of the SL Brillouin zone is 21meV, corresponding to a wavelength of 59μm. We calculated the THz gain as a function of the doping concentration (the total population density). The result is shown in Fig.13 for the temperatures of optical and acoustic phonons maintained at 300K and 77K, respectively. It can be seen from Fig.13 that the THz gain remains positive for a wide range of doping concentration from $1\times10^{17}/cm^3$ to $1.2\times10^{18}/cm^3$. The peak THz gain of 280/cm is obtained at the doping concentration of $6\times10^{17}/cm^3$. Fig.14 shows the THz gain as a function of the temperature of acoustic phonons with the optimized doping density of $6\times10^{17}/cm^3$, while maintaining the temperature of optical phonons at 300K. As the temperature of acoustic phonons increases from 10K to 140K, the THz gain monotonically decreases from 1000/cm to -100/cm, with zero THz gain occurring at T_a=115K.

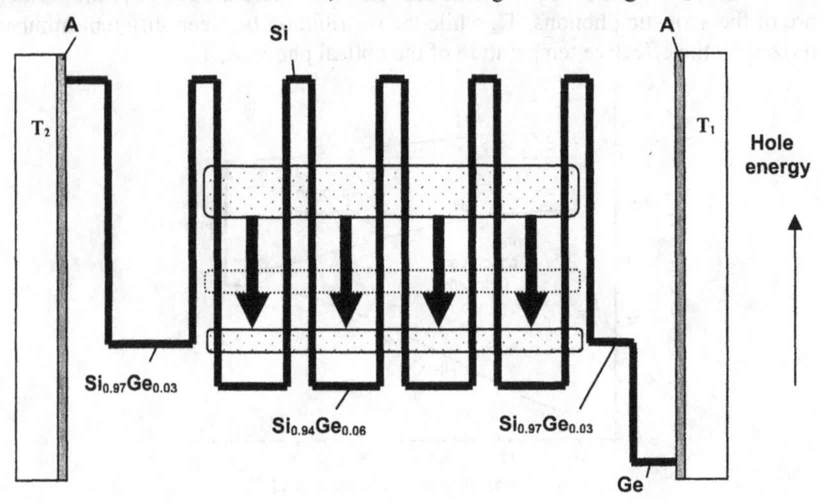

Figure 12 Valence band diagram of Fig.9 structure illustrating interminiband photon emission.

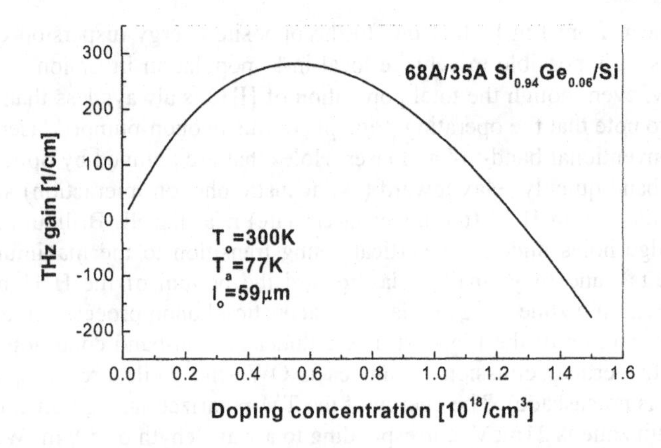

Figure 13 THz gain as a function of doping concentration for T_o=300K and T_a=77K.

A large temperature difference ΔT must be maintained across the ~10μm height dimension of the plasma-clad laser waveguide resonator. Some input power must be dissipated in order to attain and maintain ΔT. The most obvious way to do this is to attach the lower face of the SL resonator to a cold sink, e.g., to bond the face to the cold finger of a optical dewar, thereby keeping a constant 77K temperature at that face—and then to deposit a thin film electrical resistance heater on the top face of the resonator in order to heat that face up to about 300K via Joule heating of the film when a dc electric current is passed through it. This current will hold the upper-face temperature constant. To compute the required electric power dissipation in the film, the specific heat of the SL must be known along with other thermal parameters. We have made a rough estimate that ~2W input will be needed for typical SL dimensions discussed above. A second way to create the needed ΔT is essentially an optical pumping technique. Again, the lower SL face would be bonded to the cold finger, but now the upper face would be coated with a "black" film that will highly absorb visible or NIR light focused upon that film. The photon absorption will then lead to a temperature rise of that SL face, as desired. The photon source could be a laser, or if the SL is located in Space, the source could be sunlight.

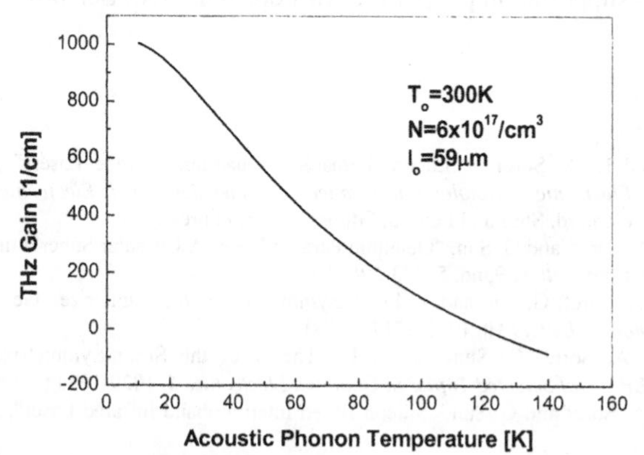

Figure 14 THz gain as a function of acoustic-phonon temperature while maintaining T_o=300K, with N=6×10^{17}/cm^3.

5. Summary and Conclusions

We have presented specific designs for strain-balanced SiGe/Si SL and QW lasers grown on SiGe-buffered-Si--devices that would be pumped electrically or optically or by phonons only. As of this writing, in September 2002, none of these designs has been demonstrated experimentally but all are promising candidates in our opinion. However, the unipolar electrically pumped SiGe/Si QCL is quite close to realization, and several cascade "emitters" are reported at this conference. For all three types of pumping, the temperature of operation is expected to be 77K or higher, and the wavelength of emission would be within the 8 to 120μm range. For electrical pumping, we see the choices as: (1) a complex cascade including SL injector sections and with perhaps two different Ge concentrations within each period, (2) a simple cascade called the quantum staircase; an electrically biased set of identical wells, (3) a QPL where the miniband in SL is in a nearly flatband condition. Our research has focused on (2) and (3). Previously, we examined LH-HH schemes for the quantum staircase. In the present paper, we give specific predictions (450/cm gain at 7.3THz) for an HH2-HH1 staircase laser in which a local-in-k_x population inversion would be produced in the HH2 inverted-mass subband. The technique proposed for the optically pumped laser is to use a tightly coupled (narrow miniband) acceptor-doped SiGe/Si SL in a ridge waveguide geometry that is end-pumped by a TM-polarized CO_2 laser which excites the HH1-HH2 interminiband transition. The 18μm lasing, TE polarized, would be HH2-LH1 at the SL zone edge in k_z space. For phonon pumping, a temperature gradient of 77 to 300K is set up along the growth axis of an acceptor-doped SiGe/Si flatband SL. Specific predictions are given for a 59μm Fabry-Perot laser embodied in a surface plasmon strip waveguide. A suggested practical geometry is to bond the waveguide to the cold finger of an optical dewar and to use an electrical resistance on the opposite face of the waveguide.

This work was supported in part by the Air Force Office of Scientific Research.

References

1. R. W. Kelsall and R. A. Soref , "Silicon-Germanium Quantum-Cascade Lasers", Chapter 8 in the book "*Sensing Science and Electronic Technology at Terahertz Frequencies, Vol I. Electronic Devices and Advanced Systems Technology*", Woolard, Shur and Leorop, Editors, (2003) in press.
2. L. Friedman, R. A. Soref, and G. Sun, "Quantum Parallel Laser: A Unipolar Superlattice Interminiband Laser", *IEEE Photonics Technology Letters* **9**, no. 5, 593-595 (1997).
3. L. Friedman, R. A. Soref, G. Sun and Y. Lu, "Asymmetric Strain-Symmetrized Ge/Si Interminiband Laser", *IEEE Photonics Technology Letters* **10**, 1715-1717 (1998)
4. L. Friedman, R. A. Soref, G. Sun and Y. Lu,"Theory of the Strain-Symmetrized Silicon-Based Ge/Si Superlattice Laser", *IEEE J. of Selected Topics in Quantum Electronics* **4**, 1029-1034 (1998)
5. L. Friedman, R. A. Soref and G. Sun, "Silicon-Based Interminiband Infrared Laser", *J. Appl. Phys*. **83**, 3480-3485 (1998).
6. G. Sun, Y. Lu, and J. B. Khurgin, "Valence intersubband lasers with inverted light-hole effective mass", *Appl. Phys. Lett*., **72**, 1481 (1998).
7. L. Friedman, G. Sun and R. A. Soref, "SiGe/Si THz laser based on transitions between inverted mass light-hole and heavy-hole subbands", *Appl. Phys. Lett*., **78**, 401-403 (2001).
8. R. A. Soref, L. Friedman, G. Sun, M. J. Noble, and L. R. Ram-Mohan, "Intersubband Quantum-Well Terahertz Lasers and Detectors", *Proc. SPIE* **3795**, Terahertz and Gigahertz Photonics, Denver, 22 July 1999, 515 (1999
9. R. A. Soref and G. Sun, "Terahertz gain in a SiGe/Si quantum staircase utilizing the heavy-hole inverted effective mass", *Appl. Phys. Lett*.**79**, 3639-3641 (2001)
10. G. Sun and R. A. Soref, "Inverted mass HH2 Intersubband Quantum-Staircase Lasers", presented at the *Sixth International Conference on Intersubband Transitions in Quantum Wells* (ITQW'01), Asilomar CA, (10-14 September 2001).
11. P. Harrison and R. A. Soref. "Room temperature population inversion in SiGe Taser designs", presented at the *Sixth International Conference on Intersubband Transitions in Quantum Wells* (ITQW'01), Asilomar CA, (10-14 September 2001)
12. Z Ikonic, P Harrison and R W Kelsall, *Proc. 8th Int. Workshop on Computational Electronics*, University of Illinois, October 15th–18th 2001, accepted for publication in Int. J. Comp. Electronics (2002)
13. G. Sun, R. A. Soref, and J. B. Khurgin, "Phonon Pumped SiGe/Si Interminiband Terahertz Laser", *IEEE J. Selected Topics in Quantum Electronics* **7**, 376-380 (2001)
14. G. Sun and R. A. Soref, "Phonon-pumped terahertz gain in *n*-type GaAs/AlGaAs superlattices", *Appl. Phys. Lett.* **78**, 3520-3522 (2001).

OPTICAL PROPERTIES OF ARRAYS OF Ge/Si QUANTUM DOTS IN ELECTRIC FIELD

A. V. DVURECHENSKII AND A. I. YAKIMOV
Institute of Semiconductor Physics, SB RAS
prospekt Lavrent'eva 13, 630090 Novosibirsk, Russia

1. Introduction

Zero-dimensional semiconductor structures or quantum dots (QDs) display many optical phenomena known from atomic physics. One of such exciting examples is the red-shift of the optical transition induced by an electric field, the quantum-confined Stark effect (QCSE). Recent theoretical [1-4] and experimental studies [5-7] for type-I InAs/GaAs and InGaAs/GaAs QDs demonstrated that the Stark effect can provide very useful information on the polarity of intra- and inter-dot electron-hole alignment and the vertical separation. The change of the potential energy of a dipole with a moment **p** in an electric field **F** is given by U=-**pF** [8]. For the electron-hole system, $\mathbf{p} = e(\mathbf{r}_h - \mathbf{r}_e)$, where $\mathbf{r}_{e,h}$ is the mean electron (hole) position. In type-II QDs, only one of the charge carriers is confined inside the dot whereas the other carrier is outside the dot. Contrary with type-I QDs, one expects that in such a system the Stark effect is extremely large because of the permanent spatial separation of electron and hole and the presence of the built-in electron-hole dipole [4]. To date, most work in the field of QCSE has concentrated on InAs/GaAs QDs, and very little is known about the influence of electric field on the excitonic properties of type-II QDs.

It is generally accepted that Ge/Si(001) quantum dots exhibit a type-II band lineup [9-11]. When an electron-hole pair is photoexcited, the hole is captured into the energy well of the Ge dot while the electron is in the low energy Si conduction band. Electron and hole interact through an attractive Coulomb potential which binds the electron at the Si/Ge interfaces [Fig. 1(a)]. This causes the formation of a spatially indirect excitons. In the present work we use photocurrent (PC) spectroscopy to study the effect of an electric field on the interband transitions in Ge/Si(001) quantum dots.

2. Experimental

To tune the electric field across the QD, these are embedded in the intrinsic region of a Si *p-i-n* diode (surface p^+ region), allowing fields up to 90 kV/cm directed long z (applying a reverse bias to a *p-i-n* structure results in an electric field pointing from the n^+ substrate to the p^+ surface). The band profile under reverse bias condition is shown schematically in Fig. 1(b).

L. Pavesi et al. (eds.), Towards the First Silicon Laser, 307–314.

To observe experimentally the QCSE by PC spectroscopy, it is necessary that: i) the size of the dots in all three dimensions should be small enough to provide actual zero-dimensional density of states; ii) the electron and hole must be well separated to ensure a large dipole moment. This last implies that the dots should be rather thick in z. However, conventional Ge/Si (001) self-assembled QDs, grown by Stranskii-Krastanov growth techniques, are always flat, i.e, they have an aspect ratio (height divided by base length) much lower than unity [12]. To fabricate thick Ge islands with small lateral size, we grow Ge dots on a Si(001) substrate covered with an ultrathin SiO_x film. Recently a similar approach has been successfully applied to form high-density ultrasmall Ge islands on Si(111) [13] and Si(001) [14] surfaces. The mechanism of Ge nanocluster formation on ultrathin SiO_x films is essentially different from that on clean Si surfaces. A possible hypothesis has been put forward by Shklyaev and co-workers [13] and takes into account a reaction between individual Ge adatoms and SiO_x followed by a local silicon oxide desorption. Reflection high-energy electron diffraction (RHEED) data show that three-dimensional Ge islands grow without the formation of a wetting layer and are almost epitaxial with the underlying silicon substrate. The latter observation implies that, similar to the case of Stranski-Krastanov islands, Ge nanoclusters fabricated on oxidized Si surface reside on *bare* Si regions.

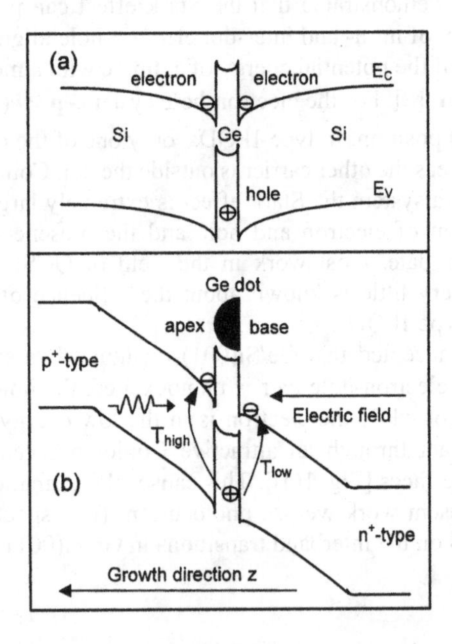

Figure 1. (a) Band structure of the type-II Si/Ge/Si heterostructure along the growth direction through the center of the Ge dot. (b) Schematic band diagram of the *p-i-n* diode under reverse bias.

The sample was grown by molecular-beam epitaxy at a temperature of 500^0C on n^+-Si(001) substrate ($7x10^{-3}$ cm^{-3} As). The growth rates were 2 ML/s for Si and 0.2 ML/s for Ge. For *p-i-n* structures, a 400 nm thick *i*-Si region was first grown. Then the surface was oxidized. Oxygen was introduced into the chamber at a pressure of 10^{-4} Pa for 10 min, which produces the growth of SiO_x with a thickness of several Å. After the oxygen was pumped out and the chamber pressure reached 10^{-7} Pa, 1 nm thick Ge layer was deposited to form the QD. This was followed by another 400 nm thick *i*-Si and 200 nm p^+-doped Si layer ($2x10^{18}$cm^{-3} B). The structure was finally capped with a 10 nm thick p^+-Si contact layer (10^{19} cm^{-3} B). The background boron concentration in the unintentionally doped Si layers was (7-8)x10^{15} cm^{-3}. The QDs formation and quality of the silicon grown to embed the dots was controlled *in situ* by RHEED. It was established that despite the presence of the silicon dioxide layer, the cap silicon is also a perfect crystal. Rectangular mesa diodes with areas ranging from 2.5x10^{-4} to 5x10^{-4} cm^2 were fabricated by standard lithography and wet chemical etching. A 1 μm SiO_2 passivation layer was deposited by chemical vapor deposition. The ohmic contacts with the p^+ and the n^+ layers were obtained by depositing 80x80 μm^2 Al contacts. The layer of QDs capped with a 10-nm-thick Si layer was examined with plan-view and cross-sectional electron microscopy (Fig. 2) [15,16]. The Ge islands have a hemispherical shape with a base diameter of 5.8±0.5nm and a height of about 3–4 nm. The apex of the dots is oriented along the growth direction. The areal density of the islands was approximately 1.8x10^{12} cm^{-2}. A reference sample was also grown under the same conditions, except that no Ge was deposited.

It is necessary to note that when the thickness of Ge reaches 1 nm, the distribution of Ge dot sizes becomes bimodal. Together with the ultra small high-density islands, a low-density (~10^8 cm^{-2}) lens-shaped Ge nanocrystals ($\approx$200 nm in diameter and $\approx$40 nm in height) appear. However, as we will argue at the end of the paper, these islands do not contribute to the measured PC spectra.

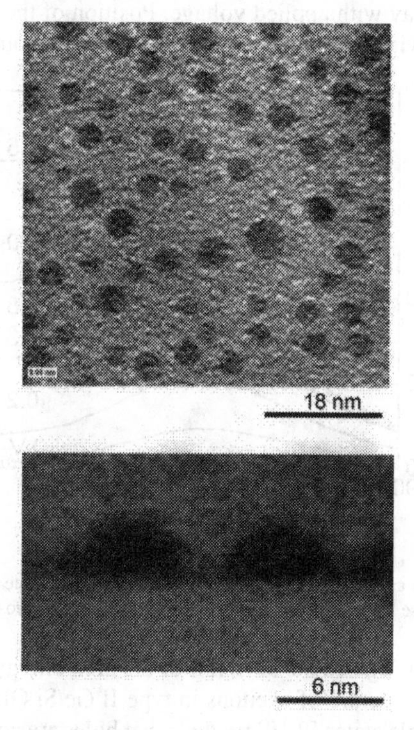

Figure 2. Plan-view (top) and cross-section (bottom) transmission electron microscopy images of a 1-nm Ge dot sample [16]. The Ge islands appear in dark contrast.

PC measurements were performed in normal-incidence geometry (incident light polarized in the plane of the samples) at room temperature. Short circuit (no bias) photocurrent was directly measured with a Keithley electrometer. For biased measurements, a lock-in amplifier was used. In the latter case, the light from globar source was mechanically chopped at the frequency of about 550 Hz. A low illumination power density of ≈ 0.1 mWcm^{-2} was employed to provide extremely low dot carrier occupancy and to avoid many-particle effects. In order to obtain the responsivity of the *p-i-n* diode, the spectral photon flux from the light source was measured using a calibrated pyroelectric detector.

3. Results and Discussion

Figure 3 shows photocurrent spectra as a function of reverse bias. There is an apparent PC peak below the silicon interband absorption edge (1126 meV) which is not seen in the reference sample (crosses in Fig. 3). At low bias, this peak at ≈ 1040 meV has a symmetric line shape. It is suggested that it originates from indirect excitonic transitions between the hole ground state in the small Ge dots and the electron ground state confined in Si near the heterojunctions. The electron-hole pairs created by interband absorption thermally escape from the dots and give rise to the measured photocurrent. As the reverse bias increases, the current maximum becomes wider and splits into two peaks which develop in a different way with applied voltage. Position of the low-energy peak T_{low} is practically unchanged with the bias while the high-energy component T_{high} apparently shifts to *higher* energies.

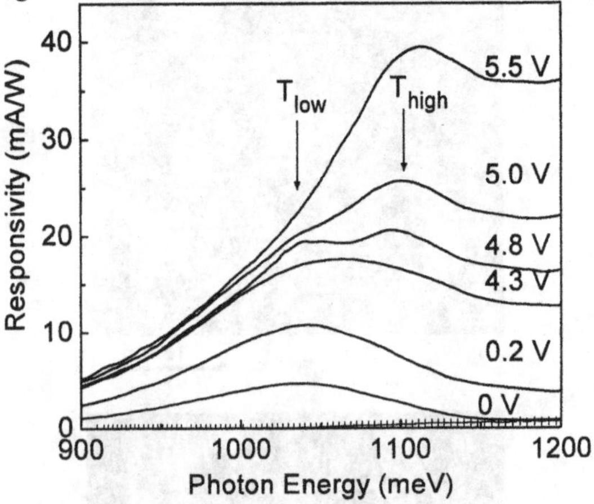

Figure 3. Photocurrent spectra as a function of applied reverse bias (lines). The nominal Ge coverage is 1 nm. The short circuit photoresponse from a reference Si photodiode is shown by crosses.

To explain splitting and the blueshift of the high-energy transition, one needs to consider the electronic structure of excitons in type-II Ge/Si QDs. The modelling of the confined electron and hole states [9,10] predicts that holes are concentrated at the bottom of the dot, and the electrons are localized in Si both on top and below the Ge island. This is the result of strain distribution and Coulomb forces around the dot. Recently the confirmation of the spatial separation of electrons in the silicon matrix surrounding the Ge islands was provided by the observation of a negative interband photoconductivity in n-type Ge/Si(001) QDs [17].

It follows from second-order perturbation theory that the field dependence of the transition energy can be described by

$$E(F) = E(0) - eF(z_h - z_e) - \beta F^2, \tag{1}$$

where $e = -|e|$ is the electron charge, $E(0)$ is the transition energy at zero field, $z_{e,h}$ is the mean electron (hole) position along the growth direction (along the nanocrystal axis), and β is the polarizability of the electron-hole system [2]. In a system possessing a non-zero dipole moment, the second order term in Eq. (1), quadratic in the applied field, must be less important than a linear one and the transition energy must vary linearly with the field.

Within this model, we interpret the high-energy maximum T_{high} as a transition between the hole ground state in the Ge dot and the electron state confined in Si near the dot apex [16]. The low-energy peak T_{low} is assigned to the transition between the same hole state and the electron state localized in Si near the dot base [see Fig. 1(b)]. Obviously, the term $eF(z_e\text{-}z_h)$ is negative for the first case and positive for the second one since the electron-hole dipoles have the opposite directions.

We can check our explanation by extracting the values of electron-hole and electron-electron separation from the observed Stark shift. First, keeping in mind that the observed PC maximum is a superposition of the two peaks, we decompose the maximum into two Gaussians. This allows us to determine the transition energies. Then we perform a self-consistent one-dimensional simulation of our *p-i-n* device to calculate the electric field near the apex and the base of the dots.

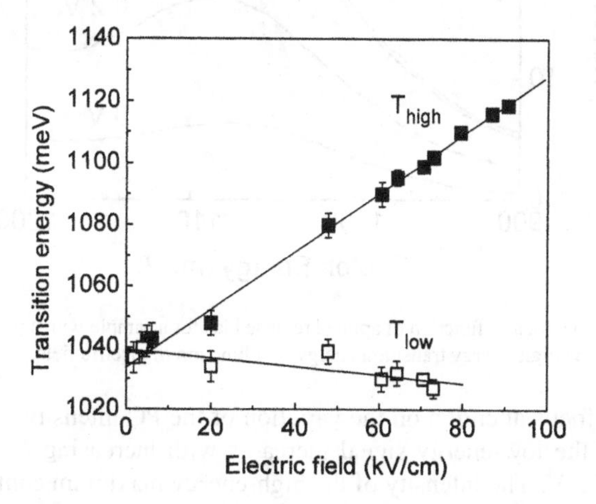

Figure 4. Transition energies as a function of electric field for 1-nm-Ge sample. The solid lines are theoretical fits to the experimental data.

The field dependence of the transition energies are plotted in Fig. 4. As expected for a system with built-in dipole moments, the Stark shift for both transitions appears to be linear. Moreover, due to the linear behaviour, the type-II Ge/Si QDs exhibit a QCSE of approximately one order-of-magnitude stronger than type-I InGaAs/GaAs QDs of similar height [7]. From a fit to the data using the Eq. (1), we find the electron-hole distance (5.1±0.2) nm for the electron near the dot apex (top electron) and −(0.8±0.3) nm for the electron near the dot base (bottom electron). It is worth to note that separation of these two electrons (≈6 nm) is somewhat larger than the mean dot height (≈4 nm), which is quite reasonable for QDs with a staggered band line-up and provides clear support for our explanation. Moreover, the small separation of the bottom electron and the hole agrees with the fact that hole is localized towards the base of the dots.

To obtain further evidence on the QD related origin of the PC maximum, we have fabricated another test sample grown under conditions similar to the previous dot sample, except that 2 nm Ge was deposited. This produced smaller Ge dots (about 4 nm in diameter and 2 nm in height) with somewhat larger areal density (2×10^{12} cm^{-2}). Due to stronger hole confinement in small dots, the ground state excitonic transition is shifted to higher energies as compared with the 1-nm Ge structure (Fig. 5). Also, since the dots are thinner, the Stark effect is rather small. From the dependence of the transition energy on the electric field (inset of Fig. 5), we deduce a top electron-hole separation of 2.5±0.8 nm, again in agreement with the dot height.

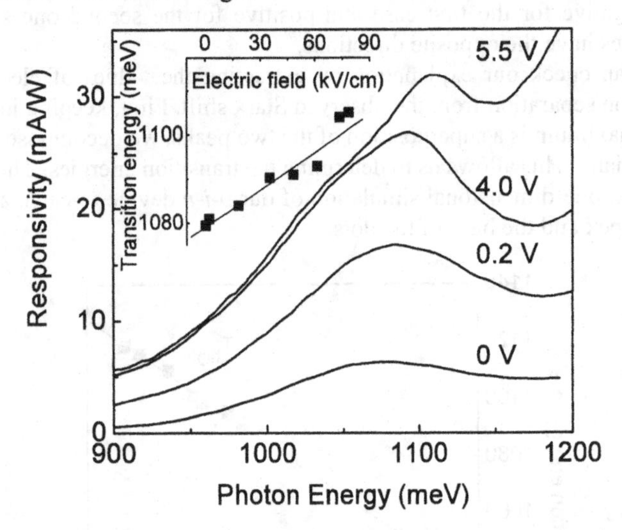

Figure 5. Photocurrent spectra as a function of applied reverse bias for a sample with a nominal Ge coverage of 2 nm. The inset shows the high-energy transition energy as a function of electric field.

We now focus attention on the variation of the PC intensity with electric field. The amplitude of the low-energy signal increases with increasing F at low fields and saturates at bias $V \gtrsim 5$V. The intensity of the high-energy maximum continues to increase even at the highest F. The increasing value of both PC peaks at low F can be related to an increasing rate of carrier escape with F. By applying a reverse bias, the electric field pushes the top electron towards the hole in the dot and pulls the bottom electron out from the hole. As a consequence, the electron-hole overlap and the corresponding absorption strength are increased for the T_{high} transition and reduced for the T_{low} transition. At highest

F, no bound state can further exist for the bottom electron and the T_{low} transition transforms into a PC tail on the low-energy side of the T_{high} absorption.

Let us discuss the role of the big Ge islands which are present in the structures due to the bimodal growth mode. We claim that these islands are of no importance for observed PC spectra due to the following arguments. First, the maximum external quantum efficiency η of the investigated photodiodes deduced from the responsivity is about 1% at 1.3 μm (at 0.954 eV). A similar value of η (2.3%) was achieved in Ge/Si quantum-dot waveguide photodetector, which contains five layers of Ge islands with a density of $3x10^9$ cm^{-2} in each layer and was designed to have *a strong optical confinement* [18]. Obviously, one layer of Ge islands having a very low density ($2x10^8$ cm^{-2} for large islands in 1-nm Ge sample and $5x10^8$ cm^{-2} for 2-nm Ge sample) cannot ensure a measurable PC, especially at normal incidence. This is possible only for an extremely high-density QD structure. Second, 100-nm-sized Ge/Si self-assembled islands usually exhibit an exciton related photoluminescence peak around ~800 meV (see Ref. [19] and references therein). An onset of interband transition at larger energies (1040–1100 meV) is possible only in presence of ultrasmall Ge QDs with enhanced size quantization of the hole energy spectrum.

4. Summary and Conclusions

In summary, the photocurrent spectroscopy of type-II Ge/Si(001) quantum dots, as a function of applied electric field, has demonstrated that the QDs possess two built-in electric dipoles of opposite orientations. We argue that this is a consequence of the spatial separation of the electrons around the dots. From the observed Stark shift, both separation of the electrons and hole at the dots and the distance between the electrons were determined. We found that, due to the linear behaviour, the type-II Ge/Si QDs exhibit a QCSE of approximately one order-of-magnitude stronger than type-I InGaAs/GaAs QDs. An external quantum efficiency of about 1% at 1.3 μm of wavelength was obtained at room temperature. This result indicates that the Ge/Si QDs are potentially applicable for Si-based 1.3–1.5 μm optical fiber communication.

The authors are much obliged to A. Milekhin and S. Schulze for TEM measurements, A. O. Govorov for helpful discussion, A. V. Nenashev for self-consistent calculations, N. P. Stepina for assistance in sample preparation, and I. B. Chistoknin for technical assistance. This work was supported by the RFBR, the Education Ministry program (grant E00-3.4-154) and INTAS-2001-0615.

References

[1] Li, S.-S. and Xia, J.-B. (2000) Quantum-confined Stark effects of InAs/GaAs self-assembled quantum dot, *J. Appl. Phys.* 88, 7171–7174.

[2] Barker, J.A. and O'Reilly, E.P. (2000) Theoretical analysis of electron-hole alignment in InAs-GaAs quantum dots, *Phys. Rev. B* 61, 13840–13851.

[3] Sheng, W. and Leburton, J.-P. (2002) Anomalous quantum-confined Stark effects in stacked InAs/GaAs self-assembled quantum dots, *Phys. Rev. Lett.* 88, 1674401.

[4] Janssens, K.L., Partoens, B., and Peeters, F.M. (2002) Stark shift in single and vertically coupled type-I and

type-II quantum dots, *Phys. Rev. B* 65, 233301.
[5] Fry, P.W., Itskevich, I.E., Mowbray, D.J., Skolnick, M.S., Finley, J.J., Barker, J.A., O'Reilly, E.P., Wilson, L.R., Larkin, I.A., Maksum, P.A., Hopkinson, M, Al-Khafaji, M., David, J.P.R., Cullis, A.G., Hill, G., and Clark, J.C. (2000) Inverted electron-hole alignment in InAs-GaAs self-assembled quantum dots, *Phys. Rev. Lett.* 84, 733–736.
[6] Patanè, A., Levin, A., Polimeni, A., Schindler, F., Main, P.C., Eaves, L., and Henini, M. (2000) Piezoelectric effects in $In_{0.5}Ga_{0.5}As$ self-assembled quantum dots grown on (311)B GaAs substrates, *Appl. Phys. Lett.* 77, 2979–2981.
[7] Findeis, F., Baier, M., Beham, E., Zrenner, A., and Abstreiter, G. (2001) Photocurrent and photoluminescence of a single self-assembled quantum dot in electric fields, *Appl. Phys. Lett.* 78, 2958–2960.
[8] Sacra, A., Norris, D.J., Murray, C.B, and Bawendi, M.J. (1995) Stark spectroscopy of CdSe nanocrystallites: The significance of transition linewidth, *J. Chem. Phys.* 103, 5236–5245.
[9] Yakimov, A.I., Stepina, N.P., Dvurechenskii, A.V., Nikiforov, A.I., Nenashev, A.V. (2000) Excitons in charged Ge/Si type-II quantum dots, *Semicond. Sci. Technol.* 15, 1125–1130; Interband absorption in charged Ge/Si type-II quantum dots, *Phys. Rev. B* 63, 045312.
[10] Schmidt, O.G., Eberl, K., and Rau, Y. (2000) Strain and band-edge alignment in single and multiple layers of self assembled Ge/Si and GeSi/Si islands, *Phys. Rev. B* 62, 16715–16720.
[11] Cusack, M.A., Briddon, P.R., North, S.M., Kitchin, M.R., and Jaros, M. (2001) Si/Ge self-assembled quantum dots for infrared applications, *Semicond. Sci. Technol.* 16, L81–L84.
[12] Pchelyakov, O.P., Bolhovityanov, Yu.B., Dvurechenskii, A.V., Nikiforov, A.I., Yakimov, A.I., and Voigtländer, B. (2000) Molecular beam epitaxy of silicon-germanium nanostructures, *Thin Solid Films* 362, 75–84.
[13] Shklyaev, A.A., Shibata, M., and Ichikawa, M. (2000) High-density ultrasmall epitaxial Ge islands on Si(111) surfaces with a SiO[Trial mode] coverage, *Phys. Rev. B* 62, 1540–1543.
[14] Barski, A., Derivaz, M., Rouvière, J.L., and Buttard, D. (2000) Epitaxial growth of germanium dots on Si(001) surface covered by a very thin silicon oxide layer, *Appl. Phys. Lett.* 77, 3541–3543.
[15] Milekhin, A.G., Nikiforov, A.I., Pchelyakov, O.P., Galzerani, J.C., Schulze, S., and Zahn, D.R.T. (2002) Resonant Raman scattering of relaxed Ge quantum dots, in *Proceedings of the 26th International Conference on the Physics of Semiconductors*. Edinburgh, Scotland.
[16] Yakimov, A.I., Dvurechenskii, A.V., Nikiforov, A.I., Ul'yanov, V.V., Milekhin, A.G., Schulze, S., and Zahn, D.R.T. (to be published).
[17] Yakimov, A.I., , Dvurechenskii, A.V., and Nikiforov, A.I. (2001) Spatial separation of electrons in Ge/Si(001) heterostructures with quantum dots, *JETP Lett.* 73, 529–531.
[18] El kurdi, M., Boucaud P., Sauvage, S., Fishman, G., Kermarrec, O., Campidelli, Y., Bensahel, D., Saint-Girons, G., Sagnes, I., and Patriarche, G. (2002) Silicon-on-insulator waveguide photodetector with Ge/Si self-assembled islands, *J. Appl. Phys.* 92, 1858–1861.
[19] Pchelyakov, O.P., Bolkhovityanov, Yu. B, Dvurechenski, A.V., Sokolov, L.V., Nikiforov, A.I., Yakimov, A.I., and Voigtlander, B. (2000) Silicon-Germanium Nanostructures with Quantum Dots: Formation Mechanisms and Electrical Properties, *Semiconductors* 34, 1229–1247.

MBE OF Si - Ge HETEROSTRUCTURES WITH Ge NANOCRYSTALS

O.P. PCHELYAKOV, A.I. NIKIFOROV, B.Z. OLSHANETSKY,
K.V. ROMANYUK, S.A. TEYS,
Institute of Semiconductor Physics SB RAS,
Novosibirsk, 630090, RUSSIA

1. Introduction

One of the most probable ways towards silicon laser runs across solution of a complicated problem of material science – development of a reproducible technology for producing heterostructures with ultimately small-sized Ge quantum dots in Si. Dramatic progress has been already made in creating Si-based light-emitting diodes with Ge nanocrystalls incorporated into Si matrix (see for example Ref. [1]). These are called "Hut" and "Dome" – clusters. A great deal of recent research pursued in this field is focused on investigation of self-assembling processes of Ge nanocrystals by Stranski-Krastanov mechanism and effect of their self-ordering during heteroepitaxy. In this investigation attempts are made to find the ways for fabricating heterosystems with the smallest size of nanocrystalls and the highest degree of self-ordering of their ensembles in a substrate plane and in their dimensional homogeneity [2-4]. Recent record in minimizing nanocrystal size using a thin silicon oxide submonolayer is reported in Ref. [5]. Some data on the degree of GeSi cluster ordering on Si are also presented in Ref. [6]. However, it should be noted that clusters of minimal size (but no less than 10 nm) turn out to be poorly ordered whereas ordered clusters are large in size.

To monitor the growth in molecular-beam epitaxy (MBE) reflection high-energy electron-diffraction (RHEED) has been used. However, only a comparative analysis of the RHEED patterns and scanning tunneling microscopy (STM) surface images allows the diffraction patterns to be interpreted correctly in the course of the film growth [7]. In early studies, the RHEED technique was used for plotting surface phase diagrams during Ge epitaxy on Si [8]. A variation in the diffraction pattern is a qualitative illustration of morphological rearrangements of the growing film and quantitative information can be obtained by recording the intensity of the diffraction pattern. Of particular interest are the data on strains in the growing layer, which are the main driving force for the observed morphological rearrangements. The strains can be estimated from variations in the lattice constant in the growing Ge film from the position of diffraction spots on the RHEED patterns. This approach was used to investigations of the elastic and plastic relaxation in the systems of Ge/Si [9] and InAs/GaAs [10]. The assumption of the island deformation during 2D growth was supported by observations of periodic changes in the size of the surface 2D cell for A_3B_5 compounds and metals [11,12].

An impressive amount of works both experimental and theoretical has been accumulated on 3D Ge nanoislands formation on Si surface. It was established, that self-organizing nanoislands with a particular shape and faceting on a silicon surface are pre-

L. Pavesi et al. (eds.), Towards the First Silicon Laser, 315–323.

ceded by the formation of a Ge wetting layer, which remains between the islands up to their coalescence (see for example, Refs. [13-17]); unclear are still the details of the formation of the wetting layer.

The present paper describes the mechanisms and structure as well as the elastic deformations of Ge nanocrystals very small in size, down to atomic units. Nanocrystals of this kind appear during the formation of Ge wetting layer by a two-dimensional layer-by-layer mechanism. This is an initial stage of heteroepitaxy preceding a morphological transition – formation of "Hut" and "Dome" clusters – the crucial moment in realization of Stranski–Krastanov mechanism. Experimental data obtained by RHEED and STM are presented. New data on atomic structure and evolution of elastic deformation of a two-dimensional unit cell of Ge nanocrystal surface are obtained during MBE. Superstructure transformation is found to effect on ordering of nanocrystal lateral arrangement. The possibility is discussed to devise a reproducible technology for manufacturing small-sized ordered Ge nanocrystal ensembles on Si surface.

2. Experimental

A MBE installation equipped with two electron-beam evaporators for Si and Ge was used for the synthesis. The analytical equipment of the chamber included a quadrupole mass spectrometer, a quartz thickness monitor and a high-energy electron (20 kV) diffractometer. Diffraction patterns were monitored during the growth using a CCD camera on-line with a PC. The software allowed both the complete at the images and chosen fragments of the diffraction patterns to be monitored at the rate of 10 frames/s. Ge was deposited at the rate of 10 BL/min, temperature was varied from 250 to 600°C. Silicon wafers misoriented by less than 0.5° were used as substrates. Before a Ge deposition, the substrate was annealed and a Si buffer layer grown to produce a clean starting surface. Scanning tunneling microscopy (STM) in ultra-high vacuum conditions (combined STM-MBE "OMICRON-RIBER") and high-energy electron diffractometry with a built - in MBE installation are used for that purpose. The surface of Si(111) samples cleaning was carried out at pressures of $(1\text{-}2)\times 10^{-10}$ Torr and consisted of a sample degassing at least for 4 hours at 600°C followed by flash at 1250°C. STM images were obtained at room temperature. A source of Ge atoms consisted of pieces of Ge welded on a W ribbon. The evaporation rate of Ge in our STM experiments was from 10^{-3} to 10^{-1} BL/min (BL or bilayer is a double layer on Ge(111) and Si(111) surfaces, 1 $BL_{Ge(111)}$ contains 1.44×10^{15}atoms/cm^2). Ge coverage was estimated by measuring a total volume of Ge islands formed at the terraces beyond the step edges and domain boundaries. The features of our experiments in more details were described in Refs. [8,14-19].

3. Results and discussion

3.1. GERMANIUM WETTING LAYER FORMATION ON Si(001) AND Si(111): RHEED INVESTIGATIONS

Several oscillations of the reflections of RHEED can be observed during the formation of a wetting layer in Ge/Si(111) by molecular beam epitaxy. Usually, the number of the oscillations is from one to four at the growth rates in the range between 0.1 and 10 BL/min. Oscillations of RHEED specular beam intensity indicate that in this range of the growth rates the formation of a Ge wetting layer occurs through layer-by-layer growth mode by nucleation and coalescence of 2D nanocrystals [15-17].

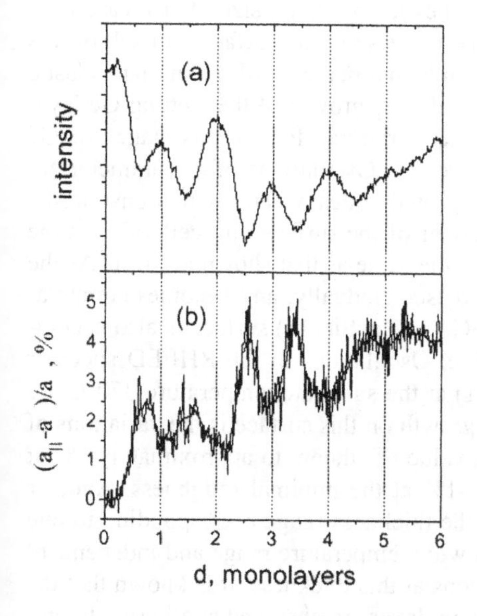

Figure 1. Variations in specular beam intensity (a) and surface 2D cell constant (b) during growth of Ge film on Si (100) at 250°C.

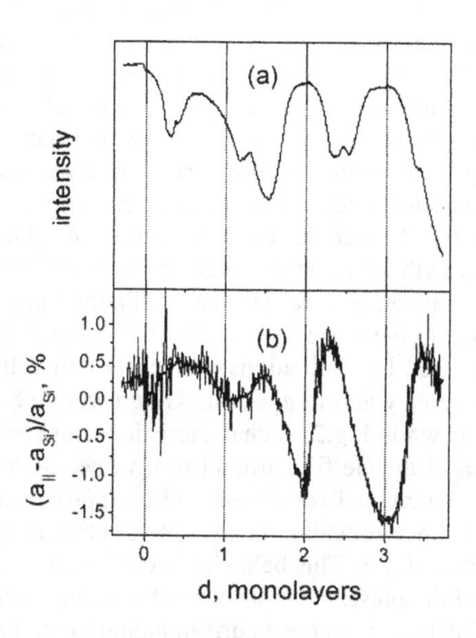

Figure 2. Variations in specular beam intensity (a) and surface 2D cell constant (b) during growth of Ge film on Si (100) at 370°C.

The variation in the RHEED during growth of wetting layer Ge on Si(111) just as on Si(001) testify that the growth of this layer is occurs by the layer-by-layer mechanism. Elastic relaxation in 2D islands results in a distortion of their unit cell. RHEED allows this deformation to be registered *in situ* as a variation in the average size ($\boldsymbol{a}_{||}$) of the surface unit cell. To accomplish this, variations in the intensity of the diffraction pattern were recorded along the line going across the streaks and bulk spots. The distance between the spots assigned to $(0\bar{1})$ and (01) reflections oscillates during growth, indicating changes in $\boldsymbol{a}_{||}$ in the (100) plane of the Ge film. Figure 1 shows variations in $\boldsymbol{a}_{||}$ (as a percentage with respect to the value a of the pure silicon surface) during growth of a Ge film on a Si(100) surface at 250°C. The corresponding intensity variations of the RHEED specular beam are also shown. The value of $\boldsymbol{a}_{||}$ is seen to change periodically during the growth and the oscillation range reaches 2%. These periodic variations are of the same nature as the specular beam intensity oscillations. Hence, the reason for variations of these quantities is the same, namely, the periodic changes in the surface roughness of the growing film due to the nucleation, growth, and coalescence of 2D islands (nanocrystals). However, 2D lattice constant oscillations are half-period-shifted with respect to the corresponding specular beam oscillations. The maximum of the lattice constant oscillations coincides with the minimum of the intensity oscillations at half-monolayer coverage. Such behavior means that the elastic deformation of 2D islands is maximal at the maximal surface roughness, and the atomic unit cell has the maximal value. When each monolayer is completed, the islands coalesce to form a smooth surface and the 2D surface unit cell of the film reaches its minimum size. A decrease in $\boldsymbol{a}_{\|}$ of the smooth surface indicates that the maximal increase in the surface unit sell occurs at the edge of the islands; island coalescence results in a decrease of the in plane elastic deformation. The mean value of the 2D unit cell during growth of the wetting Ge layer increases progressively. The whole pattern of $\boldsymbol{a}_{\|}$ variations for various stages of Ge growth on the Si(100) surface is given elsewhere [17].Oscillations of $\boldsymbol{a}_{\|}$ parameter are detected over all ranges of temperature and growth rates when oscillations of the RHEED specular beam are observed. The behavior of the surface unit cell value at the growth temperature range 250 to 500°C remains the same as that shown in Fig.1. As the temperature is raised, the oscillation range decreases gradually, and becomes comparable to the noise level at 500°C. The growth of Ge on the Si(111) surface is also accompanied by oscillations of the 2D unit cell value. Oscillations in the RHEED specular intensity and in $\boldsymbol{a}_{\|}$ during Ge growth on Si(111) at the substrate temperature 370°C are shown in Fig.2. A characteristic feature of the growth on this surface is the variations of $\boldsymbol{a}_{\|}$ of the Ge film around zero value, *i.e.* the $\boldsymbol{a}_{\|}$ value of silicon, to approximately 1% at the maximal roughness and to approximately -1% at the minimal roughness. Another distinctive feature is a slight variation in $\boldsymbol{a}_{\|}$ at the thickness range corresponding to one monolayer. This behavior is characteristic of a wide temperature range and independent of the presence or absence of intensity oscillations at this thickness. It is known that the intensity variations corresponding to the first monolayer are observed at all growth temperatures, which is accounted for by the double-level growth mode. Islands varying size are thought to occur on the growth surface at this stage. A wide-island size range results in an averaging of $\boldsymbol{a}_{\|}$ and changes in this quantity are not observable.

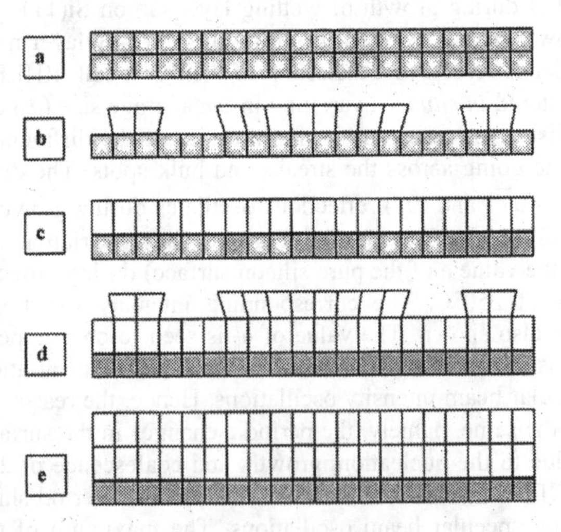

Figure 3. Schematic representation of the periodic variations of the 2D unit cells. Solid and open squares denote Si and Ge unit cell, respectively. See text for discussion.

A schematic explanation of the lattice constant variations in the Ge layer growing according to the 2D growth mode is shown in Fig.3. The cyclic nature of the roughness variations causes oscillations of the observed 2D unit cell value. The smooth surface of the continuous layer corresponds to the maximum in oscillations of the specular beam and to the minimum in oscillations of the 2D unit cell [Fig. 3(a,c,e)]. The presence of 2D islands at the growth surface results in a decrease in the specular beam intensity, and the lattice distortion at the island edges gives rise to the observed enlargement of the surface unit cell of the Ge film [Fig. 3(b.d.)]. These schematic submissions are substantiated with our STM experiments.

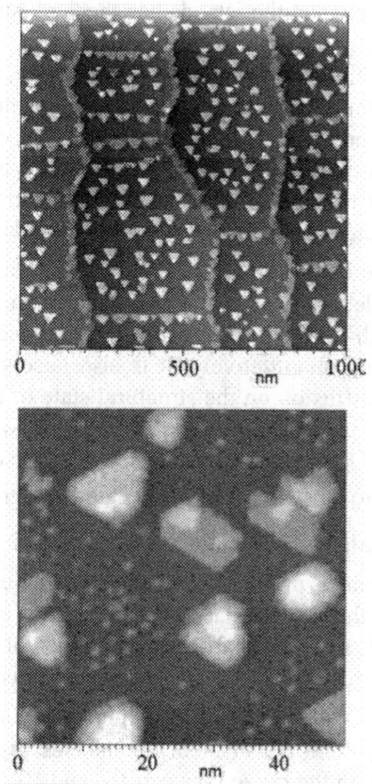

Figure 4. STM images of the Si(111)-7×7 surface after deposition of 0.5 monolayer Ge at 400°C with a rate of 1.5×10^{-3} ML/min (top). Vertical chains - islands at step edges, horizontal chains - islands at the domain boundaries; (bottom) Ge clusters in the 7×7 unit cells between islands after deposition of 0.4 monolayer Ge at 350°C with a rate of 6×10^{-3} ML/min.

3.2. GERMANIUM WETTING LAYER FORMATION ON Si(111): STM INVESTIGATIONS

At the initial stages of wetting layer formation the preferred centers of embedding of Ge atoms and of nucleation of Ge islands are different defects on Si(111) surface. Fig. 4 shows an STM image of Si(111) surface after the deposition of 0.5 BL of Ge at the temperature of 400°C with the rate of about 1.5×10^{-3} BL/min. Ge islands form at the step edges on Si surface and at the domain boundaries of the Si(111)-7×7 surface structure. They are visible as the upright and horizontal chains of triangles. The upright chains consist of the islands at step edges. The horizontal chains are of the islands at domain boundaries. Two-dimensional islands form also at flat parts of Si(111)-7×7 terraces beyond the step edges and domain boundaries. Similar observations during epitaxy of Ge on Si (111) were reported earlier [20]. Small clusters of Ge atoms form in the faulted halves of the Si (111)7x7 superstructure unit cells between large islands. Additional annealing of a surface at the growth temperature without Ge flux results in disappearance of these clusters, most probably because of the migration of Ge atoms to larger islands. Silicon surface between the islands has the 7×7 structure. We have previously considered the formation and thermostability of different superstructures on a Si(111) surface during growth of thin Ge films in Ref. [14]. Then, it was experimentally established, that the 5x5 superstructure characterizes a (111) continuous germanium pseudomorphic layer (compressed to 4%). After the breakdown of pseudomorphism, this superstructure transforms into the Ge(111)-c(2x8) superstructure that is characteristic of a clean Ge(111) surface. It is possible to make the supposition that, during the reduction of mechanical stresses, the transition from a 5x5 to 2x8 structure will happen through metastable superstructure (for example $\sqrt{3}\times\sqrt{3}$). In our case on the top surface of Ge nanoislands before coalescence some areas exist with reduced deformations and stresses. In Ref. [21] such a biaxial state of stress of strained small-area SiGe layers on Si substrate was estimated quantitatively. It is also necessary to take into the account possible influence of Si diffusion on the structural state of the Ge surface. However, all our experiments in were performed at temperatures below 400°C, when diffusion can practically be neglected. Observable Ge islands are mainly of triangular shape, and their edges are perpendicular to $[\bar{1}\bar{1}2]$ directions because the growth speed along this direction is slower than that along $[11\bar{2}]$ directions. The 7×7, 5×5 and $\sqrt{3}\times\sqrt{3}$ structures were observed on the surfaces of Ge islands (Fig. 5). At Ge coverage of 0.4 BL (growth rate of 2×10^{-3} BL/min) the 7×7 surface structure occupied from 60 to 70% of a total surface area of the islands surface regardless of their height (Fig. 5a). The 5×5 structure was observed on the surfaces of all three layers. It occupied about 30% of the total surface of Ge islands (Fig. 5b). The $\sqrt{3}\times\sqrt{3}$ structure was observed on separate islands in rare occasions (Fig. 5c). The densities of Ge islands grown at a rate of about 10^{-3} BL/min at 350-500°C are in the range between 10^9 and 10^{11} cm^{-2}. The maximum height of the islands was 3 BL. This is in agreement with the data reported in Ref. [6]. In Fig. 6 the total coverage in each of three layers is shown versus Ge deposition rate at the temperature of 400°C. At a given Ge coverage on Si(111) the relationship between the areas of different layers in islands depends on the rate of Ge deposition. At the rates of Ge deposition above 10^{-2} BL/min the growth rate of the first layer considerably exceeds that in the second and in the third layers. An epitaxial layer is filled linearly up to 100%, then the second layer starts to grow and so on. Under these conditions the formation of a wetting layer occurs through layer-by-layer growth mechanism. The differences between the growth rates of each of the three Ge layers diminish with decrease of Ge deposition rate. At Ge deposition rates of the order of 10^{-3} BL/min the values of the growth rates of all single layers in islands become almost equal. With increase of the total coverage all three layers in Ge islands are expanding in the (111) plane. The growth of successive layers proceeds simultaneously. At 400°C the transition from the layer-by-layer mode to the multilayer growth takes place, when the deposition rate changes from 10^{-2} BL/min to 10^{-3} BL/min [19]. At the coverage of about 1 BL the islands start to coalesce and form a continuous layer. The final continuous pseudomorphic layer has a thickness of 3 BL, and its surface has the 5×5 reconstruction. The dependence of the growth mode of Ge on Si(111) on the rate of Ge deposition is explained as follows. At relatively low temperatures, the density of the nucleation centers at domain boundaries is much greater than the density of surface defects on Si(111) surface.

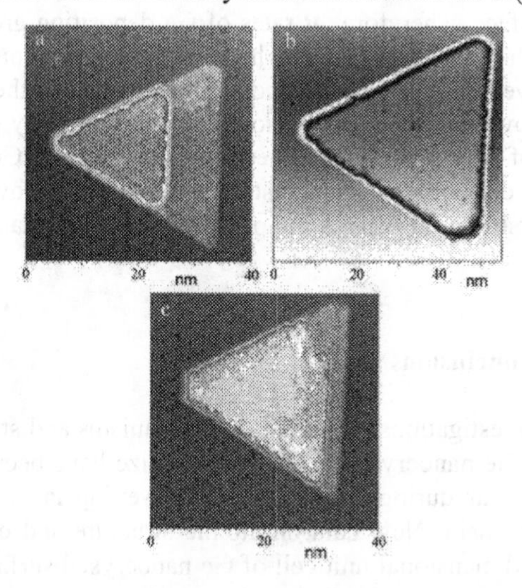

Figure 5. Surface structures of Ge islands grown with a rate of 2×10^{-3} BL /min at 400°C. Ge coverage is 0.4 BL. a) 7×7; b) 5×5; c) $\sqrt{3} \times \sqrt{3}$.

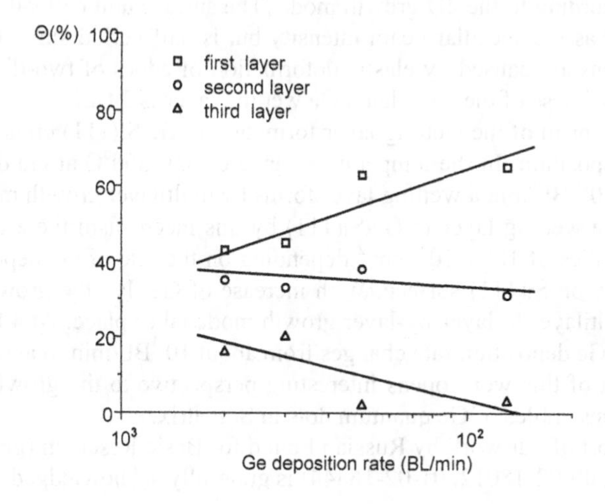

Figure 6. Total coverage Θ in Ge layers versus deposition rate, $T^{o} = 400$°C.

In our experiments the specific length of the boundaries of the surface structure domains on the surfaces of Ge islands was about thirty times greater than that on a Si(111) surface. This leads to a greater probability of the nucleation on the surfaces of Ge islands. Nucleation of Ge on the top of the first and of the second layer results in multilayer growth. At higher rates of Ge deposition the density of Ge islands increases. Therefore, the total length of the edges of Ge islands becomes large, and the density of the nucleation centers associated with step edges increases. The relative probability of the embedding of Ge atoms at the edges of Ge islands on Si surface becomes greater than that on Ge surface. Therefore, at rates of Ge deposition greater than 10^{-2} BL/min the growth rate of the first layer of Ge is high, and the formation of a wetting layer occurs through the layer-by-layer growth mode. For this reason the specular beam on the RHEED pattern shows oscillating behavior where the intensity maximum corresponds to the completion of successive monolayer. The first layer of Ge islands can represent large interest, as an ensemble of nanoclusters with extremely low sizes. The nucleation of clusters on superstructural cells is the possible way to formation of ordered systems of nanocrystals.

4. Summary and Conclusions

The results of investigations of formation mechanisms and structure as well as elastic deformations of Ge nanocrystals very small in size have been presented. Nanocrystals of this kind appear during formation of Ge wetting layer by a two-dimensional layer-by-layer mechanism. New data on atomic structure and evolution of elastic deformation of a two-dimensional unit cell of Ge nanocrystal surface are obtained during MBE. The formation and growth of several generations of ultra-small two-dimensional Ge nanocrystals were observed. The periodic variations in their surface unit cell were observed by reflection high-energy electron diffraction during the formation of a wetting layer according to the 2D growth mode. The surface unit cell value oscillates in the same manner as the specular beam intensity but is half period shifted. It is argued that these variations are caused by elastic deformation of edges of two-dimensional nanoislands. The thickness of the completed Ge wetting layer is 3 BL.

The mechanism of the wetting layer formation in Ge/Si(111) epitaxy depends on the rate of Ge deposition. In the temperature range of 350-500^{0}C at Ge deposition rates of the order of 10^{-3} BL/min a wetting layer forms by multilayer growth mechanism. During formation of a wetting layer in Ge/Si(111) by this mechanism the arrays of Ge islands with the densities of 10^{9} – 10^{11} cm^{-2} depending on the rate of Ge deposition can be obtained directly on Si(111) surface. With increase of Ge flux the growth rate the transition from multilayer to layer-by-layer growth mode takes place. At 400^{0}C the transition occurs when Ge deposition rate changes from about 10^{-3}BL/min to about 10^{-2} BL/min.

The result of this work opens interesting perspective to the growth of reproducible and ordered ensembles of Ge quantum dots in Si matrix.

The support of our work by Russian Found for Basic Research (grants 00-15-96806, 00-02-17638, 00-02-18012, 01-02-16844) is gratefully acknowledged.

References

1. Vescan, L., Stoica, T., Chretien, O., Goryll, M., Mateeva, E. and Muck A. (2000) Size distribution and electroluminescence of self-assembled Ge dots, *J. of Applied Physics*, 87, 7275-7282.
2. Bimberg, D., Heinrichsdorff, F., Ledentsov, N.N., Shchukin, V.A. (2000) Self-organized growth of semiconductor nanostructures for novel light emitters, *Applied Surface Science* **159–160**, 1–7.
3. Pchelyakov, O.P., Bolkhovityanov, Yu.B., Dvurechenskii, A.V., Nikiforov, A.I., Yakimov, A.I., Voigtlander, B. (2000) Molecular beam epitaxy of silicon-germanium nanostructures, *Thin Solid Films* **367**, 75-84.
4. Brunner, K. (2002) Si/Ge nanostructures, *Reports on Progress in Physics* **65,** 27-72.
5. Shklyaev, A.A., Shibata, M. and Ichikawa, M. (2000) High-density ultrasmall epitaxial Ge islands on Si(111) surfaces with a SiO_2 coverage, *Phisical Review B* **62**, 1540-1543.
6. Zhu, J.,Brunner, K.,Abstreiter, G.,Kienzle, O., Ernst, F. (1998) Lateral ordering of self-assembled Ge islands *Thin Solid Films,* **336,** 252-255
7. Goldfarb, I. and Briggs, G.A.D. (1999) Comparative STM and RHEED studies of Ge/Si(001) and Si/Ge/Si(001) surfaces, *Surface Science*, **433–435,** 449-454.
8. Pchelyakov, O.P., Markov, V.A., Nikiforov, A.I., Sokolov, L.V. (1997) Surface processes and phase diagrams in MBE growth of Si/Ge heterostuctures, *Thin Solid Films* **306,** 299-306.
9. Kubler, L., Dentel, D., Bischoff, J.L., Ghica, C., Ulhag-Bouillet, C., and Werckmann, J. (1998) Si adatom surface migration biasing by elastic strain gradients during capping of Ge or $Si_{1-x}Ge_x$ hut islands, *Appl.Phys.Lett.*, **73,** 1053-1055.
10. Ohtake, A., Ozeki, M., and Nakamura, J. (2000) Strain relaxation in InGa/GaAs(111)A heteroepitaxy, *Phys.Rev.Lett.*, **84,** 4665-4668.
11. Grandjian, N. and Massies, J. (1993) Epitaxial growth of highly strained InxGa1-xAs on GaAs(001) – The role of surface diffusion length, *J.Crystal Growth*, **134,** 51-62.
12. Turban, P., Hennet, L., Andrieu, S. (2000) In-plane lattice spacing oscillatory behavior during the two-dimensional hetero- and homoepitaxy of metals, *Surface Science.* **446,** 241-253
13. Vostokov, N.V., Dolgov, I.V., Drozdov, Yu.N. and Krasilnik, Z.F. (2000) The uniform nanoislands Ge on Si(001), *Izv. Acad. Nauk Fiz.* **64,** 284-287.
14. Sokolov, L.V., Stenin, S.I., Toropov, A.I., Pchelyakov, O.P. (1997) Superstructure transformation on a Ge surface during molecular beam epitaxy of Ge on Si(111), *Surface Investigation* **12,** 1151-1157.
15. Nikiforov, A.I., Markov, V.A., Cherepanov, V.A., Pchelyakov, O.P. (1998) The influence of growth temperature on the period of RHEED oscillations during MBE of Si and Ge on Si(111) surface, *Thin Solid Films* **336**, 183-187.
16. Nikiforov, A.I., Cherepanov, V.A., Pchelyakov, O.P., Dvurechenskii, A.V., Yakimov, A.I. (2000) In situ RHEED control of self-organized Ge quantum dots, *Thin Solid Films* **380**, 158-163.
17. Pchelyakov, O.P., Bolkhovityanov, Yu.B., Nikiforov, A.I., Olshanetsky, B.Z., Teys, S.A. and Voigtlander, B. (2002) Atomistic aspects of SiGe nanostructure formation by molecular-beam epitaxy, in M. Kotrla et al. (eds.), *Atomistic Aspects of Epitaxial Growth*, Kluver Academic Publishers, Dordrecht, pp. 371-381.
18. Markov, V.A., Cheng, H.H., Chih-ta Chia, Nikiforov, A.I., Cherepanov, V.A., Pchelyakov, O.P., Zhuravlev, K.S., Talochkin, A.B., McGlynn, E., Henry, M.O. (2000) RHEED studies of nucleation of Ge islands on Si(001) and optical properties of ultra-small Ge quantum dots, *Thin Solid Films* **369,** 79-83.
19. Teys, S. A. and Olshanetsky, B. Z. (2002) Formation of the wetting layer in Ge/Si(111) epitaxy at low growth rates studied with STM, *Physics of Low-Dimensional Structures* **1/2,** 37-46.
20. Koehler, U., Demuth, J.E., Hamers. R.J. (1989) Scanning tunneling microscopy study of low-temperature epitaxial growth of silicon on Si(111)-(7x7), *J. Vac.Sci.Technol.* **A7**, 2860-2867.
21. Fischer, A., Kuhne, H., Lippert, G., Richter, H. and Tillack, B. (1999) State of stress and critical thickness of strained small-area SiGe layers, *Phys. stat. sol. (a)* **171** 475-485.

STRAIN COMPENSATED SI/SIGE QUANTUM CASCADE EMITTERS GROWN ON SIGE PSEUDOSUBSTRATES

L. DIEHL, S. MENTESE, H. SIGG, E. MÜLLER, D.GRÜTZMACHER
Paul Scherrer Institut, CH-5232 Villigen, Switzerland
U. GENNSER, I. SAGNES
LPN-CNRS, F-91960 Marcoussis, France
T. FROMHERZ, J. STANGL, T.ROCH, G. BAUER
Institut für Halbleiterphysik,
Universität Linz, A-4040 Linz, Austria
Y. CAMPIDELLI, O. KERMARREC, D. BENSAHEL
STMicroelectronics, F-38926 Crolles-Cedex, France
J. FAIST
Université de Neuchâtel, CH-2000 Neuchâtel,
Switzerland

1. Introduction

Quantum cascade lasers (QCLs) based on III-V materials such as InGaAs/InAlAs [1] or GaAs/AlGaAs [2] have attracted an increasing attention since the first experimental demonstration in 1994. These convenient emitters have been successfully operated in the 3.4-24 μm range, and also recently in the terahertz range [3]. The concept of QCLs can also be applied to indirect bandgap materials since it relies on intersubband optical transitions for which the nature of the bandgap does not play any role. The Si/SiGe material system is particularly a candidate of interest, because of the possible compatibility with the well established Si integrated circuit processing.

Electroluminescence from p-type Si/SiGe cascade structures grown by MBE pseudomorphically on Si (100) substrates has been reported [4]. The non-radiative lifetime of the excited state participating to the optical transition was found to be comparable at its best to InGaAs/InAlAs based QC LEDs. However, further improvements of the devices are drastically limited by the large accumulated strain due to the 4.2 % mismatch existing between the lattice parameter of Si and Ge. The use of SiGe pseudo-substrates is a way of circumventing this drawback, since strain compensation can be achieved by alternating layers under compressive and tensile strain.

The results of preliminary experiments such as intersubband absorption performed on $Si/Si_{0.2}Ge_{0.8}$ modulation doped quantum wells (QWs) grown by molecular beam epitaxy (MBE) on $Si/Si_{0.5}Ge_{0.5}$ relaxed buffer layers gave an initial indication of the suitability of the use of relaxed buffer layers for the realization of a Si/SiGe QC laser [5]. Indeed well resolved transitions between heavy holes (HH) and light hole/split-off (LH-SO) states were observed up to room temperature as shown in Fig. 1.

L. Pavesi et al. (eds.), Towards the First Silicon Laser, 325–330.

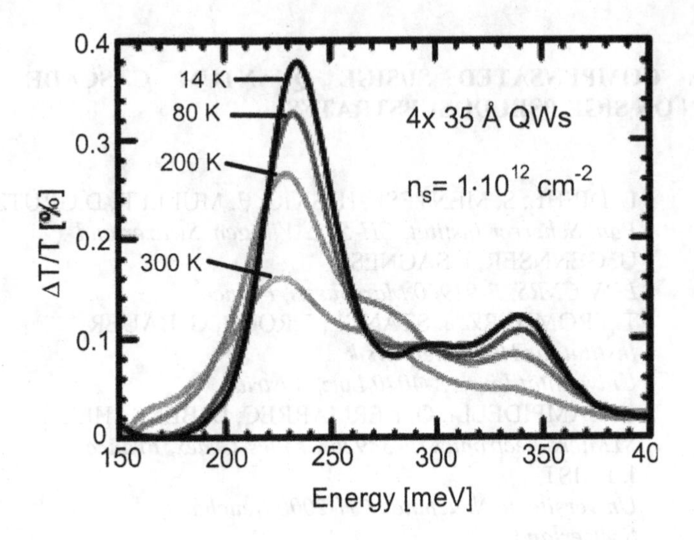

Figure 1 Transmission spectra obtained at different temperatures with 3.5 nm modulation doped QWs.

Unexpected Fano-like resonances, arising from the interplay between the light hole and split-off states together with the continuum, were also observed and well reproduced in simulations based on a 6 band k.p model. The HH-HH transition energy found for samples with different QWs width, shifts clearly as expected. This feature has been used to determine the different bandoffsets. In particular, the value of the HH discontinuity (565 meV) indicates the potential use of the SiGe material investigated for mid-infrared high-performance devices.

2. Experimental

Here we present intersubband electroluminescence (EL) resulting from optical transitions between HH states in strain compensated $Si/Si_{0.2}Ge_{0.8}$ cascade structures grown on $Si/Si_{0.5}Ge_{0.5}$ pseudosubstrate. The design of the active region follows the concept of the high-performance QCLs based on a "bound-to-continuum transition" developed in III-V material [6]. The two main features of this design are an efficient resonant tunneling injection into the upper state of the optical transition and a fast miniband extraction of the lower state. The latter is particularly of interest in Si/SiGe structures because the lifetime of the excited state of the laser transition is expected to be very short [7,8]. For this purpose, a miniband whose width is about 100 meV is created by strongly coupled HH states in the relaxation/injection region of our structure. This implies the growth of Si barriers as thin as 0.4 nm. A schematic diagram of the HH band, together with the relevant wavefunctions belonging to one active region and some of the neighboring QWs are displayed in Fig. 2. A detailed description of the layer sequence is given in the figure caption.

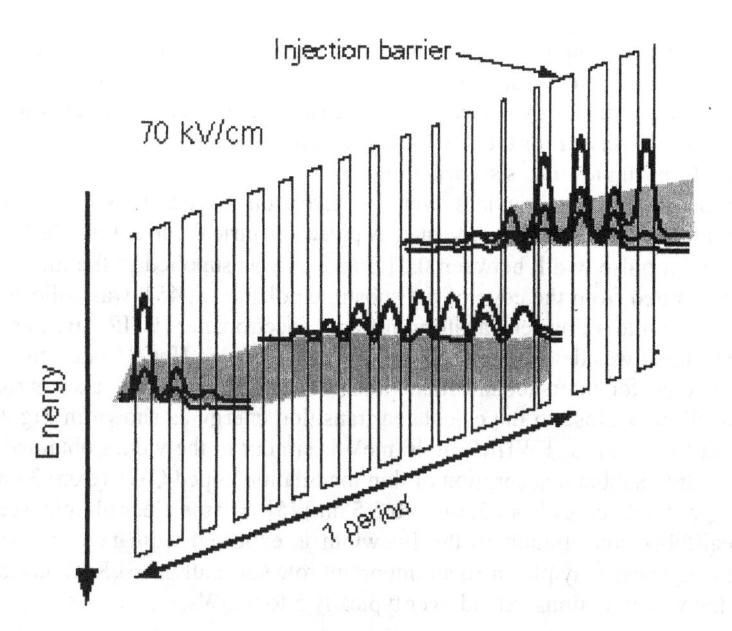

Figure 2 Schematic valence band diagram of one stage of the structure, under an applied electric field of 70 kV/cm. Only the HH band and the moduli squared of the relevant HH wave functions are shown for clarity. Note that the axis of the energy is turned upside down. Each period, starting from the injection barrier, consist of the following sequence of Si barrier (roman) and $Si_{0.2}Ge_{0.8}$ (bold) in Å: 25/**11**/4/**26**/5/**26**/6/**24**/7/**21**/8/**19**/9/**18**/10/**17**/<u>11</u>/<u>**15**</u>/<u>12</u>/<u>**15**</u>/<u>13</u>/<u>**14**</u>/<u>15</u>/**14**/16/**13**/17/**13**. The underlined numbers correspond to doped layers with a boron concentration of $5 \cdot 10^{17}$ cm^{-3}.

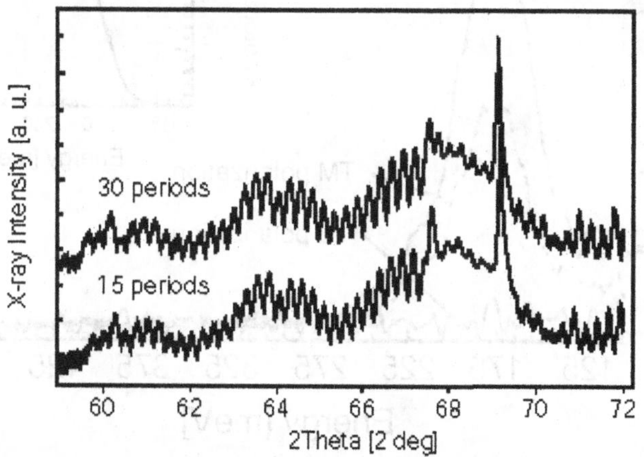

Figure 3 X-ray rocking curves measured along the 004 crystal axis, for the two samples with 15 and 30 repetitions.

Two samples, consisting of 15 and 30 periods, have been grown by MBE at a low temperature of 300°C to reduce the islanding in the Ge-rich layers and to avoid crystal relaxation. The surface of the SiGe buffer layer, deposited by low-pressure chemical vapor deposition, was chemically and mechanically polished. Transmission electron microscopy of the samples confirms the desired layer thickness, even of the thinnest (0.4 nm) Si barriers. No substantial decrease of the quality of the crystal is observed as the number of periods is increased. X-ray measurements were also performed and the same pattern is observed for the two structures, showing the excellent reproducibility of the growth from run to run, as shown on Fig 3.

After processing, the devices were soldered on copper bars, wire-bonded and mounted into a He-cooled flow cryostat. A pulsed electrical current with a frequency of 100 kHz and a pulse width between 0.01 and 5 μs was supplied to the mesa structures. The light, emitted from the edge of the substrate polished at 45°, was collected by f/0.8 optics and sent into a Fourier transform infrared spectrometer (FTIR). Behind the FTIR, the optical signal was detected with a liquid nitrogen cooled HgCdTe detector.

As expected for intersubband transitions, the emitted light reveals a strongly TM-polarized [9] peak close to the calculated transition energy as shown in Fig 4. The full width at half maximum (FWHM) of 46 meV is similar to the values obtained by investigating the intersubband absorption of thin modulation doped QWs (from 38 meV to 62 meV for QW thicknesses from 3.5 nm to 2.5 nm) [5]. For the electroluminescence peak, a non-negligible contribution to the linewidth is expected from non-parabolicity [8]. Interface roughness may play also an important role since all the Si/SiGe layers are very thin and the wavefunctions extend over typically 5 to 6 QWs.

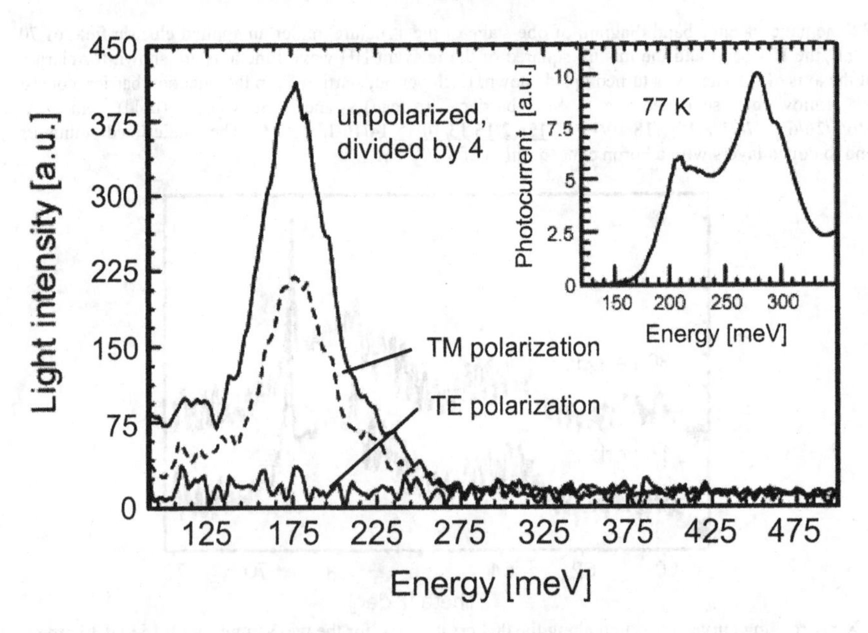

Fig. 4 Electroluminescence spectra of the sample with 15 repetitions, taken at 80K with and without a polarizer placed in the light path. The parameters are 4.7 V, 550mA, 94kHz and a duty cycle of 10%. The polarized electroluminescence is measured at 5.2V, 650mA and a 20% duty cycle. The inset shows the results of a photocurrent measurement at 77K.

Photocurrent measurements were also performed at 77K to further demonstrate the intersubband character of the electroluminescence peak (see inset of figure 4). Two features are observed and identified as the transition between the two HH states which are at the origin of the electroluminescence and as carriers excited to LH/SO states at energies just above the LH barrier potential.

On Fig. 5 is shown the current dependence of the voltage and the light intensity (referred as VI and LI curves respectively) at 80 K for the two structures investigated. In the VI curves, a clear onset is observed at 2.4 V and 5.7 V for respectively the device with 15 and 30 periods. These values scale reasonably well with the number of periods but are smaller than the calculated working biases (the conditions of Fig. 1) given by the multiplication of the length of a period (40.4 nm), the electrical field (70 kV/cm) and the number of periods. This is expected since a large current is supposed to flow already in the structure at the designed electric field. Moreover the levels in the injector tend to align already at low biases (about 40kV/cm) because of the large coupling between the different HH states, allowing carrier transport.

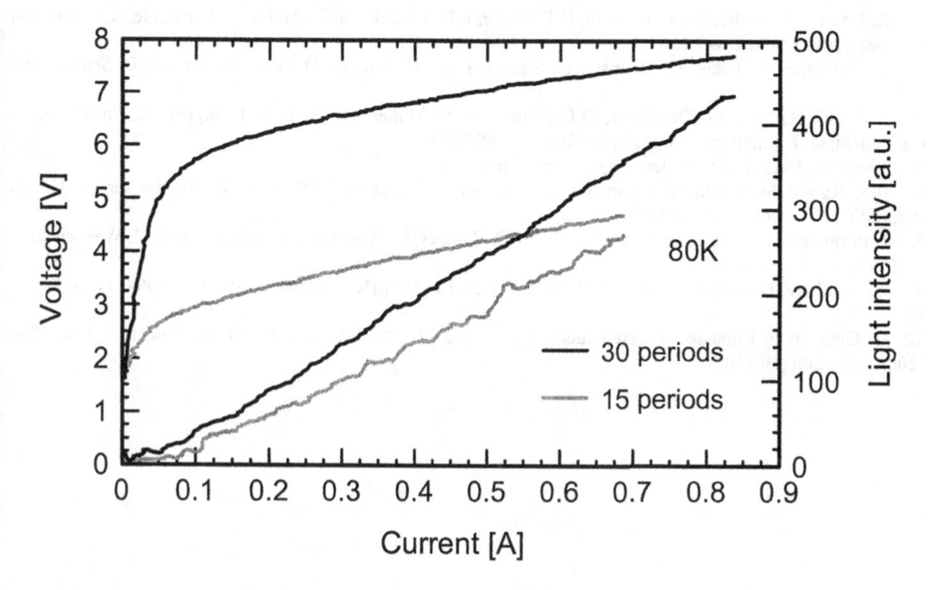

Figure 5 Voltage and light intensity vs current characteristics measured at 80K for a sample with 15 and 30 periods at 94 kHz and 10 % duty cycle.

The light intensity has a linear dependence with the drive current as shown in Fig. 5. The ratio between the slopes of the different LI curves should approach 2, since the light output power is proportional to the number of periods [10]. The factor found experimentally (1.4) deviates slightly from the expected integers, due to the standing wave pattern created at the semiconductor-metal interface on the top of the device.

3. Conclusions

In summary, electroluminescence from a p-type, strain compensated $Si/Si_{0.2}Ge_{0.8}$ cascade structure grown on $Si/Si_{0.5}Ge_{0.5}$ pseudosubstrate is demonstrated. The emitted light is fully TM polarized. A good agreement with photocurrent measurements confirmed the intersubband nature of the emitted light. The active region is based on a "bound-to-continuum transition" type of design. The successful growth of structures consisting of up to 30 periods has been confirmed by TEM images and X-ray measurements. Strain compensated structures are a solution to the strong limitations of both the number of periods and the complexity of the design encountered with the previous QC structures grown pseudomorphically on Si substrates.

References

1. J. Faist, F. Capasso, C. Sirtori, D.L. Sivco, A.L. Hutchinson and A.Y. Cho (1994) *Science* **264** 553.
2. C. Sirtori, P. Kruck, S. Barbieri, P. Collot, J. Nagle, M. Beck, J. Faist and U. Oesterle, (1998), *Appl. Phys. Lett.* **73**, 3486.
3. R. Köhler, A. Tredicucci, F. Beltram, H.E. Beere, E.H. Linfield, A.G. Davies, D.A. Ritchie, R.C. Iotti and F. Rossi, (2002) *Nature* **417** 156.
4. G. Dehlinger, L. Diehl, U. Gennser, H. Sigg, J. Faist, K. Ensslin, D. Grützmacher (2000) *Science* **290** 2277.
5. L. Diehl, H. Sigg, G. Dehlinger, D.Grützmacher, E. Müller, U. Gennser, I. Sagnes, T. Fromherz, Y. Campidelli, O. Kermarrec, (2002) *Appl. Phys. Lett.* **80** 3274.
6. J.Faist M. Beck and T. Aellen, (2001) *Appl. Phys. Lett.* **78** 147.
7. R.A. Kaindl, M. Wurm, K. Reinmann, M. Woerner, T. Elsaesser, C. Miesner, K. Brunner and G. Abstreiter, (2001) *Phys. Rev. Lett.* **86** 1122.
8. I. Bormann, K. Brunner, S. Hackenbuchner, G. Zandler, G. Abstreiter, S. Schmult and W. Wegscheider, (2002) *Appl. Phys. Lett.* **80** 2260.
9. T. Fromherz, E. Koppensteiner, M. Helm, G.Bauer, J.F. N\"{u}tzel and G. Abstreiter, (1994) *Phys. Rev. B* **50** 15073.
10. C. Gmachl, F. Capasso, A. Tredicucci, D.L. Sivco,J.N. Baillargon, A.L. Hutchinson, and A.Y. Cho, (2000) *Opt. Lett.* **50** 230.

TERAHERTZ SILICON LASERS
Intracentre optical pumping

S.G. PAVLOV, H.-W. HÜBERS and M.H. RÜMMELI
Institute of Space Sensor Technology and Planetary Exploration,
German Aerospace Centre,
Rutherfordstr. 2, 12489 Berlin, Germany
J.N. HOVENIER and T.O. KLAASSEN
Delft University of Technology,
P.O. Box 5046, 2600 GA Delft, The Netherlands
R.Kh. ZHUKAVIN, A.V. MURAVJOV and V.N. SHASTIN
Institute for Physics of Microstructures, Russian Academy of Sciences,
GSP-105, 603950 Nizhny Novgorod, Russia

1. Introduction

Silicon lasers based on the optical intracentre transitions of shallow impurities emit in the terahertz frequency range (1.2-6.6 THz or, alternatively, in wavelengths: 40-230 μm) [1]. The population inversion schemes of the lasers are realized under the conditions of optical pumping of the group V impurity centres at low Si lattice temperatures [2,3]. Two types of optical pumping have been already demonstrated successfully in order to get a laser emission from silicon. First one uses a mid-infrared CO_2 laser, which quantum energy (117-134 meV) is sufficient to photoionize carriers bound to the impurity centres from the localized ground states into the conduction band. Laser action under CO_2 laser pumping has been obtained from silicon doped by phosphor (Si:P) [4], bismuth (Si:Bi) [5,6], and antimony (Si:Sb) [7].

An alternative pump source for the silicon lasers is the free electron laser. It can tune the emission wavelength over a wide mid- and far-infrared wavelength ranges. This makes possible to realize an intracentre (from the ground state into the excited localized state) excitation of the impurities [8,9]. Although an infrared free electron laser is rather unique source, the obtained results showed the principal possibility of a more efficient way for optical pumping in comparison with the photoionization of the impurity centres. This pumping is much efficient than that for the impurity-to-band transitions. Beside this, the lower quantum energy of the pump photons allows to avoid additional losses of the laser emission inside the active medium caused by a transfer of the pump energy into non-radiative oscillations of the Si lattice (acoustic and optical phonons) and the creation of D^--centres (negatively-charged donor centres, in which two electrons are bound to a positively-charged ion [10]). These factors, as it has been proven experimentally, decrease the laser optical threshold for the Si lasers and enrich the laser emission spectra with new laser transitions.

In this work we describe experimental results on the intacentre pumping of the silicon lasers.

L. Pavesi et al. (eds.), Towards the First Silicon Laser, 331–340.

2. Intracentre Optical Pumping versus Photoionization

Schemes of population inversion for the THz bulk silicon lasers optically pumped by a CO_2 laser emission are described in detail in [1-7]. At low temperature (T < 10 K) electrons are bound to the donor neutral centres, D^0, having binding energies in the range of E_{ion}= 42-71 meV [11]. Irradiation of the Si sample by CO_2 laser emission excites the electrons high into the conduction band (Fig. 1a). The electrons lose their energy in the band by emitting optical and acoustic phonons. The following relaxation of the electrons occurs with emission of an acoustic phonon. Electrons gradually relax from the higher excited states to the lower closest impurity state. Accumulation of the electrons occurs on the $2p_0$ state in Si:P and Si:Sb and on the $2p_\pm$ state in Si:Bi, having the longest lifetimes of the other excited impurity states. Laser action occurs, therefore, on the transitions with the highest cross section between this particular level downwards to the 1s valley-orbit-split state(s) [1] (see Table 1).

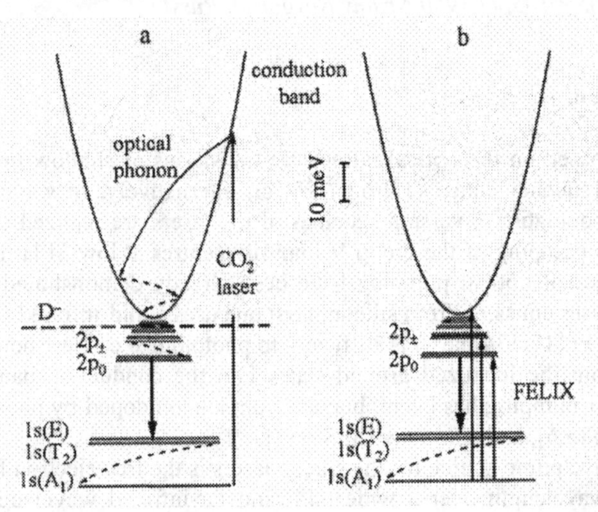

Figure 1. Schematics of the photoionization (a) and intracentre (b) optical pumping for the Si:P laser: straight arrows up indicate optical pump radiation, straight arrows down – THz emission, curved dash arrow are for electron relaxation due to emission of acoustic phonons and solid curved arrow – due to an optical phonon, correspondingly. $1s(A_1)$ is the phosphor donor ground state, D^- (dashed line) shows the level of D^--centres dynamically created after the capture of the relaxing electrons from the band bottom on the neutral donors.

Optical thresholds obtained for the Si:P and Si:Sb lasers are about 10-30 kW/cm^2 [4,6] when pumped by a 10P20 line (10.59 μm) of a CO_2 laser and for Si:Bi are about 200 kW/cm^2 [5] when pumped by a 9P20 line (9.55 μm).

Intracentre optical pumping (Fig. 1b), when the bound electrons are excited directly into the upper laser level, has several important advantages. First of all, intracentre optical transitions for n-type Si are about two orders of magnitude larger than that for the impurity-to-band transitions [12]. Pumping by a CO_2 laser into the band is followed by a non-radiative relaxation of the photoexcited electrons. This leads to heating of the Si crystal lattice. This may cause additional losses for the emitted THz light as well as affect the population inversion of the laser, depending on the acoustic phonon spectra. Additionally, the gradual relaxation of the electrons in the band and the higher excited impurity states leads to their capture to neutral donors with the creation of shallow D^--centres. These centres exhibit very efficient, broad-band ($(0.05\text{-}10)\times E_{ion}$) absorption [9] which causes additional THz light losses in the active media [4,5].

Thus, intracentre excitation of group-V donors in Si appears the most efficient optical pumping for these lasers.

TABLE I. Si laser transitions observed under photoionization and intracentre pumping

Laser	CO_2 laser pumping [1]		FELIX pumping	
	Pump line	Si emission line	Pump transition	Si emission line
Si:P	$1s(A_1)\rightarrow$band 9.3-10.6 μm	$2p_0\rightarrow 1s(T_2)$ 55.35 μm	$1s(A_1)\rightarrow 2p_\pm$ & higher 31.6 μm & shorter	$2p_0\rightarrow 1s(T_2)$ 55.35 μm
			$1s(A_1)\rightarrow 2p_0$ 36.4 μm	$2p_0\rightarrow 1s(E)$ 58.5 μm
Si:Bi	$1s(A_1)\rightarrow$band 10.6 μm	$2p_\pm\rightarrow 1s(T_2:\Gamma_7)$ 47.24 μm		
	$1s(A_1)\rightarrow$band 9.3-10.6 μm	$2p_\pm\rightarrow 1s(T_2:\Gamma_8)$ 48.57 μm	$1s(A_1)\rightarrow 2p_\pm$ & higher 19.9 μm & shorter	$2p_\pm\rightarrow 1s(T_2:\Gamma_8)$ 48.57 μm
		$2p_\pm\rightarrow 1s(E)$ 52.16 μm		$2p_\pm\rightarrow 1s(E)$ 52.16 μm
	$1s(A_1)\rightarrow$band 10.6 μm	$6p_\pm\rightarrow 2s$ 162.9 μm		
		$4p_\pm$ & $5p_0\rightarrow 2s$ 186.7 μm		
		$4p_0\rightarrow 2s$ 229.5 μm		
			$1s(A_1)\rightarrow 2p_0$ 20.8 μm	$2p_0\rightarrow 1s(E)$ 64.5 μm

3. Experimental

3.1. Si SAMPLES

A few Si samples, operating as lasers also under CO_2 laser excitation, were used in these experiments (see Table II).

TABLE II. Si laser samples

Sample	Sizes, mm^3	Net doping concentration, N_D-N_A, cm^{-3}	Compensation, N_A/N_D	Doping procedure
Si:P No. 1	5×7×1	2×10^{15}	35 %	neutron transmutation
Si:P No. 2	7×7×5	3×10^{15}	< 1 %	conventional
Si:Bi No. 1	7×6.5×5	1.2×10^{16}	~ 10 %	conventional
Si:Bi No. 2	7×6×5	8×10^{15}	~ 10 %	conventional

The Si:P sample (No.1) was cut from a neutron irradiated float zone grown Si crystal (see ref. [13]) in the form of a rectangular parallelepiped. The facets of the crystal were optically polished parallel to each other with accuracy of 1 arcmin, forming an internal reflection mode cavity. All other Si crystals were grown by the float zone technique with a simultaneous incorporation of the dopant from the melt. Then these samples were treated in the same manner.

3.2. FELIX AS A TUNABLE PUMP SOURCE FOR SI LASERS

The frequency tunable Free Electron Laser for Infrared Experiments (FELIX) at the FOM Institute in Rijnhuizen, the Netherlands, was used. The FELIX radiation consisted of 6 – 8 μs long trains of micropulses at a repetition rate of 5 Hz. The micropulses, with a 1 ns time interval, had a maximum peak power of about 0.5 MW, which corresponds to an average power of 2.5 kW during a macropulse. FELIX frequency scans covered the donor intracentre absorption bands of 25–36 μm for Si:P and 17–22 μm for Si:Bi. The scans were performed with a discrete wavelength step of 0.02 μm. FELIX wavelength bandwidth of the emitted pulse was ~ 0.5-1 % of the central emitted wavelength. The incident power was controlled by a calibrated step-attenuator and measured by a Molectron Energy Max 500 Joule meter in front of the cryostat with the sample.

3.3. SETUP

The crystals were cooled down to 5 K in a continuous flow cryostat or with a dipstick inserted into a liquid helium (LHe) transport vessel (Fig. 2). The cryostat was equipped with KRS-5 windows for the pump radiation and a cold sapphire and a room temperature polyethylene windows for the Si:P laser emission output. The dipstick had thin polyethylene windows. The THz emission from the Si crystals was registered by a liquid helium cooled Ge:Ga detector, with a maximum sensitivity in the wavelength range of 50-120 μm, protected by a cooled CaF_2 filter from leakage of the FELIX radiation. Emission spectra of the Si lasers were recorded by a Fourier transform interferometer (FTIR), equipped with another Ge:Ga photodetector, with a resolution ≈ 1 cm^{-1}.

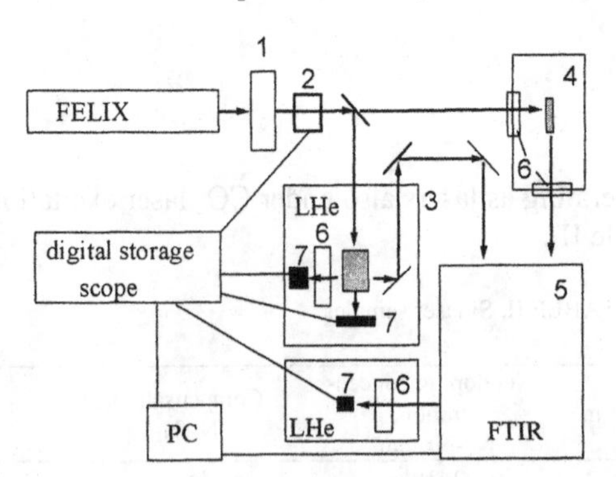

Figure 2. Layout of the experimental setup: (1) FELIX beam attenuator, (2) power meter, (3) dipstick with the Si sample (grey rectangle), (4) flow cryostat with the Si sample, (5) far-infrared Fourier interferometer, (6) mid- and far-infrared filter sets, (7) Ge:Ga photodetectors.

3.4. RESULTS

3.4.1. *Si laser action under intracentre optical pumping*

Stimulated emission in a pulsed mode (Fig. 3) was obtained from all investigated Si samples when the energy of the FELIX emission corresponded to the energy of the intracentre transitions between the ground impurity state, $1s(A_1)$, and the odd-parity excited states (Fig. 4).

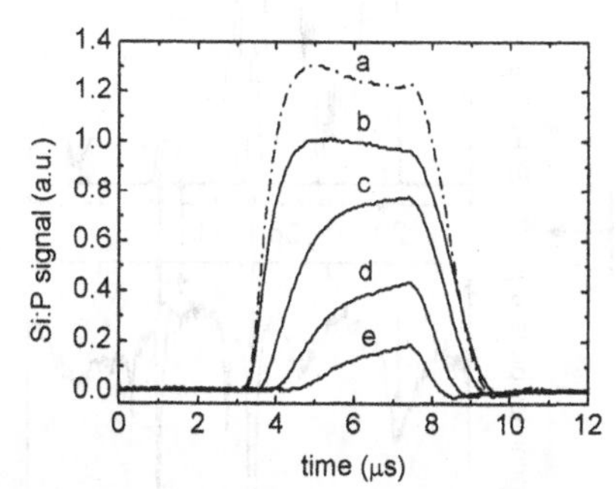

Figure 3. Typical FELIX macropulse (a, dot-dashed line), and Si:P laser emission pulses for different FELIX pump power attenuations: b) 0 dB (micropulse power ~ 160 kW/cm^2), c) 8 dB, d) 15 dB, e) 18 dB. The FELIX wavelength corresponds to direct excitation in the $2p_0$ state.

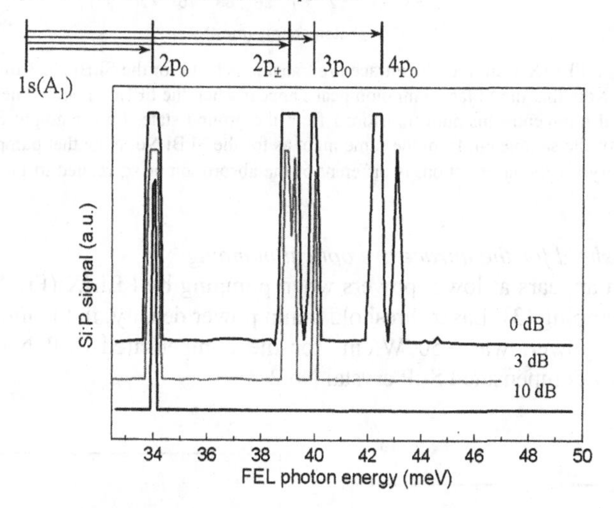

Figure 4. FELIX scan over the intracentre transitions band for the Si:P No.1 for different pump attenuations (in dB). Note that the Si laser emission peaks appear when the FELIX photon energy corresponds to the allowed optical intracentre phosphor transitions from the ground state, $1s(A_1)\rightarrow$odd-parity excited states of the donor.

In the case of the Si:Bi crystals, an additional emission line appeared when pumped into the $2p_0$ Bi state (Fig. 5). As the spectral analysis shows (see Fig. 8b below) this line corresponds to the transition from the $2p_0$ Bi state down to the 1s(E) valley-orbit-split state. This line was not previously observed when the same crystal was pumped by radiation from a CO_2 laser [1]. This lasing occurs in spite of the strong resonant-type interaction of the ground Bi state with the $2p_0$ state via an intervalley optical TO phonon [14], which leads to the fast de-excitation of the $2p_0$ state [2].

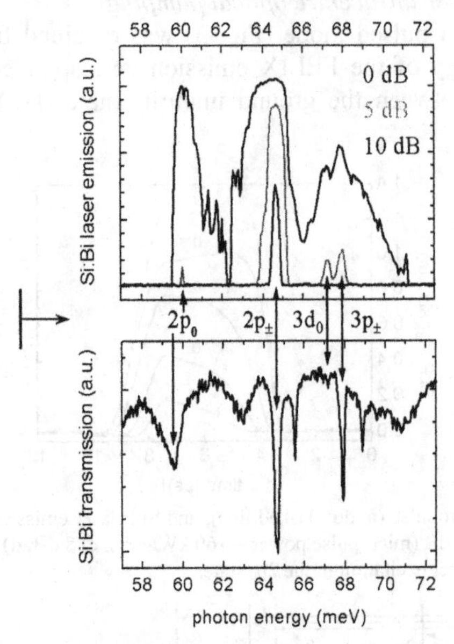

Figure 5. Upper graph: FELIX scan over the intracentre transitions band for the Si:Bi No.3 for different pump attenuations (in dB). Note that the Si laser emission peaks appear when the FELIX photon energy corresponds to the allowed optical intracentre bismuth transitions from the ground state. Lower graph: Si:Bi absorption spectrum, recorded for the sample cut from the same ingot as for the Si:Bi No. Note that pumping energy into the $2p_0$ state, necessary for the laser action, is different of the absorption peak, related to the $1s(A_1)\rightarrow 2p_0$ Bi transition.

3.4.2. *Laser threshold for the intracentre optical pumping*

Si laser emission appears at lower powers when pumping by FELIX (Fig. 6) compared to a CO_2 laser pumping [3]. Laser threshold pump power density, in the units of average over macropulse power, was ~50 W/cm^2 for the compensated Si:P No.1 and ~100 W/cm^2 for the low compensated Si:P crystal No.2.

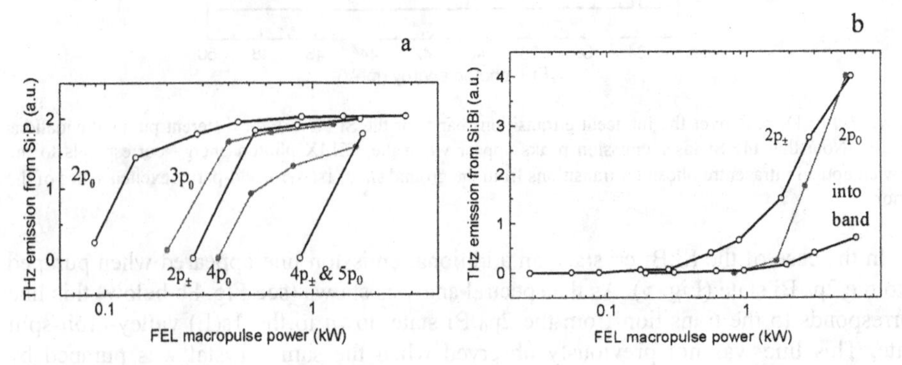

Figure 6. Dependences of the Si lasers outputs on the FELIX pump power when pumped into the Impurity states for the Si:P No.2 (a) and Si:Bi No.3 (b).

The lowest threshold, $\sim 9\times10^{21}$ photon/cm^2/s, is when the Si:P crystals are pumped by the FELIX photon with an energy of the $1s(A_1)\rightarrow 2p_0$ transition ($\approx$ 34.1 meV), i.e. into the upper Si:P laser state.

Unlike Si:P, stimulated emission from the Si:Bi crystals had a lower laser threshold when pumped into the $2p_\pm$ state (pump photon energy $\approx$ 64.6 meV) – the upper Si:Bi laser level in the schemes with a photoionization pumping [5]. Direct excitation into the $2p_0$ state of the Bi donor results also in the laser action with a pump threshold of $\sim 2\times10^{23}$ photon/cm^2/s (1.7 kW/cm^2). This is one order of magnitude higher than for the $2p_\pm$ Bi state.

3.4.3. *Laser emission spectra for the intracentre optical pumping*

Spectra of stimulated emission from the Si:P and Si:Bi crystals are strongly dependent on the FELIX pump frequency. The Si:P emission spectra obtained for different excitation wavelengths are shown in Fig 7. For all pump wavelengths, corresponding to the phosphor intracentre transitions, except into the $2p_0$ state, i.e. pumping photon energy > 38 meV, the stimulated emission has been observed on the $2p_0\rightarrow 1s(T_2)$ transition at 180 cm^{-1}. However, in a case of the direct pumping into the upper laser level, $2p_0$, the stimulated emission occurs on the $2p_0\rightarrow 1s(E)$ transition at 171 cm^{-1}.

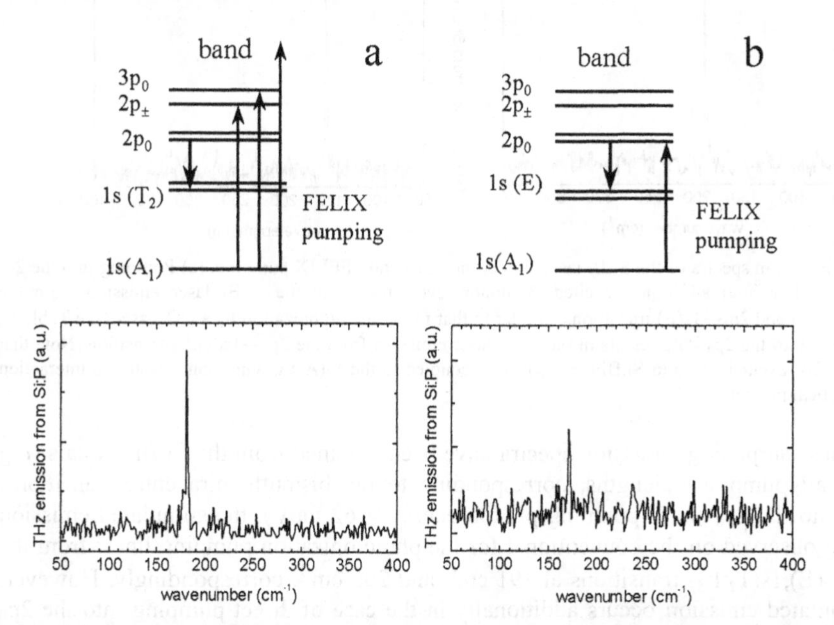

Figure 7. Emission spectra of the Si:P lasers under the intracentre FELIX pumping. (a) Pumping into the $2p_\pm$ states as well as into all higher excited P donor levels results in the Si:P laser emission from the $2p_0\rightarrow 1s(T_2)$ transition, similar to that for the photoionization (see Table I); (b) pumping into the upper Si:P laser level, $2p_0$, results in the Si:P laser emission from the $2p_0\rightarrow 1s(E)$ donor transition.

This peculiar behaviour might be related to the presence of two competing factors determining the relative gain of the $2p_0\rightarrow 1s(E)$ and of the $2p_0\rightarrow 1s(T_2)$ transitions. On the one hand, the optical cross section for the transition from the $2p_0$ state to the triplet

$1s(T_2)$ state is about 1.5 times larger than that to the doublet 1s(E) state [15]. On the other hand, the acoustic-phonon-assisted relaxation of carriers from the 1s(E) state is faster than that from the $1s(T_2)$ state because only inter-valley transitions are allowed for the $1s(T_2)$ state, whereas both the intra-valley and to the relaxation rate contribute to relaxation from the 1s(E) state [16]. In the case of direct pumping into the $2p_0$ state, apparently, stimulated emission from the $2p_0 \rightarrow 1s(E)$ transition dominates due to the lower population of the 1s(E) state resulting from a shorter lifetime [8].

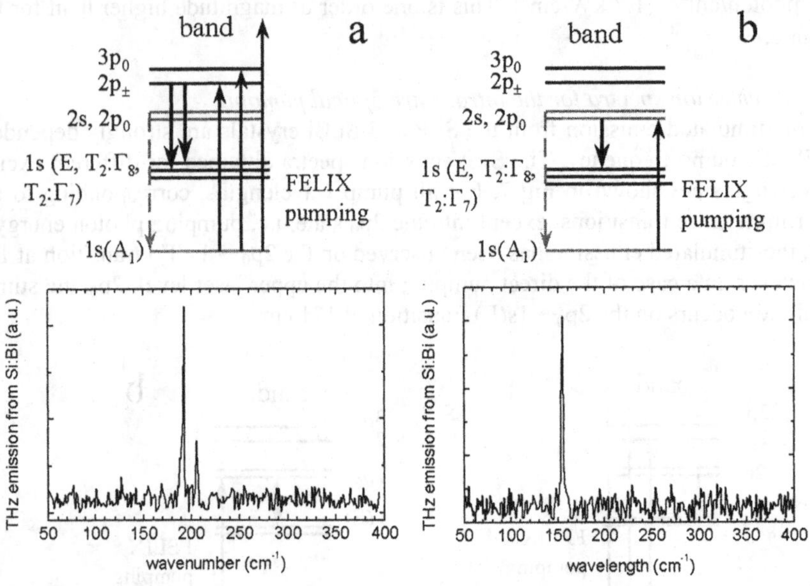

Figure 8. Emission spectra of the Si:Bi lasers under the intracentre FELIX pumping. (a) Pumping into the $2p_\pm$ states as well as into all higher excited Bi donor levels results in the Si:Bi laser emission from the $2p_\pm \rightarrow 1s(T_2{:}\Gamma_8)$ and $2p_\pm \rightarrow 1s(E)$ transitions, similar to that for the photoionization by a CO_2 laser (see Table I); (b) pumping into the $2p_0$ state results in the Si:Bi laser emission from the $2p_0 \rightarrow 1s(T_2{:}\Gamma_8)$transition. Note that the 2s and $2p_0$ excited states in Si:Bi are resonantly coupled to the $1s(A_1)$ ground donor state via interaction with an optical phonon [14].

Not less surprising, emission spectra have been obtained from the Si:Bi crystals (Fig 8). For all pump wavelengths, corresponding to the bismuth intracentre transitions, except into the $2p_0$ state (pumping photon energy > 62 meV), the stimulated emission has been observed on the conventional for the photoionization pumping lines: from the $2p_\pm \rightarrow$ 1s(E),$1s(T_2{:}\Gamma_8)$ transitions at 191 cm^{-1} and 206 cm^{-1}, correspondingly. However, the stimulated emission occurs additionally in the case of direct pumping into the $2p_0$ state on the $2p_0 \rightarrow$ 1s(E) transitions. This is in contradiction with the fact, that the $2p_0$ state is supposed to be effectively depopulated due to the resonant interaction with an optical phonon [2]. Moreover, this occurs even with a strong site-band Si lattice absorption of the FELIX pump light at the optical phonon frequency. This emission line has a relatively higher pump threshold, but still yields a significant emission signal at 155 cm^{-1} on the $2p_0 \rightarrow$ 1s(E) transition. We suppose that this emission can be achieved in the Si:Bi medium under the resonant intracentre pumping by such a powerful laser as

FELIX. The FELIX micropulse peak powers at these experiments were in the order of several MW. The corresponding photon flux density is sufficient to support the extremely fast pumping rate in the medium, up to ~ 10^{13} c^{-1}. Even considering the very short pumping time, equal to a single micropulse duration of 6-10 ps at full width on half maximum, these values might create the necessary population inversions and small gain, exceeding the absorption in the Si:Bi medium and in the laser resonator.

This explanation of the phenomenon is supported by the results of temporal measurements of the Si laser output pulse, made by using a fast Ge:Ga detector. The Si laser emission had a strong amplitude modulation with a frequency of 1 GHz, corresponding to the 1 ns intervals between the FELIX micropulses. This indicates the achievement of the laser mode when each pump micropulse creates a Si emission output. Therefore, this might support the idea about the high micropulse pump efficiency, achievable via intracentre pumping of the donors in bulk Si.

4. Summary

Intracentre optical pumping of shallow donor centres in bulk Si results in the laser action at THz frequencies. Optical thresholds of the stimulated emission are about two orders of magnitude lower than those of pumping in the conduction band. This corresponds roughly to the ratio of the cross sections of the photoionization and intracentre photoexcitation processes. Therefore it seems feasible to achieve continuous wave operation for the Si lasers under intracentre pumping. Additional laser frequencies from the Si:P and Si:Bi samples manifest another laser mechanisms compared to pumping in the conduction band.

This work was partly supported by the Deutsche Forschungsgemeinschaft and the Russian Foundation for Basic Research (RFBR) (joint grant 436 RUS 113/206/0 (R) and 00-02-04010), RFBR grants 02-02-16790. S.G. Pavlov gratefully acknowledges support through an Alexander von Humboldt Stiftung. R.Kh. Zhukavin thanks the Deutscher Akademischer Austauschdienst. Authors thank H. Riemann, M. Greiner-Bär, E.E. Orlova and A.F.G. van der Meer for technical support of the experiments and fruitful discussions of results.

References

1. Pavlov, S.G., Hübers, H.-W., Orlova, E.E., Zhukavin, R.Kh., Riemann, H., Nakata, H., and Shastin, V. N. (2003) Optically pumped terahertz semiconductor bulk lasers, *Physica Status Solidi (b)*, in press.
2. Orlova, E.E., Zhukavin, R.Kh., Pavlov, S.G., and Shastin, V.N. (1998) Far infrared active media based on the shallow impurity states transitions in silicon, *Physica Status Solidi (b)* 210, 859-863.
3. Hübers, H.-W., Pavlov, S.G., Rümmeli, M.H., Zhukavin, R.Kh., Orlova, E.E., Riemann, H., and Shastin, V.N. (2001) Terahertz emission from silicon doped by shallow impurities, *Physica B* **308-310**, 232-235.
4. Pavlov, S.G., Zhukavin, R.Kh., Orlova, E.E., Shastin, V.N., Kirsanov, A.V., Hübers, H.-W., Auen, K., and Riemann, H. (2000) Stimulated emission from donor transitions in silicon, *Phys. Rev. Lett.* **84**, 5220-5223.
5. Pavlov, S.G., Hübers, H.-W., Rümmeli, M.H., Zhukavin, R.Kh., Orlova, E.E., Shastin, V.N., and Riemann, H. (2002) Far-infrared stimulated emission from optically excited bismuth donors in silicon, *Appl. Phys. Lett.* **80**, 4717-4719.
6. Hübers, H.-W., Pavlov, S.G., Greiner-Bär, M., Rümmeli, M.H., Kimmitt, M.F., Zhukavin, R.Kh., Rie-

mann, H., and Shastin V.N. (2002) Terahertz Emission Spectra of Optically Pumped Silicon Lasers, *Physica Status Solidi (b)* **233**, 191–196.

7. Pavlov, S.G., Hübers, H.-W., Riemann, H., Zhukavin, R.Kh., Orlova, E.E., and Shastin, V.N. (2002) Terahertz optically pumped Si:Sb laser, *J. Appl. Phys.*, in press, scheduled for the issue on 1st Dec. 2002.
8. Shastin, V.N., Zhukavin, R.Kh., Orlova, E.E., Pavlov, S.G., Rümmeli, M.H., Hübers, H.-W., Hovenier, J.N., Klaassen, T.O., Riemann, H., Bradley, I.V., and van der Meer, A.F.G. (2002) Stimulated THz emission from group V donors in silicon under intracenter photoexcitation, *Appl. Phys. Lett.* **80**, 3512-3514.
9. Klaassen, T.O., Hovenier, J.N., Zhukavin, R.Kh., Gaponova, D.M., Muravjov, A.V., Orlova, E.E., Shastin, V.N., Pavlov, S.G., Hübers, H.-W., Riemann, H., and van der Meer, A.F.G. (2002) The emission spectra of optically pumped Si-based THz lasers, in J.M. Chamberlain et al. (eds) *Proc. of the 2002 IEEE Tenth Int. Conf. on Terahertz Electronics*, IEEE Cat. No. 02EX621, pp. 89-92.
10. Gershenzon, E.M., Mel'nikov, A.P., and Rabinovich, R.I. (1985) H-like impurity centers, molecular complexes and electron delocalization in semiconductors, in A.L. Efros and M. Pollak (eds.) *Electron-Electron Interactions in Disordered Systems*, Elsevier Science Publishers, Amsterdam, pp. 483-554.
11. Ramdas, A.K., and Rodriguez, S. (1981) Spectroscopy of the solid-state analogues of the hydrogen atom: donors and acceptors in semiconductors, *Reports on Progress in Physics* **44**, 1297-1387.
12. Clauws, P., Broeckx, J., Rostaert, E., and Vennik, J. (1988) Oscillator strengths of shallow impurity spectra in germanium and silicon, *Phys. Rev. B* **38**, 12377-12382.
13. Gres'kov, I.M., Smirnov, B.V., Sobolev, S.P., Stuk, A.A., and Kharchenko, V.A. (1978) Influence of growth defects on the electrical properties of radiation-doped silicon, *Soviet Physics - Semiconductors* **12**, 1118-1120.
14. Onton, A., Fisher, P., and Ramdas, A.K. (1967) Anomalous width of some photoexcitation lines of impurities in silicon, *Phys. Rev. Lett.* **19**, 781-783.
15. Mayur, A. J., Dean Sciacca, M., Ramdas, A. K., and Rodriguez, S. (1993) Redetermination of the valley-orbit (chemical) splitting of the 1s ground state of group-V donors in silicon, *Phys. Rev. B* **48**, 10893-10898.
16. Griffin, A., and Carruthers, P. (1963) Thermal Conductivity of Solids IV: Resonance Fluorescence Scattering of Phonons by Donor Electrons in Germanium, *Phys. Rev.* **131**, 1976–1995.

SILICON LASERS BASED ON SHALLOW DONOR CENTRES

Theoretical background and experimental results

V.N. SHASTIN, E.E. ORLOVA and R.Kh. ZHUKAVIN
Institute for Physics of Microstructures, Russian Academy of Sciences, GSP-105, 603950 Nizhny Novgorod, Russia
S.G. PAVLOV and H.-W. HÜBERS
Institute of Space Sensor Technology and Planetary Exploration, German Aerospace Centre, Rutherfordstr. 2, 12489 Berlin, Germany
H. RIEMANN
Institute of Crystal Growth, Max-Born-str. 2, 12489 Berlin, Germany

1. Introduction

The paper is devoted to a new class of optically pumped solid state lasers, operating in the THz frequency range, which are based on shallow donor intracentre transitions in crystalline silicon. Si-based semiconductors and semiconductor heterostructures are promising media for THz laser engineering due to the highly developed technology of silicon growth and doping, low level of lattice absorption in THz range and different ways to obtain the population inversion of charged carriers. Historically there were several attempts to make a THz laser using intraband optical transitions in silicon. In 1979 the idea to use transitions between light (l) and heavy (h) holes subbands in crossed electric and magnetic ($E\perp H$) fields was proposed [1]. Since then, hot hole Si-lasing is under discussions. Monte-Carlo calculations predicted small signal gain of 0.1 cm^{-1} on intersubband l-h transitions in $E\perp H$ fields and in the frequency range of 50-230 cm^{-1} [2]. Later it was shown that inversion in Landau levels can also provide amplification of THz radiation at the light hole cyclotron resonance (frequency of 26 cm^{-1} at H=3T) [3, 4]. In this case the required population inversion and non-equidistance of Landau levels are provided by the hybridization of light and heavy subband states. The above mentioned laser mechanism was successfully achieved in p-type doped Ge in the frequency range of 50-140 cm^{-1} [5, 6]. However, similar experimental investigations in Si have the problem of the acceptor freezing at liquid helium temperature and were not successful. Nevertheless, positive results concerning hot hole gain in $E\perp H$ fields in silicon have been obtained recently [4]. The idea was to use optically excited neutral shallow donors (D^0) for THz lasing which was proposed in 1996 [7, 8]. It was reported that both acoustic phonon (Si:P) and optical phonon (Si:Bi) assisted relaxation of non-equilibrium carriers under the photoionization of donors by CO_2 laser radiation leads to population inversion and the amplification on intracentre transitions for moderate level

L. Pavesi et al. (eds.), Towards the First Silicon Laser, 341–350.

of doping ($N_D \approx 10^{15}$ cm^{-3}) and low lattice temperatures ($T \leq$ 30-60 K). Later on, a theoretical model was refined [9, 10] which takes into account the multivalley structure of the conduction band. It was shown that the intervalley phonon assisted nonradiative transitions are very important for quantitative analysis. Up to now, THz lasing has been realized and laser transitions have been unambiguously identified for phosphorus-doped (Si:P) [11-13], bismuth-doped (Si:Bi) [13, 14] and antimony-doped (Si:Sb) [15] Si. In this article recent theoretical calculations and experimental investigations concerning THz lasing of group V shallow donor centres under their photoionization by CO_2 laser radiation are reviewed and discussed.

TABLE 1. Energy levels in meV of group V and lithium donors in silicon [16].

Level	P	As	Sb	Bi	Li	Theory
$1s(A_1)$	45.59	53.76	42.74	70.98	31.24	31.27
$1s(E)$	32.58	31.26	30.47			31.27
$1s(E+T_2)$					33.02	31.27
			32.89	32.89		
$1s(T_2)$	33.89	32.67				31.27
			32.91	31.89		
$2p_0$	11.48	11.50	11.51	11.44	11.51	11.51
$2s$		9.11		8.78		8.83
$2p_\pm$	6.40	6.40	6.38	6.37	6.40	6.40
$3p_0$	5.47	5.49	5.50	5.48	5.49	5.49
$3s$				4.70		4.75
$3d_0$	3.83	3.8		3.80		3.75
$4p_0$	3.31	3.31	3.33	3.30	3.32	3.33
$3p_\pm$	3.12	3.12	3.12	3.12	3.12	3.12
$4s$				2.89		2.85
$4f_0$	2.33			2.36		2.33
$4p_\pm$, $5p_0$	2.19	2.19	2.20	2.18	2.20	2.19-2.23
$4f_\pm$	1.90	1.90	1.94	1.91	1.90	1.89
$5f_0$	1.65		1.71	1.67	1.64	1.62
$5p_\pm$	1.46	1.46	1.48	1.46	1.47	1.44
$5f_\pm$	1.26				1.25	1.27

2. Theoretical background

2.1. LASER STATES

Shallow donor states in silicon are originating from the six equivalent conduction band valleys (see review [16]). Bound impurity states except of the lowest *1s* state can be considered using a single valley effective mass theory. In this frame, every state of the impurity exhibits at least a six-fold degeneracy originating from the different conduction band minima. However for the *1s* ground state the effective-mass theory is inadequate and the degeneracy is lifted by the crystal cell potential. The overlapping of the *1s* eigenfunctions belonging to each valley leads to intervalley interaction (hybridization of states) and as a result to valley orbit splitting of the *1s* state. The effect depends on the chemical nature of the impurity atom and is called the "chemical splitting". The binding energies of the bound states of group V donors (arsenic As, antimony Sb, phosphorus P, bismuth Bi) as well as lithium Li donor are presented in TABLE 1. One can see that the *1s* multiplet resolves into a triplet $1s(T_2)$, a doublet $1s(E)$ and a singlet $1s(A_1)$.

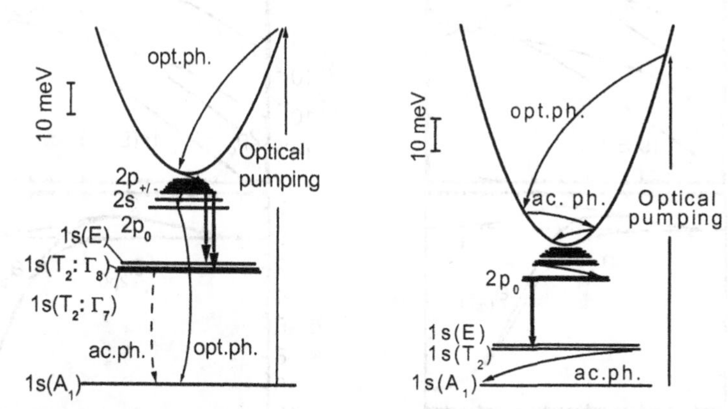

Figure 1. The energy level scheme and possible radiative and nonradiative transitions in Si:P (left) and Si:Bi (right) under optical pumping.

Moreover for the substitutional donors the $1s(A_1)$ state is the ground state. It will be shown that donor lasing arises from the $2p_0 \rightarrow \{1s(E),\ 1s(T_2)\}$ allowed optical transitions. Consequently the "chemical splitting" determines the laser states and is a major factor in the formation of the population inversion and the amplification of the donor transitions.

2.2. POPULATION INVERSION AND AMPLIFICATION

There are two mechanisms of population inversion of donor states in Si under photoexcitation. The first is connected with the low temperature intracentre acoustic phonon assisted relaxation. It is based on the accumulation of charged carriers in the long-living $2p_0$ state of P, Sb, As and perhaps Li neutral donor centres. For *n*-type Si, active optical phonons have energies of 63 meV and 59 meV, which are larger than the binding energies of P, Sb, As and Li impurity states (see TABLE 1). Therefore for these donors and at low temperature ($T< 30\ K$), the electron-phonon interaction mediated by optical lattice vibrations is negligible and the population of the donor states under the optical excitation is controlled by acoustic phonon emission (*Fig. 1*). The matrix elements of such processes decrease with increasing the energy gap ΔE between corresponding levels, provided that $qa>1$, where *a* is an effective radius of the orbit and *q* is the wavevector of the phonon required for the nonradiative transition. Note, that for the long-wavelength acoustic phonons $\hbar qs = \Delta E$, where *s* is the sound velocity. At the condition $qa>1$ the phonon-assisted transitions are ended outside the *q*-space where the wavefunctions of these states are mainly localized. Consequently, the rate of the intravalley relaxation is suppressed with increasing qa parameter. Thus, the step-by-step acoustic phonon relaxation is slowing down and transitions between adjacent levels predominate at least for the lower ($n \leq 3$) bound states.

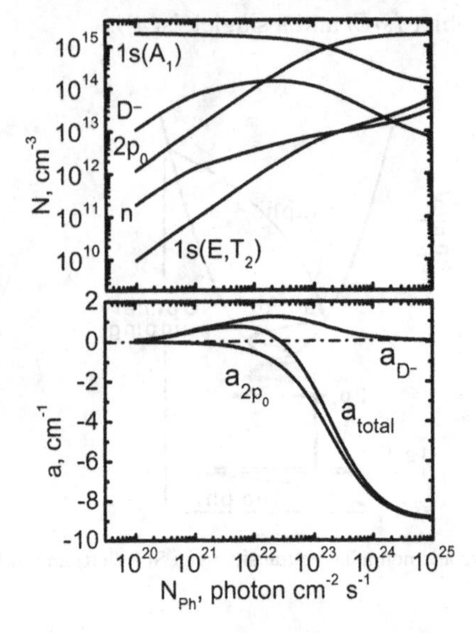

Figure 2. The populations of donor states (upper graph) and absorption/amplification coefficient (lower graph) of Si:P calculated in the frame of single valley approximation for 2×10^{15} cm^{-3} and low compensation (0.002) versus 10.6 μm photon flux density N_{Ph}.

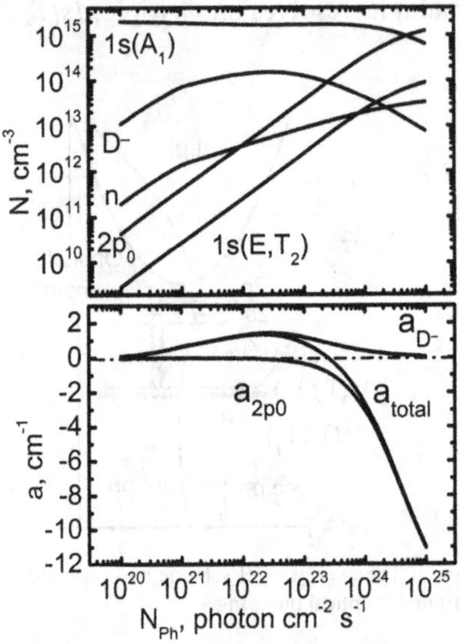

Figure 3. The populations of donor states (upper graph) and absorption/amplification coefficient (lower graph) of Si:P calculated in the frame of intervalley approximation for 2×10^{15} cm^{-3} and low compensation (0.002) versus 10.6 μm photon flux density N_{Ph}.

Moreover, the lifetime of the $2p_0$ state is the longest. An estimation gives a lifetime of 1.5×10^{-8} s. Therefore, the majority of excited carriers passes through the $2p_0$ state before relaxing to the ground state. Accumulation is possible. In comparison the lifetime of the $1s(E)$ state is much shorter (2×10^{-10}s) because of the smaller energy gap between this state and $1s(A_1)$ ground state. Thus a four-level laser scheme (*Figure 1*) with thresholdless population inversion can be realized on the $2p_0 \rightarrow 1s(E)$ transitions. However, a trapping of carriers on the $1s(T_2)$ state is expected within single valley approximation but it is ruled out on the basis of group theory analysis [17], which shows that the acoustic phonon assisted intravalley relaxation is forbidden for this state. Accumulation of carriers on the $1s(T_2)$ states terminates the inversion population and the laser effect. Fortunately, the intervalley relaxation from the $1s(T_2)$ state is allowed and makes the lifetime of the $1s(T_2)$ state comparable with the lifetime of the $1s(E)$ state. Besides this, the intervalley scattering decreases the lifetime of the $2p_0$ state to 10^{-9}s [10]. Thus, it can be concluded that both inter- and intravalley acoustic phonon assisted transitions are important and have to be taken into account in the theoretical model. The results of the calculation of the involved states populations, free carrier concentration n and the small signal gain within the single valley approximation and with the account of the intervalley phonon assisted transitions for P donors are presented in *Fig. 2* and *3* for comparison.

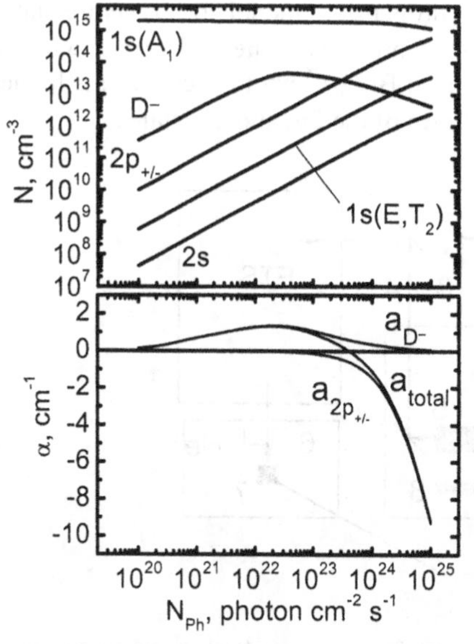

Figure 4. The $2p_0$ (solid) and D^- centre (dash) population (upper graph) and absorption/amplification coefficient (lower graph) for different levels of compensation versus 10.6 μm photon flux density N_{Ph}.

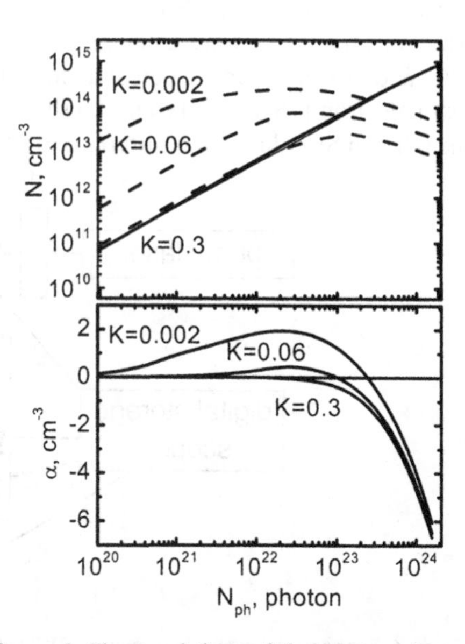

Figure 5. The populations of donor states (upper graph) and absorption/amplification coefficient (lower graph) of Si:Bi calculated within the intravalley approximation for 3×10^{15}cm^{-3} and compensation 0.1 versus 10.6 μm photon flux density N_{Ph}.

The absorption of the THz radiation by the negatively charged donor centres (D^-) [18] created by a CO_2 laser photoionization has also been taken into account. It is shown that the typical threshold flux density for laser action is 10^{23} photons/cm^{-2}s^{-1} for uncompensated samples. The optimal doping level, $10^{15}\div 3\times10^{15}$ cm^{-3}, is determined by two factors: maximum of the active centres and, minimum of impurity concentration broadening of the linewidth for the intracentre laser transition. It should be emphasised that the compensation level ($K=N_d/N_a$, where N_d and N_a is the donor and acceptor concentration correspondingly) influences the gain due the absorption changes caused by the D^- centres (*Fig*). In contrast to Si:P, in Si:Bi the inversion is caused by resonant interaction with intervalley optical phonons (*Fig. 1*). The $2p_0$ and $2s$ states in Si:Bi are coupled to the $1s(A_1)$ state via optical phonon emission [19] and have a very short lifetime, of about 1 ps. Due to this coupling, the majority of the optically excited electrons relaxes directly to the ground state and, therefore, does not reach the $1s(E)$ and $1s(T_2)$ states. As a result, the population of the $2s$ and $2p_0$ states as well as $1s(E)$ and $1s(T_2)$ states is relatively low. The lifetime of the $2p_\pm$ state (10^{-10} s) is not longer than those of $1s(E)$, $1s(T_2)$ states (10^{-9} s), controlled by intravalley acoustic phonon assisted transitions. Nevertheless, due to the fact that the probability for excited carrier to reach the $2p_\pm$ state (0.5) is much higher then that of $1s(E)$, $1s(T_2)$ states (~10^3) a population inversion is formed between the $2p_\pm \rightarrow 1s(E)$, $1s(T_2)$ as well as in both $2p_\pm \rightarrow 2p_0$ and $2p_\pm \rightarrow$

$2s$ transitions. Thus a four-level scheme can be realized. A larger gain is expected on transitions from the $2p_{\pm}$ to the $1s(E)$ and $1s(T_2)$ states (*Fig. 1*). Calculations of the level populations and gain made by using probability technique are presented on the *Figure 5*. The transition rates were estimated within an hydrogen-like model for D^0 centre states [20] and the zero-radius potential model for D^- centre states (neutral centre with an extra electron [18]). The lasing threshold value in Si:Bi (10^{24} photons cm^{-2} s^{-1}) is higher than that in Si:P due to the difference of the lifetimes of the $2p_{\pm}$ and $2p_0$ states.

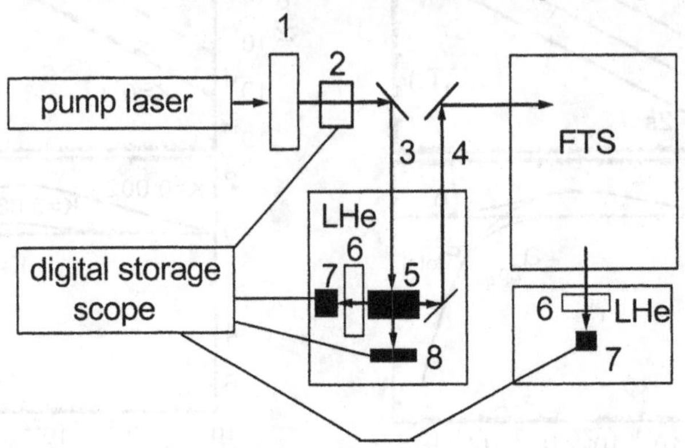

Figure 6. Experimental setup: 1) attenuator, 2)CO_2 drag detector, 3) pump beam; 4) silicon emission beam; 5) silicon sample, 6) THz filter, 7, 8) Ge detector.

3. Experiment

3.1. SAMPLES

Low compensated (K< 0.01) Si:P, Si:Sb and Si:Bi samples were grown by the float zone method with simultaneous incorporation of the doping elements from the melt. The dominant donor concentrations in these samples are in the range of $(0.1\div12)\times10^{15}$ cm^{-3}. Additionally, Si:P crystals with $K \leq 0.3$ were prepared by neutron transmutation doping [21]. Samples with different doping concentrations were cut in form of rectangular parallelepipeds (typical crystal dimensions are 7×7×5 mm^3) from the Si ingot and then polished to provide a high-Q resonator on internal reflection modes. The Si samples with the low doping concentration (ca. 10^{14} cm^{-3}) were used to identify the dominant impurity as well as the concentration of the incorporated electrically active centres by absorption spectroscopy.

3.2. EXPERIMENTAL SETUP

The samples were mounted in a holder, which was immersed in a liquid helium (LHe) vessel (*Fig. 6*). A grating tunable TEA CO_2 laser with a peak output power of 1.6 MW in the wavelength range 9.2÷10.7 μm was used as the pump source. The

THz emission from the optically pumped samples was registered by a LHe cooled Ge:Ga photodetector inside the same vessel. To prevent irradiation of the detector by the CO_2 laser, 1 mm thick sapphire filters were placed in front of the detectors. The pulses were recorded by a 500 MHz - bandwidth digital storage scope. For the spectral measurements the THz emission was guided by a stainless steel lightpipe into a Fourier transform spectrometre (FTS) and focused onto another LHe cooled Ge:Ga detector inside a separate cryostat. The spectrometre had a step-scan control, allowing averaging the emission signal over a defined number of pulses.

3.3. EXPERIMENTAL RESULTS

Spontaneous emission was detected with the Ge:Ga detector for various pump wavelengths from the silicon doped by phosphorus, antimony and bismuth. A linear dependence of the spontaneous emission signals on the pump photon flux density up to 6×10^{23} $cm^{-2}s^{-1}$ was found.

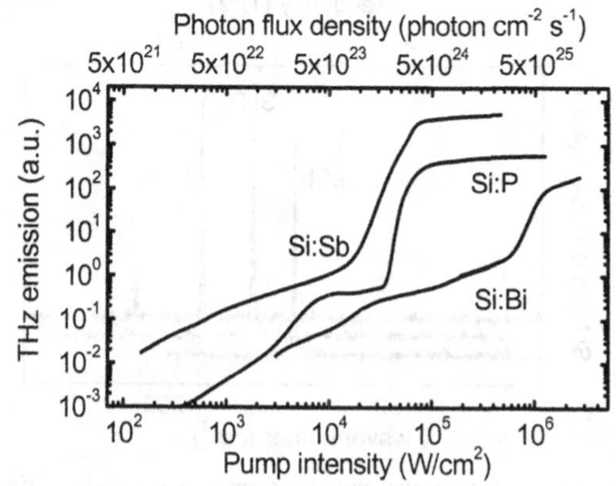

Figure 7. The dependency of THz emission from silicon doped by different impurities.

All Si:P samples doped higher than 5×10^{14} cm^{-3} showed spontaneous emission when pumped by a CO_2 laser . The spontaneous emission increases with increasing doping concentration. For these samples stimulated emission was observed for doping concentration in the range $(0.8\div5)\times10^{15}$ cm^{-3}. The neutron transmutation doped samples, which are heavily compensated, have some significant differences. First, the spontaneous emission is significantly higher than for the uncompensated samples for the same pump intensity. Second, the heavily compensated samples have lower laser thresholds. The CO_2 laser pump intensity, necessary to exceed the Si:P laser threshold for the 7 mm long sample from the best material, was about of 30 kW/cm^2 [11-13] at 10.6 μm line of the CO_2 laser (*Fig. 7*). The THz emission pulse from the Si:P laser started together with the pump pulse and had a duration of 70÷100 ns, comparable with the full width at half maximum of the pump laser pulse (*Fig. 8*). The spectrum of the stimulated emission from the Si:P sample was measured by the FTS with a resolution of 0.2 cm^{-1} (*Fig. 9*). A line at 54.1 μm was recorded, which corresponds to the $2p_0\rightarrow1s(T_2)$ intracentre transition.

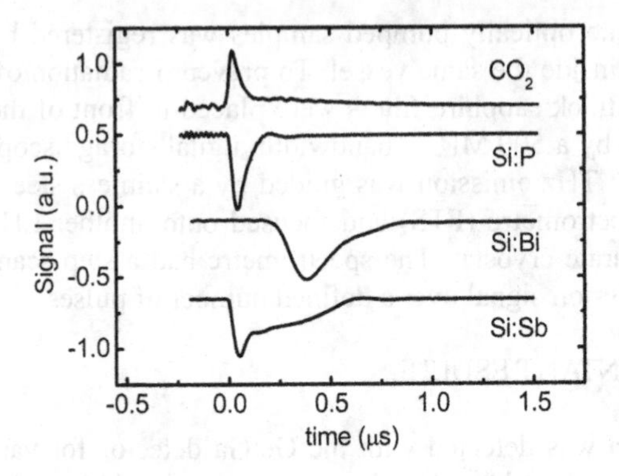

Figure 8. The stimulated emission pulses from silicon doped by different impurities (negative signals) and CO_2 laser pulse (positive signal).

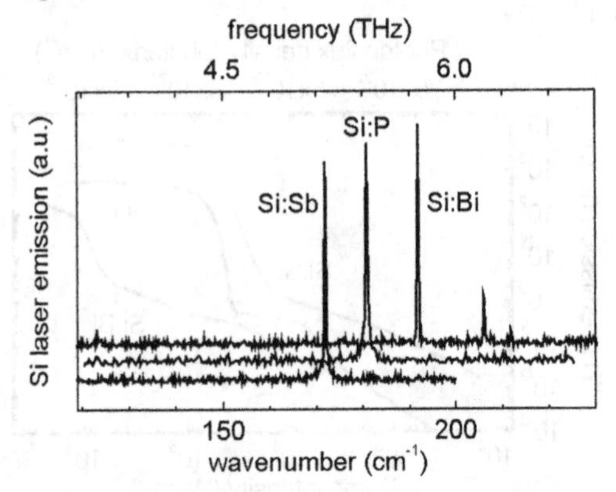

Figure 9. The spectra of stimulated emission from silicon doped by different impurities.

Stimulated emission from Si:Sb has been observed for pump photon flux density higher than 10^{24} photons cm^{-2} s^{-1} (20 kW cm^{-2}) for the 9.6 μm pump line (*Fig. 7*). Pumping by a line from the 10 μm band of CO_2 laser emission band requires a factor of 1.5÷2 higher photon flux density to reach the laser threshold. This difference is caused by the different lattice absorption of the pump emission: 0.3 cm^{-1} and 1.1 cm^{-1} for 9.6 μm and 10.6 μm pump lines [15]. The pulse shapes of the pump laser and the THz Si:Sb laser are reported in *Fig. 8*. The Si:Sb stimulated emission spectrum was measured by the FTS with a resolution of 0.2 cm^{-1} (*Fig 9*). The spectrum consists of a single line at 171.8 cm^{-1} (5.15 THz or 58.2 μm). The interesting feature of the Si:Sb in comparison with the Si:P laser is that it ends on the $1s(T_2 : \Gamma_8)$ quadruplet state. The spin-orbit coupling splits the $1s(T_2)$ state into the doublet $1s(T_2 : \Gamma_7)$ and quadruplet $1s(T_2 : \Gamma_8)$, separated with 0.3 meV energy spacing, since Sb has a higher atomic number than that of P impurity centres. In the Si:Bi samples a stimulated emission effect was observed for doping concentration in the range of N_{Bi}=(6÷12)×10^{15} cm^{-3}. The CO_2 laser pump threshold intensity for Si:Bi lasing was 100÷300 kW×cm^{-2} at the pump wavelength of 9.6 μm (photon flux density of (5÷15)×10^{25} quantum×cm^{-2}×s^{-1}) [13, 14]. Pumping by a line from the 10 μm band of the CO_2 laser required a factor ~2 higher power and exhibits a lower output signal and a longer delay than the lasing pulse. The stimulated emission pulse appears always with a time delay of 50÷200 ns after the peak of the pump laser pulse (*Fig. 8*), while the spontaneous emission has no such a delay for all pump frequencies. Two strong emission lines, having a relatively low laser thresholds, and corresponding the $2p_{\pm} \rightarrow \{1s(E),\ 1s(T_2{:}\Gamma_8)\}$ transitions were registered in the Si:Bi spectra under the pumping by of the CO_2 laser (*Fig. 9*).

4. Summary and Outlook

We reported on experimental and theoretical evidences of THz lasing in optically excited neutral donor centres embedded in crystalline silicon. The physical principles are clear and the theoretical calculations, which take into account both the intravalley and intervalley phonon-assisted captured carrier relaxation, are in good qualitative agreement with experimental data, particularly for the Si:P laser. At the same time the experiment with Si:Bi revealed an unexpected delay (3×10^{-7} s) in the temporal behavior of the stimulated emission. One can suppose that the influence of the nonequilibrium phonons on the populations of the working states in Si:Bi appears due to a bottleneck effect in the decay of *TA* (≈ 20 meV) acoustic phonons. Additionally, it should be pointed out that the lifetimes of the $1s(E)$, $1s(T_2)$ states have been estimated with a simplified hydrogen-like model. This has to be defined more exactly. Also it is important to consider reabsorption on the $2p_0 \rightarrow 1s(A_1)$ transition of the emitted *TO* intervalley optical phonons by Bi centres before they decay into the acoustic phonons. The latter can increase the life time of the $2p_0$ state. All these assumptions need further investigation. For applications frequency tunability as well as CW operation of Si donor lasers should be realized. On this way seems to be reasonable to use the effect of the uniaxial stress on Si donor states. The frequency coverage might be extended also by using other dopants such as As and Li. Si lasers covering the 5÷6 THz region with a tunability of about 1% are expected to be feasible.

This work was partly supported by the Deutsche Forschungsgemeinschaft and the Russian Foundation for Basic Research (RFBR) (joint grant 436 RUS 113/206/0 (R) and the grant of the Russian Foundation for Basic Research (02-02-16790). R.Kh. Zhukavin thanks the Deutscher Akademischer Austauschdienst (DAAD). S.G. Pavlov gratefully acknowledges support through the Alexander-von-Humboldt Stiftung.

References

1. Andronov, A.A., Kozlov, V.A., Mazov, L.S., Shastin, V.N. (1979) Amplification of Far-Infrared Radiation in Germanium during the Population Inversion of Hot Holes *JETP.Lett.* **30**, 551-555.
2. Mazov, L.S., Nefedov, I.M. (1986) Numerical Computation of Holes Population Inversion and FIR Amplification in Silicon, in A.A. Andronov (eds*) Submillimeter Wave Lasers in Semiconductors Using Hot Holes,* IAP AS ,Gorky, pp. 153-166.
3. Muravjov, A.V., Strijbos, R.C., Wenckebach, W.Th., and Shastin, V.N. (1996) Amplification of Far-Infrared Radiation on Light Hole Cyclotron Resonance in Silicon in Crossed Electric and Magnetic

Fields, in M. von Ortenberg and H.-U. Müller (eds) *Proceeding of 21th Int. Conf. on Infrared and Millimeter Waves*, Berlin, CTh11.

4. Muravjov, A.V., Strijbos, R.C., Wenckebach, W.Th., and Shastin, V.N. (1998) Population Inversion of Landau Levels in the Valence Band in Crossed Electric and Magnetic Fields, *Phys.Stat.Sol.(b)* **205**, 575-585.
5. Andronov, A.A., Zverev, I.V., Kozlov, V.A., Nozdrin, Yu.N., Pavlov, S.A., Shastin, V.N. (1984) Stimulated Emission in the Long-Wavelength IR Region from Hot Holes in Ge in Crossed Electric and Magnetic Fields, *JETP.Lett.* **40**, 804-806.
6. Ivanov, Y.L. (1991) Generation of Cyclotron Radiation by Light Holes in Germanium, *Opt.Quant.Elect.* **23**, S253-S265.
   Mitygin, Yu.A., Murzin, V.N., Stoklitsky, S.A., Chebotarev, A.P. (1991) Wide-Range Tunable Sub-Millimeter Cyclotron Resonance laser, *Opt.Quant.Elect.* **23**, S307-S311.
7. Shastin V.N. (1996) Far-Infrared Active Media Based on Inraband and Shallow Impurity States Transitions in Si, in M. von Ortenberg and H.-U. Müller (eds) *Proceeding of 21th Int. Conf. on Infrared and Millimeter Waves*, Berlin, CT2.
8. Orlova E.E., Shastin V.N. (1996) Inverse Population of Bismuth Donor Excited States and FIR Amplification in Silicon Under of Optical in Si, in M. von Ortenberg and H.-U. Müller (eds) *Proceeding of 21th Int. Conf. on Infrared and Millimeter Waves*, Berlin, CTh4.
9. Orlova, E.E. (2002) Nonequilibrium Population of Shallow Impurity States in Semiconductors and Amplification of Far-Infrared Radiation, *Ph. D. Thesis*, Institute For Physics of Microstructures RAS, Nizhny Novgorod, 128 Pages.
10. Orlova, E.E. (2002) Longliving shallow donor states in silicon-life time calculation, in J.H. Davies and A.R. Long (eds) *Proceeding of 26 international conference on the physics of semiconductors*, 29 july-2 august, Cambridge, **3**, p. 123.
11. Pavlov, S.G., Zhukavin, R.Kh., Orlova, E.E., Shastin, V.N., Kirsanov, A.V., Hübers, H.-W., Auen, K., and Riemann, H. (2000) Stimulated emission from donor transitions in silicon, *Phys. Rev. Lett.* **84**, 5220-5223.
12. Klaassen, T.O., Hovenier, J.N., Zhukavin, R.Kh., Gaponova, D.M., Muravjov, A.V., Orlova, E.E., Shastin, V.N., Pavlov, S.G., Hübers, H.-W., Riemann, H., and van der Meer, A.F.G. (2002) The emission spectra of optically pumped Si-based THz lasers, in M. Chamberlain et. al (eds) *Proceeding of 2002 IEEE Tenth International Conference on Terahertz Electronics*, pp. 89-90.
13. Hübers, H.-W., Pavlov, S.G., Greiner-Bär, M., Rümmeli, M.H., Kimmitt, M.F., Zhukavin, R.Kh., Riemann, H., and Shastin, V.N. (2002) Terahertz emission spectra of optically pumped silicon lasers, *Physica Status Solidi (b)*, **233**, 191-196.
14. Pavlov, S.G., Hübers, H.-W., Rümmeli, M.H., Zhukavin, R.Kh., Orlova, E.E., Shastin, V.N., and Riemann, H. (2002) Far-infrared stimulated emission from optically excited bismuth donors in silicon, *Appl. Phys. Lett.* **80**, 4717-4719.
15. Pavlov, S.G., Hübers, H.-W., Riemann, H., Zhukavin, R.Kh., Orlova, E.E., and Shastin, V.N. (2002) Terahertz optically pumped Si:Sb laser, *J. Appl. Phys.*, in press, scheduled for the issue on 1st Dec. 2002.
16. Ramdas, A.K. and Rodriguez, S. (1981) Spectroscopy of the solid-state analogues of the hydrogen atom: donors and acceptors in semiconductors, *Reports on Progress in Physics* **44**, 1297-1387.
17. Castner, T. G. (1963) Raman spin-lattice relaxation of shallow donors in silicon, *Phys. Rev.* **130**, 58-75.
18. Gershenzon, E.M., Mel'nikov, A.P., Rabinovich, R.I. (1985) H^- like impurity centers, molecular complexes and electron delocalization in semiconductors, in A.L. Efros and M. Pollak (eds.) *Electron-Electron Interactions in Disordered Systems*, Elsevier Science Publishers, Amsterdam, pp. 483-554.
19. Battler, N.R., Fisher, P., and Ramdas, A.K. (1975) Excitation spectrum of bismuth donors in silicon, *Phys. Rev. B* **12**, 3200-3209.
20. V.N.Abakumov, V.I. Perel, and I.N.Yassievich. (1991) *Nonradiative Recombination in Semiconductors*, North-Holland Publ. Co., Oxford.
21. Gres'kov, I.M., Smirnov, B.V., Sobolev, S.P., Stuk, A.A., and Kharchenko, V.A. (1978) Influence of growth defects on the electrical properties of radiation-doped silicon, *Soviet Physics - Semiconductors* **12**, 1118-1120.

RESONANT STATES IN MODULATION DOPED SIGE HETEROSTRUCTURES AS A SOURCE OF THZ LASING

A.A. PROKOFIEV, M.A. ODNOBLYUDOV, I.N. YASSIEVICH
A.F. Ioffe Physico-Technical Institute
RAS Politechnicheskaya 26, St. Petersburg, 194021 Russia

1. Introduction

Recent advances in semiconductor lasers considerably extended microelectronic device potentiality. In particular, significant improvements have been made in output power, stability, and tuneability of different solid state lasers with multiple quantum wells for optical and infrared ranges. A good example is the quantum cascade laser (QCL), which currently can be operated at wavelengths in excess of 10 μm [1]. The lowest frequency achieved now is 24 THz [2]. Recently, first attempts were made to fabricate QCL based on SiGe structures for the THz range (1-10 THz) [3]. However, with the increase of the emission wavelength it becomes more and more difficult to fulfil the various conditions for lasing: the energy band structure, waveguide for radiation propagation in the device, strong phonon absorption, and many other effects which deteriorate laser performance. An alternative type of THz laser source which could utilize a much simpler quantum well (QW) structure is a resonant state laser (RSL), where Ge or another semiconductor with similar band structure is employed. Recently, a tuneable continuous-wave Ge THz laser was realized under the condition of weak external electric field [4].

Since thin layers of binary alloys like, $Si_{1-x}Ge_x$, are strained internally due to the lattice mismatch, acceptor levels are split and stimulated emission can be obtained just applying an electric field. Acceptor-doped SiGe/Si heterostructures are very attractive for fabricating RSL because of their good thermal properties, of the low lattice absorption in the THz region, of the well established and relatively cheap technology, and of possible integration with Si-based electronics. Thus RSL utilizing boron delta-doped SiGe QW structures has been suggested as a THz source as soon as p-Ge lasers have been developed [5,6]. The first single QW RSL has been demonstrated in 2001 [6].

In this paper we present a detailed theoretical study of acceptor resonant states in boron doped $Si/Si_{1-x}Ge_x/Si$ structures as well as results of computer simulation of possible structures for multi quantum well resonant state lasers.

L. Pavesi et al. (eds.), Towards the First Silicon Laser, 351–358.

2. Mechanism of Resonant States Lasing

The main feature of RSL is that the population inversion as well as optical transitions takes place between the energy levels of a single impurity. The scheme of p-Ge RSL operation is shown in Fig. 1.

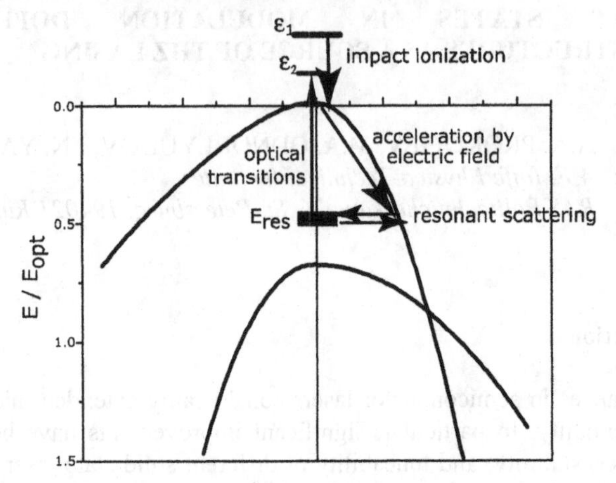

Figure 1. Strain-split valence band and holes transitions in p-Ge RSL

Population inversion in RSL is realized among the shallow acceptor states, split under external stress. If the strain is high enough, the split-off acceptor state overlaps the light-hole branch of the valence band and creates a resonant state [7]. Electric field depopulates the acceptor ground state due to impact ionisation. In the post breakdown regime, the population of the localized states of an acceptor does not exceed 1 % [8]. Free holes are accelerated towards the energy of resonant state, which acts as a trap for carriers due to a capture-emission exchange between the resonant state and the continuum. A population inversion of the resonant state with respect to the impurity states in the gap is then formed [9,10], and THz lasing occurs. The frequency range covered by the RSL is determined by the threshold stress necessary for creating the resonant state, and the maximum useful stress at which optical phonon-assisted hole transitions to the valence band depopulate the resonant state. For Ga-doped Ge, the operation frequency can be set to any value between 2 and 10 THz by simply adjusting the stress applied to the sample.

We try to apply the theory developed for p-Ge laser to RSL. Built-in electric field modifies the band structure of single QW significantly. Due to band inclination split-off acceptor states, having higher binding energy in SiGe than in Ge, overlap the continuum without being exposed to any external action like stress or electric field. Here one can also see the possibility to change frequency applying additional external electric field, i.e. even more simple than for the strained Ge. In order to achieve more intensive radiation one has to design the MQW structure like for optical frequency range lasers. Such structures are the easiest ones for applying the strained Ge RSL theory.

3. Resonant States in QW

We calculate the energy position and the width of the resonant state as well as the probabilities of resonance scattering and capture into resonant state using the configuration interaction method. The main idea is to choose two different Hamiltonians for the initial approximation: one for localized states and the other one for continuum states.

Consider the Schrödinger equation for the wavefunction of the resonant acceptor state $\Psi^{\mu}_{n,\mathbf{k}}$ overlapping the continuum:

$$\left[\hat{H}_L + U(z) - \hat{V}_c(r)\right]\Psi^{\mu}_{n,\mathbf{k}} = \varepsilon_{n,\mathbf{k}}\Psi^{\mu}_{n,\mathbf{k}}, \tag{1}$$

where V_c is the coulomb potential, *U(z)* is the QW potential and $\hat{H}_L$ is the Luttinger Hamiltonian for free holes:

$$\hat{H}_L(\mathbf{k}) = \begin{bmatrix} P+Q & -S & R & 0 \\ -S^* & P-Q & 0 & r \\ R^* & 0 & P-Q & S \\ 0 & R^* & S^* & P+Q \end{bmatrix}. \tag{2}$$

Here

$$\begin{gathered} P = \gamma_1\left(k_x^2 + k_y^2 + k_z^2 +\right), \\ R = -\gamma_2\sqrt{3}\left(k_x^2 - k_y^2\right) + i\gamma_3 2\sqrt{3}\,k_x^2 k_y^2, \\ Q = \gamma_2\left(k_x^2 + k_y^2 - 2k_z^2\right) + \zeta, \\ S = \gamma_3 2\sqrt{3}\left(k_x - ik_y\right)k_z, \end{gathered} \tag{3}$$

and the strain-split energy

$$E_{def} = \frac{\hbar^2\zeta}{2m_0}.$$

According to the standard Dirac approach to scattering:

$$\Psi^{\mu}_{n,\mathbf{k}} = \psi^{\mu}_{n,\mathbf{k}} + \sum_{\nu} a^{\mu,\nu}_{n,\mathbf{k}}\varphi^{\nu} + \sum_{n',\mathbf{k}',\mu'} \frac{t^{n\mu,n'\mu'}_{\mathbf{k},\mathbf{k}'}\psi^{\mu'}_{n',\mathbf{k}'}}{\varepsilon_{n,\mathbf{k}} - \varepsilon_{n',\mathbf{k}'} + i\gamma}, \tag{4}$$

where $a^{\mu,\nu}_{n,\mathbf{k}}$ gives the probability of capture into the resonant states, and $t^{n\mu,n'\mu'}_{\mathbf{k},\mathbf{k}'}$ is the probability of the resonant scattering.

The initial wavefunction and energy of the localized states were found as variational solutions of the acceptor problem in strained bulk $Si_{1-x}Ge_x$. They are

$$\varphi^{+1/2} = \frac{1}{\sqrt{\pi a^2 b}} e^{-\sqrt{\frac{\rho^2}{a^2}+\frac{z^2}{b^2}}} \begin{bmatrix} 0 \\ 1 \\ 0 \\ 0 \end{bmatrix}, \tag{5}$$

$$\varphi^{-\frac{1}{2}} = \frac{1}{\sqrt{\pi a^2 b}} e^{-\sqrt{\frac{\rho^2}{a^2}+\frac{z^2}{b^2}}} \begin{bmatrix} 0 \\ 0 \\ 1 \\ 0 \end{bmatrix}, \tag{6}$$

where a and b are variational parameters. These wavefunctions satisfy:

$$\left[\hat{H}_L^d - V_c(r)\right]\varphi^\nu = E_0\varphi^\nu, \tag{7}$$

where $\hat{H}_L^d$ is the diagonal part of the Luttinger Hamiltonian (2).

As zero order Hamiltonian for the 2D continuum states, we used the Luttinger Hamiltonian with appropriate boundary conditions at QW interfaces. These states are classified by subband number n, 2D wave vector $\mathbf{k}$ and parity μ, which can be either symmetric or anti-symmetric. The corresponding wavefunctions $\psi_{n,\mathbf{k}}^\mu$ and dispersion law $\varepsilon_{n,\mathbf{k}}$ are the solutions of the following equation [11]:

$$\left[\hat{H}_L + U(z)\right]\psi_{n,\mathbf{k}}^\mu = \varepsilon_{n,\mathbf{k}}\psi_{n,\mathbf{k}}^\mu. \tag{8}$$

We assume the width of QW L to be larger than the acceptor Bohr radius along the z axis: $L >> b$. So that acceptor is not affected by the quantum well potential. We also use the Luttinger Hamiltonian in cylindrical approximation which makes the angle in QW plane to contribute as a phase shift only. In Fig. 2 we provide the scheme of localized impurity levels as well as initial approximation for resonant state energy (E_0) superimposed on 2D dispersion curves for $Si/Si_{0.80}Ge_{0.20}/Si$ QW, here the variational parameters a=27 Å, b=29 Å and E_0=5.01 meV.

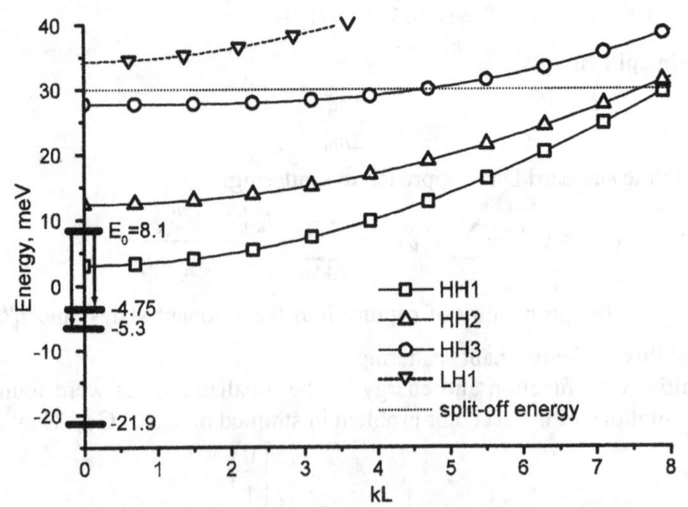

Figure 2. Energy dispersion and initial approximation for impurity levels in $Si/Si_{0.80}Ge_{0.20}/Si$ 200 Å wide QW. The kinetic energy of holes is assumed to be positive.

The dependence of variational parameters *a* and *b* of localized impurity state in Si/ $Si_{1-x}Ge_x$/Si QW on the Ge content *x* is shown in Fig. 3.

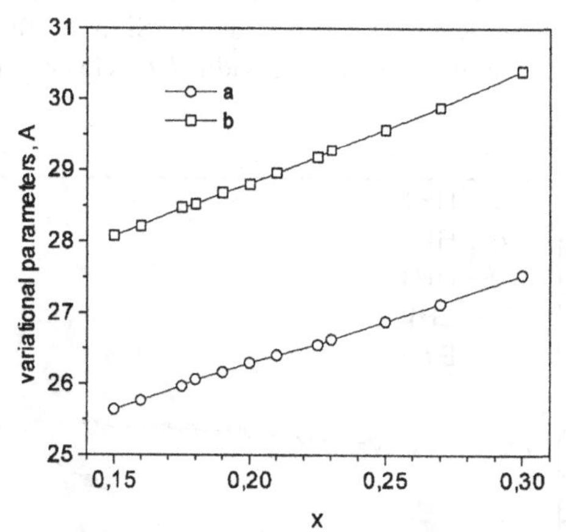

Figure 3. LH1 subband related acceptor state variational parameters *a* and *b* vs. Ge content *x* for 200 Å wide Si/ $Si_{1-x}Ge_x$/Si QW.

Substituting the wave function (4) into Schrödinger equation one can determine $a^{\mu,\nu}_{n,\mathbf{k}}$ and $t^{n\mu,n'\mu'}_{\mathbf{k},\mathbf{k}'}$ [7,8]. For the coefficient $a^{\mu,\nu}_{n,\mathbf{k}}$ the following expression containing the resonant denominator is obtained:

$$a^{\mu,\nu}_{n,\mathbf{k}} = \frac{1}{\sqrt{S}} \frac{A^{\mu,\nu}_{n,\mathbf{k}}}{\varepsilon_{n,\mathbf{k}} - (E_0 + \Delta E) + i\Gamma/2},$$

where S is the normalizing square, the resonance energy shift ΔE and width Γ should be calculated numerically, and

$$A^{\mu,\nu}_{n,\mathbf{k}} = -\left\langle \varphi^{\nu} \middle| \hat{V}_c \middle| \psi^{\mu}_{n,\mathbf{k}} \right\rangle + \sum_{n',\mathbf{k}'} \left\langle \psi^{\mu}_{n',\mathbf{k}'} \middle| \hat{V}_c \middle| \psi^{\mu}_{n,\mathbf{k}} \right\rangle \left\langle \varphi^{\nu} \middle| \psi^{\mu}_{n',\mathbf{k}'} \right\rangle.$$

For the resonant part of the scattering amplitude one can write

$$t^{n\mu,n'\mu'}_{\mathbf{k},\mathbf{k}'} = \sum_{\nu} a^{\mu,\nu}_{n,\mathbf{k}} B^{n',\mathbf{k},'\mu',\nu'}_{\varepsilon_{n,\mathbf{k}}},$$

where

$$B^{n',\mathbf{k},'\mu',\nu'}_{\varepsilon_{n,\mathbf{k}}} = \left\langle \psi^{\mu'}_{n',\mathbf{k},'} \middle| \hat{H}^{nd}_{L} \middle| \varphi^{\nu} \right\rangle - \left(\varepsilon_{n\mathbf{k}} - E_0\right) \left\langle \psi^{\mu'}_{n',\mathbf{k},'} \middle| \varphi^{\nu} \right\rangle.$$

Note, that both ΔE and Γ are functions of $\varepsilon_{n,\mathbf{k}}$ and the final resonance energy shift $\Delta E = E_r - E_0$ is calculated at the energy $\varepsilon_{n,\mathbf{k}} = E_r$. The capture probability for the hole from the *n*-th subband is given by the $a^{\mu,\nu}_{n,\mathbf{k}}$ coefficient:

$$W^{n,\mu}_{\mathbf{k}r} = \sum_{\nu} \left| a^{\mu,\nu}_{n,\mathbf{k}} \right|^2$$

The same coefficients determine the probability of elastic scattering of 2D holes.

4. The Results of Numerical Calculations

The energy of resonant state E_r vs. Ge content in the QW material together with space quantization energy levels in 200 Å width Si/ $Si_{1-x}Ge_x$ /Si QW are reported in Fig. 4. The dependence of the resonance width Γ on Ge content is presented in Fig. 5.

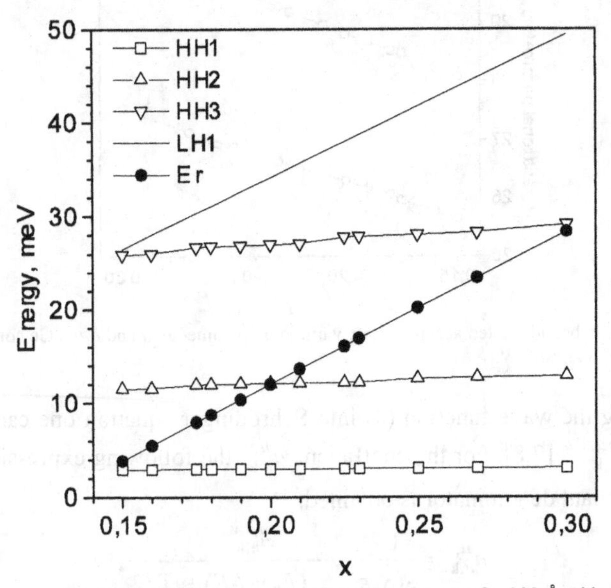

Figure 4. Resonance energy E_r and QW energy levels vs. Ge content x for 200 Å wide Si/ $Si_{1-x}Ge_x$/Si QW

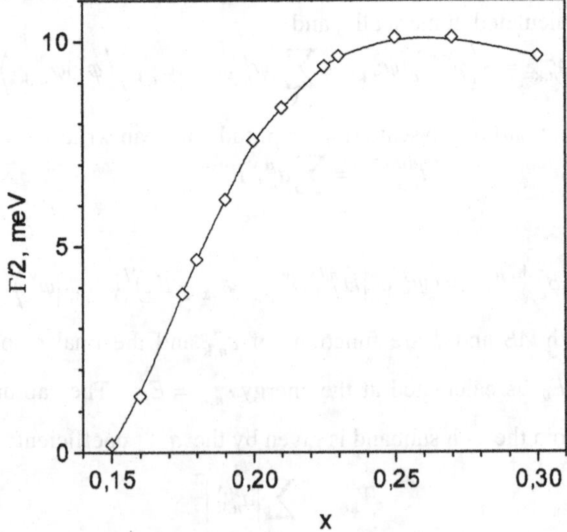

Figure 5. The width of resonant state $\Gamma/2$ vs. Ge content x for 200 Å wide Si/ $Si_{1-x}Ge_x$/Si QW

5. The Structures for Multi Quantum Well Resonant State Laser

We have performed an extensive numerical study of a $Si/Si_{1-x}Ge_x/Si$ QW laser structures. The typical structure suitable for lasing is presented in Fig. 6. A i-Si buffer layer of thickness 80 nm is deposited on a n-Si substrate (N_d= $5x10^{14}$ cm^{-3}). The delta doping of the buffer layer with boron concentration $8x10^{10}cm^{-2}$ should be formed at a distance of 10 nm from the substrate. Then four $Si_{0.8}Ge_{0.2}$ QW delta doped in the middle with boron concentration $6x10^{11}$ cm^{-2} and separated by 10 nm of undoped Si are grown. The cap layer of width 40 nm is then deposited and delta doped at a distance 25 nm from the last QW interface with boron concentration 10^{12} cm^{-2}. The purpose of the high doping concentration here is to compensate the electric field produced by charge accumulation in the surface states at Si/SiO_2 interface. In result, there is no band bending in the active region of the structure and all quantum wells are neutral with ground acceptor level populated by holes. Thus, the above described calculation of the resonant level position is equally applied to each QW. When an external electric field is applied along QW planes and conditions of impact ionisation of localized states are realized, the resonant states become populated and we expect lasing at energy $\hbar\omega$ = 15 meV corresponding to the transition between resonant (E_0) and localized states shown by arrows in Fig. 2.

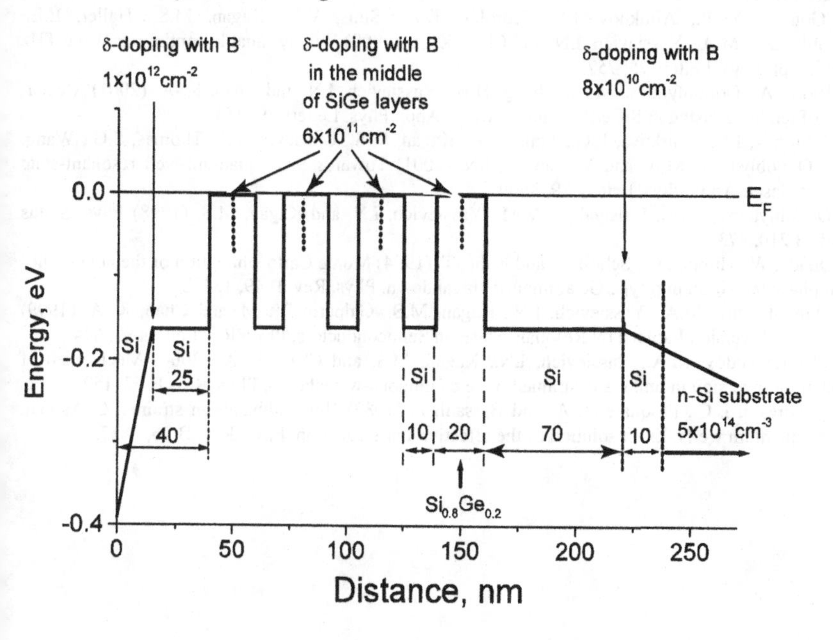

Figure 6. Possible multi QW design for laser structure

6. Conclusion

The emergence of resonant states in Si/SiGe/Si QW structures delta-doped by boron has been investigated. Energy position and width of resonant states as well as probabilities of capture on resonant state and resonant elastic scattering have been calculated. Multi QW structures suitable for THz lasing based on intracenter transitions between resonant and local states are suggested.

This work has been partially supported by the RFBR (grants 01-02-16265, 00-02-17429 and 00-15-96768), The Swedish Foundation for International Cooperation in Research and Higher Education (contract 99/527 (00)) and NorFA Grant. I. N. Yassievich thanks also Wenner-Gren Foundations for support.

References

[1] Dehlinger, G., Diehl, L., Gennser, U., Sigg, H., Faist, J., Ensslin, K., Grützmacher and D., Müller, E. (2000) Science 290, 2297.

[2] Colombelli, R., Capasso, F., Gmachl, C., Hutchinson, A.L., Sivco, D.L., Tredicucci, A., Wanke, M.C., Sergent, A.M. and Cho, A.Y. (2001), *Abstracts of International Workshop Middle Infrared Coherent Sources (St. Petersburg, Russia, 2001)*, pp. 713.

[3] Ulrich, J., Zobl, R., Schrenk, W., Strasser, G., Unterrainer, K. and Gornik, E. (2000) Appl. Phys. Letters 77, 25.

[4] Gousev, Yu.P., Altukhov, I.V., Korolev, K.A., Sinis, V.P., Kagan, M.S., Haller, E.E., Odnoblyudov, M.A., Yassievich, I.N. and Chao, K.-A. (1999) Widely tunable continuous-wave THz laser, Appl. Phys. Letters 75; 757.

[5] Blom, A., Odnoblyudov, M.A., Cheng, H.H., Yassievich, I.N. and Chao, K.-A. (2001) Mechanism of terahertz lasing in SiGe/Si quantum wells, Appl. Phys. Letters 79, 713.

[6] Altukhov, I.V., Chirkova, E.G., Sinis, V.P., Kagan, M.S., Gousev, Yu.P., Thomas, S.G., Wang, K.L., Odnoblyudov, M.A. and Yassievich, I.N. (2001) Towards SiGe quantum-well resonant-state terahertz laser, Appl. Phys. Letters 79, 3909.

[7] Odnoblyudov, M.A., Chistyakov, V.M., Yassievich, I.N. and Kagan, M.S. (1998) Phys. Status Solidi B 210, 873.

[8] Quade, W., Hupper, G., Schöll, E. and Kahn, T. (1994) Monte Carlo simulation of the nonequilibrium phase transition in p-type Ge at impurity breakdown, Phys. Rev. B 49, 13408.

[9] Odnoblyudov, M.A., Yassievich, I.N., Kagan, M.S., Galperin, Yu.M. and Chao, K.-A. (1999) Population Inversion Induced by Resonant States in Semiconductors, Phys. Rev Letters 83, 644.

[10] Odnoblyudov, M.A., Yassievich, I.N., Kagan, M.S. and Chao, K.-A. (2000) Mechanism of population inversion in uniaxially strained p-Ge continuous-wave lasers, Phys. Rev. B 62, 15291.

[11] Andreani, L.C., Pasquarello, A. and Bassani, F. (1987) Hole subbands in strained GaAs-$Ga_{1-x}Al_xAs$ quantum wells: Exact solution of the effective-mass equation, Phys. Rev. B 36, 5887.

THz LASING OF STRAINED p-Ge AND Si/Ge STRUCTURES

M. S. KAGAN
Institute of Radioengineering and Electronics of RAS
11-7 Mokhovaya, GSP-9, 101999 Moscow K-9, Russia

1. Introduction

A good example of a recent advance in semiconductor technology with impact on laser technology is the quantum cascade laser (QCL), which nowadays can be operated at wavelengths in excess of 10 μm [1]. The maximum wavelengths achieved now are about 80 μm [2]. Many effort are now spent to implement the quantum cascade THz laser scheme in SiGe/Si heterostructures [3]. Here we discuss an alternative THz laser sources which is based on a simpler quantum well (QW) structures, i.e. a resonant-state laser (RSL) [4, 5]. Population inversion in the p-Ge RSL is realized among shallow acceptor states, split under external stress (see previous chapter and [6,7]). Acceptor-doped $Si_{1-x}Ge_x$ is very attractive for fabricating RSL because of its good thermal properties, low absorption in the THz range, well established, relatively cheap technology, as well as possible integration with Si-based electronics.

2. Compressed p-Ge

Stimulated emission of THz radiation has been observed in p-Ge at strong electric and magnetic fields [8] and attributed to population inversion of the valence subbands. We have observed a stimulated emission in p-Ge without any magnetic field but under uniaxial stress [4, 5]. The gallium-doped Ge crystals with Ga concentration of 3×10^{13} to 10^{14} cm^{-3} were used in the experiment at liquid helium temperature. The samples with a square cross section of 0.5 to 1 mm^2 and 6 to 10 mm long were cut in the [111] or [100] crystallographic directions. Uniaxial pressure P and electric field E were applied along the samples. Voltage pulses of 0.2 to 1 μs duration were applied to two contacts positioned on the long (lateral) plane of the sample. The distance between the contacts was varied from 4 to 9 mm. THz luminescence was registered by a cooled Ge photodetector. For the samples with planes parallel within 4`, a steep rise in radiation intensity of up to 10^3 times was observed for some threshold stress P_c. The minimum P_c value was 3 kbar. A resonator, necessary to obtain stimulated emission, was formed by well-parallel sample planes due to total internal reflection. Indeed, the jump in emission disappeared after rough grinding of one of the long sample planes. Polishing and etching restored the resonator and the stimulated emission appeared again.

L. Pavesi et al. (eds.), Towards the First Silicon Laser, 359–366.

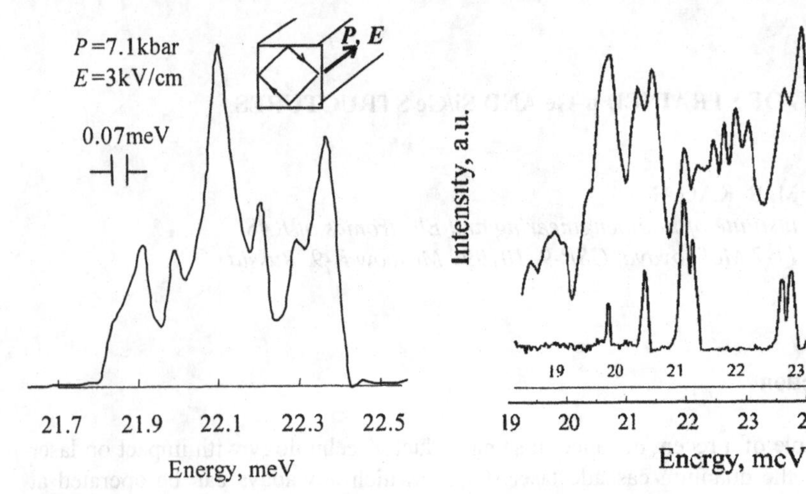

Figure 1. THz emission spectra at 14V/cm (upper curve) and 3kV/cm pulsed (0.5μsec) electric field

*Figure 2. M*odal structure of the main peak in the spectrum of Fig. 1.

The spectrum of stimulated radiation from compressed p-Ge measured by the grating monochromator [5] is shown in Fig. 1 (lower curve). It consists of several peaks. The peak energies increased with pressure. The energy of the most intensive peak (21.2 meV in Fig. 1) is varied from 21.2 to 42 meV by increasing pressure from 6.85 to 11 kbar at $P||$ [111]. The main peak measured in more details shows the structure caused by resonator modes (Fig. 2). The line spacing (≈0.11 meV) for the specimen with the cross section of 1x1 mm^2 coincides with that found from the condition $N\lambda=nL$, where λ is the radiation wavelength, n is the refractive index (n=4 for Ge), L is the optical path length (see the inset in Fig. 2), and N is an integer.

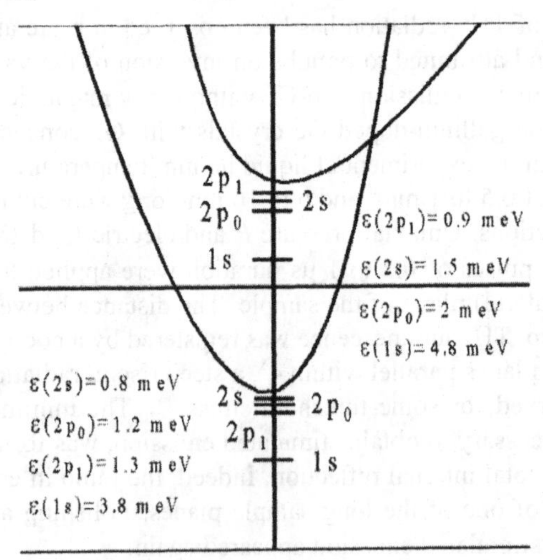

Figure 3. The scheme of energy positions of resonant and local states in uniaxially compressed Ge in the limit of strong deformation.

Uniaxial deformation removes the degeneracy of the valence band of Ge at $k=0$ and splits it into two subbands separated by an energy gap proportional to the applied pressure P (the proportionality factor is about 4 and 6 meV/kbar for P parallel to [100] and [111] crystallographic directions, respectively [9]). The degenerate ground state of an acceptor is also split into two states. The energy separation $\delta\varepsilon$ between the split-off state and the ground state increases with strain. Fig. 3 represents the p-Ge valence band structure and calculated energy spectrum of resonant and localized acceptor states [10].

At certain pressures ($P \approx 4$ kbar for $P \parallel [111]$ and $P \approx 3$ kbar for $P \parallel [100]$), the split-off state enters into the valence band continuum and becomes resonant while the ground state remains in the forbidden band. A series of excited states formed by both the ground and split-off states should exist also in the gap and the continuum, respectively. In strong electric field the acceptor ground state located in the forbidden band becomes empty due to impact ionisation while the split-off state being in the valence band is filled to some degree due to an exchange with carriers between the resonant state and the continuum. The inverted population of the resonant s-state is the result of the strong resonance scattering [6]. The calculations of the spectrum of localized and resonant acceptor levels were performed [10] by the variational method for high pressure limit (impurity states are considered in approximation of independent valence subbands). The comparison of calculated energies of optical transitions from the resonant $1s$ state to the localized $1s$ and $2p_{\pm 1}$ states and the experimental stress dependence of the energy of the main peak [5] allows us to identify the main peak as the resonant $1s$ to $2p_{\pm 1}$ transition and the peaks at 19.9 and 20.5 meV (see Fig. 1) as the transitions from resonant $1s$ to $2p_0$, $2s$, respectively. The peak of energy 23 meV is supposed to be radiative transition of hot free holes of energy equal to s-resonant level to $1s$ localized state. The linewidth for different peaks is of 0.2 to 0.5 meV. The reason for the wide spectral maximum may be the broadening of the resonant acceptor state being in the continuum [11]. The minimum pressure to observe stimulated emission is just that at which the split-off state becomes resonant. The energy splitting of the acceptor states is about 10 meV in this case. On the other hand, the stimulated radiation intensity decreases sharply [5] as the split-off acceptor state begins to depopulate via hole transitions to the valence band edge assisted by optical-phonon emission. The value of $\delta\varepsilon$ for depopulating the split-off acceptor state is about 42 meV. Thus, these data confirm that the intra-center inversion can arise as the split-off state enters into the valence band and exists until optical phonon-assisted hole transitions to the valence band edge depopulate the resonance state. The energy range of the stimulated emission is expected to be from 10 to 42 meV.

We have found that for a crystal with optical mirrors (lateral sample planes) parallel within 0.5'-4', one can only observe THz emission in the pulsed regime beginning at about 2.5 kV/cm. To obtain the lasing at low voltages, we used the dislocation-free Ge or material with small dislocation density; a resonance cavity of high enough quality is necessary, as well. In our case, a high degree of total internal reflection necessary for cw operation was achieved for the plane parallelism better than 20 arcsec. Fig. 4 shows the dependence of the THz emission intensity and the current through the sample on voltage. The THz generation appeared just after shallow impurity breakdown at a pressure of 3kbar. The threshold pressure corresponds to that at which the acceptor state, split from the ground state, becomes resonant for [100] crystallographic direction. At low voltages, due to small dissipated power in the sample, stimulated emission could be excited in a continuous regime [12]. The minimum values of bias voltage and current for lasing were 1V (2V/cm) and 3mA, respectively. The efficiency estimated by using the sensitivity of our Ge detector was higher than 10^{-4}. The stimulated emission frequency tuned by stress (from 3 to 9 kbar) corresponds to the frequency range of 3 to 10 THz (Fig. 5).

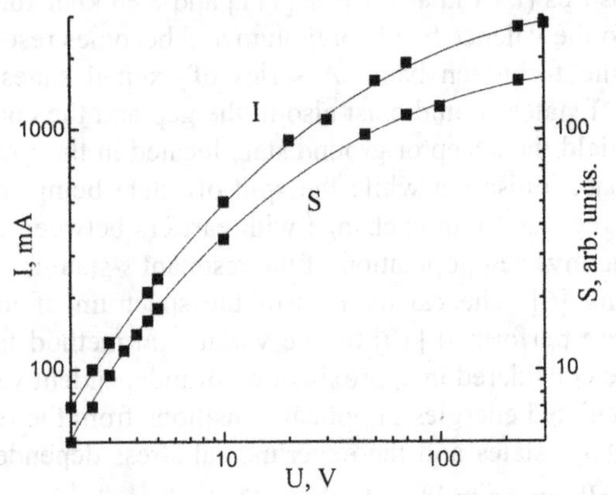

Figure 4. Signal and current vs voltage.

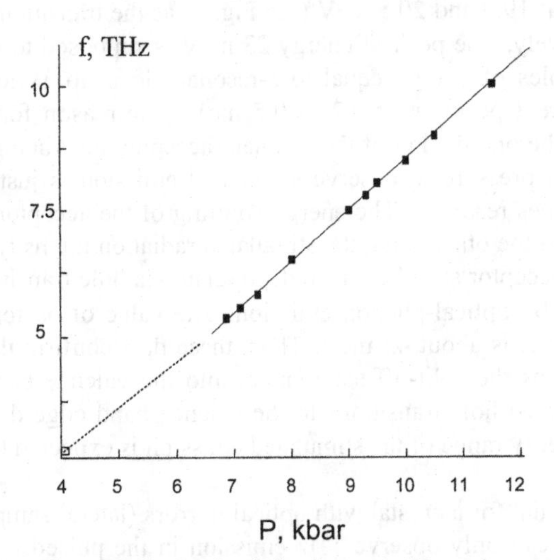

Figure 5. Pressure dependence of stimulated emission frequency. Open point corresponds to cw regime.

The spectrum of low-voltage THz radiation from compressed p-Ge was registered by a cooled InSb detector tuned by magnetic field. Voltage pulses of 1 ms duration and duty cycle of 1 were applied to contacts positioned on the long (lateral) plane of the sample on a distance 7 mm one from another. The upper curve in Fig. 1 shows the stimulated emission spectrum at 10 V/cm. It contains several peaks corresponding to direct optical transitions between the resonant and different localized acceptor states. At the pressure of 7.9 kbar, the peak energies are 20.5, 21.5, 23, 24 and 25.5 meV. The peak positions are shifted approximately by 1 meV to higher energies as compare with those observed at the pressure of 6.85 kbar (lower curve) under exitation with short (0.5 μsec) pulses of high voltage (up to 3 kV/cm). The modal structure caused by resonator was observed in this case, too, indicating that the emission is stimulated. The data obtained show that the inversion mechanism is the same at low and high electric fields. However, the heating of holes by electric field is different in these cases: at high voltages carriers accomplish so-called streaming motion while at low voltages the heating is diffusive.

3. Strained SiGe/Si structures

We studied p-type $Si_{1-x}Ge_x$ QW structures MBE-grown pseudomorphically on n-type Si substrate (Fig. 6,a). The SiGe QW was sandwiched between Si buffer and cap layers and was δ-doped with boron in the QW middle with the concentration of $6 \cdot 10^{11}$ cm^{-2}. Both the buffer and cap layers were doped with one B-δ-layer. Two kinds of structures were used. In type-I QWs, the well was 20 nm thick and Ge content x in SiGe alloy was 0.15. The thickness of cap and buffer layers was 60 and 130 nm, respectively. The boron δ-layers in the barriers, with concentration from $4 \cdot 10^{11}$ to 10^{12} cm^{-2}, were positioned each at the distance d_1=30 nm from respective QW interfaces. The QW thickness in type-II structures was about 13.5 nm. Other parameters (see Fig. 6,a) were: x=0.15, 0.1, and 0.07, and d_1=19 nm and d_2=62 nm. The concentration of B in the δ-layers was $6 \cdot 10^{11}$ cm^{-2}.

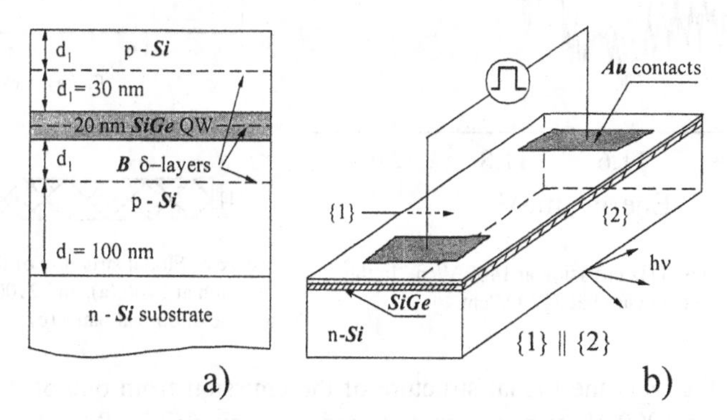

Figure 6. Scheme of structure (a) and laser design (b)

We used pulsed bias of 0.2 to 4 μs duration in order to avoid overheating. Bias was applied to the doped layer via thermal-diffusion made Au contacts as shown in Fig. 6,b. The distance between contacts was 6 mm. Emission spectrum in THz range was measured by a grating monochromator and registered with a cooled Ge photodetector. Measurements were made at liquid He temperatures.

An optical resonator was formed due to the total internal reflection between two external surfaces of a sample which were perpendicular to growth direction, and two lateral facets ({1} and {2} in Fig. 6,b) which were parallel polished. For the samples with optical resonator, we observed an intense THz emission. It came out mainly from sample ribs; that was proved using different diaphragms. Spectra of the intense emission were measured [13] in the energy range 9.8 to 15.5 meV. Shown in Fig. 7 are parts of the spectra indicating lines for the samples of type I. The peak near the wavelength 104 μm is observed. The spectral position of the peak varied from 103 to 108 μm for different samples. At higher voltages, it was possible to observe additional maximums at larger wavelengths shown in the inset in Fig. 7. They appears at the energies from 10 to 11 meV.

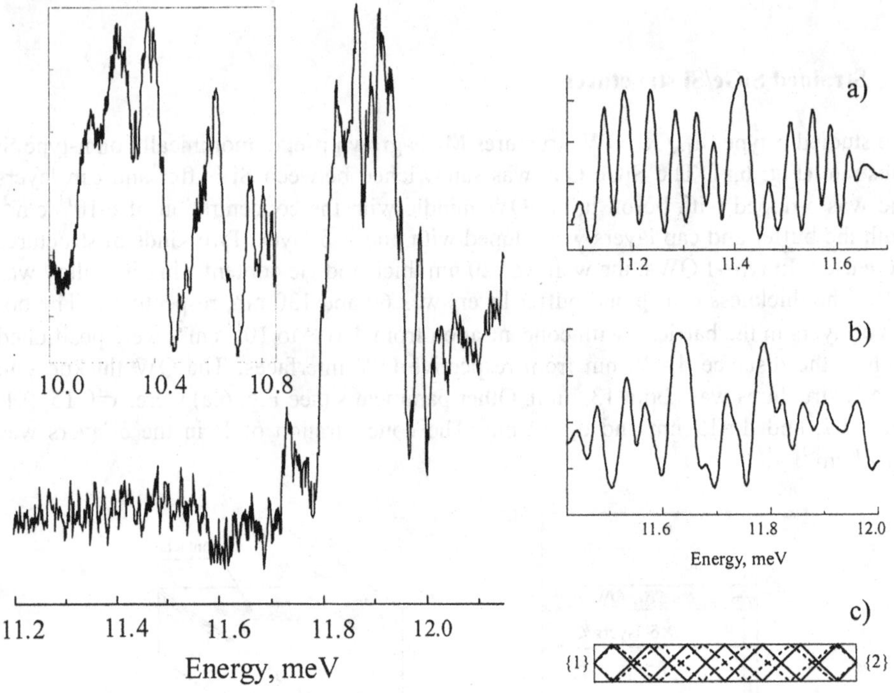

Figure 7. Spectrum of THz emission at 1400 V/cm. In the inset: additional lines appeared at 2100 V/cm.

Figure 8. Modal structure of the main line of emission at 1400 (a), and 2100 V/cm (b), and scheme of optical paths (c)

Shown in Fig. 8 is the modal structure of the emission from one of the samples at two voltages. At 1000 V, only one spatial mode is excited (Fig. 8,a). The distance between lines is 0.04 meV and corresponds to the optical path inside the structure due to total internal reflection. At higher voltages, the excitation of several spatial modes becomes possible. The modal structure shown in Fig. 8,b is more complicated and corresponds to the coexistence of two spatial modes. Optical paths for both of them are shown in Fig. 8,c. The line near 100 μm was observed also in type-II structures with 0.15 Ge content. Line widths are determined by the spectral resolution of the monochromator (the width of output slit) but allow us to estimate the value of resonator quality from below. It is not less than 300.

We believe that the lasing we have observed results from a population inversion in the QW layer due to the resonant states of boron. The resonant acceptor state is created when the energy splitting between the ground and split-off acceptor states exceeds the binding energy. In the case of $Si/Si_{0.85}Ge_{0.15}/Si$ structures, the splitting energy of the valence subbands is about 31 meV [14, 15]. According to variational calculations, the binding energy of shallow acceptors is about 27 meV; the same and the edge of higher-energy valence subband [16]. Thus, one can expect that the internal strain in the $Si_{0.85}Ge_{0.15}$ layer is sufficient for the split-off state to become resonant. The situation can be improved by an existence of built-in strong electric field, as the position of space-quantization levels is controlled by electric field. We have shown [16] that the built-in transverse electric field exists in our value was obtained for the energy difference between the split-off acceptor state structures due to surface charge. This field inclines the valence bands and gives the possibility for the resonant state to appear even if the intrinsic strain is not sufficient. The calculated energy diagram of the SiGe QW structure under study is presented in Fig. 8. The built-in electric field for given structure is about 20 kV/cm [16, 17]. The scheme of acceptor levels and the lowest space-quantization level (E_{1hh}) is presented in the inset in Fig. 8. We attribute the main line of the THz lasing observed to stimulated optical transitions between the resonant (E_{1s}) and first excited localized acceptor states (E_p); the possible transitions are shown by the arrows in the inset in Fig. 8. The energy of this transition can be found using energies given above. The resonant state is $E_{1hh} - E_{1s} = 4$ meVabove the edge of heavy-hole band. Using the value $E_g/4$ for a rough estimation of E_p ($E_g \approx 27$ meV is the energy of ground acceptor state mentioned above), we expect the energy of the transition to be about 11 meV. It is consistent with the observed quantum energy of 11.8 meV. It is natural to connect the additional maximums observed at lower energies with optical transitions from the same resonant state to higher excited localized states.

4. Conclusion.

In summary, the data obtained show that stimulated THz emission in uniaxially compressed p-Ge and strained SiGe structures doped with shallow acceptors is due to the electric field induced population inversion of strain-split acceptor states. The necessary condition for the inversion is the appearance of resonance state. For p-Ge, the possibility of a strong frequency tuning by stress is demonstrated. It is also shown that SiGe structures δ-doped with boron are promising for realization of resonant-state laser (RSL) operating in THz region.

We acknowledge financial support by Russian Foundation for Basic Research (Grants 02-02-16373, 00-02-16066, 00-02-16093), Foundation for Leading Scientific Schools (00-15-96663), RAS Program "Low-dimensional quantum structures", and European Office of Aerospace Research and Development (grant ISTC #2206).

References

1. Sistori, C., Faist, J., Capasso, F., Sivco, D.L., Hutchinson, A.L., and Cho, A.Y. (1996) Long wavelength

infrared (λ≈11 μm) quantum cascade lasers, *Appl. Phys. Lett.* **69**, 2810-2812; Colombelli, R., Capasso, F., Gmachl, C., Hutchinson, A.L., Sivco, D.L., Tredicucci, A., Wanke, M.C., Sergent, A.M., and Cho, A.Y. (2001) Far-infrared surface-plasmon quantum-cascade lasers at 21.5 μm and 24 μm wavelengths. *Appl. Phys. Lett* **79**, 2620-2622; Wanke, M.C., Capasso, F., Gmachl, C., Tredicucci, A., Sivco, D.L., Hutchinson, A.L., Chu, S.N.G., Cho, A.Y. (2001) Injectorless quantum-cascade lasers. *Appl. Phys. Lett* **79**, 3950-3952 ; Anders, S., Schrenk, W., Gornik, E., and Strasser G. (2002) Room-temperature operation of electrically pumped quantum-cascade microcylinder lasers. *Appl. Phys. Lett* **80**, 4094-4096.

2. Köhler, R., Tredicucci, A., Beltram, F., Beere, H.E., Linfield, E.H., Davies, A.G., Ritchie, D.A., Iotti, R.C., Rossi, F. (2002) Terahertz semiconductor laser. *Nature* **417**, 156; (2002) *Appl. Phys. Lett* **81,1867**.
3. Dehlinger, G., Diehl, L., Gennser, U., Sigg, H., Faist, J., Ensslin, K., Grützmacher, D., Müller, E. (2000) Intersubband electroluminescence from silicon-based quantum cascade structures. *Science*, **290**, 2277-2280.
4. Altukhov, I.V., Kagan, M.S., Korolev, K.A., and Sinis, V.P. (1994) *JETP Lett.* **59**, 476; Altukhov, I.V., Chirkova, E.G., Kagan, M.S., Korolev, K.A., Sinis, V.P., and Yassievich, I.N. (1996) *Phys. Stat. Sol.* (b) **198**, 35-40.
5. Altukhov, I.V., Kagan, M.S., Korolev, K.A., Sinis, V.P., Chirkova, E.G., Odnoblyudov, M.A., and Yassievich, I.N. (1999) Resonant acceptor states and terahertz stimulated emission of uniaxially strained germanium. *JETP* **88**, 51-60.
6. Odnoblyudov, M.A., Chistyakov, V.M., Yassievich, I.N., and Kagan, M.S. (1998) *Phys. Status Solidi (b)* **210**, 873-877.
7. Odnoblyudov, M.A., Yassievich, I.N., Kagan, M.S., Galperin, Yu.M., and Chao, K.A., (1999) *Phys. Rev. Lett.* **83**, 644-647.
8. Andronov, A.A. (1992) The highly non-equilibrium hot-hole distributions in germanium, in C.V.Shank and B.P.Zakharchenya (eds.) *Spectroscopy of Nonequilibrium Electrons and Phonons*, Modern Problems in Condensed Matter Sciences **35**, North Holland, Amsterdam.
9. Bir, G.L. and Pikus G.E. (1974) *Symmetry and Strain-Induced Effects in Semiconductors*, Wiley, New York.
10. Odnoblyudov, M.A., Chistyakov, V.M., Yassievich, I.N., and Kagan, M.S. (1998) Resonant states in strained semiconductors. *Phys. Stat. Sol.* (b) **210**
11. Ramdas, A.K., Rodriguez, S. (1981) *Rep. Prog. Phys.*, **44**, 1297.
12. Gousev, Y.P., Altukhov, I.V., Korolev, K.A., Sinis, V.P., Kagan, M.S., Haller, E.E., Odnoblyudov, M.A., Yassievich, I.N., and Chao, K.A. (1999) Widely tunable continuous-wave THz laser. *Appl. Phys. Lett.*, **75**, 757-759.
13. Altukhov, I.V., Chirkova, E.G., Kagan, M.S., Sinis, V.P., Gousev, Y.P., Odnoblyudov, M.A., and Yassievich, I.N. (2001) Towards $Si_{1-x}Ge_x$ quantum well resonant-state THz laser. *Appl. Phys. Lett.*, **79**, 3909-3911.
14. Jain, S.C. (1994) *Germanium-Silicon Strained Layers and Heterostructures.* Advances in Electronics and Electron Physics, Suppl. 24, Academic Press.
15. Schmalz, K., Yassievich, I.N., Wang, K.L., and Thomas, S.G. (1998) *Phys. Rev. B* **57**, 6579-6585.
16. Blom, A., Odnoblyudov, M.A., Cheng, H.H., Yassievich, I.N., and Chao, K.A. (2001) *Appl. Phys. Lett.* **79**, 713-716.
17. Kagan, M.S., Altukhov, I.V., Korolev, K.A., Orlov, D.V., Sinis, V.P., Thomas, S.G., Wang, K.L., and Yassievich, I.N. (1999) Lateral Transport in Strained SiGe Quantum Wells Doped with Boron. *Physica Status Solidi (b)* **211**, 495-499.

TERAHERTZ EMISSION FROM SILICON-GERMANIUM QUANTUM CASCADES

R W KELSALL[1*], Z IKONIC[1], P HARRISON[1], S A LYNCH[2], R BATES[2], D J PAUL[2], D J NORRIS[3], S L LIEW[3], A G CULLIS[3], D J ROBBINS[4], P MURZYN[5], C R PIDGEON[5], D D ARNONE[6] AND R A SOREF[7]

1. Institute of Microwaves and Photonics, The University of Leeds, U.K.
2. Cavendish Laboratory, The University of Cambridge, U.K.
3. Dept. of Electronic & Electrical Eng., The University of Sheffield, U.K.
4. QinetiQ, Malvern, UK,
5. Dept. of Physics, Heriot Watt University, U.K.
6. TeraView Ltd., Cambridge, U.K.,
7. Sensors Directorate, AFRL, Hanscom AFB, MA, USA

* Corresponding author: e-mail r.w.kelsall@leeds.ac.uk

1. Introduction

Whilst most present day efforts towards the realisation of a silicon based laser are focused on the near-infrared (telecommunications) wavelengths, one of the most promising technical approaches is that of a silicon-germanium (SiGe) quantum cascade laser (QCL) operating in the far-infrared or terahertz frequency range. Until recently, the terahertz band (1-10 THz) has proved relatively inaccessible for science and engineering applications since it lies above the present upper frequency limit of millimetre wave electronic based oscillators, and below the range of near and mid-infrared solid state lasers and detectors. However, there is currently a great deal of interest in the development of terahertz technology for imaging and sensing applications: terahertz pulsed imaging has been shown to be capable of detecting caries (the precursor of decay) in human teeth [1], and there are also promising signs that the method could be used to detect basal cell carcinoma (a common form of skin cancer)[2]. Many chemical and biological molecules have absorption lines in the THz band, and therefore applications are envisaged in chemical/biological detection and identification. THz imaging is also potentially suitable for baggage/personnel scanning for security applications, where it would provide a low-energy, non-ionising alternative to X-rays.

Most existing Terahertz pulsed imaging systems are research tools of bench-top dimensions, with the radiation usually generated by photoconductive GaAs irradiated by Ti-sapphire lasers. Consequently, there is a clear need for compact, solid state THz sources (and detectors). The quantum cascade laser has considerably extended the lowest attainable frequency limit of solid state photonic sources, with devices spanning a

L. Pavesi et al. (eds.), Towards the First Silicon Laser, 367–382.

frequency range from 86 THz down to 12.5 THz (3.5-24 μm) [3]. Very recently, laser operation has been achieved in the THz band (4.4 THz / 67 μm) in a quantum cascade device [4-5].

Although these achievements have been attained using III-V devices (InAlAs/InGaAs/InP or GaAs/AlGaAs heterostructures), the basic operating principles of the quantum cascade laser: i) the generation of radiation by intersubband, rather than interband, radiative transitions, and ii) the magnification of optical power output by 'recycling' or 'cascading' of charge carriers through a many-period heterostructure, both apply equally well to the Si/SiGe materials system. Most importantly, the fact that the quantum cascade laser does not rely at all on interband recombination means that the indirect band gap of silicon and its alloys presents no disadvantage in this device. In otherwords, the Si/SiGe quantum cascade is virtually the only candidate for a silicon based laser whose designers are not fighting against the inherently long radiative lifetimes of the bulk material.

The Si/SiGe system actually offers numerous advantages over III-V heterostructures for quantum cascade laser application. Firstly, silicon and the SiGe alloys are non-polar materials, so there is no polar electron-phonon interaction, which is the dominant non-radiative loss process in III-V QCLs. Secondly, the optical phonon energy in silicon is much higher than in GaAs (64meV compared with 36meV), providing a larger frequency window within which (non-polar) optical phonon scattering is suppressed. Thirdly, the thermal conductivity of silicon is over 3 times larger than that of GaAs, giving better prospects of CW operation at non-cryogenic temperatures. In addition to these advantageous physical properties of the Si/SiGe materials system, the main technological driver for realisation of a Si/SiGe QCL is, of course, the prospect of monolithic integration with silicon microelectronics. The laser devices may be fabricated using standard CMOS processing techniques. By using p-type heterostructures designed to emit radiation from light hole – heavy hole transitions, surface-normal THz emission can be obtained, hence the Si/SiGe quantum cascade approach is potentially amenable to the development of surface emitting lasers which could be combined in monolithically integrated arrays.

In this paper we report on the development of THz-emitting p-Si/SiGe quantum cascades. Although only spontaneous (rather than stimulated) electroluminescence has been observed to date, the achievements of this work constitute several of the key prerequisites for the ultimate realisation of a Si/SiGe THz quantum cascade laser.

2. Epitaxy

A typical mid-infrared QCL has an active region comprising 2 or 3 quantum wells, in which the laser transition occurs between two quantum-confined subbands (see figure 1). Commonly, a third subband is designed to lie one optical phonon energy below the lower of the two lasing subbands, to achieve rapid depopulation of the latter by rapid electron-phonon scattering, and hence maintain a population inversion. A QCL may have up to 100 such active regions, each joined to the next by an interstitial 'injector' region which generally takes the form of a short period superlattice. The energy spectrum of the subbands within the injector is engineered to provide extraction of carriers from the active region, carrier cooling within the injector region, and near-resonant injection into the upper subband of the following active region (figure 1). The widths of the quantum wells in the injector region are usually progressively varied (or "chirped") such that the set of confined states is closely spaced in energy (forming a pseudo-miniband) when the device is biased at the operating field.

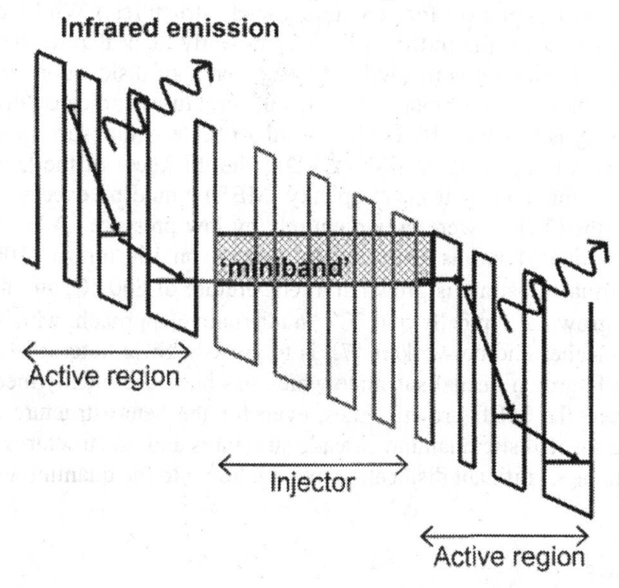

Figure 1. Schematic conduction band diagram of a quantum cascade laser, showing two identical active regions connected by an injector. The wavy arrows denote the radiative intersubband transitions.

This design concept results in a very complex heterostructure which may have over 1500 semiconductor layers, with minimum layer widths of less than 1 nm. Such a structure would be extremely difficult to grow in the Si/SiGe system: there is a 4% difference in the lattice constants of silicon and germanium, resulting in substantial lattice mismatch strain in the heterostructure. For example, for $Si_{0.7}Ge_{0.3}$ layers grown on a silicon substrate, the maximum layer thickness for stable single crystal formation, according to the Matthews-Blakeslee criterion, is approximately 8 nm [6]. The strain accumulates with increasing numbers of quantum wells, thus severely limiting the length of a Si/SiGe quantum cascade structure. If this critical thickness is exceeded, the structure becomes metastable: misfit dislocations may not form immediately upon growth, but may appear subsequently upon strain relaxation due to high temperature processing or even high current operation.

The solution to the strain problem is to grow the Si/SiGe cascade heterostructure not directly on silicon, but on a SiGe 'virtual substrate', which comprises a layer of linearly graded Ge composition, followed by a thick buffer layer at the endpoint composition value. This endpoint value is chosen to be intermediate between the composition of the quantum wells and barriers in the overlying heterostructure, such that the mismatch strain in alternate layers is of opposite sign (switching from compressive to tensile) and, for appropriately designed layer thicknesses, the net strain across the heterostructure overall is zero. This approach enables, in principle, growth of arbitrarily long heterostructures, with the only restriction being that the thickness of any single layer should not exceed the Matthews-Blakeslee stability limit. In our work, the linearly graded layer is chosen to be 3 μm thick, and the buffer layer is 1 μm thick. (Thinner virtual substrates have been recently developed for SiGe microelectronics applications, but are not required at present for quantum cascade structures.) Whilst dislocations do form in the graded layer, the buffer layer is sufficiently thick that dislocation propagation into the heterostructure is negligible. The absence of dislocations from any of the transmission electron micrographs taken of our structures indicates that the residual dislocation density is less than 10^6 cm^{-3}. Growth of SiGe virtual substrates is only practical using chemical vapour deposition (CVD): the thickness of the layers means that the growth times for molecular beam epitaxy (MBE) would be excessively long. The structures described below were grown entirely by low pressure CVD, using an industry-standard Applied Materials Epi-Centaura reactor, on 150 mm p^- (100) silicon substrates. The virtual substrate is grown at a temperature of 800 °C, and the heterostructure layers are grown at typically 580 °C. An alternative approach, which has been pursued by Grutzmacher and co-workers [7], is to grow Si/SiGe heterostructures by MBE on top of a CVD-grown virtual substrate which has been chemically/mechanically polished. However, the MBE growth times, even for the heterostructure alone, are still extremely long for realistic quantum cascade structures and, in structures grown to date, there appears to be significant dislocation propagation into the quantum well layers.

3. Design Issues

All QCLs reported to date have involved electron transport in n-type heterostructures. For III-V materials, n-type structures provide higher potential barriers and smaller effective masses (hence faster tunnelling) than in p-type. However, in the Si/SiGe system, the largest potential barriers are found in p-type systems (up to 740meV for pure Ge confined by pure Si [8]). The conduction band offset between Si and SiGe alloy layers is negligible for heterostructures grown pseudomorphically on silicon, but useable offsets can be obtained by growth on virtual substrates. Nevertheless, the main reason for using p-type heterostructures in this work is to obtain surface-normal radiation emission. The symmetry rules for radiative intersubband transition dictate that transitions between conduction subband couple only to TM mode radiation – which propagates parallel to the quantum well planes and hence is emitted through the edges of the device. In p-type heterostructures, transitions between heavy hole subbands again result only in TM emission, but light hole to heavy hole transitions couple to both TM and TE modes – and it is the latter which provides surface-normal propagation.

Even using the virtual substrate approach, simpler heterostructure designs are required for Si/SiGe QCLs, compared with the III-V structures described above. We have investigated a range of designs based on a semiconductor superlattice or "quantum staircase" structure. In the first approach, the superlattice is biased such that the lowest heavy hole (HH1) subband is resonant with the lowest light hole (LH1) subband in the next ("downstream") quantum well (see figure 2). The radiative transition then occurs between LH1 and HH1 subbands in the same quantum well. Calculations using 6-band

k.p theory show that the energy gap between these subbands at the zone centre (and hence the emitted photon energy) is determined by the lattice mismatch strain between the alloy quantum well layer and the virtual substrate (figure 3), and is virtually independent of quantum well width, except at very small well widths (figure 4).

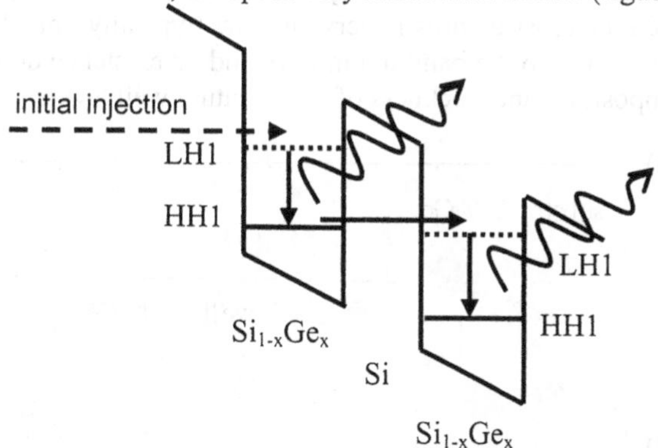

Figure 2. Schematic valence band profile of a Si/SiGe quantum staircase laser operating via radiative LH1-HH1 transitions

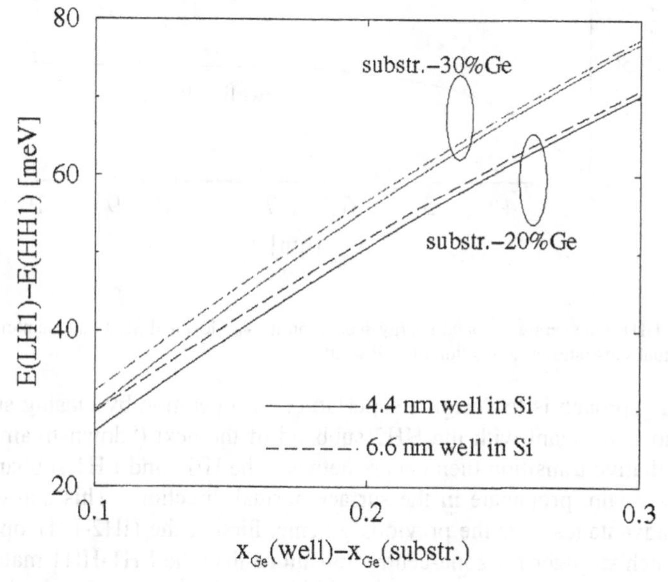

Figure 3. LH1-HH1 zone centre subband energy separation in $Si_{1-x}Ge_x$ quantum wells grown on virtual substrates, vs. the Ge mole fraction difference between the quantum well and the relaxed buffer layer of the virtual substrate. Results are shown for growth on both 20% and 30% Ge virtual substrates, and for 2 different quantum well widths.

Attainment of population inversion between LH1 and HH1 subbands in the same well requires engineering of the rates for the interwell (HH1-LH1) and intrawell (LH1-HH1) transitions. These rates are invariably dominated by non-radiative processes: phonon scattering, alloy disorder scattering, and carrier-carrier scattering. The prospects of population inversion in such a structure can be improved if an "inverted mass" feature exists in the in-plane dispersion of the LH1 subband. Carriers injected into LH1 at the zone centre then relax naturally to the off-zone-centre minimum, and carriers which make an optical ($\mathbf{k}_{\parallel}$ -conserving transition) from this point into the HH1 subband will subsequently relax to the zone centre [9,10] (see figure 5). However, the inverted mass feature in p-Si/SiGe heterostructures is very shallow (typically only 1-3meV measured from the LH1 zone centre to the band minimum), and its existence depends critically on both the alloy composition and thickness of the quantum well layers.

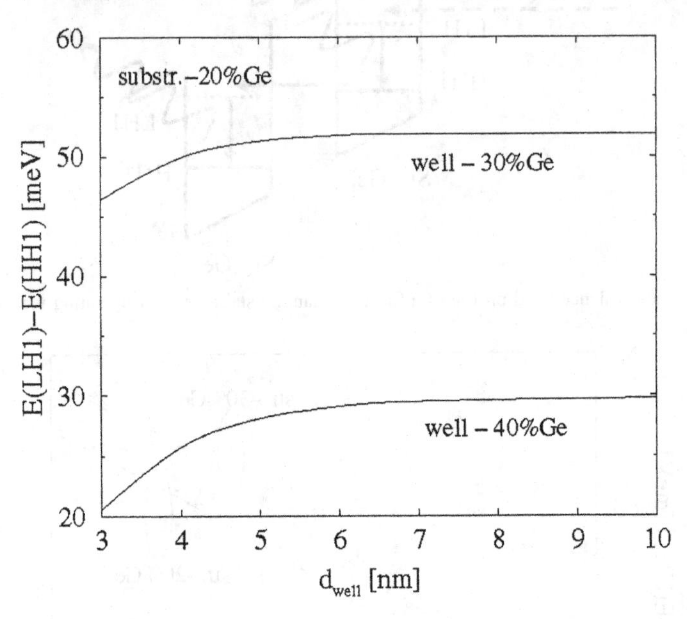

Figure 4. LH1-HH1 zone centre subband energy separation in $Si_{0.7}Ge_{0.3}$ and $Si_{0.6}Ge_{0.4}$ quantum wells grown on $Si_{0.8}Ge_{0.2}$ virtual substrates, as a function of well width.

A second approach is to design a superlattice for operation by biasing such that the HH1 subband is resonant with the HH2 subband of the next ("downstream") quantum well. The radiative transition then occurs between the HH2 and LH1 subbands, and the radiation can, again, propagate in the surface-normal direction. This approach has, in theory, two advantages over the previous scheme. Firstly, the HH2-LH1 optical matrix element is much stronger for zone-centre transitions than the LH1-HH1 matrix element, because the latter is actually symmetry-forbidden at the zone centre in the zero-field limit. Secondly, the HH1-HH2 interwell injection transition is stronger than the corresponding (HH1-LH1) carrier injection transition in the inverted mass design: transitions between pure heavy hole and pure light hole states are forbidden – hence the injection in the latter case is only made possible by the heavy/light hole mixing which occurs in non-zero-$\mathbf{k}_{\parallel}$ states.

Finally, a third approach involves designing a quantum staircase structure which emits radiation by inter-well, rather than intra-well transitions. The use of these – so-called diagonal – transitions has been reported in III-V mid-infrared QCLs [11]. In a p-type superlattice with HH1 and LH1 states confined in each well, the bias is increased *beyond* the HH1-LH1 resonance used in the inverted mass scheme, such that a radiative interwell transition may occur between the HH1 subband and the LH1 subband of the next (downstream) quantum well. The intrawell LH1-HH1 transition now serves as a non-radiative depopulation mechanism. The advantage of this approach is that, in theory, it is very easy to obtain a population inversion between the HH1 and LH1 subbands in adjacent wells. Figure 6 shows population ratios for such a device, calculated using an ensemble Monte Carlo simulation with periodic boundary conditions [12]. The matrix element for the radiative transition is obviously dependent on the barrier widths, and figure 7 shows that a trade-off exists between population inversion and optical matrix element. Whilst the matrix element is reduced due to the spatial separation of the wavefunctions, it does not suffer from the symmetry problem described above for the LH1-HH1 intra-well matrix element, since the wavefunctions in adjacent, biased quantum wells do not have the same parity relative to the two-well co-ordinate frame.

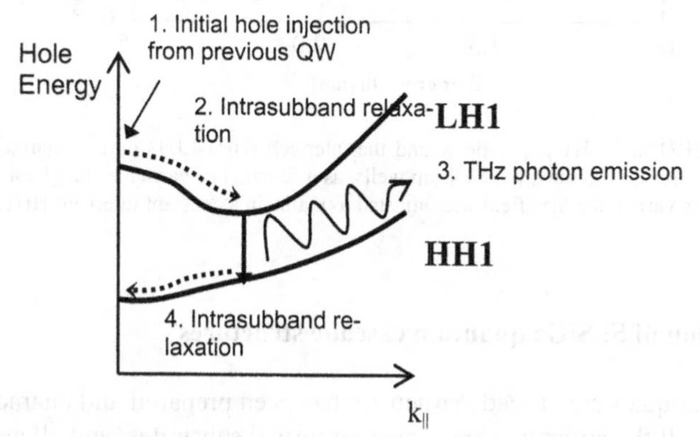

Figure 5. Use of an inverted mass feature in the LH1 subband dispersion to attain population inversion.

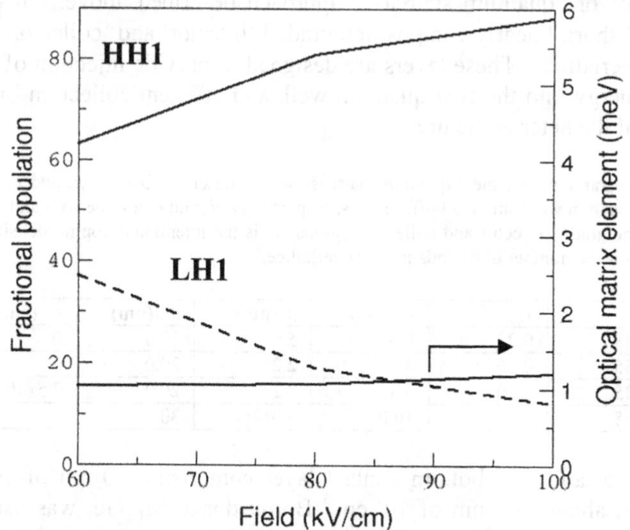

Figure 6. Simulated HH1 and LH1 populations for a $Si_{0.7}Ge_{0.3}$/Si quantum staircase structure with 4.4 nm wells and 2.2 nm barriers, grown on a $Si_{0.8}Ge_{0.2}$ virtual substrate, and biased to obtain interwell HH1-LH1 radiative transitions. The dependence of the optical matrix element on electric field is also shown.

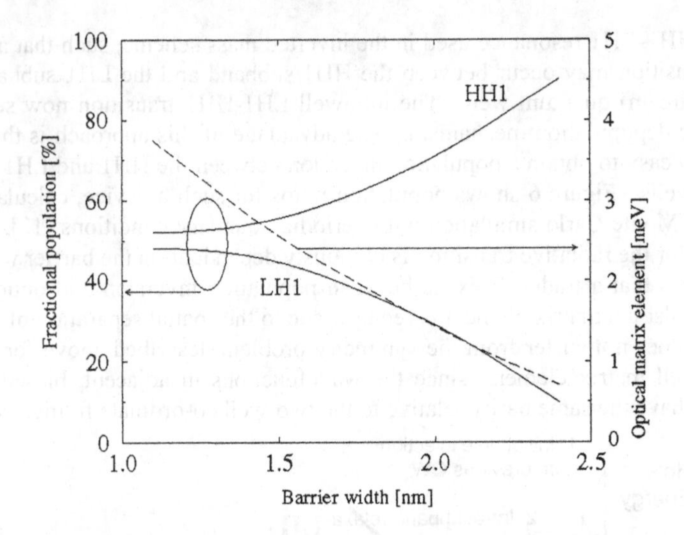

Figure 7. Simulated HH1 and LH1 populations, and the interwell (HH1-LH1) optical matrix element, for $Si_{0.7}Ge_{0.3}$/Si quantum staircase structures with 4.4 nm wells, as a function of barrier width (1 ml = 0.275 nm). As the barrier width was varied, the bias field was adjusted to maintain a constant interwell HH1-LH1 separation of 25 meV.

4. Characterisation of Si/SiGe quantum cascade structures

A range of different quantum cascade structures has been prepared and characterised, as shown in table 1. All the structures are grown on virtual substrates, and all are based on the superlattice or "quantum staircase" approach described above. In each case, we have included short, linearly composition graded 'injector' and 'collector' layers at each end of the superlattice. These layers are designed to provide injection of carriers at the appropriate energy into the first quantum well, and efficient collection of carriers from the last well of the heterostructure.

Table 1. Design parameters for the 4 quantum staircase wafers described below. x_w and x_{buf} are the Ge mole fractions in the quantum wells and the buffer layers, respectively, d_w and d_b are the well and barrier widths, d_{inj} is the width of the graded injector and collector regions, n_w is the intentional doping density in the quantum well layers and N is the number of periods in the superlattice.

	x_w	x_{buf}	d_w (nm)	d_b (nm)	d_{inj} (nm)	n_w (cm^{-3})	N
QCL3	0.28	0.23	8.0	5.0	30.0	0	30
QCL5	0.35	0.30	4.4	2.2	20.0	0	20
QCL6	0.35	0.30	4.4	2.2	20.0	7.5x10^{17}	20
QCL8	0.25	0.20	10.0	5.0	30.0		100

For QCL3,5, and 6, a bottom contact layer comprising 10 nm of 10^{18}cm^{-3} Boron doped $Si_{1-x}Ge_x$ above 400 nm of 10^{19}cm^{-3} Boron doped $Si_{1-x}Ge_x$ was used, with a top contact layer comprising 90 nm of 10^{19}cm^{-3} doped $Si_{1-x}Ge_x$ above 10 nm of 10 nm of 10^{18}cm^{-3} doped $Si_{1-x}Ge_x$. For QCL8, the bottom contact layer was reduced to 200nm thickness and the doping density increased to $3x10^{20}cm^{-3}$: the top contact layer was 40 nm thick and the doping density was $5x10^{19}cm^{-3}$.

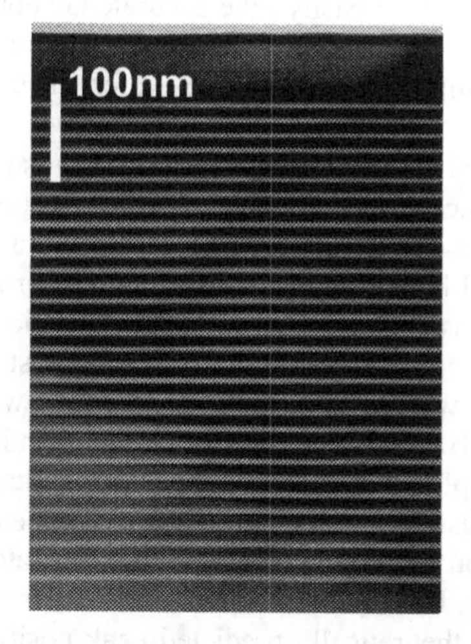

Figure 8. Bright field TEM image of the 30 period $Si/Si_{0.72}Ge_{0.28}$ wafer QCL3 [13]. The white lines are the silicon barriers

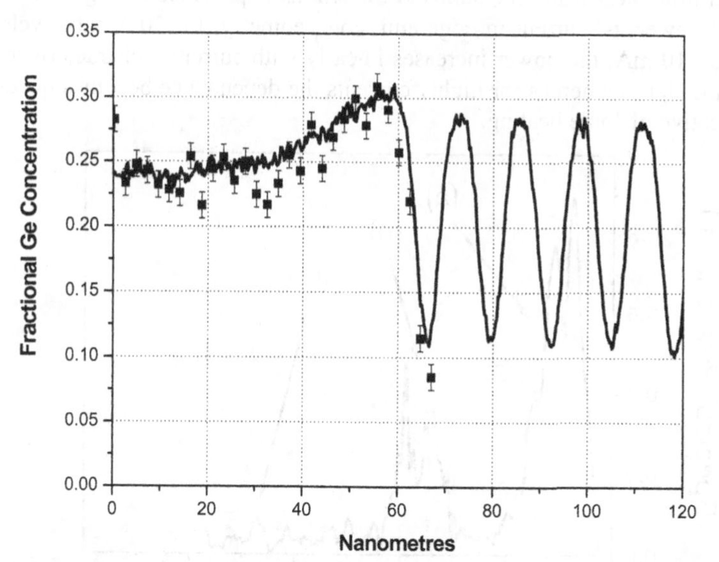

Figure 9. Germanium mole fraction through the graded injector layer and the lowest 4 quantum wells of the wafer QCL3, obtained by EELS (solid line) and EDX (squares) [13]

Following heterostructure growth, the layer thicknesses on all wafers were obtained using field-emission-gun (FEG) transmission electron microscopy (TEM). Figure 8 shows a bright field TEM image of the 30 period wafer QCL3. The planarity of the heterolayers, and the uniformity of layer thicknesses are excellent, with no evidence of strain relaxation. The germanium composition of the alloy layers was measured using both electron energy loss spectroscopy (EELS) and energy dispersive X-ray analysis (EDX). EDX is able to give absolute values of Ge composition to within ~1% Ge mole fraction, whereas EELS can most easily give accurate Ge composition values measured *relative* to a known reference. Figure 9 shows the Ge composition profile measured across the first few quantum wells of the QCL3 wafer. The linearly graded injector region is clearly visible.

Device processing involved the fabrication of 180μm square mesas by reactive ion etching, with ohmic contacts formed by sintering of evaporated aluminium (1% Si). The terahertz emission characteristics were then measured using a Bruker 66V step-scan spectrometer, with a liquid-helium-cooled silicon bolometer as a detector. The samples were placed in a continuous flow cryostat to enable variation of the measurement temperature from 4.2K up to room temperature. The entire system was evacuated to eliminate signal absorption by water vapour. A vertical bias was applied to the samples, which was pulsed at 413Hz with varying duty cycles, and the bolometer signal was measured via a lock-in amplifier, which used the pulsed voltage source as a reference.

THz emission was measured in both surface- and edge-emission geometries. Figure 10 shows the edge emission spectrum, in which three separate peaks are apparent, corresponding to the LH1-HH1, HH2-HH1 and LH2-HH1 intersubband transitions, and with good correlation with the theoretically predicted peak positions of 12 meV (2.9 THz), 40 meV (9.7 THz) and 67 meV (16.2 THz), respectively. Further confirmation that the THz radiation originates from intersubband transitions is provided by figure 11, which shows optical power vs. current in edge-emission geometry, for 50% duty cycle. For currents up to ~10 mA, the power increases linearly with current – characteristic of intersubband emission – whereas for higher currents the dependence becomes quadratic – which is indicative of Joule heating.

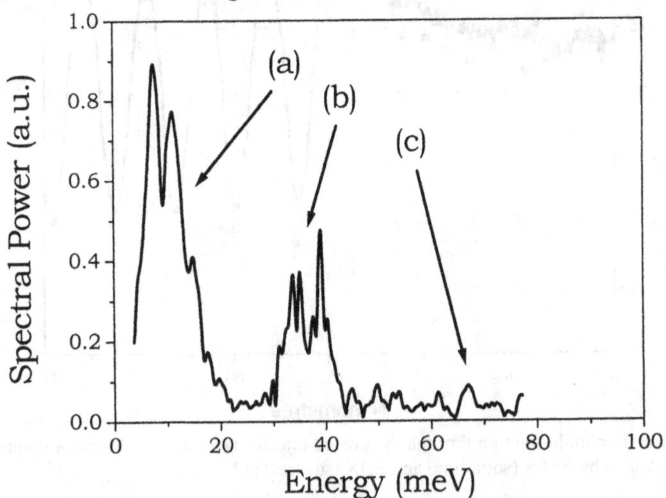

Figure 10. FTIR edge-emission spectrum for the quantum cascade structure QCL3, at a temperature of 4.2K. The pulsed bias voltage was 7 V with a 10% duty cycle [13]. The features marked (a), (b) and (c) correspond to the theoretically calculated emission peaks for the LH1-HH1, HH2-HH1 and LH2-HH1 intersubband transitions, respectively.

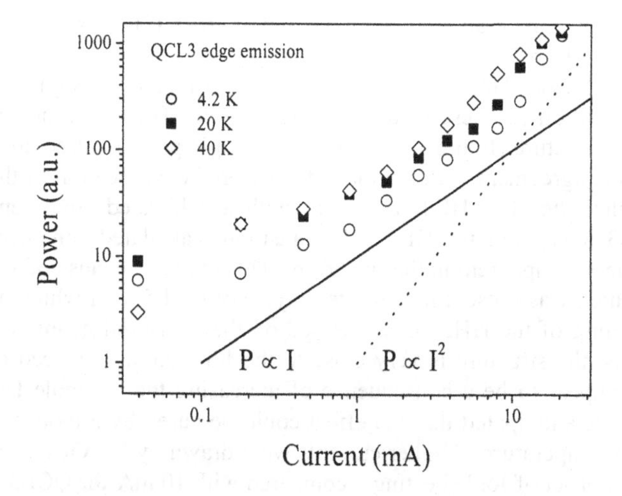

Figure 11. THz power output from a 180μm square mesa from the quantum cascade wafer QCL3 in edge emission geometry, as a function of current. [13] The full and dotted lines show the gradients for purely linear and purely quadratic dependence of power on current, respectively.

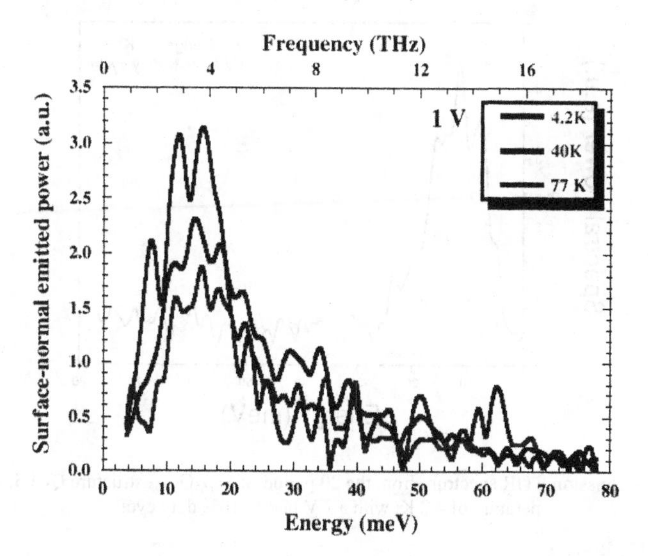

Figure 12. Surface-emission FTIR spectra for a 180μm square mesa from the wafer QCL3, for 3 different temperatures. The applied bias was 1V, with a duty cycle of 50%.

Figure 12 shows FTIR spectrum obtained from QCL3 in surface emission geometry, at a range of temperatures. No HH2-HH1 transition is observed, in accordance with the symmetry rules for intersubband transitions, but a strong peak is apparent, centred around 12 meV, corresponding once again to the LH1-HH1 transition. The intersub-band emission is still clearly visible at 77 K, although there is clearly increased black body emission at this temperature. The maximum emitted THz power in surface emission geometry was ~10 nW.

The 20 period $Si/Si_{0.65}Ge_{0.35}$ structure QCL5 was originally designed for emission via HH2-LH1 transitions, as described in section 3 above. However, the edge-emission spectrum for this structure (figure 13) shows just a single peak, centred around 13 meV (3.1 THz), in good agreement with the theoretically predicted location of the LH1-HH1 transition. Neither the HH2-HH1 transition (with a calculated zone centre subband separation of 63 meV), nor the HH2-LH1 transition (calculated zone-centre subband separation: 42 meV) is apparent in the spectrum. This may be because of a small degree of undulation which was present in the heterolayers of QCL5, and which may have resulted in broadening of the HH2 states. Based on the measured quantum well widths and compositions, the structure is very close to the fully strain-balanced condition, so the undulations appear to be a consequence of increasing the Ge mole fraction in the wells to 35%. It is anticipated that the effect could be cured by a modest reduction in the CVD growth temperature. Higher currents were drawn by the QCL5 mesas – up to 40 mA before the onset of Joule heating – compared with 10 mA for QCL3 mesas of the same area. This is probably due to the thinner barriers in QCL5 (2.2 nm compared with 5 nm), enabling increased tunnelling. However, the emitted power was much lower (~200pW).

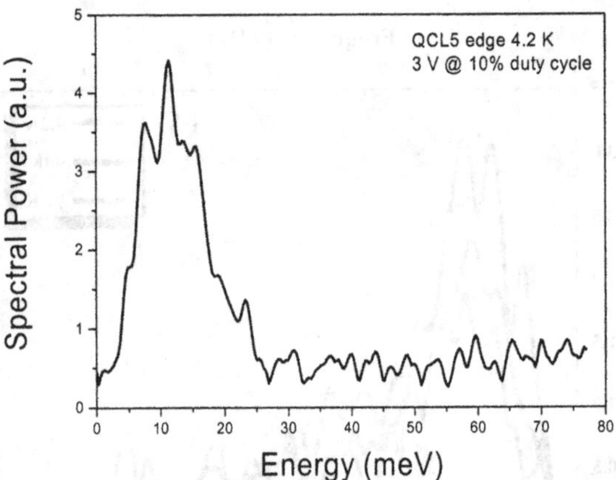

Figure 13. Edge-emission FTIR spectrum from the 20 period $Si/Si_{0.65}Ge_{0.35}$ structure QCL5, at a lattice temperature of 4.2 K, with a 3 V bias at 10% duty cycle.

The QCL6 heterostructure has the same well widths and Ge compositions as QCL5, but differs in that the quantum wells are boron doped to 7.5×10^{17} cm^{-3}, whereas, in QCL5, all the wells are undoped. The doping was introduced to facilitate pump-probe measurements of intersubband lifetimes. However, THz emission was also obtained from QCL6, in edge emission geometry. The FTIR spectra was very similar to that of QCL5, with a single peak centred around 13 meV. QCL6 showed even higher current drive than QCL6 – up to 80 mA before the onset of Joule heating – as a result of the increased carrier density. However, the emitted THz power was significantly lower for any given current, probably due to increased free carrier absorption.

Picosecond pump-probe spectroscopy was carried out on QCL6 using the free electron laser facility at Rijnhuizen, Holland. The laser can be tuned across the THz range to provide selective excitation of carriers from the lowest (HH1) subband to a higher subband. The laser beam is split to provide a probe beam by means of which the decay of sample transmission from the initial bleached state can be monitored as a function of time. The lifetime of carriers the selected subband can thus be extracted. In our experiments, the wavelength of the free electron laser was adjusted until a maximum was obtained in the initial transmission signal. This occurred at 95μm, in good agreement with the location of the peak in the FTIR spectra, thus confirming the intersubband origin of the THz emission. Pump-probe measurements carried out at a range of temperatures showed a characteristic lifetime for the LH1-HH1 transition of ~20ps [14]. Most significantly, the lifetime was essentially independent of temperature up to the maximum value (150 K) at which a signal could be obtained. There was some evidence of water absorption in the 95 μm range, so the laser beam was detuned slightly away from the peak of the LH1-HH1 transition, to 85 μm. At this wavelength, transmission relaxation was observed at temperatures ranging from 4.2 K right up to 300 K – again, with virtually no change in the measured subband lifetime. This behaviour is in marked contrast with the pump-probe response of n-type III-V heterostructures, where the intersubband lifetime decreases strongly, at temperatures above 40 K, due to polar optical phonon scattering. Our results provide clear evidence that the non-polar nature of the Si/SiGe system leads to a substantial (over an order of magnitude) reduction in the non-radiative intersubband scattering rates at high temperatures, compared to III-V systems.

Figure 14. *Bright field TEM image of the 100 period Si/$Si_{0.75}Ge_{0.25}$ quantum cascade structure QCL8*

The measurements are consistent with calculations showing that intersubband relaxation is dominated by alloy disorder scattering, which is only very weakly dependent on temperature [15]. One issue which still requires further work is that of hole-hole scattering. Scattering between carriers in light hole and heavy hole subbands is predicted to occur on a sub-picosecond timescale, due to strong mixing of the states for non-zero $\mathbf{k}_{||}$ [16]. However, the pump-probe experiments clearly show that carrier relaxation from the LH1 subband proceeds on a much longer timescale, hence hole-hole scattering does not provide any *net* transfer of carriers between subbands at such high rates.

For QCLs operating at THz frequencies, the active heterostructure must be several microns thick in order to attain sufficient overlap with the optical mode. Therefore, we increased the number of periods in our quantum staircase from the previous maximum of 30, up to 100. The wafer QCL8 was designed to have 100 10nm quantum wells, with 5nm silicon barriers, giving a total heterostructure thickness of 1.5μm. The quantum wells had 25% Ge mole fraction, and the virtual substrate buffer layer, 20%. Figure 14 shows the bright field TEM image for this wafer. The layers show a high degree of planarity, but there is some degree of variation of the layer thicknesses. (An alternative 100 period cascade structure has also been grown, with a total superlattice thickness of 600 nm, and shows excellent uniformity of layer thickness, so this may not be a recurrent problem. However, the results for QCL8 indicate that the Si/SiGe quantum cascade structures are able to tolerate some deviations from the exact strain balance condition without exhibiting any strain relaxation effects.)

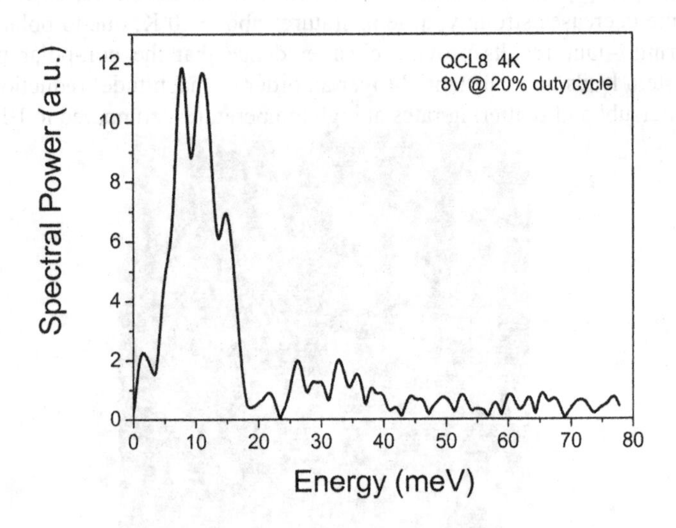

Figure 15. Surface emission FTIR spectrum for a 180μm square mesa from the wafer QCL8. The lattice temperature was 4K, with an applied bias of 8 V at 20% duty cycle.

Figures 15-16 shows the FTIR surface and edge emission spectra, respectively, for a 180 μm square mesa fabricated from this wafer. Both spectra show a distinct peak centred approximately around 12 meV, corresponding to the LH1-HH1 intersubband transition. It is notable that LH1-HH1 emission features for wafers QCL3, QCL5 and QCL8 all occur at very similar energies: this is because the Ge composition difference between quantum wells and virtual substrate is approximately the same (5%) in all three samples, and this parameter controls the strain in the quantum wells, and hence the LH1-HH1 splitting. The HH2-HH1 zone centre energy gap was calculated to be 25 meV for 10nm quantum wells; however, no feature is observed in this range in the edge emission spectra. This is a consequence of the variation of well widths, which will smear out the HH2-HH1 transition. On the other hand, these results also demonstrate clearly that the LH1-HH1 transition is not sensitive to changes in quantum well width, as discussed in section 3 above. The edge emission spectrum shows three distinct features centred on 30, 35 and 40 meV. These are the $1s\text{-}2p^1$, $1s\text{-}2p^2$, and $1s\text{-}2p^4$ interlevel transitions of the Boron dopants, and have been observed previously by us in modulation doped p-Si/SiGe heterostructures [17], and by Jagannath et al. in Boron doped bulk silicon [18].

The maximum THz power measured in surface-emission mode was approximately 50nW – well over 3 times the maximum obtained from a comparable 30 period structure - suggesting that all quantum wells in the 100 period structure were contributing to the emission. Although the superlattices in wafers QCL3, QCL5 and QCL8 are undoped, there was no sign of domain formation evident in the I/V curves for any of the devices.

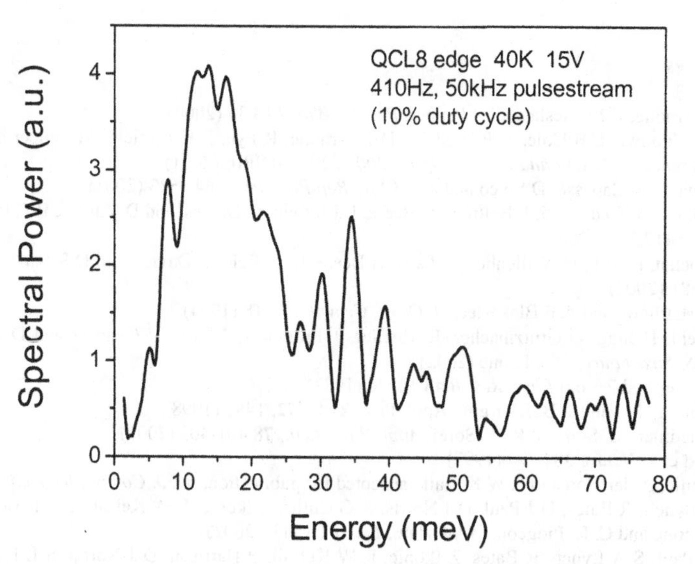

Figure 16. Edge-emission FTIR spectrum for the device described in Figure 15 above, at a temperature of 40K. The applied bias was 15V, with a 10% duty cycle.

Conclusions

Growth of p-Si/SiGe quantum cascade structures comprising up to 100 periods has been demonstrated using low pressure CVD via a strain-balanced approach on virtual substrates. The strain-balance technique enables, in principle, growth of heterostructures of arbitrary total thickness, and our results to date indicate that further increase in the number of periods is certainly possible. Intersubband THz electroluminescence from a range of Si/SiGe quantum cascade structures has been observed in both edge and surface-emission geometries. The LH1-HH1 intersubband lifetime has been measured using free electron laser pump probe spectroscopy, and was found to be virtually independent of temperature from 4.2K up to room temperature. A lifetime of ~20ps was obtained, which is over an order of magnitude longer than high temperature values in III-V heterostructures, implying that a Si/SiGe THz QCL may be capable of much higher operating temperatures than corresponding III-V devices. A maximum THz power of 50nW was measured for spontaneous emission in surface-normal geometry. This value is approximately 3 orders of magnitude higher than the power output obtained to date from mid-infrared p-Si/SiGe quantum cascade structures [19]. Most encouragingly, the power output is also 100 times higher than the spontaneous emitted power obtained for the GaAs/AlGaAs THz quantum cascade structure reported in [20] which, when subsequently fabricated into ridge-waveguide Fabry-Perot cavities, yielded THz laser action with stimulated emission of several mW [4]. In conclusion, these figures indicate that there are good prospects for realisation of a THz Si/SiGe QCL via further optimisation of the active region and appropriate cavity design.

This work is funded by DARPA on USAF contract F-19628-99-C-0074

References

1. D D Arnone, C M Ciesla and M Pepper, *Physics World* **13** 35 (2000)
2. R M Woodward, B Cole, V P Wallace, D D Arnone, R Pye, E H Linfield, M Pepper and A G Davies *Conference on Lasers and Electro-Optics 2001* 329-330 SPIE (2001)
3. C Gmachl, F Capasso, D Sivco and A Y Cho, *Rep Prog Phys* **64** 1533 (2001)
4. R Kohler, A Tredicucci, F Beltram, H Beere, E Linfield, G Davies, and D Ritchie, R C Iotti and F Rossi, *Nature* **417** 156 (2002)
5. M Rochat, L Ajili, H. Willenberg, J Faist, H Beere, E Linfield, G Davies, and D Ritchie, *Appl. Phys. Lett* **81** 1381 (2002)
6. J W Matthews and A E Blakeslee, *J. Cryst. Growth* **27** 118 (1974)
7. L Diehl, H Sigg, D Grutzmacher, E Mueller, T Fromherz, J Faist, U Gennser and D Bensahel *Proc. EMRS, Strasbourg,* 18-21 June (2002)
8. M Rieger and P Vogl *Phys. Rev. B* **48** 14276 (1983)
9. G. Sun, Y. Lu, and J. B. Khurgin, Appl. Phys. Lett., **72**, 1481 (1998).
10. L. Friedman, G. Sun and R. A. Soref, *Appl. Phys. Lett.,* **78** 401-403 (2001)
11. J Faist et al *Nature* **387** 777 (1997)
12. Z Ikonic, P Harrison and R W Kelsall, accepted for publication, *Int. J. Comp. Electronics* (2002)
13. S A Lynch, R Bates, D J Paul, D J Norris, A G Cullis, Z Ikonic, R W Kelsall, P Harrison, P.Murzyn, D D Arnone and C. R. Pidgeon, *Appl. Phys. Lett.* **81** 1543 (2002)
14. D J Paul, S A Lynch, R Bates, Z Ikonic, R W Kelsall, P Harrison, D J Norris, S L Liew, A G Cullis, P.Murzyn, C. R. Pidgeon and D D Arnone *Proc. EMRS, Strasbourg,* 18-21 June (2002)
15. Z Ikonic, P Harrison and R W Kelsall, Phys. Rev. B**64** 245311 (2001)
16. Z Ikonic, unpublished
17. S Lynch, S S Dhillon, R Bates, D J Paul, D D Arnone, D J Robbins, Z Ikonic, R W Kelsall, P Harrison, D J Norris and A G Cullis, *Mat. Sci. & Eng.* B **89** 10-12 (2002)
18. C Jagannath, Z W Grabowski and A K Ramdas, *Phys. Rev.* B**23(5)** 2082 (1981)
19. G Dehlinger, L Diehl, U. Gennser, H. Sigg, J. Faist, K. Ensslin, D. Grutzmacher, E. Muller, *Science,* **290**, 2277 (2000).
20. R Kohler, A Tredicucci, F Beltram, H Beere, E Linfield, G Davies, and D Ritchie *Appl. Phys. Lett* **80**(11) 1867 (2002)

TOWARDS AN ER-DOPED SI NANOCRYSTAL SENSITIZED WAVEGUIDE LASER – THE THIN LINE BETWEEN GAIN AND LOSS

PIETER G. KIK AND ALBERT POLMAN
F.O.M. Institute for Atomic and Molecular Physics
P.O. Box 41883, 1009 DB Amsterdam, The Netherlands

1. Introduction

Current Er-doped planar waveguide amplifiers and lasers make use of a pump laser emitting at 0.98 μm that excites Er^{3+} ions into their second excited state ($^4I_{11/2}$). From this level, the ions quickly relax non-radiatively to the long-lived first excited state ($^4I_{13/2}$) providing the population inversion necessary for gain (see Figure *1*(a)). For an Er doped waveguide laser, this means an accurately tuned semiconductor pump laser needs to be integrated with an optical waveguide. This is an expensive approach, since it requires an expensive pump laser and involves a difficult alignment procedure to obtain optimum coupling between pump laser and waveguide. If the pump laser could somehow be eliminated from this scheme, the fabrication of low-cost Si based Er doped optical amplifiers and lasers operating at 1.5 μm would become a real possibility. This is exactly what co-doping with Si nanocrystals promises to do.

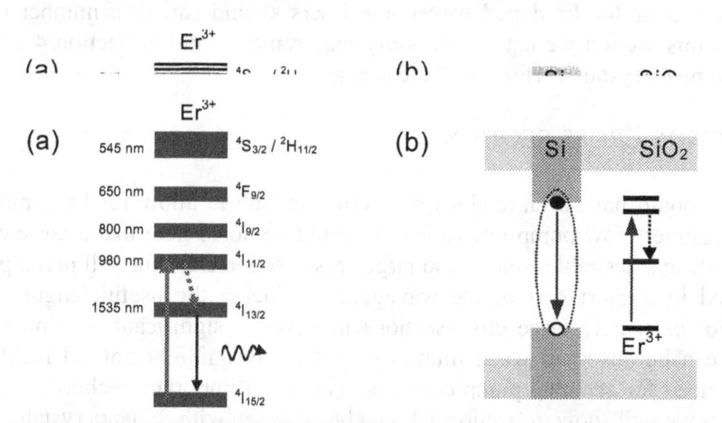

tal in SiO_2, showing the recombination of a quantum confined electron-hole pair and simultaneous Er excitation

It has been observed by several groups[1-7] that Er can be excited indirectly in Si nanocrystal doped SiO_2. The excitation occurs via optically generated electron-hole pairs inside Si nanocrystals that recombine and transfer energy to the Er ions. Figure *1*(b) shows a schematic band diagram of SiO_2 containing a Si nanocrystal and Er^{3+}. First

L. Pavesi et al. (eds.), Towards the First Silicon Laser, 383–400.

a photon is absorbed by the nanocrystal, which causes generation of an exciton inside the nanocrystal. This exciton can recombine radiatively, emitting a photon with an energy that depends on the nanocrystal size. If an Er ion is present close to the nanocrystal, the exciton can recombine non-radiatively by bringing Er into one of its excited states. Si nanocrystals can be excited in a broad wavelength range ($\lambda < \sim 1$ μm) and have a pump absorption cross-section that is several orders of magnitude higher than the cross-section for direct Er excitation. These two features make codoping with Si nanocrystals a promising route to fabricating integrated waveguide amplifiers and lasers that can be pumped with low-power broadband light sources.

Er excitation via Si nanocrystals must meet several additional requirements to be of use in an optical amplifier. In this article we show measurements that demonstrate (a) the large nanocrystal absorption cross-section $\sigma_{abs} > 10^{-16}$ cm^2, (b) a high energy transfer efficiency $\eta \geq 55\%$, (c) the broad absorption band of Si nanocrystals, (d) a high transfer rate $R \geq 10^6$ s^{-1}, (e) a limit to the excitable Er concentration of ~1-2 Er ions per nanocrystal, (f) evidence for free-carrier induced absorption in Si nanocrystal doped optical waveguides, and (g) an unexpected enhancement of the direct optical absorption cross-section of Er in a nanocrystal doped medium. We discuss how all these observations affect the potential use of Er doped Si nanocrystal co-doped SiO_2 as an optical gain medium. Finally, calculations will be presented that offer an indication of the net gain that can be expected in this type of waveguides amplifiers.

2. Er sensitizers – requirements

A potential sensitizer for Er doped waveguide lasers should satisfy a number of requirements. In this section we list all necessary requirements, and in Section 4 we will discuss how Si nanocrystals perform in these areas.

2.1. ABSORPTION CROSS-SECTION

The sensitizer should have a large absorption cross-section to allow for Er population inversion at relatively low pump intensities. It should be noted here that a large cross-section is not always desirable, since a too large cross-section can cause all pump power to be absorbed in a short part of the waveguide, reducing the useful length of the waveguide. For extremely large cross-sections however, a significant fraction of the pump power can be absorbed when illuminating the waveguide at normal incidence, removing the need for accurate pump coupling. This is an important technological advantage that, as we will show in Section 4.1, can be achieved with Si nanocrystals.

2.2. EXCITATION EFFICIENCY

Absorbed energy should be transferred to Er with high quantum efficiency and high power efficiency. In the case of nanocrystals we will show that the quantum efficiency is larger than 55%. The power efficiency is wavelength dependent: when a wavelength of 458 nm is used, the 2.7 eV pump photons generate a 0.8 eV Er excited state, with the excess 1.9 eV transformed into heat. This means that a quantum efficiency of 55% corresponds to a power efficiency of 16%. Consequently, for a high power efficiency it may be advantageous to excite the Si nanocrystals longer wavelengths. Note however that at longer wavelength the Si absorption cross-section is reduced.

2.3. ABSORPTION LINEWIDTH

The pump source is one of the main contributions to the cost of Er doped amplifiers, since it has to perfectly match the narrow Er absorption band at 980 nm. If a sensitizer is used, the pump wavelength needs to be matched to the absorption of the sensitizer. In case of a sufficiently wide sensitizer absorption band, more cost efficient broadband pump sources such as LEDs can be used for excitation, allowing for a significant cost reduction of the laser or amplifier.

2.4. TRANSFER RATE

In any laser or amplifier, the total possible output power is determined by the rate at which the amplifying transition is pumped. In the case of a sensitized Er transition the energy transfer rate from the sensitizer to Er could form an excitation bottleneck. Once the transfer rate has been determined, an estimate of the maximum output power can be made, as will be shown in Section 4.4.

2.5. EXCITABLE FRACTION

When using an Er sensitizer, it is essential that all Er ions can be excited. This is especially important for three level systems such as the 1.5 μm transition of Er. In Er^{3+} the amplifying transition occurs from the first excited state to the ground state, and consequently any fraction of non-excitable Er will efficiently absorb at the signal wavelength. Note that if the amplifying center is a four-level system in which the relevant transition occurs between two excited states, the problem of signal absorption is completely removed.

2.6. INDUCED LOSS

Any waveguide loss introduced by the sensitizer will reduce the maximum output power of the sensitized laser or amplifier. Possible loss contributions include increased scattering due to an inhomogeneous sensitizer distribution, changes in the materials processing that introduce waveguide roughness, and introduction of free charges that can lead to free carrier absorption. The latter effect is of special relevance to the case of Si nanocrystal sensitization, since the nanocrystal excitation requires the generation of electron-hole pairs, which can in fact absorb at 1.5 μm wavelength. In bulk Si, free carrier absorption cross-sections are as large as 10^{-17} cm^2 around 1.5 μm[8]. This is more than a factor 1000 higher than a typical Er emission cross-section, meaning that the presence of one exciton for every 1000 Er ions could completely suppress any amplification. It should be noted that free carrier absorption cross-sections in Si nanocrystal doped SiO_2

may be reduced compared to those in bulk Si due to quantum size effects, and that the direct Er cross-section at 1.5 μm seems to be enhanced in Si nanocrystal doped SiO_2 (see Section 4.6.).

3. Experimental

This paper contains results on two different sample types. The first part deals with thin oxide layers doped with Er and Si nanocrystals by means of ion implantation. For these samples, a 100 nm thick layer of SiO_2 was grown on a lightly B-doped Si(100) substrate by means of wet thermal oxidation. This layer was implanted with 35 keV Si to a dose of 6×10^{16} cm^{-2}. The implantation yields an approximately Gaussian depth distribution of excess Si in the SiO_2 film, with a peak concentration of 19 at.% (corresponding to a composition of $Si_{0.46}O_{0.54}$) at a depth of 45 nm. The samples were subsequently annealed at 1100 °C for 10 min. in vacuum at a base pressure below 3×10^{-7} mbar in order to induce nucleation and growth of Si nanocrystals. This treatment has been shown to produce Si nanocrystals with a diameter in the range 2-5 nm.[9] Assuming a typical nanocrystal diameter of 3 nm (~10^3 Si atoms), the nanocrystal peak concentration is estimated to be ~10^{19} cm^{-3} at the center of the SiO_2 layer. The samples were then implanted with different Er doses in the range 3.6×10^{13} cm^{-2} – 5.1×10^{15} cm^{-2} at a fixed energy of 125 keV. These implants result in an approximately Gaussian Er depth distribution, with Er peak concentrations ranging from 0.015 at.% to 1.8 at.% at a depth of 61 nm. A non Er-implanted sample was kept as a reference. After the Er implants, all samples (including the non-Er implanted) were annealed for 10 min. in vacuum at 1000 °C to remove implantation-induced damage. In order to further reduce defect-related luminescence and to saturate dangling bonds on the Si nanocrystal surface, a passivating anneal was performed at 780 °C for 30 min. in forming gas (H_2:N_2 at 1:9) at atmospheric pressure.

The second series of experiments involve the fabrication of Er doped nanocrystal sensitized waveguide samples in SiO_2. Silicon ions were implanted at 165 keV to a fluence of 1.7×10^{17} cm^{-2} into a 5 μm thick layer of SiO_2 that was grown by wet thermal oxidation of Si(100). The material was annealed at 1100 °C for 10 min. in flowing Ar in order to induce nucleation and growth of Si nanocrystals. The temperature of 1100 °C was found to yield maximum nanocrystal luminescence. This material was implanted with Er at 700 keV to a fluence of 1.2×10^{16} cm^{-2} and annealed at 1000 °C for 10 min. in flowing Ar to remove implantation induced damage. A sample containing no Er was kept as a reference. These samples were annealed at 800 °C for 10 min. in forming gas (H_2:N_2 at 1:9) to optimize the nanocrystal and Er photoluminescence intensity. The excess Si concentration profile peaks at a depth of 240 nm, and has a full width at half maximum (FWHM) of 160 nm. The Si peak concentration is 42 at.%, corresponding to an excess Si concentration of 13 at.%. Assuming an average nanocrystal diameter of 3 nm, the nanocrystal concentration at the peak of the Si profile is estimated to be 1.3×10^{19} cm^{-3}. The Er concentration profile peaks at a depth of 240 nm, with an Er peak concentration of 1.3 at.% and a FWHM of 130 nm.

A series of 3.5 μm wide ridge waveguides was formed in the implanted and annealed SiO_2 film using standard photolithography and Ar beam etching to a depth of

0.5 μm, i.e. well beyond the Si doped region. The excess Si inside the ridges locally raises the index of refraction, providing the index contrast required for transverse and lateral optical mode confinement. To reduce scattering losses, the waveguides were covered with a 1.25 μm thick SiO_2 cladding layer using microwave sputtering. The samples were subsequently cut to a length of 8 mm, and the waveguide input and output facets were mechanically polished.

Photoluminescence (PL) measurements on planar samples were taken using the 458 nm line of an Ar laser as excitation source at a peak power of 1 mW in a ~1 mm^2 laser spot. The laser beam was modulated on-off at 11 Hz using an acousto-optical modulator. The emitted light was passed through a grating monochromator and detected using standard lock-in techniques. Spectra were measured in the range 600-1150 nm using a AgOCs photomultiplier tube (PMT), and in the range 1100-1700 nm using a liquid-nitrogen cooled Ge detector. All spectra were corrected for the detector response. Photoluminescence decay traces were recorded using a multichannel photon counting system in combination with the PMT and a digitizing oscilloscope in combination with the Ge detector. The system response for the two cases was 150 ns and 160 μs, respectively. Except when noted otherwise, measurements on the planar samples were performed at 15 K using a closed-cycle He cryostat.

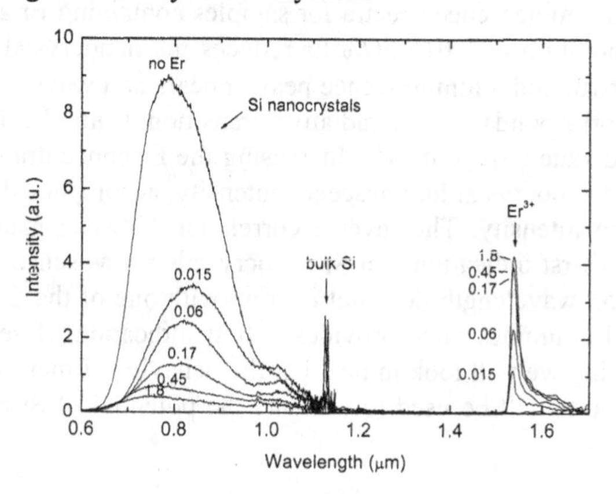

Figure 2. Photoluminescence spectra of Si nanocrystal doped SiO2 containing different Er concentrations in the range 0 - 1.8 at.%, measured at 15 K using a pump power of 1 mW at 458 nm.

Optical transmission measurements on the waveguide samples were performed using a fiber-coupled white light source emitting in the 1000-1800 nm range, which was modulated on-off at 270 Hz for lock-in detection. The white light was butt-coupled to the nanocrystal waveguides using a single mode tapered fiber. The transmitted light was collected by imaging the waveguide output facet onto a multimode fiber using a 10× microscope objective. The collected light was led to a grating monochromator and detected using a liquid nitrogen cooled Ge detector, in combination with standard lock-in techniques. Optical mode images were obtained by coupling 1.49 μm light from an InGaAsP diode laser into the waveguide, and imaging the waveguide output facet on an infrared camera using a 10× microscope objective.

For gain measurements the Si nanocrystal and Er doped waveguides were optically pumped CW using the 458 nm line of an Ar laser. The pump light was projected onto the waveguides from the top using a cylindrical lens. The elongated spot was aligned with the waveguide, and covered the full waveguide length. In this way the pump power was evenly distributed over the full length of the waveguide. In these measurements the modulated white light source was coupled into the waveguide, and the transmitted signal around 1.5 μm was detected as a function of pump power using lock-in detection.

4. Results and discussion

Figure 2 shows a PL spectrum of a SiO_2 film doped with Si nanocrystals (marked "no Er"). The sample shows a broad luminescence band peaking at 790 nm. This luminescence is caused by the radiative recombination of electron-hole pairs (excitons) confined within the Si nanocrystals. Due to quantum confinement[10] the exciton luminescence appears at energies above the bandgap energy of bulk Si (1.17 eV at 15 K). The large spectral width of the nanocrystal luminescence in these samples is caused mainly by the broad nanocrystal size distribution (2-5 nm diameter). The luminescence peak at 1.13 μm is caused by phonon-assisted electron-hole pair recombination in the Si substrate. Figure 2 also shows luminescence spectra for samples containing Er at various concentrations. The incorporation of 0.015 at.% Er reduces the nanocrystal luminescence by more than a factor two, and a luminescence peak appears at a wavelength of 1.536 μm. This wavelength corresponds to the radiative transition from the first excited state ($^4I_{13/2}$) to the ground state ($^4I_{15/2}$) of Er^{3+}. Increasing the Er concentration leads to a further reduction of the nanocrystal luminescence intensity, accompanied by an increase of the Er luminescence intensity. The inverse correlation between nanocrystal emission and Er emission is a first indication that Si nanocrystals act as sensitizers for Er. Since the applied excitation wavelength does not overlap with one of the Er absorption lines, the appearance of Er luminescence provides a first indication of sensitized luminescence. In the following we will look in detail at the behavior of these two luminescence features, and the results will be used to evaluate the potential of Si nanocrystals as Er sensitizers.

4.1. ABSORPTION CROSS-SECTION

In order to determine the nanocrystal absorption cross-section, we measured the nanocrystal luminescence $1/e$ rise time (τ_r) and decay time (τ) at 5 mW pump power (see time dependent data in Fig. *3*(a)). From these two time constants we can deduce the excitation rate $R = (1/\tau_r - 1/\tau)$, assuming that the nanocrystals effectively behave as a two-level system. At a pump intensity of ~5 mW/mm^2 we find for the nanocrystal excitation rate $R_{nc} = \sigma_{nc} \times \varphi_{phot} = 1040\ s^{-1}$ with σ_{nc} the nanocrystal absorption cross-section and φ_{phot} the photon flux. Inserting the known photon flux of $\sim 1\times10^{18}\ cm^{-2}s$ we find $\sigma_{nc} = 9\times10^{-16}\ cm^2$. This value is several orders of magnitude larger than direct Er optical absorption cross-sections, which are typically in the $10^{-21}\ cm^2$ range. Note that the value obtained is somewhat larger than other values reported at this excitation wavelength, see e.g. Künzer *et al.*[11] ($\sigma_{abs,458nm} \approx 4\times10^{-16}\ cm^2$) and Priolo *et al.*[12] ($\sigma_{abs,488nm} \approx 1\times10^{-16}\ cm^2$). This kind of discrepancy may be inherent to the type of measurement: the cross-section depends on the nanocrystal diameter to the ~3rd power, and the obtained value for the effective cross-section depends quadratically on the estimated laser beam diameter.

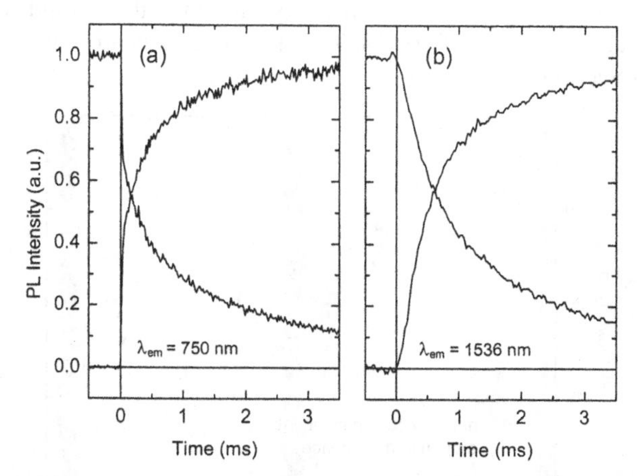

Figure 3. Luminescence rise time and decay measurements at 5 mW pump power at 15 K for (a) nanocrystal luminescence at 750 nm, and (b) Er luminescence at 1536 nm

As was mentioned in the introduction, the large absorption cross-section of Si nanocrystals may allow for unconventional pumping geometries, such as normal-incidence pumping, or side-illumination in which the whole waveguide length is illuminated by a source that is coupled into a slab-waveguide mode. For this type of geometry to be effective, a significant fraction of the pump power should be absorbed within a typical waveguide thickness. To see whether this is possible we can estimate the absorption depth in typical Si nanocrystal doped samples. The absorption depth can be obtained from the absorption coefficient $\alpha(\text{cm}^{-1})=\sigma_{abs}\times C_{nc}$. If we take a conservative absorption cross-section of 10^{-16} cm^2 and a typical nanocrystal concentration of 10^{19} cm^{-3}, we obtain an absorption depth $d_{1/e} = 1/\alpha = 10$ μm. Consequently, pumping at a wavelength of 458 nm, 63% of the pump power can be absorbed in a 10 μm thick waveguide. Pumping at shorter wavelengths will further reduce the required waveguide thickness. This shows that pumping either laterally or vertically is a realistic option in this type of material.

4.2. EXCITATION EFFICIENCY

In order to determine the Er excitation efficiency, we have also measured the Er luminescence $1/e$ rise time (τ_r) and decay time (τ) on the same sample as discussed in Section 4.1 (see Fig. 3(b)) . Under the same pump conditions we find for the Er excitation rate R_{Er} = 570 s^{-1}, which is of the same order of magnitude as the previously obtained nanocrystal excitation rate R_{nc} = 1040 s^{-1}. These numbers give an estimate of the energy transfer efficiency, or the internal quantum efficiency. Assuming each nanocrystal couples to only one Er ion, we conclude that 570/1040=55 % of the generated excitons recombine by transferring energy to the Er ion.[13] The obtained 55% is a lower estimate for the excitation efficiency. For example, if we assume each nanocrystal couples to two

Er ions, the transfer efficiency would be 2×570/1040 ≈ 100%. The requirement for efficient energy transfer is clearly satisfied. Note that it is difficult to argue that in this sample more than 2 Er ions couple to one nanocrystal, since this would lead to internal quantum efficiencies in excess of 100 %. This apparent concentration limit will be further discussed in Section 4.5.

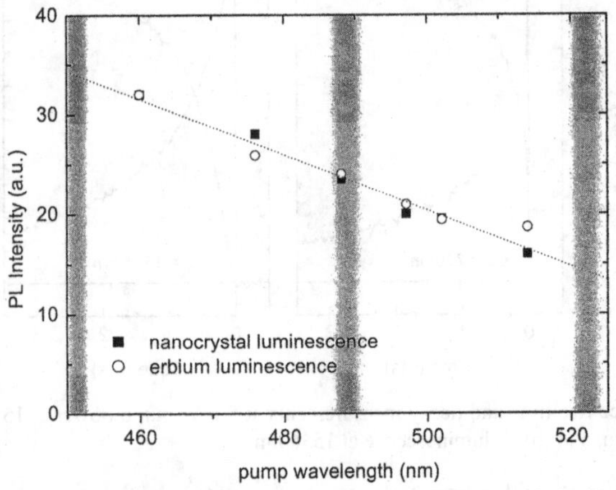

Figure 4. Excitation spectrum of the nanocrystal luminescence at 750 nm and the Er luminescence at 1536 nm. The dotted line is a guide to the eye. The vertical bands indicate the position of direct Er optical absorption lines.

4.3. ABSORPTION LINEWIDTH

Figure *4* shows photoluminescence excitation spectra of the nanocrystal luminescence at 750 nm and the Er luminescence at 1536 nm. At fixed pump power, the nanocrystal luminescence intensity and the Er luminescence intensity are seen to increase by a factor two when the excitation wavelength is decreased from 514 nm to 458 nm. The absence of absorption resonances in the Er excitation spectrum indicates that the Er is not excited directly by absorption of pump photons, but indirectly via an energy transfer process from Si nanocrystals, as was already suggested above. The observed wavelength dependence of the luminescence intensity is attributed to the increasing nanocrystal optical absorption cross-section for increasing photon energy, leading to a higher exciton generation rate. These results give direct evidence that Er in this material can be excited in a broad wavelength range, allowing for broadband pumping of amplifiers and lasers based on this materials system.

4.4. TRANSFER RATE

Figure 5 shows nanocrystal luminescence decay traces taken at 750 nm and T=15 K for samples containing different Er concentrations. In a sample containing no Er, the nanocrystal luminescence shows a 1/e lifetime of 2.0 ms. The decay is described by stretched exponential decay of the form $I(t)=\exp(-(t/\tau)^{\beta})$ with τ=2.0 ms and β=0.65 (solid line). The same curve has been overlaid on all data in order to facilitate comparison between the different traces. Incorporation of Er leads to a significant reduction of the nanocrystal luminescence intensity at 750 nm, as was already observed in Fig. 2. The decay time however varies only slightly from sample to sample, and no trend is observed with increasing Er concentration. The sample containing the highest Er concentration shows a small initial fast decay, which is attributed to luminescence from defects in the SiO_2 matrix. These lifetime measurements confirm the strong-coupling model,[6] which states that the Er induced nanocrystal PL intensity decrease in Fig. 2 is effectively due to a reduction in the number of luminescent nanocrystals. All remaining nanocrystal luminescence originates from nanocrystals that are not coupled to Er, which show their intrinsic decay characteristics. This model also explains the initially surprising observation that the Er luminescence intensity is virtually temperature independent, whereas the nanocrystal luminescence intensity varies by as much as a factor 4 in the temperature range 20 K - 300 K. This behavior is now understood: in nanocrystals that do not couple to Er, temperature dependent radiative and non-radiative exciton recombination rates affect the luminescence efficiency, while in nanocrystals that couple to Er, the non-radiative transfer to Er is the dominant recombination process, occurring at a (high) fixed efficiency over the full temperature range. These observations also allow us to obtain an upper limit to the energy transfer rate constant, since the energy transfer has to occur at a significantly higher rate than the spontaneous exciton recombination in order to be the dominant recombination mechanism. The fastest observed nanocrystal decay time in these samples is 21 μs, and consequently the energy transfer has to occur faster than ~1 μs.

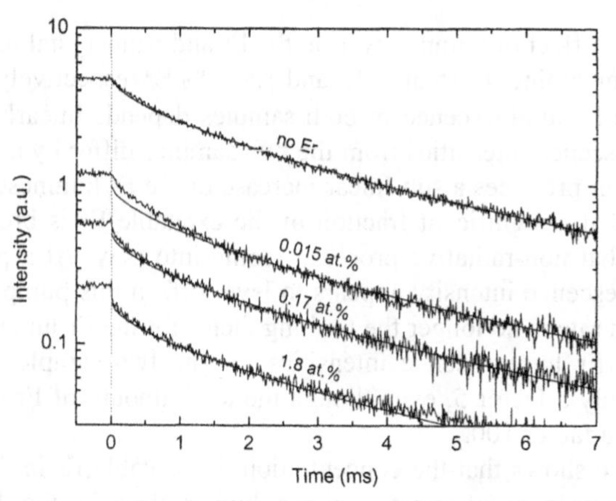

Figure 5. Photoluminescence decay traces of the nanocrystal luminescence at 750 nm for different Er concentrations in the range 0 - 1.8 at.%, measured at 15 K using 1 mW pump power at 458 nm. The pump is switched off at t=0. The drawn lines represent stretched exponential decay with τ=2.0 ms and β=0.65.

Given the obtained value for the energy transfer rate of $(1\ \mu s)^{-1} = 10^6\ s^{-1}$, we can determine the maximum output power of an Er doped Si nanocrystal sensitized waveguide amplifier. Assuming a nanocrystal concentration of $10^{19}\ cm^{-3}$ and 2 excitable Er ions per nanocrystal in a $10\ \mu m \times 10\ \mu m = 10^{-6}\ cm^2$ waveguide, we find a maximum output power of $2\times10^{19}\ cm^{-3} \times 10^{-6}\ cm^2 \times 10^6\ s^{-1} \times 0.8\ eV = 2.6$ W/cm. This shows that the transfer rate is sufficiently fast to allow for the amplification of high power signals over short waveguide lengths.

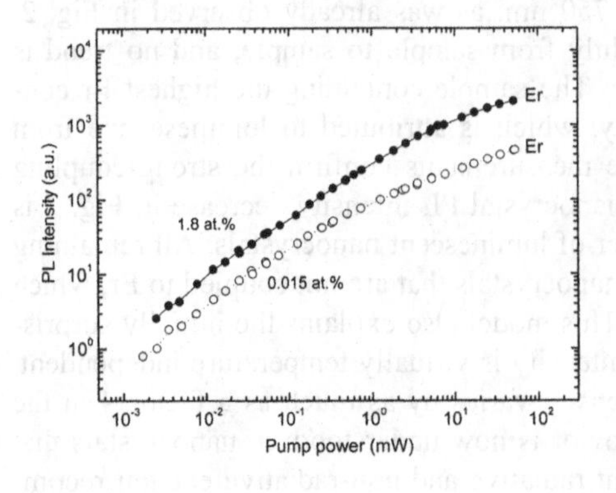

Figure 6. Photoluminescence intensity of the Er luminescence intensity as a function of pump power for samples containing 0.015 at.% and 1.8 at.% Er, measured at 15 K using 458 nm pump light in a ~1 mm² spot.

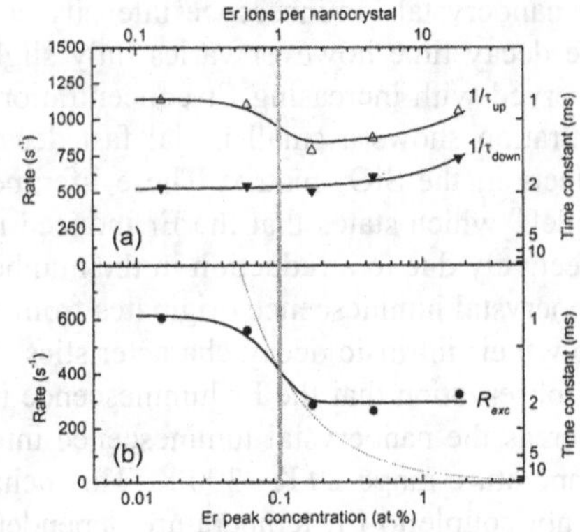

Figure 7.(a) Rise times (Δ) and decay times (▼) of the 1.536 μm Er luminescence, measured at 15 K using a pump power of 1 mW at 458 nm, and (b) the Er excitation rate derived from these data (●). The drawn lines serve as a guide to the eye. The dotted line indicates a $1/C_{Er}$ dependence.

4.5. EXCITABLE FRACTION

Figure *6* shows the effect of pump power on the Er and nanocrystal luminescence intensity for samples containing 0.015 at.% Er and 1.8 at.% Er respectively. At pump powers below 20 μW the Er luminescence in both samples depends linearly on pump power, and the Er luminescence intensities from the two samples differ by a factor ~2. Increasing the pump power produces a sub-linear increase of the Er luminescence in both samples, suggesting that a significant fraction of the excitable Er is brought into the first excited state, or that non-radiative processes come into play. At a pump power of 50 mW the Er luminescence intensity appears to level off. In this pump power regime the exciton generation rate is no longer the limiting factor for the Er luminescence intensity. Nevertheless, the Er luminescence intensities for the two samples at 50 mW pump power differ by only a factor 5, even though the total amount of Er in the samples differs by more than a factor 100.

The above shows that the concentration of excitable Er in the high concentration (1.8 at.%) sample is at most 5 times higher than in the low concentration (0.015 at.%) sample. This means that the concentration of excitable Er in these samples is less than 5×0.015 = 0.1 at.%. The existence of such a concentration limit is also observed in the Er excitation rate, as will be shown below.

When the number of Er ions coupled to a nanocrystal is increased, the Er excitation rate per ion can be expected to decrease, since several Er ions now compete for the same excitons. In order to determine the concentration dependent Er excitation rate, we performed rise time and decay time measurements of the Er luminescence at 1.536 μm. At the applied pump power of 1 mW all samples show approximately exponential time dependencies. Figure 7(a) shows the measured rates $W_{rise}=1/\tau_{rise}$ and $W_{decay}=1/\tau_{decay}$ obtained by exponential fitting of the data.

The Er decay rate increases from 500 s^{-1} to 700 s^{-1} as the Er peak concentration is increased from 0.015 at.% to 1.8 at.%. This increase is attributed to a concentration quenching effect which is known to occur when rare earth ions are spaced closely enough to allow for energy exchange between neighboring ions. As a result, excitation energy can migrate[14] to neighboring ions, which may in turn be coupled non-radiatively to quenching sites, *e.g.* defects or OH groups present in the matrix. In a simple concentration quenching model, the Er decay rate increases linearly with Er concentration, which is indeed observed in Fig. 7(a). By measuring the slope of the decay rate data in Fig. 7(a), we estimate[15] that the concentration of quenching sites in the Er implanted Si nanocrystal doped SiO_2 film is as low as 10^{18} cm^{-3}.

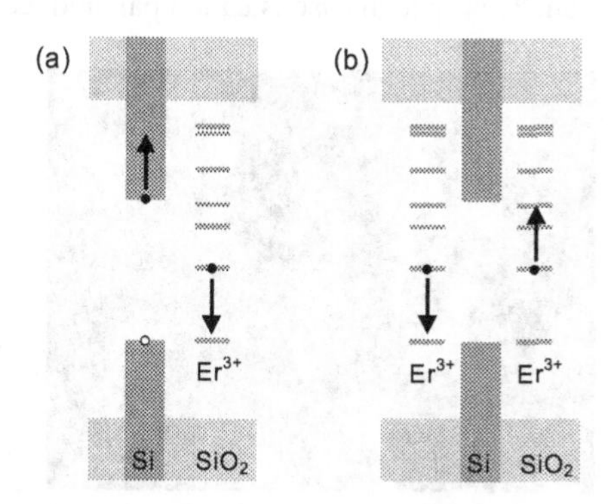

Figure 8. Schematic band diagram of Er and Si nanocrystal doped SiO_2 showing the process of (a) Auger de-excitation, and (b) pair-induced quenching. These processes can account for the observed Er concentration limit in Si nanocrystal doped SiO_2. Note that the Er energy levels represent collective states of the Er^{3+} 4f electrons, which are in fact well below the SiO_2 valence band edge.

Data for the Er excitation rate R_{exc} calculated from τ_{rise} and τ_{decay} are shown in Fig. 7(b) ( ). At an Er concentration of 0.015 at.%, the Er excitation rate is 600 s^{-1}. Increasing the Er concentration to 0.17 at.% reduces the excitation rate by a factor 2. A further increase of the Er concentration has no effect on the excitation rate, even though at these concentrations several Er ions might couple to the same nanocrystal, which would reduce the excitation rate per Er ion. The fact that such a reduction is not observed, shows that there is an upper limit to the number of Er ions that can be excited by a single nanocrystal. From the data in Fig. 7(b) it is clear that this limit is reached at an Er concentration < 0.17 at.% Er, which is consistent with the maximum value of 0.1 at.% Er found in the preceding paragraph.

The existence of an upper limit to the amount of excitable Er could indicate that there is a limit to the amount of *optically active* Er that can be incorporated in this material. This may for example be caused by Er clustering or $ErSi_2$ formation, which would prevent the Er from being in the 3+ valence state. However, measurements on Er and Si nanocrystal doped waveguides show a strong Er^{3+} related absorption,[16] suggesting that a large fraction of the Er is in the optically active state. Alternatively, the observed concentration limit could be an intrinsic property of the excitation process. The amount of excitable Er will be low when the effective Er excitation efficiency is reduced by the presence of *excited* Er. Such a situation arises when the formation of an exciton near an excited Er ion leads to

a) *Auger de-excitation*, in which an excited Er ion transfers energy to an exciton present in the nanocrystal (Fig. 8(a)). This process has been shown to occur in Er doped bulk Si.[17] After such Auger de-excitation the exciton can relax and subsequently excite an Er ion, effectively bringing the system back to the situation before the exciton was formed.

b) *pair-induced quenching.* At sufficiently high Er concentration, two excited Er ions can interact yielding one Er ion in the $^4I_{9/2}$ state, which rapidly decays to the first excited state, and one Er ion in the ground state (Fig. 8(b)). This co-operative upconversion effect usually produces a shortening of the Er decay rate at high pump powers, which has not been observed. However, if the Er-Er coupling is sufficiently strong, no effect on the lifetime is seen. This special case is called pair-induced quenching.

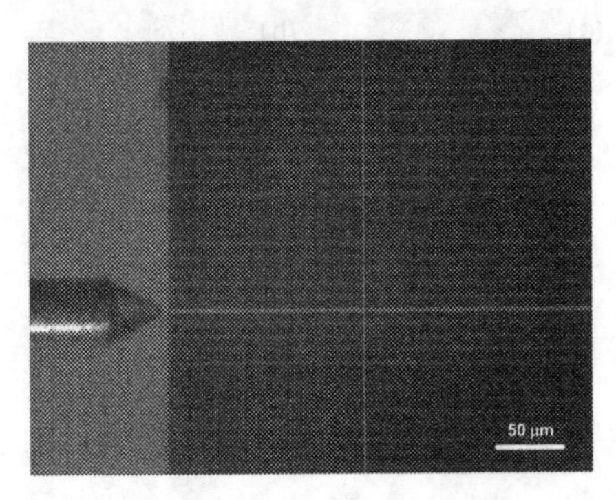

Figure 9. Top view of an array of Er doped Si nanocrystal based channel waveguides, one of which is pumped with ~10 mW at a wavelength of 1.49 µm using a tapered optical fiber. The bright (green) emission from the waveguide is caused by radiative relaxation from the $^4S_{3/2}$ and $^2H_{11/2}$ levels that are populated in a two-step co-operative upconversion process.

Evidence for upconversion has been found in Er doped Si nanocrystal sensitized optical waveguides when pumped at 1.49 µm (see Fig. 9), and in planar samples containing Er and Si nanocrystals pumped at visible wavelengths.[18] The latter observation suggests that a high concentration of excited Er can in fact be obtained in this type of material.

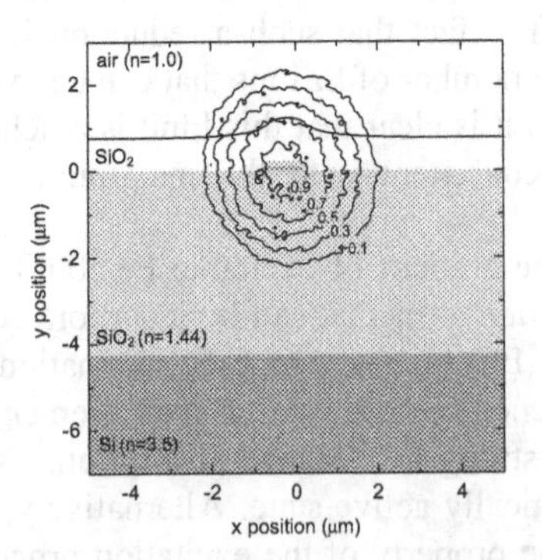

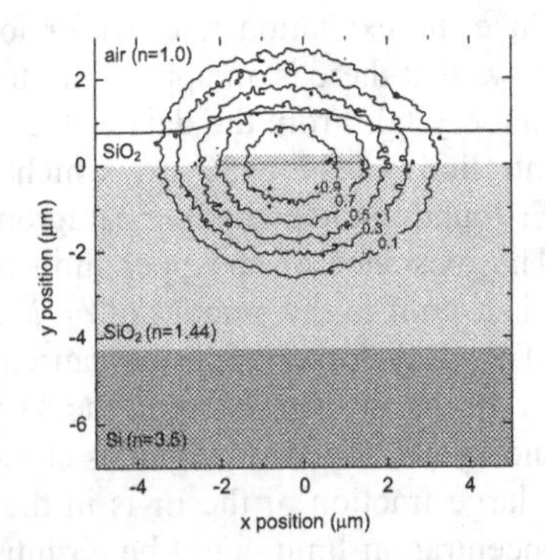

Figure 10. Optical mode images at a wavelength of 1.49 mm taken at the output facet of two Er doped Si nanocrystal based ridge waveguides with widths of 1 µm (a) and 3.5 µm (b) respectively. The contour lines indicate constant intensities. A sketch of the waveguide structure is included

4.6. INDUCED LOSSES

Figure *10* shows iso-intensity contours of two optical mode images taken using 1.49 µm light from an InGaAsP laser guided through the waveguide. The emission from the output facet was projected onto an infrared camera using a microscope objective. At this wavelength both waveguides only support the fundamental mode. For the smallest waveguide (width 1 µm) the mode is observed to be approximately circular, with a full width at half-maximum (FWHM) of 2.8 µm in the x-direction, and 3.0 µm in the y-direction. For the 3.5 µm wide waveguide, the mode is observed to be slightly elliptical, with a full width at half-maximum (FWHM) of 4.5 µm in the x-direction, and 3.5 µm in the y-direction. From these modes and the known Er depth profile, an effective mode overlap Γ with the Er doped region can be obtained. For the 3.5 µm waveguide this overlap is found to be 1%. Figure *11* shows normalized transmission spectra of Si nanocrystal waveguides containing no Er (dashed line) and 1.3 at.% Er (solid line), measured using a fiber coupled white light source butt-coupled to the nanocrystal waveguide using a single-mode tapered fiber. In the Er-doped sample a clear dip is observed around 1.53 µm, due to the $^4I_{15/2} \rightarrow {}^4I_{13/2}$ absorption transition of Er^{3+}. The peak absorption at 1.532 µm is found to be 2.7 dB/cm. From this value and the measured values for Γ and N_{Er} we obtain an absorption cross-section $\sigma_{Er}(1.532\ \mu m) = 8\times10^{-20}$ cm^2. This value is more than a factor ten larger than values found in Er doped Si (2×10^{-20} cm^{-2})[19] and SiO_2 ($\sim4\times10^{-21}$ cm^2). Evidence for such an unusually high cross-section has recently also been found by Han *et al.*[7] and Dal Negro *et al.*[20] The large cross-section may be due to the strong asymmetry of the local dielectric environment of the excitable Er^{3+} ions (note that in a perfectly symmetrical geometry the $^4I_{13/2} \rightarrow {}^4I_{15/2}$ transition is parity-forbidden). This may also explain the relatively short luminescence lifetimes typically observed in this type of material. The high absorption cross-section also translates into a high emission cross-section,[21,22] which implies that high gain could be achieved in this material over a short length.

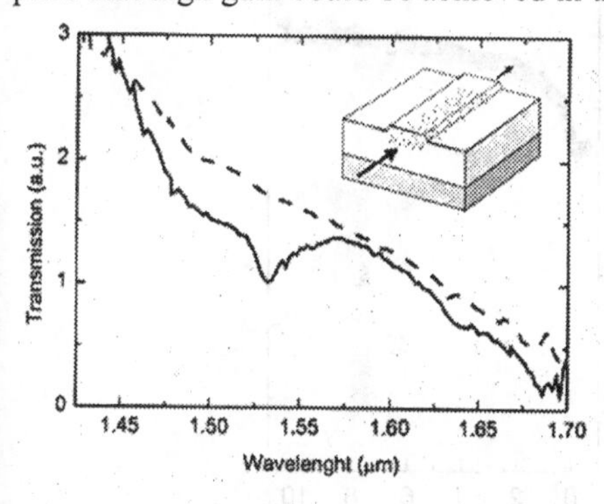

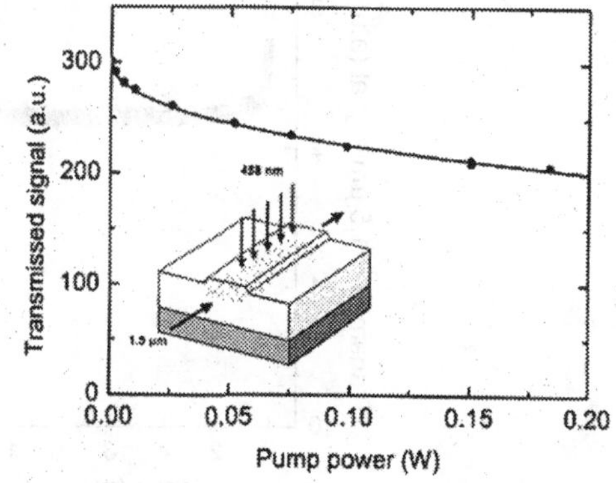

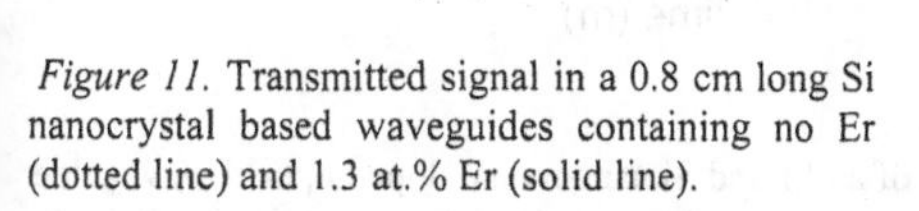

Figure 11. Transmitted signal in a 0.8 cm long Si nanocrystal based waveguides containing no Er (dotted line) and 1.3 at.% Er (solid line).

Figure 12. Transmitted signal at 1.535 µm as a function of the applied CW 458 nm pump power as obtained from transmission measurements. The transmission at zero pump power includes the waveguide loss and Er ground state absorption..

Figure *12* shows the transmitted signal at 1.535 μm as a function of applied pump power at 458 nm. The same pump power dependence of the transmitted signal was observed at all wavelengths in the range 1.4-1.7 μm. Apparently, instead of producing gain, optical pumping at 458 nm induces an absorption in the waveguide material. Figure *13* shows the time dependent transmission at 1.535 μm after pump switch-on and switch-off. Note that after pump switch-off, the signal transmission returns to approximately its original value over a period of several *minutes*.

The observed transmission change could be due to free carrier absorption, a well-known effect in bulk silicon, in which free carriers absorb sub-bandgap radiation due to intra-band transitions.[23] The free carriers causing the absorption feature are not present in the form of quantum confined electron-hole pairs, since these recombine at a rate $\sim 10^5\ s^{-1}$ at room temperature.[6,24] However, if the charge carriers are physically separated, their lifetime can increase by orders of magnitude.[25] Such separation could occur when during pumping one of the carriers in an electron-hole pair is ejected into a deep trap state near the nanocrystal surface, or when one of the carriers tunnels into a neighboring nanocrystal. The latter process is indicated schematically in Fig. *13*.

The data in Fig. *12* can be used to estimate the pump-induced free carrier concentration. The maximum transmission reduction is approximately 1.7 dB, which corresponds to α=-0.5 cm^{-1} given the waveguide length of 8 mm. Taking $\alpha(\mathrm{cm}^{-1}) = \sigma_{fc}\times N_{carr}\times\Gamma$ with σ_{fc} a typical free carrier absorption cross-section in bulk Si at 1.5 μm[8] of 7×10^{-18} cm^2, N_{carr} the concentration of pump-induced free carriers in the waveguide, and Γ=0.01 the previously determined mode overlap with the Si nanocrystal doped region, we find $N_{carr}=0.7\times10^{19}$ cm^{-3}, close to one carrier per nanocrystal. Note that this kind of pump induced absorption has not been observed in nanocrystal doped waveguides with lower excess Si concentrations,[7] possibly due to larger spacing between nanocrystals.

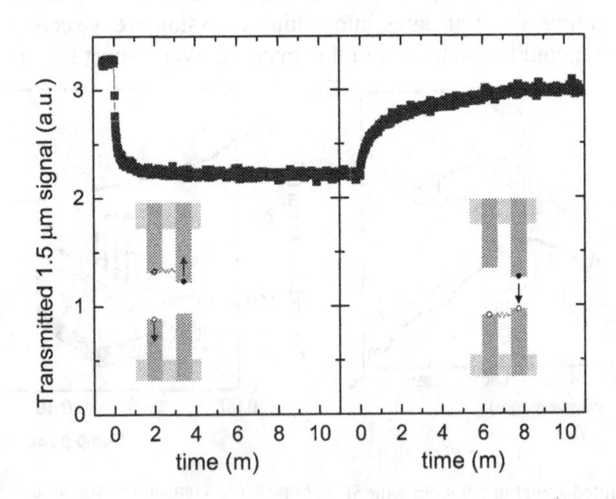

Figure 13. Time dependence of the 1.5 μm transmission of an Er and Si nanocrystal doped optical waveguide excited with a 458 nm pump laser directly after pump switch-on (left panel) and pump switch-off (right panel).

If the observed transmission reduction is indeed caused by single charges in separate nanocrystals, the recombination rate of these charges can be expected to decrease rapidly as the concentration of free charges decreases. This type of 'bimolecular recombination' can be described by the following rate equation:

$$\frac{dN_{carr}}{dt} = R - CN_{carr}^2 \tag{1}$$

with N_{carr} the concentration of *isolated* carriers, R the generation rate of isolated carriers, and $C(\mathrm{cm^3s^{-1}})$ a proportionality constant relating the recombination rate to the carrier concentration. If we assume a linear relation between the carrier generation rate and the pump power, the steady state solution gives $N_{carr} \propto P^{1/2}$. The solid line in Fig. 12 corresponds a transmission $T=\exp(-\alpha(P)L)$ with L=0.8 cm the waveguide length and $\alpha(\mathrm{cm^{-1}})=\sigma_{fc}\times N_{carr}(P)\times\Gamma$, where we have taken $N_{carr}\propto P^{1/2}$. It is clear that the simple rate equation model can produce a good fit of the observed behavior.

Equation (1) can also be used to describe the time dependence of the pump induced absorption. In Fig. 14 we have plotted -ln(T) over time after pump switch-off. Using the measured mode overlap, the bulk Si free-carrier absorption cross-section, and the known waveguide length, each value of -ln(T) can be converted to N_{carr}, as indicated on the right axis. We should be able to describe this curve with equation (1) setting the pump rate to zero. In this pump-off condition the solution is of the form $N_{carr}(t)=A/t$, with A a scaling factor indicating the initial carrier concentration and the proportionality constant relating the recombination rate to the carrier concentration. This dependency is indicated in Fig. 14 by the drawn line. Except for an initial fast drop in the concentration, the curve describes the time dependent carrier concentration quite well. Despite the shortcomings of the model (no nanocrystal size dependence, no absorption by quantum confined *excitons*) the proposed process of tunneling-limited carrier recombination can indeed describe the main features of the observed pump-induced absorption.

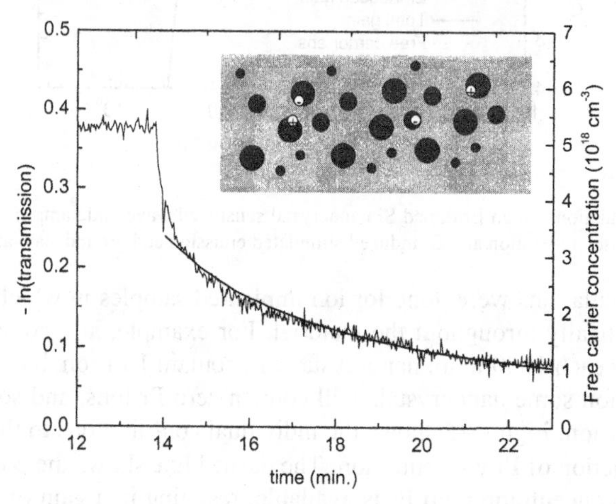

Figure 14. Time dependence of the calculated free-carrier concentration as discussed in section 4.6. The solid line is the solution of a rate equation model describing tunneling assisted bimolecular recombination of separated charges.

5. The thin line between gain and loss

In this Section we discuss a simple gain model that includes the main processes determining the gain of an Er doped Si nanocrystal sensitized waveguide amplifier. The model includes stimulated emission by excited Er, absorption by Er in the ground state, absorption by free carriers present in nanocrystals not coupled to Er, and absorption by carriers present in Si nanocrystals that *are* coupled to Er. We have made the following assumptions: 1) the Er ions, the Si nanocrystals coupled to Er, and the Si nanocrystals not coupled to Er behave as three-level systems, 2) the chosen pump rate ($R=1000\ s^{-1}$) is sufficient to keep excitable Er ions inverted, 3) each nanocrystal can keep at most 1 Er ion in the excited state, 4) excitons in Si nanocrystals not coupled to Er have a decay time of 10 μs, and excitons in Si nanocrystals that are coupled to Er have a 1 μs decay time, 5) the cross-section for free carrier absorption in Si nanocrystals is a factor 10 lower ($\sigma_{fc}=10^{-18}\ cm^2$) than in bulk Si due to a the reduced density of states in the conduction band and the valence band, 6) charge cannot tunnel to neighboring nanocrystals, 7) all nanocrystals contain 1000 atoms (diameter ~3 nm), 8) the sample contains 10^{19} nanocrystals/cm^3, and 9) the optical mode overlap is 100%. Although one can argue about the validity of some of these assumptions, the model does allow us to illustrate the main processes that influence the gain.

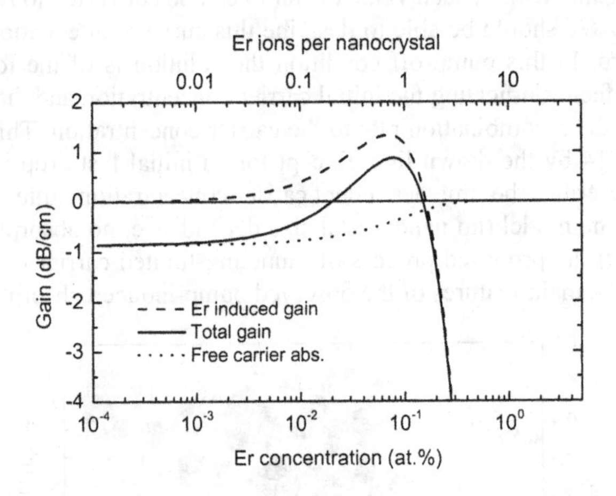

Figure 15. Gain calculations on an Er doped Si nanocrystal sensitized waveguide amplifier, showing the contributions of free-carrier absorption and Er induced stimulated emission and ground state absorption.

The calculations were done for ion implanted samples in which the Er ions are distributed statistically throughout the material. For example, at a concentration of 0.1 at.%, *on average* each 1000-atom nanocrystal will contain 1 Er ion, but due to the statistical Er distribution some nanocrystals will contain zero Er ions, and some will contain more than one Er ion. Figure 15 shows the individual contributions to the optical gain at 1.54 μm as a function of Er concentration. The dashed line shows the contribution of the Er ions. At low concentration, no Er is available, resulting in a gain of 0 dB. As the Er concentration is increased, the gain increases correspondingly. As the concentration approaches 1 Er ion per nanocrystal, in some nanocrystals the maximum excitable Er concentration is exceeded. These excess Er ions cannot be inverted, resulting in ground state absorption (i.e. negative gain) occurring at 1.54 μm. This effect becomes more important at increasing Er concentration.

The dotted line shows the contribution of absorption by excitons in nanocrystals that are not coupled to Er. At low concentration, all nanocrystals have a decay time of 10 μs. At the pump rate of 1000 s^{-1} this produces an exciton in 1% of the nanocrystals, or 10^{17} cm^{-3}. Consequently, the free carrier absorption coefficient is $\alpha_{fc}(cm^{-1})=2\times\sigma_{fc}\times10^{17}$ cm^{-3} $=0.2$ cm^{-1} or 0.87 dB/cm, where the factor 2 is due to the fact that an exciton consists of 2 free charges. As the Er concentration is increased, the chance for a nanocrystal to be coupled to Er increases, which reduces the exciton lifetime to 1 μs, resulting in reduced free carrier absorption. Since the lifetime of coupled nanocrystals is a factor ten less than that of non-coupled nanocrystals, the maximum absorption due to excitons generated in coupled nanocrystals at this pump rate is approximately 0.1×0.87 dB/cm $\approx$ 0.1 dB/cm. This contribution is negligible in Fig. 15 but could become significant at higher pump powers. The solid line shows the sum of the two gain contributions, illustrating that net gain can be achieved only in a narrow Er concentration range.

The gain calculations in Fig. 15 can be considered as a worst case scenario in several ways. For one thing, there is as yet no clear consensus on the maximum number of excitable Er ions per nanocrystal, the estimates ranging from 1 to 100. A higher concentration of excitable Er will dramatically increase the maximum attainable gain. Additionally, the free-carrier absorption observed at low Er concentration is caused by nanocrystals that are not coupled to Er. It seems feasible that a fabrication process can be developed that prevents the formation of uncoupled nanocrystals. One point of concern may be the signal power. The gain calculations were done in a relatively low pump power regime. When high intensity signals are amplified, higher pump rates will be necessary. It seems likely that given the finite lifetime of excitons in nanocrystals coupled to Er, a significant free-carrier concentration may be built up at high pump power. This could put an upper limit on the signal strength that can be amplified.

6. Conclusions

We have shown that Er excitation via Si nanocrystals can satisfy all requirements to be used in broadband pumped integrated optical waveguide amplifiers and lasers operating at 1.5 μm. In particular, it was shown that (a) the Si nanocrystal absorption cross-section at visible wavelengths is several orders of magnitude larger than direct Er optical absorption cross-sections, allowing for low pump-thresholds and unconventional pumping geometries, (b) the energy transfer quantum efficiency is larger than 55%, (c) energy transfer to Er occurs under visible excitation in a broad wavelength band, allowing for broadband pumping using cost-effective light sources such as LEDs, (d) the energy transfer rate is larger than 10^6 s^{-1}, in principle allowing for the generation of several hundreds of milliwatts 1.5 μm power in a 1 cm long waveguide, (e) concentrations of optically active Er in excess of 10^{19} cm^{-3} may be excited via nanocrystals, (f) optical pumping of Er doped nanocrystal sensitized waveguides can introduce a broad absorption in the infrared, possibly due to free-carrier absorption, and (g) the direct optical absorption cross-section of Er in a nanocrystal doped medium is enhanced by a factor

~10 compared to the absorption cross-section in typical SiO_2 based hosts. Gain calculations taking into account all these observations suggest that net gain at 1.5 μm in excess of 1 dB/cm can be achieved in Er doped Si nanocrystal sensitized optical waveguide amplifiers.

7. References

[1] S. Lombardo, S. U. Campisano, G. N. van den Hoven, A. Cacciato, and A. Polman, Appl. Phys. Lett. **63**, 1942 (1993)

[2] A. J. Kenyon, P. F. Trwoga, M. Federighi, and C. W. Pitt, J. Phys. Cond. Matter **6**, L319 (1994)

[3] M. Fujii, M. Yoshida, Y. Kanzawa, S. Hayashi, and K. Yamamoto, Appl. Phys. Lett. **71**, 1198 (1997)

[4] J. St. John, J. L. Coffer, Y. Chen, and R. F. Pinizzotto, J. Am. Chem. Soc. **121**, 1888 (1998)

[5] G. Franzò, V. Vinciguerra, and F. Priolo, Appl. Phys. A. **69**, 3 (1999)

[6] P. G. Kik, M. L. Brongersma, and A. Polman, Appl. Phys. Lett. **76**, April 24 (2000)

[7] H. S. Han, S.Y. Seo, and J. H. Shin, Appl. Phys. Lett. **79**, 4568 (2001)

[8] M. A. Green, *Silicon solar cells* (University of New South Wales, Sydney, 1995), p. 48

[9] M.L. Brongersma, A. Polman, K.S. Min, E. Boer, T. Tambo, and H.A. Atwater, Appl. Phys. Lett. **72**, 2577 (1998)

[10] C. Delerue, G. Allan, and M. Lannoo, Phys. Rev. B **48**, 11024 (1993)

[11] D. Kovalev, J. Diener, H. Heckler, G. Polisski, N. Künzner, and F. Koch, Phys. Rev. B **61**, 4485 (2000)

[12] F. Priolo, G. Franzò, F. Iacona, D. Pacifici and V. Vinciguerra, Mat. Sc. Eng. B **81**, 9 (2001)

[13] We assume that the nanocrystal absorption cross-section and spontaneous lifetime are not affected by the presence of Er

[14] W. J. Miniscalco, J. Lightwave Technol. **9**, 234 (1991)

[15] F. Auzel, in *Radiationless processes,* B. DiBartolo, Ed. (Plenum Press, New York, 1980)

[16] P. G. Kik and A. Polman, J. Appl. Phys. **91**, 534 (2002)

[17] F. Priolo, G. Franzò, S. Coffa, and A. Carnera, Phys. Rev. B **57**, 4443 (1998)

[18] D. Pacifici *et al.*, Proc. NATO Workshop on Optical Amplification and Stimulation in Silicon (2002)

[19] N. Hamelin, P.G. Kik, J.F. Suyver, K. Kikoin, A. Polman, A. Schönecker, and F.W. Saris, J. Appl. Phys. **88**, 5381 (2000).

[20] L. Dal Negro, private communication

[21] D.E. McCumber, Phys. Rev. **136**, A 954 (1964).

[22] W.J. Miniscalco, and R.S. Quimby, Opt. Lett. **16**, 258 (1991).

[23] W. Spitzer and H.Y. Fa, Phys. Rev. **108**, 268 (1957).

[24] M. Hybertsen, Phys. Rev. Lett. **72**, 1514 (1994).

[25] C. Svensson, in *The Si-SiO_2 system*, edited by P. Balk (Elsevier, 1988)

OPTICAL GAIN USING NANOCRYSTAL SENSITIZED ERBIUM: NATO-SERIES

JUNG H. SHIN, HAK-SEUNG HAN, AND SE-YOUNG SEO
Department of Physics
Korea Advanced Institute of Science and Technology (KAIST)
373-1 Kusung-dong, Yusung-Gu, Taejon, Korea

1. Introduction

The amount of information being transmitted via optical fibers has been increasing exponentially at a rate exceeding even the vaunted "Moore's Law" of silicon integrated circuits [1]. One key technology that enabled such rapid growth is erbium-doped fiber amplifiers (EDFA). By using the intra-4f transition of Er^{3+} ($^4I_{13/2} \rightarrow {}^4I_{15/2}$) to amplify optical signals near 1.54 μm, the absorption minimum of silica-based optical fibers, EDFAs made today's wideband, all-optical networks possible [2]. However, because EDFAs require expensive lasers tuned to one of the absorption bands of Er^{3+} for excitation of Er, they remain too expensive to be used widely, thereby hindering the extension of the all-optical network all the way to the individual end-users. Yet neither substituting the fiber for a waveguide nor sensitizing Er^{3+} with other rare earth ion such as Yb^{3+} can fundamentally solve this problem. In fact, the high Er^{3+} concentrations required for waveguide-based amplifiers due to the need to compress long fibers into short waveguides leads to pair-induced quenching [3], necessitating the use of even more powerful and expensive lasers.

If Er^{3+} ions are doped into Si, however, they can be excited via an Auger-type interaction with carriers instead of a direct photon absorption, as shown schematically in Fig. 1 [4,5,6,7]. Although it does not involve any resonance, this process can be surprisingly effective, with effective excitation cross section that is nearly a million times larger than that of direct optical absorption. Furthermore, since photoluminescence (PL) is essentially equivalent to electroluminescence (EL), Er doped Si has been used to fabricate Si-based light emitting diodes (LED) emitting 1.54 μm light [8,9]. However, the same carrier-Er^{3+} interactions can also lead to de-excitation of Er (dashed arrows in Fig. 1). As a consequence, the Er^{3+} luminescence efficiencies from Er-doped Si so far remain too low to be practical [6,7].

A promising and interesting method that represents a mixture of Er-doping of SiO_2 and Er-doping of Si is Er doping of silicon-rich silicon oxide (SRSO) that consists of nc-Si embedded inside an SiO_2 matrix. Such an approach essentially splits the duty of excitation and emission between nc-Si and Er^{3+} -- i.e., excitation of Er^{3+} is dominated by carriers generated inside nc-Si that now plays the role of classical sensitizer atoms such as Yb^{3+}, and emission is dominated by the intra-4f transition of the doped Er^{3+} ions

L. Pavesi et al. (eds.), Towards the First Silicon Laser, 401–420.

[10,11,12,13]. Yet despite such division, the overall Er^{3+} luminescence can be surprisingly effective, occurring within a few microseconds [14], with more than 60% efficiency and an effective excitation cross section that is several orders of magnitude greater than that of Er in pure SiO_2 [15] and a total internal quantum efficiency of greater than 15 % [16]. Furthermore, such division of labor confers several advantages over using nc-Si or Er alone. First, since nc-Si only need to excite Er, the requirement for control over the nc-Si size can be significantly relaxed. Note also that since only photocarrier generation is necessary, any low-cost light source capable of exciting nc-Si (e.g., LEDs) may be used instead of a laser, providing a possibly significant cost advantage. Second, since the emission is an atomic transition, the luminescence peak is much narrower than can be achieved with nc-Si, with a position that is insensitive to small changes in nc-Si size distribution and temperature. Finally, the emission occurs at a technologically important wavelength with great possibilities for potential applications.

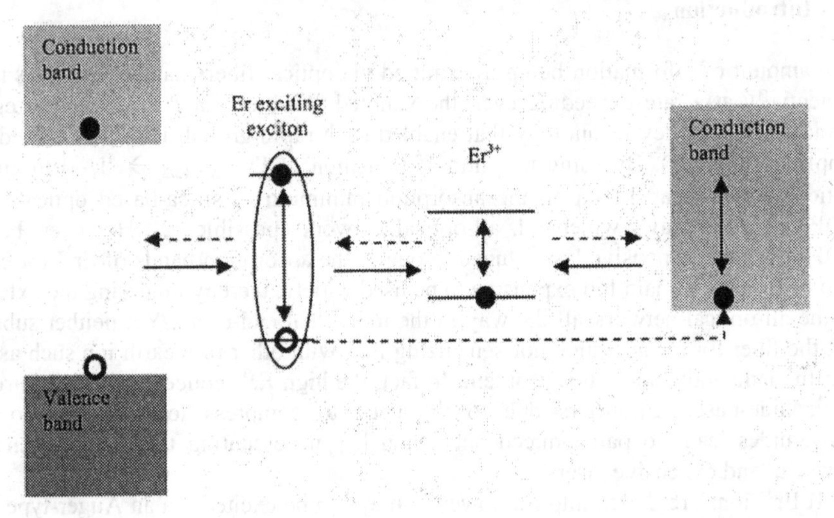

Figure 1 : Schematic description of carrier-mediated excitation processes in Si. The solid arrows represent the excitation processes, and the dashed arrows represent the de-excitation processes. Er^{3+} ions can be excited via interaction with trapped carriers, or via impact excitation with hot carriers. The de-excitation processes are in essence the reverse of the excitation processes.

Yet unlike bulk Si, SRSO is a heterogeneous mixture of nc-Si and SiO_2 with non-unique properties depending on the composition and fabrication procedures. Furthermore, there are many physically distinct possible environment for Er – e.g., inside nc-Si, at nc-Si/SiO_2 interface, or inside SiO_2 but close to nc-Si, and it is not clear which roles such different environment play in the excitation and de-excitation process of Er. Thus, many different questions need to be answered before nc-Si sensitization of Er can become a viable candidate for practical applications. In this paper, we present experimental results that aim to clarify a): how nc-Si sensitized, Er-doped SiO_2 films with high Er^{3+} luminescence efficiency may be formed; b) what the effects of the Er environment may be; and c) how nc-Si and Er interact in Er-doped SRSO. Based on the results, we show that optical gain is possible using nc-Si sensitized, Er-doped SiO_2, and practical applications may be possible. Finally, we will discuss possibilities for further research based on nc-Si sensitization.

2. Investigating nc-Si sensitization of Er

Two kinds of nc-Si sensitized films were used. One, SRSO, consists of nc-Si embedded inside a SiO_2 matrix, with random distribution of both Er and nc-Si. The other, Si/SiO_2 superlattice (SL), consists of nm-thin Si layers and SiO_2 layers. The location of Er in this case can be controlled with sub-nm precision. The schematic description of each film with a representative transmission electron microscope (TEM) images are shown in Fig. 2. SRSO films were deposited using electron-cyclotron resonance plasma enhanced chemical vapor deposition (ECR-PECVD) of SiH_4, and O_2 with concurrent sputtering of Er. The details of the SRSO deposition procedure can be found in Ref. [17]. Initially, amorphous SiO_x:H ($x<2$) films doped with Er is deposited. Post-deposition high temperature rapid thermal anneals are then used to precipitate nc-Si clusters and to activate Er. SL films are deposited either with ECR-PECVD or with ultra-high vacuum (UHV) ion sputtering. Again, the films are deposited at a low temperature, after which a high temperature anneal is used to crystallize the Si layers (for the case of ECR-PECVD deposited films) and to activate Er. The details of the SL deposition can be found in Ref. [18] and Ref. [19].

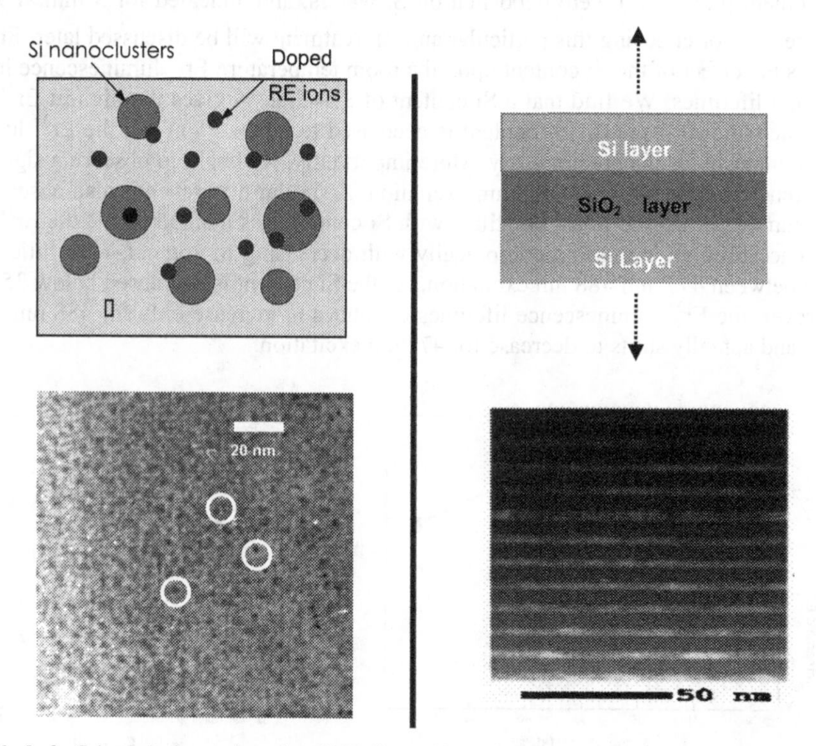

Figure 2 : Left : Schematic description of an SRSO film and its representative TEM image. Right: Schematic description of an SL film and its representative TEM image

PL measurements were performed using an Ar laser, a grating monochromator, and employing the lock-in technique. In most cases, the 477 nm line of the Ar laser was used as it is not absorbed optically by Er^{3+}, thus ensuring that excitation occurs via nc-Si only. A thermoelectrically cooled InGaAs and a photomultiplier tube was used as detectors for infra-red and visible range, respectively. For temperature-dependent measurements, a closed-cycle helium cryostat was used. For time-resolved measurements, a digitizing oscilloscope was used. a as source gases and or by ion sputtering of Si by Ar. All PL spectra are corrected for system response. The structure of the deposited films was investigated using TEM. The composition of the deposited films was investigated using Rutherford Backscattering Spectroscopy (RBS) and Medium Energy Ion Scattering (MEIS).

2.1. DEPOSITION OF ER-DOPED SRSO

2.1.1. *Composition dependence*

Er-doped SRSO films with different Si content but with similar thickness (≈ 1 μm) and Er content (0.4 at. %) were deposited on Si wafers, and annealed for 5 min at 950°C. The reason for choosing this particular anneal treatment will be discussed later. Figure 3 shows the effect of the Si content upon the room temperature Er^{3+} luminescence intensities and lifetimes. We find that a Si content of 35-45 at. % gives the highest Er^{3+} luminescence intensity. As the Si content is decreased to below 35 at. %, the Er^{3+} luminescence intensity drops precipitously. More importantly, we begin to observe a significant difference between 477 and 488 nm excitation. A similar difference is also observed for Er^{3+} luminescence lifetimes. For films with Si content less than 35 at .%, the Er^{3+} luminescence lifetime increase monotonically with decreasing Si content, with little difference between 477 and 488 nm excitation. As the Si content is decreased below 35 at. %, however, the Er^{3+} luminescence lifetimes continues to increase only for 488 nm excitation, and actually starts to decrease for 477 nm excitation.

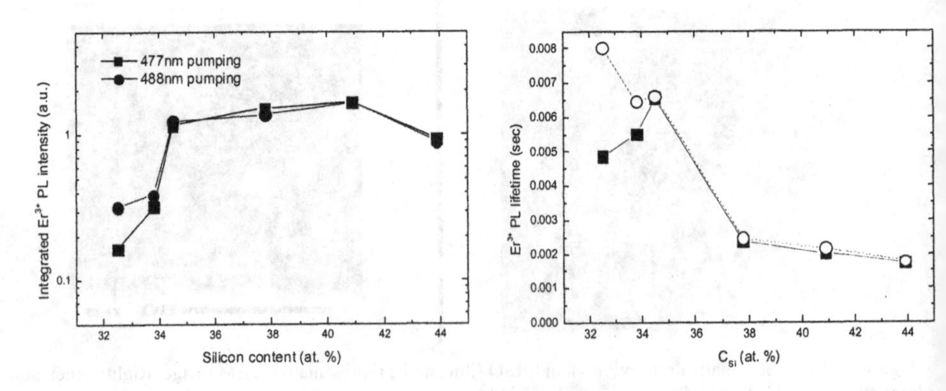

Figure 3 : The composition dependence of room temperature Er^{3+} luminescence intensities and lifetimes from Er-doped SRSO. The films were annealed for 5 min at 950°C after deposition

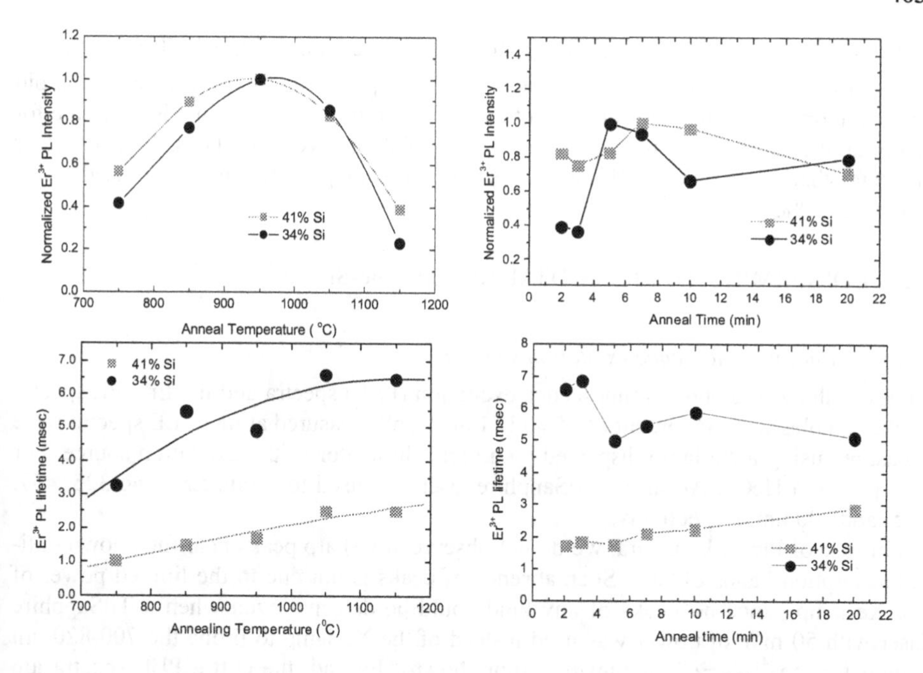

Figure 4 : The anneal temperature and time dependence of Er^{3+} luminescence intensities and lifetime of the films with Si content of 35 and 41 at. %. All measurements were performed at room temperature

2.1.2. *Annealing parameter dependence*

Figure 4 shows the anneal temperature and time dependence of Er^{3+} luminescence intensities and lifetimes from the films with Si content of 35 and 41 at. %. We find that an anneal temperature of 950-1050°C gives the highest Er^{3+} luminescence intensity. The Er^{3+} luminescence lifetime, on the other hand, continues to increase with increasing anneal temperature. An anneal of at least 5 min at 950°C seems necessary for obtaining both high luminescence intensity and long luminescence lifetime. Prolonging the anneal beyond 5 min, on the other hand, seems to have only a minor effect

2.1.3. *Discussion*

As indicated earlier, the main difference between 477 and 488 nm excitation is that 477 nm light can excite Er^{3+} via nc-Si only, while 488 nm light can excite Er^{3+} both via both nc-Si and direct optical absorption. Thus, Fig. 3 indicates that in order to obtain sufficient nc-Si for nc-Si sensitization to be effective, an Si content of at least 35 at. % is required. Such need for presence of nc-Si also explains the anneal temperature dependence see in Fig. 4. An anneal temperature of at least 900°C is necessary for precipitation of nc-Si [10]. However, annealing at too high of a temperature can lead to precipitation of Er in both Si and SiO_2 [20,21]. Thus, the temperature range of 950-1050°C represents a compromise between the need to precipitate high quality nc-Si and the need to avoid precipitation of Er ions themselves. Note a a long luminescence lifetime is desirable for many applications, since it allows for easy population inversion. Thus, Fig. 3 further indicates that for maximum efficiency, as little of Si as possible should be used – i.e., about 35 at. %.

2.2. CONFIRMING THE ESSENTIAL ROLE OF nc-Si

2.2.1. *Photoluminescence excitation spectra*

Figure 5 shows the photoluminescence excitation (PLE) spectra and the Er^{3+} PL spectra of the samples with Si content of 35 and 41 at. %, all measured at 6K. PLE spectra were measured using a Xe lamp dispersed by a monochromator as the excitation source. For PL spectra, a HeCd, Ar, and a Ti:Sapphire laser was used to excite Er using 325, 457, 515, and 700 nm, respectively.

In case of the PLE spectra, we do not observe any sharp peaks near the known optical absorption bands of Er^{3+}. Such absence of peaks is not due to the limited power of the Xe lamp, since no peaks of any kind could be detected even when a Ti:Sapphire laser with 50 mW of power was used instead of the Xe lamp to probe the 700-820 nm region for the $^4I_{15/2} \rightarrow {}^4I_{9/2}$ transition (not shown). Instead, the entire PLE spectra are rather formless, and bear striking resemblance to the absorption spectra of nc-Si [22]. Similarly, the Er^{3+} PL spectra do not show any dependence on the excitation wavelength – in fact, it is very difficult to tell that there are actually 9 PL spectra shown in Fig. 5. All spectra are broad without any sharp peaks that would indicate a well-ordered, crystalline environment for Er.

2.2.2. *Discussion*

Such lack of peaks due to direct optical absorption, coupled with the similarity of Er^{3+} PLE spectra with the absorption spectra of nc-Si, clearly indicate that for SRSO films with Si content in the excess of 35 at. %, the nc-Si:Er coupling and strong and efficient enough to completely dominate the excitation process of doped Er^{3+} ions. Furthermore, the lack of dependence of the Er^{3+} PL spectra upon the excitation wavelength indicates that the incident photons excite the same class of Er^{3+} ions irrespective of their energies – i.e., there is only one dominant class of luminescence site of Er^{3+} in SRSO.

The nature of the dominant luminescence site can be inferred from the PL spectra. They are broad without any peaks that might be expected from a crystalline environment, and are located close to that from Er doped into pure silica, indicating that the optically active Er ions are predominantly located in an amorphous, SiO_2-like environment. However, they cannot be too far away from nc-Si, since they are still able to be excited via carriers generated inside nc-Si. Therefore, Fig. 5 suggests that the optically dominant Er^{3+} ions are most likely located near the interface between nc-Si and the SiO_2 matrix.

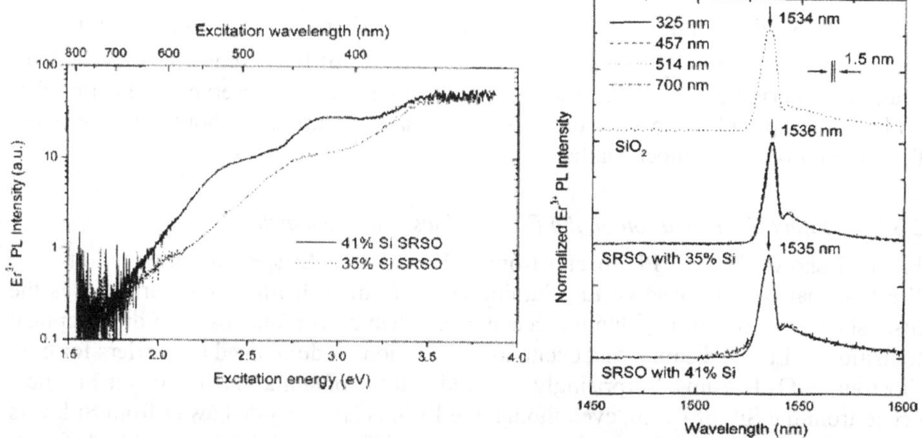

Figure 5 : Photoluminescence excitation (PLE) spectra and PL spectra of the Er-doped SRSO film with Si content of 35 and 41 at. %. Note the absence of peaks due to direct optical absorption, indicating the dominance of nc-Si mediated, carrier-induced excitation of Er. The Er^{3+} PL spectra indicate that there is only one class of dominant luminescent site, most likely the interface between nc-Si and SiO_2 matrix

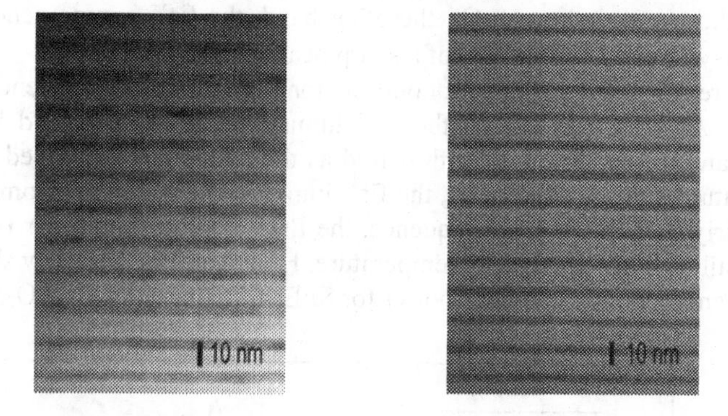

Figure 6 : The bright-field cross-section TEM images of Left : Si/SiO_2 superlattice with Er doped into the Si layers (Si :Er) , and Right : Si/SiO_2 superlattices with Er doped into the SiO_2 layers (SiO_2:Er) after anneal. The Si layers are nanocrystalline.

2.3. EFFECT OF CONTROLLING THE LOCATION OF Er

2.3.1. *Deposition of Er-doped Si/SiO_2 superlattice thin films with controlled location of Er*

To confirm the above suggestion that the dominant luminescent site is not inside nc-Si, but outside nc-Si yet still very close to nc-Si, we have deposited Er-doped Si/SiO_2 superlattice (SL) thin films with 4-5 nm-thin Si and and 8-9 nm thin SiO_2 layers. Two kinds of SL thin films were deposited, one in which Er was doped into the Si layers only (Si:Er), and one in which Er was doped into the SiO_2 layers only (SiO_2:Er). The total period of the films were adjusted to 60 and 20, respectively, in order to keep the total number of doped Er ions the same. The Er concentration was calculated from RBS to be

0.5 at. % for both films. After deposition, the films were annealed in a sequence of 20 min at 600°C → 5 min at 950°C → 5 min at 600°C in order to avoid cracking and spalling of the films. Figure 6 shows the TEM images of both films after the final anneal. Good periodicity and well-defined, planar interfaces are observed, indicating that no breaking up and agglomeration of Si layers have occurred. It should also be noted that the Si layers are nanocrystalline.

2.3.2. *Effect of Er location on the Er^{3+} luminescence properties*

Figure 7 shows the Er^{3+} PL spectra from Si:Er and SiO_2:Er spectra, measured at 20K. We find that we can observe Er^{3+} luminescence from both films. Furthermore, as the inset shows, we obtain Er^{3+} luminescence even when non-resonant with a direct optical transition of Er^{3+}, indicating that excitation of Er^{3+} ions is dominated by carriers for both Si:Er and SiO_2:Er films. Surprisingly, we find that we obtain a much stronger luminescence from the SiO_2:Er film, even though the Er ions are separated away from Si layers that provide the carriers that excite them. Such an effect is not due to possible diffusion of Er into Si, since SiO_2:Er and Si:Er films display different Er^{3+} PL spectra. The Er^{3+} luminescence peak from SiO_2:Er film is broad and symmetric, and is located at 1.533 μm, in agreement with that from pure SiO_2 film doped with Er that was deposited under identical conditions (not shown). On the other hand, the Er^{3+} luminescence peak from Si:Er film is asymmetric, with a hint of a sharp peak located at 1.537 μm

Such difference is even more pronounced for the temperature dependence of Er^{3+} luminescence. As shown in Fig. 8, the Er^{3+} luminescence intensity and lifetime from SiO_2:Er film are quenched less than threefold as the temperature is raised from 20K to room temperature. On the other hand, the Er^{3+} luminescence intensity from Si:Er film is quenched nearly tenfold. As a consequence, the Er^{3+} lifetime from Si:Er film could not be measured all the way up to room temperature. However, Fig. 8 clearly shows that the Er^{3+} luminescence lifetime is much shorter for Si:Er film than it is for SiO_2:Er film

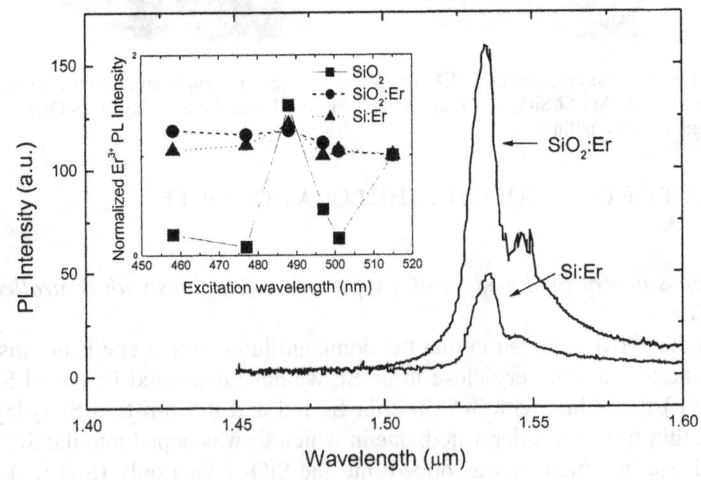

Figure 7: The PL spectra of Si :Er and SiO_2:Er superlattice films, measured at 20 K. The inset shows the dependence of the Er^{3+} PL intensity on the pump wavelength.

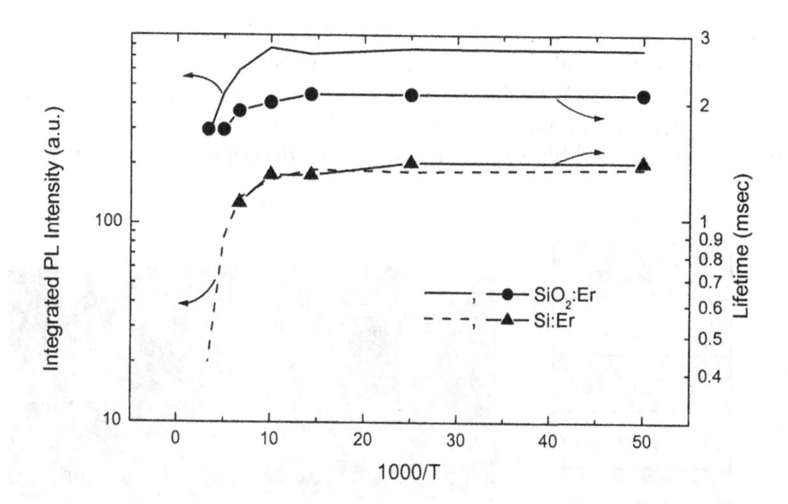

Figure 8 : The temperature dependence of Er^{3+} PL intensity and lifetimes from SiO_2:Er and Si :Er films

2.3.3. *Discussion*

Figures 7 and 8 clearly confirm the suggestion from the last section. That is, in order to obtain the best Er^{3+} luminescence properties, *Er ions must be separated from a direct contact with nc-Si*. This conclusion is somewhat paradoxical, since some sort of interaction with carriers is necessary for excitation of Er^{3+} in the first place. However, we note that the same Er-carrier interactions are also responsible for de-excitation of excited Er^{3+}. Thus, Fig. 6,7,8 indicate that there is an asymmetry between carrier-mediated excitation and carrier-mediated de-excitation, and that by carefully controlling the Er-carrier interaction, we can obtain efficient carrier-mediated excitation without the carrier-mediated de-excitation of Er^{3+}.

It is worthwhile to point out that such effects are not related to enlargement of bandgap of nc-Si due to the quantum confinement that has been correlated with suppression of quenching and increased Er^{3+} luminescence [23], since the overall film structure including the Si layer thickness is the same for both Si:Er and SiO_2:Er films. Note, however, that such separation of nc-Si and Er ions is most likely to occur for SRSO thin films with low excess Si content, since the volumetric fraction of nc-Si is very low. Therefore, Figs. 6-8 suggest that such separation of Er from direct contact with nc-Si may play as significant of a role as the enlargement of bandgap, and that with a proper engineering; we may obtain efficient Er^{3+} luminescence even without quantum confinement effects.

2.4. SEPARATING THE EFFECT OF QUANTUM CONFINEMENT AND THE EFFECT OF Er:nc-Si SEPARATION

2.4.1. *Deposition of Er-doped Si/SiO₂ superlattice thin films with controlled SiO₂ bufer layers*

In order to confirm the essential role of separation of Er^{3+} from direct contact with nc-Si, and also to obtain the interaction distance between nc-Si and Er, we have deposited SiO_2:Er thin film with nm-thin buffer layers of pure SiO_2 between nc-Si and SiO_2:Er. The thickness of Si and SiO_2:Er thin layers were kept at 4 and 3 nm, respectively, while the thickness of the SiO_2 buffer layers were varied from 0 to 2 nm. The TEM images of the deposited and annealed films with their schematic description are shown in Fig. 9.

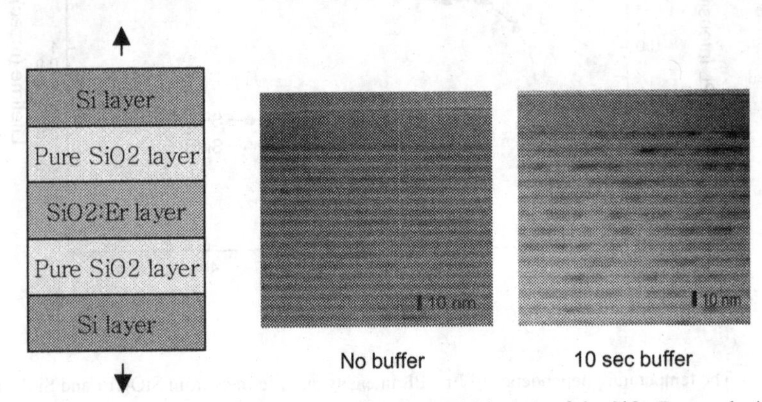

Figure 9. : The schematic description and actual cross-section TEM images of the SiO_2:Er superlattice thin films with buffer layers of pure SiO_2. The left TEM image shows the film without any buffer layers, and the right TEM image shows the film with buffer layers with a deposition time of 10 sec, corresponding to a thickness of ≈ 1.5 nm. The Si layers are nanocrystalline, and are 4-5 nm thick for both films.

2.4.2. *Effects of buffer layers on the Er^{3+} luminescence properties of SiO_2:Er superlattice films*

All films displayed the typical 1.54 μm Er^{3+} luminescence when excited with 477 nm light, confirming that carrier-mediated excitation can occur across atomic distances (not shown). The temperature dependence of the Er^{3+} luminescence intensities and luminescence lifetimes are shown in Figure 10. We find that the *Er^{3+} luminescence intensities and lifetimes increase with increasing buffer layer thickness*. With a buffer layer thickness of ≈2 nm, we obtain a several hundredfold increase in the Er^{3+} luminescence intensity. Furthermore, increasing the buffer layer thickness reduces the temperature quenching of the Er^{3+} luminescence intensity as well such that with a buffer layer thickness of ≈2 nm, we can nearly completely suppress the temperature quenching of the Er^{3+} luminescence

2.4.3. *Discussions*

Figure 10 unequivocally proves the suggestion made in previous sections that a separation of Er^{3+} ions and nc-Si is necessary for obtaining strong Er^{3+} luminescence. Furthermore, it also sets 2 nm as a lower bound for nc-Si:Er interaction distance. It should be noted here that all film investigated in Fig. 10 have nearly the same Si layer thickness, and thus should have the same quantum confinement effect. The fact that introducing a buffer layer can nearly completely suppress the temperature quenching of Er^{3+} luminescence intensity shows that quantum confinement is not essential for obtaining efficient Er^{3+} luminescence from Er-doped SRSO thin films, and that the nc-Si:Er separation that occur naturally in Er-doped SRSO films with low Si content plays a significant role in enabling efficient Er^{3+} luminescence from Er-doped SRSO.

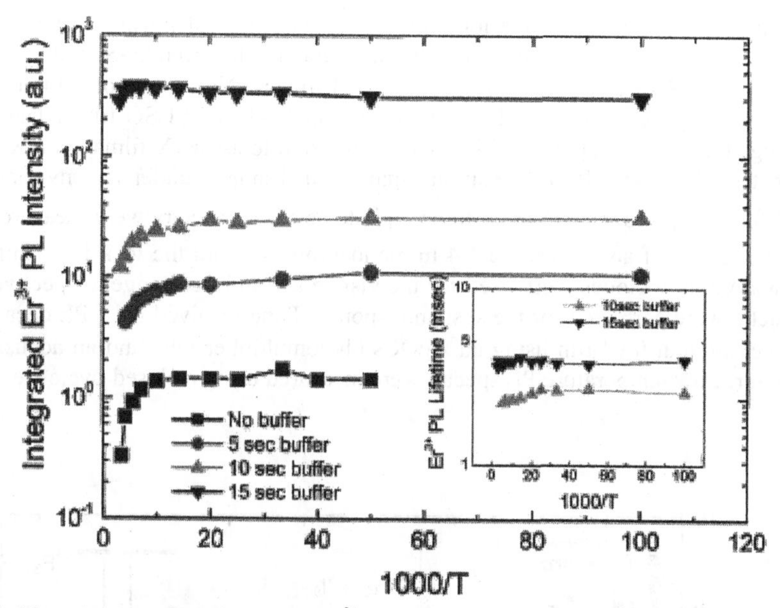

Figure 10 : The temperature dependence of the Er^{3+} luminescence intensities and lifetimes of SiO_2:Er superlattice thin films with buffer layers of pure SiO_2.

2.5. THE nc-Si/Er INTERACTION: THE NATURE OF Er-EXCITING EXCITON

So far, we have been talking rather vaguely about carrier-mediated excitation without explicitly stating how such excitation can occur. For instance, ***Figure 1*** indicates that there is an intermediate step of an exciton formation prior to excitation of Er^{3+}. In case of bulk crystalline Si, the nature of such bound excitons are well known and their interaction with Er calculated [5,24]. In case of Er-doped SRSO, the nature of Er-exciting excitons is much less clear. Since e-h pairs generated are confined inside nc-Si, they are, in a sense, always bound excitons. Furthermore, there have been recent theoretical calculations which suggest that the luminescence from nc-Si themselves are due to recombination of bound excitons [25]. As carrier-Er interaction forms the fundamental basis of nc-Si sensitization of Er^{3+}, it is imperative to develop a more detailed understanding of the nature of the excitons that excite Er^{3+} and their interactions with Er.

2.5.1. *Nd-doping of SRSO*

One drawback of studying nc-Si/Er interaction is that the intra-4f transition of Er is well separated in energy from the exciton recombination in nc-Si. Thus, it is much less sensitive to details of the nature of the exciton, since any exciton generated in nc-Si has sufficient energy to excite Er. An interesting alternative to Er is Nd. Its $^4F_{3/2} \rightarrow {}^4F_{9/2}$ and $^4F_{3/2} \rightarrow {}^4F_{11/2}$ 4f transitions luminesce at ≈0.9 and 1.1 μm, respectively. Not only are these transitions used widely in many diverse optical applications such as Nd:YAG lasers, but they can also be detected by Si diodes, thereby raising the possibility of integrating light emission and detection capabilities into one single silicon-based chip. More importantly, because these emissions fall within the range of intrinsic nc-Si luminescence, Nd^{3+} can act as a sensitive probe of the interaction between nc-Si and RE ions.

Nd-doped SRSO films with a thickness of ≈ 1 μm and Nd content of 0.14 at. % were deposited using the same procedures as the that for Er-doped SRSO films. The Si content ranged from 34 to 50 at. %. They will be referred to as SiXX films, with XX referring to the Si content. Post-deposition rapid thermal anneal under Ar environment at 950°C for 5 min was used in order to precipitate nc-Si. PL spectra were measured using the 488 nm line of an Ar laser, a 1/4 m monochromator, and the lock-in technique. Si and InGaAs photodiodes were used for the visible and infrared range, respectively. All PL spectra were corrected for the system response. Time-resolved Nd^{3+} PL decay traces were measured at 0.92 μm using an AgOCs photomultiplier tube and an acousto-optic modulator. Low temperature PL spectra were measured using a closed-cycle He cryostat.

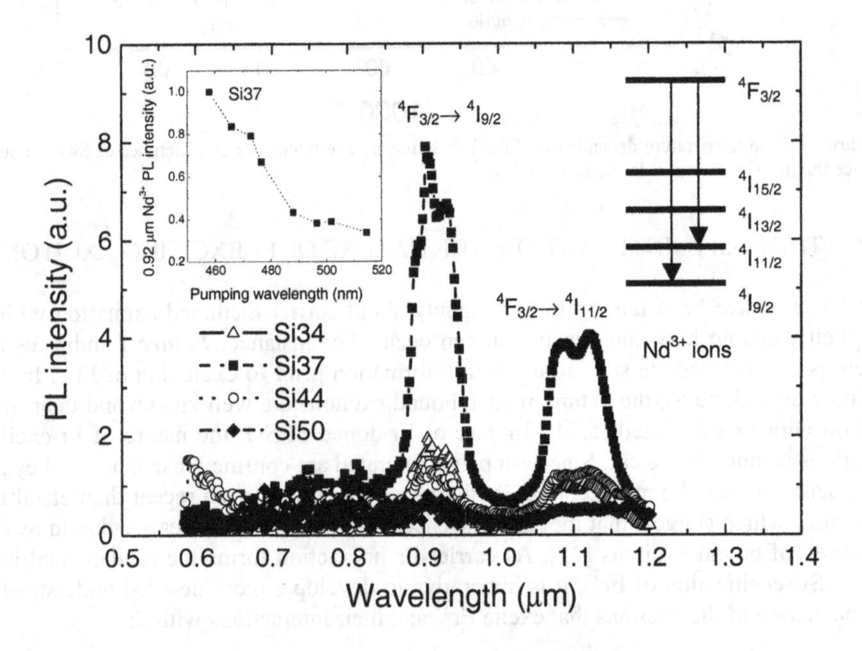

Figure 11 : Room temperature PL spectra of different Nd-doped SRSO films with different Si content. The insets show the relevant Nd^{3+} energy levels and the dependence of the Nd^{3+} PL intensity upon the pump wavelength

2.5.2. *Luminescence properties of Nd-doping of SRSO: composition dependence*

Figure 11 shows the room-temperature PL spectra of Nd-doped SRSO films and the relevant levels of Nd^{3+}. Two luminescence peaks at 0.92 μm and 1.06 μm due to the $^4F_{3/2} \rightarrow {}^4F_{9/2}$ and $^4F_{3/2} \rightarrow {}^4F_{11/2}$, respectively, can clearly be observed. On the other hand, we did not observe other Nd^{3+} luminescence peaks (e.g., at 1.3 μm due to $^4F_{3/2} \rightarrow$

$^4I_{13/2}$ transition). The maximum Nd^{3+} luminescence is observed from the Si37 film. The Nd^{3+} luminescence intensity decreases strongly as the Si content is increased, and becomes nearly undetectable for the Si50 film. The inset shows the dependence of the 0.92 μm Nd^{3+} luminescence intensity of the Si37 film on the pump wavelength. We find that Nd^{3+} luminescence can be observed irrespective of the pump wavelength.

2.5.3. *Luminescence properties of Nd-doping of SRSO: Temperature dependence*

Figure 12 shows the temperature dependence of 0.92 μm Nd^{3+} PL lifetimes of Si 37 and Si44 films. The symbols are experimental data. The line is a fit to the model discussed later. We find that the PL lifetimes show much less temperature quenching, as the Nd^{3+} luminescence lifetimes from the Si44 film decreases from 5.3 to 4.3 μsec as the temperature is raised from 15K to room temperature, while that from the Si37 film does not show any temperature quenching, remaining constant at 20 μsec.

2.5.4. *Discussion*

The composition dependence of Nd^{3+} luminescence intensity is consistent with, and may be explained by, the exciton-mediated excitation mechanism shown in Fig. 1. Since the bandgap of bulk-Si is smaller than the energy needed to excite Nd, strong quantum confinement is a prerequisite for excitation of Nd. Thus, only SRSO films with low excess Si content show Nd^{3+} luminescence as small excess Si leads to formation of small nc-Si [26].

Yet the more important result shown in Fig. 11 is the suppression of nc-Si luminescence peak by Nd doping, even in the case where excitation of Nd^{3+} is impossible. Similarly, note that Nd^{3+} doping quenches the entire nc-Si luminescence, even those in the energy range below that of Nd^{3+} intra-4f transition, and thus unable to excite Nd^{3+}. This indicates that the coupling of excitons and Nd^{3+} must occur prior to the formation of excitons that lead to nc-Si luminescence. Such a conclusion is consistent with observations by Franzò *et al* [27] and Kik *et al.*, [28] that Er doping of SRSO films, while quenching the luminescence intensity of nc-Si luminescence, have no effect on the luminescence lifetime of nc-Si luminescence, and strongly indicates that the excitons that excite Nd^{3+} (an in extension, any rare earth ions in SRSO) *are distinct from those that lead to nc-Si luminescence.*

Further insight into exciton-rare earth interaction can be obtained from the analysis of temperature dependence of Nd^{3+} luminescence lifetime, as shown in Fig. 12, using models that have been successful in describing the temperature quenching of rare earth luminescence in bulk Si [29,30]. While the physical basis differ slightly, the temperature quenching in both models can be represented as $W_o \times f(E_o,T)$, where W_o is the coupling prefactor and $f(E_o,T)$ is a model-specific function of an energy barrier E_o and temperature T. As can be seen in Fig. 12, both models fit the experimental data equally well. The values used to fit the data are also similar for both models, and are $\approx 1\times10^5$ sec^{-1} and 20 meV for W_o and E_o, respectively.

The small value of E_o is not surprising, since the Nd^{3+} transitions lie within the luminescence band of nc-Si. Note, however, that such a small energy barrier would results in a severe temperature quenching of Nd^{3+} luminescence, were it not for the fact that W_o is also very small – several orders of magnitude smaller than 10^9 - 10^{10} sec^{-1} obtained for rare earth ions doped into bulk semiconductors. In fact, the small value of W_o is the main reason for the suppression of temperature quenching of Nd^{3+}, not the main reason for the suppression of temperature quenching of Nd^{3+}, not the quantum confinement effect. We note that the the values for E_o and W_o cannot be considered to be very accurate due to large error bars and the possibility of temperature-dependent interaction with other levels of Nd^{3+}. However, the conclusion that the coupling prefactor is very small remains valid. This is shown by the dotted line in Fig. 12, which is the result of a fit using Kik's model but with W_o fixed at 1×10^9 sec^{-1}. The fit is clearly less satisfactory, and predicts a very steep drop that is not observed.

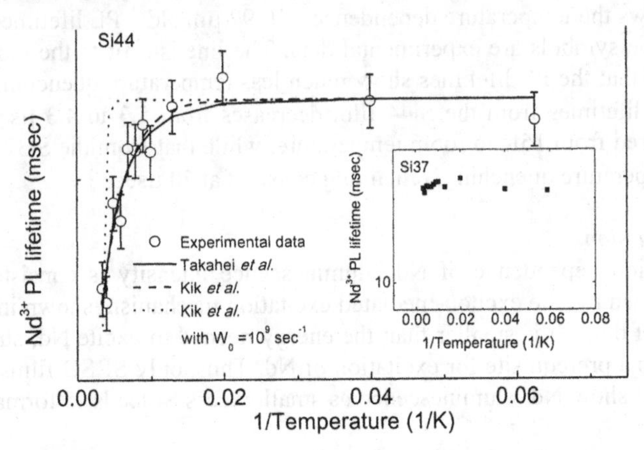

Figure 12 : Temperature dependence of Nd^{3+} PL lifetime. The symbols are eperimental data, and the line is the fit using models described in the text. The inset shows the lifetimes of Si37 sample

Note, however, that such a weak coupling prefactor is consistent with results shown in previous sections that argued that the rare earth ions that dominate luminescence are those that are separated from a direct contact with nc-Si. Furthermore, the value for the coupling constant is in good agreement with the reported excitation rate of several tens of microseconds for Er^{3+} sensitized by nc-Si [14]. Finally, it should be noted that a coupling constant of several tens of microseconds is smaller than that in bulk semiconductors, it is still much larger than the luminescence lifetimes of nc-Si, which are several hundreds of microseconds even at room temperature [15]. *Thus, the efficient Er luminescence from SRSO is in part due to a fortunate combination of coupling strengths that is strong enough to allow efficient, nc-Si mediated excitation, yet is weak enough to prevent efficient de-excitation of excited* Er^{3+} *by nc-Si.*

2.6. SUMMARY OF EXPERIMENTAL RESULTS

We summarize the results of investigation into nc-Si sensitization of Er as follows

a) We need as little of excess Si as possible
b) We need as small of nc-Si as possible
c) We need to separate Er from direct contact with nc-Si

3. Applications: nc-Si sensitized optical amplifier

3.1. DEVICE FABRICATION

Following the results obtained above, a 2.5 μm thick Er-doped SRSO thin film with 34 % Si content and 0.03 % Er was deposited on 10 μm thick thermal oxide. Following deposition and anneal, a 1 cm long, ridge-type with a 9 μm wide, 0.5 μm high ridge was defined by photolithography and wet-etching. The nc-Si provide the refractive index contrast necessary for waveguiding. No cladding layers were deposited, however. The refractive index was calculated with Maxwell-Garnett theory to be 1.46, and confirmed with ellipsometry.

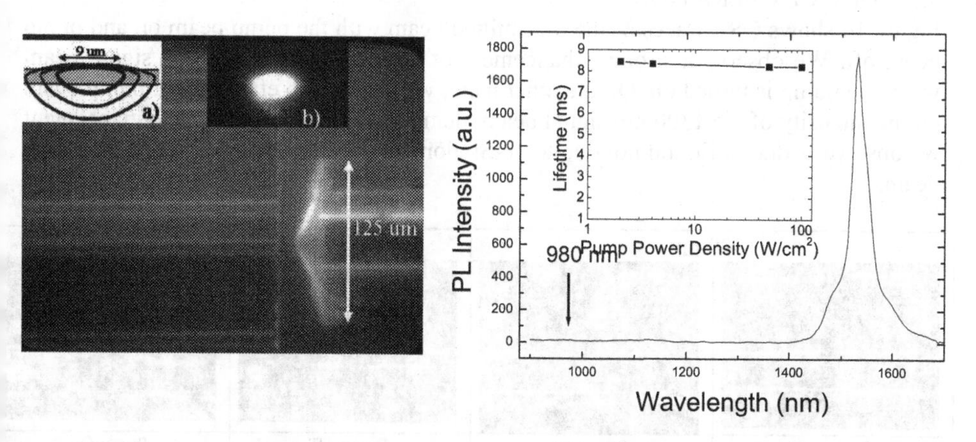

Figure 13 : Left : CCD image of the fabricated waveguide and the calculated and actual mode profile. Note the significant scatter of the waveguide. Right : the PL spectrum of the waveguide, measured from the top. The inset shows the PL lifetime as a function of the pump power.

Optical gain was measured by coupling an external signal from a DFB laser diode into the waveguide using a lensed fiber, and measuring the output using an optical spectrum analyzer (OSA). The input signal power was kept low (< 40 dBm) to measure the small signal gain. The waveguide was pumped using the 477 nm line of an Ar laser incident normally to the top surface of the waveguides, thus ensuring that all Er^{3+} are excited via Si nanoclusters and giving a good representation of a broad-band excitation source. The pump beam was focused using a cylindrical lens to a size of ≈ 0.1×2mm. Photoluminescence measurements were performed by collecting the luminescence emitted normal to the film.

Figure 13 shows the CCD image of the waveguide setup, showing the fiber and the waveguides. We note that due to the experimental nature of the waveguide fabrication process, both the film and waveguides show significant scattering. Inset a) shows the mode profile at 1535 nm, calculated with effective index method. Inset b) shows the actual shape of the transmitted beam taken with an infra-red camera, showing a good agreement with the calculated shape. From Fig. 13, we estimate the core-mode overlap to be 53 %. Also shown in Fig. 13 is the photoluminescence spectrum at a pump density of ≈100 W/cm^2, at which the Er^{3+} luminescence intensity is completely saturated. We observe the typical $^4I_{13/2} \rightarrow ^4I_{15/2}$ Er^{3+} luminescence peak near 1.54 μm, but virtually no peak near 0.98 μm due to the $^4I_{11/2} \rightarrow ^4I_{15/2}$ transition. This is in contrast to previous reports on Er-doped silica-based materials that reported many high order transitions due to upconversion [31]. Similar results are obtained from the time-resolved Er^{3+} luminescence measurements. As shown in the inset, the Er^{3+} luminescence lifetime hardly changes as the pump power is increased by nearly 2 orders of magnitude to the saturation regime, in contrast to previous reports on Er-doped silica-based materials that reported strong lifetime quenching due to cooperative upconversion [31], indicating that upconversion is nearly completely suppressed in our film.

3.2. DEVICE OPERATION

Figure 14 shows OSA traces of the transmitted beam with the pump beam on and pump beam off. We observe a strong enhancement of the transmitted 1535 nm signal beam when the pump is turned on. On the other hand, we observe no effect of the pump beam on the intensity of the 1300 nm signal beam, demonstrating that the signal enhancement we observe is due to Er and not due to nc-Si alone or other spurious effect of the pump beam.

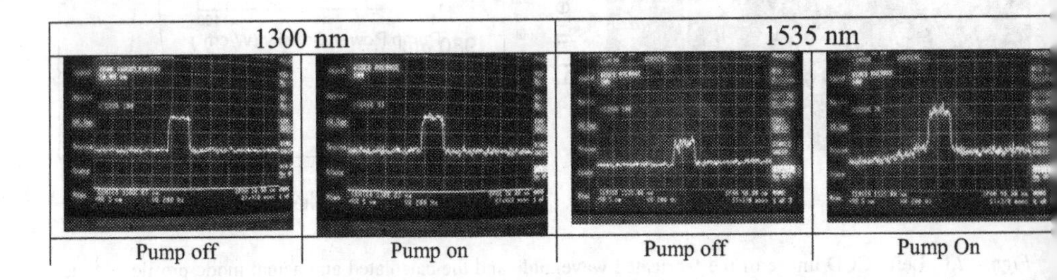

Figure 14 : OSA traces of beams after being transmitted through the waveguide

3.3. PERFORMANCE ANALYSIS

The intensity of the transmitted beam can be written as

$$I(0) = cI_o\exp[-(\sigma_{abs}N\Gamma + \alpha)L]: \qquad \text{pump off}$$
$$I(p) = cI_o\exp[-((\sigma_{abs}N_1 - \sigma_e N_2)\Gamma + \alpha)L] : \text{pump on}$$

where c is the coupling efficiency, I_o is the input signal intensity, α is the waveguide loss, σ_{abs} is the Er^{3+} absorption cross section, σ_e is the Er^{3+} emission cross section, N is total doping concentration, Γ is the core-mode overlap, and L is the illuminated length. N_1 and N_2 are the concentration of Er^{3+} ions in the ground and excited state, respectively, such that $N_1 + N_2 = N$. The coupling efficiency is difficult to estimate, but is expected to be low due to poor mode matching, as can be seen in Fig. 13. Furthermore, due to the experimental nature of the fabrication process, α is also expected to be high. These coefficients, however, are process-related and thus do not pose fundamental limitations. Therefore, we concentrate on signal enhancement (SE), defined as

$$SE \equiv I(P)/I(0) = \exp(2(\sigma N_2 \Gamma)L)$$

Note that in above equation, we have approximated that $\sigma_{abs} = \sigma_e = \sigma$, which is quite accurate for 1535 nm [32]

In general, the rate equations governing the transitions between different excited levels of Er^{3+} are coupled and quite complex. However, as can be seen from Fig. 13, there is very little upconversion in our film, which we attribute to the low Er concentration used. Thus, we model the Er^{3+} as a simple 2 level system

$$dN_2/dt = \Sigma\Phi(N-N_2) - wN_2$$

where Σ is the effective excitation cross section of nc-Si sensitized Er^{3+}, Φ is the pump flux, and w the decay rate of excited Er^{3+}. Note that we neglect the stimulated emission due to the signal beam since a very low signal power was used. Using this model, the pump power dependence of SE is simply

$$SE(P) = \exp(2(\sigma N\Sigma\Phi/(\Sigma\Phi + w))\Gamma L \quad (1)$$

Figure 15 shows the pump power dependence of SE and the fit using eq. (1). We find that we can obtain an SE of up to 14 dB/cm, implying a possible net gain of up to 7 dB/cm. From the known concentration of Er and the length of the pump beam, we obtain a value of $2\pm0.5 \times 10^{-19}$ cm^2 for the emission cross section of Er^{3+} at 1535 nm. An accurate value of the effective excitation cross section Σ is difficult to obtain due to the uncertainty of aligning a long, narrow pump beam on a long, narrow waveguide. Assuming a uniform pump intensity and optimum alignment, we obtain a value of $2\pm1\times10^{-17}$ cm^2 at 477 nm.

These values are much larger than those commonly accepted for Er^{3+} in pure silica ($\approx 10^{-21}$ cm^2 and optical gain of only about 1 dB/cm or less). The large effective excitation cross section is easily attributed to the large absorption cross section of Si nanocrystals, and is in good agreement with the value reported by Kenyon *et al.* and Priolo *et al.*, who suggested a value of 7×10^{-17} cm^2 and 2×10^{-17} cm^2, respectively [33,34]. The near hundred fold increase in the Er^{3+} emission cross section, on the other hand, is so far unexplainable, since it is that of an inner-shell transition and should be only weakly dependent upon the host material. We note, however, that our value for emission cross section is in good agreement with that reported by Kik *et al.*, who, on the basis of absorption measurement, have suggested a value of 8×10^{-20} cm^2 as the lower limit for the emission cross section of Er^{3+} in nc-Si sensitized silica [35], indicating that such enhancement in Er^{3+} emission cross section is a real effect.

It should be pointed out that such an enhancement of emission cross section is critical if nc-Si sensitized, Er doped silica is to be used for practical applications. As can be seen in eq. (1), the maximum gain achievable in a given length is completely determined by σ and N. Thus, without an increase in the emission cross section, we would still need a waveguide that is several tens of cm long to achieve practical gain even with nc-Si sensitization. Such a long waveguide, however, is impractical to pump from the top, thus negating the beneficial effect of nc-Si sensitization. However, because of the enhanced emission cross section, a waveguide that is only several cm long is sufficient, opening the possibility of a waveguide amplifier pumped from the top with a linear array of inexpensive LEDs.

4. Future research possibilities

4.1. Er DOPED Si/SiO_2 SUPERLATTICES FOR ACTIVE SILICON PHTONIC CRYSTALS

4.2.

By manipulating the dispersion relation of photons in an dielectric medium, photonic crystals allows one to control the properties of photons in a way that is similar to the control of electronic properties in a semiconductor. Furthermore, by enabling ultra-compact optical circuits, photonic crystals promise to play an important role in future, high density optical integrated circuits.

Silicon, with its high transparency in the infra-red region, and the well-established processing technologies, would be an ideal material for realization of such photonic crystal-based integrated optics. However, because of its lack of optical activity, only passive components such as waveguides and filters could be realized using silicon. Using Er-doped silica or SRSO cannot be applied in this case, however, since a high refractive index contrast is required for a photonic crystal. With Si/SiO_2 superlattices, however, it is possible to obtain both high enough of a refractive index contrast (>2.5) and high optical activity at the same time, promising the possibility of realizing an active photonic crystal circuit based on Si.

Figure 15 shows the to exist, So far, however, silicon has been EDFAs, despite their advantages, only cover the middle of the low-loss window of silica-based optical fibers (so-called C band). With increasing data transfer rate, however, there is an increasing interest in using Tm for use in amplifying the shorter (S-band) and longer (L-band) wavelength regions. However,

4.3. PHOTONIC CRYSTAL FABRICATION

In order to maximize the effect of a photonic crystal, an triangular-lattice of hole in an air-bridge configuration was fabricated using Si/SiO_2 superlattice. The Si layers were ≈ 1 nm thick and the SiO_2:Er layers were $\approx$ 2 nm thick. A total of 100 pairs of Si/SiO_2:Er layers were deposited on a Si wafer with CVD deposited Si_3N_4. The Si_3N_4 layer acts as the etch stop, and also a sacrificial layer necessary for formation of an air-bridge.

Figure 15 shows the SEM image of a fabricated photonic crystal. Good periodicity with clean features can be observed. Also shown is the calculated bandstructure of the fabricated photonic crystal. A full bandgap can be observed for the TE-like mode. The film also displayed strong Er^{3+} luminescence with lifetimes in the excess of 5 msec. A more detailed optical characterization is currently underway.

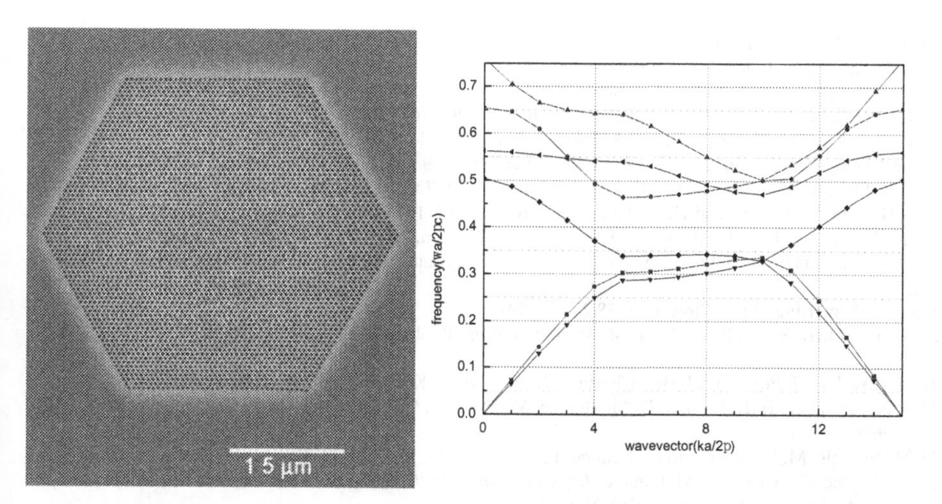

Figure 15 : SEM image and the calculated bandstructure of a photonic crystal based on Si/SiO_2 :Er superlattice thin film

5. Conclusion

In conclusion, we have investigated nc-Si sensitization of Er for optical applications. Conditions necessary to obtain strong and efficient Er^{3+} luminescence has been ascertained, and the practicality of using such nc-Si sensitization for optical amplifier applications demonstrated. We also suggest using such nc-Si sensitized, Er-doped thin films as basis for developing active, silicon-based photonic crystals.

We gratefully acknowledge Prof. Y. H. Lee for help with photonic crystals. This research was supported in part by Advanced Photonics Project and National Research Lab project from MOST, and 5th Leader Technology Development Project from MIC in Korea.

6. References

1. C. R. Giles, D. Bishop, and V. Aksyuk, *MRS Bulletinf* **26**, 328 (2001)
2. *Condensed Matter and Materials Physics}* National Research Council, National Academic Press, Washington, D.. C
3. E. Delevaque, T. Georges, M. Monerie, P. Lamouler, and J. F. Bayon, IEEE Phot. Tech. Lett. **5**, 73(1993)
4. H. Ennen, J. Schneider, G. Pomrenke, and A. Axmann, Appl. Phys. Lett. **43**, 943 (1983)
5. I. N. Yassievich and L. C. Kimerling, Semicond. Sci. Tech. **8**, 718 (1993).
6. J. Palm, F. Gan, B. Zheng, J. Michel, and L. C. Kimerling, Phys. Rev. B. **54**, 17603 (1996)
7. F. Priolo, G. Franzo, S. Coffa, and A. Carnera, Phys. Rev. B. **57**, 4443 (1998)
8 B. Zheng, J. Michel, F. Y. G. Ren, L. C. Kimerling, D. C. Jacobson, and J. M. Poate, Appl. Phys. Lett. **64** 2842 (1994)
9 G. Franzò, F. Priolo, S. Coffa, A. Polman, and A. Carnera, Appl. Phys. Lett. **64** 2235 (1994)
10. J. H. Shin, M. Kim, S. Seo, and C. Lee, Appl. Phys. Lett. **72**, 1092 (1998)
11. M. Fujii, M. Yoshida, Y. Kanazawa, S. Hayashi, and K. Yamamoto,

Appl. Phys. Lett. **71**, 1198 (1997)
12. P. G.Kik and A. Polman, J. Appl. Phys. **88**, 1992 (2000)
13. A. J. Kenyon, P. F. Trwoga, M. Federighi, and C. W. Pitt, J. Phys.: Condens. Matter **6**, L319 (1994)
14 K. Watanabe, M. Fujii, and S. Hayashi, J. App. Phys. **90** 4761 (2001)
15 P. G. Kik, M. L. Brongersma, and A. Polman, Appl. Phys. Lett. **76**, 2325 (2000)
16 Se-Young Seo and Jung H. Shin, Appl. Phys. Lett. **78** 2709 (2001)
17 J. H. Shin, M. Kim, S. Seo, and C. Lee, Appl. Phys. Lett. **72**, 1092 (1998)
18 J. H. Shin, W. H. Lee, and H. S. Han, Appl. Phys. Lett.**74**, 1573 (1999)
19 Y. H. Ha, S. H. Kim, D. W. Moon, J. H. Jhe, and J. H. Shin, Appl. Phys. Lett **79**, 287 (2001)
20 D. J. Eaglesham, J. Michel, E. A. Fitzgerald, D. C. Jacobson, J. M. Poate, J. L. Benton, A. Polman, Y.-H Xie,
and L. C. Kimerling, Appl. Phys. Lett. **58** 2797 (1991)
21 M. W. Sckerl, S. Guldberg-Kjaer, M. Rysholt Poulsen, P. Shi, and J. Chevallier, Phys. Rev. B **59** 13494 (1999)
22 D. Kovalev, G. Polisski, M. Ben-Chorin, J. Diener, and F. Koch, J. Appl. Phys. **80** 5978 (1996)
23 P. N. Favennec, H. L'Haridon, D. Moutonnet, M. Salvi, M. Ganneau, Mat. Res. Soc. Symp. Proc. **301**, 181 (1993)
24 M. Needels, M. Schlüter, and M. Lannoo, Phys. Rev. B **47**, 15533 (1993)
25 C. Delerue, G. Allan, and M. Lannoo, *Light Emission in Silicon: From Physics to Devices*, Semicond. Semimet. **49**, 253 (Academic Press,New York, 1998)
26 F. Iacona, G. Franzò, and C. Spinella, J. Appl. Phys., **87**, 1295 (2000)
27 G. Franzo`, V. Vinciguerra, and F. Priolo, Appl. Phys. A: Mater. Sci. Process. **69**, (1999)!
28 P.G. Kik, M. L. Brongersma, and A. Polman, Appl. Phys. Lett. **76** 2325 (2000)
29 P. G. Kik, M. J. A. de Dood, K. Kikoin, and A. Polman, Appl. Phys. Lett. **70**, 1721 (1997)
30 A. Taguchi and K. Takahei, J. Appl. Phys. **79**, 4330 (1996)
31 E. Snoeks, G. N. van den Hoven, A. Polman, B. Hendriksen, M. B. J. Diemeer, and F. Priolo, J. Opt. soc. Am. B 12, 1470 (1995)
32 W. J. Miniscalco, J. Light. Tech. **9** 234 (1991)
33 A. J. Kenyon, C.E. Chryssou, C. w. Pitt , T. Shimizu-lwayama, D. E. Hole, N. Sharme , and C. J. Humphreys, J. Appl. Phys. **91** 367 (2002)
34 G. Franzo, V. Vinciguerra, and F. Priolo, Appl. Phys. A **69**, 3 (1999)
35 P. G. Kik and A. Polman J. Appl. Phys. **91** 534 (2002)

EXCITATION MECHANISM OF Er PHOTOLUMINESCENCE IN BULK Si AND SiO_2 WITH NANOCRYSTALS

I.N. YASSIEVICH, A.S. MOSKALENKO, O.B. GUSEV, M.S. BRESLER
A F Ioffe Physico-Technical Institute
Politekhnicheskaya 26, 194021 St. Petersburg, Russia

1. Introduction

For about one decade, Er doped Si attracted the attention of researchers with the aim to apply it to integrated silicon-based optoelectronic devices. In fact, Er emission falls into the band of minimum absorption losses of silica glass optical fibers used in telecommunications. The advantage of Er-doped Si consists in the possibility to increase significantly the effective cross section of excitation of the rare earth ions [1,2]. This enhancement is due to a strong band-to-band absorption of the pumping light in a wide spectral range followed by an Auger process in which electron-hole pairs recombine with the transfer of energy to the 4f-shell of Er. The excess energy is taken by free electrons or phonons. However, Er luminescence in a matrix of crystalline Si is characterized by a strong temperature quenching and can be hardly observed above 200 K. Temperature quenching is caused by the thermally activated de-excitation of Er resulting from reverse Auger process accompanied by energy back transfer. Therefore, the optical medium which favors the excitation of rare earth ions is not adequate as what concerns the emission process.

In the last years, a new type of optical medium was proposed which combines the advantages of rare earth doped dielectric and rare earth doped semiconductor: a heterogeneous system in which Si nanocrystals (Si-nc) are dispersed in a SiO_2 matrix [3]. The main idea of this innovation is to separate in space the regions of efficient excitation (Er-doped Si) and emission (Er-doped SiO_2). In this medium strong Er luminescence up to room temperature was measured. Therefore excitation transfer from Si-nc, where electron-hole pairs are created by absorption, to Er in SiO_2 really occurs. Though a lot of experimental results are already obtained for this important optical medium the excitation mechanism of Er ions and transport of excitation over the dielectric matrix is understood only qualitatively. The new optical medium is promising for optical amplifiers pumped by wide band light sources. For this reason, it is interesting to revisit the mechanisms involved in excitation of photoluminescence in Er-doped Si-based media.

In this work we present theoretical considerations of the relevant mechanisms. We start from the analysis of the physical meaning of excitation cross section applied to Er ions in bulk Si where excitation is done not by direct absorption of pumping light but via semiconductor matrix. Then we discuss: i) the probability of excitation of Er ions located near Si-nc with a proper account of quantum confinement of electrons and holes; ii) transfer of excitation between Er ions (resonant and non-resonant mecha-

L. Pavesi et al. (eds.), Towards the First Silicon Laser, 421–428.

nisms); iii) diffusion of excitation over the dielectric matrix and nonradiative channels of Er de-excitation.

2. Efficient excitation cross section

The efficiency of Er excitation under direct optical absorption in the dielectric matrix is determined by the excitation cross section of Er σ , which enters the rate equation

$$\frac{dN_{Er}^{*}}{dt} = \sigma\Phi(N_{Er} - N_{Er}^{*}) - \frac{N_{Er}^{*}}{\tau} \quad , \tag{1}$$

where Φ is the photon flux, N_{Er} and N_{Er}^{*} are the total concentrations of Er ions and of excited ions, respectively, and τ is the lifetime of Er in the excited state. This equation describes the excitation and de-excitation of Er in a two-level scheme, and σ is the cross section of photon absorption by an Er ion. The same equation can be used for a three-level excitation that involves the second excited $^4I_{11/2}$ state, because the excitation relaxes quickly to the $^4I_{13/2}$ state by nonradiative processes. In this case σ is the cross section of photon absorption by Er ion via the transition $^4I_{15/2} \rightarrow {}^4I_{11/2}$. Experimental value of the excitation cross section may be deduced from the power dependence of the raise time of Er luminescence intensity after switch-on of a rectangular optical excitation pulse. Indeed, from Eq. (1) the raise time of the Er luminescence intensity τ_{on} is determined as

$$\frac{1}{\tau_{on}} = \frac{1}{\tau} + \sigma\Phi \tag{2}$$

The excitation cross section can also be obtained from the experimental power dependence of Er luminescence intensity I_{Er} [1,2]:

$$I_{Er} \propto N_{Er}\sigma\tau\Phi / (\sigma\tau\Phi + 1) \tag{3}$$

In the same way, σ can be determined in the case of Er excitation in a semiconductor mediated by band-to-band absorption. According to experimental data, σ in crystalline Si is several orders of magnitude higher than the direct Er σ in a dielectric SiO_2 and Al_2O_3 matrix [1,2]. However, here excitation of Er ions occurs by created carriers in an Auger process and σ should have another physical meaning. The problem has been studied for Er doped bulk and amorphous silicon in Ref. [4].

3. Bulk silicon

At low temperature, the most probable excitation mechanism of Er^{3+} ions in crystalline Si under optical pumping is by exciton recombination [1,5]. In the case of band-to-band absorption, free excitons are quickly formed. Neutral (at low temperatures) donors introduced by Er and/or Er-O complexes easily capture them. Excitation of an Er ion occurs via an Auger recombination of excitons in which the recombination energy is transferred by Coulomb interaction to an electron on the 4f-shell of the Er ion. The Auger process could take place by scattering of free excitons with Er induced donors.. The presence of a donor electron favors the energy conservation in the Auger process. This excitation mechanism is schematically illustrated in Fig. 1.

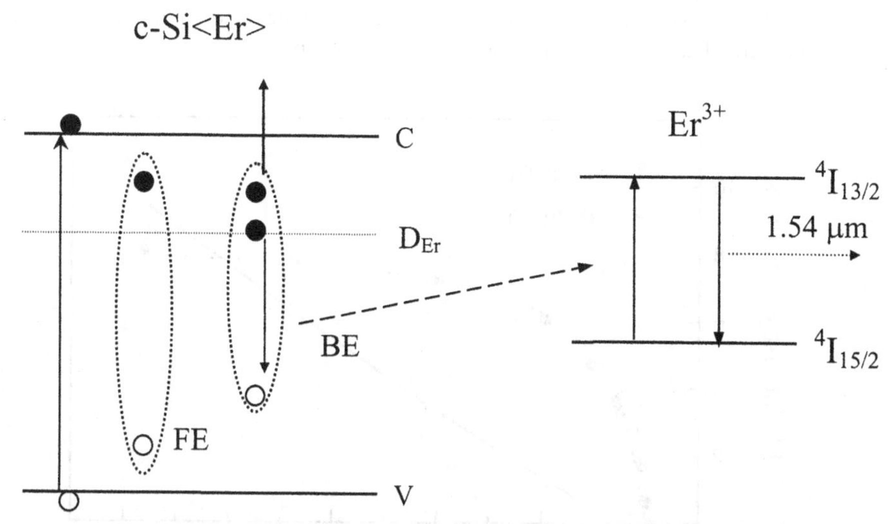

Figure 1. Model of Er^{3+} excitation in crystalline Si

The rate equation, which describes excitation of Er ion, has the form

$$\frac{dN^*_{Er}}{dt} = c_A n_{ex}\left(N_{Er} - N^*_{Er}\right) - \frac{N^*_{Er}}{\tau} \quad , \tag{4}$$

where τ is the effective lifetime of an Er ion in the $^4I_{13/2}$ state, n_{ex} is the exciton concentration, and c_A is the Auger excitation coefficient by a free exciton. The lifetime of excitons is two orders of magnitude lower than the lifetime of Er ions in the excited state. So, we can consider a stationary state and get

$$n_{ex} = \alpha\Phi\tau_{ex} \ , \tag{5}$$

where α is the absorption coefficient of photons for band-to-band transition. Eq. (4) yields

$$\frac{dN^*_{Er}}{dt} = \sigma_{eff}\Phi(N_{Er} - N^*_{Er}) - \frac{N^*_{Er}}{\tau}, \tag{6}$$

where we have introduced the effective cross-section of Er excitation σ_{eff}:

$$\sigma_{eff} = \alpha c_A \tau_{ex} \tag{7}$$

Experimental data obtained in [4] and presented in Figs. 2, 3 demonstrate a strong correlation between Er and exciton photoluminescence.

On these grounds we can conclude that at low pumping intensity the lifetime of exciton is determined mainly by the Auger process assisted by Er excitation. In this condition $\tau_{ex} = 1/c_A N_{Er}$,and Eq. (7) transforms to

$$\sigma_{eff} = \alpha / N_{Er} \, , \tag{8}$$

where N_{Er} is the concentration of optically active Er ions. Relation (8) gives a simple estimate of the number of optically active ions by the experimentally measured σ_{eff} . We found σ_{eff} = (2-8) $\times 10^{-12}$ cm^2 for a sample with an implanted Er concentration of 5 $\times$ 10^{17} cm^{-3} where only 1% of Er ions are optically active. The simultaneous measure of the exciton lifetime allows to estimate the value of the Auger excitation coefficient of

$c_A \approx 7 \times 10^{-10}$ cm^3s^{-1}. This excitation coefficient can be studied theoretically even though theoretical calculation of it is not a simple problem.

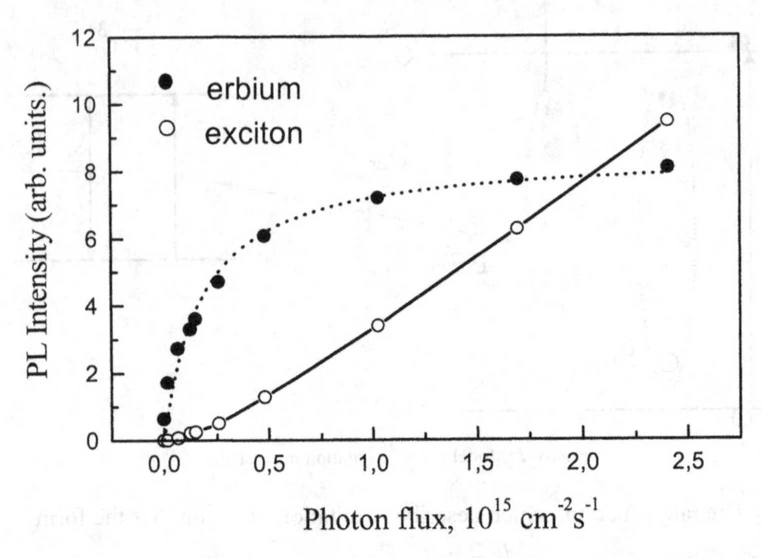

Figure 2. Intensity of Er and exciton photoluminescence as a function of pumping photon flux (Ar laser excitation). Dotted line is a fitting curve to Eq. (3) with $\sigma_{eff}\tau = 5.4$ cm^2 s, solid line is a guide for eye.

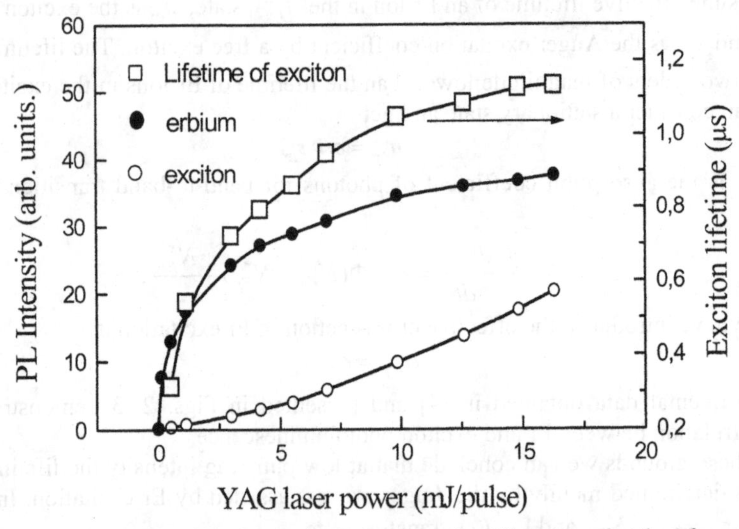

Figure 3. Intensity of Er and exciton photoluminescence and exciton lifetime as a function of power of pulsed Nd: YAG.

It should be noted that increasing pump power, i.e. the amount of excited Er, the effective cross section will increase. This result is determined by the fact that the band-to-band absorption coefficient is an intrinsic property of the semiconductor matrix, and does not depend on the concentration of Er dopant (as is the case for excitation of Er in dielectric matrices). Quite naturally, under these conditions the probability to *excite* a single ion increases when the total number of excitable ions decreases, at a given photon flux. If there are another processes controlling the lifetime of excitons, Eq. (8) is not valid and the effective cross section is less [4].

4. Optical excitation of Er^{+3} ions located in SiO_2 with Si nanocrystals

Si nanocrystals with a size of 2-4 nm strongly absorb the light because the selection rules for absorption are modified increasing the absorption by nanocrystals. Due to the effect of quantum confinement the exciton energy in nanocrystals becomes higher than the energy of the second excited state of Er $^4I_{11/2}$ (1.24 eV). Later the exciton energy is efficiently transferred to Er ions in an Auger process, the excess energy being wasted by multiphonon interaction.

Our calculations demonstrate a strong dependence of the probability of Er excitation on the location of the ions. It can be shown that de-excitation of Er in the nanocrystal is suppressed if the exciton energy sufficiently exceeds that of the second excited state of Er, since in this case the back transfer process demands large activation energy. In fact, the experimental data [3,6] show a broad band centered at ~ 0.8 μm caused by exciton emission: i) from nanocrystals and Er luminescence from the second excited state $^4I_{11/2}$ at 1.0 μm, and ii) the first excited state $^4I_{13/2}$ at 1.54 μm. This result demonstrates that excitation of Er ions occurs to the second excited state and the reverse (de-excitation) process demands activation with a significant threshold.

We shall consider the excitation of Er ions located in silicon dioxide near silicon nanocrystals by confined carriers created in result of absorption radiation by nanocrystals. The excitation probability is determined by a multiphonon-assisted transition

$$W_{ex} = \frac{2\pi}{\hbar}\sum|M|^2\sum_{NN_i}|\langle N_i + N|N_i\rangle|^2 P(N_i)\delta(E_{ex} - \Delta_{f'f} - N\hbar\omega), \quad (9)$$

where $P(N_i)$ is the probability that the number of phonons in the initial state is N_i, $N=N_f - N_i$ is the difference in the phonon number between final and initial states of the vibrational system, ω is the frequency of phonons, E_{ex} is the energy of electron-hole pair (exciton) inside the nanocrystal, $\Delta_{f'f}$ is the difference of energies of excited and ground states of an Er ion. The electronic matrix element $|M|^2$ should be calculated for dipole-dipole interaction

$$\hat{V}_{dd} = \frac{e^2}{\kappa R^3}\left[\vec{r}_1\vec{r}_2 - \frac{3(\vec{r}_1\vec{R})(\vec{r}_2\vec{R})}{R^2}\right], \quad (10)$$

where $\vec{r}_1, \vec{r}_2$ are the coordinates of an electron in a quantum dot (QD) and 4f-electron in the Er ion, R is the distance between them, κ is a dielectric constant of SiO_2. The first sum in (9) means that averaging over all the possible initial electron states and summation over their final states should be done. The result of calculation can be written in the form:

$$W_{ex} = \frac{2\pi}{\hbar}\frac{4}{3}\frac{\Im(N)}{E_{ex} - \Delta_f}\frac{e^4}{\kappa^2 R^6}d^2 d_{opt}^2, \quad (11)$$

where we introduced the notation for dipole moments: i) for the transition in the system of Er 4*f*-electrons $d^2_{opt} \approx \sum_{ff'} |\langle f' | \vec{r}_2 | f \rangle|^2$, and ii) for confined electrons going from initial states i_c to final states f_c $d^2 = \sum \left(|\langle f_c | x_1 | i_c \rangle|^2 + |\langle f_c | y_1 | i_c \rangle|^2 + |\langle f_c | z_1 | i_c \rangle|^2 \right) / 12$ (the numerical factor 12 comes from the number of valleys and two spin states). We have neglected the energy difference between states in the *f*,*f'* multiplets and used $\Delta_{f'f} \cong \Delta_f$, $N = (E_{ex} - \Delta_f)/\hbar\omega$. In Eq. (11) $\Im(N) = S^N \exp(-S)/N!$ with $S = (\varepsilon_{opt} - \varepsilon_T)/\hbar\omega$ being the Huang-Rhys factor.

In the spirit of semi-phenomenological theory it is reasonable to substitute instead of the matrix elements which are difficult to estimate, the experimentally observed quantities because the same dipole moments determine the probability of optical transitions in silicon dots and inside Er ions. The connection between the oscillator strength *P* and d^2_{opt} for a transition between the same energy levels of an Er ion is given by [7]:

$$P = \frac{\kappa^{1/2}}{3(2J+1)} \frac{2m_0 \Delta_f}{\hbar^2} d^2_{opt}, \tag{12}$$

where *2J*+1 is the number of states in the low energy multiplet $^4I_{15/2}$, m_0 is a free electron mass. According to the experimental data [3,6], the exciton energies of typical QD are in the region 1.2- 1.8 eV and therefore excitation in the state $^4I_{13/2}$ (electron transition energy 1.24 eV) and the state $^4I_{11/2}$ (electron transition energy 1.55 eV) is possible. The values of oscillator strengths for these transitions in SiO_2 are: $P = 2 - 6 \times 10^{-7}$ for the transition $^4I_{11/2} \rightarrow {}^4I_{15/2}$, and $P = 1 - 3 \times 10^{-7}$ for $^4I_{9/2} \rightarrow {}^4I_{15/2}$ [7]. From these data we get $d^{(1)}_{opt} = 5.2 \times 10^{-11}$ cm, $d^{(2)}_{opt} = 3.7 \times 10^{-11}$ cm. For QD we can use the connection between d^2 and the radiative lifetime τ_0 of confined electron-hole pairs:

$$\frac{1}{\tau_0} = W_{rad} = \frac{4d^2 e^2 n E_{ex}^3}{3c^3 \hbar^4}, \tag{13}$$

where $n = 3.42$ is the refraction index of silicon. It should be noted that E_{ex} is the quantity depending on the nanocrystal radius. (The effect of quantum confinement on exciton transition was considered in effective mass approximation and quoted in [8]). The effective probability of Er excitation is determined by competition between Auger excitation of Er and radiative transition. We shall estimate the combination $W_{ex}\tau_0$ the value of which shows what process would predominate: $W_{ex}\tau_0 = \Im(N) \cdot 25(R)^{-6}$ where *R* is measured in nanometers (for the excitation to the $^4I_{11/2}$ multiplet) and $W_{ex}\tau_0 = \Im(N) \cdot 4.3(R)^{-6}$ (for the transition to the $^4I_{9/2}$ multiplet). As the radius of nanocrystals is on the order of 2-3 nm, this result means that the energy transfer to Er occurs only for Er located efficiently at the surface of a nanocrystal.

5. Transport of excitation through Er ions

The transport of excitation energy over rare earth ions in dielectric matrix was extensively studied in solid state lasers media and organic materials [9]. There are two possible ways of excitation transport: i) the coherent transfer due to quantum beats between two adjacent Er ions with strictly coinciding energies of the ground and excited states (proposed originally by Davydov), and ii) the non-coherent transfer (the so-called Foerster mechanism) involves the ions with inhomogeneously broadened energy levels and depends on the overlap of the lineshapes of two interacting ions.

It should be stressed that in the system considered the energy levels of the ion are broadened inhomogeneously by local fluctuations of the crystal field, and the condition of strict resonance is fulfilled only for a fairly small part of the total concentration satisfying the condition $\delta E \leq W$, where δE is the energy width in some distribution, W is the dipole-dipole interaction energy in the coherent (quantum beat) approximation. For homogeneously broadened spectrum with a linewidth D_E only the fraction W/D_E of the total concentration of rare earth ions satisfies the condition of a strict resonance. This condition imposes limitation of the concentration of the "resonant" ions and, consequently, on the average distance between them $\bar{r}$.

Therefore, the Davydov case predominates quantum mechanically (the corresponding matrix element is proportional to r^{-3} for dipole-dipole interaction instead of r^{-6}) but is less important statistically in the case of inhomogeneously broadened energy states. The competition of both mechanisms could be actually resolved by an experiment. We have calculated the time of excitation transfer between two adjacent Er ions in a silicon dioxide matrix for the Davydov mechanism

$$t_{tr} = \frac{2\pi\kappa E_{n0}^3 \tau R^3}{3c^3\hbar^3}, \tag{14}$$

where R is the distance between the ions satisfying the condition of a strict resonance, κ the dielectric constant of the medium, τ is the lifetime of excitation state for an isolated ion (radiative lifetime), E_{n0} the energy of the excited state, c is the velocity of light (in the medium). Once coming out of the nanocrystal, the excitation will propagate over Er ions in the dielectric matrix. We have calculated the diffusion coefficient for propagation of excitation by Davydov mechanism. The result is

$$D = \frac{1}{2}\frac{\bar{r}^2}{t_{tr}} = \frac{3c^3\hbar^3}{2\pi\kappa E_{n0}^3\tau}\left(\frac{4\pi\tilde{N}_{Er}}{3}\right)^{1/3}, \tag{15}$$

where $\tilde{N}_{Er}$ is the concentration of Er ions being in a strict resonance. Note that $\tilde{N}_{Er}$ can be significantly less than N_{Er}, the total Er concentration.

For the Foerster mechanism

$$D = \frac{8}{3}\left(\frac{4\pi}{3}\right)^{1/3} C N_{Er}^{4/3} \tag{16}$$

where C is a constant. This is in agreement with our previous statement that statistically the Foerster mechanism is more favorable while the matrix element of the transition is higher for the Davydov case.

The lifetime τ_r of an isolated Er ion in the first excited state is controlled by radiation transition in a dielectric matrix. However, the experiment points to the existence of special traps ("black holes") which kill the excitation after it reaches the interaction radius of the "black hole" r_0. Therefore the lifetime of excitation in a dielectric matrix with the account of its diffusion is $\tau_{eff} = (4\pi D r_0 N_{Tr})^{-1}$, where $\tau_{eff} \propto N_{Er}^{-1/3} N_{Tr}^{-1}$ for the Davydov mechanism and $\tau_{eff} \propto N_{Er}^{-1} N_{Tr}^{-1}$ for the Foerster mechanism. The lifetime of excitation τ_{de} is determined by a competition between the diffusion time τ_{eff} and radiative lifetime τ_r : $\tau_{de}^{-1} = \tau_r^{-1} + \tau_{eff}^{-1}$. Thus, the character of dependence of luminescence de-excitation on Er concentration can show which mechanism of diffusion plays the main role.

6. Conclusions

In conclusion, we have clarified the physical meaning of effective excitation cross-section of Er ions in semiconductor matrix under optical pumping. The large values of excitation cross-section of Er in Si under optical pumping are due to large values of band-to-band absorption coefficient of bulk Si exceeding by several orders of magnitude the absorption coefficient of Er in dielectric SiO_2 and Al_2O_3 matrix. We have theoretically considered Auger excitation of Er ions located in SiO_2 with Si nanocrystals. Excitation probability strongly depends on the distance of an Er ion from a nanocrystal.

This work was supported by the grants from Russian Foundation of Basic Research, Nederlandse Organisatie voor Wetenschappelijk Onderzoek (NWO), Russian Ministry of Science and Technology, and Russian Academy of Sciences. I.N.Y. wishes to thank the Wenner-Gren foundations for a scholarship.

7. References

1. Palm J., Gan F., Zheng B., Michel J., and Kimerling L.C. (1996) Electroluminescence of Er-doped silicon, *Phys. Rev. B* **54**, 17603-17617.
2. Priolo F., Franzò G., Coffa S., and Carnera A. (1998) Excitation and nonradiative deexcitation processes of Er^{3+} in crystalline Si, *Phys. Rev. B* **57**, 4443-4454.
3. Franzò G., Vinciguerra V., and Priolo F. (1999) The excitation mechanism of rare-earth ions in silicon nanocrystals, *Appl. Phys. A* **69**, 3-12.
4. Gusev O.B., Bresler M.S., Pak P.E., Yassievich I.N., Forcales M., Vinh N.Q., and Gregorkiewicz T. (2001) Excitation cross section of Er in semiconductor matrices under optical pumping, *Phys. Rev. B* **64**, 075302-1-075302-7.
5. Bresler M.S., Gusev O.B., Zakharchenya B.P., and Yassievich I.N. (1996) Exciton excitation mechanism for Er ions in silicon, *Phys. Solid State* **38**, 813-817.
6. Kik P.G., Brongersma M.L., and Polman A. (2000) Strong exciton-Er coupling in Si nanocrystal-doped SiO_2, *Appl. Phys. Lett.* **76**, 2325 -2327.
7. Miniscalco W.J. (1991) Er-doped glasses for fiber amplifiers at 1500 nm, *Journ. Lightwave Techn.* **9**, 234-250.
8. Trwoga P.F., Kenyon A.J., and Pitt C.W. (1998) Modeling the contribution of quantum confinement to luminescence from silicon nanoclusters, *J. Appl. Phys.* **83**, 3789-3794.
9. Agranovich V.M. and Galanin M.D. (1982) *Electronic excitation energy transfer in condensed media*, North Holland, Amsterdam.

SiGe/Si:Er LIGHT EMITTING TRANSISTORS

W.-X. NI,[1] C.-X. DU,[1,*] G.V. HANSSON,[1] A. ELFVING,[1] A. VÖRCKEL,[2] AND Y. FU[3]
[1] *Dept. of Physics, Linköping University, SE-581 83 Linköping, Sweden* (wxn@ifm.liu.se)
[2] *AMO GmbH, D-520 74 Aachen, Germany* [3] *Dept. of Physics, Chalmers University of Technology, SE-412 96 Göteborg, Sweden* **Present address, Thin Film Electronics, Linköping, SE-582 16, Sweden*

1. Introduction

Recent research interests on the development of Si-based light emitting devices have been highly motivated by the demands of optical interconnects in the Si CMOS chip technology, in view of monolithic integration of optoelectronic devices on a Si chip to produce a new generation of high density high performance CMOS ICs with high reliability and low cost. As predicted in the SIA roadmap, accompanying with the fast reduction the device feature size, the maximum interconnect length will also increase from the present ~2 km on a logic chip to ≥20 km in 10 years. The circuit performance of CMOS will then be limited by the bandwidth of conventional electronic interconnects. It is desirable to develop fully Si-based optical interconnects. However, a highly efficient Si light emitter is so far missing, due to the inherent indirect bandgap of Si. Many different approaches have been proposed and a lot of effort has been made during the past few years. Er-doping in Si has attracted some special interest [1] due to observations of intense and sharp electroluminescence (EL) at the interesting wavelength of 1.54 μm at room temperature from structures containing layers that are doped with Er to concentrations of $0.1\text{-}2\times10^{20}$ cm^{-3} made by either low-temperature ion implantation followed by a solid phase epitaxy process, [2,3] or molecular beam epitaxy (MBE). [4,5]. Er-doped light emitting devices have several advantages in optical interconnect applications. The transition, which emits the light, is between atomic levels, so that the same output wavelength can be obtained from device to device and from wafer to wafer. Furthermore, it does not depend on the heat-sink temperature, which is a usual problem with light emission via band-to-band recombination in semiconductor homo- or hetero-structures. There are several other important issues regarding output power and modulation frequency of Er-doped light emitting sources, in particular, how the Er-ions are optically activated and being effectively pumped to obtain an optical gain. Since a long spontaneous decay time of Er emission limits Er-doped LEDs to be used as a power efficient device at high modulation frequency, to achieve stimulated emission is thus a key to making Er-doped Si materials and devices feasible for the use of optical interconnects.

In this paper, we will first give a summary of our recent studies on SiGe/Si:Er light

L. Pavesi et al. (eds.), Towards the First Silicon Laser, 429–444.

emitting transistors, which have proven to be an efficient electric pumping device for Er emission at room temperature, and a useful device module to study the fundamental Er excitation and de-excitation mechanisms as well. In the second part, we will extend the discussion to the gain and loss mechanisms in such Si:Er systems, in order to evaluate the feasibility for an Er-doped Si laser, and what the main challenges in materials and device designs and processing are.

2. Theoretical considerations and device limits

2.1 LIGHT EMISSION FROM Er^{3+} IONS AND THE ENERGY TRANSFER MODEL

As has been established, a narrow 1.54 μm light emission from Er-doped Si is originating from an intra *4f* transition of Er^{3+} ions. Since for Er^{3+} ions the partially filled inner 4f shell is well shielded by the outer 5s and 5p electrons, the atomic transitions within the *4f* shell are very weakly affected by the surrounding. Different from the optical or electrical pumping mode in Si for emitting light directly via interband recombination, excitation of the Er-doped Si system is a process of energy transfer from carriers to Er ions. There exist two main excitation processes of Er^{3+} ions: (i) excitation by electron-hole recombination-mediated energy transfer at an Er-related defect level, [3] in cases of a forward biased p-n junction or carriers generation due to laser irradiation; and (ii) hot carrier direct impact excitation [2,4,6] in case of a reverse biased p-n junction. Because the spontaneous radiative decay time of Er ions is very long (~ 1 ms), [7] non-radiative de-excitation processes strongly compete with the radiative decay and cause a significant reduction of the luminescence intensity. One major non-radiative de-excitation responsible for the luminescence intensity quenching at high temperatures is the so-called thermally activated energy back transfer process, [8] i.e., the energy stored due to the excitation of the Er ions can be transferred back to an Er-related defect level if there is enough thermal energy. That thus sets a limit for a forward biased Si:Er-LED to work at high temperature with a high emission efficiency. The luminescence intensity can persist at room temperature for a reverse biased devise, possibly due to reduced band-to-band recombination at a high bias, so that the emitted electrons can have long enough time to be recycled and accelerated for impact excitation. On the other hand, just like impact excitation of Er ions relies on the energy transfer process, any nearby carriers or Er ions can also be involved in taking energy away from excited Er ions. This process is referred to as Auger-type energy transfer. [9] The rates of Auger de-excitation due to the equilibrium free carriers by dopant ionization are usually low in a reverse biased device due to carrier depletion, as compared to the case of forward bias condition. But the non-equilibrium free carriers due to injection and/or impact ionization may play an important role for the non-radiative Auger de-excitation at reverse bias.

An energy diagram and various transfer processes for excitation and de-excitation of Er ions in Si are shown in Fig. 1, and these processes can be mathematically linked by the following differential rate equations for a first order process:

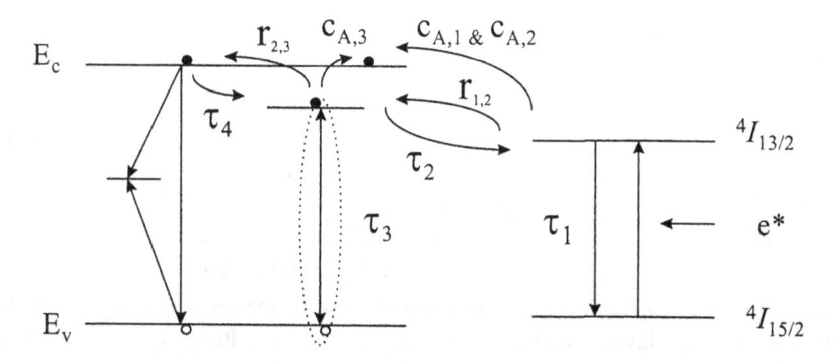

Figure 1 Schematic energy-transfer diagram for Er^{3+} ions in Si.

$$\frac{dN_{Er}^{*}}{dt}=\frac{j_{dri}}{q}\sigma_{ex}(N_{Er}-N_{Er}^{*})$$

$$-\frac{1}{\tau_1}N_{Er}^{*}-r_{1,2}N_{Er}^{*}-c_{A,1}n_{e,d}N_{Er}^{*}-c_{A,2}n_{e,i}N_{Er}^{*}+\frac{N_t}{\tau_2} \quad (1a)$$

$$\frac{dN_t}{dt}=r_{1,2}N_{Er}^{*}-\frac{1}{\tau_2}N_t-r_{2,3}N_t-c_{A,3}n_{e,d}N_t \quad (1b)$$

where, N_{Er}, N_{Er}^{*}, N_t are the densities of all the optically active Er^{3+} ions, of Er^{3+} ions in the 1st excited state ($^4I_{13/2}$), and of the occupied traps in the Er-related defect level, respectively; j_{dri} is the driving current density; σ_{ex} is the excitation cross-section; τ_1 is the spontaneous radiative decay time of the Er^{3+} ions from $^4I_{13/2}$ to $^4I_{15/2}$; τ_2 is the decay time for the energy transfer from the Er related trap state to the 1st excited state; $r_{1,2}$ is the rate to transfer the energy back to the Er-related defect level, which can be described by $r_{1,2} = \nu_1 \exp(-E_b/kT)$; and $r_{2,3}$ is the rate of the dissociation process of Er-related defect traps, described by $r_{2,3} = \nu_2 \exp(-E_d/kT)$. E_b and E_d represent the activation energy barriers for these two processes. Particularly, we list rates for three Auger-type processes, $c_{A,1}n_{e,d}$, $c_{A,2}n_{e,i}$, and $c_{A,3}n_{e,d}$, in which $c_{A,1}n_{e,d}$, and $c_{A,3}n_{e,d}$ correspond to the energy transfer to the equilibrium carriers due to the dopant ionization, while $c_{A,2}n_{e,i}$ corresponds to energy transfer to injected excess carriers. $c_{A,1}$ and $c_{A,3}$ are coefficients for Auger-free carrier transitions, and $n_{e,d}$ is the carrier density due to ionized dopants. $c_{A,2}$ is the Auger-injection carrier coefficient, and $n_{e,i}$ is the excess carrier density. For simplification, processes described by the time constants τ_3 and τ_4 are not included, but their influence can be approximated with a correction to the $r_{2,3}$-term. Moreover, it deserves to be pointed out that, according to Eqs 1, the excitation rate will be lower when most of Er ions have been excited, due to the limited number of available Er ions.

Under the steady-state excitation conditions, i.e., $dN_{Er}^{*}/dt=0$ and $dN_t/dt=0$, the temperature and pumping current dependence of the EL intensity can be calculated by using the following equation derived by combining Eqs 1a and 1b.

$$I_{EL}(T, j_{dri}) = \frac{N_{ER}^*}{\tau_1}$$

$$= \frac{I_{\max}}{1 + \frac{q}{j_{dri}\sigma_{ex}}\left\{\frac{1}{\tau_1} + c_{A,1}n_{e,d} + c_{A,2}n_{e,i} + \left(\frac{r_{1,2}\left(\frac{r_{2,3}}{c_{A,3}n_{e,d}} + 1\right)}{1 + \frac{r_{2,3}}{c_{A,3}n_{e,d}} + \frac{1}{\tau_2 c_{A,3}n_{e,d}}}\right)\right\}}, \quad (2)$$

where $c_{A,2}n_{e,i}=c_{A,2}(j_{dri}/q)(\tau_c/d)$, and τ_c is the drift time of the current carrying electrons within the active layer thickness d. I_{EL} is the resulting EL intensity and $I_{\max} = N_{Er}/\tau_1$ is the maximum EL intensity. The above equation can be written as

$$I_{EL}(T, j_{dri}) = \frac{I_{\max}}{1 + \frac{q}{j_{dri}\sigma_{ex}\hat{\tau}}}, \quad (3)$$

where $\frac{1}{\hat{\tau}} = \sum_i \frac{1}{\tau_i}$ represents an overall time constant involving all time constants related to excitation and de-excitation processes.

2.2 Er EXCITATION/DE-EXCITATION AT THE HIGH CURRENT INJECTION

In the case of a reverse biased LED, the electrons and holes in the depletion layer are driven apart by the applied electric field, so one can assume $N_t/\tau_2 = 0$, and the carriers introduced by ionized dopants are depleted in the space charge region, i.e., $n_{e,d} \approx 0$. By further neglecting the Auger effect induced by the injected carriers, the $\hat{\tau}$ can then be expressed by a simplified form,

$$\frac{1}{\hat{\tau}} = \frac{1}{\tau^*} = \frac{1}{\tau_1} + r_{1,2} \quad (4)$$

Therefore, at high injection, $j_{dri} = \infty$, $I_{EL} = I_{max}$. In fact, the situation is different when taking the Auger de-excitation due to the injected carriers into account. [10] As a consequence of Eq. 3, the maximum observable EL intensity will be saturated at the high current injection, i.e.,

$$I_{EL}(T, j_{dri} \to \infty) = I_{\max} \Big/ 1 + \frac{C_{A,2}}{\sigma_{ex} d}. \quad (5)$$

To get Eq. 5, we have used the parameter substitution $C_{A,2} = c_{A,2}\tau_c$. Making a ratio of Eqs. 2 and 5 and keeping the simplification that $N_t/\tau_2 = 0$ and $n_{e,d} = 0$, one derives

$$I_{EL}(T, j_{dri}) = \frac{I_{EL}(j_{dri} \to \infty)}{1 + \frac{q}{j_{dri}\sigma_{total}}\frac{1}{\tau^*}}, \quad (6)$$

where, $\sigma_{total} = \sigma_{ex} + C_{A,2}/d = \sigma_{ex} + \sigma_{de}$ involves both terms related to the excitation, σ_{ex}, and de-excitation, σ_{de}, by the injected carriers. τ^* can in this case be represented by Eq. 4, and is independent of the injection current. Note that the appearance of Eq. 6 is very similar to Eq. 3, but the measured maximum EL intensity is the saturation intensity rather than I_{max} that corresponds to the number of all optically active Er^{3+} ions. The measured EL saturation intensity may be only a fraction of I_{max}. [11]. Care has to be taken in this case that a high σ_{total} value may mean a strong de-excitation process due to the Auger energy transfer.We stress the following points:

(1) The driving current is composed of two parts: hot electrons (with a kinetic energy of >0.8 eV) and cold carriers (low energy electrons and holes). Cold carriers may not be used to excite Er ions, but only act as de-excitation centers via Auger processes. Therefore, it is crucial to increase the fraction of hot electrons in the driving current, in order to achieve a higher pumping efficiency.

(2) The impact ionization rate will however be dramatically increased when a larger fraction of electrons have gained a kinetic energy of >1.1 eV. This ionization process will not only consume the injected hot electron, which thus reduces the probability for impact excitation, but also produce a large density of cold electrons and holes, which only serve as Auger de-excitation centers before they can gain high enough energy through the applied electric field. Therefore, Er emission would never be efficient in an avalanche regime of the junction breakdown.

(3) It is thus a tradeoff where one should in a controlled way only inject electrons with a kinetic energy of $0.8<E_e<1.1$ eV for the impact excitation rather than for impact ionization. It is obvious that a simple diode device cannot offer the freedom to choose such an injection condition. One has to use a transfer structure to control the carrier injection through an emitter.

3. Design and manufacturing process of SiGe/Si:Er light emitting transistor

The HBT-type Si/SiGe/Si:Er:O light emitting devices have therefore been innovated, [12] aiming at further understanding of the excitation and de-excitation mechanisms and improving the excitation efficiency of Er ions by using an independent hot electron injector (e-b junction). We have then been able to separately control the applied bias across the b-c junction (the hot electron acceleration field), and the injection current density (the electron flux) during an impact excitation process. By using the advantage offered by MBE for growth of high quality n- or p-doped Si and SiGe multi-layer structures, we have designed and fabricated HBT-type Si/SiGe/Si:Er:O light emitters with an active Er/O-doped layer in the collector region and a p-type SiGe layer as the base. A calculated potential diagram and the structure design for the HBT are shown in Figs. 2a and 2b, respectively. Under the common emitter configuration, the emitter-base junction was forward biased while the base-collector junction was reverse biased. The electrons diffused from the emitter to the base and eventually to the collector, and were accelerated in the depletion region of the b-c junction, resulting in hot electron impact excitation of Er^{3+} ions. Due to the transistor current gain, the collector current passing through the Er-doped region was controlled by varying the base current. The excitation of Er ions has thus been achieved at a normal transistor working condition, i.e., in the linear regime prior to the avalanche breakdown for improving the excitation efficiency.

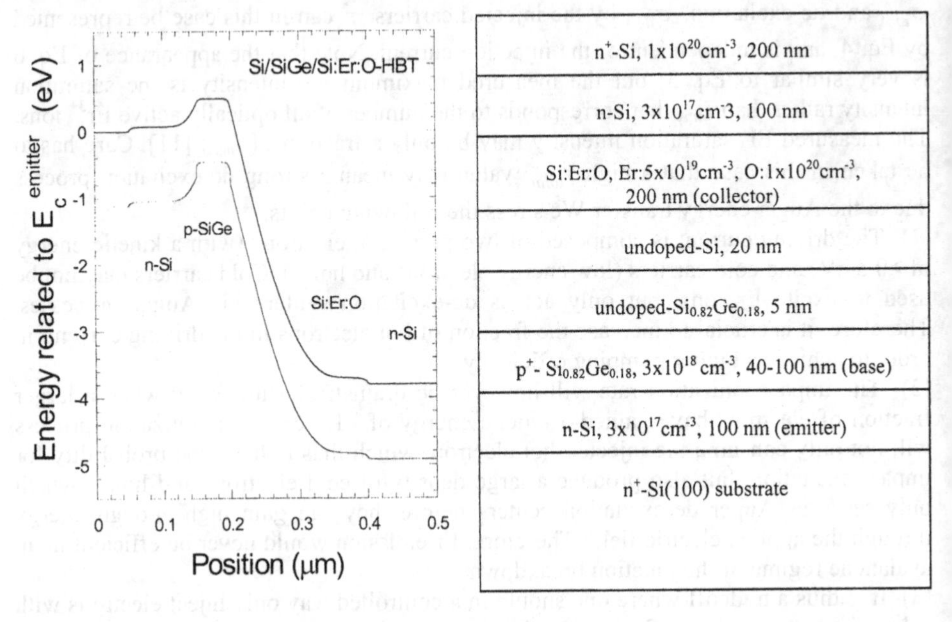

Figure 2. (a) Calculated potential diagram for an HBT-type Si/SiGe/Si:Er:O light emitter based on the layer structures shown in figure (b). For simplification, the substrate thickness has been reduced to 50 nm for the calculation.

There are several special features for this HBT device. First, an emitter-down structure was applied to ensure the growth of a high quality e-b junction. Secondly, single crystalline emitter growth was carried out through windows of an oxide layer. This uses features of the differential MBE growth, i.e., the over-layer growth on SiO_2 outside the window areas will be highly resistive for the emitter layer with a relatively low doping concentration, while the highly doped base layer is conductive. This is a key to achieving a freestanding external base contact without suffering from injection of holes from the base to the emitter. Finally, Er ions together with oxygen, supplied by sublimation of Er and SiO during the MBE growth, were incorporated in the b-c junction with the area aligned to the emitter, which permits all incorporated Er ions to be electronically pumped by injecting hot electrons from the emitter. An approximately 25 nm thick undoped region was grown between the p-SiGe and the Er:O-doped collector, to avoid the so-called dead zone effect where electrons may not have high enough energy for impact excitation.

The use of a SiGe base is two-fold. First, this enables the base engineering to achieve a thin base for hot electron injection, while reducing the base resistance and keeping a certain gain when there is a necessity to increase the base doping. Secondly, the insertion of a SiGe layer with a high refractive index is very crucial for waveguiding, which will be discuss in detail in section V.

The HBTs have been fabricated using standard device processing technology. The base contacts were achieved using KOH-based selective etching, [12] which could be stopped precisely on the surface of the SiGe base layer. A schematic cross-section view of the HBT structure and an SEM micrograph of a processed HBT for surface light emission are shown in Fig. 3 (a) and (b).

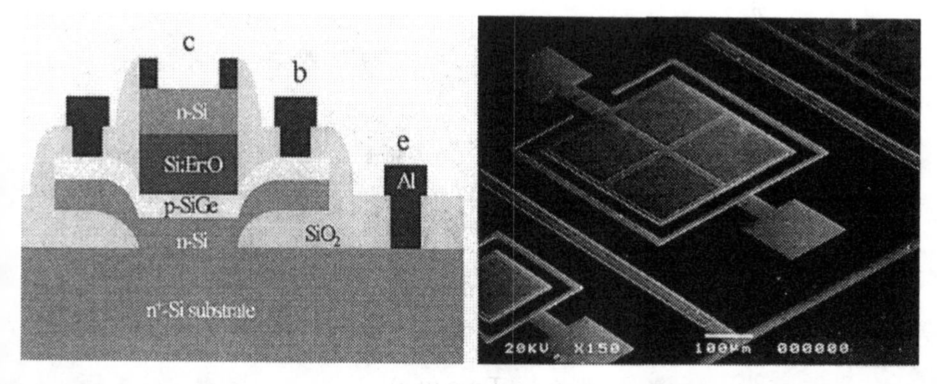

Figure 3 (a) A schematic cross-section and (b) an SEM micrograph of the Si/SiGe/Si:Er:O HBT-type light emitting device prepared by differential MBE.

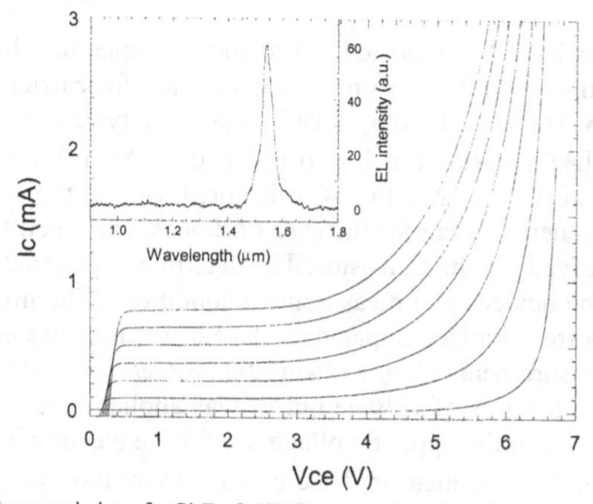

Figure 4 Typical I-V characteristics of a Si:Er:O-HBT measured at RT. The insert shows an EL spectrum of the HBT measured at I_c=0.6 mA (~0.17 Acm^{-2}) and V_{ce}=4.5 V.

4. Experimental results and discussion

Figure 4 shows typical I-V characteristics of a processed SiGe/Si:Er:O-HBT measured with the common-emitter configuration. The measured breakdown voltage with an open base circuit, BV_{ceo}, is about 6.6 V. The breakdown voltage BV_{ce} is lower than BV_{ceo} when increasing the base injection current, due to the carrier multiplication effect. In the transistor linear working regime, the small increase of I_c with V_{ce} is due to the base width modulation effect. This effect gets stronger when V_{ce} > 3.0 V. There is a much more rapid increase of I_c at high V_{ce}, which is partly due to the enhanced hot electron ballistic injection through a thin base at high bias, partly due to the onset of impact ionization and eventually avalanche breakdown.

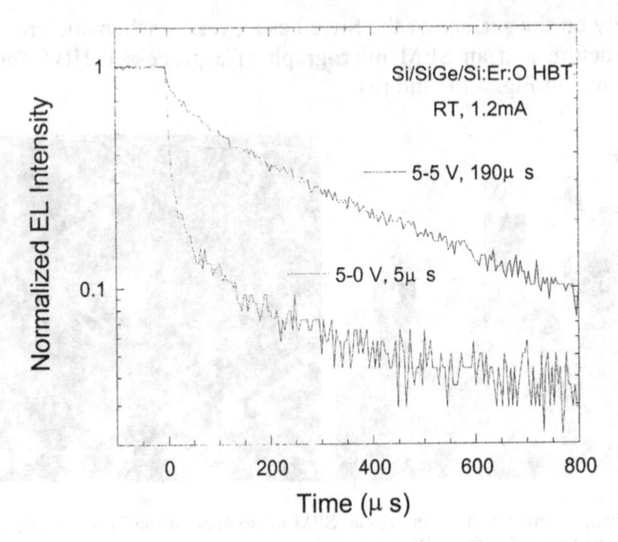

Figure 5 EL decay curves measured with two different bias conditions across the b-c junction.

EL measurements have been carried out at room temperature by applying 200 Hz, 50% duty voltage pulse modulations on the b-e junction for carrier injection, while the b-c junction was reverse biased using a DC voltage. A typical EL spectrum obtained from an Er-doped HBT measured at I_c=0.6 mA (~0.17 A/cm^2) and an applied voltage V_{ce}=4.5 V is shown in the insert of Fig. 4. It is noted that the typical broad background emission in the high-energy range for the case of diodes, attributed to hot electron luminescence, is not observed for the transistor. The determined external quantum efficiency is 3-8 x 10^{-5}, upon the devices and measurement conditions. The influence of the Auger effect due to carriers from ionized dopants on the EL intensity has been clearly revealed by the EL decay measurements (Fig. 5) using this SiGe/Si:Er:O-HBT. Under common emitter configuration, a series of voltage pulses was applied across the e-b junction for the carrier injection. When the applied voltage V_{ce} for the electron acceleration across c-e was a DC bias (5-5 V), the measured 1/e overall decay time was 190 μs for I_c = 1.2 mA. However, at the same injection current level, when V_{ce} was only applied in pulses synchronized with the V_{be} pulses, the measured 1/e decay time decreased to 5 μs. A longer decay time constant observed in the former case is due to suppression of the Auger effect because of carrier depletion in the space charge region under the constant DC bias. When V_{ce} and V_{be} were switched off simultaneously, the excited Er ions are quickly embedded in a region where the carriers due to dopant ionization act as de-excitation centers via an Auger energy transfer process, thus causing a fast decay. Even with a DC bias across the b-c junction, the EL decay time became shorter when increasing the injection current (I_c), [13] or the applied voltage (V_{ce}), which is evidence that the injected carriers may also play a role for de-excitation. As shown in Fig. 6, at a constant collector current of 1 mA during excitation, the measured 1/e decay time constant decreases monotonically from 360 μs at V_{ce}=3 V to 220 μs at 5 V. The shorter decay time at high applied bias is attributed to the occurrence of impact ionization, such that more electrons and holes are generated in the space charge region which increase Auger-carrier de-excitation. It is also expected that the back transfer process becomes more important for Er-ions in a larger field, as electrons excited to the Er-related defect level can be field-emitted.

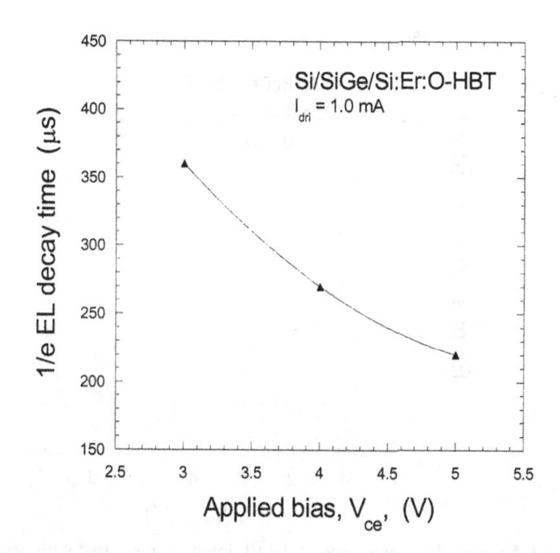

Figure 6 Measured 1/e decay time constants as a function of the bias voltage, V_{ce}.

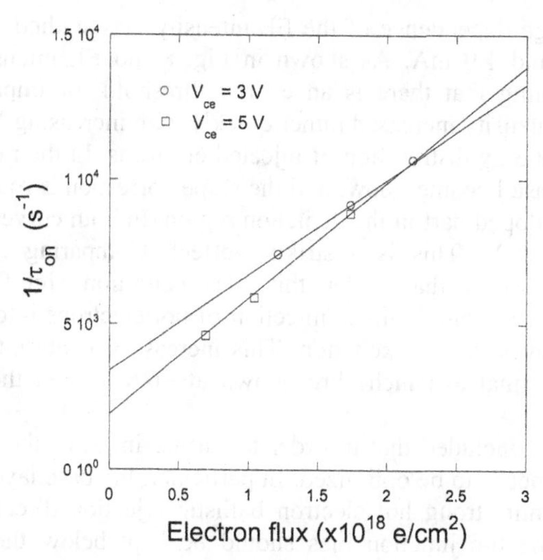

Figure 7 $1/\tau_{on}$ plotted as function of the electron flux at two bias conditions

Fig. 7 depicts the measured EL rising time constant $1/\tau_{on}$ as a function of the injected electron flux at two bias voltages, 3 and 5 V, respectively. As a consequence of Eq. 1, the slope of the linear interpolation of all data points represents the impact cross section value, while the intersection of the extrapolation to the y-axis yields an overall EL decay time constant, which is free from any carrier effect. Following this procedure, we obtain $\sigma = 4.2\times10^{-15}$ cm^2 and $\tau^* = 530$ μs at $V_{ce} = 3$ V. These values however decrease to 3.3×10^{-15} cm^2 and 290 μs, respectively, when increasing V_{ce} to 5 V. It again indicates that the occurrence of impact ionization at higher biases decreases the excitation efficiency.

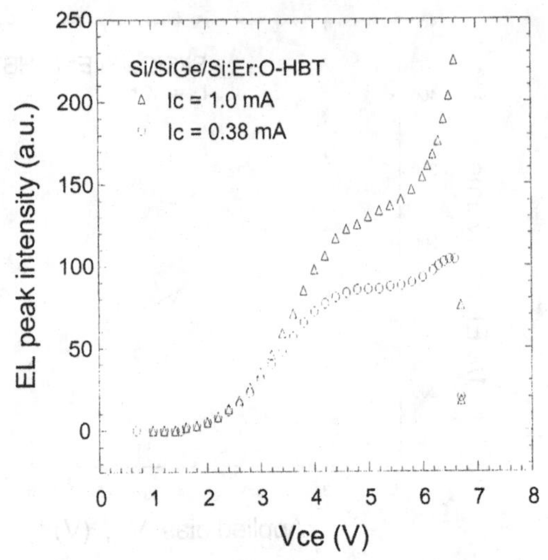

Figure. 8 Applied V_{ce} dependence of the EL intensity measured with two driving currents.

The bias voltage dependence of the EL intensity was studied for constant injection currents of 0.38 and 1.0 mA. As shown in Fig. 8, no EL intensity was observed at $V_{ce}<1.5$ V, indicating that there is an energy threshold for impact excitation. When $V_{ce}>2$ V, the EL intensity increased rather quickly with increasing V_{ce}, indicating a large sensitivity to the energy distribution of injected electrons. In the range of 4.5-5.5 V, the EL intensity increase becomes slow, and the slope corresponds roughly to the thickness increase of the Er doped part in the depletion region. In both curves, there is an increase in the slope > 5.5 V. This is a striking effect. Comparing to the measured I-V characteristics, it seems that under this bias condition significant base thickness reduction has caused strong ballistic injection of hot electrons into the collector, which thus greatly improves the Er excitation. This increase was abruptly interrupted by the occurrence of the final avalanche breakdown at ~6.6 V, and the EL intensity drops drastically

It is thus concluded that in order to further improve the excitation efficiency, the device design needs to be optimized. In particular, the base layer should be carefully engineered to permit strong hot electron ballistic injection direct from emitter to the collector, while the b-c junction bias should be kept below the onset of avalanche breakdown.

5. Perspectives for a Si:Er laser

In this section, we report a feasibility study whether stimulated emission can be observed in a Si:Er material system. The study was carried out based on some existing experimental data in combination with simulations

Already in 1991, Xie et al. [14] concluded that for laser applications the threshold inversion population of Er^{3+} ions, which is a necessary condition for observing stimulated emission, could be in principle achieved at a level of 1.5×10^{18} cm^{-3} (close to the reported solid solubility of Er in Si at that time), if one used a 300 μm long cavity with high reflectivity (R_1=R_2= 90%) mirrors at both sides, while the spectral line width was $\Delta\lambda$ = 1 Å and the overall loss coefficient was 5 cm^{-1}. As has been discussed in the preceding section, the excitation rate will be low when most Er ions are excited due to the limited number of Er sites. Therefore, the required concentration of optically active Er ions should be higher than this threshold value.

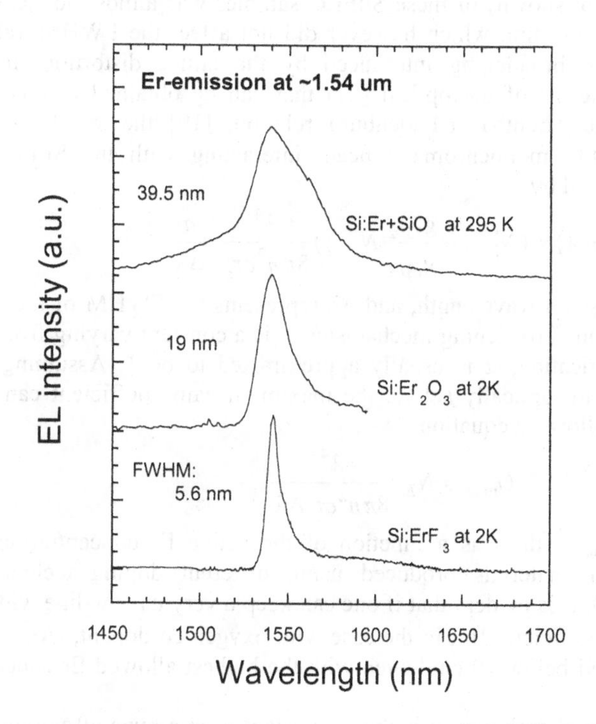

Figure 9 Typical EL spectra of three Si:Er samples prepared using different Er doping source materials during the MBE growth.

Experimentally, it has been proven that the solid solubility is no longer the limit for incorporation of Er. Both ion implantation (together with an elaborated two-step annealing procedure) [8,15] and low-temperature MBE [4,5,16] have shown to enable incorporation of Er up to ~2×10^{20} cm^{-3} with oxygen as co-dopant at the same order of magnitude. Furthermore, we have shown that a very narrow spectral line width can be achieved (FWHM = 1.8 Å, or 0.12 meV, when the Si is doped by ErF_3 and annealed at 800 °C), [17] but it may only appear at a low Er doping concentration <<1×10^{18} cm^{-3}. Because of co-dopants, there are often strong lattice disturbances, which on one hand may help Er emission but the may also broaden the emission line width significantly. Fig. 9 shows examples of EL spectra obtained from our SiGe/Si:Er diode structures with Er doping made by various techniques ($N_{Er} \approx 5x10^{19}$ cm^{-3}). As depicted in the fig. 9, the narrowest spectrum for a heavily doped Si:Er sample was obtained using ErF_3 as an MBE doping source. However, due to a p-type doping characteristic introduced by the fluorine doping, [18] it has been concluded that the Si:Er:F materials are not suitable for impact excitation using hot electrons.

The line width increases significantly, as soon as we introduce oxygen atoms. In case of using Er_2O_3, it rises up to ~20 nm. The spectrum for the sample doped using direct sublimation of Er and SiO was measured at room temperature, where the FWHM is in the range of 30-40 nm, which is similar to samples with the doping made through evaporation of Er in an O_2 ambient. The only difference observed from the low-temperature spectra (not shown) of these Si:Er:O samples was almost no EL intensity for the wavelength <1530 nm, which however did not affect the FWHM value very much. The line width broadening introduced by the lattice distortion imposes a fundamental limit in terms of the optical gain that can be obtained in this material system. Following the Füchtbauer-Ladenburg relation, [19] the small signal gain coefficient for a highly monochromatic beam interacting with the Si:Er medium material can be expressed by

$$G(\lambda) = (N_{13/2} - \frac{g_{13/2}}{g_{15/2}} N_{15/2}) \frac{\lambda^4}{8\pi n^2 c\tau_s} \cdot \frac{a}{\Delta\lambda} \tag{7}$$

where λ is the Er emission wavelength, and $\Delta\lambda$ represents the FWHM of the emission peak. Because of various broadening mechanisms, *a* is a constant varying from 0.64 to 0.94. [19] For simplification, it is usually approximated to be 1. Assuming that all incorporated Er ions are optically active, the maximum gain coefficient can then be calculated using the following equation

$$G_{max} = N_{Er} \frac{\lambda^4}{8\pi n^2 c\tau_s \Delta\lambda}. \tag{8}$$

The calculated G_{max} values as a function of the active Er concentration for the above-described Si:Er materials produced using different doping techniques are summarized in Fig. 10. It is evident that if one can keep a very narrow line width, there will be a large gain coefficient. But in the case with oxygen co-dopant, ($\Delta\lambda$ > 20 nm), the G_{max} will be limited below 10 cm^{-1} even with the highest allowed Er concentration (~$2x10^{20}$ cm^{-3}).

To achieve stimulated emission, it is thus important to minimize all possible losses so that a resulting gain may still be possible within the above limits due to Er concentration and the spectral line width broadening. In the case of a rib waveguide structure, The G_{max} value must be high enough to compensate for all losses introduced by mirror reflectance, scattering in the waveguide, and any absorptions. Especially, we address the importance of the free carrier absorption. The threshold gain required for stimulated emission is thus:

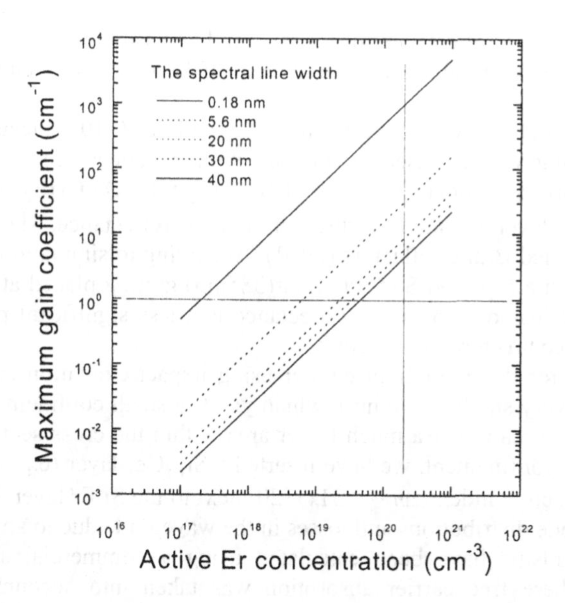

Figure. 10 Calculated maximum gain coefficient vs. the active Er concentration for several Si:Er materials with different spectral line widths. The vertical guideline indicates the maximum Er doping concentration reported in the literature.

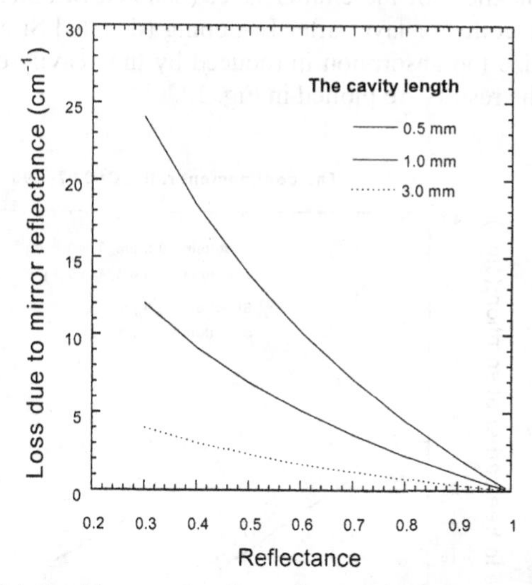

Figure 11 Calculated loses vs. the mirror reflectance for three rib lengths.

$$G_{th} = \gamma_{mirror} + \gamma_{waveguide} + \gamma_{free\ carrier\ absorption} + .. \tag{9}$$

For a rib waveguide device with a length of L, the gain to compensate the loss due the mirror reflectance can be estimated using the equation below

$$\gamma_{mirror} = \frac{1}{2L} \ln\left(\frac{1}{R_1 R_2}\right), \tag{10}$$

where, R_1 and R_2 are the reflectance for mirror 1 and 2. Assuming, $R_1 = R_2$, the losses in the waveguide have been calculated as function of the mirror reflectance and the rib length (Fig. 11).

According to our previous experience, crystalline (110) facets obtained by mechanical cleaving can be used as mirrors with a reflectance of ~0.3. Another alternative is to form a vertical sidewall by RIE using $SiF_6/O_2/CHF_3$, together with an oxide/metal double layer coating to further improve the reflectance. The best solution is to fabricate distributed Bragg reflectors (DBR). According to simulations, $R = 0.99$ can be obtained through a 5-period Si(550nm)/air(385nm) grating placed at both sides of a rib. Therefore, the loss due to mirror reflectance is a less significant part. In the best case, one can reduce this term to $< 1\ cm^{-1}$.

One difficulty for the Si:Er light emitter using impact excitation is that the active layer thickness is very small (~100 nm), which yields a small confinement factor as the modes are usually spread over a much larger area within the cross section of a rib. To improve the mode confinement, we have inserted a $Si_{1-x}Ge_x$ layer (e.g. the base) with a slightly larger refractive index ($\Delta n = 0.31x$) [20] next to the Si:Er layer, but the effect is still weak. The mode distributions and losses in the waveguide due to any scattering and absorption mechanisms have been calculated using a commercial simulation code FIMMWAVE, where free carrier absorption was taken into account based on the absorption coefficient data published by Soref et al.. [21] A waveguide device geometry with a layer structure similar to our HBT has been used in simulations. In particular, we have varied the thickness of the emitter layer (denoted Si-bottom, $N_{Sb} = 1x10^{17}\ cm^{-3}$), and the rest of the collector layer after Er-doping (denoted Si-cap, $N_{Sb} = 1x10^{17}\ cm^{-3}$), in order to minimize the absorption introduced by the heavily doped contact layer and metal contacts. The results are plotted in Fig. 12.

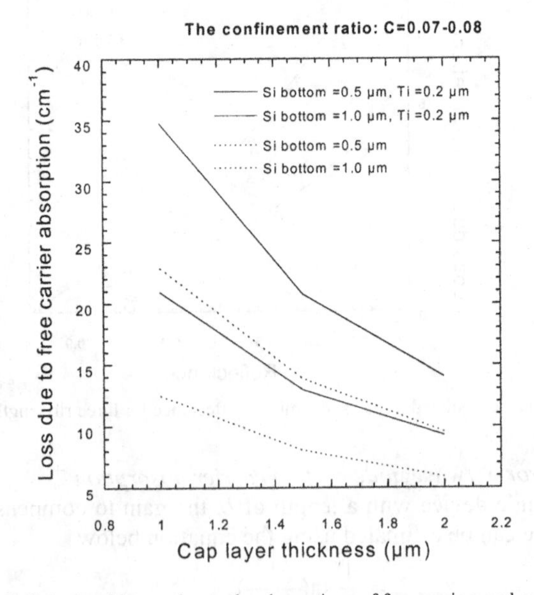

Figure 12 Calculated losses due to the absorptions of free carriers and metal contact

Top metal contact
p-Si, 2300 nm, 5x10^{15} cm^{-3}
p-Si, 100 nm, 1x10^{17} cm^{-3}
p-$Si_{0.82}Ge_{0.18}$, 80 nm, 3x10^{17} cm^{-3}
i-$Si_{0.82}Ge_{0.18}$, 5 nm, $\leq$ 1x10^{16} cm^{-3}
i-Si, 20 nm, $\leq$ 1x10^{15} cm^{-3}
n-Si:Er:O, 100 nm, 2x10^{20} cm^{-3}
n-Si, 30 nm, 3x10^{16} cm^{-3}
n-$Si_{0.82}Ge_{0.18}$, 100 nm, 3x10^{16} cm^{-3}
n-Si (sub.) 2000 nm, 3x10^{16} cm^{-3}
n-Si (back contact),150nm, 5×10^{19} cm^{-3}
Bottom metal contact

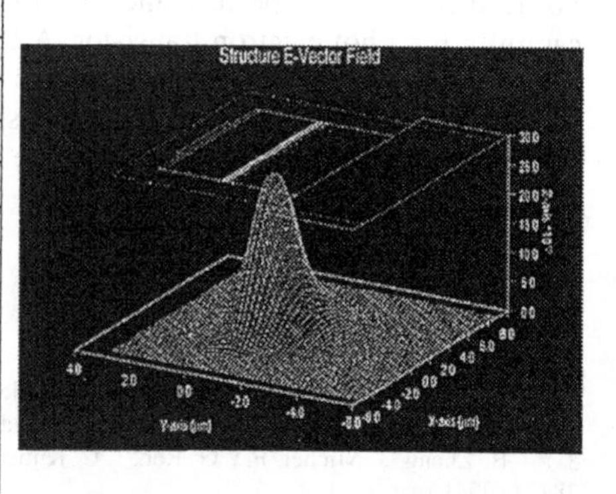

Figure 13 (a) A proposed layer structure for the Si:Er waveguide light emitter, and (b) the calculated mode distributions based on this layer structure.

A very large loss was found in this case, because of absorption of free carriers and metal contacts. Increasing the thickness of the lowly doped Si cap layer improves the situation. However, note that the confinement factor, which is limited by the active layer thickness, is very small (C = 0.07-0.08). The required gain to overcome losses should be $G = \gamma/C$. Therefore, even in the best case with thicker lowly doped layers at both sides of the active region, a maximum gain coefficient of ~80 cm^{-1} is needed for stimulated emission, which thus is impossible for this material system.

We have thus proposed a new design involving two SiGe layers at both sides of the Er-doped layer, while we also eliminate the use of a heavily doped substrate. The layer structure is depicted in Fig. 13(a). The simulation for this structure shows a significant improvement (Fig. 13(b)). The overall loss due to adsorption was reduced to 1.12 cm^{-1}. By reducing the area of top metal contact, the loss can be further decreased to 0.51. If one removes the metal contact completely, the loss was 0.34 cm^{-1}. Since the confinement factor has not yet been changed very much ($C \approx 0.08$), the required G for stimulated emission would be > 7 cm^{-1}, which is not above the limit, but very close to the maximum gain that we could ever obtain from this material system.To summarize, the observation of stimulated emission from an Er-doped structure may be feasible in particular at low temperatures. To pursue this study, there is still a need to obtain an even higher incorporation and optical activation limit of Er ions. To reduce optical losses is a key issue. Therefore, care has to be taken for the proper design of an Er-doped Si structure including a low loss waveguide, high quality mirrors, (e.g. DFB), and to reduce any absorption mechanism as much as possible.

6. Summary

To conclude, we have demonstrated that a SiGe/Si:Er:O-light emitting transistor is an efficient hot electron injector for electrically pumped Er emission, with a possibility to choose the desired acceleration potential for hot electron impact excitation. We have shown that free carriers are detrimental as de-excitation centers via Auger processes, so that the device should be operated with a bias voltage below the avalanche breakdown.

For further improved performance of Si:Er light emission, one should engineer the base carefully for a hot electron transistor. A Si:Er laser is not impossible, but demands continued efforts, i.e, (i) requiring materials studies to achieve a high and optically active Er doping concentration with a narrow spectral line width, (ii) requiring a careful device design to minimize absorptions due to free carriers and metal contacts, which are very crucial for a low-loss waveguide.

References:

1 S. Coffa, G. Franzò, and F. Priolo, MRS Bulletin **23**, 25 (1998).
2 G. Franzò, F. Priolo, S. Coffa, A. Polman, and A. Carnera, Appl. Phys. Lett. **64**, 2235 (1994).
3 B. Zheng, J. Michel, F.Y.G. Ren, L.C. Kimerling, D.C. Jacobson and J.M. Poate, Appl. Phy. Lett. **64**, 2842 (1994).
4 J. Stimmer, A. Reittinger, J.F. Nützel, G. Abstreiter, H. Holzbrecher and CH. Buchal, Appl. Phys. Lett. **68**, 3290 (1996).
5 W.-X. Ni, K.B. Joelsson, C.-X, Du, I.A. Buyanova, G. Pozina, W.M. Chen, G.V. Hansson, B. Monemar, J. Cardenas, and B.G. Svensson, Appl. Phys. Lett. **70**, 3383 (1997).
6 C.-X. Du, W.-X. Ni, K.B. Joelsson, and G.V. Hansson, Appl. Phys. Lett. **71**, 1023 (1997).
7 W.-X. Ni, C.-X. Du, K.B. Joelsson, G. Pozina, and G.V. Hansson, J. Luminescence **80**, 309 (1998).
8. F. Priolo, G. Franzò, S. Coffa, ,A. Polman, S. Libertino, R. Barklie, and D. Carey, J. Appl. Phys. **78**, 3874 (1995).
9 J. Palm, F. Gan, B. Zheng, J. Michel, and L.C. Kimerling, Phy. Rev. **B54**, 17603 (1996).
10 W.-X. Ni, C.-X. Du, F. Duteil G. Pozina, and G.V. Hansson, Thin Solid Films **369**, 414 (2000).
11 G.V. Hansson, W.-X. Ni, C.-X. Du, A. Elfving, and F. Duteil, Appl. Phys. Lett. **78**, 2104 (2001)
12 C.-X. Du, F. Duteil, G.V. Hansson, A. Elfving, and W.-X. Ni, Appl. Phys. Lett. **78**, 1697 (2001).
13 C.-X. Du, W.-X. Ni, K.B. Joelsson, F. Duteil, and G.V. Hansson, Optical Materials **14**, 259 (2000).
14 Y.H. Xie, E.A. Fitzerald, and Y.J. Mii, J. Appl. Phys. **70**, 3223 (1991).
15 A. Polman, J.S. Custer, E. Snoeks, and G.N. van der Hoven, Appl. Phys. Lett. **62**, 507 (1093).
16 W.-X. Ni, K.B. Joelsson, C.-X, Du, G. Pozina, I.A. Buyanova, W.M. Chen, G.V. Hansson, and B. Monemar, Thin Solid Films **321**, 223 (1998).
17 I.A. Buyanova, W.M. Chen, W.-X. Ni, G.V. Hansson, and B. Monemar, J. Vacuum Sci. and Technol. B**16**, 1732 (1998).
18 C.-X. Du, W.-X. Ni, K.B. Joelsson, and G.V. Hansson, Physica scripta T**79**, 155 (1999).
19 P. Bhattacharya, in Semiconductor Optoelectronic Devices, pp.249-267, (Pretice Hall, New Jersey, 1993).
20 F. Duteil, C.-X. Du, K. Järrendahl, W.-X. Ni, and G.V. Hansson, Optical Materials **17**, 131 (2001).
21 R.A. Soref and R. Bennett, IEEE J. of Quantum Electronics **23**, 123 (1987).

SMBE GROWN UNIFORMLY AND SELECTIVELY DOPED Si:Er STRUCTURES FOR LEDs AND LASERS

Z.F. KRASILNIK[1], V.Ya. ALESHKIN[1], B.A. ANDREEV[1],
O.B. GUSEV[2], W. JANTSCH[3], L.V. KRASILNIKOVA[1],
D.I. KRYZHKOV[1], V.P. KUZNETSOV[4], V.G. SHENGUROV[4],
V.B. SHMAGIN[1], N.A. SOBOLEV[2],
M.V. STEPIKHOVA[1], A.N. YABLONSKY[1]
[1] *Institute for Physics of Microstructures, Russian Academy of Sciences GSP-105, Nizhny Novgorod, 603950 Russia*
[2] *Ioffe Physico-Technical Institute Russian Academy of Sciences Politekhnicheskaya st. 26, St.-Petersburg, 194021, Russia*
[3] *Institut für Halbleiterphysik, Johannes-Kepler-Universität, A-4040 Linz, Austria*
[3] *Nizhny Novgorod State University, Gagarin Ave. 23, Nizhny Novgorod, 603600 Russia*

1. Introduction

In recent years an increasing attention has been focused on the investigation of optically active Er centers in Si in view of great application possibilities opened for this material. The intra-center emission of Er^{3+} ions occurs at a wavelength of 1.54 μm where silica based optical fibers are known to have minimum loss and low dispersion. It makes Si:Er attractive as a light emitting source for fiber optics communication systems. Moreover, the realization of efficient light emitters on Si will offer new opportunities in the application of Si-based optoelectronic devices for large-scale integrated circuits. The methods commonly used for incorporating Er into silicon are ion implantation [1-3] and molecular beam epitaxy (MBE) [4-6]. In this contribution we present an original method of sublimation MBE (SMBE) [7, 8] and describe its capabilities for growth of effective light-emitting Si:Er-based structures including light-emitting diodes operating at room temperature. Along with the SMBE grown uniform Si:Er layers, the photoluminescence (PL) efficiency of which is comparable with or even higher than that of the ion-implanted layers, we consider here a novel type of Er-doped structures - the selectively doped Si/Si:Er/Si/Si:Er.../Si multilayer structures with enhanced photo- and electroluminescence (EL) efficiency [9]. Finally, we provide the results of simulations for the parameters of real laser-type structures and discuss the prospects of achieving stimulated emission on their basis.

L. Pavesi et al. (eds.), Towards the First Silicon Laser, 445–454.

2. Experimental

SMBE is a modification of MBE where fluxes of Si and the doping impurities are produced by sublimation of appropriate current-heated sources. To grow Er-doped silicon layers we use two different impurity sources, namely: i) polycrystalline Si plates intentionally doped with Er, as a source for both the Er and Si fluxes, and ii) metallic Er plates - as an impurity (Er) source in combination with a monocrystalline Si wafer - as a source for Si. The Si:Er layers are grown on *p*- and *n*-type Si (100) substrates with resistivities of 10-20 Ω cm or on highly doped Si substrates for LED fabrication. The growth temperatures (T_{gr}) in SMBE are 430 ÷ 700°C. The growth rate can be varied from 0.3 to 5 μm/h, and the thickness of uniformly doped Si:Er layers - from 0.2 to 6 μm. Commonly, the concentration of Er atoms in the structures produced from a polycrystalline Si:Er:O source is of about $1 \div 5 \cdot 10^{18}$ cm^{-3} and can be increased up to 10^{21} cm^{-3} by applying a metallic Er source. For the selectively doped Si/Si:Er/Si/Si:Er.../Si multilayer structures that will be described separately, the thickness of Er-doped Si layers $d_{Si:Er}$ is 20 ÷ 500 Å, the thickness of pure Si space layers d_{Si} is 17 ÷ 1000 Å, and the number of periods N = 16 ÷ 400. Liquid nitrogen cooled Ge and InGaAs detectors are used for the luminescence signal detection.

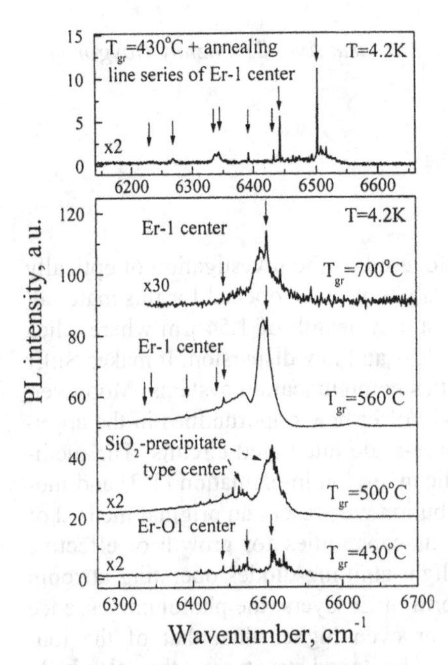

Fig. 1. PL spectra of the uniformly doped SMBE layers grown at different temperatures. The upper curve in the figure represents the spectrum of Si:Er layer grown at 430°C after subsequent annealing at 800°C for 30 min in hydrogen atmosphere.

3. Photoluminescent Si:Er structures

Fig. 1 shows the PL spectra of uniformly doped Si:Er layers grown by the sublimation MBE at different temperatures. All layers feature intense PL at the wavelength of 1.54 μm even at the "as grown" stage, without any additional heat treatment. Depending on the growth temperature the PL response of SMBE grown Si:Er layers is governed by different types of optically active Er centers. At a relatively low growth temperature of about 430°C, the series of low symmetry Er centers identified earlier in highly O- and Er-implanted Si [10] dominate in the PL spectrum. Most pronounced at this temperature is the axial symmetry Er-O complex, the well-known center Er-O1.

With an increasing growth temperature a fairly broad spectral band (FWHM = 20 - 30 cm^{-1}) is seen to develop along with the isolated low-symmetry Er centers. This is the PL band with the maximum centered at 1.538 μm and a characteristic shoulder around 1.55 μm. It was interpreted recently as pertaining to Er complexes of the so-called SiO_2-precipitate type and is believed to be primarily responsible for the light-emitting diodes operation at room temperature [3]. At the temperature of 560°C the PL spectrum, together with the dominance of the SiO_2-precipitate type Er complexes, exhibits formation of a new Er-related center, the center Er-1 with the characteristic spectral line pattern consisting of the eight well resolved PL lines at 6502, 6443, 6433, 6392, 6342, 6336, 6268 and 6231 cm^{-1}. Besides being formed at an elevated growth temperature, it emerges also as a result of the transformation of Er-oxygen complexes during the 800°C annealing treatment of Si:Er layers grown at 430°C.

Comparative low temperature PL measurements have shown much more intense PL response of SMBE grown Si:Er layers as compared with the ion implanted layers. As known, the maximal values for quantum efficiency and PL intensity in Si:Er structures are reached at low temperatures when excitation of Er^{3+} occurs through an exciton-mediated process. A theoretical analysis of the excitation processes performed for a (Si:Er/Si) structure, which also took into account free exciton generation rate and diffusion of excitons, shows that the dependence of low temperature PL intensity on Er concentration has a maximum (see Fig. 2) [11]. The position of this maximum is determined by the critical concentration of exciton capture centers in the Si:Er layer N, at which the formation of free excitons is complicated. At the concentration level N > N_{crit} the exciton generation rate f_g in a heavily Er doped layer sharply decreases (see Fig. 2). In this case excitation could occur only with participation of excitons diffusing from the neighboring undoped regions. There are the experimental results pointing to the influence of the high or low doped regions (neighboring to the Si:Er layer) on the PL intensity of Si:Er layer: we did not observe any considerable PL, either exciton- or Er- related, in SMBE samples grown on highly doped (*p*- and *n*- type) Si substrates.

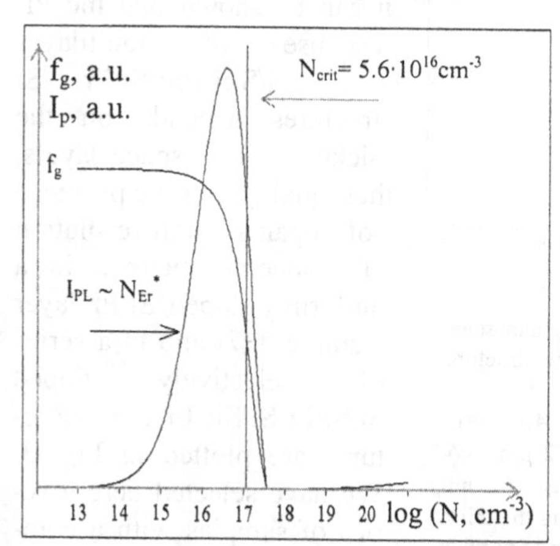

Fig. 2. Calculated dependence of PL intensity on the concentration of exciton capture centers *N* in a Si:Er layer Here f_g is the free exciton generation rate, N_{Er}^* the concentration of the excited erbium centers in the Si:Er layer. The concentration of exciton capture centers in the Si substrate, taken for simulation is 10^{15} cm^{-3}. In more detail the theoretical model and the results of the analysis were discussed in [11,12].

In the case of low-doped substrate it is no need to enhance the thickness of Si:Er layer. Because of high excitation cross-section [13,14] excitons generated in the low-doped substrate are captured by Er centers in narrow layer that adjoins to the substrate.

The above reasons resulted in a new approach which allowed us to enhance the luminescent efficiency of Si:Er structures by more than one order of magnitude [11,15]. Making use of the capabilities of the SMBE method, a novel type of Si:Er structures, namely, the selectively doped multilayer structures (Si/Si:Er/Si/Si:Er.../Si type) have been proposed. The schematic view of this kind of structures is shown in the inset in Fig. 3. The structures consist of a number of Si:Er/Si layer pairs where thin Er-doped Si layers with the thickness down to 20 Å are alternated with pure (undoped) Si space layers. PL measurements performed at low temperatures on the structures of this type have shown that their luminescence intensity exceeds considerably that obtained for the uniformly doped Si:Er layers [9]. This strong effect is connected with periodic combination of thin high-doped Si:Er layers and undoped Si spacers. Indeed, it can be shown that the PL response of multilayer Si/Si:Er/Si/Si:Er.../Si structures depends on the thickness of Si space layers, their quality and the presence of dopants. High resolution PL spectra obtained in a uniformly doped Si:Er layer (sample #37) and in a series of selectively doped Si/Si:Er/Si/Si: Er.../Si structures are plotted in Fig. 3. We have selected here a series of samples with a comparable total thickness (d) and almost the same Er concentration in Si:Er layers ~ $5\cdot10^{18}$ cm^{-3}. In all cases the absolute values of the PL intensity measured in multilayer Si/Si:Er/Si/Si:Er.../Si structures exceeded that obtained for a uniformly doped layer. To estimate the luminescent intensity per unit thickness of Er doped layers ($I_{Si:Er}$) the luminescence spectra were normalized to the total thickness of these layers $\Sigma d_{Si:Er} = d_{Si:Er}\cdot N$, which was varied from 0.1 to 1.8 μm. Dependence of the normalized PL intensity ($I_{si:Er}$) for Si/Si:Er/Si/Si:Er.../Si structures on the thickness of undoped Si space layers (d_{Si}) is shown in the inset in Fig. 3, where the point d_{Si}=0 corresponds to the $I_{Si:Er}$ value determined for a uniformly doped Si:Er layer (*sample #37*). The analysis of PL spectra in Fig. 3 shows that there is no evidence of formation in these structures of new Er centers that would be notable for their high optical efficiency and thus responsible for this strong effect. Like in the case of uniformly doped layers, the optically active Er complexes with oxygen, among which are the center *Er-O1* (transitions at 6507 cm^{-1} and 6472 cm^{-1}) and oxygen/defect related centers, dominate in these PL spectra. Moreover, the annealing behaviour of the luminescence spectra observed in multilayer Si/Si:Er/Si/Si:Er.../Si structures also has the same character as in the uniformly doped layers grown in similar conditions. It can also be shown that the co-doping of Si space layers with shallow impurities (Sb or P, the doping level $n \sim 10^{18}$ cm^{-3}) results in a decrease of the $I_{Si:Er}$ intensity in multilayer Si/Si:Er/Si/Si:Er.../Si struc-

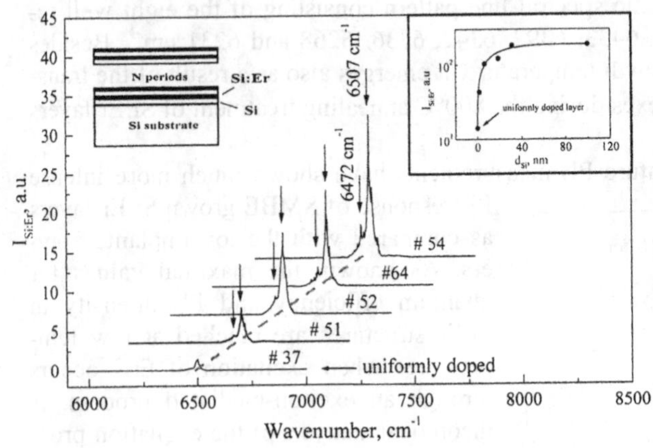

Fig. 3. PL spectra obtained at T= 4.2 K in uniformly and selectively doped Si:Er SMBE structures. Parameters of the structures are:
#37 - $d_{Si:Er}$=1.8μm; *#51* ($d_{Si:Er}$=2.3 nm, d_{Si}=1.7 nm, N= 400, d=1.6 μm); *#52* ($d_{Si:Er}$=2.3 nm, d_{Si}=6.5 nm, N= 196, d=1.72 μm); *#64* ($d_{Si:Er}$=4.5 nm, d_{Si}=19 nm, N= 94, d=2.2 μm); *#54*,($d_{Si:Er}$=6.2 nm, d_{Si}=31.3 nm, N= 44, d=1.65 μm); where $d_{Si:Er}$ and d_{Si} are the thicknesses of Er doped and undoped layers, N is the number of periods, d is the total thickness of the epylayers. Er concentration in Si:Er layers amounts to ~ $5\cdot10^{18}$ cm^{-3}. The PL intensity was normalized to the total thickness of Er doped layers ($\Sigma d_{Si:Er}=d_{Si:Er}\cdot N$). The right inset shows the dependence of the normalized PL intensity ($I_{Si:Er}$) for Si:Er structures on the thickness of undoped Si layers.

tures [11]. To complete our discussion of the PL properties of SMBE grown Si:Er structures let us consider the values for quantum efficiency determined in these structures. The measurements of the external quantum efficiency have been performed at the temperatures of 4.2 K and 77 K under excitation with an Ar^+-ion laser (λ = 514.5 nm, power 4 ÷ 150 mW), where the NOVA OPHIR instrument with the PD300-IR detector has been applied to measure a PL signal intensity at the wavelength of 1.54 μm (for more detail see [16]). The values of the external quantum efficiency obtained in the linear excitation regime (laser power ~ 4mW) for best samples amount to 0.4% and 0.04% at 4.2 K and 78 K, respectively. Taking into account the geometrical factors, the difference in the refractive indexes of Si and air, and neglecting the nonradiative losses (*i.e.*, assuming that every photon absorbed in the sample generates an e-h pair exciting one Er ion), it has been estimated that the internal quantum efficiency in our SMBE samples at 4.2 K exceeds 20%.

4. LED structures

In this chapter we will demonstrate the possibility of SMBE growth of LED structures operating at room temperature. The light-emitting diodes were grown on highly doped Si substrates with the resistivities of 0.008 - 0.01 Ωcm. These diodes have a structure of the n^+(Si substrate)/n(uniformly doped Si:Er layer)/p^+(Si layer) type, where the *p-n* junction is formed from the surface side. Generally, Er doped Si layers produced by the SMBE method have the *n*-type conductivity with the free carrier concentration at 300 K varying in the range of $5\cdot10^{15} \div 10^{17}$ cm^{-3} and being dependent on the growth temperature [17,18]. To extend the doping level of *n*-type active layers we used an additional sublimating sources cut from high doped Si:P or Si:Sb. In that way the free carrier concentration in Si:Er layers can be increased up to values of 10^{18} cm^{-3} and higher (see also

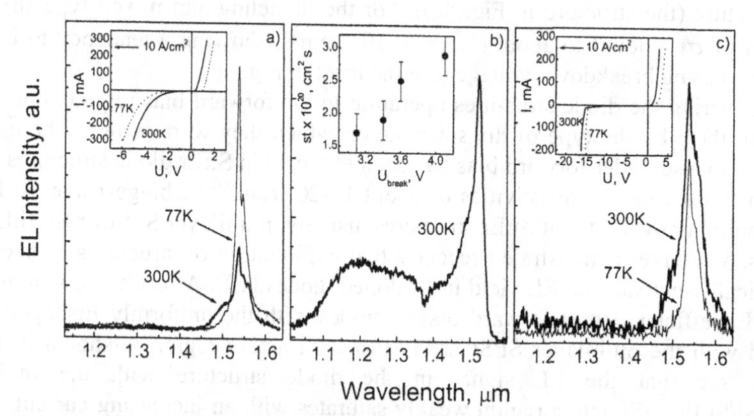

Fig. 4. EL spectra of Si:Er/Si light-emitting diodes operating in reverse bias regime at 77 and 300K. The figures correspond to diode structures with the following breakdown regimes: a) - tunneling, b) - mixed (tunneling/avalanche) and c) - avalanche breakdown. Insets show the current-voltage characteristics for the tunneling and avalanche diodes studied (the insets in Fig. a) and c), respectively) and dependence of the product $\sigma\tau$ on the breakdown voltage (U_{brek}) obtained for the series of LEDs at temperature of 300K (the inset in Fig. b)). The EL spectra presented were measured at the current densities of ~12 A/cm² (for the spectra in Fig. a) and b)), and ~8 A/cm² (for the spectrum in Fig. c)).

[17]), actually up to concentrations needed for realization of tunneling type diodes. Diode mesas with the diameters of 1.8 mm and 1.3 mm have been performed with the help of photolithography.

The EL spectra obtained for light-emitting diodes of different types - from tunneling to avalanche - are presented in Fig. 4. The parameters of the *n*-type active Si:Er layers for the tunneling and avalanche diodes, respectively, are: $n=(1-2)\cdot10^{18}$ cm^{-3}, $d = 1$ μm and $n=1\cdot10^{16}$ - $1\cdot10^{17}$ cm^{-3}, $d = 0.4$ μm. The current-voltage (I-V) plots measured at 77K and 300K clearly show the tunneling breakdown regime for sample *a* (the breakdown voltage U_{break} decreases with temperature) and avalanche regime for sample *b* (U_{break} increases with temperature). The middle curve *b* in Fig. 4 corresponds to the diode structure with a mixed (tunneling/avalanche) breakdown regime, the active Si:Er layer parameters for this structure are $n=(6-8)\cdot10^{17}$ cm^{-3}, $d = 0.5$ μm.

As one can see from Fig. 4, all of these diode structures in the breakdown regime demonstrate a strong EL response at room temperature. The EL signal at the 1.54 μm wavelength, undergoes only a weak (about 3-5 times) temperature quenching with a rising of temperature from 80 to 300 K and features a characteristic broadband spectrum with a maximum positioned at the wavelength of about 1.536 μm, closely reproducing the spectrum of the SiO_2-precipitate type Er centers described in our discussion above. Of interest are dependencies of the EL signal intensity on the reverse current density, obtained in diode structures with different types of a breakdown regime. We observed a practically linear dependence up to densities of about 12 A/cm^2 in the diode structures operating in the tunneling breakdown regime. At the same time, in the diode structure operating under avalanche breakdown, the EL signal became highly saturated at current densities as low as 2 A/cm^2 (see also [19]). This is reflected in the values of parameter $\sigma\tau$ ($\sigma\tau$ is the product of the excitation cross section of an Er^{3+} ion by its lifetime in the excited state) obtained from the saturation dependencies for EL diodes of different types. The highest value $3\cdot10^{-19}$ cm^{2}s for $\sigma\tau$ has been obtained for the avalanche type diode structure (the structure in Fig. 4c)). For the tunneling and mixed type structures the values of $\sigma\tau$ amount to about $(1.6 \div 3)\cdot10^{-20}$ cm^{2}s, showing a tendency to increase with an increasing breakdown voltage (see the inset in Fig. 4b).

Considering the diode structures operating in the forward bias regime, one should distinguish them by the type of the substrate on which they were grown. The most intense EL response under forward bias has been obtained in Si:Er diode structures grown on Si substrates with the resistivities of about 10-20 Ωcm. The biggest interest here is application of a novel type of Si:Er structures, namely, multilayer Si/Si:Er/Si/Si:Er.../Si structures. We have demonstrated recently that exploitation of structures of such kind can drastically increase the EL yield in Er doped diodes [19]. A result allowing to compare the EL efficiencies of forward biased diodes with the uniformly Er-doped active layer and with the multilayer Si/Si:Er/Si/Si:Er.../Si active region is shown in Fig. 5. One can see that the EL signal in the diode structure with the multilayer Si/Si:Er/Si/Si:Er.../Si active region weakly saturates with an increasing current density (except for the last few points on curve 1 related to overheating) and at high current densities it exceeds the signal obtained in a diode with a uniformly doped active layer by more than an order of magnitude. Similarly to the PL case, this result can be accounted for an increase in the exciton generation rate through the introduction of undoped Si space layers into the structure. So, both the PL and our latest EL studies

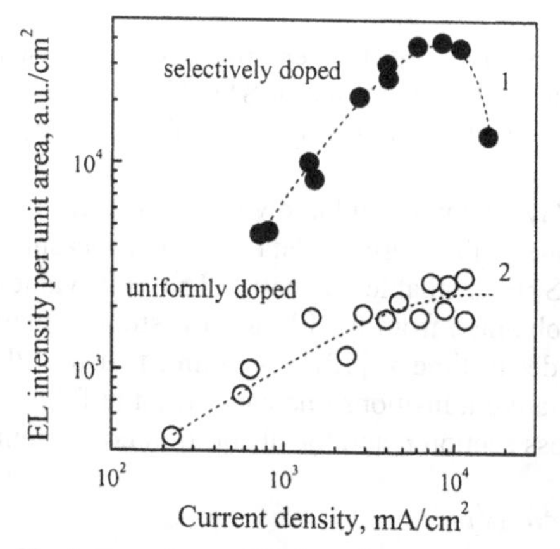

Fig. 5. Dependence of EL intensity on current density in uniformly and selectively doped LEDs at 80K in forward bias regime. Parameters of LEDs are: 1- Si:B (10 Ωcm)/ *Active region* - 52 periods of Si:Er ($d_{Si:Er}$=70Å) and the Si space layers (d_{Si}=120 Å); /n^+- Si:Sb (0.007 Ωcm), d=0.1 μm; 2- Si:B (10 Ωcm) / *Active region* - Si:Er, d=1.6 μm,/ n^+-Si:Sb (0.007 Ωcm), d=0.3 μm.

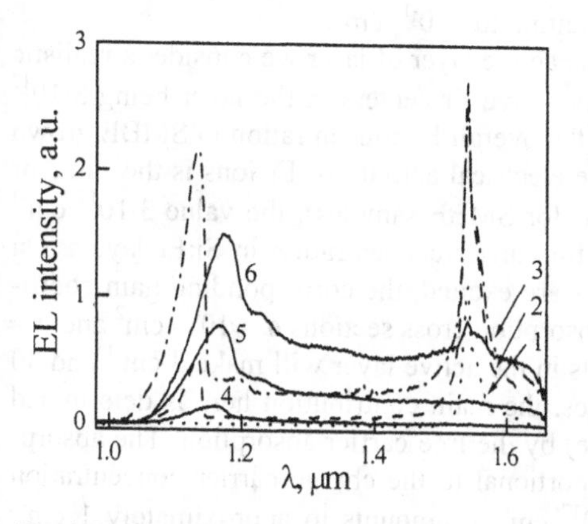

Fig. 6. EL spectra obtained in transistor-like *p-n-p* structure at 77 K (dashed curves) and 300 K (solid curves). These dependencies were measured at following current values: 1- 75 mA; 2- 175 mA; 3- 350 mA; 4- 100 mA; 5- 300 mA; 6- 700 mA.

revealed great advantages and prospects opening for these novel multilayer Si/Si:Er/Si/Si:Er…/Si structures, in particular, their application towards the development of forward biased LEDs. What concerns new approaches to realization of commercial LEDs operating at room temperature, *i.e.*, in the breakdown regime, we believe that the hopes here should be placed on development of a conceptually new design of a diode structure that will also involve multilayer Er doping as the basic principle. In particular, one can think of realization of cascade-like diode structures with multiple *p-n* junctions, in which the utilization of a specific Er doping profile will allow to considerably increase the excited volume of Er^{3+} ions and, therefore, the luminescent yield of the LEDs.

Let us finally present a new result obtained in a specific transistor-like structure where an Er-doped layer grown on the *p*-type substrate was additionally implanted with a high boron dose. Under positive bias applied to the upper implanted layer this structure emits simultaneously in two wavelength ranges: 1.13 ÷ 1.18 μm and 1.54 μm. We have observed a fairly intense EL signal both at 80 K and 300 K (see the EL spectra in Fig. 6). To understand this result we should allow the operating of two *p-n* junctions: in forward (the upper *p-n* junction) and reverse (the lower *p-n* junction) bias regimes. The reverse biased *p-n* junction determines the Er-related luminescence at 1.54 μm, whereas the other one, being forwardly biased, should be responsible for the band-edge luminescence observed in the wavelength range 1.13 ÷ 1.18 μm. Both EL signals are stable up to room temperature.

5. Si:Er laser feasibility

Realization of the laser would be a major event favouring further progress of Si-based optoelectronics. In this chapter we will provide the gain analysis for SMBE structures of interest to this paper and discuss the feasibility and possible ways of achieving stimulated emission on their basis.

To estimate the gain in the active Si:Er layers one should know the photon absorption cross section, σ, for Er^{3+} ions. Unfortunately, there are no data on direct measurements of this parameter in monocrystalline Si:Er available currently. The only values known for Er transitions in silicon are those obtained from the PL and EL studies; they are the line width $\hbar\nu$ and the luminescence decay time τ. [12]. Assuming that the PL decay time corresponds to the lifetime of radiative transitions one can write the following expressions for the phonon absorption cross section σ and the absorption coefficient α for Er transitions:

$$\sigma = \pi c^2/\epsilon\omega_0^2\nu\tau,$$

$$\alpha(\omega_0) = \sigma N,$$

here c is the light velocity, ω_0 is the photon cyclic frequency of a radiative transition, ϵ is the dielectric constant, N is the concentration of optically active Er centers.

By using the values $\tau = 10^{-3}$ s, $\epsilon_{Si} = 12.25$, $\hbar\nu = 0.125$ meV (which is the case for our structures) in the formula for the absorption cross section one can get $\sigma = 0.94\cdot10^{-17}$ cm^2. For the narrowest photoluminescence with $\hbar\nu \approx 10^{-2}$ meV [20] the σ value will be by about an order of magnitude higher amounting to $\sim 10^{-16}$ cm^2.

To estimate the gain-loss relation in the active layer of laser we consider a realistic structure with the concentration of optically active Er centers in the layer being $3\cdot10^{17}$ cm^{-3}. This corresponds to less than 10% of the overall Er concentration in SMBE grown Si:Er structures. In view of the fact that the electrical activity of Er ions is the same or less than their optical activity (in particular for SMBE samples), the value $3\cdot10^{17}$ cm^{-3} could also be regarded as the upper limit for carrier concentration in Si:Er layers. On condition that all optically active Er centers are excited, the corresponding gain coefficient is equal to $\alpha(\omega_0)$. Therefore, for the absorption cross sections $\sigma \approx 10^{-17}$ cm^2 and $\sigma \approx 10^{-16}$ cm^2 the values for the gain coefficients in the active layer will make 3 cm^{-1} and 30 cm^{-1}, respectively. As for the internal losses, the main contribution here is determined (particularly in the electrical excitation case) by the free-carrier absorption. The absorption coefficient for this mechanism is proportional to the charge carrier concentration and for the electron concentration of $3\cdot10^{17}$ cm^{-3} it amounts to approximately 1 cm^{-1} [21], which is considerably less than the estimated gain.

To realize the waveguiding effect in a laser active layer it is necessary to achieve a substantial difference in the refractive indexes (Δn) on the interfaces between the waveguiding and confinement layers. One possible candidate towards this end are structures on the basis of Ge_xSi_{1-x} and $Ge_xC_ySi_{1-x-y}$ compounds, the refraction index of which can be varied and is dependent on the Ge and C concentration.

The dependencies of the total TE mode gain in a waveguide structure on the thickness of the waveguiding layer, calculated for two values of Δn: $\Delta n = 0.1$, $\Delta n = 0.15$, are shown in Fig. 7. In the calculations, the waveguiding layer was considered as an active medium with the gain coefficients either 2 cm^{-1} (left axis) or 30 cm^{-1} (right axis), the losses in the Si confinement layers were assumed equal to 1 cm^{-1}. One can see that the total gain in the structure keeps increasing considerably within the thickness range of about 1 μm and reaches a constant value close to that of the gain in the active waveguiding layer. This tendency becomes more apparent at high values of Δn, which is related to a stronger localization of the light wave. The material commonly considered for realization of a waveguide structure is Ge_xSi_{1-x}. To obtain the refractive index difference Δn = 0.1 in Si/ Ge_xSi_{1-x} /Si structure the Ge content x should amount to the value of 0.143. However, it does not seem realistic to produce good quality waveguides in this way, because at a thickness of about 1 μm, which would be optimal for the waveguiding layer (see Fig. 7), we will have to deal with the relaxed GeSi layer. Ge_xSi_{1-x}/Si are strained heterostructures, which means that the thickness and composition of epitaxially grown layers can not be varied in a wide range. More promising in this case will be the structures on the basis of $Ge_xC_ySi_{1-x-y}$ alloy which is at x=8.3y lattice matched to Si. To provide the refractive index difference Δn = 0.1 on Si/$Ge_xC_ySi_{1-x-y}$ heterointerface, the contents x and y should amount to values of about 0.17 and 0.02, respectively. Moreover, this is a type I heterojunction that will facilitate superinjection of the minority carriers in active region thus favouring achievement of the population inversion. We have realized earlier SMBE grown Si/Si_xGe_{1-x}:Er/Si structures that demonstrate intense PL response at T = 4.2 K with the dominant lines relating to the center *Er-O1* [22].

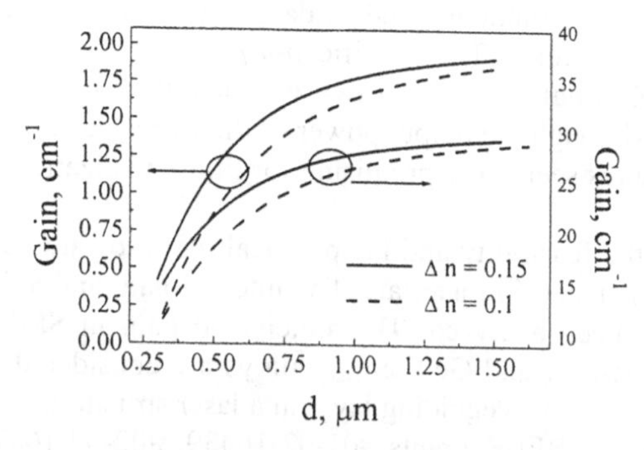

Fig. 7. The fundamental TE mode gain vs thickness of the active waveguide layer, simulated for Δn=0.1 and Δn=0.15. We have used the following values of parameters: the gain coefficients in the active wavequide layer are 2 cm^{-1} (left axis) and 30 cm^{-1} (right axis), the losses in Si confinement layers are 1 cm^{-1}.

4. Summary

In this contribution we have demonstrated the variety of light emitting Si:Er-based structures - from uniformly and selectively doped to LEDs and waveguide structures - successfully realized with the original SMBE method. We have described their PL and EL features and analyzed the origin of optically active Er centers, among which the most interesting is the center *Er-1* observed for the first time in SMBE grown structures.

The low temperature PL analysis performed within the model of the exciton-mediated excitation mechanism for Er ions shows that there are the optimal doping conditions to achieve maximal PL yield from Si:Er structures. Most promising are selectively doped Si/Si:Er/Si/Si:Er.../Si multilayer structures. The effect of a strong (by more than an order of magnitude) enhancement of PL and EL efficiency obtained in this novel type of Er doped structures has been demonstrated.

We have realized LEDs operating both in the forward and reverse bias regimes, where the latter are represented by series of tunneling and avalanche type diode structures effectively emitting at room temperature. The specific *p-n-p* type "two color" structure effectively emitting at 300 K simultaneously in two wavelength ranges 1.17 μm and 1.54 μm has been described. Although the output power of the LEDs developed to date is not high enough, these structures seem very promising for future realization of commercial devices.

Finally, we have discussed here the feasibility and the practical ways towards development of a laser on the basis of Er doped Si materials. The internal gain and a design of the waveguide structure have been analyzed. The anticipated gain in SMBE grown structures is $3 \div 30$ cm^{-1} and Ge_xSi_{1-x} and $Ge_xC_ySi_{1-x-y}$ alloys are considered as possible material for realization of the active waveguiding layer in a laser structure.

This work was supported in part by RFBR grants #01-02-16439, #02-02-16374 and #02-02-16773, INTAS grant #99-01872, RFBR-ŐAD grant #01-02-02000 BNTS and Dutch-Russian project NWO 047.009.013.

References:

[1] A. Polman, *J. Appl. Phys.*, ***82*** *(1997) 1.*
[2] S. Coffa, G. Franzo, and F. Priolo, *MRS Bulletin,* ***23*** *(1998) 25.*
[3] W. Jantsch, S. Lanzerstorfer, L. Palmetshofer et al. *J. of Lumin., 80 (1999) 9.*
[4] J. Stimmer, A. Reittinger, E. Neufeld et al. *Thin Solid Films,* ***294*** *(1996) 220.*
[5] R. Serna, Jung H. Shin, M. Lohmeier et al. *J. Appl. Phys.* ***79*** *(1996) 2658.*
[6] W.-X. Ni, K.B. Joelsson, C.-X. Du et al. *Appl. Phys. Lett.,* ***70*** *(1997) 3383.*
[7] V.P. Kuznetsov and R.A. Rubtsova *Semiconductors 34 (2000) 502.*
[8] A.Yu. Andreev, B. A. Andreev et al. *Semiconductors 33 (1999) 131*
[9] M.V. Stepikhova, B.A. Andreev et al. *Thin Solid Films,* ***381*** *(2001) 164.*
[10] H. Przybylinska, W. Jantsch et al. *Phys. Rev. B,* ***54*** *(1996) 2532.*
[11] M. Stepikhova, B. Andreev et al. *Solid State Phenomena,* ***82-84*** *(2002) 629.*
[12] B.A. Andreev, Z.F. Krasil'nik et al. *Physics of the Solid State,* ***43*** *(2001) 1012.*
[13] F. Priolo, D. Franzo et al. *Phys. Rev. B,* ***57*** *(1998) 4443*
[14] O.B.Gusev, M.S.Bresler et al. *Phys. Rev. B,* ***64*** *(2001) 075302.*
[15] M. Stepikhova, B. Andreev, et al, *Mat. Sci. & Engin. B, 81 (2001) 67.*
[16] B.A. Andreev, W. Jantsch et al. *26th Int. Conf. on Phys. Semicond. Edinburgh, Scotland, UK, 29 July - 2August 2002, Book of abstracts, Part III, p. 125, in print.*
[17] E.N. Morozova, et al. *Bulletin of RAS. Ser. Phys.,****67*** (2003) *in print (in Russian).*
[18] V.B. Shmagin et al. *Physica B: Physics of Condensed Matter,* ***308-310*** *(2001) 361.*
[19] B. Andreev, V. Chalkov, O. Gusev et al. *Nanotechnology,* ***13*** *(2002) 97.*
[20] N.Q. Vinh et al. Physica B: Physics of Condensed Matter, ***308-310*** *(2001) 340.*
[21] A. Dargys, J. Kundrotas, *Handbook on physical properties of Ge, Si, GaAs and InP*, Vilnius, Science and Encyclopedia Publishers, 1994.
[22] V. G. Shengurov, S. P. Svetlov et al. *Semiconductors 36(2002) 625.*

UV-BLUE LASERS BASED ON InGaN/GaN/Al_2O_3 AND ON InGaN/GaN/Si HETEROSTRUCTURES

G.P. YABLONSKII (a) and M. HEUKEN (b)
(a) Stepanov Institute of Physics of National Academy of Sciences of Belarus F. Skaryna Ave, 68, 220072 Minsk, Belarus, yablon@dragon.bas-net.by
(b) AIXTRON AG Kackertstr. 15-17, D-52072 Aachen, Germany M.Heuken@aixtron.com

1. Introduction

InAlGaN compounds are very promising for light emitting devices such as light emitting diodes (LED) and laser diodes (LD) due to their wide interval of the band gap energy (ultraviolet - orange spectral region), chemical, thermal and mechanical stability [1 – 4]. The GaN-based material system has gained much attention also for electronic applications due to its outstanding inherent electrical properties such as high breakdown voltages, high peak electron velocities and high sheet electron concentration especially in two-dimensional electron gas structures. A current focus of the device research are high electron mobility transistor (HEMT) devices [5] and ultraviolet detectors [6] grown on sapphire and silicon carbide. The AlInGaN based LEDs have an output power of about 14 mW with power density of about 300 kW/cm^2 at low operating voltage and current of 3 V and 20 mA [1 – 3]. AlGaInP LEDs can operate in the red region of the visible spectrum up to 620 nm with luminous intensity up to 40 lumens [1 – 3]. The GaN based LEDs grown on sapphire substrates become as bright and efficient as incandescent bulbs over the ultraviolet and visible wavelengths. If the LED efficiency would be achieved up to 50% and more, for example a 200 lm/W, white light sources two times more efficient than fluorescent lamps and ten times more efficient than incandescent lamps can be produced. The world energy saving would be more than 1,000 TWh/year. An important place among these devices belongs to lasers. The violet lasers have lifetime up to 15000 hours with power of 30 mW [4].

However, especially for automotive and other applications in high-volume low-cost market segments, the choice of substrate is still an open question. Sapphire suffers from a low thermal conductivity worsening the high-power operation of lasers, light emitting diodes and transistor devices. Silicon carbide is very expensive. Silicon is a promising alternative substrate for GaN growth because of its low cost, excellent quality, large-area availability and the possibility to integrate GaN based light emitting devices and high power electronics with Si based photodetectors and logical circuits. However, the main challenge connected with the use of silicon is the high mismatch in lattice and in thermal expansion coefficients.. Whereas the growth of GaN on sapphire requires one nucleation layer to form the crossover from sapphire to GaN, the growth of high-quality crack-free GaN on silicon requires a sophisticated growth procedure employing a combination of high and low tempera-

L. Pavesi et al. (eds.), Towards the First Silicon Laser, 455–464.

and low temperature GaN and AlGaN layers and more than one nucleation /recrystallisation step [7]. First InGaN-based blue and green LEDs on Si are developed and created now [8].

But the light emitting devices fabricated on Si show worse characteristics compared to devices grown on Al_2O_3 and SiC. The nature and reasons of the defects and electrical properties, the recombination mechanism, influence of laser and thermal effects are not completely known up to now. A comprehensive investigation of optical and electrical properties of GaN based heterostructures grown on Si is an interrelation with growth and processing technologies can lead to better understanding of physical reasons of the existing problems, to establish any feedback with technology, to optimise it and to create the UV-visible light emitting diodes and lasers with characteristics comparable to that for structures grown on Al_2O_3 and SiC.

If high brightness LEDs and LDs will be developed and fabricated out of GaN based heterostructures grown on Si substrates, it would be possible also to grow them on Si-based nanostructures appropriated for gain in the infrared region. The high intensity light of the GaN based LEDs, LDs and the vertical cavity surface emitting lasers, which where recently demonstrated [9], can be used for optical pumping of the Si-based lasers.

During the last four years, we were engaged in investigation of growth and optical characterisation of different types of the GaN epitaxial layers, InGaN/GaN single and multiple quantum wells (SQW, MQW) and electroluminescence test (ELT) heterostructures grown on sapphire substrates for a wide interval of wavelengths from 390 nm to 530 nm. The most attention was paid to optimise growth conditions, to create optically pumped lasers from the UV up to blue spectral region [10–12]. The lasing at $\lambda = 450$ - 470 nm was achieved for the first time [13]. In the course of the last year, new GaN layers and InGaN/GaN MQWs were grown on Si substrates. Laser action in the UV [14] and blue spectral regions were achieved in such heterostructures for the first time [this paper].

In this paper, the authors present the recent results on behalf of their colleagues from AIXTRON AG, Aachen, Germany (A. Alam, M. Luenenbuerger, H. Protzmann, B. Schineller) where all samples were grown, from the Institute fuer Theoretishe Elektrotechnik RWTH Aachen, Germany (Y. Dikme, H. Kalish, A. Szymakowski) where low temperature photoluminescence (PL) measurements were done and from the Institute of Physics of the National Academy of Sciences of Belarus, Minsk (A.L. Gurskii, E.V. Lutsenko, V.N. Pavlovskii, V.Z. Zubialevich) where laser and optical properties of all heterostructures under high excitation intensities were carried out.

2. Experimental

All structures were grown in AIXTRON MOVPE reactors on 2-inch (0001)-oriented Al_2O_3 and on (111)-oriented n-type Si substrates at low pressures (200 mbar or 50 mbar). Trimethylgallium (TMGa), Trimethylaluminum (TMAl), Trimethylindium (TMIn), ammonia (NH_3) and silane (SiH_4) were used as precursors. The thickness of the GaN epitaxial layers GaN/Al_2O_3 and GaN/Si was around 1 μm. The thickness of the InGaN active layer was 3 nm, the thickness of the barriers was 4 nm and the thickness of the upper cap layer was 10 - 50 nm for the MQWs grown on sapphire (MQW/Al_2O_3). The design of the ELT heterostructures grown on sapphire substrates (ELT/Al_2O_3) was the following: GaN:Mg(250 nm) / 5×(InGaN/GaN) / GaN:Si(1 μm) / GaN(1 μm) /nucleation layer /Al_2O_3.

The design of the ELT heterostructures grown on Si substrates (ELT/Si) was the following: GaN:Mg (200 nm) / AlGaN:Mg(10 nm) / 3×(InGaN/GaN:Si) / GaN:Si(600 nm) /

/ (strain reducing layer stack) / Si. The MQW heterostructures grown on Si substrate for optical excitation (MQW/Si) differed from the last one by absence of the electron blocking AlGaN:Mg layer and by a small thickness (50 nm) of the upper GaN layer. The structures were free of cracks. Photoluminescence (PL) and lasing were excited by the radiation of a N_2 laser ($h\nu = 3.68$ eV, $I_{exc} = 10^2 - 10^6$ W/cm^2, f = 1000 Hz, $\tau_p = 8$ ns), a HeCd laser ($h\nu = 3.81$ eV) and by radiation of a dye laser with tuning frequency for direct excitation of the quantum wells. The quantum energy of the dye laser was lower than the band gap value of GaN. The EL spectra were excited by voltage imposed to the ELT samples regime by stripe contacts. Reactive ion beam etching (RIBE, Ar+O_2) was used for the formation of the contacts on the n-type region of the device and for removing of a part of the cap layer.

Figure 1a shows design of the GaN/Al_2O_3, GaN/Si, InGaN MQW/Al_2O_3 and InGaN MQW/Si heterostructures. An example of the InGaN ELT/Al_2O_3 heterostructure laser and geometry of excitation by the nitrogen and dye laser radiation are given in figure 1b.

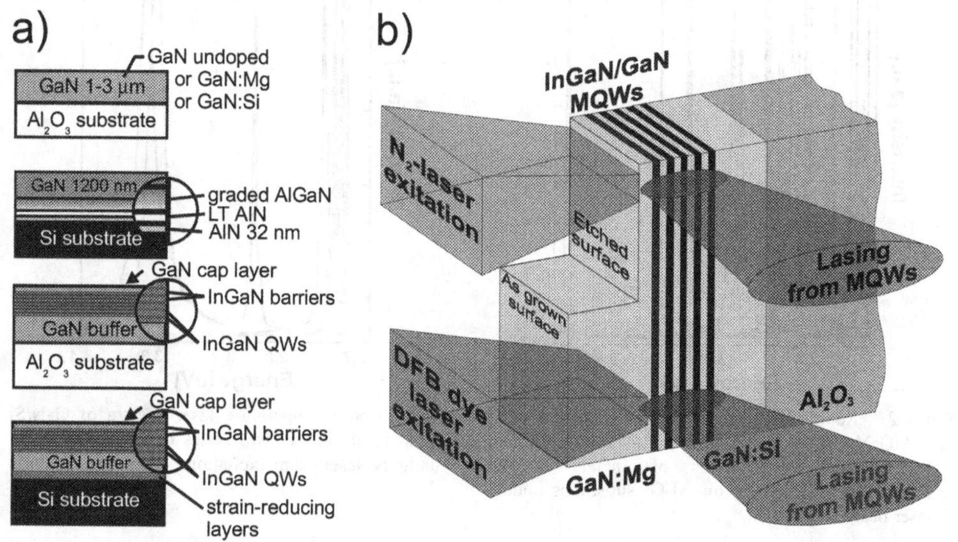

Figure 1. a) Layer design of GaN epitaxial layers and InGaN/GaN multiple quantum wells grown on Al_2O_3 and on Si substrates. b) Scheme of InGaN ELT/Al_2O_3 heterostructure laser and geometry of excitation by nitrogen and dye laser radiation.

3. Results and Discussion

3.1. OPTICALLY PUMPED LASERS BASED ON InGaN/GaN HETEROSTRUCTURES GROWN ON SAPPHIRE SUBSTRATES

Figures 2, 3 show the laser spectra of all GaN based heterostructures at room temperature excited by N_2 laser radiation over the spectral interval from the near ultraviolet up to the blue region. The lasers based on InGaN/GaN SQW and GaN/AlGaN single heterostructures covering the spectral region from $\lambda = 370$ nm up to $\lambda = 405$ nm. The best laser parameters were reached for the InGaN/GaN MQW/Al_2O_3 "violet" ($\lambda = 400 - 450$ nm) lasers. The laser action was achieved up to very high temperature $T_{max} = 580$ K. The minimal laser threshold at room temperature was $I_{thr} = 35$ kW/cm^2, the full width at half maximum (FWHM) of the laser line near the laser threshold was 0.04 nm, the pulse energy was E = 630 nJ, the pulsed power was P = 80 W and the effective temperature was T_0 = 164 K at T = 300 – 500 K. The laser parameters of the "blue" lasers were the following: λ = 450 – 470 nm, FWHM = 0.05 nm, T_{max} = 460 K, I_{thr} = 50 kW/cm^2, E = 300 nJ, P = 40 W. On the base of measurements of the laser, stimulated emission and PL spectra as functions of temperature and excitation intensity, it was concluded that at high excitation intensity (I_{exc} > 1000 kW/cm^2) a considerable thermal overheating ΔT ≈100 K of the active region takes place. It was shown that it is exclusively for the cost of the inherent InGaN/GaN laser radiation which power density on the laser mirrors was evaluated to be I > 5 mW/cm^2.

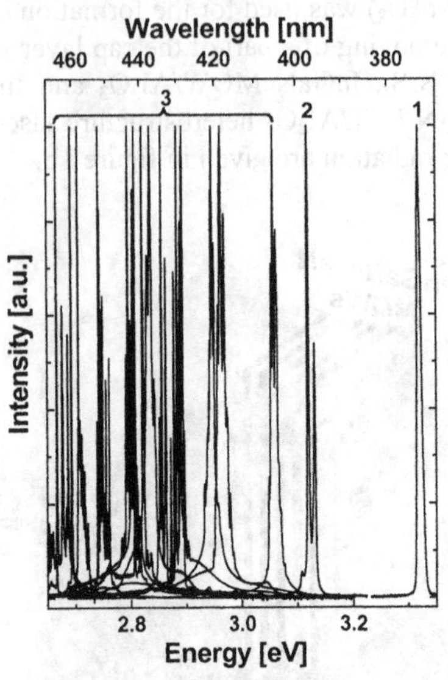

Figure 2. Room temperature laser spectra of GaN/AlGaN single heterostructure (1), InGaN/GaN SQW (2) and two series of InGaN MQW heterostructures (3) grown on Al_2O_3 substrates under N_2 laser beam excitation.

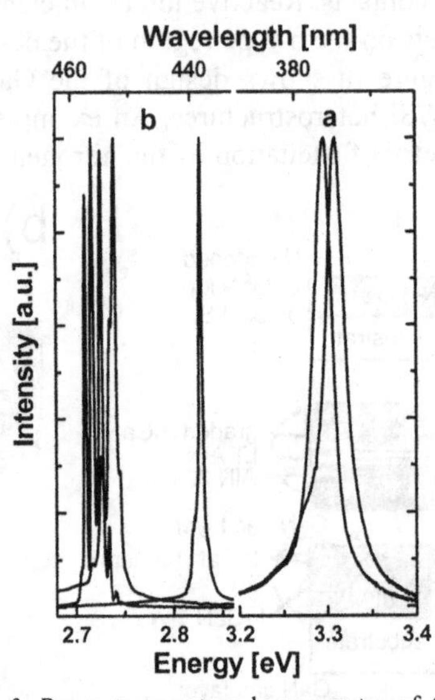

Figure 3. Room temperature laser spectra of GaN/Si epitaxial layer (a) and two series of InGaN MQW/Si (b) under N_2 laser beam excitation.

The far field patterns of the laser emission with the active layer stack d > 50 nm as a rule consist of two symmetrical spots which wavelength does not depend on the registration angle. Calculations of the electro-magnetic field distribution inside and outside the cavity showed that such lasers operate in the high order transverse mode regime. The laser wavelength depends on the registration angle for lasers with sufficiently thin stack of the active layers (d < 50 nm, figure 4) evidencing on the leaky mode operation.
The laser threshold of the MQW/Al_2O_3 depends very much on the operating wavelength (figure 5) which in turn depends on the In concentration. An increase of the In mole fraction in InGaN active layers results in a composition inhomogeneity (In rich clusters, quantum dots and discs) and apparently in increasing nonradiative defect concentrations which leads to the intensity decrease and to a broadening of the emission spectra and, thus, to an increase of the laser threshold. A considerable decrease of the In in the active layer promotes the InGaN band gap rise diminishing the band offset.

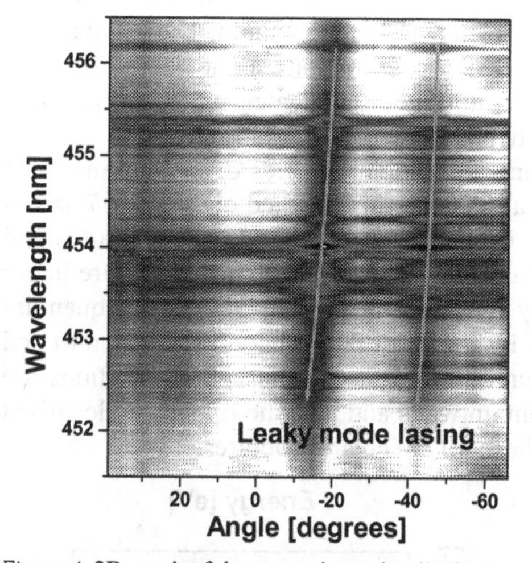

Figure 4. 2D graph of the spectral-angular distribution of the InGaN MQW/Al_2O_3 laser emission under excitation of the N_2 laser beam at room temperature.

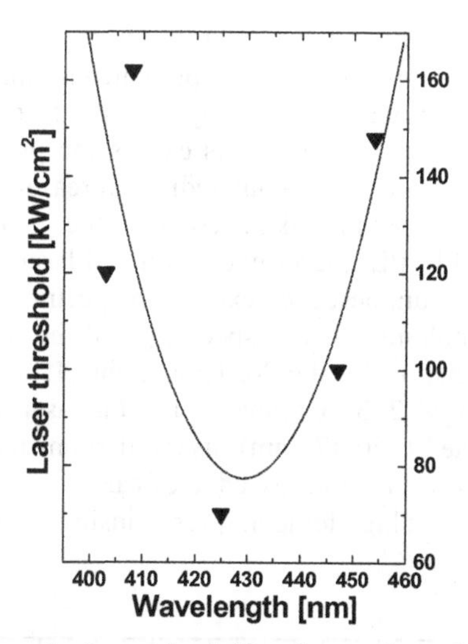

Figure 5. Laser threshold of a series of the InGaN/GaN MQW lasers as a function of operating wavelength at room temperature.

One of the most discussed problems for the GaN based lasers is the mechanism of optical gain: do localised or delocalised states play a dominant role in the mechanism?. In order to clarify a possible difference between the laser mechanisms in the short and long wavelength lasers, their laser thresholds were measured as a function of temperature. Figure 6 (curves 1, 2) presents the temperature resolved laser thresholds of both 'violet" and "blue" InGaN/GaN MQW/Al_2O_3 lasers. The temperature dependence of the laser thresholds for T = 80 - 450 K for all MQW lasers revealed two slopes with an inflection point nearby T = 200 - 250 K. The "violet" and "blue" lasers showed the characteristic temperatures T_0 = 180 K and T_0 = 530 K for the low temperature range and T_0 = 100 K and T_0 = 160 K at high temperatures correspondingly.

The alteration of the characteristic temperature values may be attributed to a change in the gain mechanism. It can be admitted that in the low temperature interval of 80 – 220 K the most important contribution to the gain mechanism is given by the localised states created by the In rich clusters (quantum dots). The characteristic temperature of the "blue" lasers T_0 = 530 K is much higher than its maximal value for the one-dimensional confinement T_0 = 285 K [15]. An evidence of the In rich clusters in the active region of the MQWs is an appearance of a red PL band with maximum near 635 nm which is observable at T = 78 K under CW excitation by the HeCd laser radiation.

It may be assumed for our heterostructures that the localised states make the most important contribution to the gain mechanism in the "blue" lasers at low temperatures and the delocalised states (recombination in the electron-hole plasma) are responsible for the gain mechanism in the «violet» lasers at high temperatures.

In order to understand the most important reasons of the laser threshold increase with temperature, a comparison of photoluminescence characteristics with threshold was made (figure 6, curve 3). These characteristics are taken into account by the equation for the laser threshold [16]:

$$I_{thr} \sim \frac{(h\nu)^2 \times \Delta h\nu}{\eta_{sp}}, \tag{1}$$

where hν - photon energy for the spectrum maximum, Δhν is the FWHM of the PL spectrum and η_{sp} is the spontaneous emission efficiency at laser threshold. The spontaneous recombination efficiency gives the most significant contribution in the laser threshold dependence despite this expression provides not quite good agreement with experimental data evidencing about additional reasons of the dependence.

In order to make account in the gain mechanism problem, a series of measurements of the PL, PLE, laser excitation and laser spectra at RT were also carried out. Figure 7 shows photoluminescence excitation spectra (1), laser excitation spectra (2), laser spectra (3) and photoluminescence spectra (4) of the InGaN/GaN MQW/Al_2O_3 at room temperature under excitation by the Xe-lamp light (1, 4) and by dye laser beam with the tuning quantum energy (2, 3). Comparison of the laser, PL and PLE spectra for both "violet" (430 nm) and "blue" (450-470 nm) lasers at room temperature showed that the laser line positions are near to the mobility edge of carriers in the quantum wells and thus the role of the localised states at high temperatures is insignificant for these MQW heterostructures.

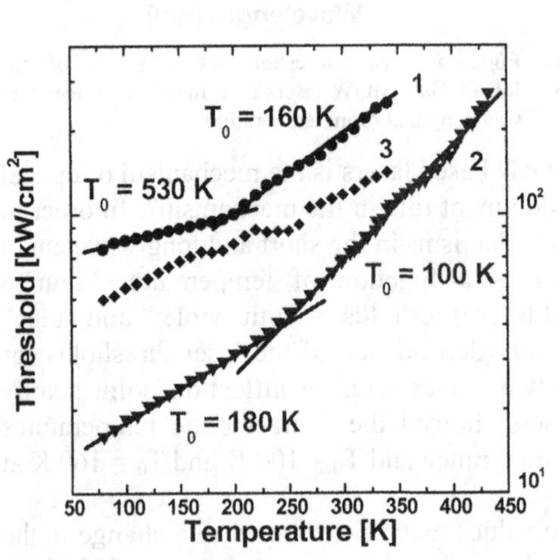

Figure 6. Laser threshold of the "blue" (curve 1, λ_{las} = 465 nm at RT) and "violet" (curve 2, λ_{las} = 430 nm at RT) InGaN MQW/Al_2O_3 lasers as a function of temperature. Curve 3 – simulation of the threshold temperature dependence of the "blue" laser by formula (1).

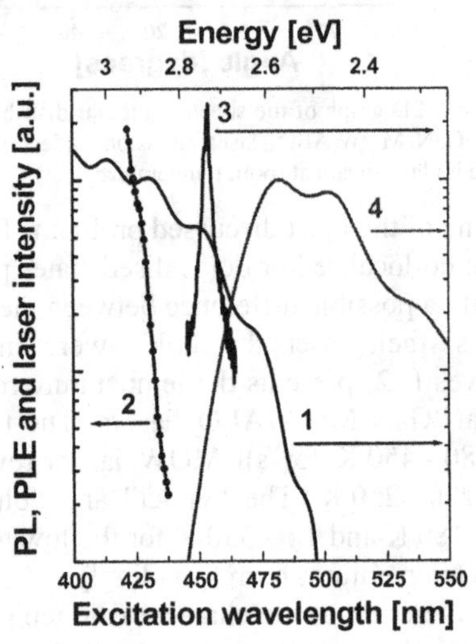

Figure 7. Part of photoluminescence excitation spectra (1), laser excitation spectra (2), laser spectra (3) and photoluminescence spectra (4) of the InGaN/GaN MQW/Al_2O_3 near the mobility edge of carriers in the QWs at room temperature under excitation of the Xe-lamp light (1, 4) and dye laser beam (2, 3).

3.2. OPTICALLY PUMPED LASERS BASED ON SILICON SUBSTRATES

Figure 8 represents low temperature PL and reflection spectra of a 1200 nm thick GaN sample grown on Si with a 32 nm AlN nucleation layer, low temperature AlN interlayer, and AlGaN buffer layer. Excitonic emission dominates in the 18 K PL spectrum, where relatively weak bands with maxima around 3.0 eV and 2.25 eV are also visible. For comparison, PL and reflection spectra of an undoped GaN grown on Al_2O_3 are also shown (inset, dashed curves). From the comparison of exciton line positions, marked as A, B, C, I_2 lines, with results of [17], it has been concluded that the value of tensile strain in the GaN/Si sample is about 7 kBar, while GaN/Al_2O_3 sample is under compressive strain of about 10.5 kBar. In figure 9, low temperature PL and reflection spectra of as grown and annealed 750 nm thick GaN/Si sample are shown. The excitonic PL spectrum of the GaN/Al_2O_3 is also given for comparison. As it is seen from figure 9, the position of the RS of GaN/Si before and after the annealing procedure did not change. These data permit us to conclude that the annealing step does not lead to the reduction of strain. It means that the layers remain non-relaxed after annealing.

In annealed samples, a strong PL band with a maximum at 3.364 eV becomes dominant in the low temperature PL spectra. This band is accompanied by relatively weak LO replica. The latter are superimposed with the well known 3.2 eV donor-acceptor pair band caused by residual impurities or native defects. The value of the electron-phonon coupling parameter S estimated from the spectrum for the recombination channel forming the 3.364 eV band is not greater than 0.17 - 0.2, which is an evidence of recombination via shallow states.

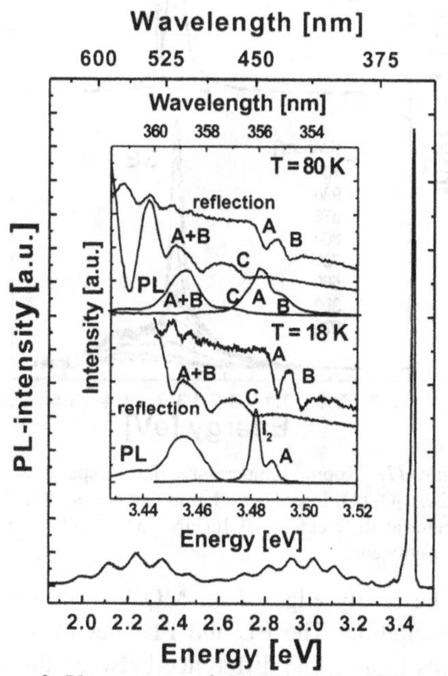

Figure 8. PL spectrum of GaN/Si at T = 18 K under CW HeCd laser beam excitation (λ = 325 nm). Inset: PL and reflection spectra of GaN/Si (solid curves) and GaN/Al_2O_3 (dashed curves) at T = 18 K and 80 K.

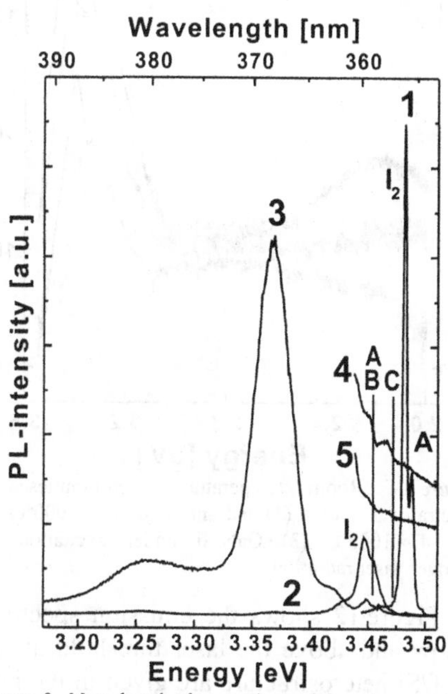

Figure 9. Near-band-edge region of low temperature (T = 18 K) luminescence (1 - 3) and reflection (4, 5) spectra of GaN/Al_2O_3 (1), GaN/Si as grown (2, 4) and annealed at T = 900°C (3, 5) under excitation by HeCd laser beam.

The thermal annealing procedure leads to an increase of near-band-edge PL efficiency of the investigated GaN/Si samples by more than an order of magnitude (about 20 - 25 times). The corresponding room temperature PL spectra are given in figure 10. Thus, the annealing promotes a significant decrease of the non-radiative centre concentration together with the increase of the number of shallow states.

The use of a low temperature AlN interlayer together with an AlGaN buffer layer reducing the strain in an epilayer, as well as the thermal annealing step increasing PL efficiency and decreasing the non-radiative recombination, allowed to achieve stimulated emission and laser action in GaN/Si epilayers. In figure 11, laser (curve 2) and stimulated emission spectra of a GaN/Si epilayer at room temperature are given. For comparison, a laser spectrum of GaN/Al_2O_3 is also given (curve 1). As it is seen from figure 11, the laser line at line at $\lambda = 377$ nm from GaN/Si is shifted to the low energy side comparing to the laser emission position in GaN/Al_2O_3.The laser threshold was 700 kW/cm^2. To our knowledge, this is the first observation of laser action in GaN/Si epitaxial layers. Previously, only lasing in GaN pyramids grown on (111) Si by selective lateral overgrowth was achieved [18].

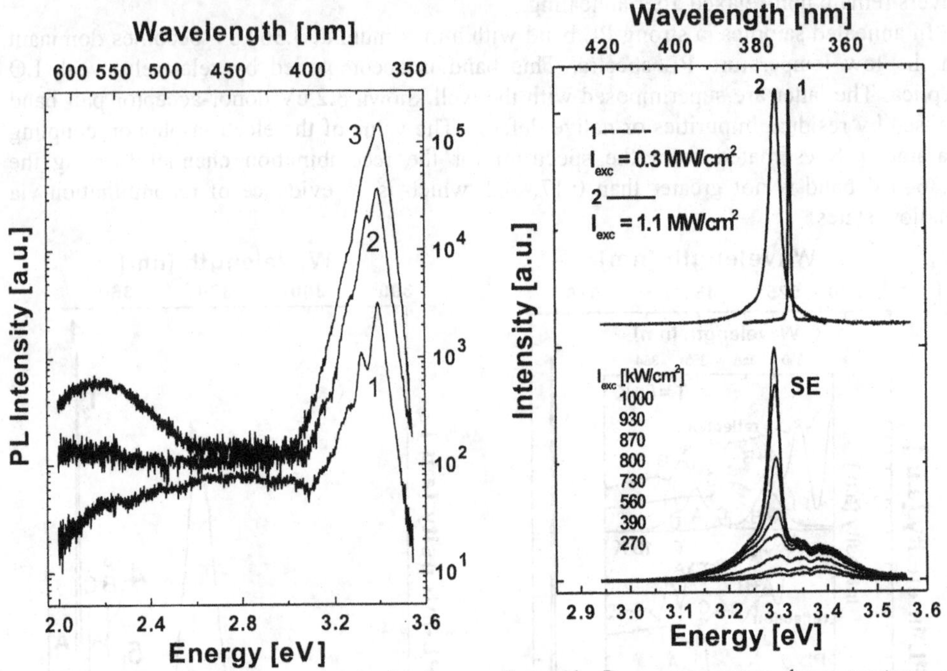

Figure 10. Room temperature photoluminescence spectra of as grown (1) and annealed at T = 900°C (2) and T = 1000°C (3) GaN/Si under excitation by nitrogen laser radiation.

Figure 11. Room temperature laser spectrum of GaN/Al_2O_3 (1), laser (2) and stimulated emission (SE) spectra at different I_{exc} of InGaN/GaN MQW/Si from the cavity end.

Figure 12 shows the emission spectra from the cavity edge of an MQW/Si structure below and above the laser threshold at optical excitation. The EL and PL spectra of the ELT/Si heterostructure are given in the inset to this figure. The difference between the EL and PL spectra position is due to higher concentration of the nonequilibrium carriers at optical excitation. The RT laser threshold was varied from 270 kW/cm^2 to 330 kW/cm^2 depending on the cavity length and quality. Laser action was achieved up to T = 420 K (figure 13) with the characteristic temperature of $T_0 = 85$ K. The energy per pulse was 60 – 80 nJ and the power was 8 - 10 W. This is the first observation of lasing in the InGaN/GaN MQWs grown on Si substrates. It is also an evidence of sufficiently good quality of the GaN templates which also showed optically pumped lasing at room temperature. These results allow hoping that fabrication of the laser diodes based on GaN quantum well heterostructures grown on Si substrates is possible in the near future.

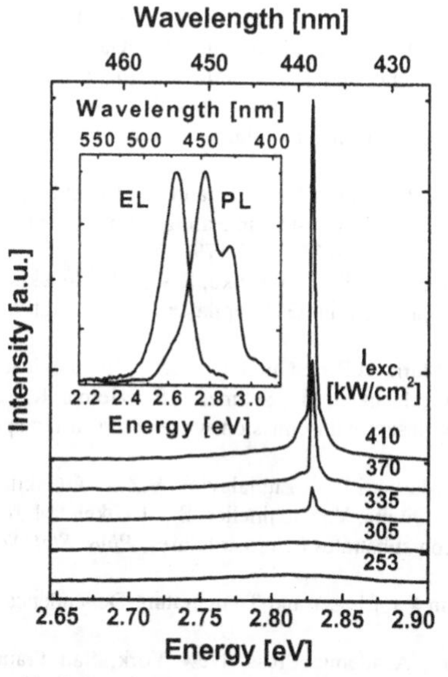

Figure 12. Room temperature emission spectra from the cavity end of the InGaN/GaN MQW/Si under N_2 laser beam excitation. Inset: normalised EL spectra at J = 100 A/cm^2 and PL spectra at I_{exc} = 300 kW/cm^2 of the InGaN/GaN ELT/Si heterostructure.

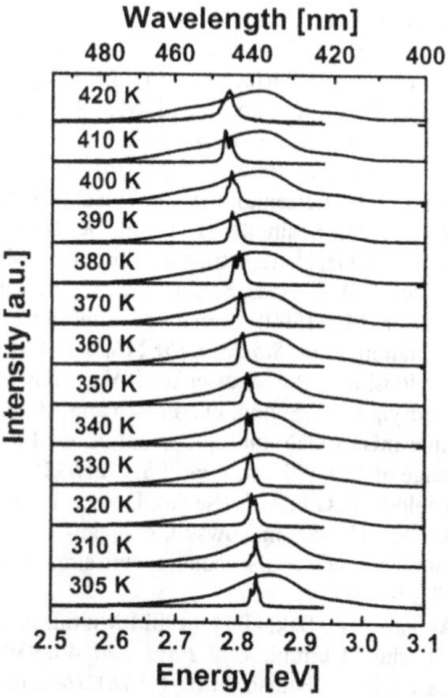

Figure 13. Photoluminescence (broad bands) and laser (narrow bands) spectra of InGaN/GaN MQW/Si at different temperatures from T = 305 K to T = 420 K under N_2 laser beam excitation.

The authors would like to acknowledge E.V. Lutsenko, A.L. Gurskii, V.N. Pavlovskii, V.Z. Zubialevich for optical measurements and discussion, H. Kalish, A. Szymakowski for low temperature PL measurements, B. Schineller, O. Schoen, H. Protzmann, M. Luenenbuerger, A. Alam and Y. Dikme for the heterostructure growth, and to acknowledge partial financial support from International Scientific and Technical Centre.

4. References

1. Muthu, S., Schuurmans, F. J., P., Pashley, M.D. (2002) Red, Green, and Blue LEDs for White Light Illumination, *IEEE J. Selected Topics in QE* **8**, 333-338.
2. Koike, M., Shibata, N., Kato, H., Takahashi Y. (2002) Development of High Efficiency GaN-Based Multiquantum-Well Light-Emitting Diodes and Their Applications, *IEEE J. Selected Topics in QE* **8**, 271-277.
3. Mukai, T. (2002) Recent Progress in Group-III Nitride Light-Emitting Diodes, *IEEE J. Selected Topics in QE* **8**, 264-270.
4. Nagahama, S., Yanamoto, T., Sano, M., Mukai, T. (2002) Characteristics of Laser Diodes Composed of GaN-Based Semiconductor, *Phys. Stat. Sol.(a)* **190**, 235–246.
5. Keller, S., Vetury, R., Parish, G., DenBaars, S.P., and Mishra, U.K. (2001) Effect of growth termination conditions on the performance of AlGaN/GaN high electron mobility transistors, *Appl. Phys. Lett.* **78**, 388-390.
6. Monroy, E., Calle, F., Pau, J.L., Munoz, E., Omnes, F., Beaumont, B., Gibart, P. (2001) AlGaN-based UV photodetectors, *J. Crystal Growth* **230**, 537-543.
7. Dadgar, A., Bläsing, J., Diez, A., Alam, A., Heuken, M., Krost, A. (2000) Metalorganic Chemical Vapour Phase Epitaxy of Crack-Free GaN on Si (111) Exceeding 1 μm in Thickness, *Jpn. J. Appl. Phys.* **39**, L1183-L1185.
8. Egawa, T., Moku, T., Ishikawa, H., Ohtsuka, K. and Jimbo, T. (2002) Improved Characteristics of Blue and Green InGaN-Based Light-Emitting Diodes on Si Grown by Metalorganic Chemical Vapor Deposition, *Jpn. J. Appl. Phys.* **41** L663–L664.

9. Krestnikov, I.L., Lundin, W.V., Sakharov, A.V., Semenov, V.A., Usikov, A.S., Tsasul'nikov, A.F., Alferov, Zh.I., Ledentsov, N.N., Hoffmann, A., Bimberg, D. (1999) Photopumped InGaN/GaN/AlGaN Vertical Cavity Surface Emitting Laser Operating at Room Temperature, *Phys. Stat.. Sol.. (b)*, **216** 511-517.
10. Marko, I.P., Lutsenko, E.V., Pavlovskii, V.N., Yablonskii, G.P., Schön, O., Protzmann, H., Lünenbürger, M., Heuken, M., Schineller B., Heime, K. (1999) High-temperature lasing in InGaN/GaN multiquantum well heterostructures, *Phys. Stat. Sol. (b)* **216**, 491-494.
11. Marko, I.P., Lutsenko, E.V., Pavlovskii, V.N., Yablonskii, G.P., Schön, O., Protzmann, H., Lünenbürger, M., Heuken, M., Schineller B., Heime, K. (1999) Influence of UV light assisted annealing on optical properties of InGaN/GaN heterostructures grown by MOVPE, *Phys. Stat. Sol. (b)* **216**, 175-179.
12. Taylor, R.A., Hess, S., Kyhm, K., Ryan, J.F., Yablonskii, G.P., Lutsenko, E.V., Pavlovskii, V.N., Heuken, M. (1999) Stimulated emission and the Mott transition in GaN epilayers under high-dencity excitation, *Phys. Stat. Sol. (b)* **216**, 465-470.
13. Yablonskii, G.P., Lutsenko, E.V., Pavlovskii, V.N., Marko, I.P., Gurskii, A.L., Zubialevich, V.Z., Mudryi, A.V., Schön, O., Protzmann, H., Lünenbürger, M., Heuken, M., Schineller B., Heime, K. (2001) Blue InGaN/GaN multiple quantum well optically pumped lasers with emission wavelength in the spectral range of 450-470 nm, *Appl. Phys. Lett.* **29**, 1953 - 1955.
14. Yablonskii, G.P., Lutsenko, E.V., Pavlovskii, V.N., Marko, I.P., Zubialevich, V.Z., Gurskii, A.L., Kalisch, H., Szymakowskii, A., Jansen, R.A., Alam, A., Dikme, Y., Schineller, B., Heuken M. (2002) Luminescence and stimulated emission from GaN on silicon substrates heterostructures, *Phys. Stat. Sol. (a)* **192**, 54-59.
15. Arakawa, Y., Sakaki, H. (1982) Multidimensional Quantum Well Laser and Temperature Dependence of its Threshold Current, *Appl. Phys. Lett.* **40,** 939-941.
16. Casey, H.C., Panish, M.B. (1978) *Heterostructure Lasers*, Academic Press, New York, San Francisco, London, 380-392.
17. Gil, B.,, Briot, O., Aulombard, R.L. (1995) Valence-band physics and the optical properties of GaN epilayers grown onto sapphire with wurtzite symmetry, *Phys. Rev. B*, **52**, R17028-R17031.
18. Bidnyk, S., Little, Y.H., Cho, B.D., Krasinski, J., Song, J.J., Yang, W., Mcpherson, S.A. (1998) Laser action in GaN pyramids grown on (111) silicon by selective lateral overgrowth, *Appl. Phys. Lett,* **73**, 2242-2244.

SILICON MICROPHOTONICS: THE NEXT KILLER TECHNOLOGY

LIONEL C. KIMERLING
MIT Microphotonics Center
Cambridge, MA 02139

1. Introduction

Silicon has become the most studied material in the history of civilization. As one of the most abundant elements on the earth's crust, and as the basis of a $250B semiconductor industry, silicon has become more pervasive than steel in its affect on quality of life. Silicon integrated circuits have provided the computation capacity that created the Information Age. The technology underlying these circuits has progressed with the help of a Roadmap [1] that coordinates materials, processing and design for successive generations. The Roadmap has defined the technological limits to progress as process yield and as feature size shrink for the past two decades, respectively. For the next decade interconnection is likely to be the major challenge. Silicon microphotonics is a potential solution to the resistance and capacitance limitations of metal conductor interconnection. This paper discusses the advantages and barriers to implementation of an integrated optical interconnection technology from the chip to the network levels of the communications interconnection hierarchy.

2. A Roadmap for Communications Technology

Point-to-point communications becomes more technology intensive as the distance x data rate product increases. Interpersonal information transfer by voice migrated to optical smoke signal technology when propagation losses for sound waves over large distances became excessive. Although optics enhanced the distance of communication (which could be further increased by serial repeaters), smoke signals were limited in data rate. Twisted pair electronic conductors provided substantially the bandwidth of interpersonal voice communication with enhanced distance. As the number of users increased it became necessary to compress bandwidth/user and multiplex multiple users on a single twisted pair line. The capacity growth has increased remarkably at a semi-logarithmic rate as coaxial cable, microwave and satellite platforms were adopted. The fibre platform for lightwave communications represented the first deviation from the trend with no concomitant change in investment or revenue. The characteristics of this designation are both a discontinuous increase in performance and a change of slope to a sustainable increase in performance/cost ratio of 100x/10 years. Figure 1 shows that both the Time Domain Multiplexed (ETDM) and the Wavelength Division Multiplexed

L. Pavesi et al. (eds.), Towards the First Silicon Laser, 465–476.

(WDM) lightwave technologies display this behaviour relative to past trends [2].

If TDM and WDM are taken together as a trend, then one must ask 'What next?' The WDM capacity increase has been enabled by the commercialisation of full spectrum optical amplification: the erbium-doped fibre amplifier (EDFA). Since 1 TB/s is only a factor of 10 away from the theoretically realizable bandwidth of the full spectrum of the EDFA, one would expect the next step to involve expansion of the optical amplification spectrum. However, in view of the recent economic difficulties in the communications component industry, one might ask instead 'Is Long Haul capacity the appropriate metric for network performance?' One might argue that only connection to the revenue-generating user can drive investment in infrastructure; and therefore, the bottleneck in network performance is more appropriately placed in long haul capacity than in broadband access.

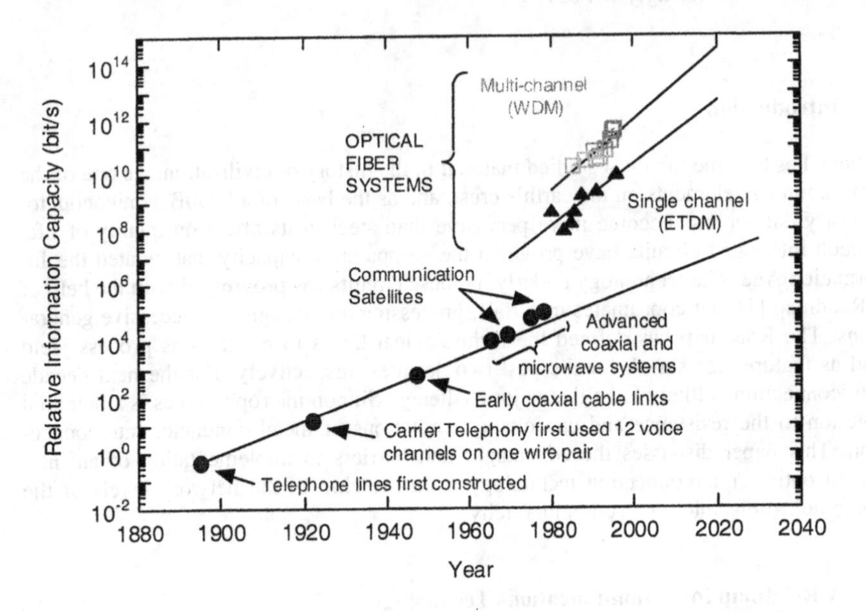

Figure 1. Growth in information carrying capacity of a single communications line [2].

3. Microphotonic Integrated Circuits

3.1 THE NETWORK

The optical network today consists of discrete components interconnected by fibre splices. This layout is reminiscent of the electronic systems of the 1950-60s. In the case of electronics performance/cost accelerated with integration, and applications proliferated as a result. The driver for integration was a scaling to smaller feature sizes that enabled higher functionality, greater speed and lower cost with the same basic set of materials and unit processes. The recent boom and bust in the network components industry has diminished both the investment by the carriers and the pricing power of the hardware manufacturers. It is clear that a return to the traditional growth rates of 4%/year for the components industry will be predicated on lower cost technology. Higher 'killer technology' growth rates will require not only a lower cost technology platform, but a scalable platform with time-dependent, exponential performance/cost enhancement.

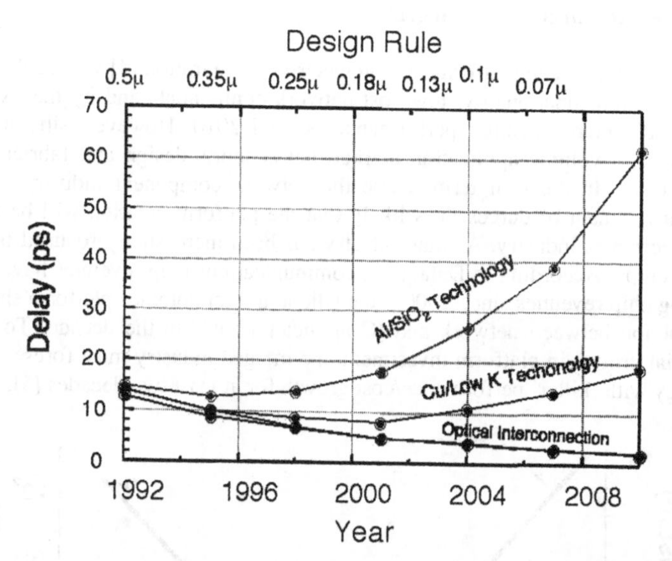

Figure 3. Trend in interconnect propagation delay (1 cm length) with technology linewidth and time for traditional aluminium metal and SiO2 insulator; projected copper and low K dielectric; and projected optical waveguide technologies [3].

3.2 ULSI ELECTRONICS

It is generally accepted that optical devices will scale to larger dimensions than electrical devices due to the size of the photon relative to that of the electron in a medium. However, this ultimate limitation should not hinder the integration of microphotonic components with current ULSI electronic circuits, and it will likely not affect the ultimate architecture and applications of optoelectronic integrated circuits (OEICs). Interconnection is the major barrier to the continued scaling of IC linewidths below 0.18μm. Designers must add complexity to maintain performance improvements, and therefore, give back some of the potential of the increased device density with linewidth reduction. Multiple layers of interconnects provide a hierarchy of conductor linewidths to reduce the RC bandwidth limit of the longest lines (~1 cm). While partially effective in maintaining performance, this multilevel architecture is a major limit to manufacturing yield. Figure 3 shows that the introduction of copper conductors to reduce resistance and low K dielectrics to reduce capacitance reaches a performance improvement limit at 0.18μm. Continued linewidth scaling with the best projected performance causes an increasing reduction in system performance by interconnection [3]. In summary, continued scaling of electronic ICs produces an 'interconnection bottleneck' to chip performance.

It is likely that optical interconnection will be implemented, at least for the longest signal lines, to extend 'Moore's Law' through the end of the decade.

4. Barriers to Large Scale Integration

The creation of a silicon platform for optoelectronic integration [4] will be driven by the demand for high functionality, low cost network components and by the extension of electronic IC interconnection performance beyond 2010. However, silicon chips are built on a high volume application market that supports design and fabrication infrastructure costs. In the near term, while the network component industry recovers to produce investment resources, it is likely that the platform creation will be led by the silicon electronics industry. As that industry has been increasingly focused on communications chips (According to Dataquest, communications chip revenues have surpassed computing chip revenues since 2000.), the silicon microphotonics platform should erase the distinction between network and IC applications within the decade. To justify an exponential growth in platform investment, the merged industry must foresee a scalable technology with 'killer' performance/cost growth for at least two decades [5].

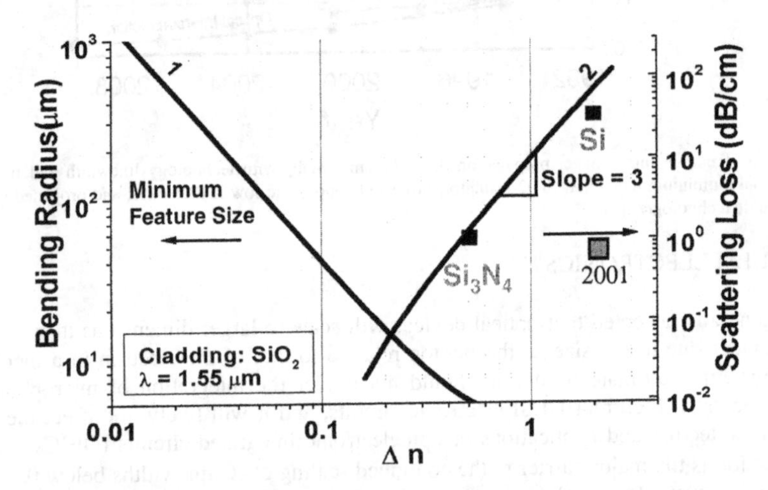

Figure 4. Dimensional scaling for microphotonic components with index difference, Δn, and the consequent increase in scattering loss with current pattern transfer technology [6].

Current Planar Lightguide Circuits (PLCs) are constructed of lightly doped SiO_2 waveguides with index contrast (Δn=0.005) closely matching that of the long haul, optical fibres that feed signals to them. The core dimension is about 10μm and the cladding is about 20μm on either side for a total film thickness approaching 50μm. While this design serves the constraint of low loss, it does not meet the critical microphotonics requirements of small bend radius and silicon CMOS compatibility. High Index Contrast (HIC) design is consistent with 1μm turn radii and total film thicknesses of <5μm. These results meet both the performance and fabrication targets for a low cost, high functionality platform. Figure 4 gives a representation of the scaling law for silicon compatible HIC design. As Δn approaches 1 the minimum device dimension falls below 10μm, and scattering loss from edge roughness becomes a dominant concern. The data for silicon nitride and silicon waveguides are taken post-lithography and match the theoretical (solid line) expectation. The green data point labelled 2001 represents the measured result achieved by post-lithography processing (oxidation smoothing) [7]. The challenge is to bring the red line down by process innovation to realize radiation limited (bending loss) performance.

4.1 OPTICAL BUS ARCHITECTURE

Microphotonics offers several architectural advantages when integrated with electronic chips. A key, first focus area is simplification of the multilevel interconnection hierarchy. While local electrical interconnects are likely to continue to offer design and process integration benefits, an optical layer could handle all high bandwidth I/O and synchronization functions.

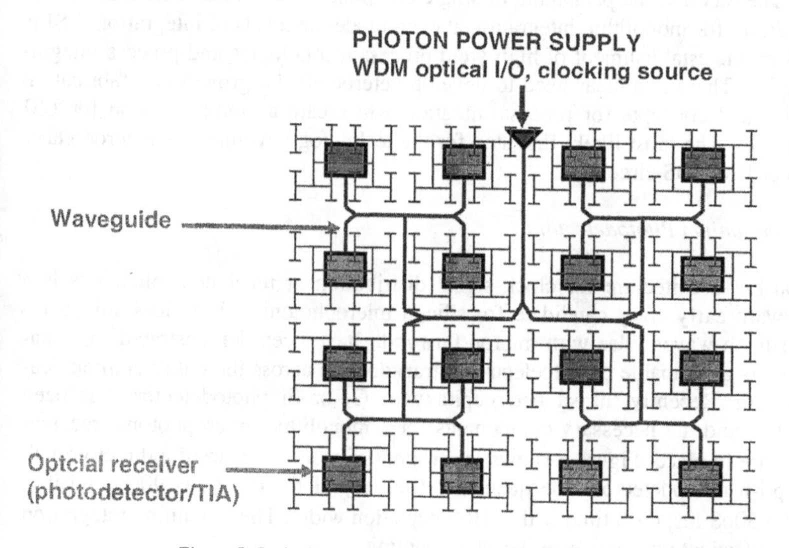

Figure 5. Optical bus architecture for electronic/photonic integration.

Figure 5 shows the optical bus architecture [8] which combines global optical signal distribution with local electrical interconnection. This architecture is applicable to inter-chip and backplane levels if interconnection as well as intrachip, as shown. The key design constraint is location of the light sources at a remote 'photon power supply' rather than integrating them locally. The key components for on-chip integration are waveguides capable of ultra-small turn radii, equal power splitters, and photodetectors or modulators to transduce between electrical and optical signals. Separating the laser source form the signal processing reduces the thermal management problem significantly. One could envision a remote mode-locked laser generating timing pulses that are distributed in optical waveguides without skew to photodetectors for local electrical distribution. As long as the optical waveguide lengths are equal, the latency or signal delay will be the same in all arms. The H-tree design shown is one version of an equal path layout.

4.2 HETEROEPITAXY

Microphotonics is inherently a materials diverse technology. Compound semiconductor sources, dielectric waveguides and ceramic isolators and modulators offer a much greater integration challenge than silicon, silicon dioxide and aluminium did for electronic circuits. The integration of microphotonic components to device densities of greater than 10^4/chip, Large Scale Integration (LSI), encounters two primary barriers: cost and complexity. At device densities of less than 10/chip, hybrid integration by pick-and-place flip-chip, solder-bumps or by wafer bonding should have the lowest barrier to technology implementation. The advantages of the hybrid approach are i) best in class optoelectronic devices, and ii) separation of yield issues to device fabrication and packaging [5]. The fact that the packaging of single components dominates cost today is the dominant driver for monolithic integration at even moderate levels of integration. LSI is predicated on the establishment of high yield process technologies and process integration strategies. There is a clear need to develop heteroepitaxial growth and fabrication technologies and concepts for process integration to create a credible vision for LSI microphotonics. The most likely first step for the technology evolution is heteroepitaxy on CMOS or BiCMOS circuits.

4.2.1 Germanium Photodetectors

An optical bus architecture for clock signal distribution or receiver applications is a straightforward early entry candidate for silicon microphotonics. Low loss silicon nitride and silicon waveguides with micron turn radii have been demonstrated[9]. Germanium detectors are capable of optoelectronic transduction across the entire communications spectrum. Recently, direct heteroepitaxy of Ge-on-Si photodetectors has been reported [10], and the necessary components for a monolithic, microphotonic receiver technology are in place. Figure 6 shows the responsivity at $\lambda=1.3\mu m$ of a direct growth Ge-on-Si p-i-n photodetector. The quantum efficiency of the device is 90%, and it is capable of a 50pS response time with a 1μm depletion width. The remaining integration challenge is efficient waveguide-to-detector coupling.

4.2.2 GaAs Lasers

Integrated optical sources are highly regarded for backplane and box-to-box interconnection at high bandwidths. Vertical Cavity Surface Emitting Laser (VCSEL) arrays are used in high capacity routers today. Using GaAs materials technology, these sources function at wavelengths of $850nm<\lambda<1310nm$. Monolithic laser integration on a silicon platform has been recently demonstrated [11], as shown in Figure 7. The envisioned applications are direct modulation electronic signals for I/O encoding and source/driver integration for 'photon power supply' architectures.

The process technology depends on lattice parameter grading from silicon to germanium and growth of GaAs on the germanium film. This method, combined with dislocation removal during growth, yields a dislocation density comparable to GaAs substrates and device performance that is nearly identical to equivalent homoepitaxial designs. Research is now focused on device reliability, and device lifetimes of greater than 4 hours have been achieved.

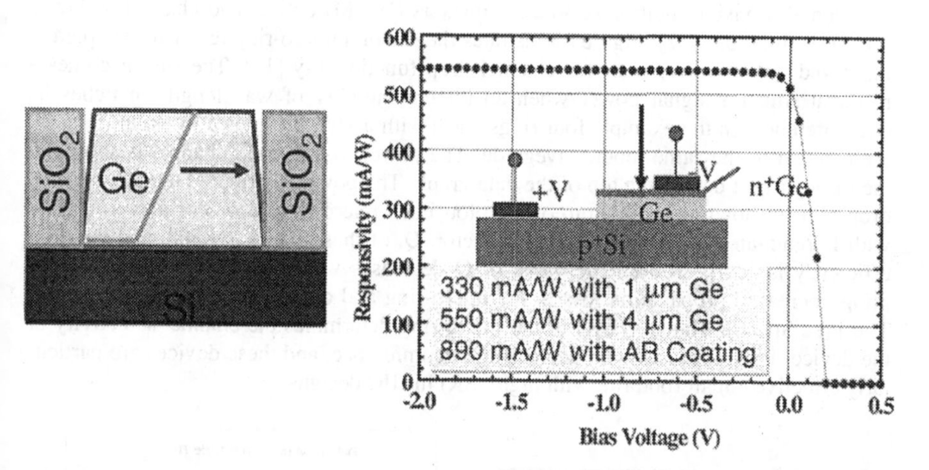

Figure 6. Responsivity curve measured for a direct growth Ge-on-Si photodetector. The process integration path could be selective epitaxial deposition of Ge through an oxide window on a Si waveguide, as shown [10].

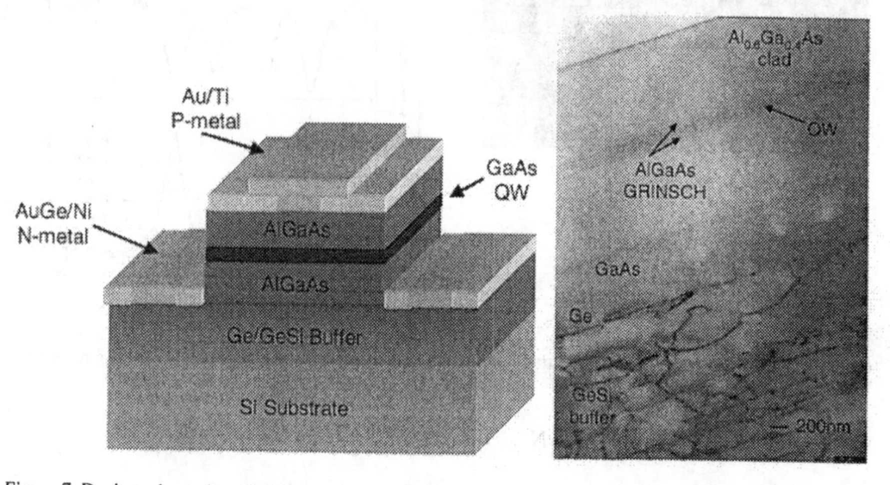

Figure 7. Device schematic and TEM micrograph of GaAs-on-graded GeSi heteroepitaxial growth technology [11].

4.3 DEVICE-WAVEGUIDE COUPLING

Coupling of optical signals is the dominant design and cost factor in optical signal distribution. For microphotonic circuits that utilize High Index Contrast (HIC) components, resonant coupling provides effective coupling with minimum footprint. A passive device that is resonant with the optical carrier frequency (wavelength) acts similarly to an inductive tank circuit in electronics as power is coupled into the device with great efficiency. Resonant devices, such as thin film microcavities (standing wave) and ring resonators (traveling wave) have immediate application as filters for high capacity, wavelength division multiplexers/demultiplexers (WDM) optical and channel add/drop (OADM) functions [12]. Figure 8 illustrates the use of a micro-ring resonator coupled to input and output waveguides for channel drop functionality [13]. The ring becomes a resonant sink for signal power when an integral number of wavelengths matches its circumference. In the example four rings, each with a 10% difference in circumference, are coupled to the same input waveguide. The power loss in each channel is shown in the through-port data at the top of the data graph. The power distributed to each channel drop port is shown at the bottom of the plot. Each micro-ring resonator acts as a filter with Lorentzian line shape. The quality factor, Q, of these filters is ~500, and the drop efficiency is nearly 100%. The value of Q decreases with increased waveguide coupling, increased propagation loss in the micro-ring and decreasing ring circumference. The Free Spectral Range (FSR) of the micro-ring determines the channel selectivity of the device. FSR increases with decreasing circumference, and these devices are particularly effective (small footprint with high FSR) in HIC designs.

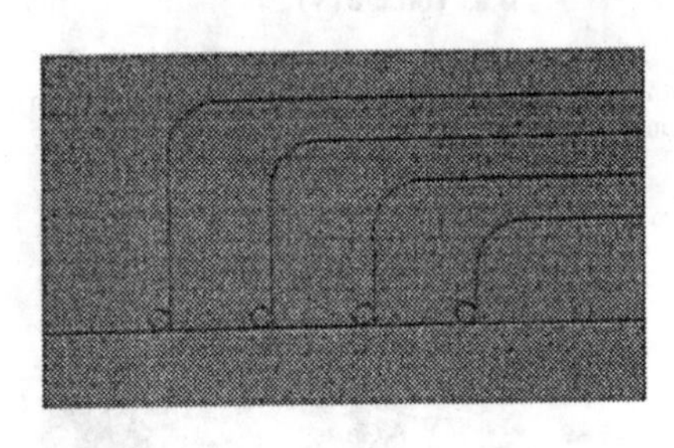

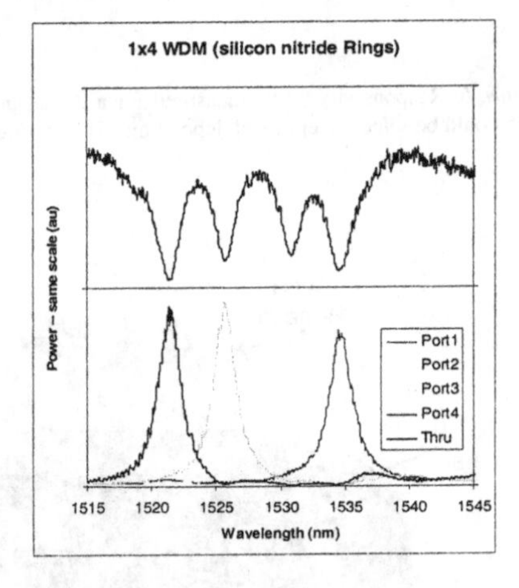

Figure 8. Four channel add/drop configuration utilizing silicon nitride waveguides and micro-ring resonators with the performance data [13].

4.4 OPTICAL AMPLIFICATION

As microphotonic circuits scale to higher complexity, the circuit fanout (number of splits) becomes a major source of signal attenuation. The WDM long haul transport architecture was enabled by the full spectrum optical amplifier that contributed >20dB of gain utilizing high power pump lasers and >20m of erbium-doped fibre. For microphotonic circuits a 3dB signal gain at each Y-split would be sufficient to compensate for fanout. To be practical these devices must occupy a negligible footprint relative to the circuit size. A set of reasonable design constraints for this compact amplifier device is a footprint of < 2 mm^2, 3 dB optical gain, noise figure of 0.02 dB and compatibility with Gbit/s data transmission. We have recently derived the scaling laws for the design of HIC waveguide optical amplifiers (HIC-WOAs) [14]. The signal and pump fluxes increase with the higher optical confinement, and allow shorter, more gain efficient amplification. The serpentine or coil wound waveguide scales dimensionally to smaller footprints with the smaller turn radius enabled by HIC design. The noise figure is invariant with HIC scaling. A Figure-of-Merit (FOM) for HIC-WOA performance is areal gain efficiency: gain/ (pump power x area). Figure 9 shows that the FOM scales dramatically as $(\Delta n)^{2.6}$, showing major benefits for microphotonic HIC-WOA design.

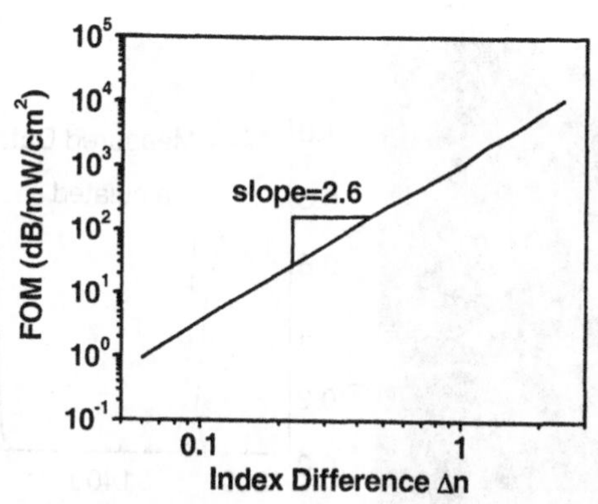

Figure 9. Dependence of the Figure-of-Merit for planar waveguide optical amplifiers on the index difference between the waveguide and its cladding [14].

4.5 PHOTONIC CRYSTALS

Photonic crystals provide the ultimate in photon control and device shrink. A photonic crystal is a periodic composite of high and low index materials that, as a medium, influence the propagation of light in a way similar to atomic potentials with the flow of electrons in a semiconductor. The fabrication of photonic crystals is difficult, and HIC design can achieve similar performance/area for many applications. Nevertheless, photonic crystals provide a dominant design paradigm for microphotonics, including HIC components. Two specialized applications are discussed below: small modal volume for in line filters and low threshold sources; and multichannel tuneable devices for ubiquitous placement in microphotonic circuits.

4.5.1 Ultrasmall Devices

An inline add/drop device can be constructed as a microcavity inserted in a waveguide [15]. One dimensional photonic crystals are inserted as dielectric stack mirrors. As the index contrast in the mirror increases, a larger spectrum of high reflectivity (photonic band gap) results. In the example of Figure 10, air sections (n=1) are separated by silicon sections (n=3.5) with the spacings chosen to give a photonic band gap overlapping the WDM gain spectrum of an Erbium Doped Fiber Amplifier (EDFA). The missing air section in the middle of the device defines the pass band (wavelength of the channel to be dropped). The simulated and measured performance on the right indicates the strong theoretical foundation behind photonic crystal design. The Q of this device is 265 at the resonant wavelength of 1564nm. Most impressive is the modal volume of V=0.055μm^3. Low threshold light sources require high values of Q/V. High Q can be designed with large numbers of sections, but low V has been the challenge until the advent of photonic crystals.

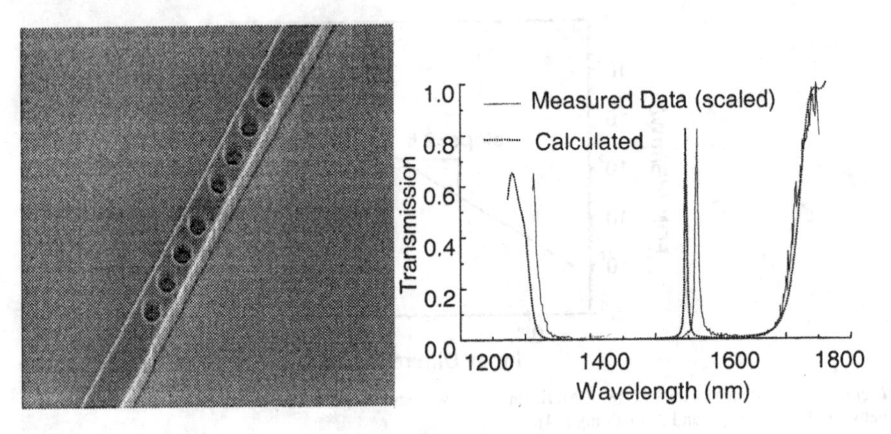

Figure 10. An inline photonic crystal add/drop filter in a silicon waveguide with measured and calculated performance.

4.5.2 Multichannel Design

The design paradigm of photonic crystals creates an analogy between channel drop frequencies of an optical filter and bound defect states in the gap of a semiconductor. As some defects can present more than one state in the gap of a semiconductor, a photonic crystal filter can be designed as a multichannel device. For microphotonic circuits to grow in complexity beyond 10 devices/chip, a ubiquitous device, such as the transistor, must perform a critical function. Since photons possess no charge, there interaction is best mediated by wavelength or frequency. A variable frequency (wavelength) filter can act as a switch, add/drop path and multiplexer/demultiplexer for alcommunications channels. If the same device can yield all functions, it can be replicated in a simplified design/process integration scheme. The photonic crystal device in Figure 11 is such a device with operation in the 1310-1600nm regime [16]. For this Microelectromechanical actuated device a 12V bias results in 100nm tuning for each channel.

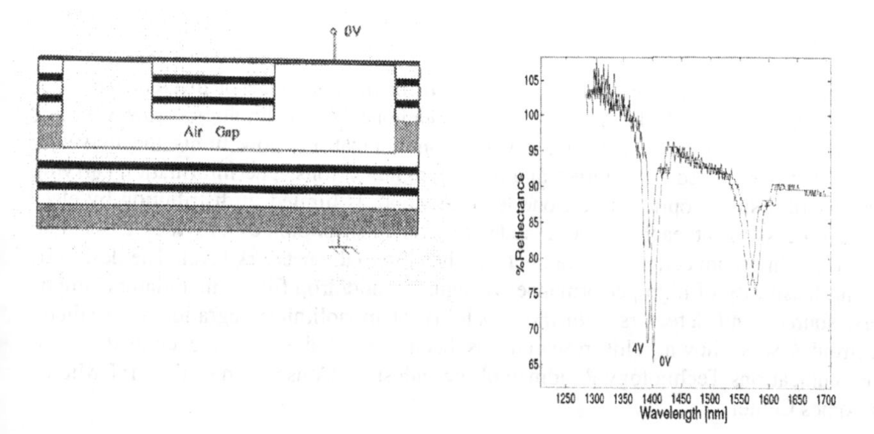

Figure 11. Device cross section and performance for a multichannel, tuneable add/drop filter [16].

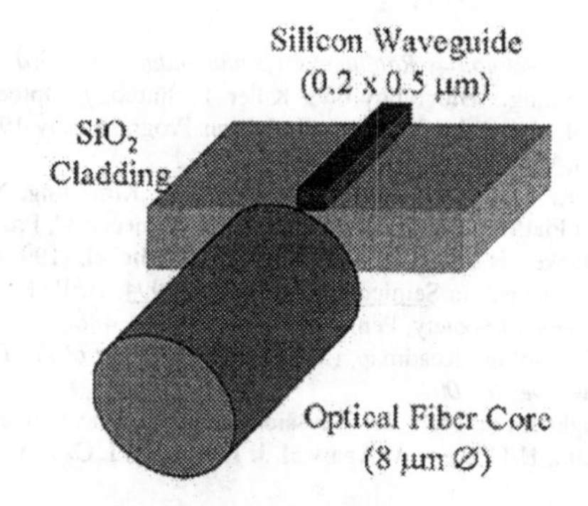

Figure 12. Schematic representation of the fiber-to-HIC-waveguide coupling challenge.

4.6 FIBRE-WAVEGUIDE COUPLING

Low index contrast fibre is designed for minimum transmission loss over long distances. High index contrast waveguides are designed for maximum routing flexibility over very short distances. Even though the primary benefit of integration is elimination of fibre splices between components, low loss coupling of fibre to HIC waveguides has been a primary barrier to research and development for microphotonics. Three problems, as shown in Figure 12, must be solved simultaneously: i) mode matching, ii) index matching, iii) and alignment tolerance. Theoretically a single solution to all three can be designed. The key constraint is that the solution be process compatible with the microphotonic circuit design and fabrication.

5. Conclusion

Photonics has provided unparalleled bandwidth for the backbone of discrete, point-to-point long distance (Wide Area Network) telephone line technology. However, this architecture that worked so well for voice communications now limits the universal access that is required for Internet and data communications. The information carrying capacity of a single optical fiber, one terabit/second, is limited in distribution by electronic processing at each network node. Silicon microphotonics provides a scalable solution to interconnection constraints from the chip to the network level. The desirability and feasibility of high performance waveguides, add/drop filters, modulators, amplifiers, sources and detectors with the capability of monolithic integration with silicon electronics was shown. This research has been presented within the context of the Communications Technology Roadmap of the Industrial Consortium of the MIT Microphotonics Center.

References

1. ITRS (2001), *International Technology Roadmap for Semiconductors : 2001,* http://public.itrs.net/.
2. C.H. Fine and L.C. Kimerling, "Biography of a Killer Technology: Optoelectronics Drives Industrial Growth with the Speed of Light," in OIDA Future Vision Program, July 1997(Optoelectronics Industry Development Association, Washington, D.C., 1997) pp. 1-22.
3. K. Wada, H.S. Luan, K.K. Lee, S. Akiyama, J. Michel, L.C. Kimerling, M. Popovic and H.A. Haus, (2002) "Silicon and Silica Platform for On-chip Optical Interconnection", Proc. LEOS Annual Meeting.
4. L. C. Kimerling, D. M. Koker, B Zheng, F. Y. G. Ren, and J. Michel, (1994) "Erbium-Doped Silicon for Integrated Optical Interconnects," in Semiconductor Silicon/1994, H. R. Huff, W. Bergholz, and K. Sumino,Eds. (The Electrochemical Society, Pennington, NJ, 1994) p. 486.
5. MIT Communications Technology Roadmap, (2002) *Interim Report of the Technical Working Group on the Silicon Platform, November 6, 2002.*
6. "Effects of Size and Roughness on Light Transmission in Si/SiO_2 Waveguides: experiments and models", (2000) K.K. Lee, D.R. Lim, H-C Luan, A. Agarwal, J. Foresi and L.C. Kimerling, *Applied Phys. Lett.* **77** 1617.
7. K.K. Lee, D.R. Lim, L.C. Kimerling, (2001) "Fabrication of ultralow-loss Si/SiO_2 waveguides by roughness reduction," *Optics Letters*, **26** (23) 1888.
8. E.A. Fitzgerald and L. C. Kimerling, (1998) 'Silicon Microphotonics and Integrated Optoelectronics', MRS Bulletin, **23** 4.
9. D.R.C. Lim, Ph. D thesis at Massachusetts Institute of Technology, 2000.
10. L. Colace, G. Masini, G. Assanto, H.C. Luan, K. Wada and L.C. Kimerling, (2000) "Efficient, High Speed Near Infrared Ge Photodetectors Integrated on a Si substrate," *Applied Phys. Letters*, **76** 1231.
11. M. Groenert, C.W. Leitz, A.J. Pitera, V. Yang, H. Lee, R. Ram and E.A. Fitzgerald, (2003) *J. Applied Physics* **93** 1.
12. B.E. Little, H.A. Haus, J.S. Foresi, L.C. Kimerling, E.P. Ippen, D.J. Ripin, (1998) "Wavelength Switching and Routing Using Absorption and Resonance," *IEEE Photonics Technology Letters,* **10**(6) 816.
13. D.R. Lim, B.E. Little, K.K. Lee, M. Morse, H.H. Fujimoto, H.A. Haus, L.C. Kimerling, (1999) "Micron-sized channel dropping filters using silicon waveguide devices" *Proceedings of Spie – the International Society for Optical Engineering* **3847** 65-71.
14. S. Saini, J. Michel and L.C. Kimerling, (2003) 'Index Contrast Scaling for Optical Amplifiers', *J. Lightwave Technology,* to be published.
15. "Photonic Bandgap Microcavities in Optical Waveguides," J.S. Foresi, P.R. Villeneuve, J. Ferrera, E.R. Thoen, G. Steinmeyer, S. Fan, J.D. Joannopoulos, L.C. Kimerling, H.I. Smith and E.P.Ippen, (1997) *Nature,* **39**, 143.
16. Y. Yi, P. Bermel, K.Wada, X. Duan, J.D. Joannopoulos, and L.C. Kimerling, 'Low Voltage Tunable One

SUBJECT INDEX

List of Participants

Aleshkin V.
Institute for Physics of Microstructures
Russian Academy of Sciences, GSP-105, Nizhny Novgorod, 603950,
Russia
aleshkin@ipm.sci-nnov.ru

Angelescu A.
National Institute for Research and Development in Microtechnologies (IMT), P.O.Box 38-160, Bucharest,
Romania
email ancaa@imt.ro

Atwater Harry
Thomas J. Watson Laboratories of Applied Physics, California Institute of Technology, Pasadena, CA 91125,
U.S.A.
email haa@daedalus.caltech.edu

Cirlin G.
Max-Planck-Institute of Microstructure Physics, Weinberg 2, 06112 Halle
Germany
email cirlin@mpi-halle.de

Dal Negro Luca
INFM and Dipartimento di Fisica, Università di Trento, via Sommarive 14, I-38050 Povo (Trento),
Italy
email dalnegro@science.unitn.it

Delerue C.
Institut d'Electronique et de Microelectronique du Nord, Dept. ISEN, 41 boulevard Vauban, 59046 Lille Cedex,
France
email delerue@isen.fr

Dvurechenskii A.V.
Institute of Semiconductor Physics, Siberian Branch of Russian Academy of Science, 630090, Novosibirsk,
Russia Lavrentiev prospect 13,
Russia
e-mail: dvurech@isp.nsc.ru

Fauchet Philippe M.
Department of Electrical and Computer Engineering Computer Studies Building Room 518 University of Rochester, PO.Box 270231, Rochester NY 14627-0231,
USA
email fauchet@ee.rochester.edu

Gaponenko Sergey
Institute of Molecular and Atomic Physics,National Academy of Science, F.Skaryna Ave. 70, Minsk 220072,
Belarus
email gaponen@imaph.bas-net.by

Gardner Donald S.
Circuits Research, Intel Labs ,
Building SC12, 2200 Mission College Blvd. Santa Clara, CA 95052-8119
California
USA
email d.s.gardner@intel.com

Garrido Blas
EME, Department d'Electrònica, Universitat de Barcelona, Martí i Franquès 1, 08028 Barcelona, Spain
email blas@el.ub.es

Goesele Ulrich
Max-Planck-Institut für Mikrostrukturphysik ,Weinberg 2 , D-06120 Halle,
Germany
email goesele@mpi-halle.de

Green Martin A.
Centre for Third Generation
Photovoltaics Electrical Engineering
Room 127, University of New South
Wales, Sydney 2052,
Australia
email m.green@unsw.edu.au

Grutzmacher Detlev
Laboratory for Micro- and
Nanotechnology Paul Scherrer Institut
CH-5232 Villigen PSI, Switzerland
e-mail:detlev.gruetzmacher@psi.ch

Gusev O. B.
A F Ioffe Physico-Technical Institute
Politekhnicheskaya 26, 194021 St.
Petersburg, Russia
email oleg.gusev@pop.ioffe.rssi.ru

Heikkila Lauri
Laboratory of Microelectronics,
Department of Information
Technology, University of Turku,
20014 Turku Finland
email lauhe@utu.fi

Homewood Kevin P.
School of Electronics and Physical
Sciences, University of Surrey,
Guildford, Surrey, GU2 7XH, United
Kingdom email
K.Homewood@eim.surrey.ac.uk

Icona Fabio
CNR-IMM, Sezione di Catania,
Stradale Primosole 50, I-95121
Catania, Italy email
iacona@imetem.ct.cnr.it

Kagan M. S.
Institute of Radioengineering and
Electronics of Russian Ac. Sci.,
Moscow, Russia
email kagan@mail.cplire.ru

Kanemitsu Yoshihiko
Nara Institute of Science and
Technology (NAIST) Graduate school
of Materials Science Quantum
Materials Science lab. 8916-5,
Takayama-cho, Ikoma, Nara 630-0101
JAPAN
email sunyu@ms.aist-nara.ac.jp

Kelsall R. W.
Institute of Microwaves and Photonics,
School of Electronic and Electrical
Engineering
The University of Leeds
LS2 9JT UK
email r.w.kelsall@leeds.ac.uk

Khriachtcev Leonid Yu
Laboratory of Physical Chemistry
University of Helsinki PO Box 55 FIN-
00014
Finland
email khriacht@rock.helsinki.fi

Kik Pieter
Thomas J. Watson Laboratories of
Applied Physics, California Institute of
Technology, Pasadena, CA 91125,
U.S.A email kik@caltech.edu

Kimerling Lionel
MIT Materials Processing Center and
MIT Microphotonics Center, 77 Mass.
Ave., Cambridge, MA 02139 ,
U.S.A.
email lckim@mit.edu

Kovalev D.
Technische Universitat Muenchen,
Physik-Department E16, 85747 Garching,
Germany
email dkovalev@physik.tu-
muenchen.de

Krasil'nik Z.
Institute for Physics of Microstructures Russian Academy of Sciences, GSP-105, Nizhny Novgorod, 603950, Russia
email zfk@ipm.sci-nnov.ru

Lazarouk Sergey
Belorusian State University of Informatics and Radioelectronics, Microelectronics Dept. P.Browky str. 6,220027 Minsk, Belarus
email serg@cit.org.by

Linnros Jan
Dept. of Microelectronics and Information Technology, Royal Institute of Technology Electrum 229, 164 40 Kista, SWEDEN
e-mail: linnros@imit.kth.se

Luterova K.
Institute of Physics, Academy of Sciences of the Czech Republic, Cukrovarnická 10, 162 53 Praha 6, Czech Republic
email luterova@fzu.cz

Mile Ivanda
Ruđer Bošković Institute, P.O.Box 180, 10002 Zagreb, Croatia,
email ivanda@rudjer.irb.hr

Miu M.
National Institute for Research and Development in Microtechnologies (IMT), P.O.Box 38-160, Bucharest, Romania
email mihaelam@imt.ro

Nayfeh M. H.
Department of Physics, University of Illinois at Urbana-Champaign 1110 W. Green Street, Urbana, Illinois 61801 USA
email m-nayfeh@staff.uiuc.edu

Ni Wei-Xin
Department Physics and Measurement Technology, Linköping University. S-581 83 Linköping, Sweden
email wxn@ifm.liu.se

Ossicini Stefano
INFM-S3 and Dipartimento di Scienze e Metodi dell'Ingegneria, Università di Modena e Reggio Emilia, Via Campi 213/A, 41100 Modena Italy
email ossicini@unimo.it

Pacifici Domenico
INFM - Unita di Catania, Dipartimento di Fisica, Universita di Catania, Corso Italia 57, I-95129 Catania ITALY
email d.pacifici@ct.infn.it

Pavesi Lorenzo
INFM and Dipartimento di Fisica, Università di Trento, via Sommarive 14, I-38050 Povo (Trento), Italy
email Pavesi@science.unitn.it

Pavlov S. G.
Institute of Space Sensor Technology and Planetary Exploration, German Aerospace Center, Rutherfordstr. 2, 12489 Berlin, Germany
email sergeij.pavlov@dlr.de

Pchelyakov O.
Institute of Semiconductor Physics SB RAS, Novosibirsk, 630090, RUSSIA
email pch@isp.nsc.ru

Pelant I.
Institute of Physics, Academy of Sciences of the Czech Republic, Cukrovarnická 10, 162 53 Praha 6, Czech Republic
email pelant@fzu.cz

Polman Albert
FOM-Institute for Atomic and Molecular Physics [AMOLF] Kruislaan 407 1098 SJ AmsterdamThe Netherlands
email polman@amolf.nl

Prokofiev A.
A F Ioffe Physico-Technical Institute Politekhnicheskaya 26, 194021 St. Petersburg, Russia
email lxpro@mail.ioffe.ru

Rebohle L.
Insitut fur Festkorperelektronik, Technische Universität Wien, A-1040 Wien, Austria
email lars.rebohle@tuwien.ac.at

Ryabtsev G.
Semiconductor Optics Laboratory, B.I. Stepanov Institute of Physics, National Academy of Sciences of Belarus, F. Skoryna Ave. 68, Minsk 220072, Belarus
e-mail: ryabtsev@dragon.bas-net.by

Shastin V.
Institute for Physics of Microstructures, Russian Academy of Sciences, GSP-105, 603950 Nizhny Novgorod, Russia
email shastin@ipm-sci-nnov.ru

Shin Jung H.
Semiconductor Physics Laboratory Department of Physics, KAIST 373-1, Kusong-dong, Yusong-Gu, Taejon 305-701, Korea
email jhs@mail.kaist.ac.kr

Sun Greg
Department of Physics, University of Massachusetts at Boston Boston, MA 02125 USA
email Greg.Sun@umb.edu

Svrcek V.
ICPMS, GONLO, UMR7504, CNRS-ULP, 23 rue du Loess, 67037 Strasbourg Cedex, France
email svrcek@phase.c-strasbourg.fr

Valenta J.
Department of Chemical Physics and Optics, Charles University, Ke Karlovu 3 CZ-121 16 Prague 2, Czech Republic
email j.valenta@mff.cuni.cz

Yablonski G. P.
Stepanov Institute of Physics of National Academy of Sciences of Belarus F. Skoryna Ave, 168, 220072 Minsk, Belarus,
email yablon@dragon.bas-net.by

Yassievich I.
A F Ioffe Physico-Technical Institute Politekhnicheskaya 26, 194021 St. Petersburg, Russia
email irinia.yassievich@mail.ioffe.ru

Zacharias M.
Max-Planck-Institute of Microstructure Physics, Weinberg 2, 06112 Halle
email zacharia@mpi-halle.de

9 781402 011948